Andrew S. Goudie

St Cross College
Oxford

The Human Impact

ON THE NATURAL ENVIRONMENT

Past, Present and Future

Seventh edition

WILEY-BLACKWELL

A John Wiley & Sons, Ltd., Publication

This edition first published 2013; © 1981, 1986, 1990, 1993, 2000, 2006 by Andrew Goudie, © 2013 by John Wiley & Sons, Ltd

Wiley-Blackwell is an imprint of John Wiley & Sons, formed by the merger of Wiley's global Scientific, Technical and Medical business with Blackwell Publishing.

Registered office: John Wiley & Sons, Ltd, The Atrium, Southern Gate, Chichester, West Sussex, PO19 8SQ, UK

Editorial offices: 9600 Garsington Road, Oxford, OX4 2DQ, UK
                       The Atrium, Southern Gate, Chichester, West Sussex, PO19 8SQ, UK
                       111 River Street, Hoboken, NJ 07030-5774, USA

For details of our global editorial offices, for customer services and for information about how to apply for permission to reuse the copyright material in this book please see our website at www.wiley.com/wiley-blackwell.

Library of Congress Cataloging-in-Publication Data
Goudie, Andrew.
    The human impact on the natural environment : past, present and future / Andrew S. Goudie. – Seventh edition.
        pages cm
    Includes bibliographical references and index.
    ISBN 978-1-118-57657-1 (cloth) – ISBN 978-1-118-57658-8 (pbk.)   1.   Nature–Effect of human beings on.   I. Title.
    GF75.G68 2013
    304.2–dc23
                                2012047946

A catalogue record for this book is available from the British Library.

Wiley also publishes its books in a variety of electronic formats. Some content that appears in print may not be available in electronic books.

Cover Image: Rice paddy in Longsheng, China. © iStockphoto.com/Rob Broek
Cover Design by Steve Thompson

Set in 10/13pt Palatino by Toppan Best-set Premedia Limited
Printed and bound in Malaysia by Vivar Printing Sdn Bhd

1   2013

The Human
Impact

ON THE NATURAL
ENVIRONMENT

CONTENTS

Colour plate section can be found between pages 196–197

PREFACE TO THE SEVENTH EDITION

It is now three decades since the first edition of this book appeared. This period has seen a remarkable transformation in interest in the impact that humans are having on the environment, together with an explosion of knowledge. In this edition, I have made substantial changes to the text, figures, tables and references, and have tried to provide updated statistical information.

A.S.G.

ABOUT THE COMPANION WEBSITE

This book is accompanied by a companion website:

www.wiley.com/go/goudiehumanimpact

The website includes:
- Powerpoints of all figures from the book for downloading
- PDFs of tables from the book

Part I

The Past and Present

1 INTRODUCTION

Chapter Overview

In this chapter the first issue that is addressed is the development of ideas over the last 200 years about the relationship between humans and their environment and in particular the development of ideas about how humans have changed their environment. The historical theme continues with a brief analysis of the changes that have taken place in human societies from prehistoric times onwards, culminating in the massive impacts that humans have achieved over the last three centuries.

The development of ideas

To what extent have humans transformed their natural environment? This is a crucial question which became very important in the seventeenth and eighteenth centuries (Grove and Damodaran, 2006) as western Europeans became aware of the ravages inflicted in the tropics by European overseas expansion. It was a theme that intrigued the eighteenth-century French natural historian, Count Buffon. He can be regarded as the first Western scientist to be concerned directly and intimately with the human impact on the natural environment (Glacken, 1967). He contrasted the appearance of inhabited and uninhabited lands: the anciently

The Human Impact on the Natural Environment: Past, Present and Future, Seventh Edition. Andrew S. Goudie.
© 2013 John Wiley & Sons, Ltd. Published 2013 by John Wiley & Sons, Ltd.

inhabited countries have few woods, lakes or marshes, many heaths and scrub, bare mountains, soils that are less fertile because they lack the organic matter which woods, felled in inhabited countries, supply, and the herbs are browsed. Buffon was also much interested in the domestication of plants and animals – one of the major transformations in nature brought about by human actions.

Studies of the torrents of the European Alps, undertaken in the late eighteenth and early nineteenth centuries, deepened immeasurably the realization of human capacity to change the environment. Fabre and Surell studied the flooding, siltation, erosion and division of watercourses brought about by deforestation in these mountains. Similarly, Horace-Bénédict de Saussure showed that Alpine lakes had suffered a lowering of water levels in recent times because of deforestation. In Venezuela, Alexander von Humboldt concluded that the lake level of Lake Valencia in 1800 (the year of his visit) was lower than it had been in previous times and that deforestation, the clearing of plains, irrigation and the cultivation of indigo, were among the causes of the gradual drying up of the basin (Cushman, 2011). Comparable observations were made by the French rural economist, Jean-Baptiste Boussingault (1845). He returned to Lake Valencia some 25 years after Humboldt and noted that the lake was actually rising. He described this reversal to political and social upheavals following the granting of independence to the colonies of the erstwhile Spanish Empire. The freeing of slaves had led to a decline in agriculture, a reduction in the application of irrigation water and the re-establishment of forest.

Boussingault also reported some pertinent hydrological observations that had been made on Ascension Island in the South Atlantic:

In the Island of Ascension there was an excellent spring situated at the foot of a mountain originally covered with wood; the spring became scanty and dried up after the trees which covered the mountain had been felled. The loss of the spring was rightly ascribed to the cutting down of the timber. The mountain was therefore planted anew. A few years afterwards the spring reappeared by degrees, and by and by followed with its former abundance. (Boussingault, 1845: 685)

Charles Lyell, in his *Principles of Geology*, one of the most influential of all scientific works, referred to the human impact and recognized that tree felling and drainage of lakes and marshes tended 'greatly to vary the state of the habitable surface'. Overall, however, he believed that the forces exerted by people were insignificant in comparison with those exerted by nature:

If all the nations of the earth should attempt to quarry away the lava which flowed from one eruption of the Icelandic volcanoes in 1783, and the two following years, and should attempt to consign it to the deepest abysses of the ocean they might toil for thousands of years before their task was accomplished. Yet the matter borne down by the Ganges and Burrampooter, in a single year, probably very much exceeds, in weight and volume, the mass of Icelandic lava produced by that great eruption. (Lyell, 1835: 197)

Lyell somewhat modified his views in later editions of the *Principles* (see e.g. Lyell, 1835), largely as a result of his experiences in the USA, where recent deforestation in Georgia and Alabama had produced numerous ravines of impressive size.

One of the most important physical geographers to show concern with our theme was Mary Somerville (1858) (who clearly appreciated the unexpected results that occurred as man 'dextrously avails himself of the powers of nature to subdue nature'):

Man's necessities and enjoyments have been the cause of great changes in the animal creation, and his destructive propensity of still greater. Animals are intended for our use, and field-sports are advantageous by encouraging a daring and active spirit in young men; but the utter destruction of some races in order to protect those destined for his pleasure, is too selfish, and cruelty is unpardonable: but the ignorant are often cruel. A farmer sees the rook pecking a little of his grain, or digging at the roots of the springing corn, and poisons all his neighbourhood. A few years after he is surprised to find his crop destroyed by grubs. The works of the Creator are nicely balanced, and man cannot infringe his Laws with impunity. (Somerville, 1858: 493)

This is in effect a statement of one of the basic laws of **ecology**: that everything is connected to everything else and that one cannot change just one thing in nature.

Considerable interest in conservation, climatic change and extinctions arose amongst European colonialists who witnessed some of the consequences of western-style economic development in tropical lands (Grove, 1997). However, the extent of human influence on the environment was not explored in detail and on the basis of sound data until George Perkins Marsh published *Man and Nature* (1864), in which he dealt with human influence on the woods, the waters and the sands. The following extract illustrates the breadth of his interests and the ramifying connections he identified between human actions and environmental changes:

Vast forests have disappeared from mountain spurs and ridges; the vegetable earth accumulated beneath the trees by the decay of leaves and fallen trunks, the soil of the alpine pastures which skirted and indented the woods, and the mould of the upland fields, are washed away; meadows, once fertilized by irrigation, are waste and unproductive, because the cisterns and reservoirs that supplied the ancient canals are broken, or the springs that fed them dried up; rivers famous in history and song have shrunk to humble brooklets; the willows that ornamented and protected the banks of lesser watercourses are gone, and the rivulets have ceased to exist as perennial currents, because the little water that finds its way into their old channels is evaporated by the droughts of summer, or absorbed by the parched earth, before it reaches the lowlands; the beds of the brooks have widened into broad expanses of pebbles and gravel, over which, though in the hot season passed dryshod, in winter sealike torrents thunder, the entrances of navigable streams are obstructed by sandbars, and harbours, once marts of an extensive commerce, are shoaled by the deposits of the rivers at whose mouths they lie; the elevation of the beds of estuaries, and the consequently diminished velocity of the streams which flow into them, have converted thousands of leagues of shallow sea and fertile lowland into unproductive and miasmatic morasses. (Marsh, 1965: 9)

More than a third of the book is concerned with 'the woods'; Marsh does not touch upon important themes like the modifications of mid-latitude grasslands, and he is much concerned with Western civilization. Nevertheless, employing an eloquent style and copious footnotes, Marsh, the versatile Vermonter, stands as a landmark in the study of environment (Thomas, 1956; Lowenthal, 2000).

Marsh, however, was not totally pessimistic about the future role of humankind or entirely unimpressed by positive human achievements (1965: 43–44):

New forests have been planted; inundations of flowing streams restrained by heavy walls of masonry and other constructions; torrents compelled to aid, by depositing the slime with which they are charged, in filling up lowlands, and raising the level of morasses which their own overflows had created; ground submerged by the encroachment of the ocean, or exposed to be covered by its tides, has been rescued

from its dominion by diking; swamps and even lakes have been drained, and their beds brought within the domain of agricultural industry; drifting coast dunes have been checked and made productive by plantation; sea and inland waters have been repeopled with fish, and even the sands of the Sahara have been fertilized by artesian fountains. These achievements are far more glorious than the proudest triumphs of war. . .

Elisée Reclus (1873), an anarchist and one of the most prominent French geographers of his generation, was an important influence in the USA and recognized that the 'action of man may embellish the earth, but it may also disfigure it; according to the manner and social condition of any nation, it contributes either to the degradation or glorification of nature' (p. 522). Reclus (1871) also displayed a concern with the relationship between forests, torrents and sedimentation.

In 1904 the German geographer Ernst Friedrich coined the term 'Raubwirtschaft', which can be translated as economic plunder, robber economy or, more simply, devastation. He believed that destructive exploitation of resources leads of necessity to foresight and to improvements, and that after an initial phase of ruthless exploitation and resulting deprivation human measures would, as in the old countries of Europe, result in conservation and improvement. This idea was opposed in the USA by Carl Sauer (1938) and Whitaker (1940), the latter pointing out that some soil erosion could well be irreversible (p. 157):

It is surely impossible for anyone who is familiar with the eroded loessial lands of northwestern Mississippi, or the burned and scarred rock hills of north central Ontario, to accept so complacently the damage to resources involved in the process of colonization, or to be so certain that resource depletion is but the forerunner of conservation.

Nonetheless Friedrich's concept of robber economy was adopted and modified by the great French geographer, Jean Brunhes, in his *Human Geography* (Brunhes, 1920). He recognized the interrelationships involved in anthropogenic environmental change (p. 332): 'Devastation always brings about, not a catastrophe, but a series of catastrophes, for in nature things are dependent one upon the other.' Moreover, Brunhes acknowledged that the 'essential facts' of human geography included 'Facts of Plant and Animal Conquest' and 'Facts of Destructive Exploitation'. At much the same time other significant studies were made of the same theme. Nathaniel Shaler of Harvard (*Man and the Earth*, 1912) was very much concerned with the destruction of mineral resources (a topic largely neglected by Marsh).

Sauer led an effective campaign against destructive exploitation, reintroduced Marsh to a wide public, recognized the ecological virtues of some so-called primitive peoples, concerned himself with the great theme of domestication, concentrated on the landscape changes that resulted from human action, and gave clear and far-sighted warnings about the need for conservation (Sauer, 1938: 494):

We have accustomed ourselves to think of ever expanding productive capacity, of ever fresh spaces of the world to be filled with people, of ever new discoveries of kinds and sources of raw materials, of continuous technical progress operating indefinitely to solve problems of supply. We have lived so long in what we have regarded as an expanding world, that we reject in our contemporary theories of economics and of population the realities which contradict such views. Yet our modern expansion has been affected in large measure at the cost of an actual and permanent impoverishment of the world.

The theme of the human impact on the environment has, however, been central to some historical geographers studying the evolution of the cultural landscape. The clearing of woodland (Darby, 1956; Williams, 1989; Williams, 2003), the domestication process (Sauer, 1952), the draining of marshlands (Williams, 1970), the introduction of alien plants and animals (McKnight, 1959), and the transformation of the landscape of North America (Whitney, 1994) are among some of the recurrent themes of a fine tradition of historical geography.

In 1956 some of these themes were explored in detail in a major symposium volume, *Man's Role in Changing the Face of the Earth* (Thomas, 1956). Kates et al. (1990: 4) write of it:

Man's role seems at least to have anticipated the ecological movement of the 1960s, although direct links between the two have not been demonstrated. Its dispassionate, academic approach was certainly foreign to the style of the movement . . . . Rather, *Man's Role* appears to have exerted a much more subtle, and perhaps more lasting, influence as a reflective, broad-ranging and multidimensional work.

In the last four decades many geographers have contributed to, and have been affected by, the phenomenon which is often called the environmental revolution or the ecological movement. The subject of the human impact on the environment, dealing as it does with such matters as environmental degradation, pollution and **desertification**, has close links with these developments, and is once again a theme in many textbooks and research monographs in geography (see Turner et al., 1990; Bell and Walker, 1992; Meyer, 1996; Mannion, 1997, 2002; Middleton, 2008).

Concerns about the human impact have become central to many other disciplines and to the public, particularly since the early 1970s, and a range of major developments in literature, legislation and international debate have taken place (Table 1.1). The concepts of global change or global environmental change have developed. These phrases are much used, but seldom rigorously defined. Wide use of the term '**global change**' seems to have emerged in the 1970s but in that period was used principally, though by no means invariably, to refer to changes in international social, economic and political systems (Price, 1989). It included such issues as proliferation of nuclear weapons, population growth, inflation and matters relating to international insecurity and decreases in the quality of life.

Since the early 1980s the concept of global change has taken on another meaning which is more geocentric in focus. This can be seen in the development of the International Geosphere–Biosphere Programme: A Study of Global Change (IBGP). This was established in 1986 by the International Council of Scientific Unions, 'to describe and understand the interactive physical, chemical and biological processes that regulate the total Earth system, the unique environment that it provides for life, the changes that are occurring in this system, and the manner in which they are influenced by human activities'. The term 'global environmental change' has in many senses come to be used synonymously with the more geocentric use of 'global change'.

In addition to the concept of global change, there is an increasing interest in the manner in which biogeochemical systems interact at a global scale and an increasing appreciation of the fact that the Earth is a single system. Earth System Science has emerged in response to this realization (see Steffen et al., 2004).

Recently, Crutzen and colleagues have introduced the term '**Anthropocene**' (e.g. Crutzen, 2002; Steffen et al., 2007; Rockström et al., 2009), as a name for a new

Table 1.1 Some environmental milestones

1864	George Perkins Marsh, *Man and Nature*
1892	John Muir founds Sierra Club in the USA
1935	Establishment of Soil Conservation Service in the USA
1956	Man's role in changing the face of the earth
1961	Establishment of World Wildlife Fund
1962	Rachel Carson's *Silent Spring*
1969	Friends of the Earth established
1971	Greenpeace established
1971	Ramsar Treaty on International Wetlands
1972	United Nations Environmental Programme (UNEP) established
1972	*Limits to Growth* published by Club of Rome
1973	Convention on International Trade in Endangered Species (CITES)
1974	F.S. Rowland and M. Molina warn about CFCs and ozone hole
1975	Worldwatch Institute established
1979	Convention on Long-Range Transboundary Air Pollution
1980	IUCN's World Conservation Strategy
1985	British Antarctic Survey finds ozone hole over Antarctic
1986	International Geosphere Biosphere Programmme (IGBP)
1986	Chernobyl nuclear disaster
1987	World Commission on Environment and Development (Brundtland Commission). *Our Common Future*
1987	Montreal Protocol on substances that deplete the ozone layer
1988	Intergovernmental Panel on Climate Change (IPCC)
1989	Global Environmental Facility
1992	Earth Summit in Rio and Agenda 21
1993	United Nations Commission on Sustainable Development
1994	United Nations Convention to Combat Desertification
1996	International Human Dimensions Programme on Global Environmental Change
1997	Kyoto Protocol on greenhouse gas emissions
2001	Amsterdam Declaration
2002	Johannesburg Earth Summit
2007	United Nations Bali Climate Change Conference
2010	United Nations Copenhagen Climate Change Conference
2010	Nagoya Biodiversity Summit and International Year of Biodiversity
2012	Rio + 20 United Nations Conference on Sustainable Development

epoch in Earth's history – an epoch when human activities have 'become so profound and pervasive that they rival, or exceed the great forces of Nature in influencing the functioning of the Earth System' (Steffen, 2010). In the last 300 years, they suggest, we have moved from the Holocene into the Anthropocene. They identify three stages in the Anthropocene. Stage 1, which lasted from *c.* 1800 to 1945, they call 'The Industrial Era'. Stage 2, which extends from 1945 to

c. 2015, they call 'The Great Acceleration', and Stage 3, which may perhaps now be starting, is a stage when people have become aware of the extent of the human impact and may thus start stewardship of the Earth System. Reviews of various aspects of the Anthropocene appear in a special issue of the *Philosophical Transactions of the Royal Society*, A, 339 (2011). However, it can be argued that the Anthropocene started more than three centuries ago, not least because of the possible effects of land-use changes on global carbon dioxide and methane budgets and thus on global climate (Ruddiman et al., 2011).

The huge increase in interest in the study of the human impact on the environment and of global change has not been without its great debates and controversies, and some have argued that environmentalists have overplayed their hand (see e.g. Lomborg's *The Skeptical Environmentalist*, 2001) and have exaggerated the amount of environmental harm that is being caused by human activities. In this book, I take a long-term perspective and seek to show the changes that mankind has caused to a wide spectrum of environmental phenomena. The current fixation with global warming should not blind us to the importance of other aspects of global change, including deforestation, desertification, salinization, pollution and the like (Slaymaker et al., 2009).

The development of human population and stages of cultural development

Some 6 or so million years ago, primitive human precursors or hominids appear in the fossil record (Wood, 2002). The earliest remains of a small, bipedal **hominin**, *Sahelanthropus tchadensis*, has been found in Chad (Brunet et al., 2002). However, the first recognizable human, *Homo habilis*, evolved about 2.5 million years ago, more or less at the time that the Pleistocene ice ages were developing in mid-latitudes. The oldest remains have been found either in sediments from the rift valleys of East Africa or in cave deposits in South Africa. Since that time the human population has spread over virtually the entire land surface of the planet (Oppenheimer, 2003) (Figure 1.1). *Homo* may have reached Asia by around 2 million years ago (Larick and Ciochon, 1996; Zhu et al., 2008) and Europe not much later (Moncel, 2010). In southern Europe

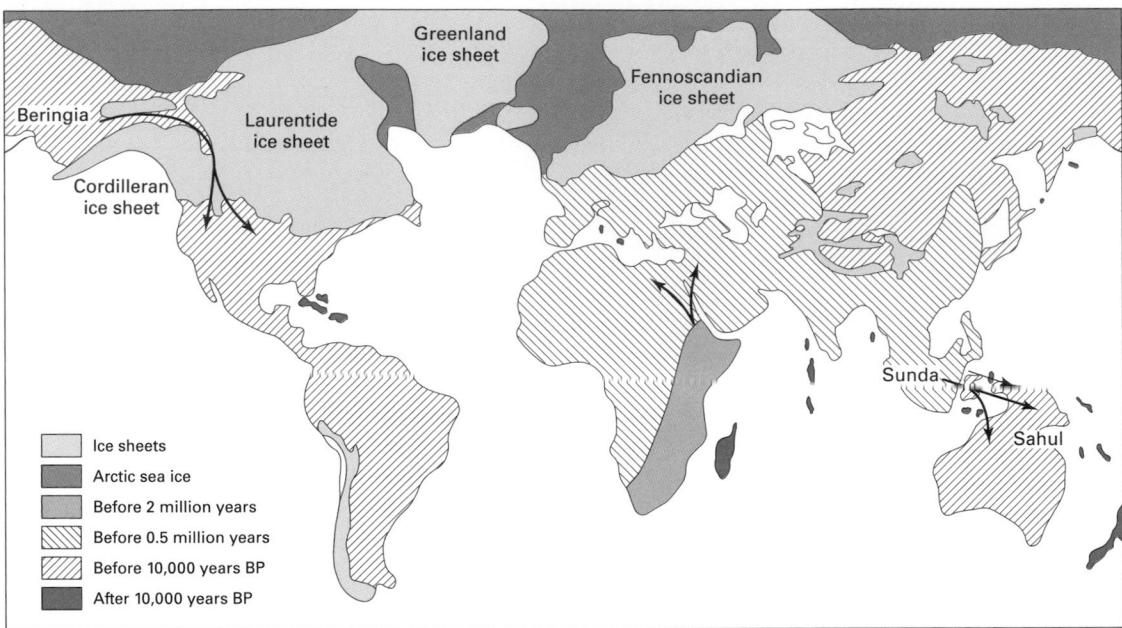

Figure 1.1 The human colonization of Ice-Age earth (after Roberts, 1989, figure 3.7).

there are stone tools in Italy associated with *Homo* that date back to 1.3–1.7 Ma (Arzarello et al., 2007) and also in Spain (Carbonell et al., 2008). In northwest Europe and Britain the earliest dates for human occupation are >0.78 Ma (Parfitt et al., 2010). Modern humans, *Homo sapiens*, appeared in Africa around 160,000 years ago (Stringer, 2003; White et al., 2003) and then spread 'out of Africa' to other parts of the world.

Table 1.2 gives data on recent views of the dates for the arrival of humans in selected areas. Some of these dates are controversial, and this is especially true of Australia, where they range from *c.* 40,000 years to as much as 150,000 years (Kirkpatrick, 1994: 28–30), but with a date of *c.* 50,000 years ago being widely accepted (Balme, 2011). There is also considerable uncertainty about the dates for humans arriving in the Americas (Goebel et al., 2008). Many authorities have argued that the first colonizers of North America, equipped with so-called Clovis spears, arrived via the Bering land bridge from Asia around 12,000 years ago. However, some earlier dates exist for the Yukon (Yesner, 2001) and for South America, and these perhaps imply an earlier phase of colonization (Dillehay, 2003). The settlement of Oceania occurred relatively late, with colonization of the western archipelagos of Micronesia and eastern Melanesia taking place at *c.* 3500–2800 BP, of

Table 1.2 Dates of human arrivals

Area	Source	Date (years BP)
Africa	Klein (1983)	2,700,000–2,900,000
China	Huang et al. (1995)	1,900,000
Georgian Republic	Gabunia and Vekua (1995)	1,600,000–1,800,000
Java	Swisher et al. (1994)	1,800,000
Europe	Moncel (2010)	c. 1,500,000
Britain	Parfitt et al. (2010)	c. 790,000
Japan	Ikawa-Smith (1982)	c. 50,000
New Guinea	Bulmer (1982)	c. 50,000
Australia	Bowler et al. (2003)	c. 40,000–50,000
North America	Goebel et al. (2008)	15,000
Peru	Keefer et al. (1998)	12,500–12,700
Ireland	Edwards (1985)	9000
Caribbean	Morgan and Woods (1986)	4500
Polynesia	Kirch (1982)	2000
Madagascar	Crowley (2010)	2500
New Zealand	Lowe (2008)	750

central and eastern Micronesia at 2200–2000 BP, and of eastern and southern Polynesia at 1100–700 BP (Anderson, 2009).

Estimates of population levels in the early stages of human development are difficult to make with any degree of certainty (Figure 1.2a). Before the agricultural 'revolution' some 10,000 years ago, human groups lived by hunting and gathering in parts of the world where this was possible. Population densities were low, and the optimum territory for a band of hunter–gatherers in the Middle Eastern woodland–parkland belt would have been 300–500 km², while in the drier regions it would have been 500–2000 km² (Bar-Yosef, 1998). At that time the world population *may* have been of the order of 5 million people (Ehrlich et al.,

1977: 182), and large areas would only recently have witnessed human migration. The Americas and Australia, for example, were probably virtually uninhabited until about 15,000 and 40,000 years ago, respectively. Human population estimates for the Holocene are diverse and controversial (Boyle et al., 2011).

However, the agricultural revolution probably enabled an expansion of the total human population to about 200–300 million by the time of Christ, and to 500 million by AD 1650. It is since that time, helped by the

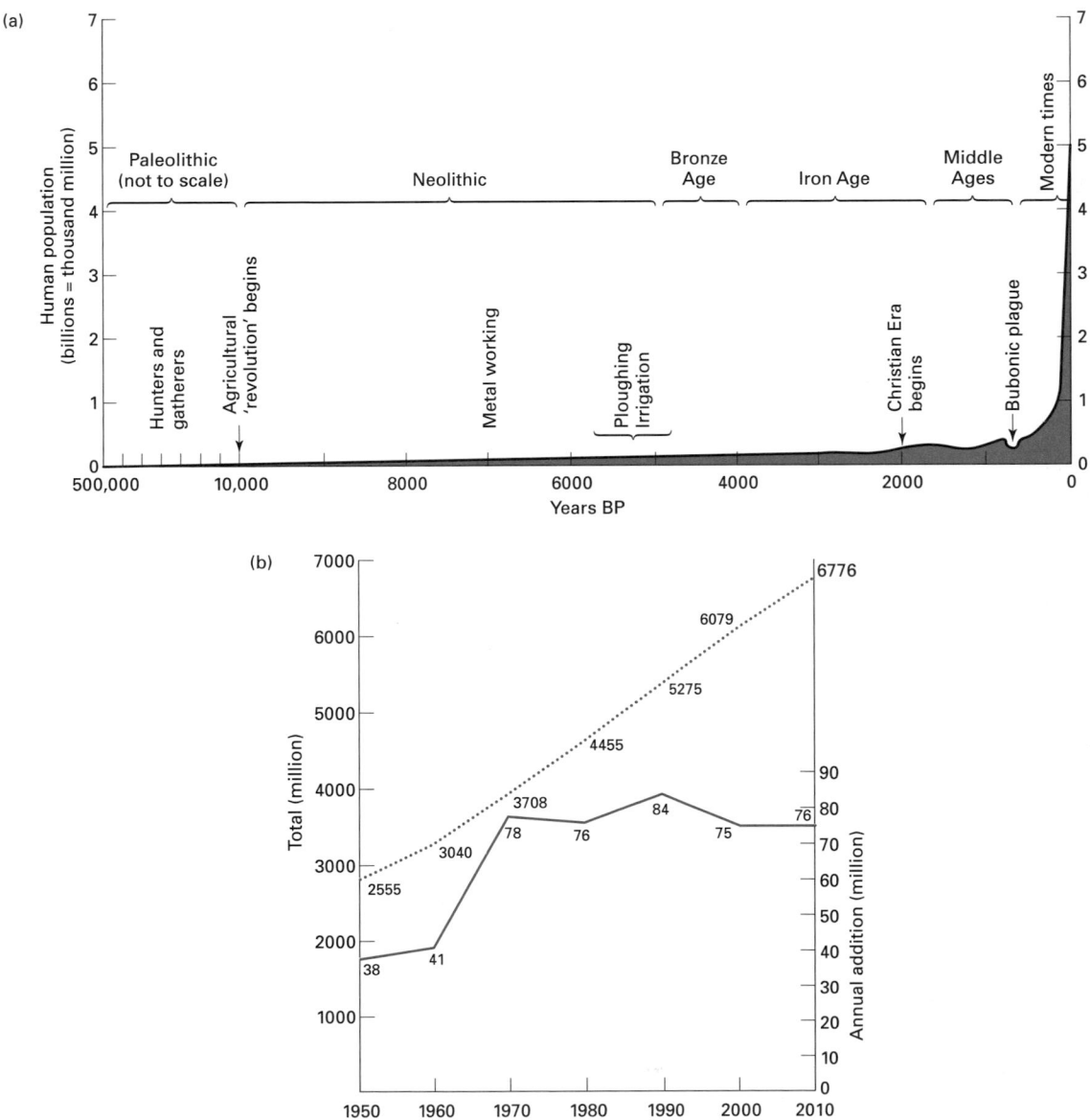

Figure 1.2 (a) The growth of human numbers for the past half million years (modified from Ehrlich et al., 1977, figure 5.2). (b) Annual growth of population since 1950.

medical and industrial revolutions and developments in agriculture and colonization of new lands, that human population has exploded, reaching about 1000 million by AD 1850, 2000 million by AD 1930 and 4000 million by AD 1974. The figure had reached over 6000 million by the end of the millennium and in 2012 exceeded 7000 million. Victory over malaria, smallpox, cholera and other diseases has been responsible for marked decreases in death rates throughout the non-industrial world, but death-rate control has not in general been matched by birth control. Thus the annual population growth rate in the late 1980s in South Asia was 2.64%, Africa, 2.66% and Latin America (where population increased sixfold between 1850 and 1950), 2.73%. In the period from 2005–2010 these rates had slowed down substantially, with Latin America down to 1.2% and Africa to 2.2%. The global annual growth in population has over the last decade been around 75–77 million people (Figure 1.2b).

The history of the human impact, however, has not been a simple process of increasing change in response to linear population growth over time, for in specific places at specific times there have been periods of reversal in population growth and ecological change as cultures have collapsed, wars occurred, disease struck and habitats were abandoned. Denevan (1992), for example, has pointed to the decline of native American populations in the new world following European entry into the Americas. This created what was 'probably the greatest demographic disaster ever'. The overall population of the western hemisphere in 1750 was perhaps less than a third of what it may have been in 1492, and the ecological consequences were legion.

Clearly, this growth of the human population of the Earth is in itself likely to be a highly important cause of the transformation of nature. Of no lesser importance, however, has been the growth and development of culture and technology. Sears (1957: 51) has put the power of humankind into the context of other species:

Man's unique power to manipulate things and accumulate experience presently enabled him to break through the barriers of temperature, aridity, space, seas and mountains that have always restricted other species to specific habitats within a limited range. With the cultural devices of fire, clothing, shelter, and tools he was able to do what no other organism could do without changing its original character. Cultural change was, for the first time, substituted for biological evolution as a means of adapting an organism to new habitats in a widening range that eventually came to include the whole earth.

The evolving impact of humans on the environment has often been expressed in terms of a simple equation:

$$I = P\,A\,T$$

where I is the amount of pressure or impact that humans apply on the environment, P is the number of people, A is the affluence (or the demand on resources per person), and T is a technological factor (the power that humans can exert through technological change). P, A and T have been seen by some as 'the three horsemen of the environmental apocalypse' (Meyer, 1996: 24). There may be considerable truth in the equation and in that sentiment; but as Meyer points out, the formula cannot be applied in too mechanistic a way. The 'cornucopia view', indeed, sees population not as the ultimate depleter of resources but as itself the ultimate resource capable of causing change for the better (see e.g. Simon, 1996). There are cases where strong population growth has appeared to lead to a reduction in environmental degradation (Tiffen et al., 1994). Likewise, there is debate about whether it is poverty or affluence that creates deterioration in the environment. On the other hand many poor countries have severe environmental problems and do not have the resources to clear them up, where as affluent countries do. Conversely it can be argued that affluent countries have plundered and fouled less fortunate countries and that it would be environmentally catastrophic if all countries used resources at the rate that the rich countries do. Similarly, it would be naïve to see all technologies as malign, or indeed benign. Technology can be a factor either of mitigation and improvement or of damage. Sometimes it is the problem (as when ozone depletion has been caused by a new technology – the use of **chlorofluorocarbons**) and sometimes it can be the solution (as when renewable energy sources replace the burning of polluting lignite in power stations).

In addition to the three factors of population, affluence and technology, environmental changes also depend on variations in the way in which different societies are organized and in their economic and social structures (see Meyer, 1996: 39–49 for an elaboration of this theme). For example, the way in which land is owned is a crucial issue.

The controls of environmental changes caused by the human impact are thus complex and in many cases contentious, but all the factors discussed play a role of some sort, at some places and at some times.

We now turn to a consideration of the major cultural and technical developments that have taken place during the past 2–3 million years. Takács-Sánta (2004) argued that there have been six major transformations in the history of the human transformation of the environment: the use of fire, the development of language, the birth of agriculture, the development of cities and states, European conquests since the fifteenth century AD and the Technological-Scientific Revolution, with the emergence of fossil fuels as primary energy sources. In this book, three main phases will form the basis of the analysis: the phase of hunting and gathering; the phase of plant cultivation, animal keeping and metal working; and the phase of modern urban and industrial society. These developments are treated in much greater depth by Simmons (1996) and Ponting (2007).

Hunting and gathering

The oldest records of human activity and technology are pebble tools (crude stone tools which consist of a pebble with one end chipped into a rough cutting edge). These have been found with human bone remains in various parts of Africa (Gosden, 2003). At Dikika in Ethiopia there is evidence for stone-tool-assisted consumption of meat at 3.42–3.24 Ma (McPherron et al., 2010). At Lake Turkana in northern Kenya and the Omo Valley in southern Ethiopia, a tool-bearing bed of volcanic material called tuff has been dated by isotopic means at about 2.6 million years old, another from Gona in the north-east of Ethiopia at about 2.5 million years old (Semaw et al., 1997), while another bed at the Olduvai Gorge in Tanzania (Figure 1.3) has been dated by similar means at 1.75 million years. Indeed, these very early tools are generally termed 'Oldowan'.

As the Stone Age progressed the tools became more sophisticated, varied and effective, and Figure 1.4 shows some beautiful Palaeolithic hand axes from Olorgesailie in East Africa. Greater exploitation of plant and animal resources became feasible. Stone may not, however, have been the only material used. Sticks and animal bones, the preservation of which is less

Figure 1.3 A view of the Olduvai Gorge in Tanzania – one of the great sites for the investigation of early man. (See Plate 1)

Figure 1.4 A cluster of Palaeolithic hand axes from Olorgesailie in East Africa. Courtesy of Jean-Marc Jancovici, www.manicore.com, 2012. (See Plate 2)

likely than stone, are among the first objects that may have been used as implements, although the sophisticated utilization of antler and bone as materials for weapons and implements appears to have developed surprisingly late in prehistory. There is certainly a great deal of evidence for the use of wood throughout the Palaeolithic Age, for ladders, fire, pigment (charcoal), the drying of wood and digging sticks. Tyldesley and Bahn (1983: 59) went so far as to suggest that 'The Palaeolithic might more accurately be termed the "Palaeoxylic" or "Old Wood Age".'

The building of shelters and the use of clothing became a permanent feature of human life as the Palaeolithic period progressed and permitted habitation in areas where the climate was otherwise not congenial.

European sites from the Mousterian of the Middle Palaeolithic have revealed the presence of purposefully made dwellings as well as caves, and by the Upper Palaeolithic more complex shelters were in use, allowing people to live even in the tundra lands of Central Europe and Russia.

Another feature of early society which seems to have distinguished humans from the surviving non-human primates was their seemingly omnivorous diet. In the Palaeolithic Age humans secured a wide range of animal meats, whereas the great apes, through not averse to an occasional taste of animal food, are predominately vegetarian. One consequence of enlarging the range of their diet was that, in the long run, humans were able to explore a much wider range of environment (Clark, 1977: 19). Another major difference that set humankind above the beasts was the development of communicative skills such as speech. Until hominids had developed words as symbols, the possibility of transmitting, and so accumulating, culture hardly existed. Animals can express and communicate emotions, never designate or describe objects.

At an early stage humans discovered the use of fire (Figure 1.5). This, as we shall see (Chapter 2), is a major agent by which humans have influenced their environment. The date at which fire was first deliberately employed is a matter of ongoing controversy (Bogucki, 1999: 51–54; Caldararo, 2002). It may have been employed very early in South Africa, where Beaumont (2011) and Berna et al. (2012) found some traces of repeated burning events from Acheulean cave sediments dating back to more than a million years ago, and from East Africa, where Gowlett et al. (1981) claimed to find evidence for deliberate manipulation of fire from over 1.4 million years ago. However, it is not until after around 400,000 years ago that evidence for the association between human and fire becomes compelling. Nonetheless, as Pyne (1982: 3) has written:

It is among man's oldest tools, the first product of the natural world he learned to domesticate. Unlike floods, hurricanes or windstorms, fire can be initiated by man; it can be combated hand to hand, dissipated, buried, or 'herded' in ways unthinkable for floods or tornadoes.

He goes on to stress the implications that fire had for subsequent human cultural evolution (p. 4):

It was fire as much as social organisation and stone tools that enabled early big game hunters to encircle the globe and to begin the extermination of selected species. It was fire that assisted hunting and gathering societies to harvest insects, small game and edible plants; that encouraged the spread of agriculture outside the flood plains by allowing for rapid landclearing, ready fertilization, the selection of food grains, the primitive herding of grazing animals that led to domestication, and the expansion of pasture and grasslands against climate gradients; and that, housed in machinery, powered the prime movers of the industrial revolution.

Overall, compared with later stages of cultural development, early hunters and gatherers had neither the numbers nor the technological skills to have a very substantial effect on the environment. Besides the effects of fire, early cultures may have caused some diffusion of seeds and nuts, and through hunting activities (see Chapter 3) may have had some dramatic effects on animal populations, causing the extinction of many great mammals (the so-called Pleistocene overkill). Locally some **eutrophication** may have occurred, and around some archaeological sites phosphate and nitrate levels may be sufficiently raised to make them an indicator of habitation to archaeologists today (Holliday, 2004). Equally, although we often assume that early humans were active and effective hunters, they may well have been dedicated scavengers of carcasses of animals which had either died natural deaths or been killed by carnivores like lion.

Figure 1.5 Fire was one of the first and most powerful tools of environmental transformation employed by humans. The high grasslands of southern Africa may owe much of their character to regular burning, as shown here in Swaziland. (See Plate 3)

It is salutary to remember, however, to remember just how significant this stage of our human cultural evolution has been. As Lee and DeVore (1968: 3) wrote:

Of the estimated 80,000,000,000 men who have ever lived out a life span on earth, over 90 per cent have lived as hunters and gatherers, about 6 per cent have lived by agriculture and the remaining few per cent have lived in industrial societies. To date, the hunting way of life has been the most successful and persistent adaptation man has ever achieved.

Figure 1.6 indicates the very low population densities of hunter/gatherer/scavenger groups in comparison with those that were possible after the development of pastoralism and agriculture.

Humans as cultivators, keepers and metal workers

Humans have been foragers rather than farmers for around 95% of their history, but during the end of the Pleistocene major changes were afoot. It is possible to identify some key stages of economic development that have taken place since the end of the Pleistocene (Table 1.3). For example around 14,000–15,000 years ago, in the Middle Eastern region, now consisting of Jordan, Syria, Israel, Palestine and Lebanon, the hunting folk – the Natufians – in addition to their hunting, began to build permanent houses of stone

and wood, they buried their dead in and around them with elaborate rituals, gathered in communities of up to several hundred people, ground up wild cereals with pestles and mortars, and made tools and art objects from animal bones (Bar-Yosef, 1998; Barker,

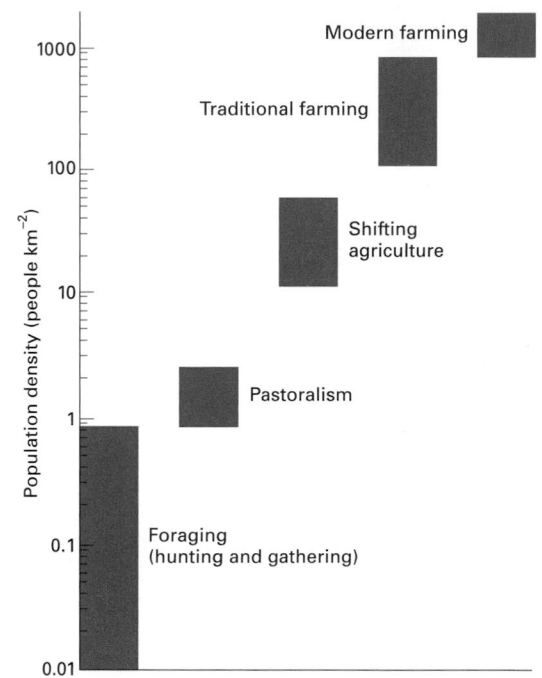

Figure 1.6 Comparison of carrying capacities of foraging, pastoralist and agricultural societies.

Table 1.3 Five stages of economic development. Source: Adapted from Simmons (1993: 2–3)

Economic stage	Dates and characteristics
Hunting–gathering and early agriculture	Domestication first fully established in south-western Asia around 7500 BC; hunter–gatherers persisted in diminishing numbers until today. Hunter–gatherers generally manipulate the environment less than later cultures and adapt closely to environmental conditions.
Riverine civilizations	Great irrigation-based economies lasting from c. 4000 BC to first century AD in places such as the Nile Valley and Mesopotamia. Technology developed to attempt to free civilizations from some of the constraints of a dry season.
Agricultural empires	From 500 BC to around 1800 AD a number of city-dominated empires existed, often affecting large areas of the globe. Technology (e.g. terracing and selective breeding) developed to help overcome environmental barriers to increased production.
The Atlantic-industrial era	From c.1800 AD to today a belt of cities from Chicago to Beirut, and around the Asian shores to Tokyo, form an economic core area based primarily on fossil fuel use. Societies have increasingly divorced themselves from the natural environment, through air conditioning for example. These societies have also had major impacts on the environment.
The Pacific-global era	Since the 1960s there has been a shifting emphasis to the Pacific Basin as the primary focus of the global economy, accompanied by globalization of communications and the growth of multinational corporations.

2006; Balter, 2010). Then, round the beginning of the Holocene, about 10,000 years ago, the Natufians and other groups in various other parts of the world began to domesticate rather than to gather food plants and to keep, rather than just hunt, animals. This phase of human cultural development is well reviewed in Roberts (1998). By taking up farming and domesticating food plants, they reduced enormously the space required for sustaining each individual by a factor of the order of 500 at least (Sears, 1957: 54) and population densities could thus become progressively greater (Figure 1.6). As a consequence, we see shortly thereafter, notably in the Middle East, the establishment of the first major settlements – towns. So long as man had

> to subsist on the game animals, birds and fish he could catch and trap, the insects and eggs he could collect and the foliage, roots, fruits and seeds he could gather, he was limited in the kind of social life he could develop; as a rule he could only live in small groups, which gave small scope for specialization and the subdivision of labour, and in the course of a year he would have to move over extensive tracts of country, shifting his habitation so that he could tap the natural resources of successive areas. It is hardly to be wondered at that among communities whose energies were almost entirely absorbed by the mere business of keeping alive, technology remained at a low ebb. (Clark, 1962: 76)

It is now recognized that some hunters and gathers had considerable leisure and did not need to develop agriculture to avoid drudgery and starvation. Moreover, some believe that the mobile hunter–gatherer lifestyle was far more attractive than a sedentary one, which creates problems of refuse disposal, hygiene and social conflict (Mithen, 2007). However, there is no doubt that through the controlled breeding of animals and plants humans were able to develop a more reliable and readily expandable source of food and thereby create a solid and secure basis for cultural advance, an advance which included civilization and the 'urban revolution' of Childe (1936) and others. Indeed, Isaac (1970) termed **domestication** 'the single most important intervention man had made in his environment'; and Harris (1996) termed the transition from foraging to farming as 'the most fateful change in the human career'. Diamond (2002) termed it 'the most momentous change in Holocene human history', while Mithen (2007: 705) has said that 'The origins of farming is the defining event of human history – the one turning point that has resulted in modern humans having a

quite different type of lifestyle and cognition to all other animals and past types of humans'.

A distinction can be drawn between cultivation and domestication. Whereas cultivation involves deliberate sowing or other management, and entails plants which do not necessarily differ genetically from wild populations of the same species, domestication results in genetic change brought about through conscious or unconscious human selection. This creates plants that differ morphologically from their wild relatives and which may be dependent on humans for their survival. Domesticated plants are thus necessarily cultivated plants, but cultivated plants may or may not be domesticated. For example, the first plantations of *Hevea* rubber and quinine in the Far East were established from seed which had been collected from the wild in South America. Thus at this stage in their history, these crops were cultivated but not yet domesticated.

The origin of agriculture remains controversial (Harris, 1996; Scarre, 2005; Barker, 2006). Some early workers saw agriculture as a divine gift to humankind, while others thought that animals were domesticated for religious reasons. They argued that it would have been improbable that humans could have predicted the usefulness of domestic cattle before they were actually domesticated. Wild cattle are large, fierce beasts, and no one could have foreseen their utility for labour or milk until they were tamed – tamed perhaps for ritual sacrifice in connection with lunar goddess cults (the great curved horns being the reason for the association). Another major theory – the demographic hypothesis – was that domestication was produced by crowding, possibly brought on by a combination of climatic deterioration (alleged post-Glacial progressive desiccation) and population growth. Gordon Childe's 'oasis propinquity hypothesis' held that increasing desiccation brought wild animals and plants into ever closer relationships, from which symbiosis and ultimately domestication emerged (Renfrew, 2006). Such pressure may have forced communities to intensify their methods of food production. Current palaeoclimatological research tends not to support this interpretation, but that is not to say that other severe climatic changes could not have played a role (Sherratt, 1997).

Sauer (1952) believed that plant domestication was initiated in Southeast Asia by fishing folk, who found that lacustrine and riverine resources would underwrite a stable economy and a sedentary or semi-sedentary lifestyle. He surmises that the initial domesticates would be multi-purpose plants set around small fishing villages to provide such items as starch foods, substances for toughening nets and lines and making them water-resistant, and drugs and poisons. He suggested that 'food production was one and perhaps not the most important reason for bringing plants under cultivation.'

Yet another model was advanced by Jacobs (1969) which turned certain more traditional models upside down. Instead of following the classic pattern whereby farming leads to village which leads to town which leads to civilization, she proposed that one could be a hunter–gatherer and live in a town or city, and that agriculture originated in and around such cities rather than in the countryside. Her argument suggests that even in primitive hunter–gatherer societies particularly valuable commodities such as fine stones, pigments and shells could create and sustain a trading centre which would possibly become large and stable. Food would be exchanged for goods, but natural produce brought any distance would have to be durable, so meat would be transported on the hoof for example, but not all the animals would be consumed immediately; some would be herded together and might breed. This might be the start of domestication. Indeed, settlements may have been a cause of agriculture rather than a consequence (Watkins, 2010).

Another hypothesis – the feasting hypothesis – is based on the idea that in many societies, those wishing to achieve rank and status do so by throwing feasts. The adoption of cultivation and the husbanding of domestic animals made it possible for ambitious individuals to produce increasing amounts of food which would give them an advantage in social competition (Hayden, 1995). It is also possible that as humans developed art and equipment to process plants, they developed new ideas and saw cultivation and domestication as a means of social prestige (Mithen, 2007). In other words, the origins of agriculture 10,000 years ago may perhaps be explained by a fundamental change in the way in which the human mind conceived of nature.

The process of domestication and cultivation was also once considered a revolutionary system of land procurement that had evolved in only one or two hearths and diffused over the face of the earth, replac-

Table 1.4 Dates which indicate that there may have been some synchroneity of plant domestication in different centres

Centre	Dates (000 years BP)	Plant
Mesoamerica	10.7–9.8	Squash-pumpkin
	9.0	Bottle gourd
		Maize
Near East	11.0–9.3	Fig tree
		Emmer wheat
		Two-rowed barley
		Einkorn wheat
		Pea
		Lentil
		Flax
Far East	11.0–7.0	Broomcorn millet
		Rice
		Gourd
		Water chestnut
Andes	9.4–8.0	Chile pepper
		Common bean
		Ullucu
		White potato
		Squash and Gourd

ing the older hunter–gatherer systems by stimulus diffusion. It was felt that the deliberate rearing of plants and animals for food was a discovery or invention so radical and complex that it could have developed only once (or possibly twice) – the so-called Eureka model. In reality, however, the domestication of plants occurred at approximately the same time in widely separated areas (Table 1.4). As Barker (2006: 412) has written:

... probably many more societies than commonly envisaged, in all parts of the world, started to engage in different kinds of animal and/or plant husbandry at or soon after the transition to the Holocene – in South-West Asia, South Asia, East Asia, Island South-East Asia, several parts of the Americas, and North Africa (and who knows when in tropical West Africa?). Independent of one another (at the regional scale, that is) and in many different ways, very many societies arrived at solutions to living in the transformed landscapes they were encountering which we can recognize as the beginnings of systematic husbandry.

So, the balance of botanical and archaeological evidence seems to suggest that humans started experimenting with domestication and cultivation of different plants at different times in different parts of the world (Figure 1.7) (Mithen, 2007).

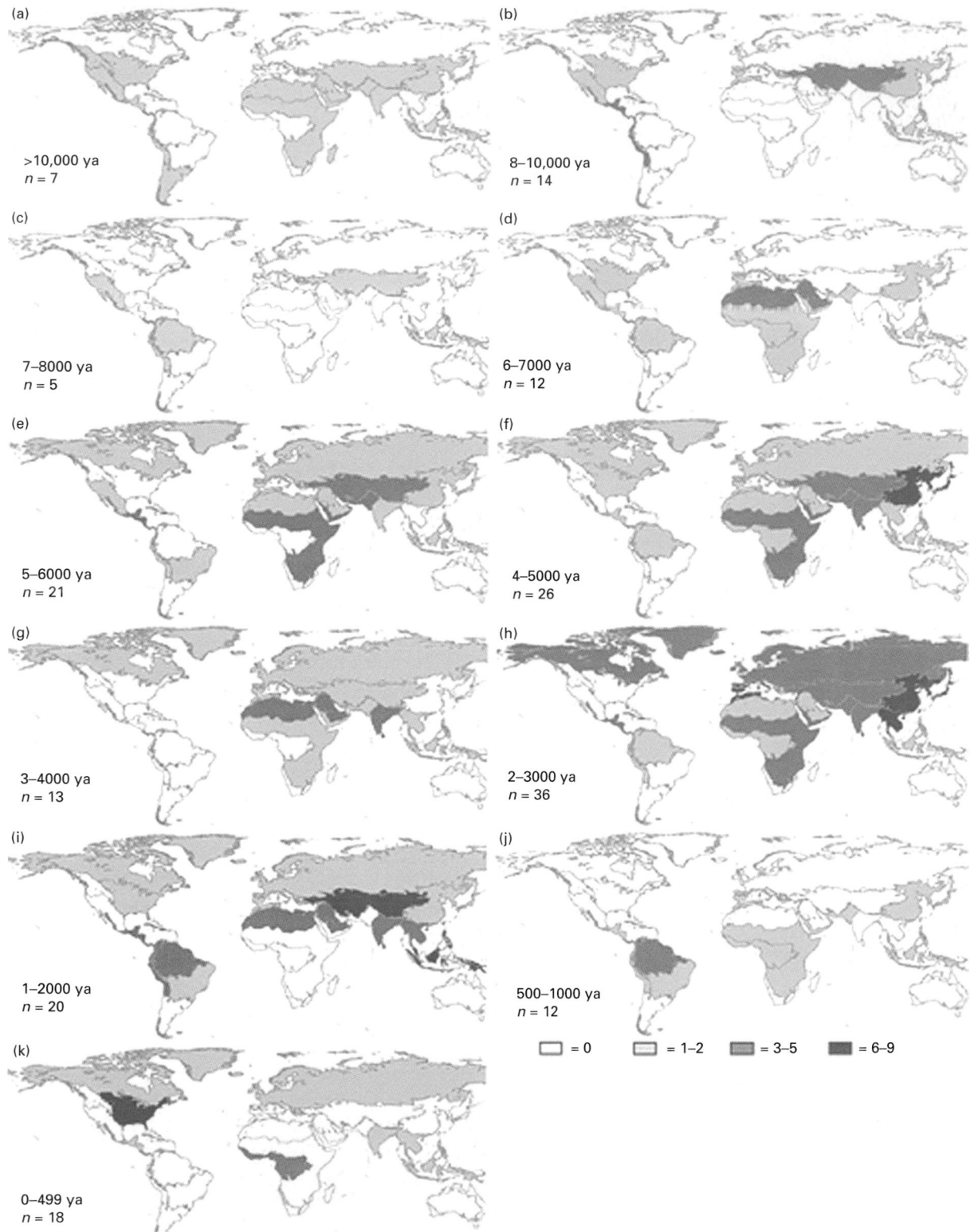

Figure 1.7 The geographical locations of new domesticated food crops worldwide grouped into 2000-year time intervals from >10,000 years ago (ya) to the present. Higher numbers of domestication events are represented by darker shading (from Meyer et al., 2012, figure 4). Reproduced with permission.

The dates for the first evidence of domestication tend to have become earlier in recent years, and the first steps towards domestication of crops in the eastern Mediterranean lands and the so-called Hilly Flanks of the Middle East may have taken place around 12,000 years ago (Zeder, 2008; Özkan et al., 2011), and the domestication of millet in China has been extended back to 10,000 years ago (Lu et al., 2009). In Mesoamerica maize, beans and squashes may have been domesticated as early as 10,000–9000 years ago (Zizumbo-Villarreal and Colunga-GarcíaMarín (2010).

The locations and dates for domestication of some important domestic animals are shown in Figure 1.8. Note that the domestication of sheep, goats, pigs and cattle took place in the Near East and neighbouring areas round about 10,500–9000 years ago, with the cat following shortly thereafter (Driscoll et al., 2009). However, the animal that was first domesticated, probably because of its utility for hunting, was the dog, *Canis familiaris*, which was probably domesticated by *c.* 16,000 BP (Galibert et al., 2011), possibly in southern China (Pang et al., 2009). The horse was first domesticated in the steppes of Kazakhstan at *c.* 5500 BP (Outram et al., 2009), while the chicken was domesticated at several centres in South Asia and Southeast Asia (Tixier-Boichard et al., 2011), including the Indus Valley and China (Liu et al., 2006). The donkey was

domesticated from the wild ass in north-east Africa about 5000 years ago (Kimura et al., 2011), and the water buffalo was domesticated during the Neolithic in various regions in south and eastern Asia (Yang et al., 2008). The pig may have been independently domesticated not only in the Middle East but also in China (Amills, 2011).

However, it was the Near East, and in particular the Hilly Flanks, that was especially important for both plant and animal domestication (Bar-Yosef, 1998; Lev-Yadun et al., 2000; Abbo et al., 2010; Zohary et al., 2012), and wild progenitors were numerous in the area, including those of wheat, barley, flax, lentils, peas, sheep, goats, cows and pigs – a list that includes what are still the most valuable crops and livestock of the modern world (Diamond, 1997, 2002; Morris, 2010). The grape was also domesticated in the Middle East at *c.* 6000–8000 years ago (Myles et al., 2011), and the date palm was domesticated in southern Mesopotamia around 7000 years ago (Tengberg, 2012). Nonetheless, recent syntheses of sites of plant domestication on a global basis (Meyer et al., 2012) have shown the importance of such centres as China, Southeast Asia and the Americas (Dillehay et al., 2007; Pickersgill, 2007). The earliest domestication of rice occurred in China (though the exact locations are a matter of debate) after *c.* 9000 years ago (Liu et al., 2007; Fuller et al., 2009) while

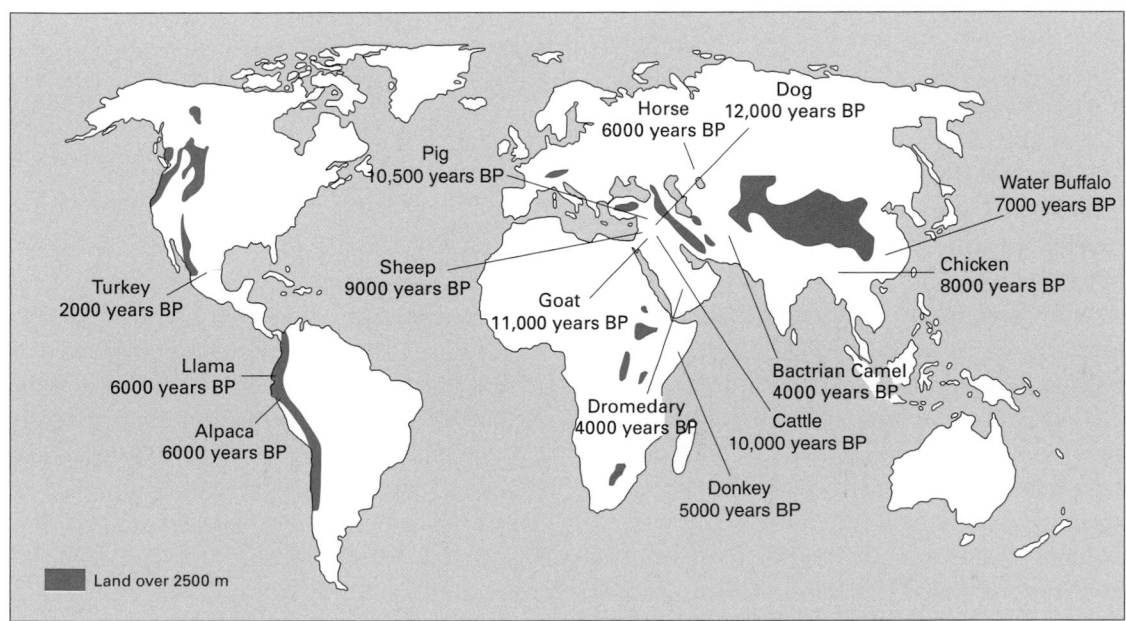

Figure 1.8 The places of origin, with approximate dates, for the most common domesticated animals.

Figure 1.9 Irrigation using animal power, as here in Rajasthan, India, is an example of the use of domesticated stock to change the environment. (See Plate 4)

Figure 1.10 The development of ploughs provided humans with the ability to transform soils. This simple type is in Pakistan. (See Plate 5)

evidence for cotton domestication occurs in Baluchistan (Pakistan) at *c.* 8000 years ago (Moulherat et al., 2002).

One highly important development in agriculture, because of its rapid and early effects on environment, was irrigation (Figure 1.9) and the adoption of riverine agriculture. This came rather later than domestication. Amongst the earliest evidence of artificial irrigation is the mace-head of the Egyptian Scorpion King which shows one of the last pre-dynastic kings ceremonially cutting an irrigation ditch around 5050 years ago (Butzer, 1976), although it is possible that irrigation at Sumer in Iraq started even earlier.

A major difference has existed in the development of agriculture in the Old and New Worlds; in the latter there were few counterparts to the range of domesticated animals which were an integral part of the former (Sherratt, 1981). A further critical difference was that in the Old World the secondary applications of domesticated animals were explored. This has been termed 'The Secondary Products Revolution' (Greenfield, 2010). The plough was particularly important in this process (Figure 1.10) – the first application of animal power to the mechanization of agriculture. Closely connected to this was the use of the cart, which both permitted more intensive farming and enabled the transportation of its products. Furthermore, the development of textiles from animal fibres afforded, for the first time, a commodity which could be produced for exchange in areas where arable farming was not the optimal form of land

use. Finally, the use of animal milk provided a means whereby large herds could use marginal or exhausted land, encouraging the development of the pastoral sector with transhumance or nomadism.

This Secondary Products Revolution therefore had radical effects, and the change took place over quite a short period. The plough was invented some 5000 years ago and was used in Mesopotamia, Assyria and Egypt. The wheeled cart was first produced in the Near East in the fourth millennium BC and rapidly spread from there to both Europe and India (Figure 1.11) during the course of the third millennium. The development of other means of transport preceded the wheel. Sledge-runners found in Scandinavian bogs have been dated to the Mesolithic period (Cole, 1970: 42).

The origins of seafaring are difficult to establish. On the one hand, by *c.* 800,000 years ago, people seem to have crossed a deep water strait between Bali and Lombok in Southeast Asia, and humans crossed a 100-km wide strait to reach Australia by *c.* 50,000 years ago (Broodbank, 2006; Erlandson, 2010; Balme, 2011). However, in the Mediterranean world, there is very little firm evidence of any maritime activity until the late Pleistocene (Knapp, 2010), and this is also the time when humans may have navigated down the coast of the Americas (Erlandson et al., 2011). Neolithic people undertook a 130-km open sea voyage to reach Taiwan by *c.* 5000 years ago (Rolett et al., 2011). Indeed, by the Neolithic era, humans had developed boats, floats and rafts that were able to cross to Mediterranean islands and sail the Irish Sea. Dugout canoes could hardly have been common before polished stone axes and adzes came into general use during Neolithic times, although some paddle and canoe remains are recorded from Mesolithic sites in northern Europe. The middens of the hunter–fishers of the Danish Neolithic contain bones of deep-sea fish such as cod, showing that these people certainly had seaworthy craft with which to exploit ocean resources.

Both the domestication of animals and the cultivation of plants have been among the most significant causes of the human impact (see Mannion, 1995). Pastoralists have had many major effects – for example, on soil erosion – nomadic pastoralists are probably more conscious than agriculturalists that they share the earth with other living things. Agriculturalists, on the other hand, deliberately transform nature in a sense which nomadic pastoralists do not. Their main role has been to simplify the world's ecosystems. Thus in the prairies of North America, by ploughing and seeding the grasslands, farmers have eliminated a hundred species of native prairie herbs and grasses, which they replace with pure stands of wheat, corn or alfalfa. This simplification may reduce stability in the ecosystem (but see Chapter 13, section on 'The susceptibility to change'). Indeed, on a world basis (see Harlan, 1976), such simplification is evident. Whereas people once enjoyed a highly varied diet, and have used for food several thousand species of plants and several hundred species of animals, with domestication their sources are greatly reduced. For example, today, four crops (wheat, rice, maize and potatoes) at the head of the list of food supplies contribute more tonnage to the world total than the next 26 crops combined. Simmonds (1976) provides an excellent account of the history of most of the major crops produced by human society.

The spread of agriculture has transformed land cover at a global scale. As Table 1.5 shows, there have been great changes in the area covered by particular biomes

Figure 1.11 In the Secondary Products Revolution, cattle, such as these in Haryana, India, were used as beasts of burden, for pulling ploughs, and as a source of products such as leather. (See Plate 6)

Table 1.5 Estimated changes in the areas of the major land cover types between pre-agricultural times and the present[a]. Source: From J.T. Matthews (personal communication), in Meyer and Turner (1994). With permission from Cambridge University Press

Land cover type	Pre-agricultural area	Present area	Percent change
Total forest	46.8	39.3	−16.0
Tropical forest	12.8	12.3	−3.9
Other forest	34.0	27.0	−20.6
Woodland	9.7	7.9	−18.6
Shrubland	16.2	14.8	−8.6
Grassland	34.0	27.4	−19.4
Tundra	7.4	7.4	0.0
Desert	15.9	15.6	−1.9
Cultivation	0.0	17.6	+1760.0

[a]Figures are given in millions of square kilometres.

since pre-agricultural times. Even in the last 300 years the areas of cropland and pasture have increased by around five- to sixfold (Goldewijk, 2001). It is possible (Ruddiman, 2003) that Holocene deforestation and land cover change modified global climates by releasing carbon dioxide into the atmosphere.

One further development in human cultural and technological life which was to increase human power was the mining of ores and the smelting of metals (Roberts et al., 2009). Neolithic cultures used native copper from the eighth millennium BC onwards, but evidence for its smelting occurs at Catal Hüyük in Turkey and in Jordan from the sixth millennium BC (Craddock, 2000; Grattan et al., 2007) and in Serbia from the fifth millennium BC (Radivojević et al., 2010). The spread of metal working into other areas was rapid particularly in the second half of the fifth millennium (Muhly, 1997) and by 2500 BC, bronze products were in use from Britain in the West to northern China in the East. The smelting of iron ores may date back to the late third millennium BC.

In recent decades fossil-fuelled machinery has allowed mining activity to expand to such a degree that in terms of the amount of material moved its effects are reputed to rival the natural processes of erosion. Taking overburden into account, the total amount of material moved by the mining industry globally is probably at least 28 billion tonnes – about 1.7 times the estimated amount of sediment carried each year by the world's rivers (Young, 1992). The environmental impacts of mineral extraction are diverse but extensive and relate not only to the process of excavation and removal, but also to the processes of mineral concentration, smelting and refining (Table 1.6). Metal working required enormous amounts of wood and so led to deforestation.

Modern industrial and urban civilizations

Although modest communal settlements may have occurred before the adoption of domestication (Watkins, 2010), it was within a few thousand years of the adoption of cereal agriculture that people began to gather into ever larger settlements (cities) and into more institutionalized social formations (states). After around 6000 years ago, cities developed in the basin of the Tigris and Euphrates, and more followed by c. 5000

Table 1.6 Environmental impacts of mineral extraction. Source: Young (1992, table 5). Reproduced with permission

Activity	Potential impacts
Excavation and ore removal	• Destruction of plant and animal habitat, human settlements and other features (surface mining) • Land subsidence (underground mining) • Increased erosion: silting of lakes and streams • Waste generation (overburden) • Acid drainage (if ore or overburden contain sulfur compounds) and metal contamination of lakes, streams and groundwater
Ore concentration	• Waste generation (tailings) • Organic chemical contamination (tailings often contain residues of chemicals used in concentrators) • Acid drainage (if ore contains sulfur compounds) and metal contamination of lakes, streams and groundwater
Smelting/refining	• Air pollution (substances emitted can include sulfur dioxide, arsenic, lead, cadmium and other toxic substances • Waste generation (slag) • Impacts of producing energy (most of the energy used in extracting minerals goes into smelting and refining)

years ago in the coastal Mediterranean, the Nile valley, the Indus plain and coastal Peru. In due course cities had evolved which had considerable human populations. It has been estimated that Nineveh may have had a population of 700,000, that Augustan Rome may have had a population of around 1 million and that Carthage (Figure 1.12), at its fall in 146 BC, had 700,000 (Thirgood, 1981). Such cities would have already exercised a considerable influence on their environs, but this influence was never as extensive as that of the last few centuries; for the modern era, especially since the late seventeenth century, has witnessed the transformation of, or revolution in, culture and technology – the development of major industries. This, like domestication, has reduced the space required to sustain each individual and has increased the intensity with which resources are utilized. Modern science and modern medicine have compounded these effects, leading to accelerating population increase even in non-industrial societies. Urbanization has gone on apace (Figure 1.13), and it is now recognized that large

Figure 1.12 Carthage, in northern Tunisia, was one of the great cities of the ancient world. (See Plate 7)

Figure 1.13 Urbanization (and, in particular, the growth of large conurbations such as Kuwait) is an increasingly important phenomenon. Urbanization causes and accelerates a whole suite of environmental problems. (See Plate 8)

cities have their own environmental problems (Cooke et al., 1982) and environmental effects (Douglas, 1983). As Table 1.7 shows, the world now has some enormous urban agglomerations. These, in turn, have large **ecological footprints**. They suck in resources and materials and export vast amounts of waste.

It is now also possible to use such terms as 'urban hydrology', 'urban ecology' and 'urban geomorphology' (Gurnell et al., 2007). Cities have their own range

Table 1.7 World's urban agglomerations of 10 million or more inhabitants 2010

City and country	Population (million)
Tokyo, Japan	36.7
Delhi, India	22.2
Sao Paulo, Brazil	20.3
Mumbai, India	20.0
Mexico City, Mexico	19.5
New York-Newark, USA	19.4
Shanghai, China	16.6
Kolkata, India	15.6
Dhaka, Bangladesh	14.7
Karachi, Pakistan	13.1
Buenos Aires, Argentina	13.1
Los Angeles, Long Beach, Santa Ana, USA	12.8
Beijing, China	12.4
Rio de Janeiro, Brazil	12.0
Manila, Philippines	11.6
Osaka-Kobe, Japan	11.4
Cairo, Egypt	11.0
Lagos, Nigeria	10.6
Moscow, Russia	10.6
Istanbul, Turkey	10.5
Paris, France	10.5

of internal environmental impacts and characteristics, and these are referred to elsewhere in this book (Table 1.8), and some of their wildlife impacts are discussed by Goddard et al. (2010).

The perfecting of sea-going ships in the sixteenth and seventeenth centuries was part of this industrial and economic transformation, and this was the time when mainly self-contained but developing regions of the world coalesced so that the ecumene became to all intents and purposes continuous. The invention of the steam engine in the late eighteenth century, and the internal combustion engine in the late nineteenth century, massively increased human access to energy and lessened dependence on animals, wind and water.

Modern science, technology and industry have also been applied to agriculture, and in recent decades some spectacular progress has been made through, for example, the use of fertilizers and the selective breeding of plants and animals.

The twentieth century was a time of extraordinary change (McNeill, 2003). Human population increased from 1.5 to 6 billion, the world's economy increased 15-fold, the world's energy use increased 13- to 14-fold,

Table 1.8 Urban environments

Phenomenon	Discussion in this book (chapter)
Accelerated salinization	4
Areas of land reclamation and offshore islands	6
Biodiversity	3
Channelization and burying of streams	6
Ground levelling for airports, etc.	6
Ground subsidence caused by fluid extraction	6
Groundwater augmentation	5
Groundwater extraction	5
Increased sediment loads during construction	4
Lanslides produced by building on slopes	6
Pits produced by excavation of building materials	6
Pollution of air	7
Pollution of soil (e.g. by heavy metals)	4
Pollution of water	5
Runoff generation and floods from storm drains, sewers and impermeable surfaces	5
Soil transformation (compaction, addition of rubble, loss of organics, etc)	4
Sulfation and other forms of weathering	6
Urban animals	3
Urban heat island effect	7
Urban vegetation – parks and gardens	2
Waste dumps and landfill	6

freshwater use increased ninefold and the irrigated area by fivefold. In the hundred centuries from the dawn of agriculture to 1900, McNeill calculates that humanity only used about two-thirds as much energy (most of it from biomass) than it used in the twentieth century. Indeed, he argued that humankind used more energy in the twentieth century than in all preceding human history put together. In addition he suggests that the seas surrendered more fish in the twentieth century than in all previous centuries and that the forest and woodland area shrank by about 20%, accounting for perhaps half the net deforestation in world history.

One measure of the extent of the human impact is the human appropriation of net primary production (HANPP) as a percentage of the total amount of net primary productivity (NPP) generated on land. There is a considerable range of estimates of HANPP as a percentage of NPP, but Imhoff et al. (2004) give a global figure of around 20%, but also point to great differences between the continents, with values ranging from *c.* 6.1% for Latin America to 80.4% for south central Asia.

To conclude, we can recognize certain trends in human manipulation of the environment which have taken place in the modern era. The first of these is that the ways in which humans are affecting the environment are proliferating, so that we now live on what some people have argued is a human-dominated planet (Vitousek et al., 1997). For example, nearly all the powerful pesticides post-date the Second World War, and the same applies to the construction of nuclear reactors. Secondly, environmental issues that were once locally confined have become regional or even global problems. An instance of this is the way in which substances such as dichlorodiphenyltrichloroethane (DDT), lead and sulfates are found at the poles, far removed from the industrial societies that produced them. This is one aspect of increasing globalization. Thirdly, the complexity, magnitude and frequency of impacts are probably increasing; for instance, a massive modern dam like that at Aswan in Egypt or the Three Gorges Dam in China has very different impacts from a small Roman one. Finally, compounding the effects of rapidly expanding populations is a general increase in per capita consumption and environmental impact (Myers and Kent, 2003; McNeill, 2005) (Table 1.9). Energy resources are being developed at an ever-increasing rate, giving humans enormous power to transform the environment, not least in China (see Chapter 13 and Figure 13.5). Figure 1.14 shows worldwide energy consumption since 1860 on a per capita basis. Nonetheless, it is important to recognize that there are huge differences in the likely environmental impacts of different economies in different parts of the world. As Table 1.10 indicates, the environmental impact, as measured by the so-called ecological footprint, is 12 times greater, for example, for the average American than for the average Indian (Wackernagel and Rees, 1995). On a global basis, our ecological footprint, according to the 2010 WWF *Living Planet Report*, doubled between 1961 and 2007.

Modern technologies have immense power output. A pioneer steam engine in AD 1800 might rate at 8–16 kW. Modern railway diesels top 3.5 MW and a large aero engine 70 MW. Figure 1.15 shows how the human impact on six 'component indicators of the **biosphere**' has increased over time. This graph is based on work by Kates et al. (1990). Each component

Table 1.9 Indicators of global change (Modified from McNeill, 2005, tables 1 and 2)

(a) Some indicators of change in the global economy from 1950–2005

World indicator	1950	2005	Change (xn)
Grain production (million tonnes)	631	2008	3.18
Meat production (million tonnes)	44	264	6.0
Coal consumption (million tonnes of oil equivalent)	1074	2597	2.42
Oil consumption (million tonnes)	470	3861	8.21
Natural gas consumption (million tonnes of oil equivalent)	171	2512	14.69
Car production (million)	8.0	46	5.75
Bike production (million)	11	124	11.27
Human population (million)	2555	6469	2.53

(b) Scale of changes between 1890s and 1990s (modified from McNeill, 2005, tables 1 and 2)

Indicator	Increase
Freshwater use	9-fold
Marine fish catch	35
Cropland	2
Irrigated area	5
Pasture area	1.8
Forest area	0.8 (i.e. 20% reduction)
Carbon dioxide emissions	17
Sulfur dioxide emissions	13
Lead emissions	8
Cattle population	4
Goat population	5
Pig population	9
World industrial output	40
World energy use	13
Coal production	7
Urban population	13
World economy	14

Table 1.10 Comparing people's average consumption in Canada, the USA, India and the world. Sources: Wackernagel and Rees (1995) and Ewing et al. (2009)

Consumption per person in 1991	Canada	USA	India	World
CO_2 emission (metric tonnes per year)	15.2	19.5	0.81	4.2
Purchasing power ($US)	19,320	22,130	1150	3800
Vehicles per 100 persons	46	57	0.2	10
Paper consumption (kilograms per year)	247	317	2	44
Fossil energy use (gigajoules per year)	250 (234)	287	5	56
Freshwater withdrawal (cubic metres per year)	1688	1868	612	644
Ecological footprint[a] (hectares per person) (1991)	4.3	5.1	0.4	1.8
Ecological footprint (2006)	5.8	9.0	0.8	2.6

[a]An ecological footprint is an accounting tool for ecological resources in which various categories of human consumption are translated into areas of productive land required to provide resources and assimilate waste products. It is thus a measure of how sustainable the lifestyles of different population groups are.

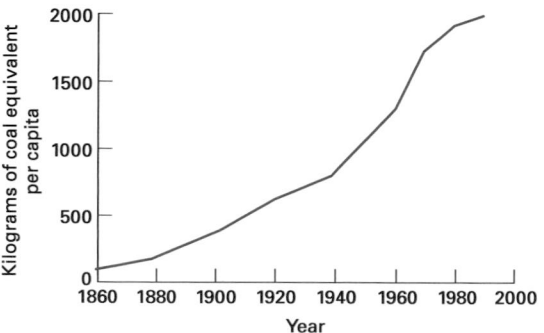

Figure 1.14 World per capita energy consumption since 1860, based on data from the United Nations.

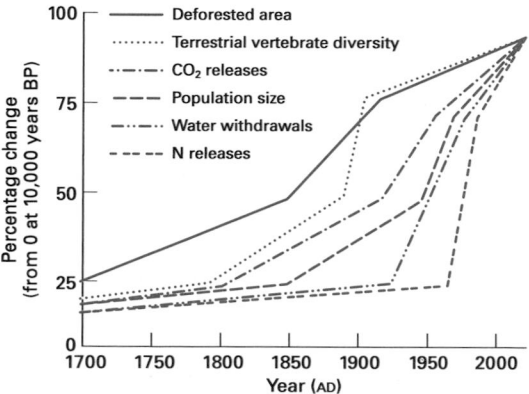

Figure 1.15 Percentage change (from assumed zero human impact at 10,000 BP) of selected human impacts on the environment.

indicator was taken to be 0% for 10,000 years ago (before the present = BP) and 100% for 1985. They then estimated the dates by which each component had reached successive quartiles (that is, 25, 50 and 75%) of its total change at 1985. They believe that about half of the components have changed more in the single generation since 1950 than in the whole of human history before that date. McNeill (2000) provides an exceptionally fine picture of all the changes in the environment that humans achieved in the twentieth century, and Carpenter (2001) examines the issue of whether civil engineering projects are environmentally sustainable.

Likewise, we can see stages in the pollution history of the earth. Mieck (1990), for instance, has identified a sequence of changes in the nature and causes of pollution: *pollution microbienne* or *pollution bacterielle,* caused by bacteria living and developing in decaying and putrefying materials and stagnant water associated with settlements of growing size; *pollution artisanale,* associated with small-scale craft industries such as tanneries, potteries and other workshops carrying out various rather disagreeable tasks, including soap manufacture, bone burning and glue-making; *pollution industrielle,* involving large-scale and pervasive pollution over major centres of industrial activity, particularly from the early nineteenth century in areas like the Ruhr or the English 'Black Country'; *pollution fondamentale,* in which whole regions are affected by pollution, as with the desiccation and subsequent salination of the Aral Sea area; *pollution foncière,* in which vast quantities of chemicals are deliberately applied to the land as fertilizers and biocides; and finally, *pollution accidentale,* in which major accidents can cause pollution which is neither foreseen nor calculable (e.g. the Chernobyl nuclear disaster of 1986).

Humans now play a very substantial role in major biogeochemical cycling, including the nitrogen cycle. For example, human activities now convert more N_2 from the atmosphere into reactive forms than all of the Earth's terrestrial processes combined (Galloway et al., 2008; Rockström et al., 2009). This is achieved through four processes: industrial fixation of N_2 to ammonia, agricultural fixation of N_2 via cultivation of leguminous crops, fossil-fuel combustion and biomass burning. Much of this reactive N_2 eventually ends up in the environment, polluting waterways and coastal zones.

Above all, as a result of the escalating trajectory of environmental transformation, it is now possible to talk about *global* environmental change. There are two components to this (Turner et al., 1990): systemic global change and cumulative global change. In the systemic meaning, 'global' refers to the spatial scale of operation and comprises such issues as global changes in climate brought about by atmospheric pollution. This is a topic discussed at length in Chapter 7, Chapter 8, Chapter 9, Chapter 10, Chapter 11, Chapter 12. In the cumulative meaning, 'global' refers to the areal or substantive accumulation of localized change, and a change is seen to be 'global' if it occurs on a worldwide scale or represents a significant fraction of the total environmental phenomenon or global resource. Both types of change are closely intertwined. For example, the burning of vegetation can lead to systemic change through such mechanisms as carbon dioxide release and albedo modification, and to cumulative change through its impact on soil and biotic diversity (Table 1.11). It is for this reason that we now talk of Earth System Science and recognize the complex interactions that take place at a multitude of scales on our planet (Steffen et al., 2004).

We can conclude this introductory chapter by quoting from Kates et al. (1990: 1):

Table 1.11 Types of global environmental change. Source: From Turner et al. (1990, table 1) with permission from Elsevier

Type	Characteristic	Examples
Systemic	Direct impact on globally functioning system	(a) Industrial and land-use emissions of 'greenhouse' gases (b) Industrial and consumer emissions of ozone-depleting gases (c) Land cover changes in albedo
Cumulative	Impact through worldwide distribution of change	(a) Groundwater pollution and depletion (b) Species depletion/genetic alteration (biodiversity)
	Impact through magnitude of change (share of global resource)	(a) Deforestation (b) Industrial toxic pollutants (c) Soil depletion on prime agricultural lands

Most of the change of the past 300 years has been at the hands of humankind, intentionally or otherwise. Our ever-growing role in this continuing metamorphosis has itself essentially changed. Transformation has escalated through time, and in some instances the scales of change have shifted from the locale and region to the earth as a whole. Whereas humankind once acted primarily upon the visible 'faces' or 'states' of the earth, such as forest cover, we are now also altering the fundamental flows of chemicals and energy that sustain life on the only inhabited planet we know.

Points for review

What have been the main stages in the development of ideas about the human impact on the environment?

How have human population levels changed over the last few millions of years?

To what extent did early humans change their environment?

What have been the main changes in the environment wrought by humans over the last 300 years?

What do you think is meant by the term Earth System Science?

Guide to reading

Baker, A.R.H. (2003) *Geography and History, Bridging the Divide*. Cambridge: Cambridge University Press. Chapter 3 contains a valuable and perceptive discussion of environmental geographies and histories.

Barker, G. (2006) *The Agricultural Revolution in Prehistory. Why Did Foragers Become Farmers?* Cambridge: Cambridge University Press. An impressive assessment at a global scale of what drove foragers to become farmers near the transition to the Holocene.

Cuff, D.J. and Goudie, A.S. (eds) (2009) *The Oxford Companion to Global Change*. New York: Oxford University Press. A multi-author encyclopedia of global change.

Diamond, J. (1997) *Guns, Germs and Steel*. London: Chatto and Windus. A lively review of human history over the last 13,000 years.

Goudie, A.S. (ed.) (1997) *The Human Impact Reader. Readings and Case Studies*. Oxford: Blackwell. A collection of key papers on many of the themes discussed in this book.

Goudie, A.S. and Viles, H. (1997) *The Earth Transformed*. Oxford: Blackwell. An introductory treatment of the human impact, with many case studies.

Govorushko, S.M. (2011) *Natural Processes and Human Impacts*. Heidelberg: Springer. A highly comprehensive and well-illustrated survey.

Kemp, D.D. (2004) *Exploring Environmental Issues: An Integrated Approach*. London: Routledge. A balanced, accessible and comprehensive analysis of many environmental issues.

Meyer, W.B. (1996) *Human Impact on the Earth*. Cambridge: Cambridge University Press. A good point of entry to the literature that brims over with thought-provoking epigrams.

Middleton, N.J. (2008) *The Global Casino*. London: Hodder Education. The fourth edition of an introductory text, by a geographer, which is well illustrated and clearly written.

Morris, I. (2010) *Why the West Rules – For Now*. London: Profile Books. A massive survey of the patterns of development since prehistory and how they have varied in the west and the east.

Oppenheimer, S. (2002) *Out of Eden. Peopling of the World*. London: Constable. A very accessible account of human development in prehistory.

Ponting, C. (2007) *A New Green History of the World*. London: Penguin. The second edition of an engaging and informative treatment of how humans have transformed the Earth through time.

Scarre, C. (ed.) (2005) *The Human Past: World History and the Development of Human Societies*. London: Thames & Hudson. A splendid account of current ideas about the evolution of humans and human societies.

Simmons, I.G. (1996) *Changing the Face of the Earth: Culture, Environment and History* (2nd edn). Oxford: Blackwell. A characteristically amusing and perceptive review of many facets of the role of humans in transforming the Earth, from a essentially historical perspective.

Simmons, I.G. (1997) *Humanity and Environment: A Cultural Ecology*. Harlow: Longman. A broad account of some major themes relating to humans and the environment.

Steffen, W. 10 others, 2004, *Global Change and the Earth System*. Berlin: Springer. A multi-author, high-level, earth system science-based overview of environmental change at a global scale.

Turner, B.L. II (ed.) (1990) *The Earth as Transformed by Human Action*. Cambridge: Cambridge University Press. A great analysis of global and regional changes over the past 300 years.

2 THE HUMAN IMPACT ON VEGETATION

Chapter Overview

Humans have done much to transform the vegetation cover of the Earth. From very early times, they have used fire to modify the environment. Also important has been grazing by domestic livestock. However, it is deforestation that has been the most potent cause of change. Humans have also modified or contributed to the character of some major biomes, including secondary forests, desert margins, savannas, prairies, lowland heaths and Mediterranean shrublands. Other major changes have been caused by the introduction of alien species around the world, species that sometimes invade explosively and by pollution episodes. Finally, attention is drawn to changes in genetic composition.

Introduction

In any consideration of the human impact on the environment, it is probably appropriate to start with vegetation, for humankind has possibly had a greater influence on plant life than on any of the other components of the environment (Ellis, 2011). We have appropriated a large amount of the world's biomass for our own use (Haberl et al., 2007), and Smil (2011) has estimated that through harvesting, deforestation, and conversion of grasslands and wetlands, humans have reduced the stock of global terrestrial plant mass by as much as 45% in the last 2000 years, with a third of this being achieved in the twentieth century. Through

The Human Impact on the Natural Environment: Past, Present and Future, Seventh Edition. Andrew S. Goudie.
© 2013 John Wiley & Sons, Ltd. Published 2013 by John Wiley & Sons, Ltd.

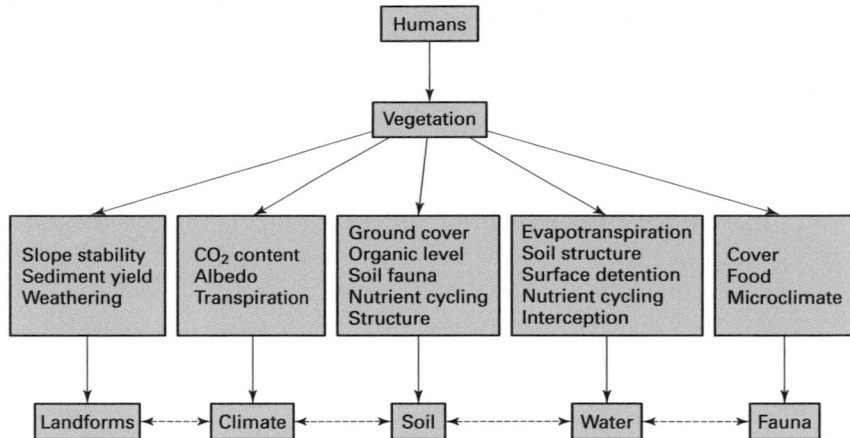

Figure 2.1 Some ramifications of human-induced vegetation change.

the many changes humans have brought about in land use and land cover, they have modified soils (Meyer and Turner, 1994) (see Chapter 4), changed the quality and quantity of some natural waters (see Chapter 5), affected geomorphic processes (see Chapter 6) and influenced climates (see Chapter 7). Indeed, the nature of whole landscapes has been transformed by human-induced vegetation change (Figure 2.1). Hannah et al. (1994) attempt to provide a map and inventory of human disturbance of world ecosystems, but many of their criteria for recognizing undisturbed areas are naïve. Large tracts of central Australia are classed as undisturbed, when plainly they are not (Fitzpatrick, 1994). Following Hamel and Dansereau (1949, cited by Frenkel, 1970) we can recognize five principal degrees of interference – each one increasingly remote from pristine conditions. These are

1 *Natural habitats*: those that develop in the absence of human activities.
2 *Degraded habitats*: those produced by sporadic, yet incomplete, disturbances; for example, the cutting of a forest, burning and the non-intensive grazing of natural grassland.
3 *Ruderal habitats*: where disturbance is sustained but where there is no intentional substitution of vegetation. Roadsides are an example of a ruderal habitat.
4 *Cultivated habitats*: when constant disturbance is accompanied by the intentional introduction of plants.
5 *Artificial habitats*: which are developed when humans modify the ambient climate and soil, as in greenhouse cultivation.

An alternative model for classifying the extent of human influence on vegetation is provided by Westhoff (1983), who adopts a four-part scheme:

1 *Natural*: a landscape or an ecosystem not influenced by human activity.
2 *Subnatural*: a landscape or ecosystem partly influenced by humans, but still belonging to the same (structural) formation type as the natural system from which it derives (e.g. a wood remaining a wood).
3 *Semi-natural*: a landscape or ecosystem in which flora and fauna are largely spontaneous, but the vegetation structure is altered so that it belongs to another formation type (e.g. a pasture, moorland or heath deriving from a wood).
4 *Cultural*: a landscape or ecosystem in which flora and fauna have been essentially affected by human agency in such a way that the dominant species may have been replaced by other species (e.g. arable land).

In this chapter we shall be concerned mainly with degraded and ruderal habitats, or sub-natural and semi-natural habitats. However, first we need to consider some of the processes that human societies employ, notably, fire, grazing and the physical removal of forest.

The use of fire

Humans are known to have used fire since Palaeolithic times (see Chapter 1). As Sauer, one of the great proponents of the role of fire in environmental change, put it (1969: 10–11):

Through all ages the use of fire has perhaps been the most important skill to which man has applied his mind. Fire gave to man, a diurnal creature, security by night from other predators. . . . The fireside was the beginning of social living, the place of communications and reflection.

People have utilized fire for a great variety of reasons: to clear forest for agriculture, to improve grazing land for domestic animals or to attract game; to deprive game of cover; to drive game from cover in hunting; to kill or drive away predatory animals, ticks, mosquitoes and other pests; to repel the attacks of enemies, or burn them out of their refuges; for cooking; to expedite travel; to burn the dead and raise ornamental scars on the living; to provide light; to transmit messages via smoke signalling; to break up stone for tool-making; to protect settlements or encampments from great fires by controlled burning; to satisfy the sheer love of fires as spectacles; to make pottery and smelt ores; to harden spears; to provide warmth; to make charcoal; and to assist in the collection of insects such as crickets for eating. Indeed fire is still much used, especially by pastoralists, such as the cattle-keepers of Africa and by shifting agriculturalists. For example, the Malaysian and Indonesian *ladang* and the *milpa* system of the Maya in Latin America involved the preparation of land for planting by felling or deadening forest, letting the debris dry in the hot season and burning it before the commencement of the rainy season. With the first rains, holes were dibbled in the soft ash-covered earth with a planting stick. This system was suited to areas of low population density with sufficiently extensive forest to enable long intervals of 'forest fallow' between burnings. Land which was burned too frequently became overgrown with perennial grasses, which tended to make it difficult to farm with primitive tools. Land cultivated for too long rapidly suffered a deterioration in fertility, while land recently burned was temporarily rich in nutrients.

The use of fire, however, has not been restricted to primitive peoples in the tropics. Remains of charcoal are found in Holocene soil profiles in Britain (Moore, 2000); large parts of North America appear to have suffered fires at regular intervals prior to European settlement (Parshall and Coster, 2002); and in the case of South America, the 'great number of fires' observed by Magellan during the historical passage of the Strait that bears his name resulted in the toponym, 'Tierra

del Fuego'. Indeed, says Sternberg (1968: 718): 'for thousands of years, man has been putting the New World to the torch, and making it a "land of fire".' Given that the native American population may have been greater than once thought, their impact, even in Amazonia, may have been appreciable (Denevan, 1992). North American Indians were also not passive occupants of the land. They managed forests extensively, creating 'a dynamic mosaic containing patches of young and old trees, interspersed with patches of grass and shrubs' (Bonnichsen, 1999: 442). As a result, some North American forests, prior to European settlement, were open and park-like, not dark and dense.

Fire (Figure 2.2) was also central to the way of life of the Australian aboriginals, including those of Tasmania (Hope, 1999), and the carrying of fire sticks was a common phenomenon. As Blainey (1975: 76) has put it, 'Perhaps never in the history of mankind was there a people who could answer with such unanimity the question: "have you got a light, mate?" There can have been few if any races who for so long were able to practise the delights of incendiarism.' That is not to say that aboriginal burning necessarily caused wholesale

Figure 2.2 The pindan bush of Australia, composed of *Eucalyptus*, *Acacia* and grasses, is frequently burned. The pattern of the burning shows up clearly on Landsat imagery. (See Plate 9)

modification of Australian vegetation in pre-European times. That is a hotly debated topic (Kohen, 1995).

In neighbouring New Zealand, Polynesians carried out extensive firing of vegetation in pre-European settlement times after their arrival in *c.* 1280 AD (McWethy et al., 2009), and hunters used fire to facilitate travel and to frighten and trap a major food source – the flightless moa (Cochrane, 1977). The fires continued over a period of about a thousand years up to the period of European settlement (Mark and McSweeney, 1990). The changes in vegetation that resulted were substantial. The forest cover was reduced from about 79% to 53%, and fires were especially effective in the drier forests of central and eastern South Island in the rain shadow of the Southern Alps. Pollen analyses (Figure 2.3) show dramatically the reduction in tall trees and the spread of open shrubland following the introduction of the widespread use of fire by the Maoris (McGlone and Wilmshurst, 1999).

Fires: natural and anthropogenic

Although people have used fire for all the reasons that have been mentioned, before one can assess the role of this facet of the human impact, one must ascertain how important fires started by human action are in comparison with those caused naturally, especially by lightning (Bowman et al., 2011). Some natural fires may result from spontaneous combustion, for in certain ecosystems heavy vegetal accumulations may become compacted, rotted and fermented, thus generating heat. Other natural fires can result from sparks produced by falling boulders and by landslides (Booysen and Tainton, 1984). In the forest lands of the western USA about half the fires are caused by lightning. Lightning starts over half the fires in the pine savanna of Belize, Central America and about 8% of the fires in the bush of Australia, while in the south of France, nearly all the fires are caused by people.

One method of gauging the long-term frequency of fires (Clark et al., 1997; Marlon et al., 2012) is to look at fire scars in tree rings and charcoal in lake sediments. In the USA, fires occurred with sufficient frequency over wide areas to have effects on annual tree rings and lake cores every 7–80 years in pre-European times. The frequency of fires in different environments tends to show some variation. Rotation periods may be in excess of a century for tundra, 60 years for boreal pine forest, 100 years for spruce-dominated ecosystems,

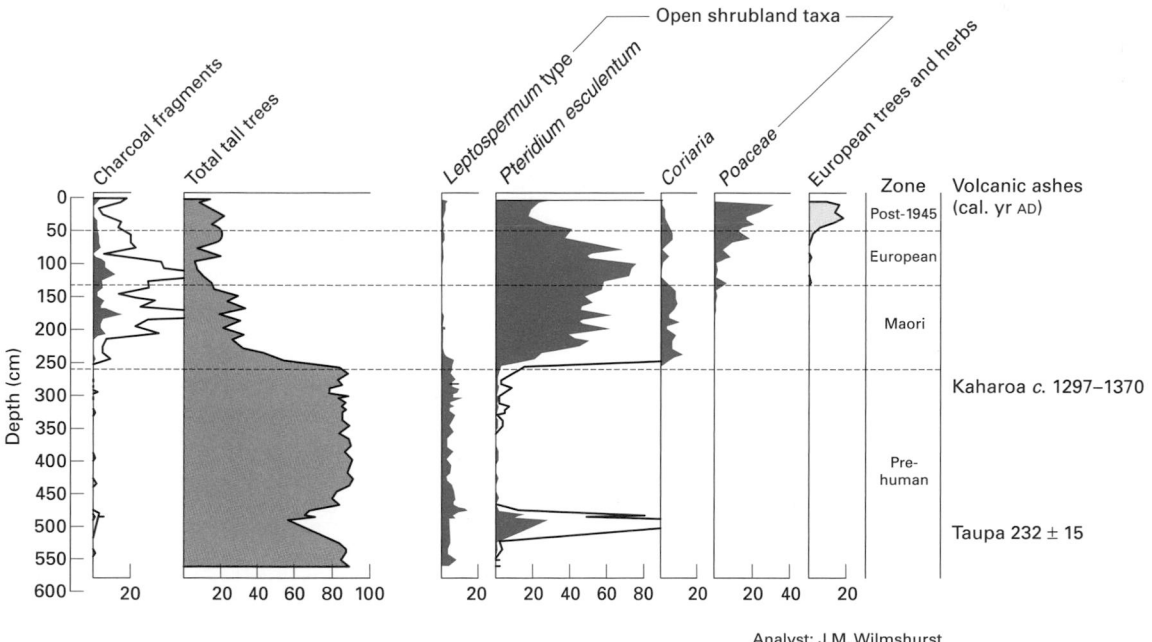

Analyst: J.M. Wilmshurst

Figure 2.3 Summary percentage pollen diagram, Lake Rotonuiaha, eastern North Island showing the changes in vegetation brought about in Maori and European times (from McGlone and Wilmshurst, 1999, figure 1. Reproduced with permission from Elsevier).

5–15 years for savanna, 10–15 years for chaparral and less than 5 years for semi-arid grasslands (Wein and Maclean, 1983). Fire prevalence can be increased by vegetation change, and in the Great Basin and Mojave Desert of the USA, the invasion of non-native annual grasses such as the drooping brome, *Bromus rubens*, has created a more or less continuous fuel bed following years of heavy rainfall (Brooks and Matchett, 2006; Pierson et al., 2011). Similarly, in northern Australia, the invasion by a big African grass called gamba (*Andropogon gayanus*) has increased fire intensity in recent decades because of its great biomass in comparison with native grasses (Setterfield et al., 2010). Fire frequency can also be increased by changes in land use and land cover, as appears to be the case in Patagonia, where a substantial increase followed the arrival of European-Chilean farmers during recent centuries (Holz and Veblen, 2011). At a global scale, biomass burning showed a marked increase in the nineteenth century, and at that time direct anthropogenic activities became the dominant driver of global fire activity trends (Pechony and Shindell, 2010). On the other hand, in the western USA, because of human activities that include fire suppression, fire activity has declined since the late 1800s so that given the present climate there is what they call a 'fire deficit'.

Some consequences of fire suppression

Given that fire has long been a feature of many ecosystems, and irrespective of whether the fires were or were not caused by people, it is clear that any deliberate policy of fire suppression might have important consequences for vegetation (Sherriff and Veblen, 2006; Kasischke et al., 2010). Fire exclusion in the Ponderosa pine forests of the Cascade Mountains in the USA over the last hundred years caused compositional and structural changes (Taylor, 2010). Rigid fire-protection policies have often had undesirable results (Bonnicksen et al., 1999) and as a consequence many foresters stress the need for 'environmental restoration burning'. One of the best ways to prevent the largest forest fires is to allow the small and medium fires to burn (Malamud et al., 1998). In Alaska, fire suppression policies, particularly adjacent to urban settlements, meant that the area burned from human-ignited fires

fell from 26% for the 1950s and 1960s to 5% for the 1990s and 2000s (Kasischke et al., 2010).

Fire suppression, as has already been suggested, can magnify the adverse effects of fire. The position has been well stated by Sauer (1969: 14):

The great fires we have come to fear are effects of our civilization. These are the crown fires of great depths and heat, notorious aftermaths of the pyres of slash left by lumbering. We also increase fire hazard by the very giving of fire protection which permits the indefinite accumulation of inflammable litter. Under the natural and primitive order, such holocausts, that leave a barren waste, even to the destruction of the organic soil, were not common.

In the coniferous forests of the middle upper elevations in the Sierra Nevada Mountains of California, fire protection since 1890 has made the stands denser, shadier and less park-like, and sequoia seedlings have decreased in number. In the high-altitude mountains of the western USA the whitebark (*Pinus albicaulis*) has suffered from the expansion of fire-intolerant species (Larson and Kipfmueller, 2012), and the same has happened to Douglas fir in southwestern Oregon (Messier et al., 2012). Likewise, at lower elevations, the Mediterranean semi-arid shrubland, called chaparral, has had its character changed. The vegetation has increased in density, the amount of combustible fuels has risen, fire-intolerant species have encroached, and vegetation diversity has decreased, resulting in a monotony of old-age stands, instead of a mosaic of different successional stages. Unfavourable consequences of fire suppression have also been noted in Alaska (Oberle, 1969), where it has been found that when fire is excluded from many lowland sites, an insulating carpet of moss tends to accumulate and raise the permafrost level. **Permafrost** close to the surface encourages the growth of black spruce (*Picea mariana*), a low-growing species with little timber or food value. In the Kruger National Park, in South Africa, fires have occurred less frequently after the establishment of the game reserve, when it became uninhabited by natives and hunters. As a result, bush encroachment has taken place in areas that were formerly grassland, and the carrying capacity for grazing animals has declined. Controlled burning has been reinstituted as a necessary game-management operation.

However, following the severe fires that ravaged America's Yellowstone National Park in 1988, there

THE HUMAN IMPACT ON VEGETATION

has been considerable debate as to whether the inferno, the worst since the park was established in the 1870s, was the result of a policy of fire suppression. Without such a policy the forest would burn at intervals of 10–20 years because of lightning strikes. Could it be that the suppression of fires over long periods of, say, 100 years or more, allegedly to protect and preserve the forest, led to the build-up of abnormal amounts of combustible fuel in the form of trees and shrubs in the understory? Should a programme of prescribed burning be carried out to reduce the amount of available fuel?

Fire suppression policies at Yellowstone did indeed lead to a critical build-up in flammable material. However, other factors must also be examined in explaining the severity of the fire. One of these was the fact that the last comparable fire had been in the 1700s, so that the Yellowstone forests had had nearly 300 years in which to become increasingly flammable. In other words, because of the way vegetation develops through time (a process called succession), very large fires may occur every 200–300 years as part of the natural order of things (Figure 2.4). Another crucial factor was that the weather conditions in the summer

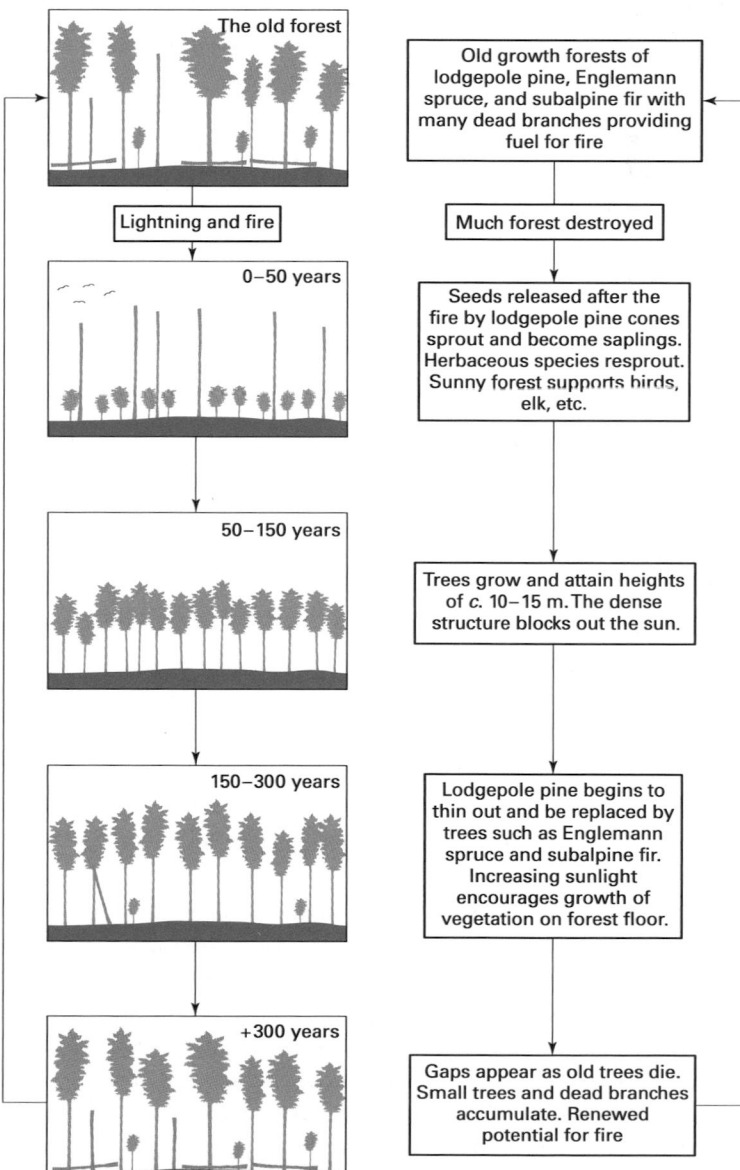

Figure 2.4 Ecological succession in response to fire in Yellowstone National Park (after Romme and Despain, 1989: 24–25, heavily modified).

of 1988 were abnormally dry, bringing a great danger of fire. As Romme and Despain (1989: 28) remarked:

It seems that unusually dry, hot and windy weather conditions in July and August of 1988 coincided with multiple ignition in a forest that was at its most flammable stage of succession. Yet it is unlikely that past suppression efforts greatly exacerbated the Yellowstone fire. If fires occur naturally at intervals ranging from 200 to 400 years, then 30 or 40 years of effective suppression is simply not enough for excessive quantities of fuel to build up. Major attempts at suppression in the Yellowstone forests may have merely delayed the inevitable.

Certainly, studies of the long-term history of fire in the western USA have shown the great importance of climatic changes at the decennial and centennial scales in controlling fire frequency during the Holocene (Meyer and Pierce, 2003; Whitlock et al., 2003; Marlon et al., 2012). There is also evidence that an increase in the severity of forest fires in the mountains of California and Nevada has taken place since the mid 1980s, partly because of the availability of combustible materials following fire suppression but also possibly because of a more favourable climate for the generation of large fires (Miller et al., 2009).

Some effects of fire on vegetation

The effects which fires have on the environment depend very much on their size, duration and intensity. Some fires are relatively quick and cool, and only destroy ground vegetation. Other fires, crown fires, affect whole forests up to crown level and generate very high temperatures. In general, forest fires are hotter than grassland fires. Perhaps more significantly in terms of forest management, fires which occur with great frequency do not attain too high a temperature because there is inadequate inflammable material to feed them.

There is evidence that fire has played an important role in the formation of various major types of vegetation (e.g. savanna, pine forests) and in influencing the operation of ecosystems (Crutzen and Goldammer, 1993; O'Connor et al., 2011). Fire may assist in seed germination. For example, the abundant germination of dormant seeds on recently burned chaparral sites has been reported by many investigators, and it seems that some seeds of chaparral species require scarifica-

tion by fire. The better germination of those not requiring scarification may be related to the removal by fire of competition, litter and some substances in the soils which are toxic to plants. Fire alters seedbeds. If litter and humus removal are substantial, large areas of rich ash, bare soil, or thin humus may be created. Some trees, such as Douglas fir (*Pseudotsuga*) and the giant sequoia (*Sequoiadendron giganteum*), benefit from such seedbeds. Fire sometimes triggers the release of seeds (as with the Jack pine, *Pinus bankdiana*) and seems to stimulate the vegetative reproduction of many woody and herbaceous species. Fire can control forest insects, parasites and fungi – a process termed 'sanitization'. It also seems to stimulate the flowering and fruiting of many shrubs and herbs, and to modify the physicochemical environment of the plants. Mineral elements are released both as ash and through increased decomposition rates of organic layers. Above all, areas subject to fire often show greater species diversity, which is a factor that tends to favour stability. Fire may also influence climate by changing surface albedo and releasing greenhouse gases into the atmosphere and also cause the degradation of permafrost.

One can conclude by quoting at length from Pyne (1982: 3), who provides a detailed and scholarly analysis of the history of cultural fires in America:

Hardly any plant community in the temperate zone has escaped fire's selective action, and, thanks to the radiation of *Homo sapiens* throughout the world, fire has been introduced to nearly every landscape on earth. Many biotas have consequently so adapted themselves to fire that, as with biotas frequented by floods and hurricanes, adaptation has become symbiosis. Such ecosystems do not merely tolerate fire, but often encourage it and even require it. In many environments fire is the most effective form of decomposition, the dominant selective force for determining the relative distribution of certain species, and the means for effective nutrient recycling and even the recycling of whole communities.

The role of grazing

Many of the world's grasslands have long been grazed by wild animals like the bison of North America or the large game of East Africa, but the introduction of pastoral economies also affects their nature and productivity (Figure 2.5).

Light grazing may increase the productivity of wild pastures. Nibbling, for example, can encourage the

Figure 2.5 Intensive grazing by domestic stock, such as these cattle near Belo Horizonte in Brazil, has a major influence on many biomes, including grasslands and savannas. (See Plate 10)

vigour and growth of plants, and in some species, such as the valuable African red grass, *Themeda triandra*, the removal of coarse, dead stems permits succulent sprouts to shoot. Likewise the seeds of some plant species are spread efficiently by being carried in cattle guts and then placed in favourable seedbeds of dung or trampled into the soil surface. Moreover, the passage of herbage through the gut and out as faeces modifies the nitrogen cycle, so that grazed pastures tend to be richer in nitrogen than ungrazed ones. Also, like fire, grazing can increase species diversity by opening out the community and creating more niches.

On the other hand, heavy grazing may be detrimental. Excessive trampling when conditions are dry will reduce the size of soil aggregates and break up plant litter to a point where they are subject to removal by wind. Trampling, by puddling the soil surface, can accelerate soil deterioration and water erosion as infiltration capacity is reduced. Heavy grazing can kill plants or lead to a marked reduction in their level of photosynthesis. In addition, when relieved of competition from palatable plants or plants liable to trampling damage, resistant and usually unpalatable species expand their cover. Thus in the western USA, poisonous burroweed (*Haploplappius* spp.) has become dangerously common, and many woody species have intruded. These include the mesquite (*Prosopis juliflora*), the big

sagebrush (*Artemisia tridentata*), the one-seed juniper (*Juniperus monosperma*) and the Pinyon pine. Grover and Musick (1990) saw shrubland encroachment by creosote bush (*Larrea tridentata*) and mesquite as part and parcel of desertification and indicate that in southern New Mexico the area dominated by them has increased several fold over the last century. This has been as a result of a corresponding decrease in the area's coverage of productive grasslands. They attributed both tendencies to excessive livestock overgrazing at the end of the nineteenth century, but pointed out that this was compounded by a phase of rainfall regimes that were unfavourable for perennial grass growth.

It is evident then that the semi-arid grasslands of southwestern North America have changed dramatically over the last 150 years as a result of the encroachment of native woody species such as mesquite. Van Auken (2000: 207) concluded: 'The major cause of the encroachment of these woody species seems to be the reduction of grass biomass (fire fuel) by chronic high levels of domestic herbivory coupled to a reduction of grassland fires, which would have killed or suppressed the woody plants to the advantage of the grasses.'

There are many reasons why excessive grazing can lead to shrub dominance (Archer et al., 1999):

1 Livestock may effectively disperse woody plant seeds, particularly those of some leguminous shrubs and arborescents.
2 Utilization of grasses increases chances for germination and early establishment of woody seedlings.
3 Concomitant reductions in transpirational leaf area, root biomass and root activity associated with grazing of grasses can:
   (a) increase superficial soil moisture to enhance woody seedling establishment and growth of shallow-rooted woody species;
   (b) increase the amount of water percolating to deeper depths and benefit established woody species with deep root systems;
   (c) increase nutrient availability to woody plants; and
   (d) release suppressed populations of established tree or shrub "seedling reserves."
4 Grazing increases mortality rates and decreases plant basal area, seed production and seedling establishment of palatable grasses.

5 Grazing may also increase susceptibility of grasses to other stresses such as drought. These factors would combine to increase the rate of gap formation and available area for woody plant seedling establishment, especially in post-drought periods.

6 Herbaceous species may be replaced by assemblages that compete less effectively with woody plants.

7 Reduction of the biomass and continuity of fine fuel may reduce fire frequency and intensity.

8 Invading woody species are often unpalatable relative to grasses and forbs and are thus not browsed with sufficient regularity or severity to limit establishment or growth.

9 Lower soil fertility and alterations in physicochemical properties occur with loss of ground cover and subsequent erosion. This favours N_2-fixing woody plants (e.g. *Prosopis*, *Acacia*) and evergreen woody plant growth forms that are tolerant of low nutrient conditions.

In Australia the widespread adoption of sheep grazing after European settlement led to significant changes in the nature of grasslands over extensive areas. In particular, the introduction of sheep led to the removal of kangaroo grass (*Themeda australis*) – a predominantly summer-growing species – and its replacement by essentially winter-growing species such as *Danthonia* and *Stipa*. Also in Australia, not least in the areas of tropical savanna in the north, large herds of introduced feral animals (e.g. cattle, horse, dromedary and sambar deer) have resulted in overgrazing and alteration of native habitats. As they appear to lack significant control by predators and pathogens, their densities, and thus their effects, became very high (Freeland, 1990).

Similarly in Britain many plants are avoided by grazing animals because they are distasteful, hairy, prickly or even poisonous (Tivy, 1971). The persistence and spread of bracken (*Pteridium aquilinium*) on heavily grazed rough pasture in Scotland is aided by the fact that it is slightly poisonous, especially to young stock. The success of bracken is furthered by its reaction to burning, for with its extensive system of underground stems (rhizomes) it tends to be little damaged by fire. The survival and prevalence of shrubs such as elder (*Sambucus nigra*), gorse (*Ulex* spp.), broom (*Sarothamus scoparius*) and the common weeds ragwort (*Senecio jacobaea*) and creeping thistle (*Cirsium arvense*), in the face of grazing, can be attributed to their lack of palatability.

The role of grazing in causing marked deterioration of habitat has been the subject of further discussion in the context of upland Britain. In particular, Darling (1956) stressed that while trees bring up nutrients from rocks and keep minerals in circulation, pastoralism means the export of calcium phosphate and nitrogenous organic matter. The vegetation gradually deteriorates, the calcicoles disappear, and the herbage becomes deficient in both minerals and protein. Progressively more xerophytic plants come in: *Nardus stricta* (mat grass), *Molinia caerulaea* (purple moor grass), *Erica tetralix* (cross-leaved heath) and then *Scripus caespitosu* (deer grass). However, this is not a view that has received universal support (Mather, 1983). Studies in several parts of upland Britain have shown that mineral inputs from precipitation are very much greater than the nutrient losses in wool and sheep carcasses. During the 1970s the idea of upland deterioration was gradually undermined.

In general terms it is clear that in many parts of the world the grass family is well equipped to withstand grazing. Many plants have their growing points located on the apex of leaves and shoots, but grasses reproduce the bulk of fresh tissue at the base of their leaves. This part is least likely to be damaged by grazing and allows regrowth to continue at the same time that material is being removed.

Communities severely affected by the treading of animals (and indeed people) tend to have certain distinctive characteristics. These include diminutiveness (since the smaller the plant is, the more protection it will get from soil surface irregularities); strong ramification (the plant stems and leaves spread close to the ground); small leaves (which are less easily damaged by treading); tissue firmness (cell-wall strength and thickness to limit mechanical damage); a bending ability; strong vegetative increase and dispersal (e.g. by stolons); small hard seeds which can be easily dispersed; and the production of a large number of seeds per plant (which is particularly important because the mortality of seedlings is high under treading and trampling conditions).

Deforestation

Deforestation is one of the great processes of landscape transformation (Williams, 2003). However, controversy surrounds the meaning of the word '**defor-estation**' and this causes problems when it comes to assessing rates of change and causes of the phenom-enon (Williams, 1994). It is best defined (Grainger, 1992) as 'the temporary or permanent clearance of forest for agriculture or other purposes'. According to this definition, if clearance does not take place then deforestation does not occur. Thus, much logging in the tropics, which is selective in that only a cer-tain proportion of trees and only certain species are removed, does not involve clear felling and so cannot be said to constitute deforestation. However, how 'clear' is clear? Many shifting cultivators in the humid tropics leave a small proportion of trees standing (perhaps because they have special utility). At what point does the proportion of trees left standing permit one to say that deforestation has taken place? This is not a question to which there is a simple answer. There is not even a globally accepted definition of forest, and the UN's Food and Agriculture organiza-tion (FAO) defines forest cover as greater than 10% canopy cover, while the International Geosphere Biosphere programme (IGBP) defines it as greater than 60%.

The deliberate removal of forest is one of the most long-standing and significant ways in which humans have modified the environment, whether achieved by fire or cutting. Pollen analysis shows that temperate forests were removed in Mesolithic and Neolithic times and at an accelerating rate thereafter. The example of Easter Island in the Pacific is salutary, for its deforesta-tion at an unsustainable rate caused the demise of the culture that created its great statues (Flenley et al., 1991). Since pre-agricultural times, world forests have declined approximately one-fifth, from 5 to 4 billion hectares. The highest loss has occurred in temperate forests (32–5%), followed by subtropical woody savan-nas and deciduous forests (24–5%). The lowest losses have been in tropical evergreen forests (4–6%) because many of them have for much of their history been inac-cessible and sparsely populated (World Resources Institute, 1992: 107).

In Britain, Birks (1988) analysed dated pollen sequences to ascertain when and where tree pollen values in sediments from upland areas drop to 50% of their Holocene maximum percentages, and takes this as a working definition of 'deforestation'. From this work he identified four phases:

1 *3700–3900 years BP*: north-west Scotland and the eastern Isle of Skye;
2 *2100–2600 years BP (the pre-Roman Iron Age)*: Wales, England (except the Lake District), northern Skye and northern Sutherland;
3 *1400–1700 years BP (post-Roman)*: Lake District, south-ern Skye, Galloway and Knapdale-Ardnamurchan;
4 *300–400 years BP*: the Grampians and the Cairngorms.

Sometimes forests are cleared to allow agriculture; at other times to provide fuel for domestic purposes, or to provide charcoal or wood for construction; some-times to fuel locomotives, or to smoke fish; and some-times to smelt metals. The Phoenicians were exporting cedars as early as 4600 years ago (Mikesell, 1969) both to the Pharaohs and to Mesopotamia. Attica in Greece was laid bare by the fifth century BC, and classical writers allude to the effects of fire, cutting and the destructive nibble of the goat. The great phase of defor-estation in central and western Europe, described by Darby (1956: 194) as 'the great heroic period of recla-mation', occurred from *c.* AD 1050 onwards for about 200 years. In particular, the Germans moved eastward: 'What the new west meant to young America in the nineteenth century, the new east meant to Germany in the Middle Ages' (Darby, 1956: 196). The landscape of Europe was transformed (Kaplan et al., 2009), just as that of North America, Australia, New Zealand and South Africa was to be as a result of the European expansions, especially in the nineteenth century.

Temperate North America underwent particularly brutal deforestation (Williams, 1989) and lost more woodland in 200 years than Europe did in 2000. The first colonialists arriving in the *Mayflower* found a con-tinent that was wooded from the Atlantic seaboard as far as the Mississippi River (Figure 2.6). The forest originally occupied some 170 million hectares. Today only about 10 million hectares remain.

Fears have often been expressed that the mountains of High Asia (e.g. in Nepal) have been suffering from a wave of deforestation that has led to a whole suite of environmental consequences, which include accel-erated landsliding, flooding in the Ganges Plain and sedimentation in the deltaic areas of Bengal. Ives and Messerli (1989) doubt that this alarmist viewpoint is soundly based and argue that 'the popular claims about catastrophic post-1950 deforestation of the Middle Mountain belt and area of the high mountains of the

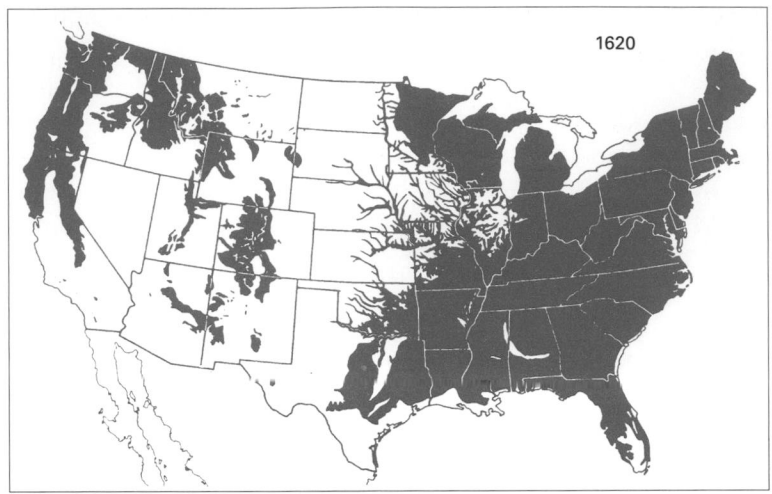

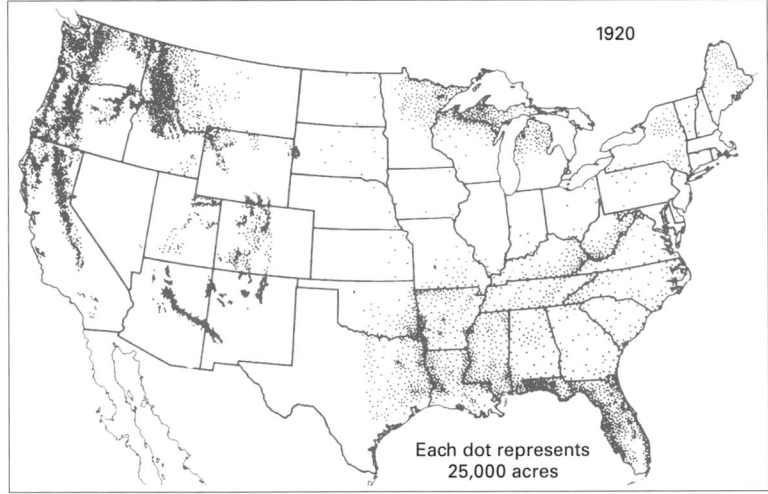

Figure 2.6 The distribution of American natural forest in 1620 and 1920 (modified from Williams, 1989).

Himalayas are much exaggerated, if not inaccurate' (p. 67).

Likewise, some workers in West Africa have interpreted 'islands' of dense forest in the savanna as the relics of a once more extensive forest cover that was being rapidly degraded by human pressure. More recent research (Fairhead and Leach, 1996) suggests that this is far from the truth and that villagers have in recent decades been extending rather than reducing forest islands (Table 2.1). There are also some areas of Kenya and Ethiopia (Nyssen et al., 2009) in East Africa where, far from recent population pressures promoting vegetation removal and land degradation, there has been an increase in woodland since the 1930s.

With regard to the equatorial rain forests, the researches of Flenley and others (Flenley, 1979) have indicated that forest clearance for agriculture has been

Table 2.1 Estimates of deforestation since 1900 (millions of hectares)

Country	Orthodox estimate of forest area lost	Forest area lost according to World Conservation Monitoring Centre	Forest area lost according to Fairhead and Leach (1996, table 9.1)
Côte d'Ivoire (Ivory Coast)	13	20.2	4.3–5.3
Liberia	4–4.5	5.5	1.3
Ghana	7	12.9	3.9
Benin	0.7	12.9	3.9
Togo	0	1.7	0
Sierra Leone	0.8–5	6.7	c.0
Total	25.3–30.2	48.6	9.5–10.5

Figure 2.7 A satellite image of the spread of agriculture and settlement through part of Amazonia (courtesy of NASA). (See Plate 11)

Table 2.2 The causes of deforestation. (Source: From Grainger, 1992). Reproduced with permission

A. Immediate causes – land use
1 Shifting agriculture
  (a) Traditional long-rotation shifting cultivation
  (b) Short-rotation shifting cultivation
  (c) Encroaching cultivation
  (d) Pastoralism
2 Permanent agriculture
  (a) Permanent staple crop cultivation
  (b) Fish farming
  (c) Government sponsored resettlement schemes
  (d) Cattle ranching
  (e) Tree crop and other cash crop plantations
3 Mining
4 Hydro-electric schemes
5 Cultivation of illegal narcotics

B. Underlying causes
1 Socio-economic mechanisms
  (a) Population growth
  (b) Economic development
2 Physical factors
  (a) Distribution of forests
  (b) Proximity of rivers
  (c) Proximity of roads
  (d) Distance from urban centres
  (e) Topography
  (f) Soil fertility
3 Government policies
  (a) Agriculture policies
  (b) Forestry policies
  (c) Other policies

going on since at least 3000 BP in Africa, 7000 BP in South and Central America and possibly since 9000 BP or earlier in India and New Guinea. Recent studies by archaeologists and palaeoecologists have tended to show that prehistoric human activities were rather more extensive in the tropical forests than originally thought (Willis et al., 2004).

The causes of the present spasm of deforestation in the tropics (Figure 2.7) are complex and multifarious and are summarized in Grainger (1992) and Table 2.2. He also provides a review of the problems of measuring and defining loss of tropical forest area (Grainger, 2008). Indeed, there are very considerable difficulties in estimating rates of deforestation in part because different groups of workers use different definitions of what constitutes a forest (Allen and Barnes, 1985) and what distinguishes rainforest from other types of forest. There is thus some variability in views as to the present rate of rainforest removal and this is brought out in a debate (see Achard et al., 2002; Fearnside and Laurance, 2003). FAO estimates (Lanly et al., 1991) show that the total annual deforestation in 1990 for 62 countries (representing some 78% of the tropical forest area of the world) was 16.8 million hectares, a figure significantly higher than the one obtained for these same countries for the period 1976–1980 (9.2 million hectares per year). Myers (1992) suggests that there was an 89% increase in the tropical deforestation rate

during the 1980s (compared with an FAO estimate of a 59% increase). He believed that the annual rate of loss in 1991 amounted to about 2% of the total forest expanse.

The change in forest area on a global basis in the 1990s is shown in Table 2.3a. These FAO data indicate the ongoing process of deforestation in Africa, Asia and the Pacific, Latin America and the Caribbean, and the stable or slightly reducing amount of deforestation in Europe, West Asia and North America. The rate of deforestation in Brazilian Amazonia, while still high, may also have become reduced by almost three quarters since 2004 (Regalado, 2010; Malingreau et al., 2011).

There is, however, a considerable variation in the rate of forest regression in different areas (Figure 2.8), with some areas under relatively modest threat (e.g.

Table 2.3 Changes in extent of forests

(a) Change in forested land 1990–2000 by region. Source: Compiled from FAO (2001)

	Total land area (million hectares)	Total forest 1990 (million hectares)	Total forest 2000 (million hectares)	% of land forested in 2000	Change 1990–2000 (million hectares)	% change per year
Africa	2963.3	702.5	649.9	21.9	−52.6	−0.7
Asia and the Pacific	3463.2	734.0	726.3	21.0	−7.7	−0.1
Europe	2359.4	1042.0	1051.3	44.6	9.3	0.1
Latin America and the Caribbean	2017.8	1011.0	964.4	47.8	−46.7	−0.5
North America	1838.0	466.7	470.1	25.6	3.9	0.1
West Asia	372.4	3.6	3.7	1.0	0.0	0.0
World	13,014.1	3960.0	3866.1	29.7	−93.9	−0.24

Note: numbers may not add due to rounding.

(b) Annual deforestation rates as a percentage of the 1990 forest cover, for selected areas of rapid forest cover change (hot spots) within each continent. Source: Achard et al. (2002). Reprinted with permission from AAAS

Hot-spot areas by continent	Annual deforestation rate of sample sites within hot-spot area (range), %
Latin America	0.38
Central America	0.8–1.5
Brazilian Amazonian belt	
Acre	4.4
Rondônia	3.2
Mato Grosso	1.4–2.7
Pará	0.9–2.4
Columbia-Ecuador border	~1.5
Peruvian Andes	0.5–1.0
Africa	0.43
Madagascar	1.4–4.7
Côte d'Ivoire	1.1–2.9
Southeast Asia	0.91
Southeastern Bangladesh	2.0
Central Myanmar	~3.0
Central Sumatra	3.2–5.9
Southern Vietnam	1.2–3.2
Southeastern Kalimantan	1.0–2.7

western Amazonia, the forests of Guyana, Surinam and French Guyana, and much of the Zaire Basin in central Africa; Myers, 1983, 1984). Some other areas are being exploited so fast that minimal areas will soon be left, for example, the Philippines, peninsular Malaya, Thailand, Australia, Indonesia, Vietnam, Bangladesh, Sri Lanka, Central America, Madagascar, West Africa and eastern Amazonia. The current annual rate of loss exceeds 5% in the eastern lowlands of Sumatra and the peatlands of Sarawak (Miettingen et al., 2011). In Indonesia and Malaysia, one of the main causes of forest disappearance is the rapidly expanding removal of forest for palm oil cultivation (Wicke et al., 2011). Production and demand for palm oil has soared in recent years (Figure 2.9).

Myers (1992) refers to particular 'hot spots' where the rates of deforestation are especially threatening and presents data for certain locations where the percentage loss of forest is more than three times his global figure of 2%: southern Mexico (10%), Madagascar (10%), northern and eastern Thailand (9.6%), Vietnam (6.6%) and the Philippines (6.7%). A more

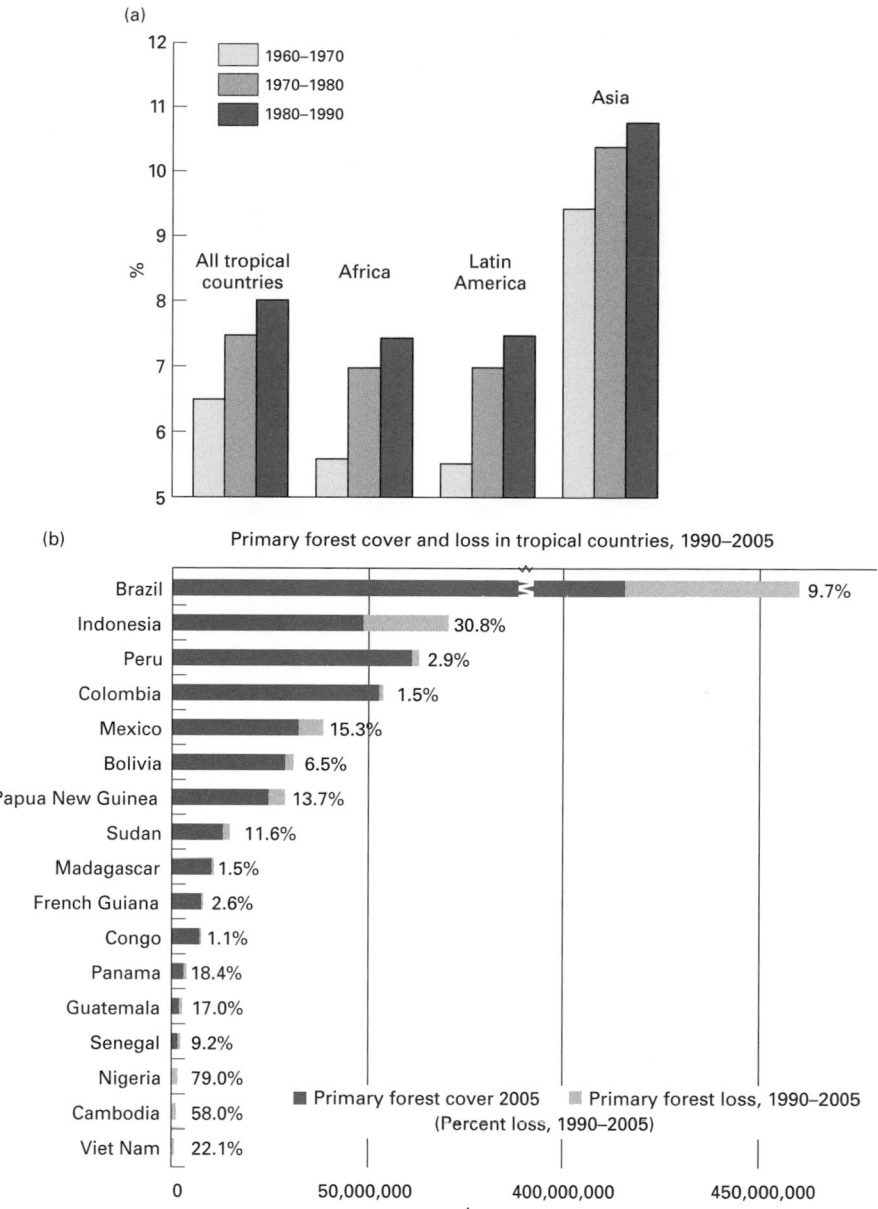

Figure 2.8 (a) Estimated rate of tropical deforestation, 1960–1990, showing the particularly high rates in Asia (modified from *World Resources, 1996–1997*). (b) Primary forest cover and loss in selected tropical countries, 1990–2005, based on FAO data.

recent examination of rates of deforestation in selected hot spots is provided by Achard et al. (2002) (Table 2.3b). They suggest that the annual deforestation rate for rainforests is 0.38% for Latin America, 0.43% for Africa and 0.91% for Southeast Asia, but that the hot spot areas can have rates that are much greater (e.g. up to 5.9% in central Sumatra and up to 4.7% in Madagascar). These rates are not as high as those of Myers, but are still extremely rapid.

The rapid loss of rain forest is potentially extremely serious, because as Poore (1976: 138) stated, these forests are a source of potential foods, drinks, medicines, contraceptives, abortifacients, gums, resins, scents, colorants, specific pesticides and so on. Their removal may contribute to crucial global environmental concerns (e.g. climatic change and loss of biodiversity), besides causing regional and local problems, including lateritization, increased rates of erosion and accelerated

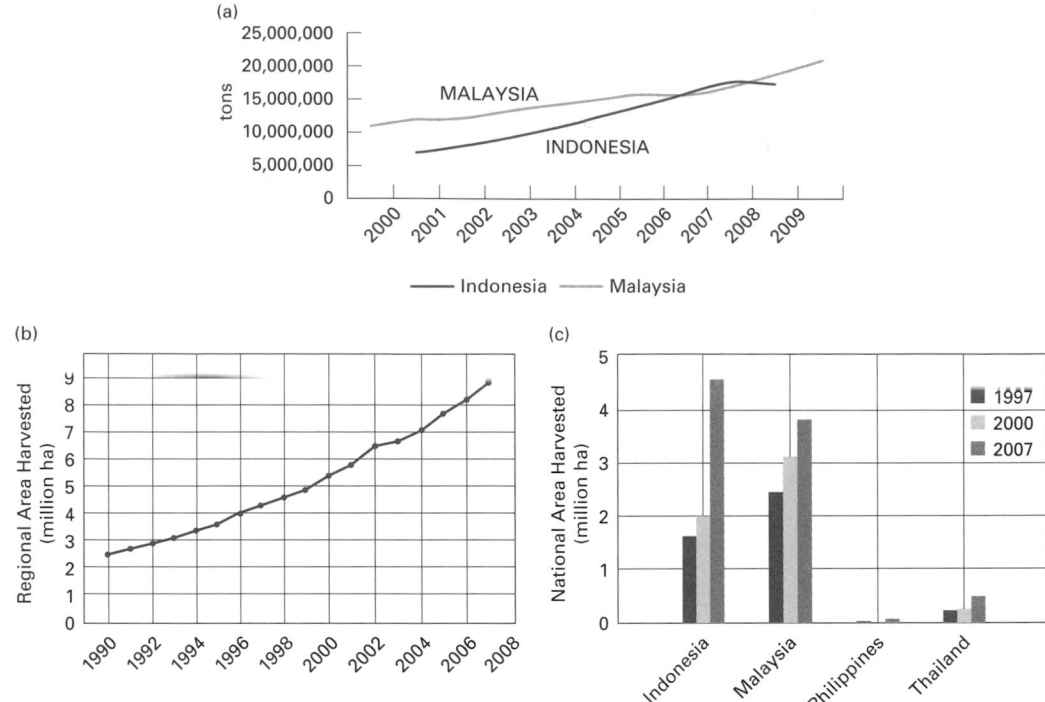

Figure 2.9 Palm oil trends. (a) Palm oil production in Malaysia and Indonesia, 2000–2009. (b) Area of palm oil harvested in Southeast Asia, 1990–2007. (c) National area of palm oil harvested in selected major producers in Southeast Asia (based on FAO data).

mass movements. The great range of potential impacts of tropical deforestation is summarized in Table 2.4.

Some traditional societies have developed means of exploiting the rain forest environment which tend to minimize the problems posed by soil fertility deterioration, soil erosion and vegetation degradation. Such a system is known as shifting agriculture (*swidden*). As Geertz (1963: 16) remarked:

In ecological terms, the most distinctive positive characteristic of swidden agriculture . . . is that it is integrated into and, when genuinely adaptive, maintains the general structure of the pre-existing natural ecosystem into which it is projected, rather than creating and sustaining one organised along novel lines and displaying novel dynamics.

The tropical rain forest and the swidden plots have certain common characteristics. Both are closed cover systems, in part because in swidden some trees are left standing, in part because some tree crops (such as banana, papaya, areca, etc.) are planted, but also because food plants are not planted in an open field, crop-row manner, but helter-skelter in a tightly woven, dense botanical fabric. It is, in Geertz's words (1963:

Table 2.4 The consequences of tropical deforestation. Source: From Grainger (1992). Reproduced with permission

1 Reduced biological diversity
  (a) Species extinction
  (b) Reduced capacity to breed improved crop varieties
  (c) Inability to make some plants economic crops
  (d) Threat to production of minor forest products
2 Changes in local and regional environments
  (a) More soil degradation
  (b) Changes in water flows from catchments
  (c) Changes in buffering of water flows by wetland forests
  (d) Increased sedimentation of rivers, reservoirs, etc.
  (e) Possible changes in rainfall characteristics
3 Changes in global environments
  (a) Reduction in carbon stored in terrestrial biota
  (b) Increase in carbon dioxide content of atmosphere
  (c) Changes in global temperature and rainfall patterns by greenhouse effects
  (d) Other changes in global climate due to changes in land surface processes

25) 'a miniaturized tropical forest'. Secondly, swidden agriculture normally involves a wide range of cultigens, thereby having a high diversity index like the rain forest itself. Thirdly, both swidden plots and the rain forest have high quantities of nutrients locked up in the biotic community compared to that in the soil. The primary concern of 'slash-and-burn' activities is not merely the clearing of the land, but rather the transfer of the rich store of nutrients locked up in the prolific vegetation of the rain forest to a botanical complex whose yield to people is a great deal larger. If the period of cultivation is not too long and the period of fallow is long enough, an equilibrated, non-deteriorating and reasonable productive farming regime can be sustained in spite of the rather impoverished soil base upon which it rests.

Unfortunately this system often breaks down, especially when population increase precludes the maintenance of an adequately long fallow period. When this happens, the rain forest cannot recuperate and is replaced by a more open vegetation assemblage ('derived savanna') which is often dominated by the notorious *Imperata* savanna grass which has turned so much of Southeast Asia into a green desert. *Imperata cyclindrica* (also called cogongrass) is a tall grass which springs up from rhizomes and is highly invasive. Because of its rhizomes it is fire-resistant, but since it is a tall grass it helps to spread fire (Gourou, 1961).

Given the perceived and actual severity of tropical deforestation it is evident that strategies need to be developed to reduce the rate at which it is disappearing. These include:

- *research, training and education* to give people a better understanding of how forests work and why they are important, and to change public opinion so that more people appreciate the uses and potential of forests;
- *land reform* to reduce the mounting pressures on landless peasants caused by inequalities in land ownership;
- *conservation of natural resources* by setting aside areas of rain forest as National Parks or nature reserves;
- *restoration and reforestation* of damaged forests;
- *sustainable development*, namely development which, while protecting the habitat, allows a type and level of economic activity that can be sustained into the future with minimum damage to people or forest

(e.g. selective logging rather than clear felling; promotion of non-tree products; small-scale farming in plots within the forest);
- *control of the timber trade* (e.g. by imposing heavy taxes on imported tropical forest products and outlawing the sale of tropical hardwoods from non-sustainable sources);
- *'debt-for-nature' swaps* whereby debt-ridden tropical countries set a monetary value on their ecological capital assets (in this case forests) and literally trade them for their international financial debt;
- *improvement of local peoples* in managing the remaining rain forests;
- *careful control of international aid* and development funds to make sure that they do not inadvertently lead to forest destruction;
- *reducing demand for wood products*.

One particular type of tropical forest ecosystem coming under increasing pressure from various human activities is the mangrove forest characteristic of inter-tidal zones (Polidoro et al. 2010) (Figure 2.10). These ecosystems constitute a reservoir, refuge, feeding ground and nursery for many useful and unusual plants and animals (Mercer and Hamilton, 1984). In particular, because they export decomposable plant debris into adjacent coastal waters, they provide an important energy source and nutrient input to may tropical estuaries. In addition they can serve as buffers against the erosion caused by tropical storms – a crucial consideration in low-lying areas like Bangladesh. In spite of these advantages, mangrove

Figure 2.10 Mangrove swamps are highly productive and diverse ecosystems that are being increasingly abused by human activities. (See Plate 12)

forests are being degraded and destroyed on a large scale in many parts of the world, either through exploitation of their wood resources or because of their conversion to single-use systems such as agriculture, aquaculture, salt-evaporation ponds or housing developments. Data from FAO suggest that the world's mangrove forests, which covered 19.8 million hectares in 1980, have now been reduced to only 14.7 million hectares, with the annual loss running at about 1% per year (compared to 1.7% a year from 1980 to 1990) (http://www.fao.org/DOCREP/005/Y7581E/y7581e04.htm). However, some areas are now being relatively well protected, including the Sundarbans of Bangladesh.

On a global basis Richards (1991: 164) calculated that since 1700 about 19% of the world's forests and woodlands have been removed. Over the same period the world's cropland area has increased by over four and a half times, and between 1950 and 1980 it amounted to well over 100,000 km^2 per year.

Deforestation is not an unstoppable or irreversible process. What is called the '**forest transition**' has taken place in some parts of the world. This is a shift from net deforestation to net reforestation The forest transition is now being recognized in some developing countries including Vietnam, China and India (Mather, 2007; Meyfroidt and Lambin, 2008), but has also been a feature of many European and other developed countries (Meyfroidt and Lambin, 2009; Rudel et al., 2005; 2010). Around the Mediterranean Sea, for instance, farmland abandonment has been widely reported, caused by national and international migrations and because market conditions made farming unprofitable in areas with steep slopes, small field sizes, and with difficulties for access and mechanization (García-Ruiz et al., 2011). In the UK, the area of woodland has increased from 1.2 million hectares in 1924 to 2.8 million hectares in 2009 (Figure 2.11). In the USA the forested area has increased substantially since the 1930s and 1940s, though it is possible that some authors have exaggerated the extent of abandoned agricultural land and the amount of forest regrowth that has occurred (Ramankutty et al., 2010). Nevertheless, this 'rebirth of the forest' (Williams, 1988) has a variety of causes: new timber growth and planting was rendered possible because the old forest had been removed, forest fires have been suppressed and controlled, farmland has been abandoned and reverted to forest, and there has been a falling demand for lumber and lumber derived

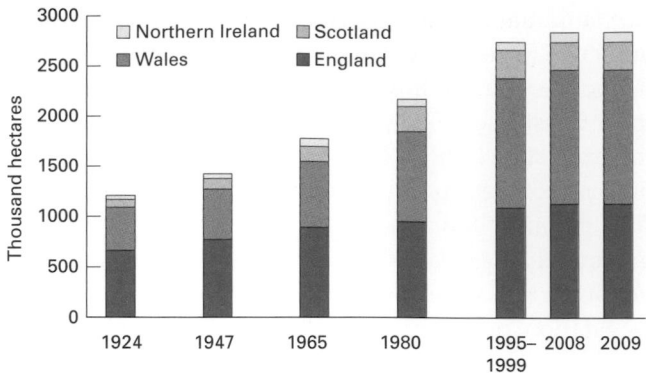

Figure 2.11 The area of woodland in the UK from 1924 to 2009 (From Forestry Commission Statistics (DEFRA, 2011)).

products. It is, however, often very difficult to disentangle the relative importance of grazing impacts, fire and fluctuating climates in causing changes in the structure of forested landscapes. In some cases, as for example the Ponderosa pine forests of the American south-west, all three factors may have contributed to the way in which the park like forest of the nineteenth century has become significantly denser and younger today (Savage, 1991). Even when forest recovery occurs, the type of forest that develops may not compositionally be the same as it was prior to clearance. Rhemtulla et al. (2009), compare forests for three time slices in Wisconsin (mid-1800s, 1930s-2000s) and document the recovery of forests since the peak agricultural extent that occurred in the 1930s. They found that in the south of the state, maples and other shade-tolerant trees were thriving in areas which prior to clearance had been dominated by oak forests and savannas. In southern Wisconsin the original savanna ecosystems were maintained through a frequent low-intensity fire regime. Such fires are now suppressed so that the forests are succeeding towards species that are less fire resistant and more shade tolerant.

Having considered the importance of the three basic processes of fire, grazing and deforestation, we can now turn to a consideration of some of the major changes in vegetation types that have taken place over extensive areas as a result of such activities.

Secondary rain forest

When an area of rain forest that has been cleared for cultivation or timber exploitation is abandoned by humans, the forest begins to regenerate; but for an extended period of years the type of forest that occurs – **secondary forest** – is very different in character from the virgin one it replaces. The features of such secondary forest, which is widespread in many tropical regions and accounts for as much as 40% of the total forest area in the tropics, have been summarized by Brown and Lugo (1990) and Corlett (1995).

First, secondary forest is lower and consists of trees of smaller average dimensions than those of primary forest; but since it is comparatively rare that an area of primary forest is clear-felled or completely destroyed by fire, occasional trees much larger than average are usually found scattered through secondary forest. Secondly, very young secondary forest is often remarkably regular and uniform in structure, though the abundance of small climbers and young saplings gives it a dense and tangled appearance which is unlike that of primary forest and makes it laborious to penetrate. Thirdly, secondary forest tends to be much poorer in species than primary, and is sometimes, though by no means always, dominated by a single species or a small number of species. Fourthly, the dominant trees of secondary forest are light-demanding and intolerant of shade, most of the trees possess efficient dispersal mechanisms (having seeds or fruits well adapted for transport by wind or animals), and most of them can grow very quickly. The rate of net primary production of secondary forests exceeds that of the primary forests by a factor of 2 (Brown and Lugo, 1990). Some species are known to grow at rates of up to 12 m in 3 years, but they tend to be short-lived and to mature and reproduce early. One consequence of their rapid growth is that their wood often has a soft texture and low density.

There is an increasing number of studies of the nature of secondary forests of differing ages and of the rates at which regrowth can occur (Chazdon et al., 2007). For example, Gehring et al. (2005), working in central Amazonia, found that 25 years of regrowth restored half of the mature forest biomass, and they estimated that 75% would be restored after 175 years. Bonnell et al. (2011), working in East Africa, found that it would take 95–112 years for structural recovery of logged areas to approach that of adjacent mature forest, but considerably longer for the original species composition to occur. Liebsch et al. (2008) estimated that it might take about one to four thousand years for

a secondary forest to reach the endemism levels that exist in mature forests. This was a finding similar to that of Chai and Tanner (2011) working in Jamaica. They found that it would take one or two centuries for tree density and basal area to match that of old growth forests but that even longer periods would be required to fully recover original species composition. However, secondary forests can be species rich, though the composition of species will not be the same (Finegan, 1996). They become more species rich with age and in eastern Brazil Piotto et al. (2009) found that more than half the species found in a 40-year-old secondary forest were shared with the neighbouring old growth forests.

Secondary forests should not be dismissed as useless scrub. All but the youngest secondary forests are probably effective at preventing soil erosion, regulating water supply and maintaining water quality. They provide refuge for some flora and fauna and can provide a source of timber, albeit generally of inferior quality than derived from primary forest. In addition, secondary forest regrowth can have a profound effect on water resources (see e.g. Wilby and Gell, 1994).

The human role in the creation and maintenance of savanna

The **savannas** (Figure 2.12) can be defined, following Hills (1965: 218–219), as:

a plant formation of tropical regions, comprising a virtually continuous ecologically dominant stratum of more or less

Figure 2.12 A tropical savanna in the Kimberley District of north-west Australia. (See Plate 13)

xeromorphic plants, of which herbaceous plants, especially grasses and sedges, are frequently the principal, and occasionally the only, components, although woody plants often of the dimension of trees or palms generally occur and are present in varying densities.

They are 'unique among the world's biomes because they are the only major one defined by the coexistence of two major plant forms – grasses and trees' (Laris, 2011: 1067). They are extremely widespread in low latitudes (Harris, 1980), covering about 18 million km² (an area about 2.6 million km² greater than that of the tropical rain forest). They occupy about 65% of the African continent and are home to a fifth of the global human population and a large portion of the world's ungulates. Their origin has been the subject of great contention in the literature of biogeography, though most savanna research workers agree that no matter what savanna origins may be, the agent which seems to maintain them is intentional or inadvertent burning (Scott, 1977). As with most major vegetation types a large number of interrelated factors are involved in causing savanna (Mistry, 2000), and too many arguments about origins have neglected this fact (Figure 2.13). Confusion has also arisen because of the failure to distinguish clearly between predisposing, causal, resulting and maintaining factors (Hills, 1965). It appears, for instance, that in the savanna regions around the periphery of the Amazon basin, the climate *predisposes* the vegetation towards the development of savanna rather than forest. The geomorphic evolution of the landscape may be a *causal* factor; increased laterite

development a *resulting* factor; and fire, *a maintaining* factor.

Originally, however, savanna was envisaged as a predominantly natural vegetation type of climatic origin (see e.g. Schimper, 1903). According to supporters of this theory, savanna is better adapted than other plant formations to withstand the annual cycle of alternating soil moisture: rain forests could not resist the extended period of extreme drought, while dry forests could not compete successfully with perennial grasses during the equally lengthy period of large water surplus (Sarmiento and Monasterio, 1975). Strong correlations have been established between the degree of woody cover in African savannas and mean annual precipitation (Sankaran et al., 2008).

Other workers have championed the importance of edaphic (soil) conditions, including poor drainage, soils which have a low water-retention capacity in the dry season, soils with a shallow profile due to the development of a lateritic crust and soils with a low nutrient supply (either because they are developed on a poor parent rock such as quartzite, or because the soil has undergone an extended period of leaching on an old land surface). Associated with soil characteristics, ages of land surfaces and degree of drainage is the geomorphology of an area. This may also be an important factor in savanna development. Soil moisture is certainly a major control of vegetation patterns in the savannah landscapes of West Africa, as is bedrock lithology (Duvall, 2011).

Some other researchers – for example, Eden (1974) – found that savannas are the product of former drier

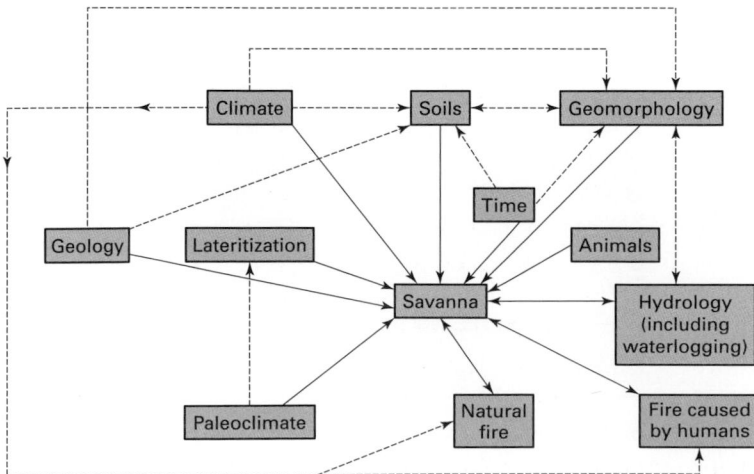

Figure 2.13 The interrelated factors involved in the formation of savanna vegetation.

conditions (such as late Pleistocene aridity) and that, in spite of a moistening climate they have been maintained by fire. He pointed to the fact that the patches of savanna in southern Venezuela occur in forest areas of similar humidity and soil infertility, suggesting that neither soil, nor drainage, nor climate was the cause. Moreover, the present 'islands' of savanna are characterized by species which are also present elsewhere in tropical American savannas and whose disjunct distribution conforms to the hypothesis of previous widespread continuity of that formation.

The importance of fire in maintaining and originating some savannas is suggested by the fact that many savanna trees are fire-resistant. Periodic fires maintain the structure of a savanna by preventing the transition to dense woodland or forest, and fire differentiates between areas with a high and a low tree cover, especially in regions with rainfall between 1000 and 2000 mm per year (Staver et al., 2011). There are also many observations of the frequency with which, for example, African herdsmen and agriculturists burn over much of tropical Africa and thereby maintain grassland. Savannas are the most frequently burned ecosystems in the world (Beerung and Osborne, 2006). Certainly the climate of savanna areas is conductive to fire for, as Gillon (1983: 617) put it, 'Large scale grass fires are more likely to take place in areas having a climate moist enough to permit the production of a large amount of grass, but seasonally dry enough to allow the dried material to catch fire and burn easily.' On the other hand Morgan and Moss (1965) expressed some doubts about the role of fire in western Nigeria and point out that fire is not itself necessarily an independent variable:

The evidence suggests that some notions of the extent and destructiveness of savanna fires are rather exaggerated. In particular the idea of an annual burn which affects a large proportion of the area in each year could seem to be false. It is more likely that some patches, peculiarly susceptible to fire, as a result of edaphic or biotic influences upon the character of the community itself, are repeatedly burned, whereas others are hardly, if ever, affected . . . It is also important to note that there is no evidence anywhere along the forest fringe . . . to suggest that fire sweeps into the forest, effecting notable destruction of forest trees.

Some savannas are undoubtedly natural, for pollen analysis in South America shows that savanna vegeta-

tion was present before the arrival of humans. The origin of modern savannah biomes dates back to the late Miocene, *c.* 8 million years ago (Beerung and Osborne, 2006). Many modern savannas, however, are 'nature-society hybrids' (Duvall, 2011). Even natural savannas, when subjected to human pressures, change their characteristics. For example, the inability of grass cover to maintain itself over long periods in the presence of heavy stock grazing may be documented from many of the warm countries of the world (Johannessen, 1963). Heavy grazing tends to remove the fuel (grass) from much of the surface. The frequency of fires is therefore significantly reduced, and tree and bush invasion take place. As Johannessen (p. 111) wrote: 'Without intense, almost annual fires, seedlings of trees and shrubs are able to invade the savannas where the grass sod has been opened by heavy grazing . . . the age and size of the trees on the savannas usually confirm the relative recency of the invasion.' In the case of the savannas of interior Honduras, he reports that they only had a scattering of trees when the Spaniards first encountered them, whereas now they have been invaded by an assortment of trees and tall shrubs.

Elsewhere in Latin America changes in the nature and density of population have also led to a change in the nature and distribution of savanna grassland. C.F. Bennett (1968: 101) noted that in Panama the decimation of the Indian population saw the re-establishment of trees:

Large areas which today are covered by dense forest were in farms, grassland or low second growth in the early sixteenth century when the Spaniards arrived. At that time horses were ridden with ease through areas which today are most easily penetrated by river, so dense has the tree growth become since the Indians died away.

Such bush or shrub encroachment is a serious cause of rangeland deterioration in savanna regions (Roques et al., 2001), not least in South Africa (Wigley et al., 2009), though some communal land users recognize that bush encroachment can be advantageous by providing increased woody resources for building and firewood and increased browse availability (Wigley et al., 2009). It can be defined as 'The suppression of palatable grasses and herbs by encroaching woody species often unpalatable to domestic livestock' (Ward, 2005: 101). It is extremely expensive to remove established dense scrub cover by mechanical means, though some success has been achieved by introducing animals that have a browsing or bulldozing effect, such as goats, giraffes and elephants. A prime cause of bush encroachment, it is often asserted, is overgrazing, especially around cattle watering points and near settlements (Moleele et al., 2002; Gil-Romera et al., 2011). This reduces the vigour of favourable perennial grasses, which tend to be replaced, as we have seen, by less reliable annuals and by woody vegetation. Annual grasses do not adequately hold or cover the soil, especially early in the growing season. Thus runoff increases and topsoil erosion occurs. Less water is then available in the topsoil to feed the grasses, so that the woody species, which depend on deeper water, become more competitive relative to the grasses which can only use water within their shallow root zones. The situation is exacerbated when there are few browsers in the herbivore population and, once established, woody vegetation competes effectively for light and nutrients. Increasing levels of grazing also reduce fuel load, making fire less intense and thus less damaging to trees. Deliberate reductions in fire use can have a similar effect and enable trees and shrubs to flourish. Conversely, fire frequency can be increased because of the spread of invasive grasses (Foxcroft et al., 2010; Setterfield et al., 2010). Tree densities in some savannas appear to have increased in recent decades (Buitenwerf et al., 2012) as a possible response to increased atmospheric carbon dioxide levels (Bond and Midgley, 2012).

Another example of human interventions modifying savanna character is through their effects on savanna-dwelling mammals such as the elephant. They are what is known as a '**keystone species**' because they exert a strong influence on many aspects of the environment in which they live (Waithaka, 1996). They diversify the ecosystems which they occupy and create a mosaic of habitats by browsing, trampling and knocking over of bushes and trees. They also disperse seeds through their eating and defecating habits and maintain or create water holes by wallowing. All these roles are beneficial to other species. Conversely, where human interference prevents elephants from moving freely within their habitats and leads to their numbers exceeding the carrying capacity of the savanna, their effect can be environmentally catastrophic. Equally, if humans reduce elephant numbers in a particular piece of savanna, the savanna may become less diverse and

less open, and its water holes may silt up. This will be to the detriment of other species.

The spread of desert vegetation on desert margins

One of the most contentious and important environmental issues of recent years has been the debate on the question of the alleged expansion of deserts (Middleton and Thomas, 1997).

The term 'desertification' was first used but not formally defined by Aubréville (1949), and for some years the term 'desertization' was also employed, for example, by Rapp (1974: 3) who defined it as: 'The spread of desert-like conditions in arid or semi-arid areas, due to man's influence or to climatic change'.

There has been some variability in how 'desertification' itself is defined. Some definitions stress the importance of human causes (e.g. Dregne, 1986: 6–7):

Desertification is the impoverishment of terrestrial ecosystems under the impact of man. It is the process of deterioration in these ecosystems that can be measured by reduced productivity of desirable plants, undesirable alterations in the biomass and the diversity of the micro and macro fauna and flora, accelerated soil deterioration, and increased hazards for human occupancy.

Others admit the possible importance of climatic controls but give them a relatively inferior role (e.g. Sabadell et al., 1982: 7):

The sustained decline and/or destruction of the biological productivity of arid and semi arid lands caused by man made stresses, sometimes in conjunction with natural extreme events. Such stresses, if continued or unchecked, over the long term may lead to ecological degradation and ultimately to desert-like conditions.

Yet others, more sensibly, are even-handed or open-minded with respect to natural causes (e.g. Warren and Maizels, 1976: 1):

A simple and graphic meaning of the word 'desertification' is the development of desert like landscapes in areas which were once green. Its practical meaning . . . is a sustained decline in the yield of useful crops from a dry area accompanying certain kinds of environmental change, both natural and induced.

It is also by no means clear how extensive desertification is or how fast it is proceeding. Indeed, the lack of agreement on the former makes it impossible to determine the latter. As Grainger (1990: 145) remarked in a well-balanced review, 'Desertification will remain an ephemeral concept to many people until better estimates of its extent and rate of increase can be made on the basis of actual measurements.' He continued (p. 157): 'The subjective judgements of a few experts are insufficient evidence for such a major component of global environmental change . . . Monitoring desertification in the drylands is much more difficult than monitoring deforestation in the humid tropics, but it should not be beyond the ingenuity of scientists to devise appropriate instruments and procedures.'

The United Nations Environment Programme (UNEP) has played a central role in the promotion of desertification as a major environmental issue, as is made evident by the following remark by Tolba and El-Kholy (1992: 134):

Desertification is the main environmental problem of arid lands, which occupy more than 40 per cent of the total global land area. At present, desertification threatens about 3.6 billion hectares – 70% of potentially drylands, or nearly one-quarter of the total land area of the world. These figures exclude natural hyper-arid deserts. About one sixth of the world's population is affected.

However, in their book *Desertification: exploding the myth*, Thomas and Middleton (1994) have discussed UNEP's views on the amount of land that is desertified. They state:

The bases for such data are at best inaccurate and at worst centred on noting better than guesswork. The advancing desert concept may have been useful as a publicity tool but it is not one that represents the real nature of desertification processes. (Thomas and Middleton, 1994: 160)

Their views have been trenchantly questioned by Stiles (1995), but huge uncertainties do indeed exist.

There are relatively few reliable studies of the rate of supposed desert advance. Lamprey (1975) attempted to measure the shift of vegetation zones in the Sudan (see Figure 2.14) and concluded that the Sahara had advanced by 90–100 km between 1958 and 1975, the average rate of about 5.5 km per year. However, on the basis of analysis of remotely sensed data and ground observation, Helldén (1985) found sparse evidence that

— ·· — ·· Limit of Khartoum's exploitation of wood and charcoal
— — —   Approximate desert boundary

Figure 2.14 Desert encroachment in the northern Sudan 1958–1975, as represented by the position of the boundary between sub-desert scrub and grassland in the desert (modified from Rapp et al., 1976, figure 8.5.3). Also shown is the expanding wood and charcoal exploitation zone south of Khartoum, Sudan (modified from Johnson and Lewis, 1995, figure 6.2).

this had in fact happened. One problem is that there may be very substantial fluctuations in biomass production from year to year. This has been revealed by meteorological satellite observations of green biomass production levels on the south side of the Sahara (Dregne and Tucker, 1988).

The spatial character of desertification is also the subject of some controversy (Helldén, 1985). Contrary to popular rumour, the spread of desert-like conditions is not an advance over a broad front in the way that a wave overwhelms a beach. Rather, it is like a 'rash' which tends to be localized around settlements (Figure 2.15). It has been likened to Dhobi's itch – a ticklish problem in difficult places. Fundamentally, as Mabbutt (1985: 2) has explained, 'the extension of desert-like conditions tends to be achieved through a process of

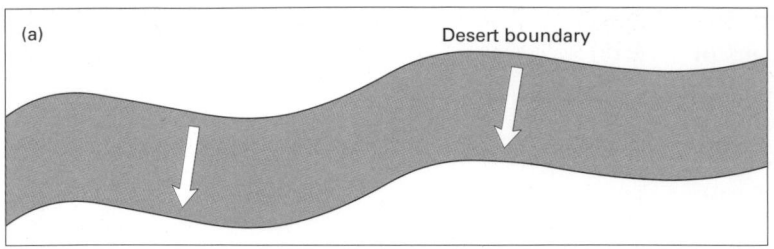

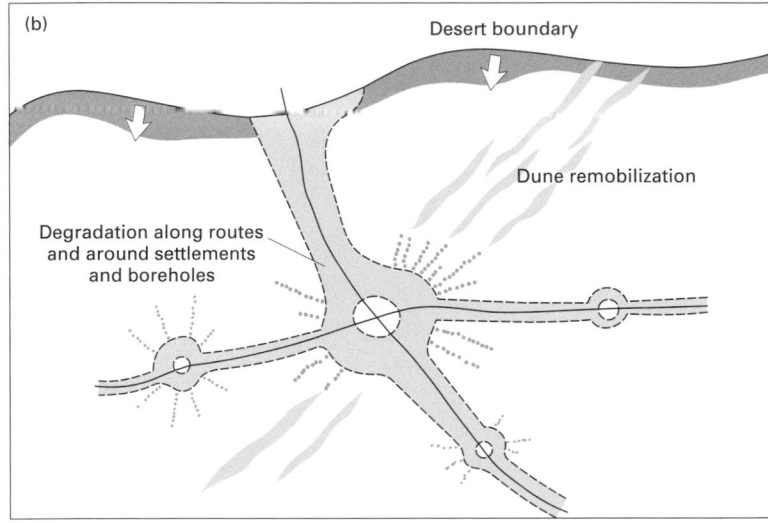

Figure 2.15 Two models of the spatial pattern of desertification: (a) The myth of desert migration over a wide front; (b) The more local development of degradation around routes, settlement and boreholes (modified from Kadomura, 1994, figure 9).

accretion from without, rather than through expansionary forces acting from within the deserts.' This distinction is important in that it influences perceptions of appropriate remedial or combative strategies, which are discussed in Goudie (1990). Figure 2.15 shows two models of the spatial pattern of desertification. One represents the myth of a desert spreading outward over a wide front, while the other shows the more realistic pattern of degradation around routes and settlements.

Woodcutting is an extremely serious cause of vegetation decline around almost all towns and cities of the Sahelian and Sudanian zones of Africa. Many people depend on wood for domestic uses (cooking, heating, brick manufacture, etc.), and the collection of wood for charcoal and firewood is an especially serious problem in the vicinity of urban centres. This is illustrated for Khartoum in Sudan in Figure 2.14. Likewise, the installation of modern boreholes has enabled rapid multiplication of livestock numbers and large-scale destruction of the vegetation in a radius of 15–30 km around boreholes. Given this localization of degradation, amelioration schemes such as local tree-planting

may be partially effective, but ideas of planting green belts as a 'cordon sanitaire' along the desert edge (whatever that is) would not halt deterioration of the situation beyond this Maginot line (Warren and Maizels, 1976: 222). The deserts are not invading from without; the land is deteriorating from within. Nonetheless, proposals for the creation of 'green walls' are still made, the latest being for a strip of forest 15 km wide and 7675 km long stretching from Dakar to Djibouti (Minassian, 2011).

There has been considerable debate as to whether the vegetation change and environmental degradation associated with desertization is irreversible. In many cases, where ecological conditions are favourable because of the existence of such factors as deep sandy soils or beneficial hydrological characteristics, vegetation recovers once excess pressures are eliminated. There is evidence of this in arid zones throughout the world where temporary or permanent enclosures have been set up (Le Houérou, 1977). The speed of recovery will depend on how advanced deterioration is, the size of the area which is degraded, the nature of the soils and moisture resources, and the character of local vegetation. It needs to be remembered in this context that much desert vegetation is adapted to drought and to harsh conditions, and that it often has inbuilt adaptations which enable a rapid response to improved circumstances.

Nonetheless, in certain specific circumstances recovery is slow and so limited that it may be appropriate to talk of 'irreversible desertization'. In North Africa, decades after the end of hostilities, wheel tracks and degraded vegetation produced in World War II are still present in the desert. Some desert surfaces, such as those covered by lichens, are easily destroyed by vehicular traffic but take a long time to recover (Figure 2.16).

The causes of desertification are also highly controversial and diverse (Table 2.5). The question has been

Figure 2.16 A lichen field, near Swakopmund in Namibia, shows the effects of vehicular traffic. Such scars can take a long time to recover. (See Plate 14)

Table 2.5 Some examples of causes and consequences of desertification. Source: Williams (2000, table 2)

Trigger factor	Consequences
• Direct land use • Overcultivation (decreased fallows, mechanized farming) • Overgrazing • Mismanagement of irrigated lands • Deforestation (burning, fuel and fodder collection) • Exclusion of fire	• Physical processes affected • Decline in soil structure, water permeability, depletion of soil nutrients and organic matter, increases susceptibility to erosion, compaction of soil, sand dune mobilization • Loss of biodiversity and biomass, increased soil erosion via wind and water, soil compaction from trampling, increased runoff, sand dune mobilization • Causes water logging and salinization of soil, hence lower crop yields, possible sedimentation of water reservoirs • Promotes artificial establishment of savanna vegetation, loss of soil-stabilizing vegetation, soil exposed and eroded, increases aridity, increased frequency of dust storms, sand dune mobilization (e.g. Ethiopia is losing approx. 1000 million tonnes of topsoil per year) • Promotes growth of unpalatable woody shrubs at the expense of herbage
Indirect government policies Failed population planning policies Irrigation subsidies Settlement policies/land tenure Improved infrastructure (e.g. roads, large-scale dams, canals, boreholes) Promotion of cash crops and push towards nation/international markets Price increases on agricultural produce War High interest rates	• Drives overexploitation land-use practices • Increases need for food cultivation, hence overexploitation • Exacerbates flooding and salinization • Forces settlement of nomads, promotes concentrated use of land which often exceeds the carrying capacity • Although beneficial, can exacerbate the problem by attracting increased livestock and human populations or increasing risk from salinization, possible lowering of groundwater table below dams, problems of silting up reservoirs, water-logging; promotes large-scale commercial activity with little local benefit, flooding may displace people and perpetuate cycles of poverty • Displaces subsistence cropping, pushes locals into marginal areas to survive, promotes less resilient monocultures, and promotes expansion and intensification of land use • Incentive to crop on marginal lands • Valuable resources, both human and financial, are expended on war at the expense of environmental management and the needs of the people, large-scale migration with resultant increased pressure on receiving areas • Forcing grazing or cultivation to levels beyond land capacity
• Natural • Extreme drought • Ecological fragility	• Decreased vegetation cover and increases land vulnerability for soil erosion. Creates an environment which exacerbates overexploitation • Impact of land-use practices – impact also depends on resilience of environment

asked whether this process is the result of temporary drought periods of high magnitude, whether it is due to long-term climatic change towards aridity (either as alleged post-glacial progressive desiccation or as part of a 200-year cycle), whether it is caused by anthropogenic climatic change, or whether it is the result of human action degrading the biological environments in arid zones. There is little doubt that severe droughts do take place and that their effects become worse as human and domestic animal populations increase. The devastating drought in the African Sahel from the mid 1960s caused greater ecological stress than the droughts of 1910–1915 and 1944–1948, largely because of the increasing anthropogenic pressures.

It is evident that it is largely a combination of human activities with occasional series of dry years that leads to presently observed desertization. The process also seems to be fiercest not in desert interiors, but on the less arid marginal areas around them. It is in semi-arid areas – where biological productivity is much greater than in extremely arid zones, where precipitation is frequent and intense enough to cause rapid erosion of unprotected soils, and where humans are prone to mistake short-term economic gains under temporarily favourable climatic conditions for long-term stability – that the combination of circumstances particularly conductive to desert expansion can be found. It is in these marginal areas that dry farming and cattle rearing can be a success in good years, so that susceptible areas are ploughed and cattle numbers become greater than the vegetation can support in dry years. In this way, a depletion of vegetation occurs which sets in train such insidious processes as water erosion and deflation. The vegetation is removed by clearance for cultivation, by the cutting and uprooting of woody species for fuel (Figure 2.17), by over-grazing and by the burning of vegetation for pasture and charcoal. Cow dung may be used for fuel rather than for fertilizing the ground (Figure 2.18).

These tendencies towards bad land-use practices result in part from the restrictions imposed on many nomadic societies through the imposition of national boundaries across their traditional migration routes, or through various schemes implemented for political and social reasons to encourage their establishment in settled communities. Some of their traditional grazing lands have been taken over by cash-crop farmers. In Niger, for example, there was a sixfold increase in the

Figure 2.17 Deforestation for charcoal production and for firewood is a major cause of environmental change in the dry savanna woodlands of Swaziland. (See Plate 15)

Figure 2.18 In the drylands of north-west India, the use of buffalo dung as fuel rather than as fertilizer, may have a deleterious effect on soil fertility. (See Plate 16)

acreage of peanuts grown between 1934 and 1968. The traditional ability to migrate enabled pastoral nomads and their cattle to emulate the natural migration of such wild animals as wildebeest and kob, and thereby to make flexible use of available resources according to season and according to yearly variations in rainfall. They could also move away from regions that had become exhausted after a long period of use. As soon as migrations are stopped and settlements imposed, such options are closed, and severe degradation occurs (Sinclair and Fryxell, 1985).

The suggestion has sometimes been made that not only are deserts expanding because of human activity, but that deserts themselves are created by human activity. There are authors, who have suggested, for example, that the Thar Desert of India is a post-glacial and possibly post-medieval creation (see Allchin et al., 1977, for a critique of such views), while Ehrlich and Ehrlich (1970) wrote: 'The vast Sahara desert itself is largely man-made, the result of over-grazing, faulty irrigation, deforestation, perhaps combined with a shift in the course of a jet stream.' Nothing could be further from the truth. The Sahara, while it has fluctuated greatly in extent, is many millions of years old, pre-dates human life and is the product of the nature of the general atmospheric circulation, occupying an area of dry descending air.

One of the fascinating developments in the study of desertification in recent years has been the realization that there has been an increase in the greenness of the Sahel and some other areas. This 'recovery' may partially be due to some increase in rainfall, but is also possible that some of the greening is due to improved land management (including the protection of trees) and migration from marginal areas (Rasmussen et al., 2001; Olsson et al., 2005). Repeat photography in northern Ethiopia (Nyssen et al., 2009) has demonstrated that over the last 140 years, in spite of a 10-fold increase in rural populations, there has been some regeneration of natural vegetation, the spread of new eucalypt woodlands, and improved soil and water husbandry. Land degradation in drylands is not irreversible.

The maquis of the Mediterranean lands

Around much of the Mediterranean basin there is a plant formation called **maquis** (alternative names include phrygana, matorral, macchia and garrigue). This consists of a stand of xerophilous non-deciduous bushes and shrubs which are evergreen and thick and whose trunks are normally obscured by low-level branches. It includes such plants as holly oak (*Quercus ilex*), kermes oak (*Quercus coccifera*), tree heath (*Erica arborea*), broom heath (*Erica scoparia*) and strawberry trees (*Arbutus unedo*).

Holocene pollen analyses indicate that some maquis pre-dates substantial human impact (Collins et al., 2012), and some may represent a stage in the evolution towards true forest in places where the climax has not yet been reached; but in large areas it may represent the degeneration of forest caused by intensive land use over millennia (Noti et al., 2009). Considerable concern has been expressed about the speed with which degeneration to, and degeneration beyond, maquis is taking place as a result of human influences (Tomaselli, 1977), of which cutting, grazing and fire are probably the most important and long continued. Charcoal burners, goats and frequent outbreaks of fires among the resinous plants in the dry Mediterranean summer have all taken their toll. Such aspects of degradation in Mediterranean environments are discussed in Conacher and Sala (1998).

On the other hand, in some areas, particularly marginal mountainous and semi-arid portions of the Mediterranean basin, agricultural uses of the land have declined in recent decades, as local people have sought easier and more remunerative employment Puddu et al. 2011). In such areas scrubland, sometimes termed 'post-cultural shrub formations', may start to invade areas of former cultivation (May, 1991), while maquis has developed to become true woodland. Mediterranean vegetation appears to be very resilient (Grove and Rackham, 2001).

There is considerable evidence that maquis vegetation is in part adapted to, and in part a response to, fire (Malkisnon et al., 2011). One effect of fire is to reduce the frequency of standard trees and to favour species which after burning send up a series of suckers from ground level. Both *Quercus ilex* and *Quercus coccifera* seem to respond in this way. Similarly, a number of species (e.g. *Cistus albidus*, *Erica arborea* and *Pinus halepensis*) seem to be distinctly advantaged by fire, perhaps because it suppresses competition or perhaps because (as with the comparable chaparral of the south-west USA) a short burst of heat encourages germination (Wright and Wanstall, 1977).

The prairies and other mid-latitude and high-altitude grasslands

In mid-latitudes there are extensive grasslands. Those in western Europe are for the most part secondary and have been caused by forest clearance, but the large grasslands of the Eurasian steppes, the Prairies of North America and the Pampas of Argentina are generally regarded as primary features that are a response to particular climatic conditions (Bredenkamp et al., 2002). They generally occupy areas where mean annual rainfall is about 300–1000 mm and where the length of the dry season ranges from 0 to 8 months.

The human role in the development of the North American prairies is the subject of controversy (Whitney, 1994). It was once fairly widely believed that the prairies were essentially a climatically related phenomenon (see Changnon et al., 2003, for a historical review). Workers like J.E. Weaver (1954) argued that under the prevailing conditions of soil and climate the invasion and establishment of trees was significantly hindered by the presence of a dense sod. High evapotranspiration levels combined with low precipitation were thought to give a competitive advantage to herbaceous plants with shallow, densely ramifying root systems, capable of completing their life cycles rapidly under conditions of pervasive drought. There is ample evidence that grasslands pre-date humans, with, for example, a great expansion of grasslands during the Miocene-Pliocene transition, 8–6 million years ago (Anderson, 2006).

An alternative view was, however, put forward by Stewart (1956: 128): 'The fact that throughout the tall-grass prairie planted groves of many species have flourished and have reproduced seedlings during moist years and, furthermore, have survived the most severe and prolonged period of drought in the 1930s suggests that there is no climatic barrier to forests in the area.'

Other arguments along the same lines have been advanced. Wells (1965), for example, pointed out that in the Great Plains a number of woodland species, notably the junipers, are remarkably drought-resistant and that their present range extends into the Chihuahua Desert where they often grow in association with one of the most xerophytic shrubs of the American deserts, the creosote bush (*Larrea divaricata*). He remarks (p. 247): 'There is no range of climate in the vast grassland climate of the central plains of North America which can be described as too arid for all species of trees native to the region.' Moreover, confirming Stewart, he points out that numerous plantations and shelterbelts have indicated that trees can survive for at least 50 years in a 'grassland' climate. One of his most persuasive arguments is that, in the distribution of vegetation types in the plains, a particularly striking vegetational feature is the widespread

but local occurrence of woodlands along escarpments and other abrupt breaks in topography remote from fluvial irrigation. A probable explanation for this is the fact that fire effects are greatest on flat, level surfaces, where there are high wind speeds and no interruptions to the course of fire. It has also been noted that where burning has been restricted there has been extension of woodland into grassland.

The reasons why fire tends to promote the establishment of grassland have been summarized by Cooper (1961: 150–151).

In open country fire favours grass over shrubs. Grasses are better adapted to withstand fire than are woody plants. The growing point of dormant grasses from which issues the following years growth, lies near or beneath the ground, protected from all but the severest heat. A grass fire removes only one year's growth, and usually much of this is dried and dead. The living tissue of shrubs, on the other hand, stands well above the ground, fully exposed to fire. When it is burned, the growth of several years is destroyed. Even though many shrubs sprout vigorously after burning, repeated loss of their top growth keeps them small. Perennial grasses, moreover, produce seeds in abundance one or two years after germination; most woody plants require several years to reach seed-bearing age. Fires that are frequent enough to inhibit seed production in woody plants usually restrict the shrubs to a relatively minor part of the grassland area.

Thus, as with savanna, anthropogenic fires may be a factor which maintains, and possibly forms, the prairies, and there is now some palaeoecological information to support this view (Boyd, 2002). Though again, following the analogy with savanna, it is possible that some of the American prairies may have developed in a post-glacial dry phase, and that with a later increase in rainfall re-establishment of forest cover was impeded by humans through their use of fire and by grazing animals. The grazing animals concerned were not necessarily domesticated, however, for Larson (1940) has suggested that some of the short-grass plains were maintained by wild bison. These, he believed, stocked the plains to capacity so that the introduction of domestic livestock, such as cattle, after the destruction of the wild game was merely a substitution so far as the effect of grazing on plants is concerned. There is indeed pollen analytical evidence that shows the presence of prairie in the western mid-west over 11,000 years ago, prior to the arrival of human settlers

(Bernabo and Webb, 1977). Therefore some of it, at least, may be natural.

Comparable arguments have attended the origins of the great Pampa grassland of Argentina. The first Europeans who penetrated the landscape were much impressed by the treeless open country, and it was always taken for granted, and indeed became dogma, that the grassland was a primary **climax** unit. However, this interpretation was challenged, notably by Schmieder (1927a, b), who pointed out that planted trees thrived, that precipitation levels were quite adequate to maintain tree growth and that, in topographically favourable locations such as the steep gullies (*barrancas*) near Buenos Aires, there were numerous endemic representatives of the former forest cover (*monte*). Schmieder believed that Pampa grasslands were produced by a pre-Spanish aboriginal hunting and pastoral population, the density of which had been underestimated, but whose efficiency in the use of fire was proven. On the other hand, pollen analysis has shown that pampas grassland was present in the Late Pleistocene and probably prior to the arrival of substantial human populations. Moreover, severe seasonal drought and the presence of fine textured soils favour grassland over trees in the region (Bredenkamp et al., 2002). Some of the grassland may also be derived from the formerly drier conditions of the Late Pleistocene and the drier phases of the Holocene, when dune, dust and deflation activity were greater than today.

In addition to its lowland savannas, Africa has a series of high altitude grasslands called 'Afromontane grasslands'. They extend as a series of 'islands' from the mountains of Ethiopia to those of the Cape area of South Africa. Are they the result mainly of forest clearance by humans in the recent past? Or are they a long-standing and probably natural component of the pattern of vegetation (Meadows and Linder, 1993)? Are they caused by frost, seasonal aridity, excessively poor soils or an intensive fire regime? This is one of the great controversies of African vegetation studies.

Almost certainly a combination of factors has given rise to these grasslands. On the one hand current land management practices, including the use of fire, prevent forest from expanding. There has undoubtedly been extensive deforestation in recent centuries. On the other hand, pollen analysis from various sites in southern Africa suggests that grassland was present in the area as long ago as 12,000 BP. This would mean that much grassland is not derived from forest through very recent human activities.

In recent centuries mid-latitude grasslands have been especially rapidly modified by human activities. Notable has been the role of excessive grazing in the western USA in causing grasslands to be invaded by dense covers of shrubs, such as creosote bush (*Larrea tridentata*) and mesquite (*Prosopis glandulosa*) (Eldridge et al., 2011). As Whitney (1994: 257) points out, these grasslands have been altered in ways that can be summarized in three phrases: 'ploughed out', 'grazed out' and 'worn out'. Very few areas of natural North America prairies remain. The destruction of 'these masterpieces of nature' took less than a century. The preservation of remaining areas of North American grasslands is a major challenge for the conservation movement (Joern and Keeler, 1995). On a global basis Graetz (1994) estimated that grasslands have been reduced by about 20% from their pre-agricultural extent, but a more recent study by Boakes et al. (2010) suggest that about half the world's grasslands have been converted to agriculture since 1700, a greater decline than that experienced by the world's forests (Figure 2.19).

Post-glacial vegetational change in Britain and Europe

The classic interpretation of the vegetational changes of post-glacial (Holocene) times – that is, over the past 11,000 or so years – has been in terms of climate. That changes in the vegetation of Britain and other parts of Western Europe took place was identified by pollen analysts and palaeo-botanists, and these changes were used to construct a model of climatic change – the Blytt-Sernander model. There was thus something of a chicken-and-egg situation: vegetational evidence was used to reconstruct past climates, and past climates were used to explain vegetational change. The importance of humanly induced vegetation changes in the Holocene has been described thus by Behre (1986: vii):

Human impact has been the most important factor affecting vegetation change, at least in Europe, during the last 7,000 years. With the onset of agriculture, at the so-called Neolithic revolution, the human role changed from that of a passive

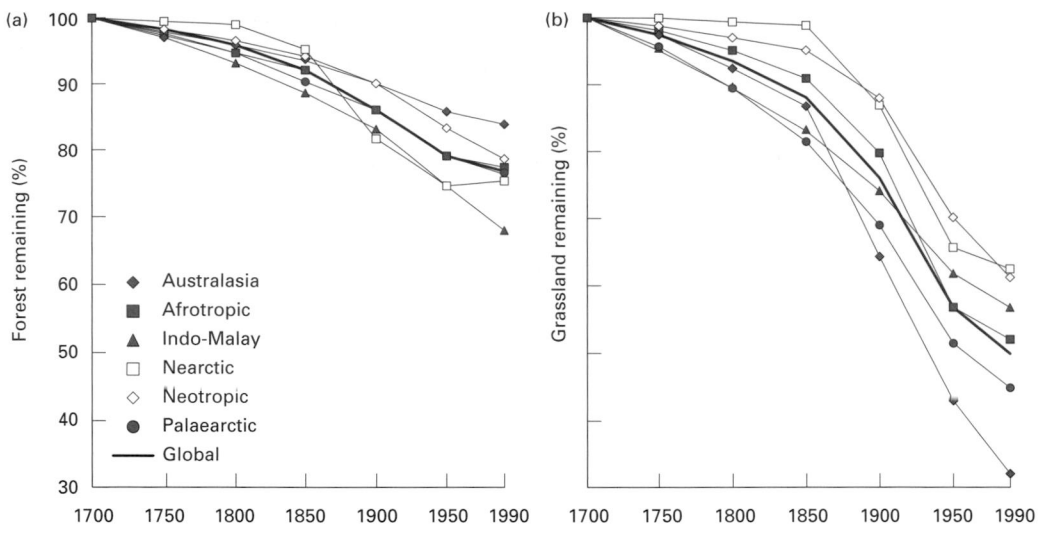

Figure 2.19 Percentage losses of (a) forest and (b) grassland, both globally and by biological realm, over the period 1700–1990 (modified from Boakes et al., 2010, figure 1).

component to an active element which impinged directly on nature. This change had dramatic consequences for the natural environment and landscape development. Arable and pastoral farming, the actual settlements themselves and the consequent changes in the economy significantly altered the natural vegetation and created the cultural landscape with its many different and varying aspects.

Recent palaeo-ecological research has indicated that certain features of the post-glacial pollen record in Europe and Britain can perhaps be attributed to the action of Mesolithic and, especially, Neolithic peoples (Innes et al., 2010). It is, for example, possible that the expansion of the alder (*Alnus glutinosa*) was not so much a consequence of supposed wetness in the Atlantic period as a result of Mesolithic colonizers. Their removal of natural forest cover helped the spread of alder by reducing competition, as possible did the burning of reed swamp. It may also have been assisted in its spread by the increased runoff of surface waters occasioned by deforestation and burning of catchment areas. Furthermore, the felling of alder itself promotes vegetative sprouting and cloning, which could result in its rapid spread in swamp forest areas (Moore, 1986).

In the Yorkshire Wolds of northern England pollen analysis suggests that Mesolithic peoples may have caused forest disturbance as early as 8900 BP. They may even have so suppressed forest growth that they permitted the relatively open landscapes of the early

post-glacial to persist as grasslands even when climatic conditions favoured forest growth (Bush, 1988). Also, late Mesolithic hunter–gatherers made extensive use of fire in the North York Moors (Innes et al., 2010) and in the Pennines (Ryan and Blackford, 2010).

One vegetational change which has occasioned particular interest is the fall in *Ulmus* pollen – the so-called elm decline – which appears in all pollen diagrams from north-west Europe, though not from North America. A very considerable number of radiocarbon dates have shown that this decline was approximately synchronous over wide areas and took place at about 5000 BP. It is now recognized that various hypotheses can be advanced to explain this major event in vegetation history: the original climatic interpretation; progressive soil deterioration; the spread of disease; or the role of people.

The climatic interpretation of the elm decline as being caused by cold, wet conditions has been criticized on various grounds (Rackham, 1980: 265):

A deterioration of climate is inadequate to explain so sudden, universal and specific a change. Had the climate become less favourable for elm, this would not have caused a general decline in elm and elm alone; it would have wiped out elm in areas where the climate had been marginal for it, but would not have affected elm at the middle of its climatic range unless the change was so great as to affect other species also. A climatic change universal in Europe ought to have some effect on North American elms.

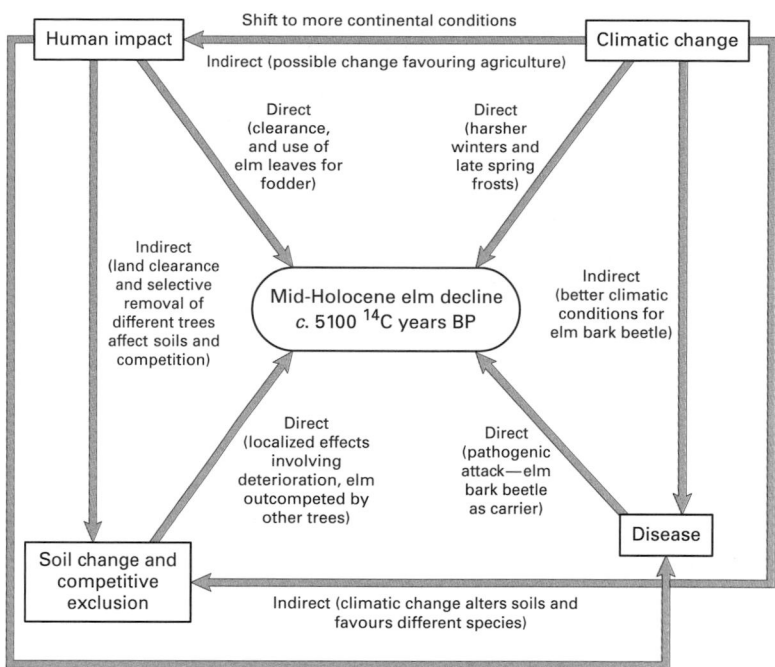

Figure 2.20 Relationships between factors influencing the mid-Holocene elm decline (after Parker et al., 2002, figure 9). With permission of Sage Publications.

Humans may have contributed to soil deterioration and affected elms thereby. Troels-Smith (1956), however, postulated that around 5000 years ago a new technique of keeping stalled domestic animals was introduced by Neolithic peoples and that these animals were fed by repeated gathering of heavy branches from those trees known to be nutritious – elms. This, it was held, reduced enormously the pollen production of the elms. In Denmark it was found that the first appearance of the pollen of a weed, Ribwort plaintain (*Plantago lanceolata*), always coincided with the fall in elm pollen levels, confirming the association with human settlements.

Experiments have also shown that Neolithic peoples, equipped with polished stone axes, could cut down mature trees and clear by burning a fair-sized patch of established forest within about a week. Such clearings were used for cereal cultivation. In Denmark a genuine chert Neolithic axe was fitted into an ash wood shaft. Three men managed to clear about 600 m² of birch forest in four hours. Remarkably, more than 100 trees were felled with one axe head, which had not been sharpened for about 4000 years (Cole, 1970: 38). Rackham, however, doubts whether humans alone could have achieved the sheer extent of change in such a short period (see also Peglar and Birks, 1993) and

postulates that epidemics of elm disease may have played a role, aided by the fact that the cause of the disease, a fungus called *Ceratocystis*, is particularly attracted to pollarded elms (Rackham, 1980: 266).

In reality the elm decline may have resulted from a complex cocktail of causes relating to climate change, soil deterioration, disease and human activity. A model of the interacting factors that could be involved (Parker et al., 2002) is shown in Figure 2.20.

The elm was not the only tree to show a decline in lowland Britain during the mid-Holocene. The lime (*Tilia*) disappeared as a result of clearance activity in the Late Neolithic to the Late Bronze Age (Grant et al., 2011).

Lowland heaths

Heathland is characteristic of temperate, oceanic conditions on acidic substrates and is composed of low shrubs, which form a closed canopy at heights which are usually less than 2 m. Dominant species are heathers (*Calluna, Erica,* etc.), woody legumes such as gorse (*Ulex*), broom (*Genista*) and various grasses (Loidi et al., 2010). Trees and tall shrubs are absent or scattered. Some heathlands are natural: for example, communities at altitudes above the forest limit on mountains

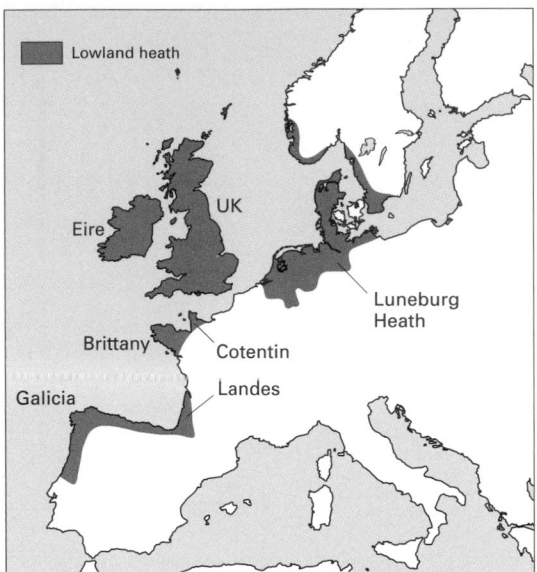

Figure 2.21 The lowland heath region of Western Europe (modified from Gimingham and de Smidt, 1983, figure 2).

and those on exposed coasts. There are also well-documented examples of heath communities which appear naturally in the course of plant succession as, for example, where *Calluna vulgaris* (heather) colonizes *Ammophila arenaria* (marram grass) and *Carex arenaria* (sand sedge) on coastal dunes.

However, at low and medium altitudes on the western fringes of Europe and North Africa between Morocco, Iberia and Scandinavia (Figure 2.21) extensive areas of heathland occur. The origin of these is strongly disputed (Gimingham and de Smidt, 1983). Some areas were once thought to have developed where there were appropriate edaphic conditions (e.g. well-drained loess or very sandy, poor soils), but pollen analysis showed that most heathlands occupy areas which were formerly tree-covered. This evidence alone, however, did not settle the question whether the change from forest to heath might have been caused by Holocene climatic change. However, the presence of human artefacts and buried charcoal, and the fact that the replacement of forest by heath has occurred at many different points in time between the Neolithic and the late nineteenth century, suggest that human actions established, and then maintained, most of the heathland areas. In particular, fire is an important management tool for heather in locations such as upland Britain, since the value of *Calluna* as a source

of food for grazing animals increases if it is periodically burned.

The area covered by heathland in Western Europe reached a peak around 1860, but since then there has been a very rapid decline. For example, by 1960 there had been a 60–70% reduction in Sweden and Denmark (Gimingham, 1981). Reductions in Britain averaged 40% between 1950 and 1984, and this was a continuation of a more long-term trend which saw a decline of 3.6 times between 1830 and 1980 in southern England (Nature Conservancy Council, 1984). The reasons for this fall are many and include unsatisfactory burning practices, peat removal, drainage fertilization, replacement by improved grassland, conversion to forest, and sand and gravel abstraction. In England, the Dorset heathlands that were such a feature of Hardy's Wessex novels are now a fraction of their former extent. The nature of lowland heaths is also currently being modified by the effects of increasing levels of nitrate additions from the atmosphere, which effects the growth of lichens and *Calluna* (Southon et al., 2012).

Thus far we have considered the human impact on general assemblages of vegetation over broad zones. However, in turning to questions such as the range of individual plant species, the human role is no less significant.

Introduction, invasion and explosion

People are important agents in the spread of plants and other organisms. Some plants are introduced deliberately by humans to new areas; these include crops, ornamental and miscellaneous landscape modifiers (trees for reafforestation, cover plants for erosion control, etc.). Indeed, some plants, such as bananas and breadfruit, have become completely dependent on people for reproduction and dispersal, and in some cases they have lost the capacity for producing viable seeds and depend on human-controlled vegetation propagation. Most cultigens are not able to survive without human attention, partly because of this low capacity for self-propagation, but also because they cannot usually compete with the better-adapted native vegetation.

However, some domesticated plants have, when left to their own devices, shown that they are capable of at least ephemeral colonization, and a small number have successfully naturalized themselves in areas other than their supposed region of origin (Gade, 1976). Examples of such plants include several umbelliferous annual garden crops (fennel, parsnip and celery) which, though native to Mediterranean Europe, have colonized wastelands in California. The Irish potato, which is native to South America, grows unaided in the mountains of Lesotho. The peach (in New Zealand), the guava (in the Philippines), coffee (in Haiti) and the coconut palm (on Indian Ocean island strands) are perennials that have established themselves as wild-growing populations, though the last-named is probably within the hearth region of its probable domestication. In Paraguay, orange trees (originating in Southeast Asia and the East Indies) have demonstrated their ability to survive in direct competition with natural vegetation.

Plants that have been introduced deliberately because they have recognized virtues (Jarvis, 1979) can be usefully divided into an economic group (e.g. crops, timber trees) and an ornamental or amenity one. In the British Isles, Jarvis believes that the great bulk of deliberate introductions before the sixteenth century had some sort of economic merit, but that only a handful of the species introduced thereafter were brought in because of their utility. Instead, plants were introduced increasingly out of curiosity or for decorative value.

A major role in such deliberate introductions was played by European botanic gardens (Figure 2.22) and those in the colonial territories from the sixteenth century onwards. Many of the 'tropical' gardens (such as those in Calcutta, Mauritius and Singapore) were often more like staging posts or introduction centres than botanic gardens in the modern sense (Heywood, 1989).

Many plants, however, have been dispersed accidentally as a result of human activity: some by adhesion to moving objects, such as individuals themselves or their vehicles; some among crop seed; some among other plants (like fodder or packing materials); some among minerals (such as ballast or road metal); and some by the carriage of seeds for purposes other than planting (as with drug plants). As illustrated in Figure 2.23, based on California, the establishment of alien species has proceeded rapidly over the last 150 years. In the Pampa of Argentina, Schmieder (1927b) estimates that the invasion of the country by European plants has taken place on such a large scale that at

Figure 2.22 A monument in the botanic gardens in Mauritius, drawing attention to the value of introduced plants. (See Plate 17)

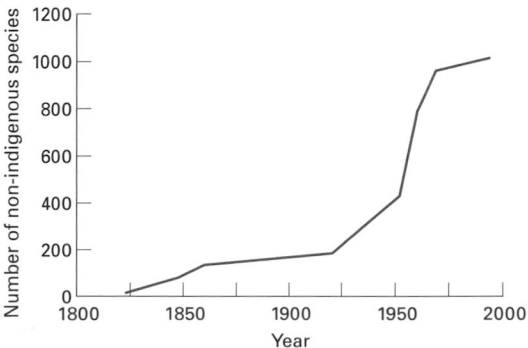

Figure 2.23 Number of non-indigenous plant species by date as reported for California (from Schwartz et al., 1996, figure 47.1. Reproduced with permission).

present only one-tenth of the plants growing wild in the Pampa are native. Many trees have invaded, particularly along roads (Ghersa et al., 2002), and introduced conifers are a menace in many Southern Hemisphere countries (Simberloff et al. 2010).

The accidental dispersal of such plants and organisms can have serious ecological consequences (see Williamson, 1996, and Henderson et al., 2006, for recent analyses) and can lead to loss of biodiversity. In Britain, for instance, many elm trees died in the 1970s because of the accidental introduction of the Dutch elm disease fungus which arrived on imported timber at certain ports, notably Avonmouth and the Thames Estuary ports (Sarre, 1978). There are also other examples of the dramatic impact of some introduced plant pathogens (von Broembsen, 1989). The American chestnut *Castanea dentata* was, following the introduction of the chestnut blight fungus *Cryphonectria parasitica* in ornamental nursery material from Asia late in the 1890s, almost eliminated throughout its natural range in less than 50 years. Two North American hemlock species, *Tsuga Canadensis* and *Tsuga caroliniana*, have been threatened by another accidental Asian invader, a small sap-sucking insect, *Adelges tsugae* (Nuckolls et al., 2009). In western Australia the great jarrah forests have been invaded and decimated by a root fungus, *Phytophthora cinnamomi*. This was probably introduced on diseased nursery material from eastern Australia, and the spread of the disease within the forests was facilitated by road building, logging and mining activities that involved movement of soil or gravel containing the fungus. More than three million hectares of forest have been affected. In South Africa (Figure 2.24), the unique *fynbos* heathland, with its remarkable biodiversity, has been threatened by the spread of introduced pines and various types of Australian acacia. Significant declines have taken place in native species richness (Gaertner et al., 2009).

In the USA, Kudzu (*Pueraria montana*) has been a particularly difficult invader. It was introduced into the USA from Japan in 1876 and was promoted as an ornamental and forage crop plant. From 1935 to the mid-1950s it was recommended as a good means of reducing soil erosion. Unfortunately this large vine, with massive tap roots and the ability to spread at *c.* 30 cm per day, now smothers large expanses of the south-eastern USA. It is fiendishly difficult to eradicate because of the nature of its root system. No less diffi-

Figure 2.24 The invasion of Australian acacia (background) is a major threat to the Fynbos (foreground) of the Cape region of South Africa. (See Plate 18)

cult has been *Imperata cylindrica*. Originating from east Asia, it was introduced to the Mobile Bay area of Alabama in the mid-1900s, as a forage crop and soil stabilizer. It has since spread throughout much of the outer coastal plain region of the south-eastern USA and because of its fire resistance has the power to dominate natural ecosystems such as the long leaf pine (*Pinus palustris*) savanna (Brewer, 2008). It has invaded nearly 500,000 hectares within the USA and over 500 million hectares worldwide (Holzmueller and Jose, 2010). Another fiendishly bad invading weed in many parts of the world is the Spanish flag (*Lantana camara* L.). A native of the American tropics, it was introduced as an ornamental garden plant to Africa, Asia and Oceania in the mid-nineteenth century. It has invaded more than 13 million hectares in India, 5 million hectares in Australia and 2 million hectares in South Africa (Bhagwat et al., 2012).

The invasibility (vulnerability of a habitat and associated biological community to invasion) of particular types of environment is still an imperfectly understood area of study (Moles et al., 2012; Catford et al., 2012). However, ocean islands have often been particularly vulnerable to plant invasions (Sax and Gaines, 2008) (Figure 2.25). The simplicity of their ecosystems inevitably leads to diminished stability, and introduced species often find that the relative lack of competition enables them to broaden their ecological range more easily than on the continents. Moreover, because the natural species inhabiting remote islands have been selected primarily for their dispersal capacity, they

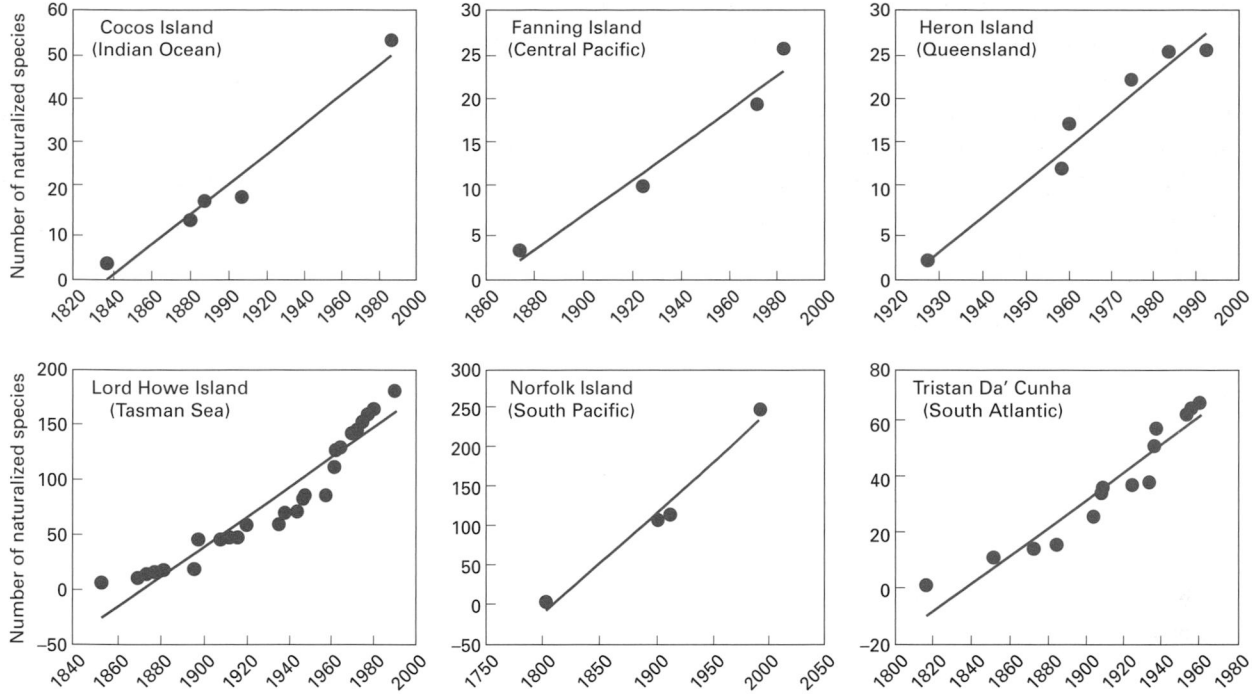

Figure 2.25 Naturalized plant richness on these six oceanic islands has increased in approximately linear fashion over the past 200 years (modified from Sax and Gaines, 2008, figure 3).

have not necessarily being dominant or even highly successful in their original continental setting. Therefore, introduced species may prove more vigorous and effective. There may also be a lack of indigenous species to adapt to conditions such as bare ground caused by humans. Many successful invasive plants have escaped the natural enemies that hold them in check, freeing them to utilize their full competitive potential (Callaway and Aschehoug, 2000). Thus introduced weeds may catch on.

Table 2.6 illustrates clearly the extent to which the floras of selected islands now contain alien species, with the percentage varying between about one-quarter and two-thirds of the total number of species present. Figure 2.25 shows the trends in the spread of naturalized species of plants for six oceanic islands over the last two hundred years.

There are a number of major threats that invasive plants pose to natural ecosystems (Levine et al., 2003). These have been discussed by Cronk and Fuller (1995) and Holzmueller and Jose (2010):

(i)   Replacement of diverse systems with single species stands of aliens, leading to a reduction in biodi-

Table 2.6 Alien plant species on oceanic islands. Source: Data from Moore (1983)

Island	Number of native species	Number of alien species	% of alien species in flora
New Zealand	1200	1700	58.6
Campbell Island	128	81	39.0
South Georgia	26	54	67.5
Kerguelen	29	33	53.2
Tristan da Cunha	70	97	58.6
Falklands	160	89	35.7
Tierra del Fuego	430	128	23.0

versity, as for example, where Australian acacias have invaded the fynbos heathlands of South Africa (Le Maitre et al., 2000);

(ii)   Direct threats to native faunas by producing a change of habitat – the invasional meltdown hypothesis;

(iii)   Alteration of soil chemistry. For example, the African *Mesembryanthemum crystallinum* (ice plant) accumulates large quantities of salt. In this way it salinizes invaded areas in Australia and may prevent the native vegetation from establishment.

Some plants exude allelopathic chemicals (weapons) that are inhibitory to native species;

(iv)   Alteration of geomorphological processes, especially rates of sedimentation and movement of mobile land forms (e.g. dunes and salt marshes);

(v)   Plant extinction by competition for light, nutrients, etc.;

(vi)   Alteration of fire regime. For example, in Florida, USA, the introduction of the Australasian myrtle (*Melaleuca quinquenervia*) has increased the frequency of fires because of its flammability and has damaged the native vegetation which is less well adapted to fire;

(vii)   Alteration of hydrological conditions (e.g. reduction in groundwater levels caused by some species having high rates of transpiration).

The introduction of new animals can have an adverse effect on plant species. A clear demonstration of this comes from the atoll of Laysan in the Hawaii group. Rabbits and hares were introduced in 1903 in the hope of establishing meat cannery. The number of native species of plants at this time was 25; by 1923 it had fallen to four. In that year all the rabbits and hares were systematically exterminated to prevent the island turning into a desert, but recovery has been slower than destruction. By 1930 there were nine species, and by 1961 sixteen species, on the island (Stoddart, 1968). Pigs are another animal that have been introduced extensively to the islands of the Pacific. Long-established feral populations are known on many islands. Like rabbits they have caused considerable damage, not least because of their non-fastidious eating habits and their propensity for rooting into the soil. This is a theme that is reviewed by Nunn (1991).

It has often been proposed that the introduction of exotic terrestrial mammals has had a profound effect on the flora of New Zealand. Among the reasons that have been put forward for this belief are that the absence of native terrestrial mammalian herbivores permitted the evolution of a flora highly vulnerable to damage from browsing and grazing, and that the populations of wild animals (including deer and opossums) that were introduced in the nineteenth century grew explosively because of the lack of competitors and predators. It has, however, proved difficult to determine the magnitude of the effects which the introduced mammals had on the native forests (Veblen and Stewart, 1982).

Overall, invasive species may be one of the prime causes of loss of **biodiversity** because of their role in competition, predation and hybridization, though this is a view that has been challenged by Theodoropoulos (2003). They also have enormous economic costs. On a global basis plant and animal invaders may cost as much as $1.4 trillion a year, representing 5% of the world economy (Pimentel, 2003).

There are many other examples of ecological explosions – bioinvasions – caused by humans creating new habitats. Some of the most striking are associated with the establishment of artificial lakes in place of rivers. Riverine species which cannot cope with the changed conditions tend to disappear, while others that can exploit the new courses of food, and reproduce themselves under the new conditions, multiply rapidly in the absence of competition (Lowe-McConnell, 1975). Vegetation on land flooded as the lake waters rise decomposes to provide a rich supply of nutrients which allow explosive outgrowth of organisms as the new lake fills. In particular, floating plants may form dense mats of vegetation, which in turn support large populations of invertebrate animals, may cause fish deaths by deoxygenating the water, and can create a serious nuisance for turbines, navigators and fishermen. On Lake Kariba in Central Africa there were dramatic growths in the communities of the South American water fern (*Salvinia molesta*), bladder-wort (*Utricularia*) and the African water lettuce (*Pistia stratiotes*); on the Nile behind the Jebel Aulia Dam there was a huge increase in the number of water hyacinths (*Eichhornia crassipes*); and in the Tennessee Valley lakes there was a massive outbreak of the Eurasian water-millfoil (*Myriophyllum*).

Roads have been of major importance in the spread of plants. As Frenkel (1970) has pointed out in a valuable survey of this aspect of anthropogenic biogeography:

By providing a route for the bearers of plant propagules – man, animal and vehicle – and by furnishing, along their margins, a highly specialized habitat for plant establishment, roads may facilitate the entry of plants into a new area. In this manner roads supply a cohesive directional component, cutting across physical barriers, linking suitable habitat to suitable habitat.

Roadsides tend to possess a distinctive flora in comparison with the natural vegetation of an area. As Frenkel (1970: 1) has again written:

Roadsides are characterized by numerous ecological modification including: treading, soil compaction, confined drainage, increased runoff, removal of organic matter and sometimes additions of litter or waste material of frequently high nitrogen content (including urine and faeces), mowing or crushing of tall vegetation but occasionally the addition of wood chips or straw, substrate maintained in an ecologically open condition by blading, intensified frost action, rill and sheetwash erosion, snow deposition (together with accumulated dirt, gravel, salt and cinder associated with winter maintenance), soil and rock additions related to slumping and rock-falls, and altered microclimatic conditions associated with pavement and right-of-way structures. Furthermore, road rights-of-way may be used for driving stock in which case unselective, hurried but often close grazing may constitute an additional modification. Where highway landscaping or stability of cuts and fills is a concern, exotic or native plants may be planted and nurtured.

The speed with which plants can invade roadsides is impressive. A study by Helliwell (1974) demonstrated that the M1 motorway in England, less than twelve years after its construction, had on its cuttings and embankments not only the thirty species which had been deliberately sown or planted, but more than 350 species that had not been introduced there.

Railways have also played their role in plant dispersal. The classic example of this is provided by the Oxford ragwort, *Senecio squalidus*, a species native to Sicily and southern Italy. It spread from the Oxford Botanical Garden (where it had been established since at least 1690) and colonized the walls of Oxford. Much of its dispersal to the rest of Britain (Usher, 1973) was achieved by the Great Western Railway, in the vortices of whose trains and in whose cargoes of ballast and iron ore the plumed fruits were carried. The distribution of the plant was very much associated with railway lines, railway towns and waste ground.

Indeed, by clearing forest, cultivating, depositing rubbish and many other activities, humans have opened up a whole series of environments which are favourable to colonization by a particular group of plants. Such plants are generally thought of as weeds. In fact it has often been said that the history of weeds is the history of human society (though the converse might equally be true) and that such plants follow people like flies follow a ripe banana or a gourd of unpasteurized beer (see Harlan, 1975b).

One weed which has been causing especially severe problems in upland Britain in the 1980s is bracken (*Pteridium aquilinum*), although the problems it poses through rapid encroachment are of wider geographical significance (Pakeman et al., 1996). Indeed, Taylor (1985: 53) maintains that it 'may justifiably be dubbed as the most successful international weed of the twentieth century', for 'it is found, and is mostly expanding, in all of the continents.' This tolerant, aggressive opportunist follows characteristically in the wake of evacuated settlement, deforestation or reduced grazing pressure, and estimated encroachment rates in the upland parts of the UK average 1% per annum. This encroachment results from reduced use of bracken as a resource (e.g. for roofing) and from changes in grazing practices in marginal areas. As bracken is hostile to many other plants and animals, and generates toxins, including some carcinogens, this is a serious issue.

Air pollution and its effects on plants

Air pollutants exist in gaseous or particulate forms. The gaseous pollutants may be separated into primary and secondary forms. The primary pollutants, such as sulfur dioxide, most oxides of nitrogen and carbon monoxide, are those directly emitted into the air from, for example, industrial sources. Secondary air pollutants, like ozone and products of **photochemical** reactions, are formed as a consequence of subsequent chemical processes in the atmosphere, involving the primary pollutants and other agents such as sunlight. Particulate pollutants consist of very small solid – or liquid-suspended droplets (e.g. dust, smoke and aerosolic salts), and contain a wide range of insoluble components (e.g. quartz) and of soluble components (e.g. various common cations, together with chloride, sulfate and nitrate).

Some of the air pollutants humans have released into the atmosphere have had detrimental impacts on plants (Yunus and Iqbal, 1996): sulfur dioxide, for example, is toxic to them. Increasing levels of atmospheric nitrogen deposition appear to have had a marked effect on the biodiversity of grasslands in Britain (Maskell et al., 2010) and Europe (Stephens et al., 2010, 2011). Some species become crowded out by more robust plants which benefit from the fertilizing effect of this pollutant.

Lichens are also sensitive to air pollution and have been found to be rare in central areas of cities such as

Bonn, Helsinki, Stockholm, Paris and London. Overall it was calculated by Rose (1970) that more than one-third of England and Wales, extending in a belt from the London area to Birmingham, broadening out to include the industrial Midlands and most of Lancashire and West Yorkshire, and reaching up to Tyneside, has lost nearly all its epiphytic lichen flora, largely because of sulfur dioxide pollution. By 1970 there were very few lichens present on trees in central London. Hawksworth (1990: 50), indeed, believes 'the evidence that sulfur dioxide is the major pollutant responsible for impoverishment of lichen communities over wide areas of Europe is now overwhelming', but he also points out that sulfur dioxide is not necessarily always the cause of impoverishment. Nitrate pollution also can have an adverse effect (van Herk et al., 2003). Some lichens were, however, tolerant of sulfur dioxide (e.g. *Lecanora conizaeoides*) or of nitrate and so survived or flourished (Fenn et al., 2003). With the lower atmospheric sulfate levels of recent years, various cities have seen a dramatic recovery in the diversity of lichen species, as is the case in Tampere, Finland (Ranta, 2001) and in London (Larsen et al., 2007).

Local concentrations of industrial fumes also kill vegetation (Freedman, 1995), as has been the case with the smelters of the Sudbury mining district of Canada (Figure 2.26). In Norway a number of the larger Norwegian aluminium smelters built immediately after the Second World War were sited in deep, narrow, steep-sided valleys at the heads of fjords. The relief has not proved conductive to the rapid dispersal of fumes, particularly of fluoride. In one valley a smelter with a production of 1,110,000 tonnes per year causes the death of Scots pines (*Pinus sylveststris*) for over 13 km in each direction up and down the valley. To about 6 km from the source all pines are dead. Birch, however, seems to be able to withstand these conditions and to grow vigorously right up to the factory fence (Gilbert, 1975).

Photochemical smog is also known to have adverse effects on plants both within cities and on their outskirts. In California, Ponderosa pines (*Pinus ponderosa*) in the San Bernadino Mountains as much as 129 km to the east of Los Angeles have been extensively damaged by smog and ozone (Diem, 2003). Fumigation experiments in the USA show that plant injury can occur at levels only marginally above the natural maximum and well within the summertime levels now known to

Figure 2.26 The great smelter at Sudbury in Canada. Pollution derived from the fumes from such sources can have many deleterious impacts on vegetation in their neighbourhood. (See Plate 19)

be present in Britain and the USA. Tropospheric ozone has a number of impacts on forests (Karnosky et al., 2007), including leaf damage, needle damage to conifers, decreasing photosynthesis, reduced productivity and increasing susceptibility to pest attack. Vegetation will also be adversely affected by excessive quantities of suspended particulate matter in the atmosphere. The particles, by covering leaves and plugging plant stomata, reduce both the absorption of carbon dioxide from the atmosphere and the intensity of sunlight reaching the interior of the leaf. Both tendencies may suppress the growth of some plants. This and other consequences of air pollution are well reviewed by Elsom (1992).

The adverse effects of pollution on plants are not restricted to air pollution: water and soil pollution can also be serious. Excessive amounts of heavy metals may prove toxic to them and, as a consequence, distinctive patterns of plant species may occur in areas contaminated with the waste from copper, lead, zinc and nickel mines (Cole and Smith, 1984). Heavy metals in soils may also be toxic to microbes and especially to fungi which may in turn change the environment by reducing rates of leaf-litter decomposition. In many areas it has proved extremely difficult to undertake effective re-establishment of plant communities on mining spoil tips, though toxicity is only one of the problems (Kent, 1982).

Other types of industrial effluents may smother and poison some species. Salt marshes, mangrove swamps and other kinds of wetlands are particularly sensitive to oil spills, for they tend to be anaerobic environments in which the plants must ventilate their root systems through pores or openings that are prone to coating and clogging (Lugo et al., 1981). The situation is especially serious if the system is not subjected to flushing by, for example, frequent tidal inundation.

Forest decline

Forest decline, often called *Waldsterben* or *Waldschäden* (the German words for 'forest death' and 'forest decline'), is an environmental issue that attained considerable prominence in the 1980s. The common symptoms of this phenomenon (modified from World Resources Institute, 1986, table 12.9) are:

1 *Growth-decreasing symptoms*:
   discoloration and loss of needles and leaves;
   loss of feeder-root biomass (especially in conifers);
   decreased annual increment (width of growth rings);
   premature ageing of older needles in conifers;
   increased susceptibility to secondary root and foliar pathogens;
   death of herbaceous vegetation beneath affected trees;
   prodigious production of lichens on affected trees;
   death of affected trees.
2 *Abnormal growth symptoms*:
   active shedding of needles and leaves while still green, with no indication of disease;
   shedding of whole green shoots, especially in spruce;
   altered branching habit;
   altered morphology of leaves.

3 *Water-stress symptoms*:
   altered water balance;
   increased incidence of wet wood disease.

The decline appeared to be widespread in much of Europe and to be particularly severe in Poland and the Czech Republic (see Table 2.7). The process was also thought to be undermining the health of North America's high elevation eastern coniferous forests (World Resources Institute, 1986, Chapter 12). In Germany it was the white fir, *Abies alba*, which was afflicted initially, but since then the symptoms spread to at least 10 other species in Europe, including Norway spruce (*Picea abies*), Scotch pine (*Pinus sylvestris*), European larch (*Larix decidua*) and seven broad-leaved species.

Table 2.7 Results from forest damage surveys in Europe: percentage of trees with >25% defoliation (all species). Source: Data from *Acid News* (1995: 7)

	Mean of 1993/1994
Austria	8
Belarus	33
Belgium	16
Bulgaria	26
Croatia	24
Czech Republic	56
Denmark	35
Estonia	18
Finland	14
France	8
Germany	24
Greece	22
Hungary	21
Italy	19
Latvia	33
Lithuania	26
Luxembourg	29
Netherlands	22
Norway	26
Poland	52
Portugal	7
Romania	21
Slovak Republic	40
Slovenia	18
Spain	16
Switzerland	20
UK	15

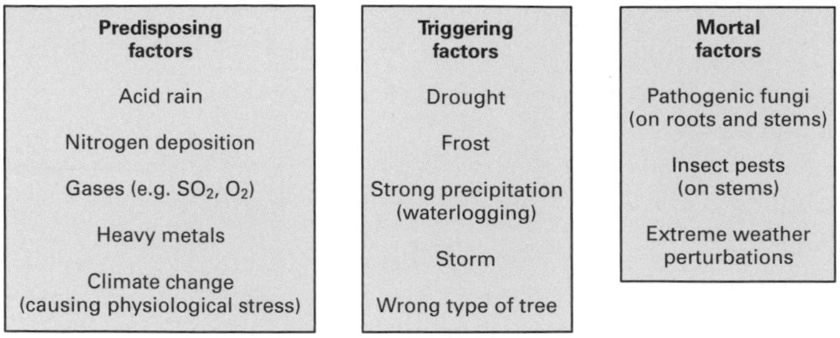

Predisposing factors	Triggering factors	Mortal factors
Acid rain	Drought	Pathogenic fungi (on roots and stems)
Nitrogen deposition	Frost	Insect pests (on stems)
Gases (e.g. SO_2, O_2)	Strong precipitation (waterlogging)	Extreme weather perturbations
Heavy metals	Storm	
Climate change (causing physiological stress)	Wrong type of tree	

Figure 2.27 Some of the different stress variables used to explain forest decline (modified from Nihlgård, 1997, figure 24.1).

Trees in North America were also adversely affected, probably because of acid rain. Likens (2010) reported that since the 1960s, more than half of large canopy red spruce in the Adirondack Mountains of New York and the Green mountains of Vermont and approximately one quarter of large canopy red spruce in the White Mountains of New Hampshire have died. Extensive mortality among sugar maples in Pennsylvania has also occurred. In the mountains of the north-west USA *Pinus albicaulis*, the whitebark, has suffered from a range of possible factors: fire suppression policies, blister rust, mountain pine beetle infestation and warmer temperatures (Larson and Kipfmueller, 2012).

Indeed, many hypotheses have been put forward to explain forest dieback (Wellburn, 1988; Federal Research Centre for Forestry and Forest Products, 2000): poor forest management practices, ageing of stands, climatic change, severe climatic events (such as the severe droughts in Britain during 1976), nutrient deficiency, viruses, fungal pathogens and pest infestation (Hicke et al., 2012). The last of these has been accelerated by the introduction of invasive pests in plant imports (Liebhold et al., 2012). However, particular attention is being paid to the role of pollution, either by gaseous pollutants (sulfur dioxide, nitrous oxide or ozone), acid deposition on leaves and needles, soil acidification and associated aluminium toxicity problems and excess leaching of nutrients (e.g. magnesium), over-fertilization by deposited nitrogen, and trace metal or synthetic organic compound (e.g. pesticide, herbicide) accumulation as a result of atmospheric deposition.

The arguments for and against each of these possible factors have been expertly reviewed by Innes (1987), who believes that in all probability most cases of forest decline are the result of the cumulative effects of a number of stresses. He draws a distinction between predisposing, inciting and contributing stresses (p. 25):

Predisposing stresses are those that operate over long time scales, such as climatic change and changes in soil properties. They place the tree under permanent stress and may weaken its ability to resist other forms of stress. Inciting stresses are those such as drought, frost and short-term pollution episodes that operate over short time scales. A fully healthy tree would probably have been able to cope with these, but the presence of predisposing stresses interferes with the tree's mechanisms of natural recovery. Contributing stresses appear in weakened plants and are frequently classed as secondary factors. They include attack by some insect pests and root fungi. It is probable that all three types of stress are involved in the decline of trees.

An alternative categorization of the stresses leading to forest decline has been proposed by Nihlgård (1997). He also has three classes of stress that have a temporal dimension: predisposing factors (related primarily to pollution and long-term climatic change); triggering factors (including droughts, frosts and inappropriate management or choice of trees); and mortal factors such as pests, pathogenic fungi and extreme weather events, which can lead to plant death (Figure 2.27).

As with many environmental problems, interpretation of forest decline is bedevilled by a paucity of long-term data and detailed surveys. Given that forest condition oscillates from year to year in response to variability in climatic stress (e.g. drought, frost, wind throw) it is dangerous to infer long-term trends from short-term data. There may also be differences in causation in different areas. Thus while widespread forest death in eastern Europe may result from high concentrations of sulfur dioxide combined with extreme winter stress, this is a much less likely explanation in Britain, where sulfur dioxide concentrations have shown a marked decrease in recent years. Indeed, in Britain Innes and Boswell (1990: 46) suggest that the direct effects of gaseous pollutants appear to be very limited.

It is also important to recognize that some stresses may be particularly significant for a particular tree species. Thus, in 1987 a survey of ash trees in Great Britain showed extensive dieback over large areas of the country. Almost one-fifth of all ash trees sampled showed evidence of this phenomenon. Hull and Gibbs (1991) indicated that there was an association between dieback and the way the land is managed around the tree, with particularly high levels of damage being evident in trees adjacent to arable land. Uncontrolled stubble burning, the effects of drifting herbicides and the consequences of excessive nitrate fertilizer applications to adjacent fields were seen as possible mechanisms. However, the prime cause of dieback was seen to be root disturbance and soil compaction by large agricultural machinery. Ash has shallow roots and if these are damaged repeatedly the tree's uptake of water and nutrients might be seriously reduced, while broken root surfaces would be prone to infection by pathogenic fungi.

Innes (1992: 51) also suggests that there has been some modification in views about the seriousness of the problem since the mid-1980s:

The extent and magnitude of the forest decline is much less than initially believed. The use of crown density as an index of tree health has resulted in very inflated figures for forest 'damage' which cannot now be justified . . . If early surveys are discounted on the basis of inconsistent methodology . . . then there is very little evidence for a large-scale decline of tree health in Europe.

. . . the term 'forest decline' is rather misleading in that there are relatively few cases where entire forest ecosystems are declining. Forest ecosystems are dynamic and may change through natural processes.

He also suggests that the decline of certain species has been associated with climatic stress for as long as records have been maintained.

Removal of some stresses, such as reduction in atmospheric sulfur dioxide levels in recent decades, has enabled some forests to show signs of marked recovery (e.g. Elling et al., 2009; Hauck et al., 2012), though longer term nutrient imbalances may limit the amount of recovery that occurs (Jonard et al., 2012).

Miscellaneous causes of plant decline

Some of the main causes of plant decline have already been referred to: deforestation, grazing, fire and pollution. However, there are many records of species being affected by other forms of human interference. For example, casual flower-picking has resulted in local elimination of previously common species and has been held responsible for decreases in species such as the primrose (*Primula vulgaris*) on a national scale in England. In addition, serious naturalists or plant collectors using their botanical knowledge to seek out rare, local and unusual species can cause the eradication of rare plants in an area. More significantly, agricultural 'improvements' mean that many types of habitat are disappearing or that the range of such habitats is diminishing. Plants associated with distinctive habitats suffer a comparable reduction in their range. Other plants have suffered a reduction in range because of drainage activities.

The introduction of pests, either deliberately or accidentally, can also lead to a decrease in the range and numbers of a particular species (Liebhold et al., 2012). Reference has already been made to the decline in the fortunes of the elm in Britain because of the unintentional establishment of Dutch Elm Disease. There are, however, many cases where 'pests' have been introduced deliberately to check the explosive invasion of a particular plant. One of the most spectacular examples of this involves the history of the prickly pear (*Opuntia*) imported into Australia from the Americas (Figure 2.28). It was introduced some time before 1839 and spread dramatically. By 1900, it covered 4 million hectares and by 1925 more than 24 million. Of the latter figure approximately one-half was occupied by dense growth (1200–2000 tonnes per ha) and other more useful plants were excluded. To combat this menace one of *Opuntia's* natural enemies, a South American moth, *Cactoblastus*, was introduced to remarkable effect. By the year 1940 not less than 95% of the former 50 million acres (20 million hectares) of prickly pear in Queensland had been wiped out (Dodd, 1959).

A further good example also comes from Australia. By 1952 the aquatic fern, *Salvinia molesta*, which originated in south-eastern Brazil, appeared in Queensland and spread explosively. Significant pests and parasites appear to have been absent. In June 1980 possible

Figure 2.28 The prickly pear (*Opuntia*) is a plant that has been introduced from the Americas to Africa and Australia. It has often spread explosively. Recently it has been controlled by the introduction of moths and beetles, an example of biological control. (See Plate 20)

control agents from the *Salvinia's* native range in Brazil – the black, long-snouted weevil (*Cyrobagous salviniae*) – were released on to Lake Moondarra (which carried an infestation of 50,000 tonnes fresh weight of *Salvinia*, covering an area of 400 ha). By August 1981 there was estimated to be less than 1 tonne of the weed left on the lake.

Finally, the growth of leisure activities is placing greater pressure on increasingly fragile communities, notably in **tundra** and high-altitude areas. These areas tend to recover slowly from disturbance, and both the trampling of human feet and the actions of vehicles can be severe.

The change in genetic and species diversity

The application of modern science, technology and industry to agriculture has led to some spectacular progress in recent decades through such developments as the use of fertilizers and the selective breeding of plants and animals. The latter has caused some concern, for in the process of evolution domesticated plants have become strikingly different from their wild progenitors. Plant species that have been cultivated for a very long time and are widely distributed demonstrate this particularly clearly. Crop evolution through the millennia has been shaped by complex interactions reflecting the pressures of both artificial and natural selection. Alternate isolation of stocks followed by migration and seed exchanges brought distinctive stocks into new environments and permitted new hybridizations and the recombination of characteristics. Great genetic diversity resulted.

There are fears, however, that since the Second World War the situation has begun to change (Harlan, 1975a). Modern plant-breeding programmes have been established in many parts of the developing world in the midst of genetically rich centres of diversity. Some of these programmes, associated with the so-called **Green Revolution**, have been successful, and new, uniform high-yielding varieties have begun to replace the wide range of old, local strains that have evolved over the millennia. This may lead to a serious decline in the genetic resources which could potentially serve as reservoirs of variability. Ehrlich et al. (1977: 344) have warned:

Aside from nuclear war, there is probably no more serious environmental threat than the continued decay of the genetic variability of crops. Once the process has passed a certain point, humanity will have permanently lost the coevolutionary race with crop pests and diseases and will no longer be able to adapt crops to climatic change.

New, high-yielding crop varieties need continuous development if they are to avoid the effects of crop pests, and Ehrlich and Ehrlich (1982: 65) have summarized the situation thus:

The life of a new cultivated wheat variety in the American Northwest is about five years. The rusts (fungi) adapt to the strain, and a new resistant one must be developed. That development is done through artificial selection: the plant breeder carefully combines genetic types that show promise of giving resistance.

The impact of human activities on species diversity, while clearly negative on a global scale (as evidenced by extinction rates) is not so cut-and-dried on the local scale. Under certain conditions chronic stress caused by humans can lead to extremely high numbers of coexisting species within small areas. By contrast, site enrichment or fertilization can result in a decline of species density (Peet et al., 1983).

Certain low productivity grasslands, which have been grazed for long periods have high species densities in Japan, the UK and the Netherlands. The same applies to Mediterranean scrub vegetation in Israel, and to savanna in Sri Lanka and North Carolina (USA). Studies have confirmed that species densities may increase in areas subject to chronic mowing, burning, domestic grazing, rabbit grazing or trampling. It is likely that in such ecosystems humans encourage a high diversity of plant growth by acting as a 'keystone predator', a species which prevents competitive exclusion by a few dominant species.

Grassland enrichment experiments employing fertilizers have suggested that in many cases high growth rates result in the competitive exclusion of many plants. Thus the tremendous increase in fertilizer use in agriculture has had with in retrospect is the predictable result of a widespread decrease in the species diversity of grasslands.

One other area in which major developments will take place in the coming years, which will have implications for plant life is the field of **genetic engineering**. This involves the manipulation of **DNA** the basic chromosomal unit that exists in all cells and contains genetic information that is passed on to subsequent generations. Recombinant DNA technology (also known as *in vitro* genetic manipulation and gene cloning) enables the insertion of foreign DNA (containing the genetic information necessary to confer a particular target characteristic) into a vector. The advent of modern techniques of genetic modification (GM) enables the removal of individual genes from one species and their insertion into another, without the need for sexual compatibility. Once the new gene has been inserted, offspring that will contain copies of that new gene can be produced in the traditional manner. For example, a bacterial gene can be inserted into maize to give it resistance to certain insect pests. Indeed, genetic engineering is vastly different from conventional plant breeding methods because it allows scientists to insert a gene or genes from virtually any organism into any other organism. Fears have been expressed that this technology could produce pathogens that might interact detrimentally with naturally occurring species (see e.g. Beringer, 2000; Letourneau and Burrows, 2001; Hails, 2002).

It has also been feared that GM could make a plant more invasive, leading to a new threat from superweeds. Equally, some genetically modified crops might be highly effective at controlling or eliminating weeds, thereby disrupting essential food sources for a wide range of organisms. Conversely, genetically modified herbicide-tolerant crops would bring greater economy and feasibility in weed control, so that fewer environmentally damaging chemical herbicides would be used than in conventional farming (Squire et al., 2003). Other beneficial traits from GM technology include resistance to insects or pathogens, drought resistance and salt tolerance, the ability to fix nitrogen and so forth (Jones, 2011). There is an urgent need to evaluate the costs and benefits of this new technology, for already GM crops are being widely grown. The number of hectares planted to them in the USA increased from 1.4 m hectares in 1996 to around 40 million hectares in 2003. An estimated 38 trillion GM plants have now been grown in American soil! Globally, between 1997 and 2010 the total surface area of land cultivated with GM crops had increased from $17,000 \, km^2$ to $1,480,000 \, km^2$.

Conclusion: threats to plant life

In 2010, the Royal Botanic Gardens at Kew produced a report on 'Plants at Risk' (http:www.kew.org/science/plants-at-risk/) in which an attempt was made to summarize what were the world's most endangered plant habitats and to identify what the main threats were to plant life. As Figure 2.29a shows, the most endangered habitats are tropical wet forests, with almost two thirds of plant species endangered. Areas least suitable for conversion to agriculture, such as wetlands and deserts, contain the smallest proportions of threatened plant species, with 4% and 0.5% of threatened species respectively. Some of the threats to plant

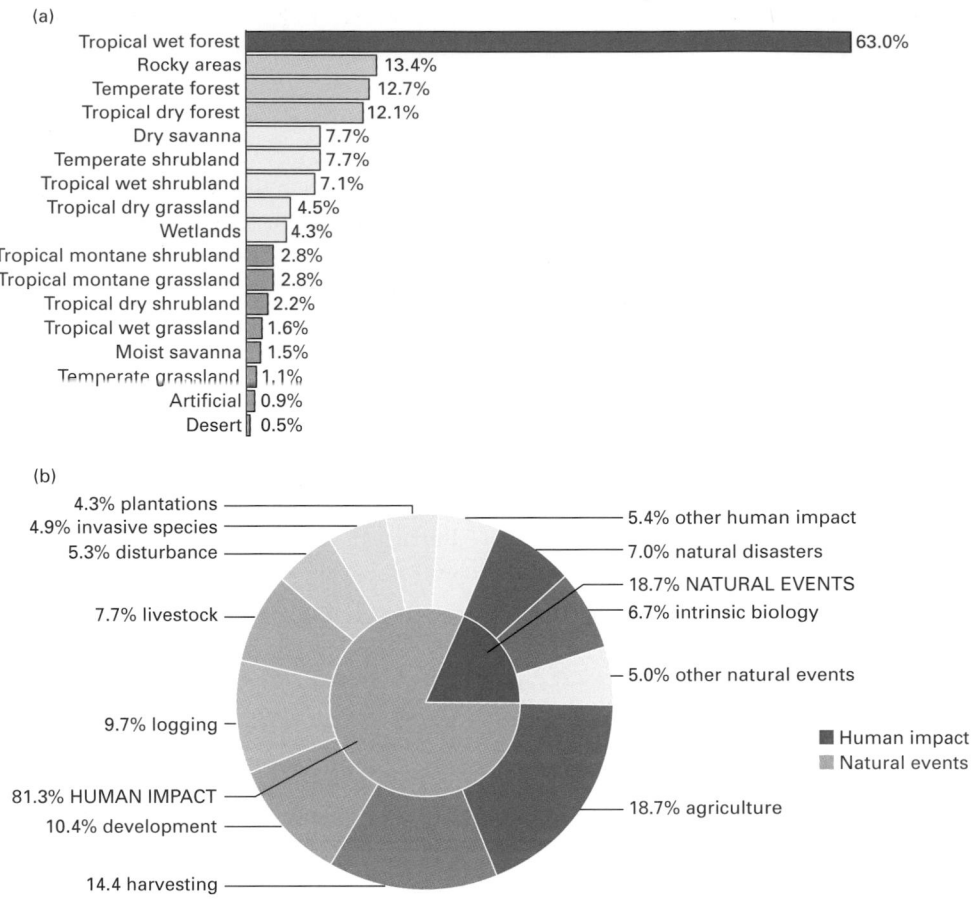

Figure 2.29 (a) Percentage of plant species at risk for different biomes and (b) the main threats, both natural and human (modified from Kew, *Plants at Risk*, 2010).

life are natural, accounting for 18.7% of the total (Figure 2.29b). These include natural disasters as well as the intrinsic characteristics of some species, such as limited dispersal ability, restricted range or poor reproduction. The biggest human threats, which account for 81.3% of the total, are agriculture, harvesting and other types of development. However, there is some evidence that ecosystems can recover quite rapidly from many human impacts if sustainable practices are introduced (Jones and Schmitz, 2009). What is clear overall is that humans have transformed the biosphere and particularly over the last century have transformed the majority of the terrestrial biosphere into intensively used 'anthromes' with predominantly novel anthropogenic processes (Ellis, 2011).

Points for review

How does the use of fire impact upon the environment?
In what ways does grazing by domestic stock modify vegetation?
Where is present day deforestation taking place and why?
What controversies surround desertification?
What role have humans played in the development of savannas and mid-latitude grasslands?
What are ecological explosions?
How does pollution affect plants?

Guide to reading

Crutzen, P.J. and Goldammer, J.G. (1993) *Fire in the Environment*. Chichester: Wiley. A particularly useful study of fire's importance.

Douglas, I., Goode, D., Houck, C. and Wang, R. (eds) (2011) *The Routledge Handbook of Urban Ecology*. New York: Routledge. A large edited compendium on the environment of cities.

Grainger, A. (1990) *The Threatening Desert: Controlling Desertification*. London: Earthscan. A very readable and wide-ranging review of desertification.

Grainger, A. (1992) *Controlling Tropical Deforestation*. London: Earthscan. An introduction with a global perspective.

Grove, A.T. and Rackham, O. (2002) *The Nature of Mediterranean Europe: An Ecological History*. New Haven and London: Yale University Press. A well illustrated account of natural and human changes in the Mediterranean lands.

Holzner, W., Werger, M.J.A., Werger, I. and Ikusima, I. (eds) (1983) *Man's Impact on Vegetation*. Hague: Junk. A wide-ranging edited work with examples from many parts of the world.

Imeson, A. (2012) *Desertification, Land Degradation and Sustainability*. Chichester: Wiley-Blackwell. A review of research on desertification and on landscape degradation more generally.

Manning, R. (1995) *Grassland: The History, Biology, Politics and Promise of the American Prairie*. New York: Viking Books. A 'popular' discussion of the great grasslands of North America.

Meyer, W.B. and Turner, B.L. (eds) (1994) *Changes in Land Use and Land Cover: A Global Perspective*. Cambridge: Cambridge University Press. An excellent edited survey of human transformation of the biosphere.

Pimentel, D. (ed.) (2003) *Biological Invasions*. Washington, DC: CRC Press. An edited, international account of invasive species.

Stewart, C.N. (2004) *Genetically Modified Planet. Environmental Impacts of Genetically Engineered Plants*. Oxford: Oxford University Press. A readable review of the costs and benefits of GM plants.

Williams, M. (2003) *Deforesting the Earth. From Prehistory to Global Crisis*. Chicago: University of Chicago Press. A magisterial and scholarly overview of the global history of deforestation.

3 HUMAN INFLUENCE ON ANIMALS

Chapter Overview

Humans now have the greatest **biomass** or anthropomass of any species and it is an order of magnitude greater than the mass of all terrestrial mammals alive today (Smil, 2011). The impact that they have had on animals can be grouped conveniently into five main categories: domestication, dispersal, extinction, expansion and contraction. Humans have domesticated many animals, to the extent that, as with many plants, those animals depend on humans for their survival and, in some cases, for their reproduction. As with plants, people have helped to disperse animals deliberately, though many have also been dispersed accidentally, for the number of animals that accompany people without their leave is enormous, especially if we include the clouds of micro-organisms that infest their land, food, clothes, shelter, domestic animals and their own bodies. The extinction of animals by human predators has been extensive over the past 20,000 years, and in spite of recent interest in conservation continues at a high rate. In addition, the presence of humans has led to the contraction in the distribution and welfare of many animals (because of factors like pollution), though in other cases human alteration of the environment and modification of competition has favoured the expansion of some species, both numerically and spatially.

The Human Impact on the Natural Environment: Past, Present and Future, Seventh Edition. Andrew S. Goudie.
© 2013 John Wiley & Sons, Ltd. Published 2013 by John Wiley & Sons, Ltd.

Domestication of animals

We have already referred in Chapter 1 to one of the great themes in the study of human influence on nature: domestication. This has been one of the most profound ways in which humans have affected animals, for during the ten or eleven millennia that have passed since this process was initiated the animals that human societies have selected as useful to them have undergone major changes. Relatively few animal species have been domesticated in comparison with plant species, but for those that have been the consequences are so substantial that the differences between breeds of animals of the same species often exceed those between different species under natural conditions (Figure 3.1). A cursory and superficial comparison of the tremendous range of shapes and sizes of modern dog breeds (as, for example, between a wolfhound and a chihuahua) is sufficient to establish the extent of alteration brought about by domestication and the speed at which domestication has accelerated the process of evolution. In particular humans have changed and enhanced the characteristics for which they originally chose to domesticate animals. For example, the wild ancestors of cattle gave no more than a few hundred millilitres of milk; today the best milk cow can yield up to 15,000 L of milk during its lactation period. Likewise, sheep have changed enormously (Ryder, 1966). Wild sheep have short tails, while modern domestic sheep have long tails which may have arisen during human selection of a fat tail. Wild sheep also have an overall brown colour, whereas domestic sheep tend to be mainly white. Moreover, the woolly undercoats of wild sheep have developed at the expense of bristly outer coats. With the ancestors of domestic sheep, wool (which served as protection for the skin and as insulation) consisted mainly of thick rough hairs and a small amount of down; the total weight of wool grown per year probably never reached 1 kg. The wool of present-day fine-fleeced sheep consists of uniform, thin down fibres and the total yearly weight may now reach 20 kg. Wild sheep also undergo a complete spring moult, while domestic sheep rarely shed wool.

Indeed, one of the most important consequences or manifestations of the domestication of animals consists of a sharp change in their seasonal biology. Whereas wild ancestors of domesticated beasts are often characterized by relatively strict seasonal reproduction and moulting rhythms, most domesticated species can reproduce at almost any season of the year and tend not to moult to a seasonal pattern.

The most important centre of animal domestication, shown in Figure 1.8, was south-west Asia (Diamond, 1997) (cattle, sheep, goat and pig) but other centres were important (e.g. the chicken in Southeast Asia, the turkey in the central America and the horse in the Ukraine). Curiously, in spite of its large number of mammals, Africa was not a major centre of animal domestication, except for the donkey in the north-east part of that continent (Beja-Pereira et al., 2004).

Dispersal and invasions of animals

Zoogeographers have devoted much time to dividing the world into regions with distinctive animal life – the great 'faunal realms' of Wallace and subsequent workers. This pattern of wildlife distribution evolved slowly over geological time. The most striking dividing line between such realms is probably that between Australasia and Asia – **Wallace's Line**. Because of its isolation from the Asian land mass, Australasia developed a distinctive fauna characterized by its relative absence of placental mammals, its well-developed marsupials and the egg-laying monotremes (echidna and the platypus).

Figure 3.1 The consequences of domestication for the form of animals is dramatically illustrated by this large-horned but dwarf cow being displayed in a circus in Chamonix, France. (See Plate 21)

Modern societies, by moving wildlife from place to place, consciously or otherwise, are breaking down these classic distinct faunal realms and are facilitating what are called ecological or biological invasions (Davis, 2009). Highly adaptable and dispersive forms are spreading, perhaps at the expense of more specialized organisms. In terms of the number of countries where they are considered to be invasive, the Global Native Species Information Network (2011) (http://www.niss.org) listed rats (*Rattus rattus* and *Rattus norvegicus*), cats (*Felis catus*), goats (*Capra hircus*), American mink (*Mustela vison*) and the house mouse (*Mus musculus*) as the most widespread invasive animals. Humans have introduced a new order of magnitude into distances over which dispersal takes place, and through the transport by design or accident of seeds or other propagules, through the disturbance of native plant and animal communities and of their habitat, and by the creation of new habitats and niches, the invasion and colonization by adventive species is facilitated. The effective management of invasive alien species is a major concern for biological conservation worldwide (Clout and Williams, 2009), for invasive species are the major threat, for example, to seabird populations, including penguins and albatrosses/petrels (Croxall et al., 2012).

Di Castri (1989) identified three main stages in the process of biological invasions stimulated by human actions. In the first stage, covering several millennia up to about AD 1500, human historical events favoured invasions and migrations primarily within the Old World. The second stage commenced about AD 1500, with the discovery, exploration and colonization of new territories, and the initiation of 'the globalization of exchanges'. It was characterized by flows of invaders from, to and within the Old World. The third stage, which only covers the last 100 to 150 years, sees an even more extensive 'multifocal globalization' and an increasing rate of exchanges. The Eurocentric focus has diminished.

Deliberate introductions of new animals to new areas have been carried out for many reasons (Roots, 1976): for food, sport, revenue, sentiment, control of other pests and for aesthetic purposes. Such *deliberate* actions probably account, for instance, for the widespread distribution of brown trout (*Salmo trutta*) (Elliott, 1989) (see Figure 3.2). This was originally an essentially European species but has now been deliberately introduced for commercial and sport fishing to at least 24 countries outside Europe. Likewise the rainbow trout (*Oncorhynchus mykiss*) has spread from its natural habitats on the Pacific margins of North America and Russia to much of the rest of the world (Crawford and Muir, 2008).

There have, however, been many *accidental* introductions, especially since the development of ocean-going vessels. The rate is increasing, for whereas in the eighteenth century there were few ocean-going vessels of more than 300 tonnes, today there are thousands. Sea-borne trade now exceeded 8.4 billion tonnes per year in 2010, with much of it carried in millions of containers. Many vessels dump ballast containing potentially invasive exotic species. The spread of the highly invasive fish, the goby (*Neoglobius melanostomus*) into the Great Lakes region of North America from the Black Sea is due to this cause (Kornis et al., 2012). Because of this, in the words of Elton (1958: 31), 'we are seeing one of the great historical convulsions in the world's fauna and flora.' Indeed, many animals are introduced with vegetable products, for 'just as trade followed the flag, so animals have followed the plants.' One of the most striking examples of the accidental introduction of animals given by Elton is the arrival of some chafer beetles, *Popillia japonica* (the Japanese beetle), in New Jersey in a consignment of plants from Japan. From an initial population of about one dozen beetles in 1916, the centre of population grew rapidly outwards to cover many thousands of square kilometres in only a few decades. It now extends as far as parts of Wisconsin, Minnesota, Iowa, Missouri, Nebraska, Kansas, Arkansas and Oklahoma.

A more recent example of the spread of an introduced insect in the Americas is provided by the Africanized honey bee which is popularly known as the 'killer bee' (Rinderer et al., 1993). A number of these were brought to Brazil from South Africa in 1957 as an experiment and some escaped. Since then (Figure 3.3) they have moved northwards to Central America and Texas, spreading at a rate of 300–500 km per year and competing with established populations of European honey bees. In 1998 they reached as far north as Nevada and in 2009 as far north as southern Utah. Goulson (2003) reviews some of the adverse environmental effects of introduced bees, including their pollination of exotic weeds.

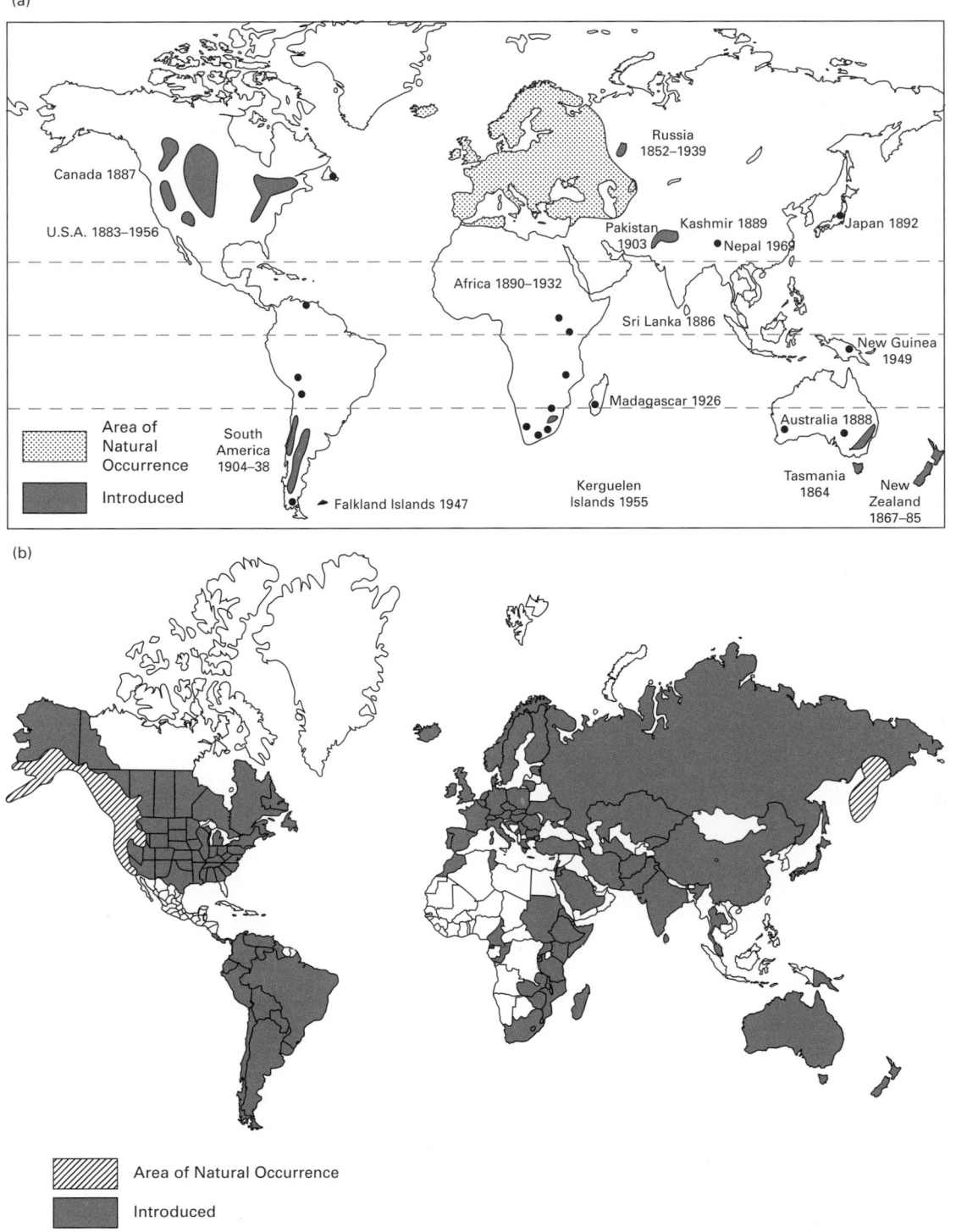

Figure 3.2 (a) The original area of distribution of the brown trout, *Salmo trutta*, areas where it has been artificially naturalized (modified from MacCrimmon in Illies, 1974, figure 3.2 and Elliott, 1989, figure 2) and the dates when first introduced. (b) Native distribution of rainbow trout, *Oncorhynchus mykiss*, and the countries, states or provinces into which this species has been introduced (modified from Crawford and Muir, 2008, figure 7).

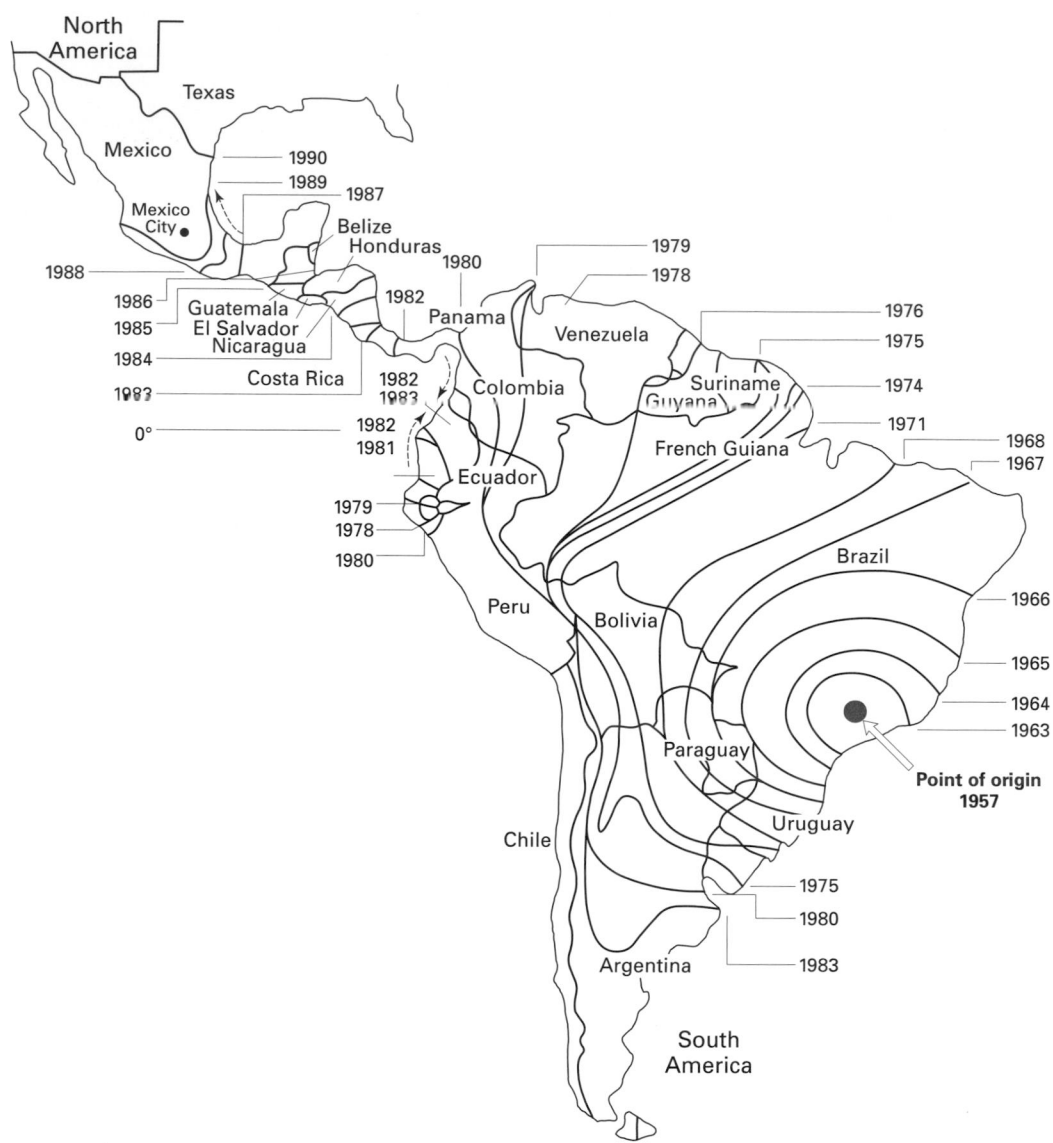

Figure 3.3 The spread of the Africanized honey bee in the Americas between 1957 (when it was introduced to Brazil) and 1990 (modified from Texas Agricultural Experiment Station in *Christian Science Monitor*, September 1991).

Some animals arrive accidentally with other beasts that are imported deliberately. In northern Australia, for instance, water buffalo (*Bubalus bubalis*) were introduced (McKnight, 1971) and brought their own blood-sucking fly, a species which bred in cattle dung and transmitted an organism sometimes fatal to cattle. Australia's native dung beetles, accustomed only to the small sheep-like pellets of the grazing marsupials, could not tackle the large dung pats of the buffalo. Thus untouched pats abounded and the flies were able

to breed undisturbed. Eventually African dung beetles were introduced to compete with the flies (Roots, 1976).

While domesticated plants have, in most cases, been unable to survive without human help, the same is not so true of domesticated animals. There are a great many examples of cattle, horses (see e.g. McKnight, 1959), donkeys and goats which have effectively adapted to new environments and have virtually become wild (feral). Frequently feral animals have both ousted native animals and, particularly in the case of goats on ocean islands, caused desertification. Indeed, animal invaders have had tragic consequences for the native Australian fauna and flora. Among the worst offenders are the rabbit, the fox, the feral pigs and goats, dogs, cats, mice, buffalo, horses, camels, rats, cane toads and various birds (Bomford and Hart, 2002). In neighbouring New Zealand, Australian brushtail possums (*Trichosaurus vulpecula*) have become major pests, and its population is estimated to be around 30 million.

The story of the feral camel in Australia is especially striking. The one-humped dromedary (*Camelus dromadarius*) was introduced in 1840 and between 1880 and 1907 some 20,000 were imported. However, with the advent of motor transport they became largely redundant, and between 1920 and 1941, between 5000 and 10,000 were released into the wild. There are now probably around one million feral camels spread over 3.3 million km² (Saalfield and Edwards, 2010). They have a wide range of adverse environmental effects (Edwards et al., 2010). The numbers of other feral animals are even greater, with 2.6 million goats, 5 million donkeys, and between 13 and 23 million pigs. In addition there are over 7 million red foxes and more than 200 million rabbits.

Sometimes, however, introduced animals have spread so thoroughly and rapidly, and have led to such a change in the environment to which they were introduced, that they have sown the seeds of their own demise. Reindeer (*Rangifer tarandus*), for example, were brought from Lapland to Alaska in 1891–1902 to provide a new resource for the Eskimos, and the herds increased and spread to 640,000 animals in 1932. By the 1950s, however, there was less than a twentieth of that number left, since the reindeer had been allowed to eat the lichen supplies that are essential to winter survival; as lichen grows very slowly their food supply was drastically reduced (Elton, 1958: 129). Now there may be only around 17,650 reindeer in Alaska.

The accidental dispersal of animals can be facilitated by means other than transport on ships or introduction with plants. This applies particularly to aquatic life which can be spread through human alteration of waterways by methods such as canalization. The opening of the Suez Canal in 1869 enabled the exchange of animals between the Red Sea and the eastern Mediterranean. Initially, the high salinity of the Great Bitter Lake acted to prevent movement, but the infusion of

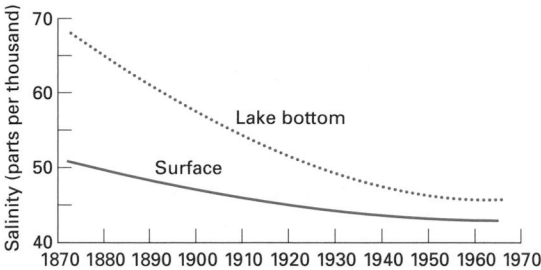

Figure 3.4 Decrease in the salinity of the Great Bitter Lake, Egypt, resulting from the intrusion of fresher water by way of the Suez Canal (after Wooster, 1969, with subsequent modification).

progressively fresher waters (Figure 3.4) through the Suez Canal has meant that this barrier has gradually become less effective. Some 63 Red Sea fish immigrants have now been identified in the Mediterranean, and they are especially important in the Levant basin (Mavruk and Avsar, 2008). They have been less successful in invading the western Mediterranean but now are regularly encountered off Sicily. Menacing jellyfish (*Rhopilema nomadica*) have invaded Levantine beaches (Spanier and Galil, 1991). This type of movement has been termed 'Lessepsian migration'.

Similarly, the construction of the Welland Canal, linking the Atlantic with the Great Lakes of North America, has permitted similar movements with disastrous consequences. Much of the native fish fauna has been displaced by alewife (*Alosa pseudoharengus*) through competition for food, and by the sea-lamprey (*Petromyzon marinum*) as a predator, so that once common Atlantic salmon (*Salmo salar*), lake trout (*Salvelinus namaycush*) and lake herring (*Leucichtys artedi*) have been nearly exterminated (Aron and Smith, 1971). The whole question of the presence and effects of exotic species in large lakes is reviewed by Hall and Mills (2000).

Can one make any generalizations about the circumstances that enable successful invasion by some exotic vertebrates and less successful invasion by others (Duncan et al., 2003)? Brown (1989) suggests that there may by 'five rules of biological invasions':

Rule 1
'Isolated environments with a low diversity of native species tend to be differentially susceptible to invasion.'

Rule 2
'Species that are successful invaders tend to be native to continents and to extensive, non-isolated habitats within continents.'
Rule 3
'Successful invasion is enhanced by similarity in the physical environment between the source and target areas.'
Rule 4
'Invading exotics tend to be more successful when native species do not occupy similar niches.'
Rule 5
'Species that inhabit disturbed environments and those with a history of close association with humans tend to be successful in invading man-modified habitats.'

Human influence on the expansion of animal populations

Although most attention tends to be directed towards the decline in animal numbers and distribution brought about by human agency, there are many circumstances where alterations of the environment and modification of competition has favoured the expansion of some species. Such expansion is not always welcome or expected.

Human actions, however, are not invariable detrimental, and even great cities may have effects on animal life which can be considered desirable or tolerable (Faeth et al., 2011). This has been shown in the studies of bird populations in several urban areas. For example, Nuorteva (in Jacobs, 1975) examined the bird fauna in the city of Helsinki (Finland), in agricultural areas near rural houses and in uninhabited forests (see Table 3.1). The city supported by far the highest

Table 3.1 Dependence of bird biomass and diversity on urbanization. Source: Data from Nuorteva (modified from Jacobs 1975, table 2)

	City (Helsinki)	Near rural houses	Uninhabited forest
Biomass ($kg\,km^{-2}$)	213	30	22
No. of birds (km^{-2})	1089	371	297
Number of species	21	80	54
Diversity	1.13	3.40	3.19

biomass and the highest number of birds, but exhibited the lowest number of species and the lowest diversity. In the artificially created rural areas the number of species, and hence diversity, were much higher than in the uninhabited forest, and so was biomass. Altogether, human civilization appeared to have brought about a very significant increase of diversity in the whole area: there were 37 species in city and rural areas that were not found in the forest. Similarly, after a detailed study of suburban neighbourhoods in west-central California, Vale and Vale (1976) found that in suburban areas the number of bird species and the number of individuals increased with time. Moreover, when compared to the pre-suburban habitats adjacent to the suburbs, the residential areas were found to support a larger number of both species and individuals. Horticultural activities appear to provide more luxuriant and more diverse habitats than do pre-suburban environments. However, a later review of 51 bird studies found that 61% showed lower species richness compared with more natural rural areas (McKinney, 2002). In America, the huge population of the domestic cat (*Felis catus*) has been a serious problem for many birds. There are no less than 77 million pet cats in the USA and possibly an additional 60–100 million homeless strays.

Some other types of beast are also favoured by urban expansion. Animals that can tolerate disturbance, are adaptable, utilize patches of open or woodland-edge habitat, creep about inside buildings, tap people's food supply surreptitiously, avoid recognizable competition with humans, or attract human appreciation and esteem, may increase in the urban milieu. For these sorts of reasons the north-east megalopolis of the USA hosts thriving populations of squirrels, rabbits, racoons, skunks and opossums, while some African cities are now frequently blessed with the scavenging attention of hyenas (*Crocuta crocuta*).

There is, of course, considerable variability in the diversity of species and richness of populations in different parts of cities. Most central urban areas, heavily covered by pavements and buildings, have low biodiversity, but suburban areas with many gardens may be rich in wildlife, though not invariably richer than neighbouring rural areas (McKinney, 2002). However, urban wildlife may be composed of many exotic species rather than native ones (Kühn and Klotz, 2006) and a big question is whether the addition of non-native species associated with urbanization exceeds the loss of native species to produce a net gain in species richness with urbanization. McKinney (2008), in a review of a large number of previous studies in many parts of the world, found that whereas 65% of studies showed increasing species richness with moderate urbanization, only a minority of invertebrate studies (about 30%) and a very small minority of non-avian vertebrate studies (about 12%) showed increasing species-richness.

Human economic activities may lead to a rise in the number of examples of a particular **habitat** which can lead to an expansion in the distribution of certain species, though often humans have prompted a contraction in the range of a particular species because of the removal or modification of its preferred habitat. In Britain the range of the little ringed plover (*Charadrius dubius*), a species virtually dependent on anthropogenic habitats, principally wet gravel and sand pits, greatly expanded as mineral extraction accelerated (see Figure 3.5). Indeed, it needs to be stressed that the changes of habitat brought about by urbanization, industry (Ratcliffe, 1974) and mining need not be detrimental.

One of the most remarkable examples of the consequences of creating new environments is provided by the European rabbit (*Oryctolagus cuniculus*). Introduced into Britain in early medieval times, and originally an inhabitant of the western Mediterranean lands, it was kept for food and fur in carefully tended warrens. Agricultural improvements, especially to grassland, together with the increasing decline in the numbers of predators such as hawks and foxes, brought about by game-guarding landlords (Sheail, 1971), enabled the rabbit to become one of the most numerous of mammals in the British countryside. By the early 1950s there were 60–100 million rabbits in Britain. Frequently, as many contemporary reports demonstrated, it grazed the land so close that in areas of light soil, like the Breckland of East Anglia or in coastal dune areas, wind erosion became a serious problem. Similarly, the rabbit flourished in Australia, especially after the introduction of the merino sheep which created favourable pasture-lands. Erosion in susceptible lands like the Mallee was severe. Both in England and Australia an effective strategy developed to control the rabbit was the introduction of a South American virus, *Myxoma*.

Some of the familiar British birds have benefitted from agricultural expansion. Formerly, when Britain was an extensively wooded country, the starling (*Sturnus vulgaris*) was a rare bird. The lapwing (*Vanellus vanellus*) is yet another component of the grassland fauna of Central Europe which has benefited from agriculture and the creation of open country with relatively sparse vegetation (Murton, 1971). There is also a large class of beasts which profit so much from the environmental conditions wrought by humans that they become very closely linked to them. These animals are often referred to as **synanthropes**. Pigeons and sparrows now form permanent and numerous populations in almost all the large cities of the world; human food supplies support many synanthropic rodents (rats and mice); a once shy forest-bird, the blackbird (*Turdus merula*), has in the course of a few generations become a regular and bold inhabitant of many gardens in the UK; and the squirrel (*Sciurus vulgaris*) in many places now occurs more frequently in parks than in forests (Illies, 1974: 101). The English house sparrow (*Passer domesticus*), a familiar bird in towns and cities, currently exists over approximately one-quarter of Earth's surface, and over the past 100 years it has doubled the area that it inhabits as settlers, immigrants and others have carried it from one continent to another (Figure 3.6), though recently in Britain, for reasons that are not clear, it has suffered a decline in numbers. It is found around settlements both in Amazonia and the Arctic Circle (Doughty, 1978). The marked increase in many species of gulls over recent decades is largely attributable to their growing utilization of food scraps on refuse tips (Croxall et al., 2012). This is, however, something of a mixed blessing, for most of the gull species which feed on urban rubbish dumps breed in coastal areas, and there is now good evidence to show that their greatly increased numbers are threatening other less common species, especially the terns (*Sterna* spp.). It may seem incongruous that the niches for scavenging birds which exploit rubbish tips should have been filled by sea-birds, but the gulls have proved ideal replacements for the kites which humans had removed, as they have the same ability to watch out for likely food sources from aloft and then to hover and plunge when they spy a suitable meal.

Most of the examples given so far to illustrate how human actions can lead to expansion in the numbers and distributions of certain animal species, have been

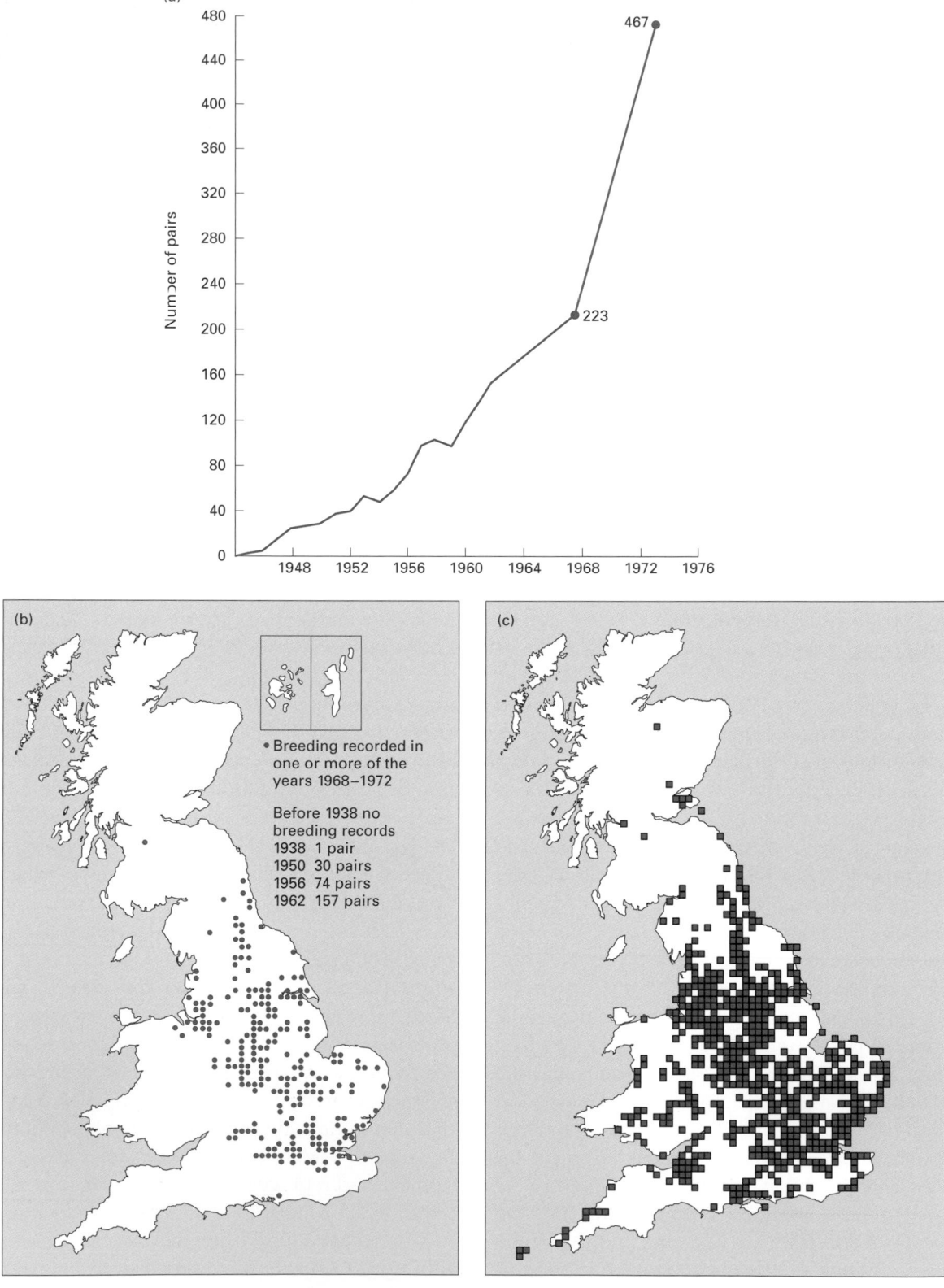

Figure 3.5 The changing range of the little ringed plover (*Charadrius dubius*): (a) The increase in the number summering in Britain (Modified after Murton, 1971, figure 6); (b) The increasein the range, related to habitat change, especially as a result of the increasing number of gravel (after Sharrock, J. T. R., 1976. Reproduced with permission from the British Trust for Ornithology); (c) The range of *Charadrius dubius* in 2004 pits (Data courtesy of the National Biodiversity Network Gateway with thanks to all the data contributors).

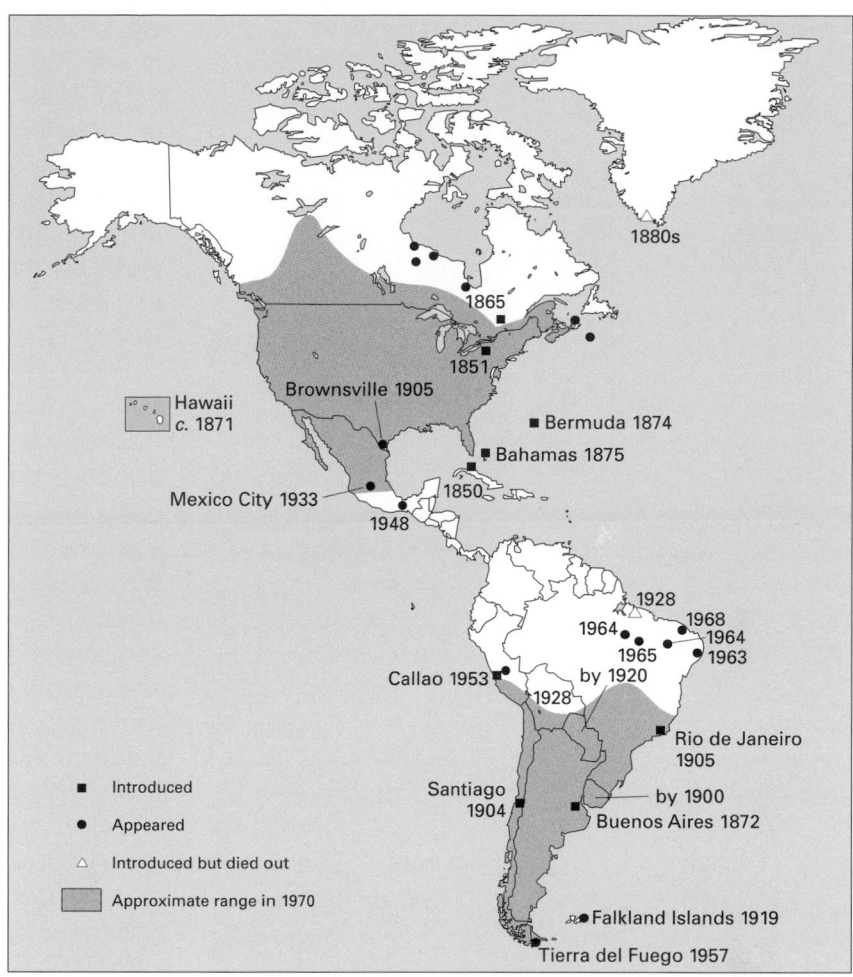

Figure 3.6 The spread of the English house sparrow in the New World (modified from Doughty, 1978: 14).

used to make the point that such expansion frequently occurs as an incidental consequence of human activities. There are, of course, many ways in which people have intentionally and effectively promoted the expansion of particular species (Table 3.2 presents data for some introduced mammals in Britain). This may sometimes be done deliberately to reduce the numbers of a species which has expanded as an unwanted consequence of human actions. Perhaps the best known example of this is biological control using introduced predators and parasites. Thus an Australian insect, the cottony-cushion scale, *Icerya purchasi*, was found in California in 1868. By the mid-1880s it was effectively destroying the citrus industry, but its ravages were quickly controlled by deliberately importing a parasitic fly and a predatory beetle (the Australian ladybird) from Australia. Because of the uncertain ecological effects of synthetic pesticides, such biological control

has its attractions but the importation of natural enemies is not without its own risks. For example, the introduction into Jamaica of mongooses to control rats led to the undesirable decimation of many native birds and small land mammals.

One further calculated method of increasing the numbers of wild animals, of conserving them and of gaining an economic return from them, is game cropping. In some circumstances, because they are better adapted to, and utilize more components of, the environment, wild ungulates provide an alternative means of land use to domestic stock. The exploitation of the saiga antelope, *Saiga tartarica* in the former Soviet Union was once thought to be the most successful story of this kind (Edington and Edington, 1977). Hunting, much overdone because of the imagined medical properties of its horns, had led to its near demise, but this was banned in the 1920s and a system

Table 3.2 Introduced mammals in Britain. Source: Reproduced with permission from Jarvis (1979: 188, table 1)

Species	Date of introduction of present stock	Reason for introduction
House mouse *Mus musculus*	Neolithic?	Accidental
Wild goat *Capra hircus*	Neolithic?	Food
Fallow deer *Dama dama*	Roman or earlier	Food, sport
Domestic cat *Felis catus*	Early Middle Ages	Ornament, pest control
Rabbit *Oryctolagus cuniculus*	Mid- to late-twelfth century	Food, sport
Black rat *Rattus rattus*	Thirteenth century?	Accidental
Brown rat *Rattus norvegicus*	Early eighteenth century (1728–1729)	Accidental
Sika deer *Cervus nippon*	1860	Ornament
Grey squirrel *Sciurus carolinensis*	1876	Ornament
Indian muntjak *Muntiacus muntjak*	1890	Ornament
Chinese muntjak *Muntiacus reevesi*	1900	Ornament
Chinese water deer *Hydropetes inermis*	1900	Ornament
Edible dormouse *Glis glis*	1902	Ornament
Musk rat *Ondatra zibethica*	1929–1937	Fur
Coypu *Myocastor coypus*	1929	Fur
Mink *Mustela vison*	1929	Fur
Bennett's wallaby *Macropus rufogriseus bennetti*	1939 or 1940	Ornament
Reindeer *Rangifer tarandus*	1952	Herding, ornament
Himalayan (Hodgson's) porcupine *Hystrix hodgsoni*	1969	Ornament
Crested porcupine *Hystrix cristata*	1972	Ornament
Mongolian gerbil *Meriones unguiculatus*	1973	Ornament

of controlled cropping was instituted. Under this regime (which produces appreciable quantities of meat and leather) total numbers rose from about 1000 to over 2 million in the 1970s, and in the process the herds reoccupied most of their original range. Since the break-up of the Soviet Union, however, previously successful conservation schemes have suffered from a shortage of funding, and high levels of poverty and unemployment have encouraged poaching (Neronov

et al., 2012). WWF now estimates that only 50,000 remain.

The global warming of recent decades, which may be anthropogenic in origin, has also facilitated the expansion of some species. One example of this is the southward expansion of a sea urchin, *Centrostephanus rodgersii*, along the coast of Tasmania (Ling, 2008). Its grazing activities have had a massive and deleterious impact on the macro-algal beds of the region, causing what are termed 'sea urchin barrens'.

Causes of animal contractions and decline: pollution

The extreme effect of human interference with animals is extinction, but before that point is reached humans may cause major contractions in both animal numbers and animal distribution. This decline may be brought about partly by intentional killings for subsistence and commercial purposes, but much wildlife decline occurs indirectly (Doughty, 1974), for example, through pollution, habitat change or loss, and competition from invaders.

Particular concern has been expressed about the role of veterinary drugs and certain chemical pesticides in creating undesirable and unexpected changes (Mitra et al., 2011). Recently, for example, it has been noted that there has been a catastrophic decline in the numbers of Gyps vultures in both Africa and India, and this has been attributed to the fact that the diclofenac given to cattle for treating inflammation, pain and fevers, is toxic to the vultures so that when they eat cattle carcases they are in effect poisoned (Naidoo et al., 2009).

There has also been a catastrophic decline in bumble and honey bees in some parts of the world and this has been attributed to the use of neonicotinoid pesticides, which are often used as a seed dressing for crops (Henry et al., 2012; Whitehorn et al., 2012). This could have highly detrimental effects on pollination both of crops and of wild plants.

The classic case, however, is that of DDT and related substances. These were introduced on a worldwide basis after the Second World War and proved highly effective in the control of insects such as malarial mosquitoes. However, evidence accumulated that DDT was persistent, capable of wide dispersal and reached high levels of concentration in certain animals at the top trophic levels. The tendency for DDT to become more concentrated as one moves up the food chain is illustrated further by the data in Table 3.3c. DDT levels in river and estuary waters may be low, but the zooplankton and shrimps contain higher levels, the fish that feed on them higher levels still, while fish-eating birds have the highest levels of all.

DDT had a major effect on sea-birds. Their egg shells become thinned to the extent that reproduction fails in fish-eating birds. Similar correlations between eggshell thinning and DDT residue concentrations have been demonstrated for various raptorial land-birds. The bald eagle (*Haliaeetus leucocephalus*), peregrine falcon (*Falco peregrinus*) and osprey (*Pandion haliaetus*) all showed decreases in eggshell thickness and population decline from 1947 to 1967, a decline which correlated with DDT usage and subsequent reproductive and metabolic effects upon the birds (Johnston, 1974).

However, appreciation of the undesirable side-effects of DDD and DDT brought about by ecological concentration led to severe controls of their use. For example, DDT reached a peak in terms of utilization in the USA in 1959 (35×10^6 kg) and by 1971 was down to 8.1×10^6 kg. It was banned for general use in the USA in 1972. The monitoring of birds in Florida over the same period indicated a parallel decline in the concentration of DDT and its metabolites (DDD and DDE) in their fat deposits (Johnston, 1974).

Other substances are also capable of concentration. For example, heavy metals and methyl mercury may build up in marine organisms, and filter feeders like shellfish have a strong tendency to concentrate the metals from very dilute solutions, as is clear in Table 3.3a and b.

It is possible, however, that the importance of biological accumulation and magnification has been overstated in some textbooks. Bryan (1979) reviewed the situation and noted that, although the absorption of pollutants from foods is often the most important route for bioaccumulation and transfer along food chains certainly occurs, this does not automatically mean that predators at high trophic levels will always contain the highest levels. He wrote (p. 497):

although, for a number of contaminants, concentrations in individual predators sometimes exceed those of their prey, when the situation overall is considered only the more persistent organochlorine pesticides, such as DDT and its metabolites and methyl mercury, show appreciable signs of being biologically magnified as a result of food-chain transfer.

Table 3.3 Examples of biological magnification

(a) Enrichment factors for the trace element compositions of shellfish compared with the marine environment. Source: Merlini (1971) in King (1975: 303, table 8.8)

Element	Enrichment factor		
	Scallops	*Oysters*	*Mussels*
Ag	23,002	18,700	330
Cd	260,000	318,000	100,000
Cr	200,000	60,000	320,000
Cu	3000	13,700	3000
Fe	291,500	68,200	196,000
Mn	55,000	4000	13,500
Mo	90	30	60
Ni	12,000	4000	14,000
Pb	5300	3300	4000
V	4500	1500	2500
Zn	28,000	110,300	9100

(b) Mean methyl mercury concentrations in organisms from a contaminated salt marsh. Source: Gardner et al. (in Bryan 1979)

Organism	Parts per million
Sediments	<0.001
Spartina	<0.001–0.002
Echinoderms	0.01
Annelids	0.13
Bivalves	0.15–0.26
Gastropods	0.25
Crustaceans	0.28
Fish muscle	1.04
Fish liver	1.57
Mammal muscle	2.2
Bird muscle	3.0
Mammal liver	4.3
Bird liver	8.2

(c) The concentration of DDT in the food chain. Source: King (1975: 301, table 8.7)

Source	Parts per million
River water	0.000003
Estuary water	0.00005
Zooplankton	0.04
Shrimps	0.16
Insects – Diptera	0.30
Minnows	0.50
Fundulus	1.24
Needlefish	2.00
Tern	2.80–5.17
Cormorant	26.40
Immature gull	75.50

Oil pollution is a serious problem for marine and coastal fauna and flora, although some of it derives from natural seeps. Sea-birds are especially vulnerable since oil clogs their feathers; while the ingestion of oil when birds attempt to preen themselves leads to enteritis and other complaints. Local bird populations may be seriously diminished (Guitard et al. 2010). Fortunately, though oil is toxic it becomes less so with time, and the oil spilt from the *Torrey Canyon* in March 1967 was almost biologically inert when it was stranded on the Cornish beaches (Cowell, 1976). There are 'natural' oil-degrading organisms in nature. Much of the damage caused to marine life as a result of this particular disaster was caused not by the oil itself but by the use of 2.5 million gallons of detergents to disperse it (Smith, 1968). Recovery from oil spills can be rapid and in most cases is complete within 2–10 years (Kingston, 2002).

It is possible that oil spills pose a particular risk in cold areas, for **biodegradation** processes achieved by microbes appear to be slow. This is probably because of a combination of low temperatures and limited availability of nitrogen, phosphorus and oxygen. The last of these is a constraint because compared to temperate ecosystems arctic tundra and coastal ecosystems are relatively stagnant – ice dampens re-aeration due to wave action in marine ecosystems, while standing water in tundra soils limits inputs of oxygen to them. Detailed reviews are provided in Engelhardt (1985).

There is far less information available on the effects of oil spills in freshwater environments, though they undoubtedly occur. Freshwater bodies have certain characteristics which, compared with marine environments, tend to modify the effects of oil spills. The most important of these are the smaller and shallower dimensions of most rivers, streams and lakes, which means that dilution and spreading effects are not as vigorous and significant in reducing surface slicks as would be the case at sea. In addition, the confining dimensions of ponds and lakes are likely to cause organisms to be subjected to prolonged exposure to dissolved and dispersed hydro-carbons. Information on this problem is summarized in Vandermeulen and Hrudey (1987).

One particular type of aquatic ecosystem where pollution is an increasingly serious problem is the coral reef (see Kuhlmann, 1988; Sheppard et al., 2009). Coral reefs are important because they are among the most diverse, productive and beautiful communities in the world. Although they only cover an area of less than 0.2% of the world's ocean beds, up to one-quarter of all marine species and one-fifth of known marine fish species live in coral reef ecosystems (World Resources Institute, 1998: 253). They also provide coastal protection, opportunities for recreation and potential sources of substances like drugs. Accelerated sedimentation resulting from poor land management (Nowlis et al., 1997; Fabricius, 2005), together with dredging, is probably responsible for more damage to reef communities than all the other forms of human insult combined, for suspended sediments create turbidity which restricts the light penetration necessary for coral growth. Also, the soft, shifting sediments may not favour colonization by reef organisms. Sewage is the second worst form of stress to which coral reefs are exposed, for oxygen-consuming substances in sewage result in reduced levels of oxygen in the water of lagoons. The detrimental effects of oxygen starvation are compounded by the fact that sewage may cause nutrient enrichment to stimulate algal growth and feed various predators, which in turn can overwhelm coral.

Runoff from the land can also cause nutrient enrichment, because it may contain nitrates and other nutrients derived from fertilizers. Equally, poor land management can also cause salinity levels to be reduced below the level of tolerance of reef communities as a consequence of accelerated runoff of freshwater from catchments draining into lagoons. An interesting example of the role of sedimentation in damaging reef health is provided by geochemical studies of long-lived corals from Australia's Great Barrier Reef (McCulloch et al., 2003). These showed that since 1870 and the start of European settlement in its catchment, the Burdekin River has carried five to ten times more sediment to the reef than it did previously. Reefs may also be threatened by accelerated coral bleaching associated with recent global warming. One of the reasons why all these stresses may be especially serious for reefs is that corals live and grow for several decades or more, so that it can take a long time for them to recover from damage. Whatever may be the causes, severe declines in coral cover on reefs have been identified in recent decades, though there are debates as to just how serious the degradation has been, not least in the context of the Great Barrier Reef (GBR) of Australia. On the one hand Hughes et al. (2011) have suggested that there has been a close to 40% loss of coral cover on the inner reefs of the GBR, whereas Sweatman et al. (2011) believe that losses of GBR corals over the last 40 years have probably been exaggerated. In the Indo-Pacific region coral cover declined from 42.5% in the early 1980s to only 22.1% by 2003, an average annual rate of 1%. In the Caribbean between 1977 and 2001 the annual rate was even higher – 1.5% (Bruno and Selig, 2007). Heavily degraded reefs are now little more than a mix of rubble, seaweed and slime (Pandolfi et al. 2005).

Industrial air pollution has had an adverse effect on wild animals (Newman, 1979). Arsenic emissions from silver foundries are known to have killed deer and rabbits in Germany; sulfur emissions from a pulp mill in Canada are known to have killed many song-birds; industrial fluorosis has been found in deer living in the USA and Canada; asbestosis has been found in baboons and rodents in the vicinity of asbestos mines in South Africa; and oxidants from air pollution are recognized causes of blindness in bighorn sheep (*Ovis canadensis*) in the San Bernadino Mountains near Los Angeles in California.

In many parts of the world considerable concern has arisen over the effects of lead poisoning on wildlife, particularly on wildfowl. Poisoning occurs particularly in raptors, pigeons, partridge and waterfowl, for example, when they feed and seek grit, since they ingest spent shotgun pellets or discarded anglers' weights in the process. Such pellets are eroded in the bird's gizzard and the absorbed lead causes a variety of adverse physiological effects that can result in death: damage to the nervous system, muscular paralysis, anaemia and liver and kidney damage (Fisher et al., 2006; Guitard et al. 2010).

Soil erosion, by increasing stream **turbidity** (another type of pollution), may adversely affect fish habitats. The reduction of light penetration inhibits photosynthesis, which in turn leads to a decline in food and a decline in carrying capacity. Decomposition of the organic matter, which is frequently deposited with sediment, uses dissolved oxygen, thereby reducing the oxygen supply around the fish; sediment restricts the emergence of fry from eggs; and turbidity reduces the ability of fish to find food (though conversely it may also allow young fish to escape predators).

Turbidity has also been increased by mining waste, construction and dredging. In the case of some of the Cornish rivers in china-clay mining areas, river turbidities reach 5000 mg per litre and trout are not present in streams so affected. Indeed, the non-toxic turbidity tolerances of river fish are much less than this figure. Alabaster (1972), reviewing data from a wide range of sources, believes that in the absence of other pollution, fisheries are unlikely to be harmed at chemically inert suspended sediment concentrations of less than 25 mg per litre, that there should be good or moderate fisheries at 25–80 mg per litre, that good fisheries are unlikely at 80–400 mg per litre and that at best only poor fisheries would exist at more than 400 mg per litre.

Habitat change and animal decline

Land use and **land cover** changes, including the conversion of natural ecosystems to cropland (Ramankutty and Foley, 1999), have so modified habitats that animal declines have ensued. Elsewhere ancient and diverse anthropogenic landscapes have themselves been modified by modern agricultural intensification. Good reviews of changes in biodiversity are provided by Heywood and Watson (1995) and Chapin et al. (2001). Certainly, many species have lost a very large part of their natural ranges as a result of human activities, and a study of historical and present distributions of 173 declining mammal species from six continents has indicated that these species have collectively lost over 50% of their historic range area (Ceballos and Ehrlich, 2002). A particularly stark example of a species whose range is shrinking, because of habitat loss and intense poaching, is the wild tiger, *Panthera tigris*. Its historical range was from the Caspian to southern China, but now, in less than a century, it occupies only 7% of this range (Dinerstein et al., 2007) (Figure 3.7) and now there may only be a few thousand individuals left in the wild. In Britain, currently declining species include red squirrels, Scottish wildcats, mountain hares, harvest mice, dormice, and, in rural areas, hedgehogs (Macdonald and Burnham, 2011).

One of the most important habitat changes in Britain has been the removal of many of the hedgerows that form the patchwork quilt so characteristic of large tracts of the countryside (Sturrock and Cathie, 1980). Their removal has taken place for a variety of reasons:

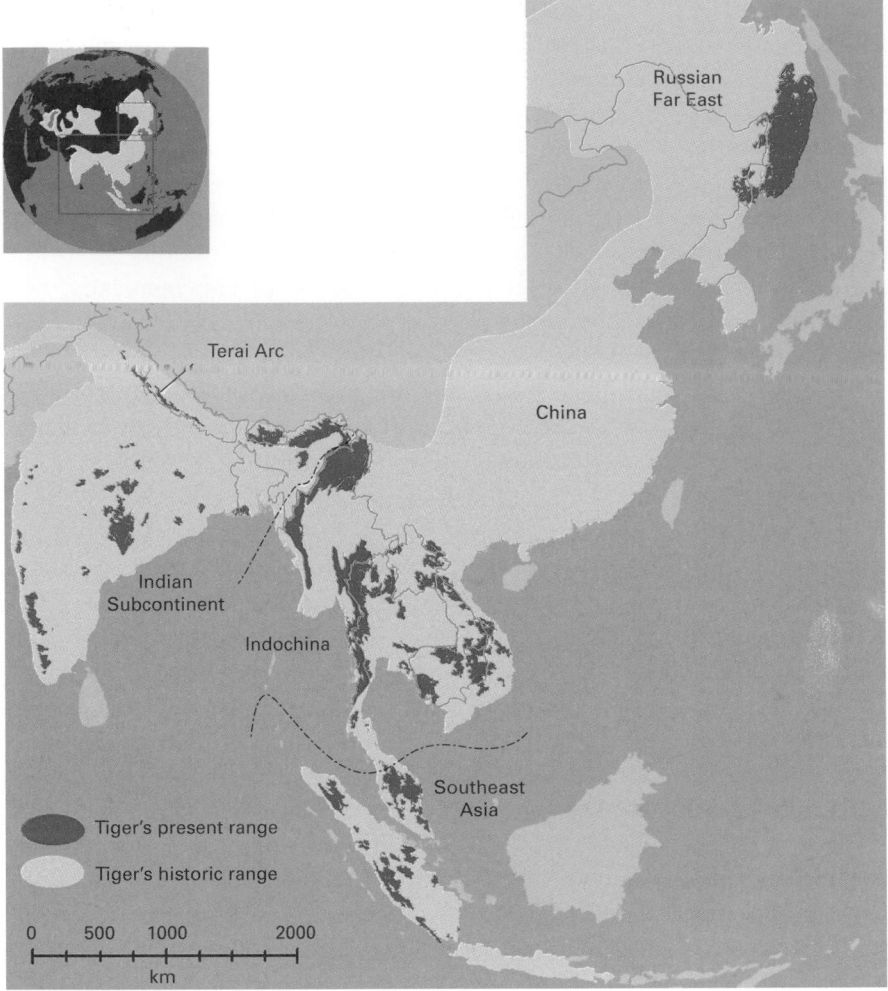

Figure 3.7 The historic and present distribution of tigers. They once occupied portions of central Asia around the Caspian Sea (see inset globe) (modified from Dinerstein et al., 2007, figure 1). (See Plate 22)

to create larger fields, so that larger machinery can be employed and so that it spends less time turning; because as farms are amalgamated boundary hedges may become redundant; because as farms become more specialized (e.g. just arable) hedges may not be needed to control stock or to provide areas for lambing and calving; and because drainage improvements may be more efficiently executed if hedges are absent. In 1962 there were almost 1 million km of hedge in Britain, probably covering the order of 180,000 hectares, an area greater than that of the national nature reserves. Bird numbers and diversity are likely to be severely reduced if the removal of hedgerows continues, for most British birds are essentially of woodland type and, since woods only cover a small percentage of the land surface of Britain, depend very much on hedgerows (Moore et al., 1967).

Other types of habitat have disappeared at a quick rate with agricultural changes (Nature Conservancy Council, 1977). For example, the botanical diversity of old grasslands is often reduced by replacing the lands with grass leys or by treating them with selective herbicides and fertilizers; this can mean that the habitat does not contain some of the basic requirements essential for many species. One can illustrate this by reference to the larva of the common blue butterfly (*Polyommatus icarus*) which feeds upon bird-foot trefoil (*Lotus corniculatus*). This plant disappears when pasture is ploughed and converted into a ley, or when it is treated with a selective herbicide. Once the plant has gone the butterfly vanishes too because it is not adapted to feeding on the plants grown in the ley of improved pasture. Overall, in the UK, since 1976, the populations of butterflies associated strongly with semi-natural

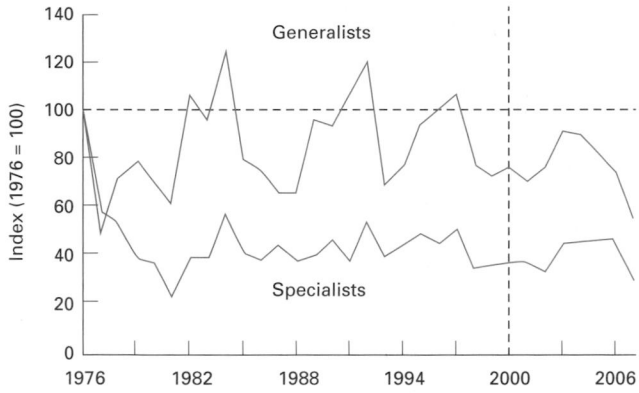

Figure 3.8 The population of butterflies in the UK from 1976 to 2007 (based on data from DEFRA, 2010).

Table 3.4 Decline in farmland birds in Britain between 1972 and 1996 as shown by the Common Birds Census. Based on data from the RSPB at http://www.rspb.org.uk/images/decline_tcm9-133467.pdf (accessed 01 February 2013) and from JNCC 1998

Species	% decline
Corn bunting	74
Grey partridge	78
Tree sparrow	87
Lapwing	42
Bullfinch	62
Song thrush	66
Turtle dove	62
Linnet	40
Skylark	60
Spotted flycatcher	78
Blackbird	33

habitats (specialists) show a 70% decline, and those found in the wider countryside (generalists) show a 45% decline (Figure 3.8).

Over large parts of western Europe the number of brown hares (*Lepus europaeus*) has declined sharply since the 1960s (Edwards et al., 2000). While this may have been partially due to hunting, to poisoning by pesticides such as paraquat, or to diseases such as European Brown Hare Syndrome, the prime cause appears to have been loss of crop and landscape diversity, which in turn has reduced the nutrition available to hares at certain times of year.

Agricultural intensification on British farms also appears to have had a deleterious effect on species of farmland birds, many of which have shown a downward trend in numbers over the last three decades (Table 3.4) (Gregory et al., 2004b). Three farmland birds, the grey partridge (*Perdix perdix*), the turtle dove (*Streptopelia turtur*) and the corn bunting (*Milaria calandra*) have declined by over 70% relative to 1970 (DEFRA, 2011). This is because of habitat changes – lack of fallows, less mixed farming, new crops, modern farm management, biocide use, hedgerow removal, etc. (Fuller et al., 1991). However, in Europe, both farmland and woodland specialist birds have shown a decline since the 1980s (Gregory et al., 2007) whereas some more generalist species have flourished. In the USA the greatest declines over the last 40 years have been in grassland and arid land birds. This results from habitat loss from urban development, overgrazing and the spread of invasive plants (State of the Birds, United States of America, 2009).

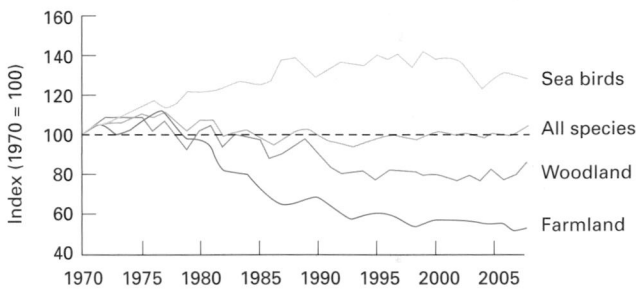

Figure 3.9 Population of wild birds in the UK, 1970–2008 (based on data from UK DEFRA, 2011).

The replacement of natural oak-dominated woodlands in Britain by conifer plantations, another major land-use change of recent decades, also has its implications for wildlife. It has been estimated that where this change has taken place the number of species of birds found has been approximately halved (Figure 3.9). Likewise the replacement of upland sheep walks with conifer plantations led to a sharp decline in raven numbers in southern Scotland and northern England. The raven (*Corvus corax*) obtains much of its carrion from open sheep country (Marquiss et al., 1978). Other birds that have suffered from planting moorland areas with forest plantations are miscellaneous types of wader, golden eagles, peregrine falcons and buzzards (Grove, 1983).

On a global basis the loss of **wetland** habitats (marshes, bogs, swamps, fens, mires, etc.) is a cause of considerable concern (Williams, 1990). In all, wetlands cover about 6–9% of the earth's surface (not far short of the total under tropical rainforest), and so they are far from being trivial, even though they tend to occur in relatively small patches. However, they also account for about one-quarter of the earth's total net primary productivity, have a very diverse fauna and flora, sequester carbon, help to abate floods, and provide crucial wintering, breeding and refuge areas for wild-life. According to some sources, the world may have lost half of its wetlands since 1900 (Zedler and Kercher, 2005), and the USA alone lost 53% of its original wetland area, primarily because of agricultural devel-opments, by the 1980s. There are, however, other threats, including drainage, dredging, filling, peat removal, urbanization, pollution and **channelization**. In Europe and the USA 90% of riverine flood plains have been gravely and extensively modified as habitats and in ecological terms have been described as 'func-tionally extinct' (Tockner and Stanford, 2002). The rate of wetland loss in the USA is now somewhat lower than it was, partly because of a growing awareness of the value of wetlands, the range of ecological services they provide and also because of an important inter-national treaty, the Ramsar Convention of 1971. Man-grove swamps are another wetland type that is under great pressure, and along the Atlantic and Pacific coasts of Central America, as many as 40% of mangrove species are threatened with extinction (Polidoro et al. 2010). Techniques for wetland restoration are discussed by Zedler and Kercher (2005).

Other causes of animal decline

One could list many other indirect causes of wildlife decline. For example, hunting and poaching of the African black rhino (*Diceros bicornis*), particularly for its horn, have caused a crash in numbers, from several hundreds of thousands in the early 1900s, to 70,000 in the late 1960s, and to a mere 2410 in 2004. Likewise, African elephant (*Loxondonta africana*) numbers were 3–5 million in the 1930s and 1940s, but are now only 600,000. Purposeful hunting, particularly when it involves modern firearms and snares, means that the distribution of many animal species has become smaller very rapidly (Figure 3.10). This clearly illus-trated by a study of the North American bison (*Bison

Figure 3.10 A major cause of animal decline is trapping. Here we see a sample of the many snares that have been found in a game reserve in Swaziland, southern Africa. (See Plate 23)

bison), which on the eve of European colonization still had a population of some 60 million in spite of the presence of a small Indian population. By 1850 only a few dozen examples of bison still survived, though conscious protective and conservation measures have saved it from extinction. By the late 1990s the total number of bison in North America has crept up to 250,000.

Animals native to remote ocean islands have been especially vulnerable to hunters since many have no flight instinct. They also provided a convenient source of provisions for seafarers. Thus the last example of the dodo (*Raphus cucullatus*) was slaughtered on Mauritius in 1681 and the last Steller's sea-cow (*Hydrodamalis gigas*) was killed on Bering Island in 1768 (Illies, 1974: 96). The land-birds, which are such a feature of coral atoll ecosystems, rapidly become extinct when cats, dogs and rats are introduced. The endemic flightless rail of Wake Island, *Rallus wakensis*, became extinct during the prolonged siege of the island in the Second World War, and the flightless rail of Laysan, *Porzanula palmeri*, is also now extinct. Sea-birds are more numer-ous on the atolls, but many colonies of terns, noddies and boobies, have been drastically reduced by vegeta-tion clearance and by introduced rats and cats in recent times (Stoddart, 1968; Sax and Gaines, 2008).). Other atoll birds, including the albatross, have been culled because of the threat they pose to aircraft using mili-tary airfields on the islands. Some birds have declined because they are large, vulnerable and edible.

The fish resources of the oceans are still exploited by hunting and gathering techniques, and over 90 million tonnes of fish are caught in the world each year, a more than fivefold increase since 1950. The world catch may have peaked, however, with only modest expansion since 1990 (Food and Agriculture Organization, 2009) (Figure 3.11) for some major fisheries have collapsed as a result of unsustainable exploitation (Hilborn et al., 2003). A dramatic illustration of a collapse is provided by the Atlantic cod (*Gadhus morhua*)off Newfoundland, Canada (Figure 3.12). Catches of this fish plummeted catastrophically after *c.* 1970. Chronic over-fishing is a particular problem in shallow marginal seas (e.g. the North Sea) and in the vicinity of coral reefs, where the removal of herbivorous fish can enable algal bloom to develop catastrophically (Hughes, 1994). Marine

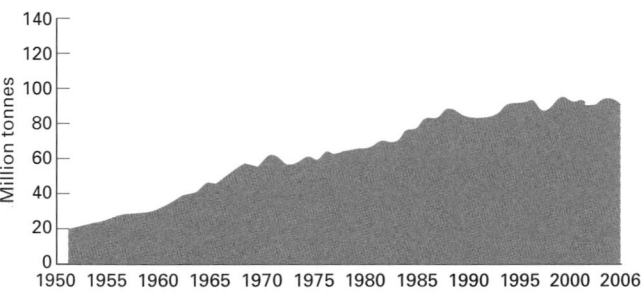

Figure 3.11 World capture fisheries production (source, FAO, 2009, figure 3).

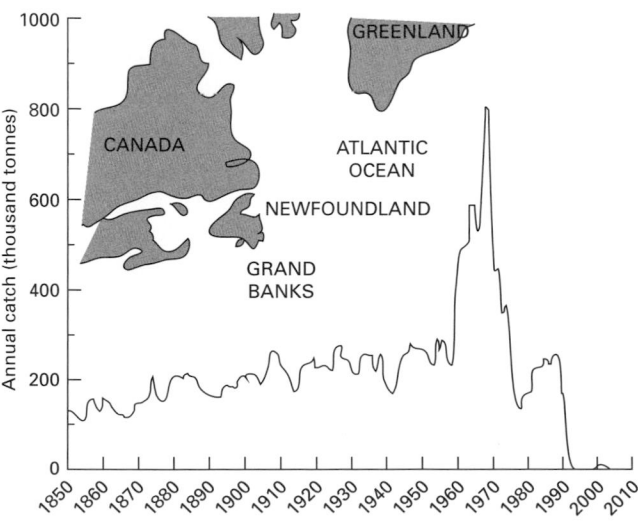

Figure 3.12 Catch history of Atlantic cod off Newfoundland since 1850 (modified from Millennium Ecosystem Assessment, 2005).

mammals are also at risk of extinction (Davidson et al., 2012). For instance, populations of the large varieties of whales, those that have been most heavily exploited, have been cut to a tiny fraction of their former sizes, and the hunting, especially by the Russians and the Japanese, still goes on though since the 1980s, most nations have banned whaling.

It is extremely difficult to estimate whale stocks before whaling began and therefore to quantify the magnitude of hunting effects that have taken place. However, genetic techniques have recently been used to establish pre-whaling stocks (Roman and Palumbi, 2003). These techniques have indicated that previous estimates of natural stocking levels were too low and that current populations are a fraction of past numbers. Pre-whaling populations in the North Atlantic of humpback, fin and minke whales were perhaps 865,000 in total, whereas current populations are 215,000. With bans on whaling that now exist, a number of large whale species have shown signs of recovery, though the status of some baleen whale populations continues to cause concern. In addition even if commercial whaling was banned in perpetuity, whales would face other threats of mankind's making, including acoustic disturbance from seismic exploration, sonar and motor-vessel traffic.

Vehicles and roads are another threat. Vehicle speed, noise and mobility upset remote and sensitive wildlife populations, and this problem has intensified with the rapid development in the use of off-road recreation vehicles. The effects of roads are becoming ever more pervasive as a result of increasing traffic levels, vehicle speed and road width and a new sub-discipline of road ecology has emerged (Sherwood et al., 2002; Forman et al., 2003). Roads affect the abiotic components of neighbouring ecosystems through pollution (e.g. by deicing salts) by altering local hydrological conditions (e.g. through ponding) and by causing localized erosion and sedimentation. They also act as barriers to movement, as cause of death (roadkill) as corridors and conduits for migration and as causes of habitat fragmentation (Coffin, 2007). A fine-meshed road network is a highly effective cause of habitat isolation, acting as a series of barriers to movement, particularly of small, cover-loving animals. As Oxley et al. (1974) have put it, 'a four-lane divided highway is as effective a barrier to the dispersal of small forest mammals as a body of fresh water twice as wide.' This barrier effect results from a variety of causes (Mader, 1984): roads interrupt microclimatic conditions; they are a broad band of emissions and disturbances; they are zones of instability due to cutting and spraying, etc.; they provide little cover against predators; and they subject animals to death and injury by moving wheels. Roads can also prevent animal migrations, and a proposed highway across the Serengeti in Tanzania has been opposed on the grounds that the area's huge herds of wildebeest (*Connochaetes*) would suffer a severe population collapse as a consequence (Dobson et al., 2010). The construction of fences also poses a threat to aggregated migrations of such large terrestrial mammals (Harris et al., 2009).

Leisure activities in general may create problems for some species (Speight, 1973). A survey of the breeding status of the little tern (*Sterna albifrons*) in Britain gave a number of instances of breeding failure by the species, apparently caused by the presence of fishermen and bathers on nesting beaches. The presence of even a few people inhibited the birds from returning to their nests. In like manner, species building floating nests on inland waters, such as great crested grebes (*Podiceps cristatus*), are very prone to disturbance by water-skiers and power-boats.

New types of construction can pose threats. As Doughty (1974) remarked '"wirescapes", tall buildings, and towers can become to birds what dams, locks, canals and irrigation ditches are to the passage of fish.' The accelerating construction of wind turbines is regarded by many as a threat to birds, not least in estuaries and shallow coastal waters (Powlesland, 2009). The construction of canals can cause changes in aquatic communities. Likewise turbines associated with hydroelectric schemes can cause numerous fish deaths, either directly through ingesting them, or through gas-bubble disease. This resembles the 'bends' in divers and is produced if a fish takes in water supersaturated with gases. The excess gas may come out of solution as bubbles which can lodge in various parts of the fish's body causing injury or death. Supersaturation of water with gas can be produced in turbines or when a spillway plunges into a deep basin (Baxter, 1977).

It might also be thought that human use of fire could be directly detrimental to animals, though the evidence is not conclusive. Many forest animals appear to be able to adapt to fire, and fire also tends to maintain habitat diversity. Thus, after a general review of the available ecological literature Bendell (1974) found that fires did not seem to produce as much change in birds or small mammals as one might have expected. Some of his data are presented in Table 3.5. The amount of change is fairly small, though some increase in species diversities and animal densities is evident. Indeed Vogl (1977) has pointed to the diversity of fauna associated with fire-affected ecosystems (p. 281):

Birds and mammals usually do not panic or show fear in the presence of fire, and are even attracted to fires and smoking or burned landscapes . . . The greatest arrays of higher animal species, and the largest numbers per unit area, are associated with fire-dependent, fire-maintained, and fire-initiated ecosystems.

The introduction of a new animal species can cause the decline of another, whether by predation, competition, the introduction of disease, or by hybridization. The British native red squirrel (*Sciurus vulgaris*), which used to be widespread throughout Britain, has contracted in range very greatly since the introduction of the North American grey squirrel (*Sciurus carolinensis*) at the turn of the twentieth century and is now largely restricted to parts of Scotland. One cause of this is that red squirrels have been killed by a disease that was introduced with the grey squirrels, squirrel parapoxvirus (Tomkins et al., 2003). **Hybridization** has been found to be a major problem with fish in California, where many species have been introduced by humans. This results when fish are transferred from one basin to a neighbouring basin, since closely related species of the type likely to hybridize usually exist in adjacent basins. One species can eliminate another closely related to it through genetic swamping (Moyle, 1976).

Many species that have suffered serious contraction in numbers and range have done so for a range of reasons, rather than in response to just one pressure. This was the case of the European otter, *Lutra lutra*, which in Britain had by the mid 1970s been reduced to such an extent that it only survived in Scotland, parts of Wales and western England, with a few remnant populations in parts of England. This was caused by a combination of hunting, a general decline in water quality which reduced fish stocks upon which otter populations depended, and the introduction of the persistent organochlorine pesticides dieldrin and aldrin. Since the 1970s, however, because of new con-

Table 3.5 Environmental effects of burning

(a) Change in number of species of breeding birds and small mammals after fire

	Foraging zone	Before fire	After fire	Gained (%)	Lost (%)
Birds	Grassland and shrub	48	62	38	8
	Tree trunk	25	26	20	16
	Tree	63	58	10	17
	Total	136	146	21	14
Small mammals	Grassland and shrub	42	45	17	10
	Forest	16	14	13	25
	Total	58	59	16	14

(b) Change in density and trend of population of breeding birds and small mammals after fire

	Foraging zone	Density (%)			Trend (%)		
		Increase	Decrease	No change	Increase	Decrease	No change
Birds	Grassland and shrub	50	9	41	24	10	66
	Tree trunk	28	16	56	4	8	88
	Tree	24	19	57	6	6	88
	Total	35	15	50	12	8	80
Small mammals	Grassland and shrub	24	13	63	20	5	75
	Forest	23	42	35	0	11	89
	Total	23	25	52	14	1	80

trols on water quality, restrictions on the use of organo-chlorine pesticides, legal protection of the species and active re-introduction programmes, the otter has now returned to most of England (Environment Agency, 2010; Macdonald and Burnham, 2011).

The complexity of possible causes of decline and the difficulty of establishing the precise causes of such decline is exemplified by the case of the house sparrow (*Passer domesticus*). This bird, which used to be a prominent and cheeky inhabitant of cities in the UK and western Europe, has suffered from near catastrophic crashes in numbers in the last few decades (Shaw et al., 2008). The possible explanations include:

1 Changes in the urban habitat because of the conversion of front gardens into parking spaces, the fashion of replacing lawns and flower beds with decking, infill between houses, the building of modern houses at high densities, a reduction in the rearing of chickens in gardens and the demise of horse-drawn transport. All these factors have reduced the food supply.

2 The use of pesticides in gardens and parks has led to a decline in the availability of insects and other food sources.
3 Pollution from growth in cars and from electromagnetic sources.
4 Reduction in nesting spaces as new houses have less nooks and crannies than old ones
5 Increased predation by cats and sparrow hawks.
6 Increased competition for food from magpies, gulls, feral pigeons.

Another example of the complex causes of animal decline is provided by the hedgehog (*Erinaceus europaeus*) in the UK. Its numbers have declined by a half over the last quarter of a century. Possible reasons include:

1 Loss of hedgerows
2 Presence of less rough field edges
3 Pesticide use
4 Fragmented habitats
5 Predation by badgers

6 Loss of suitable refuges in the urban habitat because of greater garden tidiness and the construction of impenetrable boundaries.

Animal extinctions in prehistoric times

Between 50,000 and 10,000 years ago, most large mammals became extinct everywhere except Africa, and a range of hypotheses can be erected to account for this phenomenon (Table 3.6). Some workers, notably Martin (1967, 1974), believed that the human role in these late Quaternary extinctions was crucial. They believed that extinction closely followed the chronology of the spread of prehistoric civilization and the development of big-game hunting. They would also maintain that there are no known continents or islands in which accelerated extinction definitely pre-dates the arrival of human settlements. An alternative interpretation, however, is that the Late Pleistocene extinctions of big mammals were caused by the rapid and substantial changes of climate at the termination of the last glacial (Martin and Wright, 1967; Martin and Klein, 1984).

Martin (1982) argued that the global pattern of extinctions of the large land mammals follows the footsteps of Palaeolithic settlements. He suggested that Africa and parts of southern Asia were affected first, with substantial losses at the end of the Acheulean, around 200,000 years ago. Europe and northern Asia were affected between 20,000 and 10,000 years ago, while North and South America were stripped of large herbivores between 12,000 and 10,000 years ago. Megafaunal extinctions in Australia may have occurred around 41,000 years ago, shortly after human arrivals (Roberts et al., 2001). Extinctions continued into the Holocene on oceanic islands (Turvey, 2009) and in the Galapagos Islands, for example, virtually all extinctions took place after the first human contact in AD 1535 (Steadman et al., 1991). Likewise the complete deforestation of Easter Island in the Pacific between 1200 and 800 years BP was an ecological disaster that led to the demise of much of the native flora and fauna and also precipitated the decline of the megalithic civilization that had erected the famous statues on the islands (Flenley et al., 1991). As Burney (1993: 536) has written:

Table 3.6 Hypotheses to explain the late Quaternary extinctions (modified from Koch and Barnosky 2006, table 1)

Type or name	Description
Environmental hypotheses	
Catastrophies	Mega-drought, rapid cooling, bolide impact
Habitat loss	Preferred habitat types lost or to fragmented
Mosaic-nutrient hypothesis	Loss of floras with high local diversity
Co-evolutionary disequilibrium	Disruption of coevolved plant–animal interactions due to flora rearrangements
Self-organized instability	Collapse of systems due to intrinsic dynamics
Human impacts other than hunting	
Habitat alteration	Loss or fragmentation of viable habitat due to human impacts, including fires
Introduced predators	Direct predation by dogs, rats, cats, pigs, etc.
Hyper-disease	Introduction of virulent diseases
Overkill hypotheses	
Blitzkrieg	Rapid loss of prey due to overhunting
Protracted overkill	Loss of prey after prolonged interaction with predator
Combined hypotheses	
Keystone mega-herbivores	Ecosystem collapse due to loss of landscape altering mega-herbivores, perhaps with increase in fire
Prey-switching	Nonhuman carnivores switch prey as human usurp preferred prey
Predator avoidance	Herbivores restricted to in viable refugia

For millennia after the late-Pleistocene extinctions, these seemingly fragile and ungainly ecosystems apparently thrived throughout the world. This is one of the soundest pieces of evidence against purely climatic explanations for the passing of Pleistocene faunas. In Europe, for instance, the last members of the elephant family survived climatic warming not on the vast Eurasian land mass, but on small islands in the Mediterranean.

The dates of the major episodes of Pleistocene extinction are crucial to the discussion. Martin (1967: 111–115) wrote:

Radiocarbon dates, pollen profiles associated with extinct animal remains, and new stratigraphic and archaeological evidence show that, depending on the region involved, late-Pleistocene extinction occurred either after, during or somewhat before worldwide climatic cooling of the last maximum . . . of glaciation . . . Outside continental Africa and South East Asia, massive extinction is unknown before the earliest known arrival of prehistoric man. In the case of Africa, massive extinction coincides with the final development of Acheulean hunting cultures which are widespread throughout the continent . . . The thought that prehistoric hunters ten to fifteen thousand years ago (and in Africa over forty thousand years ago) exterminated far more large animals than has modern man with modern weapons and advanced technology is certainly provocative and perhaps even deeply disturbing.

The nature of the human impact on animal extinctions can conveniently be classified into three types (Marshall, 1984): the 'blitzkrieg effect', which involves rapid deployment of human populations with big-game hunting technology so that there is very rapid demise of animal populations (Faith and Surovell, 2009); the 'innovation effect', where-by long-established human population groups adopt new hunting technologies and erase fauna that have already been stressed by climatic changes; and the 'attrition effect', whereby extinction takes place relatively slowly after a long history of human activity because of loss of habitat and competition for resources. The effects of habitat loss, fire usage and of introduced predators may be applicable in some circumstances (Koch and Barnosky, 2006).

The arguments that have been used in favour of an anthropogenic interpretation can be summarized as follows. First, massive extinction in North America seems to coincide in time with the arrival of humans in sufficient quantity and with sufficient technological

Figure 3.13 A reconstruction of the demise of one member of the mega-fauna, a mastodon, at the La Brea Tar Pits, Los Angeles (ASG). (See Plate 24)

skill in making suitable artefacts (the bifacial Clovis blades) to be able to kill large numbers of animals around 13,800–11,400 years ago (Faith and Surovell, 2009) (Figure 3.13). Second, in Europe the efficiency of Upper Palaeolithic hunters is attested by such sites as Solutré in France, where a late-Perigordian level is estimated to contain the remains of over 100,000 horses. Third, many beasts unfamiliar with people are remarkably tame and stupid in their presence, and it would have taken them a considerable time to learn to flee or seek concealment at the sight or scent of human life. Moreover, some, such as birds on oceanic islands, may have lost the ability to fly away from predators. Fourth, humans, in addition to hunting animals to death, may also have competed with them for particular food or water supplies or may have introduced predators, much as the dingo in Australia, which then competed with native fauna (Johnson and Wroe, 2003). Fifth, Coe

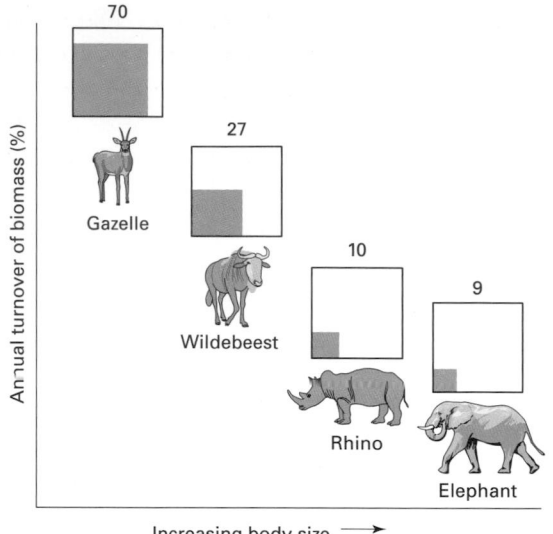

Figure 3.14 Decreasing rate of population biomass turnover with increasing body size in warm- and cold-blooded vertebrates (modified from Coe, 1982).

(1981, 1982) has suggested that the supposed preferential extinction of the larger mammals could also lend support to the role of human actions. He argued that while large body size has certain definite advantages, especially in terms of avoiding predation and being able to cover vast areas of savanna in search of food, large body size also means that these herbivores are required to feed almost continuously to sustain a large body mass. Furthermore, as the size and generation time of a mammal increases, the rate at which they turn over their biomass decreases (Figure 3.14). Hence a population of Thompson's gazelle (*Eudorcas thomsonii*) will turn over up to 70% of their biomass each year, a wildebeeste over 27%, but a rhino only 10% and the elephant just 9%. The significance of this is that, since large mammals can only turn over a small percentage of their population biomass each year, the rate of slaughter that such a population can sustain in the face of even a primitive hunter is very low indeed.

Certain objections have been levelled against the climatic change model and these tend to support the anthropogenic model. It has been suggested, for instance, that changes in climatic zones are generally sufficiently gradual for beasts to be able to follow the shifting vegetation and climatic zones of their choice. Similar environments are available in North America today as were present, in different locations and in

different proportions, during Late Pleistocene times. Second, it can be argued that the climatic changes associated with the multiple glaciations, interglacials, pluvials and interpluvials do not seem to have caused the same striking degree of elimination as those in the Late Pleistocene. The last glacial-interglacial transition was not so different from previous ones. A third difficulty with the climatic-cause theory is that animals like the mammoth occupied a broad range of habitats from arctic to tropical latitudes, so that is unlikely all would perish as a result of a climatic change (Martin, 1982).

This is not to say, however, that the climatic hypothesis is without foundation or support. The migration of animals in response to rapid climatic change could be halted by geographical barriers such as high mountain ranges or seas. The relatively rich state of the African big mammalian fauna is due, according to this point of view, to the fact that the African biota is not, or was not, greatly restricted by an insuperable geographical barrier. Another way in which climatic change could cause extinction is through its influence on disease transmission. It has been suggested that during glacials animals would be split into discrete groups cut off by ice sheets but that, as the ice melted (before 11,000 BP in many areas), contacts between groups would once again be made enabling diseases to which immunity might have been lost during isolation to spread rapidly. Large mammals, because of their low reproduction rates, would recover their numbers only slowly, and it was large mammals (according to Martin) that were the main sufferers in the Late Pleistocene extinctions.

The detailed dating of the European megafauna's demise lends further credence to the climatic model (Reed, 1970). The Eurasiatic **boreal** mammals, such as mammoth, woolly rhinoceros, musk ox and steppe bison, were associated with and adapted to the cold steppe which was the dominant environment in northern Europe during the glacial phases of the last glaciation. Each of these forms, especially the mammoth and the steppe bison, had been hunted by humans for tens of thousands of years, yet managed to survive through the last glacial. They appear to have disappeared, according to Reed, within the space of a few hundred years when warm conditions associated with the Allerød interstadial led to the restriction and near disappearance of their habitat. Late Pleistocene climate change has also been implicated in the demise of the Siberian and Japanese mammoths (Iwase et al., 2011; Nikolskiy et al., 2011)

Grayson (1977) added to the doubts expressed about the anthropogenic overkill hypothesis. He suggested that the overkill theory, because it states that the end of the North American Pleistocene was marked by extraordinary high rates of extinction of mammalian genera, required terminal Pleistocene mammalian generic extinctions to have been relatively greater than generic extinctions within other classes of vertebrates at this time. When he examined the extinction of birds he found that an almost exactly comparable proportion became extinct at the end of the Pleistocene as one finds for the megafauna. Moreover, as the radiocarbon dates for early societies in countries like Australia are pushed back, it becomes increasingly clear that humans and several species of megafauna were living together for quite long periods, thereby undermining the notion of rapid overkill (Wroe and Field, 2006).

Further arguments can be marshalled against the view that humans as predators played a critical role in the Late Pleistocene extinctions in North America (Butzer, 1972: 509–510). There are, for example, relatively few Palaeo-Indian sites over an immense area, and the majority of these have a very limited cultural inventory. In addition there is no clear evidence that Palaeo-Indian subsistence was necessarily based, in the main part, on big-game hunting; if it was, only two genera were hunted intensively: mammoth and bison. A final point that militates against the argument that humans were primarily responsible for the waves of extinction is the survival of many big-game species well into the Holocene (Johnson, 2005) and even to the nineteenth century, despite much larger and more efficient human populations.

Thus the human role in the great Late Pleistocene extinction is still a matter of debate (Field and Wroe, 2012). The problem is complicated because certain major cultural changes in human societies may have occurred in response to climatic change. The cultural changes may have assisted in the extinction process (Koch and Barnosky, 2006), and increasing numbers of technologically competent humans may have delivered in the final coup de grace to isolated remnants already doomed by rapid post-glacial environmental changes. In this context it is worth noting that Haynes (1991) suggested that at around 11,000 years ago conditions were dry in the interior of the USA and that

Clovis hunters may as a result have found large game animals easier prey when concentrated around waterholes and under stress. Humans may have been assisted in their hunting efficiency as a result of their early domestication of hunting dogs (Fiedel, 2005). Humans may also inadvertently have caused animal declines because of the introduction of diseases (sometimes called 'hyperdiseases') to new environments (Lyons et al., 2004).

Actual extinction may have occurred after or concurrently with a dwarfing in the size of animals, and this too is a subject of controversy (Guthrie, 2003). On the one hand, there are those who believe that the dwarfing could result from a reduction in food availability brought about by climatic deterioration. On the other, it can be maintained that this phenomenon derives from the fact that small animals, being more adept at hiding and being a less attractive target for a hunter, are more likely to survive human predators, so that a genetic selection towards reduced body size takes place (Marshall, 1984).

Although the debate about the cause of this extinction spasm has now persisted for a long time, there are still great uncertainties. This is particularly true with regard to the chronologies of extinction and of human colonization (Grayson, 1988; Brook and Bowman, 2002). It may well be that in many cases it was a combination of technological and climate changes that caused the extinctions to occur (Pushkina and Raia, 2008; Yule, 2009).

It is probable that the Late Pleistocene extinctions themselves had major ecological consequences. As Birks (1986: 49) has put it:

The ecological effects of rapid extinction of over 75% of the New World's large herbivores . . . must have been profound, for example on seed dispersal, browsing, grazing, trampling and tree regeneration . . . large grazers and browsers such as bison, mammoth and woolly rhinoceros may have been important in delaying or even inhibiting tree growth.

This is a theme that has been reviewed by Johnson (2009), who argues that in particular the extermination of herbivores would have had profound influences on the structure of vegetation, by reduction the amount of heavy grazing and trampling. Vegetation may have become less open, habitat mosaics may have become less varied, and with accumulation of uncropped plant material and increased vegetation density, fire may have increased. This has been indicated as having occurred in Australia by Rule et al. (2012).

Modern-day extinctions

As the size of human populations has increased and technology has developed, humans have been responsible for the extinction of many species of animals and a reduction in biodiversity – 'the only truly irreversible global environmental change the Earth faces today' (Dirzo and Raven, 2003). As Lynas (2011: 30) put it, 'We have poisoned, outcompeted or simply eaten so many other species that the Earth is currently in the throes of its most severe mass extinction event for 65 million years. . . .' That said, it is very difficult to estimate current rates of extinction (Stork, 2010), not least because we have a very imperfect idea of how many species exist. Nonetheless, it has often been maintained that there has been a close correlation between the curve of population growth since the mid-seventeenth century and the curve of the number of species that have become extinct (Figure 3.15a and b). However, although it is likely that European expansion overseas during that time has led to many extinctions, the dramatic increase in the rate of extinctions over the last few centuries that is shown in the figure may in part be a consequence of a dramatic increase in the documentation of natural phenomena.

A more recent attempt to look at the historical trends in extinction rates (Figure 3.15c) for 20-year intervals from the year 1600 shows a rather different pattern,

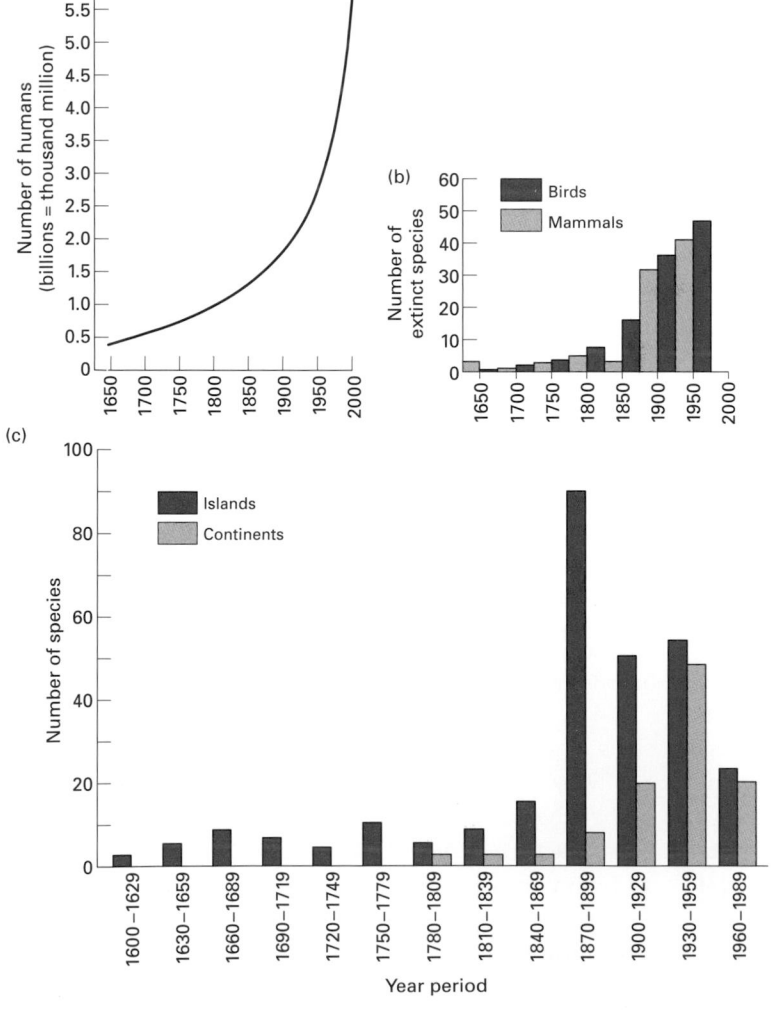

Figure 3.15 Time series of animal extinctions since the seventeenth century, in relation to human population increase (a). Figure (b) shows an early attempt to trace the number of extinctions in birds and mammals (modified from Ziswiler in National Academy of Sciences, 1972, figure 3.2), while figure (c) is a more recent attempt and displays a decreasing rate of animal extinctions in recent decades (modified from World Conservation Monitoring Centre, 1992, figure 16.5).

with a decline in rate for 1960–1990. This may partly be because of a time-lag in the recording of the most recent extinction events, but it may also be the result of successful conservation efforts and the spread of the areal extent of protected areas (see Chapter 13). What is clear, however, is just what a large proportion of total extinctions have been recorded from islands.

When considering modern-day extinctions the role of human agency is much less controversial. Nonetheless, some modern extinctions of species are natural. Extinction is a biological reality: it is part of the process of evolution. In any period, including the present, there are naturally doomed species, which are bound to disappear, either through over-specialization or an incapacity to adapt themselves to climatic change and the competition of others, or because of natural cataclysms such as earthquakes, eruptions and floods. Fisher et al. (1969) believed that probably one-quarter of the species of birds and mammals that have become extinct since 1600 may have died out naturally. In spite of this, however, the rate of animal extinctions brought about by humans in the last 400 years is of a very high order when compared to the norm of geological time. In 1600 there were approximately 4226 living species of mammals: since then 36 (0.85%) have become extinct; and at least 120 of them (2.84%) are presently believed to be in some (or great) danger of extinction. A similar picture applies to birds. In 1600 there were about 8684 living species; since then 94 (1.09%) have become extinct; and at least 187 of them (2.16%) are presently, or have very lately been, in danger of extinction (Fisher et al., 1969). At the present time, of the 5487 mammal species on Earth, at least one fifth are at risk of extinction in the wild (Hoffmann et al., 2011).

Some beasts appear to be more prone to extinction than others, and a distinction is now often drawn between r-selected species and k-selected species. These are two ends of a spectrum. The former have a high rate of increase, short gestation periods, quick maturation and the advantage that they have the ability either to react quickly to new environmental opportunities or to make use of transient habitats (such as seasonal ponds). The life duration of individuals tends to be short, populations tend to be unstable and the species may overexploit their environment to their eventual detriment. Many pests come into this category. The other end of the spectrum, the k-selection species are those which tend to be endangered or to become extinct. Their prime

characteristics are that they are better adapted to physical changes in their environment (such as seasonal fluctuations in temperature and moisture) and live in a relatively stable environment (Miller and Botkin, 1974). These species tend to have much greater longevity, longer generation times, fewer offspring, but a higher probability of survival of young and adults. They have traded a high rate of increase and the ability to exploit transient environments for the ability to maintain more stable populations with low rates of increase, but correspondingly low rates of mortality and a closer adjustment to the long-term capacity of the environment to support their population.

Table 3.7, which is modified after Ehrenfeld (1972), attempts to bring together some of the characteristics of species which affect their survival in the human world. This is also a theme discussed in Jeffries (1997).

Table 3.7 Characteristics of species affecting survival

Endangered	Safe
Large size	Small size
Predator	Grazer, scavenger, insectivore
Narrow habitat tolerance	Wide habitat tolerance
Valuable fur, oil, hide, etc.	Not a source of natural products
Restricted distribution	Broad distribution
Lives largely in international waters	
Migrates across international boundaries	Lives largely in one country
Reproduction in one or two vast aggregates	Reproduction by solitary pairs or in many small aggregates
Long gestation period	Short gestation period
Small litters	Big litters and quick maturation
Behavioural idiosyncracies that are non-adaptive today	Adaptive
Intolerance to the presence of humans	Tolerance of humans
Dangerous to humans, livestock, etc.	Perceived as harmless

There are at least nine ecological or life history traits that have been proposed as factors which determine the sensitivity of an animal species to a reduction and fragmentation in habitat (World Conservation Monitoring Centre, 1992: 193):

Rarity
Several studies have found that the abundance of a species prior to habitat fragmentation is a significant predictor of

extinction . . . This is only to be expected, since fewer individuals of a rare species than a common species are likely to occur in habitat fragments, and the mechanisms of extinction mean that small populations are inherently more likely to become extinct than large.

Dispersal ability

If animals are capable of migrating between fragments or between 'mainland' areas and fragments, the effects of small population size may be partly or even greatly mitigated by the arrival of 'rescuers'. Species that are good dispersers may therefore be less prone to extinction in fragmented habitats than poor dispersers.

Degree of specialisation

Ecological specialists often exploit resources which are patchily distributed in space and time, and therefore tend to be rare. Specialists may also be vulnerable to successional changes in fragments and to the collapse of coevolved mutualisms or food webs.

Niche location

Species adapted to, or able to tolerate, conditions at the interface between different types of habitats may be less affected by fragmentation than others. For example, forest edge species may actually benefit from habitat fragmentation.

Population variability

Species with relatively stable populations are less vulnerable than species with pronounced population fluctuations, since they are less likely to decline below some critical threshold from which recovery becomes unlikely.

Trophic status

Animals occupying high **trophic** levels usually have small populations: e.g. insectivores are far fewer in number than their insect prey and, as noted above, rarer species are more vulnerable to extinction.

Adult survival rate

Species with naturally low adult survival rates may be more likely to become extinct. . .

Longevity

Long-lived animals are less vulnerable to extinction than short-lived.

Intrinsic rate of population increase

Populations which can expand rapidly are more likely to recover after population declines than those which cannot.

In addition, there seems to be a strong positive association between body size and extinction risk for mammals. The most threatened families are dominated by large species, such as primates and ungulates, whereas the least threatened include small mammals such as rodents and bats. The explanation for this is that larger species tend to have lower population densities, slower life histories, larger home ranges and a greater risk of being hunted (Schipper et al., 2008).

Possibly one of the most fundamental ways in which humans are causing extinction is by reducing the area of natural habitat available to a species (Turner, 1996) and by fragmenting it (Fahrig, 2003). Habitat fragmentation divides once continuous, large populations into many smaller ones, which can be more or less isolated. Small population size and strong isolation of populations is associated with various negative consequences, which include susceptibility to natural disasters, genetic drift and inbreeding (Lienert, 2004). Even wildlife reserves tend to be small 'islands' in an inhospitable sea of artificially modified vegetation or urban sprawl.

We know from many of the classic studies in true island biogeography that the number of species living at a particular location is related to area; islands support fewer species than do similar areas of mainland, and small islands have fewer species than do large ones. Thus it may well follow that if humans destroy the greater part of a vast belt of natural forest, leaving just a small reserve, initially it will be 'supersaturated' with species, containing more than is appropriate to its area when at equilibrium (Gorman, 1979). Since the population sizes of the species living in the forest will now be greatly reduced, the extinction rate will increase and the number of species will decline towards equilibrium. For this reason it may be a sound principle to make reserves as large as possible; a larger reserve will support more species at equilibrium by allowing the existence of larger populations with lower extinction rates. Several small reserves will plainly be better than no reserves at all, but they will tend to hold fewer species at equilibrium than will a single reserve of the same area. If it is necessary to have several small reserves, they should be placed as close to each other as possible so that each may act as a source area for the others. Connectivity of reserves is an important issue, and if it can be achieved their equilibrial number of species will be raised due to increased immigration rates.

There are situations when small reserves may have advantages over a single large reserve: they will be less prone to total decimation by some natural catastrophe such as fire; they may allow the preservation of a range of rare and scattered habitats; and they may allow the survival of a group of competitors one of which would exclude the others from a single reserve.

Reduction in area leads to reduction in numbers, and this in turn can lead to genetic impoverishment through inbreeding (Frankel, 1984). The effect on reproductive performance appears to be particularly marked. Inbreeding degeneration is, however, not the only effect of small population size for, in the longer term the depletion of genetic variance is more serious since it reduces the capacity for adaptive change. Space is therefore an important consideration, especially for those animals that require large expanses of territory. For example, the population density of the wolf (*Canis lupus*) is about one adult per 20 km², and it has been calculated that for a viable population to exist, one might need 600 individuals ranging over an area of 12,000 km². The significance of this is apparent when one realizes that most nature reserves are small: 93% of the world's national parks and reserves have an area less than 5000 km² and 78% less than 1000 km².

Equally, range loss, the shrinking of the geographical area in which a given species is found, often marks the start of a downward spiral towards extinction. Such a contraction in range may result from habitat loss or from such processes as hunting and capture. Particular concern has been expressed in this context about the pressures on primates, notably in Southeast Asia. Of 44 species, 33 have lost at least half their natural range in the region. In two cases, those of the Javan leaf monkey (*Presbytis comata*) and the Javan (*Hylobates moloch*) and grey (*Hylobates muelleri*) gibbons, the loss of range is no less than 96%. Recent figures produced by the International Union for the Conservation of Nature and the United Nations Environment Programme for wildlife habitat loss show the severity of the problem (see *World Resources*, 1988–1989, tables 6.4 and 6.5). In the Indomalayan countries 68% of the original wildlife habitat has been lost, and the comparable figure for tropical Africa is 65%. In these regions, only Brunei and Zambia have lost less than 30% of their original habitat, while at the other end of the spectrum, Bangladesh, the most densely populated large country in the world, has suffered a loss of 94%. Figure 3.16 shows the fragmentation and reduction in area of rainforest within three areas: Sumatra, Costa Rica and Borneo.

Although habitat change and destruction is clearly a major cause of extinction in the modern era, a remark-able important cause is the introduction of competitive species (Figure 3.17). When new species are deliberately or accidentally introduced to an area, they can cause the extermination of local fauna by preying on them or out-competing them for food and space. As we have seen elsewhere island species have proved to be especially vulnerable. The World Conservation Monitoring Centre, in its analysis of the known causes of animal extinctions since 1600 AD, believed that 39% were caused by species introduction, 36% by habitat destruction, 23% by hunting and 2% for other reasons (*World Resources*, 1994–1995: 149). With respect to land mammals, Schipper et al. (2008) suggest that habitat loss and degradation is the prime threat (affecting 40% of the species assessed) followed by harvesting (i.e. hunting or gathering for food, medicine, materials, etc.), which is the threat to 17% of land mammals. Among marine mammals the threats are very different. Here the dominant threat is accidental mortality, which affects 78% of species and is caused by fishing and shipping. Pollution is the second most prevalent threat to marine mammals. The main threat to seabirds is competition from invasive species (Croxall et al., 2012). Figure 3.18, based on the work of Baillie et al. (2010), shows the threats posed to various types or organism and also population trends in fish, amphibians, reptiles, birds and mammals since 1970. The most serious decline is that of amphibians.

Islands have been especially subject to faunal extinctions. Since 1500, 95% of extinctions of birds and terrestrial mammals have been on islands, and on a per unit basis the extinction rate was 177 times higher for mammals and 187 times higher for birds than on continents (Loehle and Eschenbach, 2012).

There are certainly some particularly important environments in terms of their species diversity (Myers, 1990; Myers et al., 2000) and such biodiversity 'hotspots' need to be made priorities for conservation. They include coral reefs, tropical forests (which support well over half the planet's species on only about 6% of its land area) and some of the Mediterranean climate ecosystems (including the extraordinarily diverse Fynbos shrublands of the Cape region of South Africa). Some environments are crucial because of their high levels of species diversity or endemic species, others are crucial because their loss would have consequences elsewhere. This applies for example, to wetlands which provide habitats for migratory birds and produce the

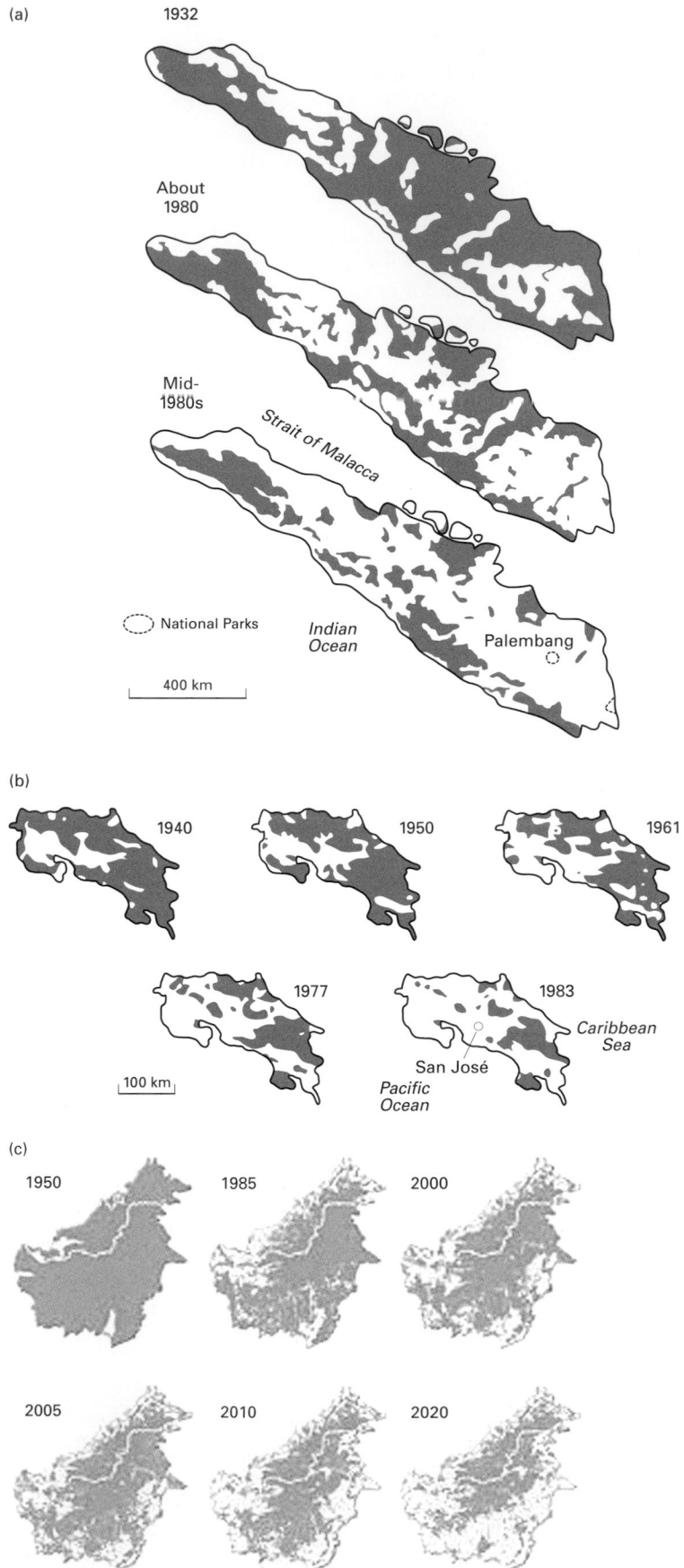

Figure 3.16 Progressive habitat fragmentation in the rainforest environments of (a) Sumatra and (b) Costa Rica (c) Borneo (modified from Whitten et al., 1987 and Terborgh, 1992 and with data from UNEP).

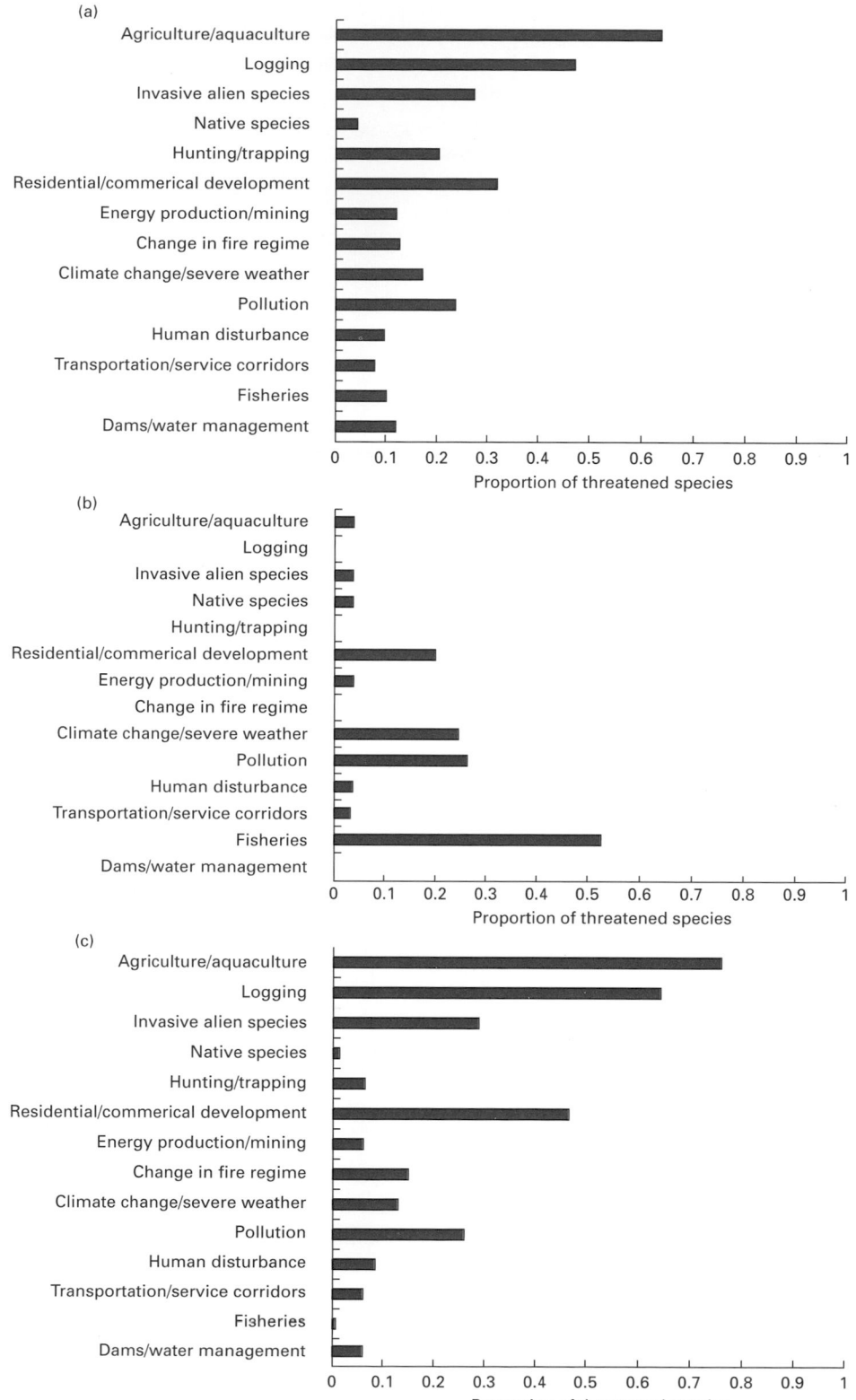

Figure 3.17 Global threats to (a) vertebrates, (b) marine fish, (c) amphibians, (d) reptiles, (e) mammals and (f) birds (modified from Baillie et al., 2010).

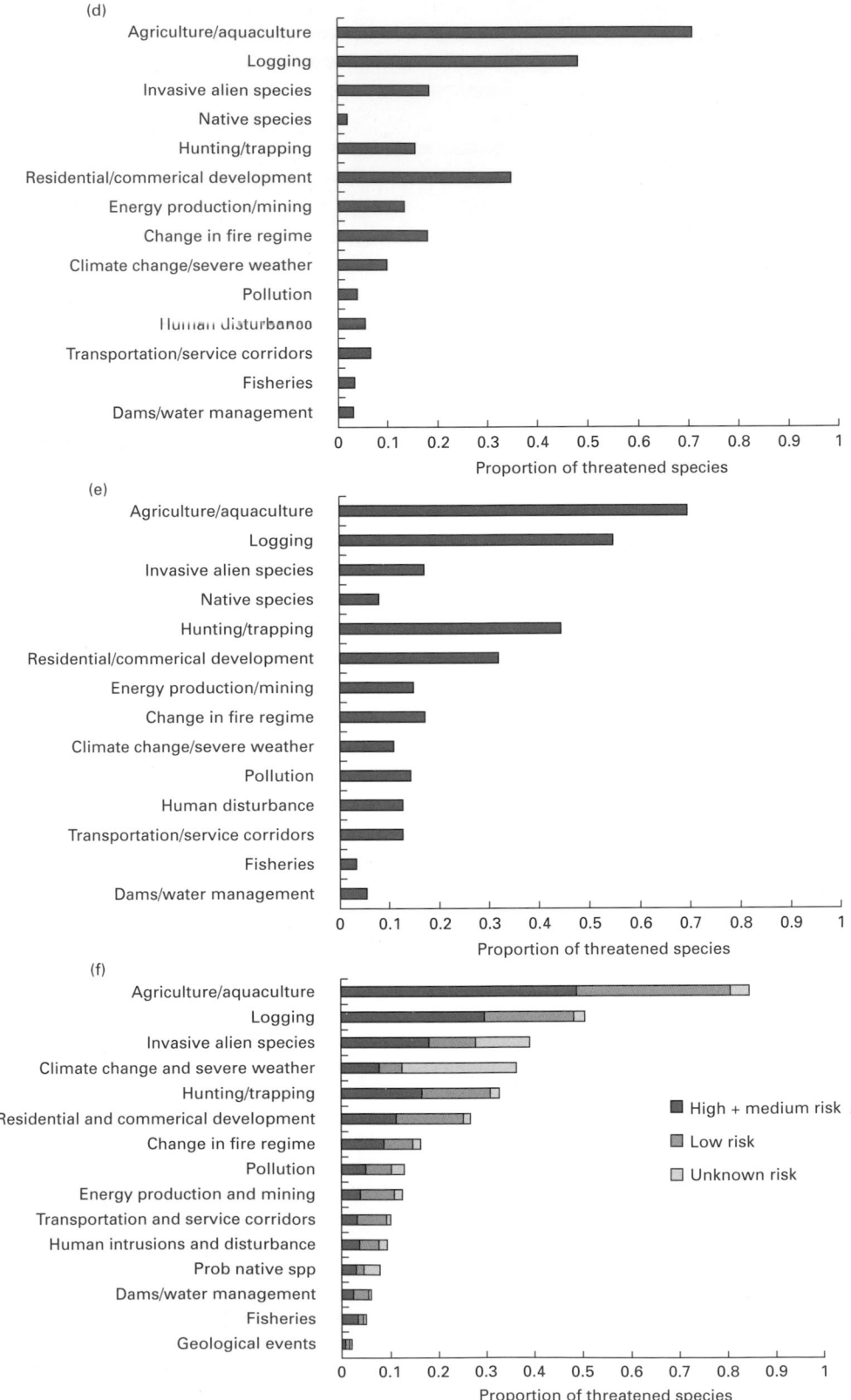

Figure 3.17 (*Continued*)

nutrients for many fisheries. Myers et al. (2000) have argued that as many as 44% of all species of vascular plants and 35% of all species in four groups of vertebrate animals are confined to just 25 hotspots (Figure 3.19) that comprise a mere 1.4% of the earth's land surface. These are major locations for conservation.

Finally, future climate change could act as a major cause of extinctions in the coming decades and could be a major threat to biodiversity (Thomas et al., 2004). As we will see in Chapters 8 to 13, many habitats will change markedly, and as a result many nature reserves will be in the wrong place for the species they are meant to protect. For instance, high altitude habitats at the tops of mountains may simply disappear, and ice-dependent species like the polar bear (*Ursus maritimus*) may be particularly at risk.

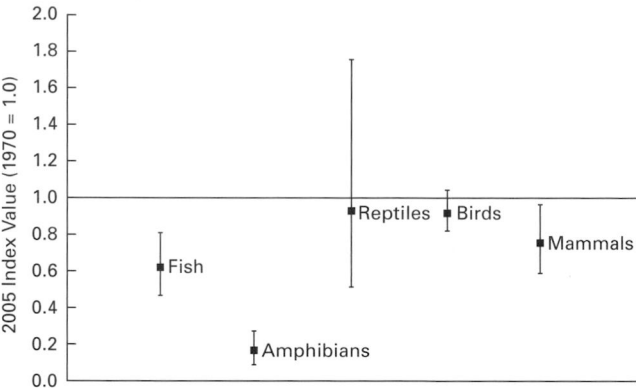

Figure 3.18 Population trends in fish, amphibians, reptiles, birds and mammals since 1970. The change in the size of these populations relative to 1970 (1970 = 1.0) is plotted over time. Squares are index values in 2005, and bars are 95% confidence limits (modified from Baillie et al., 2010).

Points for review

What do you understand by plant and animal domestication?
What are the ecological effects of invasive animals?
What are the main factors that lead to declines in animal populations?
Why did some animals become extinct in prehistoric times?

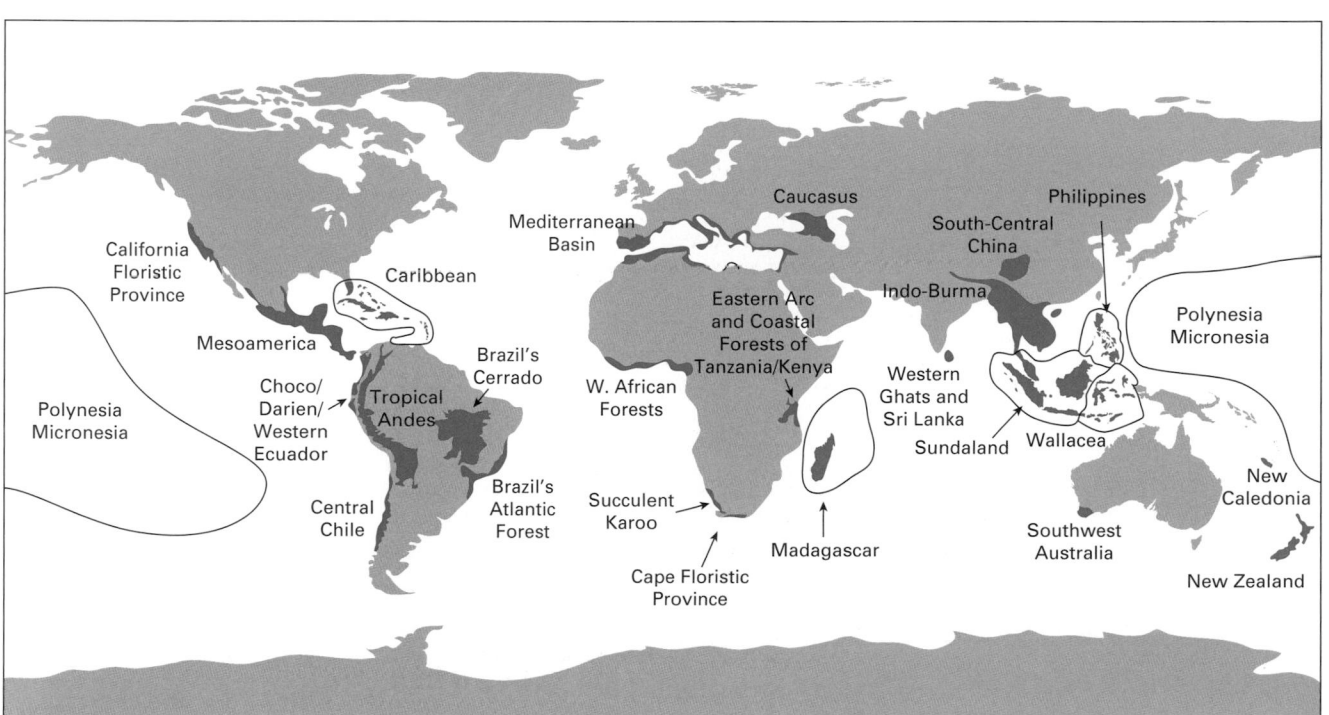

Figure 3.19 Conservation hot spot areas. Hot spots are habitats with many species found nowhere else and in greatest danger of extinction from human activity (after Myers et al., 2000, figure 1). Reproduced with permission.

Guide to reading

Baillie, J.E.M., Griffiths, J., Turvey, S.T., Loh, J. and Collen, B. (2010) *Evolution Lost. Status and Trends of the World's Vertebrates*. London: Zoological Society of London. An authoritative and well illustrated discussion of the current threats to the world's mammals, fish, birds, amphibians, and reptiles.

Clout, M.N. and Williams, P.A. (eds) (2009) *Invasive Species Management*. New York: Oxford University Press. An edited review of how to try and counter the threat of invasive species.

Davis, M.A. (2009) *Invasion Biology*. Oxford: Oxford University Press. A concise, comprehensive and controversial account of all aspects of biological invasions.

Martin, P.S. and Klein, R.G. (1984) *Pleistocene Extinctions*. Tucson: University of Arizona Press. An enormous survey of whether or not late Pleistocene extinctions were caused by humans.

Sodhi, N.S. and Ehrlich, P.R. (eds) (2010) *Conservation Biology for All*. Oxford: Oxford University Press. An edited collection of papers on threats to habitats and ways of conserving them.

Turvey, S.T. (ed.) (2009) *Holocene Extinctions*. Oxford: Oxford University Press. A wide-ranging survey of extinctions over the last 11,500 years.

Wilson, E.O. (1992) *The Diversity of Life. Harvard*: Belknap Press. A beautifully written and highly readable discussion of biodiversity.

4 THE HUMAN IMPACT ON THE SOIL

Chapter Overview

Humans modify soils in many ways. Among the most important changes are those involving accelerated salinization, lateritization, podzolization and acidification. Humans also affect soils by such processes such as land drainage and the addition of fertilizers. Of very great importance is accelerated erosion brought about primarily by land cover changes.

Introduction

Humans live close to and depend on the soil. It is one of the thinnest and most vulnerable human resources and is one upon which, both deliberately and inadvertently, humans have had major impacts (Richter, 2007). Moreover, such impacts can occur with great rapidity in response to land-use change, new technologies or waves of colonization (see Russell and Isbell, 1986, for a review in the context of Australia). Soils modified by human activities are called anthrosols.

Natural soil is the product of a whole range of factors, and the classic expression of this is that of Jenny (1941):

$$s = f(cl, o, r, p, t \dots),$$

where s denotes any soil property, cl is the regional climate, o the biota, r the topography, p the parent

The Human Impact on the Natural Environment: Past, Present and Future, Seventh Edition. Andrew S. Goudie.
© 2013 John Wiley & Sons, Ltd. Published 2013 by John Wiley & Sons, Ltd.

material, t the time (or period of soil formation), and the dots represent additional, unspecified factors. In reality soils are the product of highly complex interactions of many interdependent variables, and the soils themselves are not merely a passive and dependent factor in the environment. Nonetheless, following Jenny's subdivision of the classic factors of soil formation, one can see more clearly the effects humans have had on soil be it detrimental or beneficial. These can be summarized as follows (adapted from the work of Bidwell and Hole, 1965):

1 *Parent material*
Beneficial: adding mineral fertilizers; accumulating shells and bones; accumulating ash locally; removing excess amounts of substances such as salts.

Detrimental: removing through harvest more plants and animal nutrients than are replaced; adding materials in amounts toxic to plants or animals; altering soil constituents in a way to depress plant growth.

2 *Topography*
Beneficial: checking erosion through surface roughening, land forming and structure building; raising land level by accumulation of material; land leveling.

Detrimental: causing subsidence by drainage of wetlands and by mining; accelerating erosion; excavating.

3 *Climate*
Beneficial: adding water by irrigation; rainmaking by seeding clouds; removing water by drainage; diverting winds, etc.

Detrimental: subjecting soil to excessive insolation, to extended frost action, to wind, etc.

4 *Organisms*
Beneficial: introducing and controlling populations of plants and animals; adding organic matter including 'night-soil'; loosening soil by ploughing to admit more oxygen; fallowing; removing pathogenic organisms, for example, by controlled burning.

Detrimental: removing plants and animals; reducing organic content of soil through burning, ploughing, over-grazing, harvesting, etc.; adding or fostering pathogenic organisms; adding radioactive substances.

5 *Time*
Beneficial: rejuvenating the soil by adding fresh parent material or through exposure of local parent material by soil erosion; reclaiming land from under water.

Detrimental: degrading the soil by accelerated removal of nutrients from soil and vegetation cover; burying soil under solid fill or water.

Space precludes, however, that we can follow all these aspects of anthropogenic soil modification or, to use the terminology of Yaalon and Yaron (1966), of *meta-pedogenesis*. We will therefore concentrate on certain highly important changes which humans have brought about, especially chemical changes (such as salinization and lateritization), various structural changes (such as compaction), some hydrological changes (including the effects of drainage and the factors leading to peat-bog development), and, perhaps most important of all, soil erosion.

Salinity: natural sources

Increasing salinity has a whole series of consequences that include a reduction in the availability of potable water (for humans and/or their stock), deterioration in soil structure, reduction in crop yields and decay of engineering structures and cultural treasures. It is, therefore, a major environmental issue.

Many semi-arid and arid areas are, however, naturally salty. By definition they are areas of substantial water deficit where evapotranspiration exceeds precipitation. Thus, whereas in humid areas, there is sufficient water to percolate through the soil and to leach soluble materials from the soil and the rocks into the rivers and hence into the sea, in deserts this is not the case. Salts therefore tend to accumulate. This tendency is exacerbated by the fact that many desert areas are characterized by closed drainage basins which act as terminal evaporative sumps for rivers.

The amount of natural salinity varies according to numerous factors, one of which is the source of salts. Some of the salts are brought in to the deserts by rivers. A second source of salts is the atmosphere – a source which in the past has often been accorded insufficient importance. Rainfall, coastal fogs and dust storms all transport significant quantities of soluble salts. Further soluble salts may be derived from the weathering and solution of bedrock. In the Middle East, for example, there are extensive salt domes and evaporite beds within the bedrock which create locally high groundwater and surface-water salinity levels. In other areas, such as the Rift Valley of East Africa, volcanic rocks may provide a large source of sodium carbonate to groundwater, while elsewhere the rocks in which groundwater occurs may contain salt because they are themselves ancient desert sediments. Even in the absence of such localized sources of highly saline ground water it needs to be remembered that over a period of time most rocks will provide soluble products to groundwater, and in a closed hydrological system such salts will eventually accumulate to significant levels.

A further source of salinity may be marine transgressions. At times of higher sea levels, it has sometimes been proposed that salts would have been laid down by the sea. Likewise in coastal areas, salts in groundwater aquifers may be contaminated by contact with seawater.

Human agency and increased salinity

Human activities cause enhanced or secondary salinization in drylands in a variety of ways (Goudie and Viles, 1997). In Table 4.1 these mechanisms are grouped into five main classes: irrigation salinity; dryland salinity; urban salinity; salinity brought about by interbasin water transfers; and coastal zone salinity.

Human-induced salinization affects about 77 million hectares on a global basis, of which 48 million hectares are in susceptible drylands (Middleton and Thomas, 1997) (Table 4.2).

Irrigation salinity

In recent decades there has been a rapid and substantial spread of irrigation across the world (Table 4.3). The irrigated area in 1900 amounted to less than 50 million hectares. By 2000 the total area amounted to five times that figure. During the 1950s the irrigated area was increasing at over 4% annually, though this

Table 4.1 Enhanced salinization

1. Irrigation salinity
   (a) Rise in groundwater
   (b) Evaporation of water from fields
   (c) Evaporation of water from canals and reservoirs
   (d) Waterlogging produced by seepage losses
2. Dryland salinity
   (a) Vegetation clearance
3. Urban salinity
   (a) Water importation and irrigation
   (b) Faulty drains and sewers
4. Interbasin water transfers
   (a) Mineralization of lake waters
   (b) Deflation of salts from desiccating lakes
5. Coastal zone salinity
   (a) Overpumping
   (b) Reduced freshwater recharge
   (c) Sea-level rise
   (d) Ground subsidence

Table 4.2 Global extent of human-induced salinization in the susceptible drylands (million hectares). Source: GLASOD; Middleton and Thomas (1997, table 4.17)

Continent	Light	Moderate	Strong	Extreme	Total
Africa	3.3	1.9	0.6	–	5.8
Asia	10.7	8.1	16.2	0.4	35.4
South America	0.9	0.1	–	–	1.0
North America	0.3	1.2	0.3	–	1.8
Europe	0.8	1.7	0.5	–	3.0
Australasia	–	0.5	–	0.4	0.9
Global total	16.0	13.5	17.6	0.8	47.9

Table 4.3 Estimates of the increasing area of irrigated land on a global basis. Source: United Nations, World Population Prospects, the 2004 Revision and CIA World Factbook

Year	Irrigated area (10^6 hectares)
1900	44–48
1930	80
1950	94
1955	120
1960	168
1980	211
1990	245
2000	276
2008	325

figure has now dropped to only about 1%. This spread of irrigation has brought about a great deal of salinization and waterlogging (Figure 4.1).

The amount of salinized irrigated land varies from area to area (Table 4.4), but in general ranges between 10 and 50% of the total. This is the case, for example, in Central Asia, where more than 50% of irrigated soils are salt affected and /or waterlogged (Qadir et al., 2009). However, there is a considerable range in these values according to the source of the data (compare Table 4.5), and this may in part reflect differences in the definition of the terms 'salinization' and 'waterlogging' (see Thomas and Middleton, 1993).

Irrigation causes secondary salinization in a variety of ways (Rhoades, 1990). Firstly, the application of irrigation water to the soil leads to a rise in the water table

Figure 4.1 The extension of irrigation in the Indus valley of Pakistan by means of large canals has caused widespread salination of the soils. Waterlogging is also prevalent. The white efflorescence of salt in the fields has been termed 'a satanic mockery of snow'. (See Plate 25)

Table 4.4 Salinization of irrigated cropland. Source: FAO data as summarized in *World Resources* (1987 and 1988, table 19.3)

Country	Percentage of irrigated lands affected by salinization
Algeria	10–15
Australia	15–20
China	15
Colombia	20
Cyprus	25
Egypt	30–40
Greece	7
India	27
Iran	<30
Iraq	50
Israel	13
Jordan	16
Pakistan	<40
Peru	12
Portugal	10–15
Senegal	10–15
Sri Lanka	13
Spain	10–15
Sudan	<20
Syria	30–35
USA	20–25

Table 4.5 Global estimate of secondary salinization in the world's irrigated lands. Source: Ghassemi et al. (1995, table 18). Reproduced with permission from CAB International and the University of New South Wales Press

Country	Cropped area (Mha)	Irrigated area (Mha)	Share of irrigated to cropped area (%)	Salt-affected land in irrigated area (%)	Share of salt-affected to irrigated land (%)
China	96.97	44.83	46.2	6.70	15.0
India	168.99	42.10	24.9	7.00	16.6
CIS	232.57	20.48	8.8	3.70	18.1
USA	189.91	18.10	9.5	4.16	23.0
Pakistan	20.76	16.08	77.5	4.22	26.2
Iran	14.83	5.74	38.7	1.72	30.0
Thailand	20.05	4.00	19.9	0.40	10.0
Egypt	2.69	2.69	100.00	0.88	33.0
Australia	47.11	1.83	3.9	0.16	8.7
Argentina	35.75	1.72	4.8	0.58	33.7
South Africa	13.17	1.13	8.6	0.10	8.9
Subtotal	852.80	158.70	18.8	29.62	20.0
World	1473.70	227.11	15.4	45.4	20.0

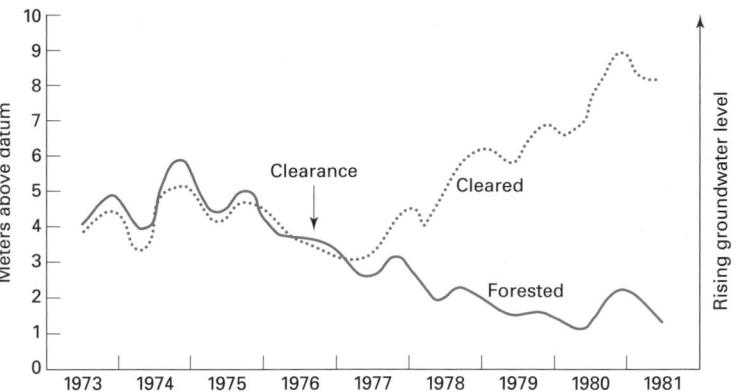

Figure 4.2 Comparison of hydrographs recorded from the boreholes in Wights (——) and Salmon (-----) catchments in Western Australia. Both catchments were forested until late in 1976 when Wights was cleared (modified from Peck, 1983, figure 1). Reproduced with permission.

so that it may become near enough to the ground surface for capillary rise and subsequent evaporative concentration to take place. When groundwater comes within 3 m of the surface in clay soils, and even less for silty and sandy soils, capillary forces bring moisture to the surface where evaporation occurs. There is plenty of evidence that irrigation does indeed lead to rapid and substantial rises in the position of the water table. Rates typically range between 0.2 and 3 m per year.

Secondly, many irrigation schemes, being in areas of high temperatures and rates of evaporation, suffer from the fact that the water applied over the soil surface is readily concentrated in terms of any dissolved salts it may contain. This is especially true for crops with a high water demand (e.g. rice) or in areas where, for one reason or another, farmers are profligate in their application of water.

Thirdly, the construction of large dams and barrages creates extensive water bodies from which further evaporation can take place, once again leading to the concentration of dissolved solutes.

Fourthly, notably in sandy soils with high permeability, water seeps both laterally and downwards from irrigation canals so that waterlogging may occur. Many irrigation canals are not lined, with the consequence that substantial water losses can result.

Dryland salinity

A prime cause of dryland salinity extension is vegetation clearance (Peck and Halton, 2003). The removal of native forest, by reducing interception and evapotranspirational losses, allows a greater penetration of rainfall into deeper soil layers which causes groundwater levels to rise,

thereby creating conditions for the seepage of sometimes saline water into low-lying areas. This is a particularly serious problem in the wheatbelt of Western Australia and in some of the prairies of North America. In the case of the former area it is the clearance of *Eucalyptus* forest that has led to the increased rate of groundwater recharge and to the spreading salinity of streams and bottomlands. Salt 'scalds' have developed. The speed and extent of groundwater rise following such forest clearance is shown in Figure 4.2. Until late 1976 both the Wights and Salmon catchments were forested. Then the Wights catchment was cleared. Before 1976 both catchments showed a similar pattern of groundwater fluctuation, but after that date there was a marked divergence of 5.7 m (Peck, 1983). The process can be reversed by afforestation (Bari and Scholfield, 1992). Re-vegetation policies could also provide increased carbon sinks and so could provide synergistic value (Pittock and Wratt, 2001: 603).

Groundwater levels have increased some tens of metres since clearance of the natural vegetation began. They have increased by up to 30 m since the 1880s in south-eastern Australia and by about 20 m in parts of south-west Australia. In some of the upland areas of New South Wales, groundwater levels have increased by up to 60 m over the last 70–80 years. In New South Wales, the area of land affected by dryland salinity in currently reported to be about 120,000 hectares. However, if current land-use trends and groundwater rise continue, this figure has the potential to increase to as much as 7.5 million hectares by 2050. In Western Australia there is already an estimated 1.8 million hectares of farmland that is salt-affected. This area could double in the next 15–20 years and in all some 6.1 million hectares have the potential to be affected by dryland salinity.

Dryland salinity is also a major problem on the Canadian prairies. In Alberta, approximately 0.65 million hectares are affected by secondary salinity, with an average crop yield reduction of 25%. In Saskatchewan 1.3 million hectares are affected and in Manitoba 0.24 million hectares.

Rising water tables resulting from land use changes are now being identified in other areas. For example, there has been a marked rise in the water table of the Continental Terminal in south-west Niger (Leduc et al., 2001). The rates have been between 0.01 and 0.45 million per year. The reason for this is the replacement of natural woodland savanna with millet fields and associated fallows. This has promoted increased surface runoff which concentrates in temporary endoreic ponds and then infiltrates to the water table.

Urban salinity

Drylands have seen rapid rates of urbanization. There are now some enormous urban agglomerations. The growth of cities has also been substantial. As Table 4.6 suggests, the average size of major dryland cities expanded 7.9 times between 1950 and 2000. Urbanization can cause a rise in groundwater levels by affecting the amount of moisture lost by evapotranspiration. Many elements of urbanization, and in particular the spread of impermeable surfaces (roads, buildings, car parks, etc.), interrupt the soil evaporation process so that groundwater levels in *sabkha* (salt plain) areas along the coast of the Arabian Gulf rise at a rate of 40 cm per year until a new equilibrium condition is attained; the total rise from this cause may be 1–2 m (Shehata and Lotfi, 1993). This can require the construction of horizontal drains.

Urbanization can lead to other changes in groundwater conditions that can aggravate salinization. In some large desert cities the importation of water, its usage, wastage and leakage, can produce the ingredients to feed this phenomenon. This has, for example, been identified as a problem in Cairo and its immediate environs (Hawass, 1993). The very rapid expansion of Cairo's population has outstripped the development of an adequate municipal infrastructure. In particular, leakage losses from water pipes and sewers have led to a substantial rise in the groundwater level and have subjected many buildings to attack by sulfate-

Table 4.6 Population of selected dryland cities in 1950 and 2010 (millions). Source: UN data processed by author

City and country	1950	2010
Cairo, Egypt	2.41	12.66
Ouagadougou, Burkina Faso	0.03	2.55
Ndjamena, Chad	0.04	1.58
Lanzhou, China	0.32	2.10
Alexandria, Egypt	1.04	5.53
Jodhpur, India	0.18	1.22
Tehran, Iran	1.04	8.71
Esfahan, Iran	0.18	3.92
Alma Ata, Kazakhstan	0.32	1.20
Kuwait City, Kuwait	0.09	1.51
Bamako, Mali	0.06	2.13
Karachi, Pakistan	1.03	16.61
Lima, Peru	0.97	8.84
Riyadh, Saudi Arabia	0.11	4.59
Jeddah, Saudi Arabia	0.12	2.75
Damascus, Syria	0.37	3.10
Las Vegas, USA	0.04	1.12
Phoenix, USA	0.22	2.86
Los Angeles, USA	4.05	13.86
Dubai, UAE	0.02	1.70
Abu Dhabi, UAE	0.01	1.09
Baghdad, Iraq	0.58	5.44
Windhoek, Namibia	0.02	0.23
Total	13.37	105.38

and chloride-rich water. There are other sites in Egypt, such as those of Luxor, where urbanization and associated changes in ground water levels has been identified as a major cause of accelerated salt weathering of important monuments.

Interbasin water transfers

A further reason for increases in levels of salinity is the changing state of water bodies caused by interbasin water transfers. The most famous example of this is the shrinkage of the Aral Sea, the increase in its mineralization, and the deflation of saline materials from its surface and their subsequent deposition downwind (Saiko and Zonn, 2000; Kravtsova and Tarasenko, 2010) (Figure 4.3). The sea itself has had its mineral content increased more than threefold since 1960. Another illustration of the effects of interbasin water transfers is the desiccation of the Owens Lake in California. Diversion of water to feed the insatiable demands of

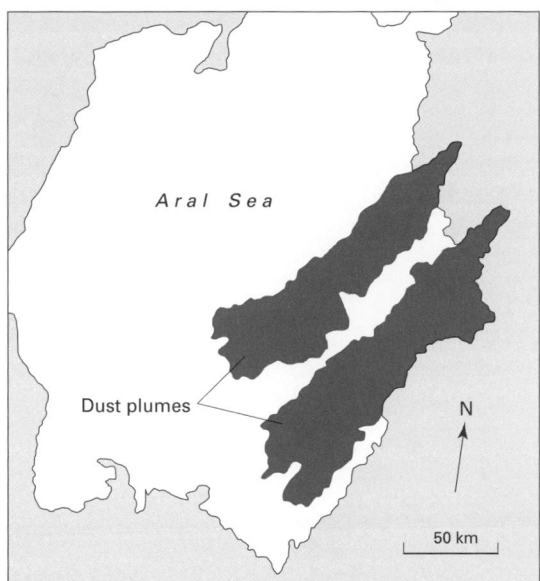

Figure 4.3 Dust plumes caused by the deflation of salty sediments from the drying floor of the Aral Sea as revealed by a satellite image (153/Metero-Priroda, 18 May 1975) (modified from Mainguet, 1995, figure 4).

Table 4.7 Enhanced salinity of lake basins. Source: Data from Williams (1999) with permission from Blackwell

Lake	Period	Change (g L^{-1})
Mono (California)	1941–1992	48–90
Pyramid Lake (Nevada)	1933–1980	3.75–5.5
Dead Sea	1910–1990s	200–300
Aral Sea	1960–1991	10–30
Qinghai (China)	1950s–1990s	6–12
Corangamite (Australia)	1960–1990s	35–50

Los Angeles caused the lake to dry out, so that highly saline dust storms have become increasingly serious (Gill, 1996; Tyler et al., 1997). Data on recent salinity enhancement in lakes from the USA, Asia and Australia are provided by Williams (1999) (Table 4.7).

Coastal zone salinity

Another prime cause of the spread of saline conditions is the incursion of seawater brought about by the over-pumping of groundwater (Ferguson and Gleeson, 2012). Saltwater displaces less saline groundwater through a mechanism called the Ghyben–Herzberg

principle. The problem presents itself on the coastal plain of Israel, on the island of Bahrain and in some of the coastal aquifers of the United Arab Emirates. It is also a widespread phenomenon in the coastal regions of North America (Barlow and Reichard, 2010). Figure 4.4 shows the spread of saltwater into the Biscayne aquifer in Miami-Dade County in Florida between 1904 and 1995. A comparable situation has also arisen in the Nile Delta (Kotb et al., 2000), though here the cause is not necessarily solely groundwater over-pumping, but may also be due to changes in water levels and freshwater recharge caused by the construction of the Aswan High Dam. Figure 4.5 shows the way in which chloride concentrations have increased and spread in the Llobregat Delta area of eastern Spain because of the incursion of seawater.

The Ghyben–Herzberg relationship (Figure 4.6) is based on the fact that freshwater has a lower density than saltwater, such that a column of seawater can support a column of freshwater approximately 2.5% higher than itself (or a ratio of about 40:41). So where a body of freshwater has accumulated in a reservoir rock or sediment which is also open to penetration from the sea, it does not simply lie flat on top of the saltwater but forms a lens, whose thickness is approximately 41 times the elevation of the piezometric surface above sea-level. The corollary of this rule is that if the hydrostatic pressure of the freshwater falls as a result of overpumping in a well, then the underlying saltwater will rise by 40 units for every unit by which the freshwater table is lowered. A rise in sea-level can cause a comparably dramatic alteration in the balance between fresh and saltwater bodies. This is especially serious on low islands (Figure 4.7) (Broadus, 1990).

An example of concern about salinization following on from sea-level rise is the Shanghai area of China (Chen and Zong, 1999). The area is suffering sub-sidence as a result of delta sedimentation (at 3 mm per year) and groundwater extraction (now less than 10 mm per year). It is very low-lying and in some counties the groundwater table is at 1.0–1.2 m. Thus a rise in sea level will raise the groundwater level, prolong waterlogging and cause a greater dominance of saline water in the deltaic area.

In any one location the causes of seawater incursion may be complex. A particularly good exemplification of this is provided by the coastal plains of northern Australia. Here Mulrennan and Woodroffe (1998)

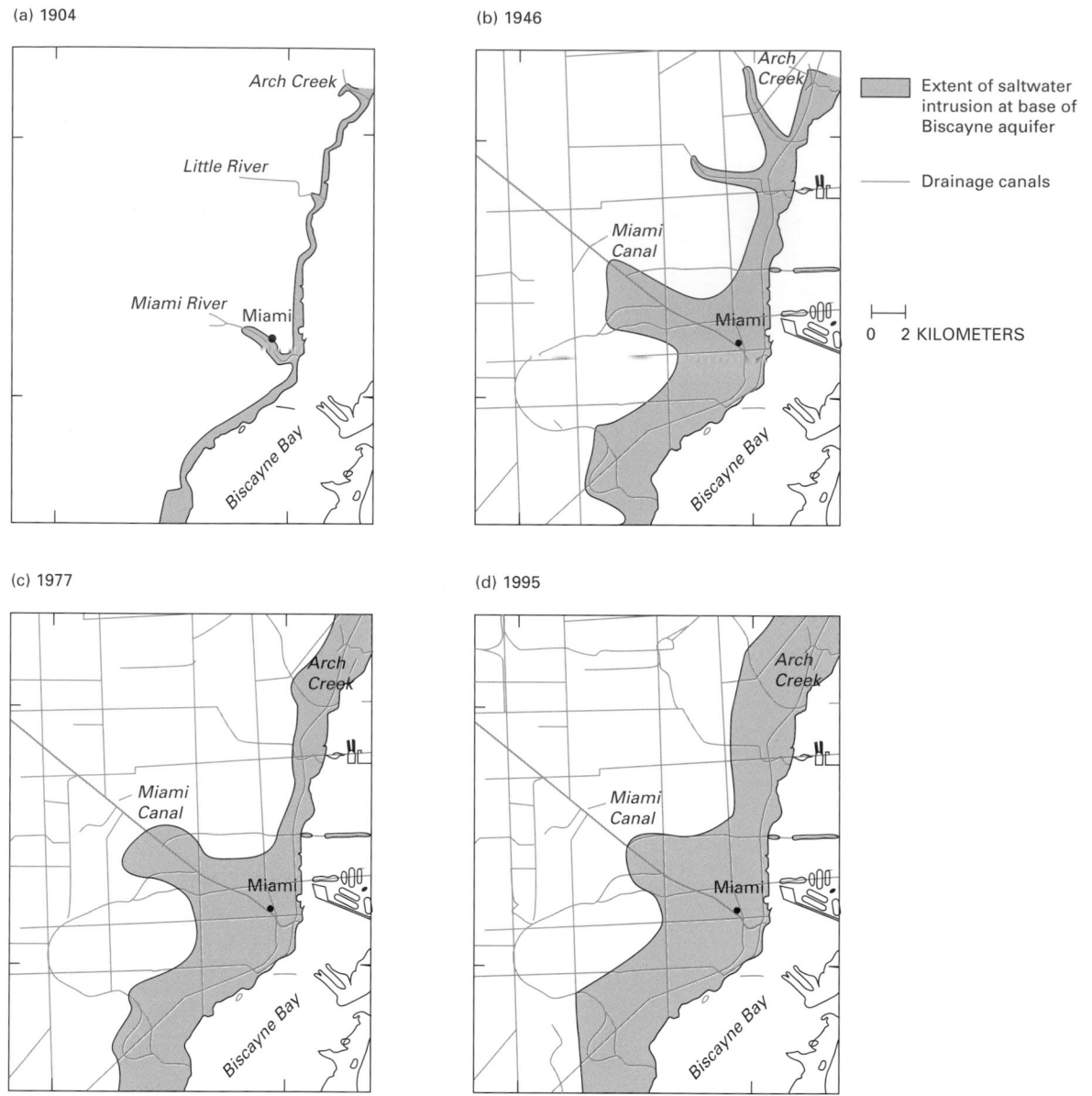

Figure 4.4 Saltwater intrusion in the Biscayne aquifer near Miami, Florida, between 1904 and 1995 (modified from Barlow and Reichard, 2010, figure 5).

have assessed the potential role of such factors as sea-level change, rainfall variability, boat erosion of creeks, the activities of feral water buffalo and sediment compaction.

Consequences of salinity

One consequence of the evaporative concentration of salts, and the pumping of saline waters back into rivers and irrigation canals from tubewells and other sources, is that river waters leading from irrigation areas show higher levels of dissolved salts. These, particularly when they contain nitrates, can make the water undesirable for human consumption.

A further problem is that, as irrigation water is concentrated by evapotranspiration, calcium and magnesium components tend to precipitate as carbonates, leaving sodium ions dominant in the soil solution. The sodium ions tend to be absorbed by colloidal clay par-

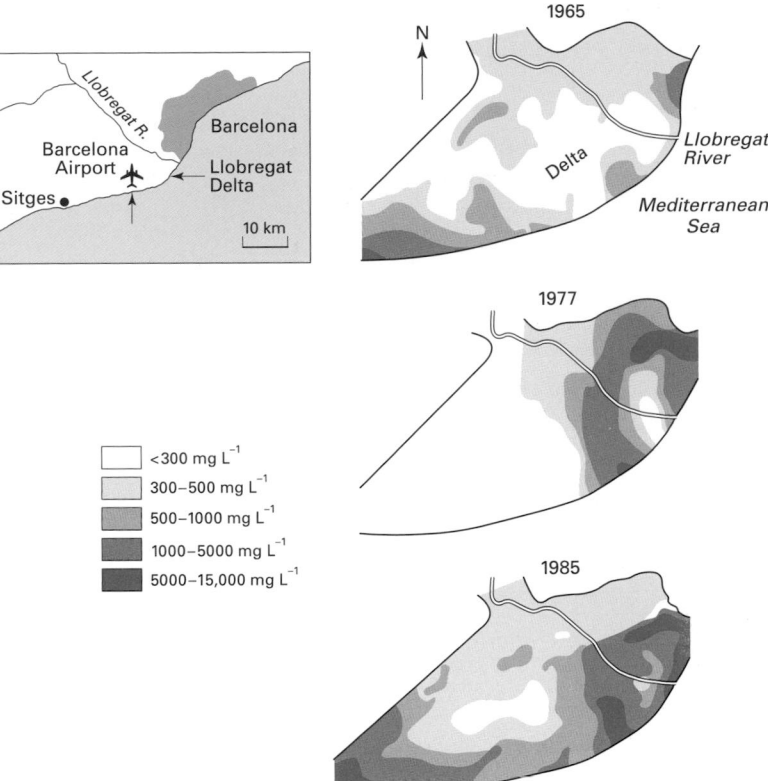

Figure 4.5 Changes in the chloride concentration of the Llobregat delta aquifer, Barcelona, Spain as a result of seawater incursion caused by the over-pumping of groundwater (modified from Custodio et al., 1986).

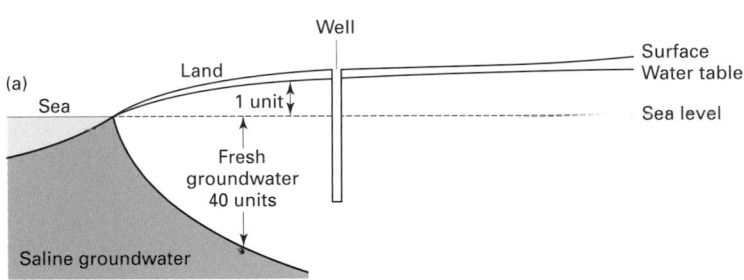

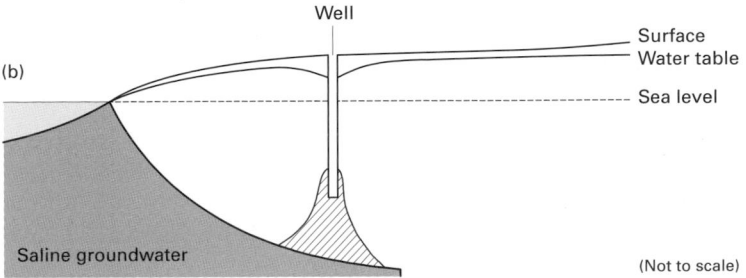

Figure 4.6 (a) The Ghyben–Herzberg relationship between fresh and saline groundwater; (b)The effect of excessive pumping from the well. The diagonal hatching represents the increasing incursion of saline water (modified from Goudie and Wilkinson, 1977, figure 63).

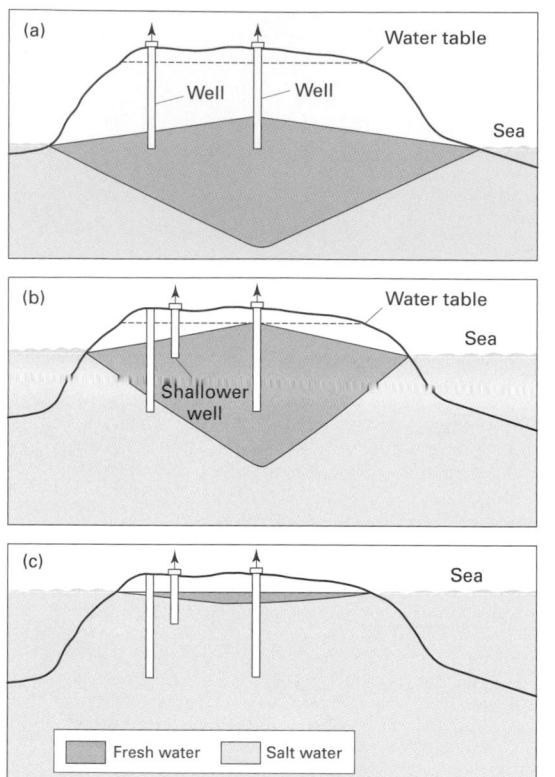

Figure 4.7 Impact of sea-level rise on an island water table. Note: The freshwater table extends below sea level 40 cm for every 1 cm by which it extends above sea level. (A) For islands with substantial elevation a 1-m rise in sea level simply shifts the entire water table up 1 m, and the only problem is that a few wells will have to be replaced with shallower wells (B). However, for very low islands the water table cannot rise due to runoff, evaporation and transpiration. A rise in sea level would thus narrow the water table by 40 cm for every 1 cm that the sea level rises (C) effectively eliminating groundwater supplies for the lowest islands (modified from Broadus, 1990).

ticles, deflocculating them and leaving the resultant structureless soil almost impermeable to water and unfavourable to root development.

The death of vegetation in areas of saline patches, due both to poor soil structure and toxicity, creates bare ground which becomes a focal point for erosion by wind and water. Likewise, **deflation** from the desiccating surface of the Aral Sea causes large amounts of salt to be blown away in **dust storms** and to be deposited downwind. Some tens of millions of tonnes are being translocated by such means and their plumes are evident on satellite images.

Probably the most serious impact of salination is on plant growth. This takes place partly through its effect on soil structure, but more significantly through its effects on osmotic pressures and through direct toxicity. When a water solution containing large quantities of dissolved salts comes into contact with a plant cell it causes a shrinkage of the protoplasmic lining. The phenomenon is due to the osmotic movement of the water, which passes from the cell towards the more concentrated soil solution. The cell collapses and the plant succumbs. Crop yields fall.

The toxicity effect varies with different plants and different salts. Sodium carbonate, by creating highly alkaline soil conditions, may damage plants by a direct caustic effect; while high nitrate may promote undesirable vegetative growth in grapes or sugar beets at the expense of sugar content. Boron is injurious to many crop plants at solution concentrations of more than 1 or 2 ppm.

Anthropogenic increases in salinity are not new. Jacobsen and Adams (1958) have shown that they were a problem in Mesopotamian agriculture after about 2400 BC. Individual fields, which in 2400 BC were registered as salt-free, can be seen in the records of ancient temple surveyors to have developed conditions of sporadic salinity by 2100 BC. Further evidence is provided by crop choice, for the onset of salinization strongly favours the adoption of crops which are most salt-tolerant. Counts of grain impressions in excavated pottery from sites in southern Iraq dated at about 3500 BC suggest that at that time the proportions of wheat and barley were nearly equal. A little more than 1000 years later the less salt-tolerant wheat accounted for less than 20% of the crop, while by about 2100 BC it accounted for less than 2% of the crop. By 1700 BC the cultivation of wheat had been abandoned completely in the southern part of the alluvial plain. These changes in crop choice were accompanied by serious declines in yield which can also probably be attributed to salinity. At 2400 BC the yield was 2537 L per hectare, by 2100 BC it was 1460, and by 1700 BC it was down to 897. It seems likely that this played an important part in the break-up of Sumerian civilization, though the evidence is not conclusive. Moreover, the Sumarians appear to have understood the problem and to have had coping strategies (Powell, 1985).

Reclamation of salt-affected lands

Because of the extent and seriousness of salinity, be the causes natural or anthropogenic, various reclamation techniques have been initiated. These can be divided into three main types: eradication, conversion and control.

Eradication predominantly involves the removal of salt either by improved drainage or by the addition of quantities of freshwater to leach the salt out of the soil. Both solutions involve considerable expense and pose severe technological problems in areas of low relief and limited freshwater availability. Improved drainage can either be provided by open drains or by the use of tubewells (as at Mohenjo Daro, Pakistan) to reduce groundwater levels and associated salinity and waterlogging. A minor eradication measure, which may have some potential, is the biotic treatment of salinity through the harvesting of salt-accumulating plants such as *Suaeda fruticosa* (shrubby seablite), or through bioremediation with crops such as licorice (*Glycyrrhiza glabra*) (Qadir et al., 2009).

Conversion involves the use of chemical methods to convert harmful salts into less harmful ones. For example, gypsum is frequently added to sodic soils to convert caustic alkali carbonates to soluble sodium sulfate and relatively harmless calcium carbonate:

$$Na_2CO_3 + CaSO_4 \leftrightarrow CaCO_3 + Na_2SO_4 \downarrow \text{leachable}$$

Some of the most effective ways of reducing the salinity hazard involve miscellaneous control measures, such as less wasteful and lavish application of water through the use of sprinklers rather than traditional irrigation methods; the lining of the canals to reduce seepage (Figure 4.8); the realignment of canals through less permeable soil; and the use of more salt-tolerant plants. As salinity is a particularly serious threat at the time of germination and for seedlings, various strategies can be adopted during this critical phase of plant growth: plots can be irrigated lightly each day after seeding to prevent salt build-up; major leaching can be carried out just before planting; and areas to be seeded can be bedded in such a way that salts accumulate at the ridge tops with the seed planted on the slope between the furrow bottom and the ridge top. Useful

Figure 4.8 A lined irrigation canal in Haryana, India. Such lining is an important way in which to reduce waterlogging and salinization. (See Plate 26)

general reviews of methods for controlling soil salinity are given by Rhoades (1990) and Qadir et al. (2000).

Lateritization

In some parts of the tropics are extensive sheets of a material called **laterite**, an iron and/or aluminium-rich duricrust (see Maignien, 1966 or Macfarlane, 1976). These iron-rich sheets result naturally, either because of a preferential removal of silica during the course of extensive weathering (leading to a *relative* accumulation of the sesquioxides of iron and aluminium), or because of an *absolute* accumulation of these compounds.

One of the properties of laterites is that they harden on exposure to air and through desiccation. Once hardened they are not favourable to plant growth. One particular way in which exposure may take place is by accelerated erosion, while forest removal may so cause a change in microclimate that desiccation of the laterite surface can take place. Indeed, one of the main problems with the removal of humid tropical rain forest is that soil hardening may occur. The phenomenon may occur in some, but by no means in all so-called tropical soils (Richter and Babbar, 1991). Should hardening occur, it tends to limit the extent of successful soil utilization and severely retards the re-establishment of forest. although Vine (1968: 90) and Sanchez and Buol (1975) have rightly warned against exaggerating this difficulty in agricultural land use, there are records from many parts of the tropics of accelerated induration brought about by forest removal (Goudie, 1973). In the Cameroons, for example, around 2 m of complete induration can take place in less that a century. In India, foresters have for a long time been worried by the role that plantations of teak (*Tectona grandis*) can play in lateritization. Teak is deciduous, demands light, likes to be well spaced (to avoid crown friction), dislikes competition from undergrowth and is shallow-rooted. These characteristics mean that teak plantations tend to expose the soil surface to erosive and desiccative forces more than does the native vegetation cover.

One of the main exponents of the role that human agency has played in lateritization in the tropical world has been Gourou (1961: 21–22). although he may be guilty of exaggerating the extent and significance of laterite, Gourou gives many examples from low latitudes of falling agricultural productivity resulting from the onset of lateritization. It is worth quoting him at length:

Laterite is a pedological leprosy. Man's activities aggravate the dangers of laterite and increase the rate of the process of lateritization. To begin with, erosion when started by negligent removal of the forest simply wears away the friable and relatively fertile soil which would otherwise cover the laterite and support forest or crops. . . . The forest checks the formation of the laterite in various ways. The trees supply plenty of organic matter and maintain a good proportion of humus in the soil. The action of capillary attraction is checked by the loosening of the soil; and the bases are retained through the absorbent capacity of humus. The forest slows down evaporation from the soil . . . it reduces percolation and consequently leaching. Lastly, the forest may improve the composition of the soil by fixing atmospheric dust.

Accelerated podzolization and acidification

There is an increasing amount of evidence that the introduction of agriculture, deforestation and pastoralism to parts of upland Western Europe promoted some major changes in soil character: notably an increase in the development of acidic and **podzolized** conditions, associated with the development of peat bogs. Climatic changes may have played a role, as could pro-

gressive leaching of last glacial drifts during the passage of the Holocene. But the association in time and space of human activities with soil deterioration has become increasingly clear (Evans et al., 1975).

Replacing the natural forest vegetation with cultivation and pasture, human societies set in train various related processes, especially on base-poor materials. First, the destruction of deep-rooting trees curtailed the enrichment of the surface of the soil by bases brought up from the deeper layers. Second, the use of fire to clear forest may have released nutrients in the form of readily soluble salts, some of which were inevitably lost in drainage, especially in soils poor in colloids (Dimbleby, 1974). Third, the taking of crops and animal products depleted the soil reserves to an extent probably greater than that arising from any of the manuring practices of prehistoric settlements. Fourth, as the soil degraded, the vegetation which invaded – especially bracken and heather – itself tended to produce a more acidic humus type of soil than the original mixed deciduous forest, and so continued the process.

The development of podzols and their associated ironpans (Cunningham et al., 2001), by impeding downward percolating waters, may have accelerated the formation of peats, which tend to develop where there is waterlogging through impeded drainage. Another fact which would have contributed to their development is that when a forest canopy is removed (as by deforestation) the transpiration demand of the vegetation is reduced, less rainfall is intercepted, so that the supply of groundwater is increased, aggravating any waterlogging. It has been argued that many peat bogs in highland Britain coincide broadly in age with the first major land-clearance episodes (Moore, 1973), but this has been disputed by Tipping (2008) who argued that peat development was essentially controlled by climatic conditions, that it was widespread in highland Scotland very early in the Holocene and that there is very little evidence that it is the result of anthropogenic activities. Indeed, the role of natural processes must not be totally forgotten, and Ball's (1975: 26) assessment would seem judicious:

It seems to be on balance that the highland trends in soil formation due to climate, geology and relief have been clearly running in the direction of leaching, acidity, podzolization, **gleying** and peat formation. For the British highlands generally, man has only intervened to hasten or slow the rate of these trends, rather than being in a position to alter the whole trend from one pedogenetic trend to another.

One serious type of soil acidification is that which produces acid sulfate soils. As Dent and Pons (1995: 263) wrote:

Acid sulphate soils are the nastiest soils in the world. They generate sulphuric acid that brings their pH as low as 2 and leaks into drainage and floodwaters. In this acid environment, aluminium and other toxic elements kill vegetation and aquatic life or, in sub-lethal doses, render many species stunted and sickly. Generations of people depending on these soils have been impoverished and, probably, poisoned by their drinking water.

The reason for the development of such extremely acid soils is that originally they accumulated as sediments under severely reducing conditions in environments like tidal (e.g. mangrove) swamps, or brackish lakes. Large amounts of sulfitic mud accumulated. When such materials are drained reduction is replaced by oxidation, and sulfuric acid is produced. Infamous examples are known from the drained polders of the Netherlands and from the drained coastal swamps of Southeast Asia.

Some soils are currently being acidified by air pollution and the deposition of acid precipitation (Grieve, 2001) (see also Chapter 7). Many soils have a resistance to acidification because of their buffering capacity, which enables them to neutralize acidity. However, this resistance very much depends on soil type and situation, and soils which have a low buffering capacity because of their low calcium content (as, for example, on granite), and which are subjected to high levels of precipitation, may build up high levels of acidity. The concept of **critical loads** has been developed. They are defined as exposures below which significant harmful effects on sensitive elements of the environment do not occur according to currently available knowledge. The critical load for sensitive forest soils on gneiss, granite or other slow-weathering rocks is often less than 3 kg of sulfur per hectare per year. In some of the more polluted parts of central Europe, the rates of sulfur deposition may be between 20 and 100 kg per hectare per year (Ågren and Elvingson, 1996). Areas where crucial loads are being exceeded include the heavily industrialized regions of eastern USA, Europe, Russia and large parts of Asia, but risks are increasing in parts of Brazil, Argentina and South Africa (Bouwman et al., 2002).

The immediate impact of high levels of acid input to soils is to increase the exchange between hydrogen ($H+$) ions and the nutrient cations, such as potassium ($K+$), magnesium ($Mg++$) and calcium ($Ca++$). As a result of this exchange, such cations can be quickly leached from the soil, along with the sulfate from the acid input. This leaching leads to nutrient deficiency. Acidification also leads to a change in the rate at which dead organic matter is broken down by soil microbes. It can also render some ions, such as aluminium, more mobile and this has been implicated in the phenomenon of forest decline (see Chapter 2).

On the other hand, the recent reduction in acid rain in some areas as a result of pollution controls and a reduction in the burning of sulfur-rich fossil fuels has seen some reversal of soil acidity (RoTap, 2012). There has, for example, been an increase in soil pH is UK forested areas in recent years (Vanguelov et al., 2010).

Changing levels of acid precipitation are not, however, the sole reason for acidity trends in soils. A recent study in China (Guo et al. 2010) has indeed shown widespread acidification of agricultural soils since the 1980s, and this has been attributed to high nitrogen fertilizer application rates together with the uptake and removal of base cations by plants.

Some anthrosols resulting from agriculture and urbanization

It would be plainly misleading to stress only the deleterious effects of human actions on European soils. Traditional agricultural systems have often employed laborious techniques to augment soil fertility and to reduce such properties as undesirable acidity. In Britain, for example, the addition of chalk to light sandy land goes back at least to Roman times and the marl pits from which the chalk was dug are a striking feature of the Norfolk landscape, where Prince (1962) has identified at least 27,000 hollows. Similarly, in the sandy lowlands of the Netherlands, Germany, Belgium and western Russia there are soils which for centuries (certainly back to the Bronze Age) have been built up (often over 50 cm) and fertilized with a mixture of manure, animal bedding, sods, litter or sand. Such humiferous soils are called Plaggen soils (Pope, 1970; Blume and Leinweber, 2004). Plaggen soils also occur in Ireland and Scotland, where the addition of sea-sand to peat was carried out in pre-Christian times. Likewise, before European settlement in New Zealand, the Maoris used thousands of tonnes of gravel and sand, carried in flax baskets, to improve soil structure (Cumberland, 1961).

Another type of soil which owes much to human influence is the category called 'paddy soil'. Long-continued irrigation, levelling and manuring of terraced land in China and elsewhere have changed the nature of the pre-existing soils in the area. Among the most important modifications that have been recognized (Gong, 1983; Zhang and Gong, 2003) are an increase in organic matter and increase in base saturation, and the translocation and reduction of iron and manganese.

In Amazonia, recent studies have revealed the existence of some distinctive 'Amazonian Dark Earths' under what was previously thought to be pristine rain forest. In pre-European times, before their numbers were decimated, there appear to have been shifting cultivator communities in the region who over a few hundreds or thousands of years deliberately or non-deliberately modified soils with various types of rubbish such as food remains, ash, human excrement and collapsed houses (Neves et al., 2003).

Many soils in urban areas have distinctive characteristics, and some pedologists now talk about 'Urban Soils' (Jim, 1998). Soils in cities, because of the limited vegetation cover, may suffer from a decline in organic content through time. In addition they may be compacted by traffic, churned up and eroded during construction, contaminated with large amount of rubble, polluted with heavy metals such as cadmium, lead and mercury.

Soil carbon

One important consequence of land use and land cover changes is changes in the carbon content of soils. This has potential significance in terms of carbon release to the atmosphere and hence for the enhanced greenhouse effect (Lal et al., 1995). Soil organic carbon includes plant, animal and microbial residues.

Under prolonged cultivation there is strong evidence for loss of soil carbon. Agricultural practices that contribute to this are deforestation and biomass burning, drainage of wetlands, ploughing and other forms of soil disturbance, and removal of crop residues. Soil carbon is depleted by oxidation or mineralization due to breakdown of aggregate leading to exposure of carbon, leaching and translocation of dissolved organic carbon or particulate organic carbon, and accelerated erosion by runoff or wind (Lal, 2002). Losses of soil organic carbon of as much as 50% in surface soils (20 cm) have been observed after cultivation for 30–50 years. Reductions average around 30% of the original amount in the top 100 cm (Post and Kwon, 2000).

Much depends, however, on the nature of land management. In recent decades, for example, the carbon stock in agricultural land in the USA may have increased because of the adoption of conservation tillage practices on cropland and a reduction in the use of bare fallow (Eve et al., 2002). The use of nitrogenous fertilizers may also explain some of the increase (Buyanovsky and Wagner, 1998).

Conversely, reafforestation and conversion of cropland to grassland can cause substantial gains in soil organic carbon. In general, after afforestation there may be a period when soil organic carbon declines, because of low rates of litter fall and continuing decomposition of residues from the preceding agricultural phase. As the forest cover develops, inputs of carbon exceed outputs. The rates of building of carbon vary, and tend to be greater in cool and humid regions and under hardwoods and softwoods rather than under *Eucalyptus* or *Pinus radiata* (Paul et al., 2002).

Soil structure alteration

One of the most important features of a soil, in terms of both its suitability for plant growth and its inherent erodibility, is its structure. There are many ways in which humans can alter this, especially by compacting it with agricultural machinery (Horn et al., 2000), by the use of recreation vehicles, by changing its chemical character through irrigation and by trampling (Grieve, 2001; Batey, 2009). Soil compaction, which involves the compression of a mass of soil into a smaller volume, tends to increase the resistance of soil to penetration

Table 4.8 Change to soil properties resulting from the passage of 100 motorcycles in New Zealand. Source: Crozier et al. (1978, table 1)

Soil property	Total no. of sites	No. of sites with significant change[a]	Mean percentage at significant sites	No. of sites with significant increase	No. of sites with significant decrease	Main direction of change	Mean percentage change in direction
Infiltration capacity	16	16 (100%)	84.3	3	13	Decrease	78.1
Bearing capacity	21	10 (48%)	22.8	2	8	Decrease	18.6
Soil moisture	20	14 (70%)	15.5	5	9	Decrease	16.7
Dry bulk density	19	11 (58%)	13.6	9	2	Decrease	13.3

[a]Change is significant when greater than: 10% for infiltration capacity; 10% for bearing capacity; 5% for soil moisture; 5% for bulk density.

by roots and emerging seedlings, and limits oxygen and carbon dioxide exchange between the root zone and the atmosphere Moreover, it reduces the rate of water infiltration into the soil, which may change the soil moisture status and accelerate surface runoff and soil erosion (Nawaz et al., 2012).

For example, the effects of the passage of vehicles on some soil structural properties are shown in Table 4.8. Excessive use of heavy agricultural machinery is perhaps the major cause of soil compaction, and most procedures in the cropping cycle, from tillage and seedbed preparation, through drilling, weeding and agrochemical application to harvesting, are now largely mechanized, particularly in the developed world. Most notable of all is the reduction that is caused in soil infiltration capacity, which may explain why vehicle movements can often lead to gully development.

Grazing is another activity that can damage soil structure through trampling and compaction. Heavily grazed lands tend to have considerably lower infiltration capacities than those found in ungrazed lands. Particular fears have been expressed that the replacement of Amazonian rainforest to cattle-trampled pasture could lead to great increases in the frequency and volume of stormflow. One study showed that the frequency of storm flow in such grazed areas increased twofold, while its volume increased 17-fold (Germer et al., 2010). Trimble and Mendel (1995) have drawn particular attention to why it is that cows have the ability to cause soil compaction. Given their large mass, their small hoof area and the stress that may be imposed on the ground when they are scrambling up a slope, they are probably remarkably effective in compacting soils. The removal of vegetation cover and

associated litter also changes **infiltration capacity**, since cover protects the soil from packing by raindrops and provides organic matter for binding soil particles together in open aggregates. Soil fauna that live on the organic matter assist this process by churning together the organic material and mineral particles. Dunne and Leopold have ranked the relative influence of different land-use types on infiltration (1978, table 6.2, after US Soil Conservation Service):

Highest infiltration	Woods, good
	Meadows
	Woods, fair
	Pasture, good
	Woods, poor
	Pasture, fair
	Small grains, good rotation
	Small grains, poor rotation
	Legumes after row crops
	Pasture, poor
	Row crops, good rotation (more than one quarter in hay or sod)
	Row crops, poor rotation (one quarter or less in hay or sod)
Lowest infiltration	Fallow

An interesting study of the effect of grazing on soil infiltration capacities has been undertaken in the Nama-Karoo rangelands of South Africa (du Toit et al., 2009). There no simple linear relationship was found between grazing pressure and soil infiltration rates. Ungrazed areas had biological crusts which meant that infiltration rates were low. Moderately grazed areas had their crusts disrupted by trampling and so had a higher infiltration rates. Heavily grazed areas became compacted and so had lower infiltration capacities.

In general, experiments show that reafforestation improves soil structure, especially the pore volume of the soils (see e.g. Challinor, 1968). Ploughing is also known to produce a compacted layer at the base of the zone of ploughing. This layer has been termed the 'plough sole'. The normal action of the plough is to leave behind a loose surface layer and a dense subsoil where the soil aggregates have been pressed together by the sole of the plough. The compacting action can be especially injurious when the depth of ploughing is both constant and long term, and when heavy machinery is used on wet ground (Greenland and Lal, 1977).

Attempts have been made to reduce soil crusting by applying municipal and animal wastes to farm land, by adding chemicals such as phosphoric acid and by adopting a cultivation system of the no-tillage type. The last practice is based on the idea that the use of herbicides has eliminated much of the need for tillage and cultivation in row crops; seeds are planted directly into the soil without ploughing, and weeds are controlled by the herbicides. With this method less bare soil is exposed and heavy farm machinery is less likely to create soil compaction problems (see Carlson, 1978, for some of these methods).

Soil structures may also be modified to increase water runoff, particularly in arid zones where the runoff obtained can augment the meagre water supply for crops, livestock, industrial and urban reservoirs, and groundwater recharge projects. In the Negev farming was practised in this way, especially in the Nabatean and the Romano-Byzantine periods (about 300 BC to AD 630), and attempts were made to induce runoff by clearing the surface gravel of the soil and heaping it into thousands of mounds. This exposed the finer silty soil beneath, facilitating soil crust formation by raindrop impact, decreasing infiltration capacity and reducing surface roughness, so that runoff increased (Evenari et al., 1971).

Soil drainage and its impact

Soil drainage 'has been a gradual process and the environmental changes to which it has led have, by reason of that gradualness, often passed unnoticed' (Green, 1978: 171). To be sure, the most spectacular feats of drainage – arterial drainage – involving the construction of veritable rivers and large dike systems, as seen in the Netherlands and the Fenlands of eastern England, have received attention. However, more widespread than arterial drainage, and sometimes independent of it, is the drainage of individual fields, either by surface ditching or by underdrainage with tile pipes and the like. Green (1978) mapped the areas of drained agricultural land in Europe and found that in Finland, Denmark, Great Britain, the Netherlands and Hungary, the majority of agricultural land is drained. Such drainage is also widespread in the USA Corn Belt and so, for example, nearly a half of Indiana was artificially drained by 1960s (Kumar et al., 2009).

In Britain underdrainage was promoted by government grants and reached a peak of about 1,000,000 hectares per year in the 1970s in England and Wales. More recently government subsidies have been cut and the uncertain economic future of farming has led to a reduction in farm expenditure. Both tendencies have led to a reduction in the rate of field drainage, which by 1900 was only 40,000 hectares per year (Robinson, 1990).

The drainage conditions of the soil have also frequently been altered by the development of ridge and furrow patterns created by ploughing. Such patterns are a characteristic feature of many of the heavy soils of lowland England where large areas, especially of the Midland lowlands, are striped by long, narrow ridges of soil, lying more or less parallel to each other and usually arranged in blocks of approximately rectangular shape. They were formed by ploughing with heavy ploughs pulled by teams of oxen; the precise mechanism has been described with clarity by Coones and Patten (1986: 154).

Soil drainage has been one of the most successful ways in which communities have striven to increase agricultural productivity; it was practised by Etruscans, Greeks and Romans (Smith, 1976). Large areas of marshland and floodplain have been drained to human advantage. By leading water away, the water table is lowered and stabilized, providing greater depth for the root zone. Moreover, well-drained soils warm up earlier in the spring and thus permit earlier planting and germination of crops. Farming is easier if the soil is not too wet, since the damage to crops by winter freezing may be minimized, undesirable salts carried away, trafficability of farm equipment improved and the general physical condition of the soil ameliorated. In addition, drained land may have certain inherent virtues: tending to be flat it is less prone to erosion and more amenable to mechanical cultivation. It will also be less prone to drought risk than certain other types of land (Karnes, 1971). Paradoxically, by reducing the area of saturated ground, drainage can alleviate flood risk in some situations by limiting the extent of a drainage basin that generates saturation excess overland flow.

Conversely, soil drainage can have quite undesirable or unplanned effects. For example, some drainage systems, by raising drainage densities, can increase flood risk by reducing the distance over which unconcentrated overland flow (which is relatively slow) has to travel before reaching a channel (where flow is relatively fast). In central Wales, for instance, the establishment of drainage ditches in peaty areas to enable afforestation increased flood peaks in the rivers Wye and Severn (Howe et al., 1966) and this has also been a problem in northern England (Holden et al., 2006). Stream flow trends have also been influenced by subsoil drainage in Indiana, USA (Kumar et al., 2009). In addition, the quality of stream flow may be impacted by drainage, leading to the export of dissolved organic matter (Dalzell et al., 2011), nitrates (Cuadra and Vidon, 2011) and phosphates (Gardner et al., 2002).

Drainage can also cause long-term damage to soil quality. A fall in water level in organic soils can lead to the oxidation and eventual disappearance of peaty materials (Leifeld et al., 2011), which in the early stages of post-drainage use may be highly productive. This has occurred, for instance, in the English Fenland (Dawson et al., 2010), the Sacramento-San Joaquin Delta of California (Drexler et al., 2009) and the Everglades of Florida, where drainage of peat soils has led to a subsidence in the soil of 32 mm per year (Stephens, 1956).

Soil moisture content can also determine the degree to which soils are subjected to expansion and contraction effects, which in turn may affect engineering structures in areas with expansive soils (Holtz, 1983). Soils containing sodium montmorillonite type clays, when drained or planted with large trees, may dry out and cause foundation problems (Driscoll, 1983).

Soil fertilization

The chemistry of soils has been changed deliberately by the introduction of chemical fertilizers. The employment of chemical fertilizers on a large scale is little more than 150 years old. In the early nineteenth century, nitrates were first imported from Chile, and sulfate of ammonia was produced only after the 1820s as a by-product of coal gas manufacture. In 1843 the first fertilizer factory was established at Deptford Creek (London), but for a long time superphosphates were the only manufactured fertilizers in use. In the twentieth century, synthetic fertilizers, particularly nitrates, were developed, notably by Scandinavian countries which used their vast resources of water power. Potassic fertilizers came into use much later than the phosphatic and nitrogenous; nineteenth-century farmers hardly knew them. The rise in nitrogen fertilizer consumption for the world as a whole between 1950 and 1990 is shown in Figure 4.9. Global fertilizer consumption in 2000 was around 141 million tonnes of nutrient of which around 61% was nitrogen, 23% phosphate and 16% potash. The use of synthetic fertilizers has greatly increased agricultural productivity in many parts of the world, and remarkable increases in yields have been achieved. It is also true that in some circumstances proper fertilizer use can help minimize erosion by ensuring an ample supply of roots and plant residues, particularly on infertile or partially degraded soils (Bockman et al., 1990).

On the other hand, the increasing use of synthetic fertilizers can create environmental problems such as water pollution, while their substitution for more

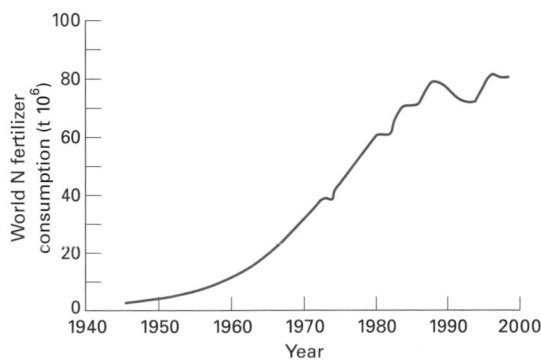

Figure 4.9 Global consumption of nitrogen fertilizer over the last seven decades.

traditional fertilizers may accelerate soil structure deterioration and soil erosion. One effect has been the increase in the water repellency of some surface soil materials. This in turn can reduce soil infiltration rates and a consequential increase in erosion by overland flow (Conacher and Conacher, 1995: 39). Fertilizers can also promote soil acidity and may lead to deficiencies or toxic excesses of major nutrients and trace elements. They may also contain impurities, such as fluoride, lead, cadmium, zinc and uranium. Some of these heavy metals can inhibit water uptake and plant growth. They may also become concentrated in food crops, which can have important implications for human health.

Fires and soil quality

The importance and antiquity of fire as an agency through which the environment is transformed requires that some attention be given to the effects of fire on soil characteristics.

Fire has often been used intentionally to change soil properties, and both the release of nutrients by fire and the value of ash have long been recognized, notably by those involved in shifting agriculture based on slash-and-burn techniques. Following cultivation, the loss of nutrients by leaching and erosion is very rapid (Nye and Greenland, 1964), and this why after only a few years the shifting cultivators have to move on to new plots. Fire rapidly alters the amount, form and distribution of plant nutrients in ecosystems, and, compared to normal biological decay of plant remains, burning rapidly releases some nutrients into a plant-available form. Indeed, the amounts of P, Mg, K and Ca released by burning forest and scrub vegetation are high in relation to both the total and available quantities of these elements in soils (Raison, 1979). In particular phosphate loss can be very detrimental to soil fertility (Thomas et al., 2000). In forests, burning often causes the pH of the soil to rise by three units or more, creating alkaline conditions where formerly there was acidity. Burning also leads to some direct nutrient loss by volatilization and convective transfer of ash, or by loss of ash to water erosion or wind deflation. The removal of the forest causes soil temperatures to increase because of the absence of shade, so that humus is often lost at a faster rate than it is formed.

Soil erosion: general considerations

The scale of accelerated soil erosion that has been achieved by human activities has been well summarized by Myers (1988: 6):

Since the development of agriculture some 12,000 years ago, soil erosion is said by some to have ruined 4.3 million km^2 of agricultural lands, or an area equivalent to rather more than one-third of today's crop-lands . . . the amount of agricultural land now being lost through soil erosion, in conjunction with other forms of degradation, can already be put at a minimum of 200,000 km^2 per year.

That soil erosion is a major and serious aspect of the human role in environmental change is not to be doubted (Sauer, 1938). There is a long history of weighty books and papers on the subject (see e.g. Marsh, 1864; Bennett, 1938; Jacks and Whyte, 1939; Morgan, 1995; Imeson, 2012). although many techniques have been developed to reduce the intensity of the problem (see Hudson, 1987) it appears to remain intractable. As L. J. Carter (1977: 409) has reported of the USA:

Although nearly $15 billion has been spent on soil conservation since the mid-1930s, the erosion of croplands by wind and water . . . remains one of the biggest, most pervasive environmental problems the nation faces. The problem's surprising persistence apparently can be attributed at least in part to the fact that, in the calculation of many farmers, the hope of maximizing short-term crop yields and profits has taken precedence over the longer term advantages of conserving the soil. For even where the loss of topsoil has begun to reduce the land's natural fertility and productivity, the effect is often masked by the positive response to heavy application of fertilizer and pesticides, which keep crop yields relatively high.

Although construction, urbanization, war, mining and other such activities are often significant in accelerating the problem, the prime causes of soil erosion are deforestation and farming (Figure 4.10). Land with a permanent vegetation cover is characterized by soil losses which are more than an order of magnitude lower than those on agricultural land (Cerdan et al., 2010). Pimentel (1976) estimated that in the USA soil erosion on agricultural land operates at an average rate of about 30 tonnes per hectare per year, which is approximately eight times quicker than topsoil is formed. He calculated that water runoff delivers around 4 billion tonnes of soil each year to the rivers of the 48 contiguous states and that three-quarters of this comes from agricultural land. He estimated that another billion tonnes of soil is eroded by the wind, a process which created the Dust Bowl of the 1930s. More recently, Pimentel et al. (1995) argued that about 90% of US cropland is losing soil above the sustainable rate, that about 54% of US pasture land is overgrazed and subject to high rates of erosion, and that erosion costs about $44 billion each year. They

argued that on a global basis soil erosion costs the world about $400 billion each year. However, as Trimble and Crosson (2000) and Boardman (1998) point out, determination of general rates of soil erosion is fraught with uncertainties.

One serious consequence of accelerated erosion is the sedimentation that takes place in reservoirs, shortening their lives and reducing their capacity. Many small reservoirs, especially in semi-arid areas and in areas with erodible sediments in their catchments such as the loess lands of China, appear to have an expected life of only 30 years or even less (see e.g. Rapp et al., 1972). Soil erosion also has serious implications for soil productivity. A reduction in soil thickness reduces available water capacity and the depth through which root development can occur. Loss of soil humus, whether as a result of fire, drainage, deforestation or ploughing, is an especially serious manifestation of human alteration of soil. As Table 4.9 indicates, humus has many beneficial effects on both the chemical and the physical properties of soil. Its removal by human activity can be a potent contributory cause of soil erosion.

The water-holding properties of the soil may be lessened as a result of the preferential removal of organic material and fine sediment. Hardpans and duricrusts

Figure 4.10 Soil erosion near Baringo in Kenya has exposed the roots of a tree, thereby indicating the speed at which soil can be lost. (See Plate 27)

Table 4.9 The beneficial properties of humus. Source: Swift and Sanchez (1984, table 2)

Property	Explanation	Effect
(a) Chemical properties		
Mineralization	Decomposition of humus yields CO_2, NH^{+4}, NO^{-3}, PO^{3-4} and SO^{2-4}	A source of nutrient elements for plant growth
Cation exchange	Humus has negatively charged surfaces which bind cations such as Ca^+ and K^+	Improved cation exchange capacity (CEC) of soil. From 20–70% of the CEC of some soils (e.g. Mollisols) is attributable to humus
Buffer action	Humus exhibits buffering in slightly acid, neutral and alkaline ranges	Helps to maintain a uniform pH in the soil
Acts as matrix for biochemical action in soil	Binds other organic molecules electrostatically or by covalent bonds	Affects bioactivity persistence and biodegradability of pesticides
Chelation	Forms stable complexes with Cu^{2+}, Mn^{2+}, Zn^{2+}, and other polyvalent cations	May enhance the availability of micronutrients to higher plants
(b) Physical properties		
Water retention	Organic matter can hold up to 20 times its weight in water	Helps prevent drying and shrinking. May significantly improve moisture-retaining properties of sandy soils
Combination with clay	Cements soil particles into structural units called aggregates	Improves aeration. Stabilizes structure. Increases permeability
Colour	The typical dark colour of many soils is caused by organic matter	May facilitate warming

may become exposed at the surface and provide a barrier to root penetration. Furthermore, splash erosion may cause soil compaction and crusting, both of which may be unfavourable to germination and seedling establishment. Erosion also removes nutrients preferentially from the soil. Some damage may be caused by associated excessive sedimentation, while wind erosion may lead to the direct sandblasting of crops. Finally, extreme erosion may lead to wholesale removal of both seeds and fertilizer.

Soil erosion associated with deforestation and agriculture

Forests protect the underlying soil from the direct effects of rainfall, generating what is generally an environment in which erosion rates tend to be low. The canopy plays an important role by shortening the fall of raindrops, decreasing their velocity and thus reducing kinetic energy. There are some examples of certain types (e.g. beech) in certain environments (e.g. maritime temperate) creating large raindrops, but in general most canopies reduce the erosion effects of rainfalls. Possibly more important than the canopy in reducing erosion rates in forest is the presence of humus in forest soils (Trimble, 1988). This both absorbs the impact of raindrops and has an extremely high permeability. Thus forest soils have high infiltration capacities. Another reason that forest soils have an ability to transmit large quantities of water through their fabrics is that they have many macropores produced by roots and their rich soil fauna. Forest soils are also well aggregated, making them resistant to both wetting and water drop impact. This superior degree of aggradation is a result of the presence of considerable organic material, which is an important cementing agent in the formation of large water-stable aggregates. Furthermore, earthworms also help to produce large aggregates. Finally, deep-rooted trees help to stabilize steep slopes by increasing the total shear strength of the soils.

It is therefore to be expected that with the removal of forest, for agriculture or for other reasons, rates of soil loss will rise (Figure 4.11), rock will be exposed to create a phenomenon called 'rocky desertification' (Yang et al., 2011a), and mass movements will increase in magnitude and frequency. This has been the case in

Figure 4.11 The removal of vegetation in Swaziland creates spectacular gully systems, which in southern Africa are called *dongas*. The smelting of local iron ores in the early nineteenth century required the use of a great deal of firewood which may have contributed to the formation of this example. (See Plate 28)

New Zealand, where the clearance of indigenous forest caused actively eroding gullies to develop, particularly from the 1880s to the 1920s (Marden et al., 2012). The rates of erosion that result will be particularly high if the ground is left bare; under crops the increase will be less marked. Furthermore, the method of ploughing, the time of planting, the nature of the crop and size of the fields will all have an influence on the severity of erosion.

One particular type of erosion associated with agriculture is called 'Tillage Erosion'. Tillage is responsible for the movement of soil material, particularly on slopes, and leads to a net soil loss from convex landscape positions and a net soil gain in concave landscape positions. Erosion rates for mechanized agriculture are often of the order of 400–800 kg/m/year and are of the same order of magnitude or larger than water erosion rates (Van Oost et al., 2006).

However, it is seldom that we have reliable records of rates of erosion over a sufficiently long time-span to show just how much human activities have accelerated these effects, and it is important to try and isolate the role of human impacts from climatic changes (Wilby et al., 1997). Recently, however, techniques have been developed which enable rates of erosion on slopes to be gauged over a lengthy time-span by means of dendrochronological techniques that date the time of root exposure for suitable species of tree. In Colorado,

USA, Carrara and Carroll (1979) found that rates over the last 100 years have been about 1.8 mm per year, whereas in the previous 300 years rates were between 0.2 and 0.5 mm per year, indicating an acceleration of about sixfold. This great jump has been attributed to the introduction of large numbers of cattle to the area about a century ago.

Another way of obtaining long-term rates of soil erosion is to look at rates of sedimentation on continental shelves and on lake floors. The former method was employed by Milliman et al. (1987) to evaluate sediment removal down the Yellow River in China during the Holocene. They found that, because of accelerated erosion, rates of sediment accumulation on the shelf over the last 2300 years have been 10 times higher than those for the rest of the Holocene (i.e. since around 10,000 BP).

A further example of using long-term sedimentation rates to infer long-term rates of erosion is provided by Hughes et al.'s (1991) study of the Kuk Swamp in Papua New Guinea (Figure 4.12). They identify low rates of erosion until 9000 BP, when, with the onset of the first phase of forest clearance, erosion rates increased from 0.15 cm per thousand years to about 1.2 cm per thousand years. Rates remained relatively stable until the last few decades when, following European contact, the extension of anthropogenic grasslands, subsistence gardens and coffee plantations has produced a rate that is very markedly higher: 34 cm per thousand years.

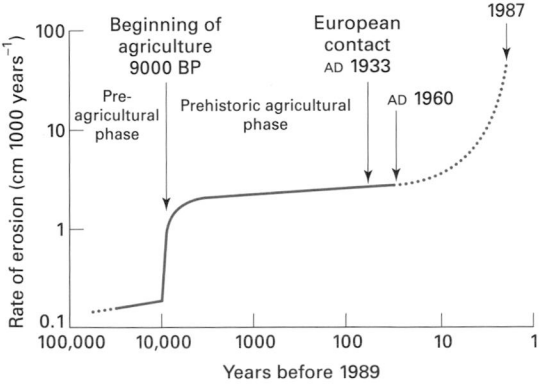

Figure 4.12 Rates of erosion in Papua New Guinea in the Holocene derived from rates of sedimentation in Kuk Swamp (after Hughes et al., 1991, figure 5, with modifications).

There have been a number of studies of the history of erosion in the Holocene in Europe and Britain (Foulds and Macklin, 2006; Notebaert and Verstraeten, 2010; Dusar et al., 2011). For example, Macklin et al. (2010) have shown that in British catchments there was a very marked increase in erosion and associated floodplain sediment accumulation after c. 1000 years ago, and this can be related to the agricultural revolution of the Middle Ages. Similarly, Heine et al. (2005), working in Bavaria, showed how agricultural intensification led to both slope colluviation and floodplain sedimentation. also working in Germany, Dreibrodt et al. (2010) suggested that erosion was at a maximum in the late Bronze Age and pre-Roman iron age (c. 1600 BC–1000 AD), high and late Medieval times (c. 1000–1350 AD) and late modern times (from c. 1500 until today). The combination of slope clearance for agriculture with times of intense storms would be an especially powerful stimulus to soil erosion, and catastrophic soil erosion in Central Europe was identified for the first half of the fourteenth century and in the mid-eighteenth to the early nineteenth century by Dotterweich (2008).

Some of the sediment produced by accelerated erosion accumulates as colluvium downslope, some on floodplains (Knox, 2002) and some in lake basins (see Chapter 6).

Another good long-term study of the response rates of erosion to land use changes is provided by a study undertaken on the North Island of New Zealand by Page and Trustrum (1997). During the last 2000 years of human settlement their catchment has undergone a change from indigenous forest to fern/scrub following Polynesian settlement (c. 560 BP) and then a change to pasture following European settlement (AD 1878). Sedimentation rates under European pastoral land use are between five and six times the rates that occurred under fern/scrub and between eight and seventeen times the rate under indigenous forest. In a broadly comparable study, Sheffield et al. (1995) looked at rates of infilling of an estuary fed by a sheepland catchment in another part of New Zealand. In pre-Polynesian times rates of sedimentation were 0.1 mm per year, during Polynesian times the rate climbed to 0.3 mm per year, while since European land clearance in the 1880s the rate has shot up to 11 mm per year (see also Nichol et al., 2000). Other examples of such trends in New Zealand are provided by Glade (2003), while a

Table 4.10 Runoff and erosion under various covers of vegetation in parts of Africa. Source: After Charreau (table 5.5, p. 153) in Greenland and Lal (1977). With permission from John Wiley & Sons Ltd

Locality	Average annual rainfall (mm)	Slope (%)	Annual runoff (%)			Erosion (t/hectare/year)		
			A	B	C	A	B	C
Ouagadougou (Burkina Faso)	850	0.5	2.5	2–32	40–60	0.1	0.6–0.8	10–20
Sefa (Senegal)	1300	1.2	1.0	21.2	39.5	0.2	7.3	21.3
Bouake (Ivory Coast)	1200	4.0	0.3	0.1–26	15–30	0.1	1–26	18–30
Abidjan (Ivory Coast)	2100	7.0	0.1	0.5–20	38	0.03	0.1–90	108–170
Mpwapwa[a] (Tanzania)	c. 570	6.0	0.4	26.0	50.4	0	78	146

Note: A = forest or ungrazed thicket; B = crop; C = barren soil.
[a]From Rapp et al. (1972: 259, figure 5).

good case study of the effect of European settlement on soil erosion rates in neighbouring Australia is given by Olley and Wasson (2003).

In a more general sense there are plainly huge difficulties in estimating erosion rates in pre-human times, but in a recent analysis McLennan (1993) has estimated that the pre-human suspended sediment discharge from the continents was about 12.6×10^{15} g per year, which is about 60% of the present figure.

Table 4.10, which is based on data from tropical Africa, shows the comparative rates of erosion for three main types of land use: trees, crops and barren soil. It is very evident from these data that under crops, but more especially when ground is left bare or under fallow, soil erosion rates are greatly magnified. At the same time, and causally related, the percentage of rainfall that becomes runoff is increased.

In some cases the erosion produced by forest removal will be in the form of widespread surface stripping. In other cases the erosion will occur as more spectacular forms of mass movement, such as mudflows, landslides and debris avalanches. Some detailed data on debris-avalanche production in North American catchments as a result of deforestation and forest road construction are presented in Table 4.11. They illustrate the substantial effects created by clear-cutting and by the construction of logging roads. The significance of logging roads has also been highlighted in Borneo (Walsh et al., 2006). It is indeed probable that a large proportion of the erosion associated with forestry operations is caused by road construction, and care needs to be exercised to minimize these effects. Accelerated soil erosion can also be associated with grazing and trampling by domestic animals (Figure 4.13).

The digging of drainage ditches in upland pastures and peat moors to permit tree-planting in central Wales has also been found to cause accelerated erosion (Clarke and McCulloch, 1979), while the elevated sediment loads can cause reservoir pollution (Burt et al., 1983). In general, the greater the deforested proportion of a river basin the higher the sediment yield per unit area will be. In the USA the rate of sediment yield appears to double for every 20% loss in forest cover.

García-Ruiz (2010) has provided a full analysis of how land use changes in Spain have affected soil erosion rates. One change that has taken place in recent decades is the abandonment of farmland because of rural depopulation and the problems of mechanization on small packets of land on steep slopes. On the one hand one would expect vegetation recolonization to cause erosion rates to be reduced, but on the other the lack of maintenance to field terraces on steep slopes can cause gully erosion to occur. Overall, however, the effects of farmland abandonment and reductions in grazing pressures appear to have been to reduce overland flow and sediment yield from hillslopes (Nadal-Romero et al., 2012). Nonetheless, soil erosion has become severe in some vineyards because their soils are often left bare for much of the year, and many vineyards are on steep slopes. Olive and almond cultivation, some of which has been encouraged because of European Union subsidies, can also be a major cause of accelerated erosion on susceptible soils.

Soil erosion resulting from deforestation and agricultural practice is often thought to be especially serious in tropical areas or semi-arid areas (see Moore, 1979, for a good case study), but it is also a problem in some humid parts of Europe (Fuller, et al., 2009),

Table 4.11 Debris-avalanche erosion in forest, clear-cut and roaded areas. Source: After Swanston and Swanson (1976, table 4)

Site	Period of records (years)	Area (%)	(km²)	No. of slides	Debris-avalanche erosion (m³/km/year)	Rate of debris-avalanche erosion relative to forested areas
Stequaleho Creek, Olympic Peninsula						
Forest	84	79	19.3	25	71.8	×1.0
Clear-cut	6	18	4.4	0	0	0
Road	6	3	0.7	83	11,825	×165
Total	–	–	24.4	108	–	–
Alder Creek, western Cascade Range, Oregon						
Forest	25	70.5	12.3	7	45.3	×1.0
Clear-cut	15	26.0	4.5	18	117.1	×2.6
Road	15	3.5	0.6	75	15,565	×344
Total	–	–	17.4	100	–	–
Selected drainages, Coast Mountains, south-west British Columbia						
Forest	32	88.9	246.1	29	11.2	×1.0
Clear-cut	32	9.5	26.4	18	24.5	×2.2
Road	32	1.5	4.2	11	282.5	×25.2
Total	–	–	276.7	58	–	–
H.J. Andrews Experimental Forest, western Cascade Range, Oregon						
Forest	25	77.5	49.8	31	35.9	×1.0
Clear-cut	25	19.3	12.4	30	132.2	×3.7
Road	25	3.2	2.0	69	1772	×49
Total	–	–	64.2	130	–	–

Figure 4.13 Linear erosion furrows along domestic animal tracks developed in vertisols in the highlands of Ethiopia. (See Plate 29)

Russia (Sidorchuk and Golosov, 2003) and in the UK. Measurements by Morgan (1977) on sandy soils in the English East Midlands near Bedford indicate that rates of soil loss under bare soil on steep slopes can reach 17.69 tonnes per hectare per year, compared with 2.39 under grass and nothing under woodland (Table 4.12), and subsequent studies have demonstrated that water-induced soil erosion is a substantial problem, in spite of the relatively low erosivity of British rainfall. Walling and Quine (1991: 123) have identified the following farming practices as contributing to this developing problem:

1 Ploughing up of steep slopes that were formerly under grass, in order to increase the area of arable cultivation.
2 Use of larger and heavier agricultural machinery which has a tendency to increase soil compaction.
3 Removal of hedgerows and the associated increase in field size. Larger fields cause an increase in slope length with a concomitant increase in erosion risk.

Table 4.12 Annual rates of soil loss (tonnes per hectare) under different land-use types in eastern England. Source: From Morgan (1977). Reproduced with permission

Plot	Splash	Overland flow	Rill	Total
1. Bare soil				
Top slope	0.33	6.67	0.10	7.10
Mid-slope	0.82	16.48	0.39	17.69
Lower slope	0.62	14.34	0.06	15.02
2. Bare soil				
Top slope	0.60	1.11		1.71
Mid-slope	0.43	7.78	–	8.21
Lower slope	0.37	3.01	–	3.38
3. Grass				
Top slope	0.09	0.09	–	0.18
Mid-slope	0.09	0.57	–	0.68
Lower slope	0.12	0.05	–	0.17
4. Woodland				
Top slope	–	–	–	0.00
Mid-slope	–	0.012	–	0.012
Lower slope	–	0.008	–	0.008

4 Declining levels of organic matter resulting from intensive cultivation and reliance on chemical fertilizers, which in turn lead to reduced aggregate stability.

5 Availability of more powerful machinery which permits cultivation in the direction of maximum slope rather than along the contour. Rills often develop along tractor and implement wheelings and along drill lines.

6 Use of powered harrows in seedbed preparation and the rolling of fields after drilling.

7 Widespread introduction of autumn-sown cereals to replace spring-sown cereals. Because of their longer growing season, winter cereals produce greater yields and are therefore more profitable. The change means that seedbeds are exposed with little vegetation cover throughout the period of winter rainfall.

Accelerated wind erosion

Human activities may have had an important effect on soil erosion and as a consequence generate dust storms in the world's drylands. Von Suchodoletz et al. (2010) have even speculated that humans intensified dust storm activity in the north-west Sahara as early as 7–8 ka ago. The situation becomes less speculative as we move towards the present, and Neff et al. (2008), for instance, used analyses of lake cores in the San Juan Mountains of south-western Colorado, USA, to show that dust levels increased by 500% above the late Holocene average following the increased western settlement and livestock grazing during the nineteenth and early twentieth centuries. The USA Dust Bowl of the 1930s was caused by a combination of a major drought and adverse land management, with the latter having a feedback effect on the drought itself (Cook et al., 2009). A dust core from the Antarctic Peninsula (McConnell et al., 2007) showed a doubling in dust deposition in the twentieth century, and this is explained by increasing temperatures, decreasing relative humidity and widespread desertification in the source region – Patagonia and northern Argentina. Finally, analysis of a 3200-year marine core off West Africa shows a marked increase in dust activity at the beginning of the nineteenth century, which was a time which saw the advent of commercial activity (including groundnut production) in the Sahel (Mulitza et al., 2010). Ginoux et al. (2012) have calculated that around one quarter of global dust emissions come from anthropogenic sources (primarily agricultural land), while Mahowald et al. (2010) estimated that global dust loads doubled in the twentieth century due to anthropogenic activities.

The Dust Bowl of the 1930s in the USA (Figure 4.14) was in part caused by a series of hot, dry years which depleted the vegetation cover and made the soils dry enough to be susceptible to wind erosion. The effects of this drought were gravely exacerbated by years of over-grazing and unsatisfactory farming techniques. However, perhaps the prime cause of the event was the rapid expansion of wheat cultivation in the Great Plains. The number of cultivated hectares doubled during the First World War as tractors (for the first time) rolled out on to the plains by the thousands. In Kansas alone wheat hectarage increased from under 2 million hectares in 1910 to almost 5 million in 1919. After the war, wheat cultivation continued apace, helped by the development of the combine harvester and government assistance. The farmer, busy sowing wheat and reaping gold, could foresee no end to his land of milk and honey, but the years of favourable climate were not to last, and over large areas the tough

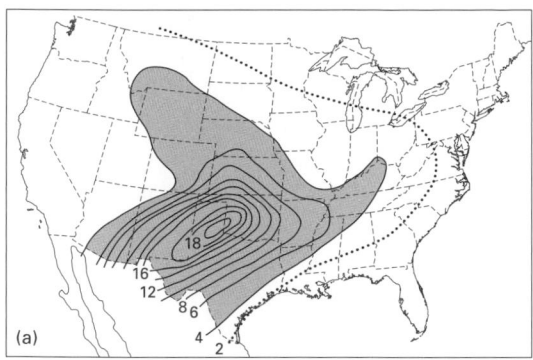

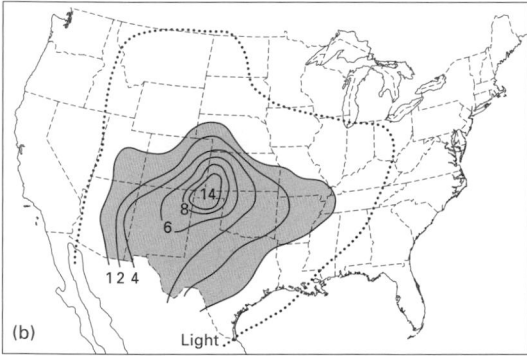

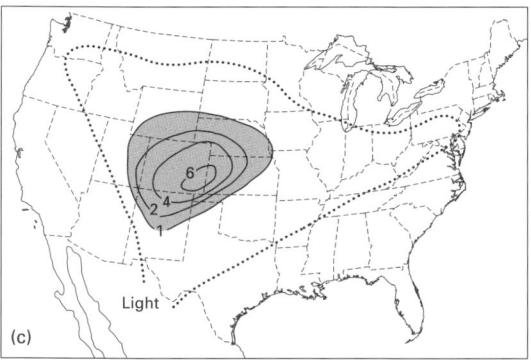

Figure 4.15 Surface degradation on a stone pavement caused by off-road driving in the Libyan Desert, Egypt. (See Plate 30)

Figure 4.14 The concentration of dust storms (number of days per month) in the USA in 1939, illustrating the extreme localization over the High Plains of Texas, Colorado, Oklahoma and Kansas. (a) March; (b) April; (c) May; (modified from Goudie, 1983).

sod which exasperated the earlier homesteaders had given way to friable soils of high erosion potential. Drought, acting on damaged soils, created the 'black blizzards' which have been so graphically described by Coffey (1978).

Dust storms are still a serious problem in various parts of the USA; the Dust Bowl was not solely a feature of the 1930s. Thus, for example, in the San Joaquin Valley area of California in 1977 a dust storm

caused extensive damage and erosion over an area of about 2000 km². More than 25 million tonnes of soil were stripped from grazing land within a 24-h period. While the combination of drought and a very high wind provided the predisposing natural conditions for the stripping to occur, over-grazing and the general lack of windbreaks in the agricultural land played a more significant role. In addition, broad areas of land had recently been stripped of vegetation, levelled or ploughed up prior to planting. Other quantitatively less important factors included stripping of vegetation for urban expansion, extensive denudation of land in the vicinity of oilfields and local denudation of land by vehicular recreation (Wilshire et al., 1981). One interesting observation made in the months after the dust storm was that in subsequent rain storms runoff occurred at an accelerated rate from those areas that had been stripped by the wind, exacerbating problems of flooding and initiating numerous gullies. Elsewhere in California dust yield has been considerably increased by mining operations in dry lake beds (Wilshire, 1980) and by disturbance of playas (Gill, 1996).

In many drylands, vehicles can disturb desert pavement surfaces, exposing the fine material under the surface lag to wind deflation (Figure 4.15).

A comparable acceleration of dust storm activity occurred in the former Soviet Union. After the 'Virgin Lands' programme of agricultural expansion in the 1950s, dust storm frequencies in the southern Omsk region increased on average by a factor of 2.5 and locally by factors of 5 to 6. Data on trends elsewhere are evaluated by Goudie (1983: 520) and Goudie and Middleton (1992). A good review of wind erosion of agricultural land is provided by Warren (2002).

Soil erosion produced by fire

Many fires are started by humans, either deliberately or non-deliberately, and because fires remove vegetation and expose the ground, they tend to increase rates of soil erosion (see Table 4.13). Fire causes a loss of vegetation and leaf litter cover which reduces the interception of rainfall. Overland flow is increased in volume and speed as a result of the increased percentage of rainfall available and the reduction in surface roughness caused by the removal of the vegetation and litter. The resulting exposure of the soil leaves it prone to rainsplash detachment (Shakesby et al., 2007). Most, though not all, studies indicate that fires can lead to a decrease in soil resistance because of a reduction in soil organic content and in aggregate stability. If the fire does not succeed in destroying the organic-rich surface layer (sometimes called 'duff'), then the effect of burning on soil erosion may be limited, as in parts of the Canadian Rockies (Martin et al., 2011).

However, in general, the burning of forests can, especially in the first years after the fire event, lead to high rates of soil loss. Burnt forests often have rates a whole order of magnitude higher than those of protected areas. Comparably large changes in soil erosion rates have been observed to result from the burning of heather in the Yorkshire moors in northern England (see Table 4.14), and the effects of burning may be felt for the 6 years or more than may be required to regenerate the heather (*Calluna*).

Likewise, watershed experiments in the chaparral scrub of the southwest USA have shown marked increases in runoff and erosion after burning (Warrick et al., 2012). The causes of this are particularly interesting. There is normally a distinctive 'non-wettable' layer in the soils supporting chaparral. This layer, composed of soil particles coated by hydrophobic substances leached from the shrubs or their litter, is normally associated with the upper part of the soil profile (Mooney and Parsons, 1973) and builds up through time in the unburned chaparral. The high temperatures which accompany chaparral fires cause these hydrophobic substances to be distilled so that they condense on lower soil layers. This process results in a shallow layer of wettable soil overlying a non-wettable layer. Such a condition, especially on steep slopes, can result in severe surface erosion (DeBano, 2000; Shakesby et al., 2000; Letey, 2001; Ravi et al., 2009; Shakesby, 2011). However, the importance of fire induced water repellency varies between different vegetation types and observations made in chapparal may not necessarily apply to Australian eucalyptus forest (Shakesby et al., 2007). Similarly, the time taken for forests to recover from a severe burn also varies between different plant types, and eucalypts, for example, may quickly re-sprout whereas conifers may rely upon seed germination to recover.

Nonetheless, in chaparral terrain it is possible to envisage a fire-induced sediment cycle (Graf, 1988:

Table 4.13 Recent examples of fire-induced soil erosion

Source	Location
Wilson (1999)	Tasmania
Cerda (1998); Shakesby (2011)	Mediterranean region
Moody and Martin (2001)	Colorado Front Range, USA
Dragovich and Morris (2002)	Eastern Australia
Pierson et al. (2002)	Idaho, USA
Martin et al. (2011)	Canadian Rockies
Kean et al. (2011); Warrick et al. (2012)	California
Neris et al. (2012)	Canary Islands

Table 4.14 Soil erosion associated with *Calluna* (heather) burning on the North Yorkshire moors. Source: Data with permission from Imeson (1971). With permission from John Wiley & Sons Ltd

Condition of Calluna or ground surface	Mean rate of litter accumulation (+) or erosion (−) (mm/year)	No. of observations
(a) *Calluna* 30–40 cm high. Complete canopy	+3.81	60
(b) *Calluna* 20–30 cm high. Complete canopy	+0.25	20
(c) *Calluna* 15–20 cm high. 40–100% cover	−0.74	20
(d) *Calluna* 5–15 cm high. 10–100% cover	−6.4	20
(e) Bare ground. Surface of burnt *Calluna*	−9.5	19
(f) Bare ground. Surface of peaty or mineral subsoil	−45.3	25

243). It starts with a fire that destroys the scrub and the root net, and changes surface soil properties in the way already discussed. After the fire, a precipitation event of low magnitude (with a return interval of around 1 or 2 years) is sufficient to induce extensive sheet and rill erosion, which removes enough soil to retard vegetation recovery. Eventually, a larger precipitation event occurs (with a return interval of around 5–10 years) and, because of limited vegetation cover, produces severe debris slides (Kean et al., 2011). Slowly the vegetation cover re-establishes itself, and erosion rates diminish. However, in due course enough vegetation grows to create a fire hazard, and the whole process starts again.

Soil erosion associated with construction and urbanization

There are now a number of studies which illustrate clearly that urbanization can create significant changes in erosion rates and sediment production.

The highest rates of erosion are produced in the construction phase, when there is a large amount of exposed ground and much disturbance produced by vehicle movements and excavations. In pioneer studies, Wolman and Schick (1967) and Wolman (1967) have shown that the equivalent of many decades of natural or even agricultural erosion may take place during a single year in areas cleared for construction. In Maryland they found that sediment yields during construction reached 55,000 tonnes per km^2 per year, while in the same area rates under forest were around 80–200 tonnes per km^2 per year and those under farm 400 tonnes per km^2 per year. New road cuttings in Georgia were found to have sediment yields up to 20,000–50,000 tonnes per km^2 per year. Likewise, in Devon, England, Walling and Gregory (1970) found that suspended sediment concentrations in streams draining construction areas were 2–10 times (occasionally up to 100 times) higher than those in undisturbed areas. In Virginia, USA, Vice et al. (1969) noted equally high rates of erosion during construction and reported that they were ten times those from agricultural land, 200 times those from grassland and 2000 times those from forest in the same area. Rates of sediment production may be especially high in humid tropical cities, where there is highly intense rainfall (Chin, 2006).

Table 4.15 Rates of erosion associated with construction and urbanization

Location	Land use	Source	Rate (t/km/year)
1. Maryland, USA	Forest	Wolman (1967)	39
	Agriculture		116–309
	Construction		38,610
	Urban		19–39
2. Virginia, USA	Forest	Vice et al. (1969)	9
	Grassland		94
	Cultivation		1876
	Construction		18,764
3. Detroit, USA	General non-urban	Thompson (1970)	642
	Construction		17,000
	Urban		741
4. Maryland, USA	Rural	Fox (1976)	22
	Construction		37
	Urban		337
5. Maryland, USA	Forest and grassland	Yorke and Herb (1978)	7–45
	Cultivated land		150–960
	Construction		1600–22,400
	Urban		830
6. Wisconsin, USA	Agricultural	Daniel et al. (1979)	<1
	Construction		19.2
7. Tama New Town, Japan	Construction	Kadomura (1983)	c. 40,000
8. Okinawa, Japan	Construction	Kadomura (1983)	25,000–125,000

However, construction does not go on forever, and once the disturbance ceases, roads are surfaced, and gardens and lawns are cultivated. The rates of erosion fall dramatically and may be of the same order as those under natural or pre-agricultural conditions (Table 4.15). Moreover, even during the construction phase, several techniques can be used to reduce sediment removal, including the excavation of settling ponds, the seedings and mulching of bare surfaces, and the erection of rock dams and straw bales (Reed, 1980).

Even small rural villages can produce substantial amounts of erosion, especially in the tropics. For example, De Meyer et al. (2011) have shown that in Uganda village compounds and associated unpaved footpaths and roads generate sediment that is a significant source of pollution in neighbouring Lake Victoria. Indeed, a scarcity of paved roads, especially in areas with communal land and easily motorable terrain, encourages motorists to make multiple tracks, and this can accelerate land erosion and other forms of environmental degradation (Keshkamat et al., 2011).

Soil conservation

Because of the adverse effects of accelerated erosion a whole array of techniques has now been widely adopted to conserve soil resources (Hudson, 1987) (Table 4.16) (Figure 4.16). Some of the techniques such as hill slope terracing may be of some antiquity, and traditional techniques have both a wide range of types and have many virtues (see Critchley et al., 1994; Reij et al., 1996). The following are some of the main ways in which soil cover may be conserved:

1 *Revegetation*:
  (a) Deliberate planting;
  (b) Suppression of fire, grazing, etc., to allow regeneration.
2 *Measures to stop stream bank erosion.*
3 *Measures to stop gully enlargement*:
  (a) Planting of trailing plants, etc.;
  (b) Weirs, dams, gabions, etc.
4 *Crop management*:
  (a) Maintaining cover at critical times of year;
  (b) Rotation;
  (c) Cover crops.
5 *Slope runoff control*:
  (a) Terracing;
  (b) Deep tillage and application of humus;
  (c) Transverse hillside ditches to interrupt runoff;
  (d) Contour ploughing;
  (e) Preservation of vegetation strips (to limit field width).
6 *Prevention of erosion from point sources like roads, feedlots*:
  (a) Intelligent geomorphic location;
  (b) Channelling of drainage water to non-susceptible areas;
  (c) Covering of banks, cuttings, etc., with vegetation.
7 *Suppression of wind erosion*:
  (a) Soil moisture preservation;
  (b) Increase in surface roughness through ploughing up clods or by planting windbreaks.

Table 4.16 Examples of the evaluation of methods to control water erosion on slopes in drylands

Techniques	Source
Soil compaction	Poesen et al. (2003)
Hedges	Kiepe (1996); Smolikowski et al. (2001); Poesen et al. (2003)
Control of trampling by sheep	Eldridge (1998)
Fallow cropping	Valentin et al. (2004)
Plant Strips	Martinez Raya et al. (2006)
Geotextiles	Rickson (2006)
Trash and stone lines	Quinton et al. (1997); Wakindiki and Ben-Hur (2002)
Vegetation cover	Rogers and Schumm (1991); Snelder and Bryan (1995); Durán Zuazo et al. (2004)
Engineering trenches	Marston and Dolan (1999)
Bench terracing	Ternan et al. (1996)
Matorral species	Bochet et al. (1998)
Mulching	Bautista et al. (1996); G.D. Smith et al. (1992)
Juniper control	Belski (1996)
Blade ploughing and exclosure	Eldridge and Robson (1997)
Rock mulches	Poesen et al. (1994)
Crop residue management	Unger et al. (1991)
Addition of fertilizer	Lasanta et al. (2000)
No tillage	Kabakci et al. (1993)
Afforestation	Cao et al. (2010); Romero-Diaz et al., (2010)
Synthetic polymers and biopolymers	Orts et al. (2007)

Figure 4.16 Soil conservation banks across a small ephemeral channel (wadi) in the Matmata area of Tunisia. (See Plate 31)

Table 4.17 Effect of various soil conservation practices on the detachment (D) and transport (T) phases of erosion. Source: Morgan (1995, table 7.1)

Practice	Rainsplash		Runoff		Wind	
	D	T	D	T	D	T
Agronomic measures						
Covering soil surface	*	*	*	*	*	*
Increasing surface roughness	–	–	*	*	*	*
Increasing surface depression storage	+	+	*	*	–	–
Increasing infiltration	–	–	+	*	–	–
Soil management						
Fertilizers, manures	+	+	+	*	+	*
Subsoiling, drainage	–	–	+	*	–	–
Mechanical measures						
Contouring, ridging	–	+	+	*	+	*
Terraces	–	+	+	*	–	–
Shelterbelts	–	–	–	–	*	*
Waterways	–	–	–	–	*	–

–no control; +moderate control; *strong control.
D = detachment, T = transport.

An alternative way of classifying soil conservation is provided by Morgan (1995). He identifies three main types of measure: agronomic, soil management and mechanical. The effect of these in relation to the main detachment and transport phases of erosion are shown in Table 4.17.

There are some parts of the world where terraces (a mechanical measure) are one of the most prominent components of the landscape. This applies to many wine-growing areas, to some arid zone regions (such as southwest USA, Yemen, western Saudi Arabia and Peru) and to a wide selection of localities in the more humid tropics (Luzon, Java, Sumatra, Assam, Ceylon, Uganda, Cameroons, the Andes, etc.). However, much traditional terracing only had soil erosion control as a secondary motive. More often these were constructed to provide level planting surfaces, to provide deeper soil and to manage the flow of water (Doolittle, 2000). In areas subject to wind erosion other strategies may be necessary. Since soil only blows when it is dry, anything which conserves soil moisture is beneficial. Another approach to wind erosion is to slow down the wind by physical barriers either in the form of an increased roughness of the soil surface brought about by careful ploughing, or by placing palm frond fences on the ground surface (Figure 4.17), or by planting vegetative barriers, such as windbreaks and shelterbelts.

Figure 4.17 Palm frond fences to reduce sand movement and dune migration in Erfoud, Morocco. (See Plate 32)

Various attempts have been made to control the occurrence of dust storms, and these include the array of techniques that have been used for wind erosion control, most of them developed to protect cultivated fields from soil loss (Sterk, 2003; Nordstrom and Hotta, 2004). These techniques are normally classified into three categories: (1) crop management practices, (2) mechanical tillage operations and (3) vegetative barriers. All of these aim to decrease wind speed at the soil surface by increasing surface roughness and/or increasing the threshold velocity that is required to initiate particle movement by wind. The numerous crop management practices, also commonly referred to as agronomic measures, can influence both the detachment and transport phases of soil particle movement, particularly when combined with good soil management. Mechanical methods, by contrast, effectively do little to prevent soil detachment, but tend to be more effective in preventing soil transport (Morgan, 2005).

Some attempts at soil conservation have been particularly successful. In New Zealand, gully systems produced by the removal of indigenous forest in the 1880s–1920s have been stabilized as a result of replanting with exotic plantations (Marden, 2012). In Wisconsin, USA, a study by Trimble and Lund (1982) showed that in the Coon Creek Basin, erosion rates declined fourfold between the 1930s and the 1970s. One of the main reasons for this was the progressive adoption of contour-strip ploughing.

However, it is important to stress that such techniques are not always successful. For example, in semi-arid Tunisia, where approximately 1 million hectares of agricultural land have been installed with anti-erosive contour benches, the life of the benches has proved to be limited, with those on gypsum clays being especially prone to failure (Baccari et al., 2008). It was also found that soil conservation was in part achieved at the cost of reduced runoff and thus a reduction in available water resources for irrigation. Indeed, soil erosion counter-measures can be counter-productive, as has been made evident by reviews of large-scale afforestation schemes in China (Cao et al., 2010, 2011) and Spain (Romero-Diaz et al., 2010). This latter study showed that aggressive land sculpting and the subsequent planting of inappropriate trees on scrublands, actually accelerated rates of erosion on marl slopes by between 1 and 2 orders of magnitude. Bulldozing can be a curse in Mediterranean environments. One also needs to be aware that some species of exotic plant that have been introduced for control of water erosion can prove to be highly invasive, as with mesquite, *Prosopis juliflora*, in Kenya (Muturi et al., 2009).

Moreover, attempts at soil conservation have not always been without their drawbacks. For example, the establishment of ground cover in dry areas to limit erosion may so reduce soil moisture because of acceler-

ated evapotranspiration that the growth of the main crop is adversely affected. On a wider scale major afforestation schemes can cause substantial runoff depletion in river catchments. Likewise, the provision of mulching is sometimes detrimental: in cool climates, reduced soil temperature shortens the growing season, whilst in wet areas, higher soil moisture may induce gleying and anaerobic conditions (Morgan, 1979: 60). Similarly strip cropping, because it involves the farming of small areas, is incompatible with highly mechanized agricultural systems, and insect infestation and weed control are additional problems which it has posed.

Soil conservation measures may not always be appropriate, for, paradoxically, soil erosion can be a useful phenomenon. Sanchez and Buol (1975), for instance, pointed out that in recent volcanic areas soil erosion has enabled removal of the more weathered base-depleted material from the soil surface, exposing the more fertile, less weathered, base-rich material beneath. Likewise, in the Nochixtland area of southern Mexico, soil erosion has been utilized by local farmers to *produce* agricultural land. Severe gullies have cut in to steep valley-side slopes, and since the Spanish Conquest an average depth of 5 m has been stripped from the entire surface area. The local Mixtec farmers, far from seeing this high rate of erosion as a hazard to be feared, have directed the flow of the eroded material to feed their fields with fertile soil and to extend their land. Over the past 1000 years (Whyte, 1977), they have managed to use gully erosion to double the width of the main valley floors with flights of terraces and to convert poor hilltop fields into the rich alluvial farmland below. On some Pacific islands, accelerated erosion led to the creation of large coastal plains on which extensive settlement and agricultural exploitation occurred, and erosion on steep slopes may have been deliberately encouraged to increase the area of valuable land (Spriggs, 2010).

Nonetheless it is undoubtedly true that manipulation of the soil is one of the most significant ways in which humans change the environment and one in which they have had some of the most detrimental effects. Soil deterioration has led to many cases of what W.C. Lowdermilk once termed 'regional suicide', and the overall situation is at least as bleak as it was in the post-Dust Bowl years when Jacks commented: 'The organization of civilized societies is founded upon the measures taken to wrest control of the soil from wild Nature, and not until complete control has passed into human hands can a stable super-structure of what we call civilization be erected on the land' (Jacks and Whyte, 1939: 17).

Points for review

What are the causes of accelerated salinization?

Do humans contribute to lateritization and podzolization?

How does the removal of forest cover affect soil erosion?

What effects do fires have on soils?

How might you seek to reduce rates of soil erosion by (a) water and by (b) wind?

How serious a risk is soil erosion to agricultural sustainability?

Guide to reading

Conacher, A.J. and Sala, M. (eds) (1998) *Land Degradation in Mediterranean Environments of the World*. Chichester: Wiley. An edited discussion of land degradation processes in Mediterranean environments.

Fullen, M.A. and Catt, J.A. (2004) *Soil management: problems and solutions*. A review of some of the problems involved in the management of soils in the face of human impacts.

Holliday, V.T. (2004) *Soils in Archaeological Research*. New York: Oxford University Press. This book includes a treatment of long-term human impacts on soils.

Johnson, D.L. and Lewis, L.A. (1995) *Land Degradation: Creation and Destruction*. Oxford: Blackwell. A broadly-based study of intentional and unintentional causes of many aspects of land degradation.

McTainsh, G. and Boughton, W.C. (1993) *Land Degradation Processes in Australia*. Melbourne: Longman Cheshire. An Australian Perspective on soil modification.

Morgan, R.P.C. (2005) *Soil Erosion and Conservation* (3rd edn). Oxford: Blackwell. A revised edition of a fundamental work.

Russell, J.S. and Isbell, F.R. (eds) (1986) *Australian Soils: the Human Impact*. St. Lucia: University of Queensland Press. A detailed collection of studies on land degradation in Australia.

5 THE HUMAN IMPACT ON THE WATERS

Chapter Overview

The chapter starts with a consideration of the various ways in which humans have deliberately modified river systems, by the creation of dams, reservoirs, embankments and through channelization. It then considers how stream flows and lake levels have been transformed by land cover changes and how groundwater levels have been modified by water abstraction. The second part of the chapter concentrates on changes in water quality, whether caused by pollution or by changes in river catchment conditions. Finally, there is a consideration of the ways in which our oceans and seas are being modified by pollution.

Introduction

Because water is so important, humans have sought to control water resources in a whole variety of ways (Carpenter et al., 2011). Also, because water is such an important part of so many natural and human systems its quantity and quality have undergone major changes as a consequence of human activities. We can quote Gleick (1993: 3):

. . .we must now acknowledge that many of our efforts to harness water have been inadequate or misdirected. . . Rivers, lakes, and groundwater aquifers are increasingly contaminated with biological and chemical wastes. Vast numbers of people lack clean drinking water and rudimen-

The Human Impact on the Natural Environment: Past, Present and Future, Seventh Edition. Andrew S. Goudie.
© 2013 John Wiley & Sons, Ltd. Published 2013 by John Wiley & Sons, Ltd.

tary sanitation services. Millions of people die every year from water-related diseases such as malaria, typhoid, and cholera. Massive water developments have destroyed many of the world's most productive wetlands and other aquatic habitats.

As Lynas (2011: 139) has graphically remarked,

Whole natural drainage basins, which once responded to the grand seasonal cycles of winter flood and summer drought, now react meekly to the whims of water managers seated in the control rooms that govern sluice gates in tens of thousands of large dams. The Colorado River may have gouged out the most spectacular cutting in the world – the Grand Canyon – but today the flow of this powerful torrent is as much a product of human hydrological engineering as it is of any natural force.

In recent decades human demand for freshwater has increased rapidly. Global water use has more than tripled since 1950, and annual irretrievable water losses increased about sevenfold in the twentieth century (Table 5.1a). Trends in the interception of runoff and in the capacity of reservoirs are shown in Figure 5.1. Many rivers now have highly modified flow

regimes as a result of a variety of human interventions (Table 5.2).

Deliberate modification of rivers

Although there are many ways in which humans influence water quantity and quality in rivers and streams – for example, by direct channel manipulation, modification of basin characteristics, urbanization and pollution – the first of these is of particularly great importance. Indeed, there are a great variety of methods of direct channel manipulation and many of them have a long history. Perhaps the most widespread of these is the construction of dams and reservoirs (Figure 5.2). The first recorded dam was constructed in Egypt some 5000 years ago, but since that time the adoption of this technique has spread, variously to improve agriculture, prevent floods, generate power or to provide a reliable source of water. Some dams – impoundment dams – are of immense size, but others are much smaller – 'run-of-river dams' (Csiki and Rhoads, 2010) – and the implications of these two

Table 5.1 Major changes in the hydrological environment. Source: Data provided by UNEP and World Commission on Dams

(a) Irretrievable water losses (km³/year)

Users	1900	1940	1950	1960	1970	1980	1990	2000
Agriculture	409	679	859	1180	1400	1730	2050	2500
Industry	3.5	9.7	14.5	24.9	38.0	61.9	88.5	117
Municipal supply	4.0	9.0	14	20.3	29.2	41.1	52.4	64.5
Reservoirs	0.3	3.7	6.5	23.0	66.0	120	170	220
Total	417	701	894	1250	1540	1950	2360	2900

(b) Number of large dams (>15 m high) constructed or under construction, 1950–1986

Continent	1950	1982	1986	Under construction 31.12.86
Africa	133	665	763	58
Asia	1562	22,789	22,389	613
Australasia/Oceania	151	448	492	25
Europe	1323	3961	4114	222
North and Central America	2099	7303	6595	39
South America			884	69
TOTAL	5268	35,166	36,327[a]	1036

[a]The figure by the end of the twentieth century was *c.* 45,000.

(a)

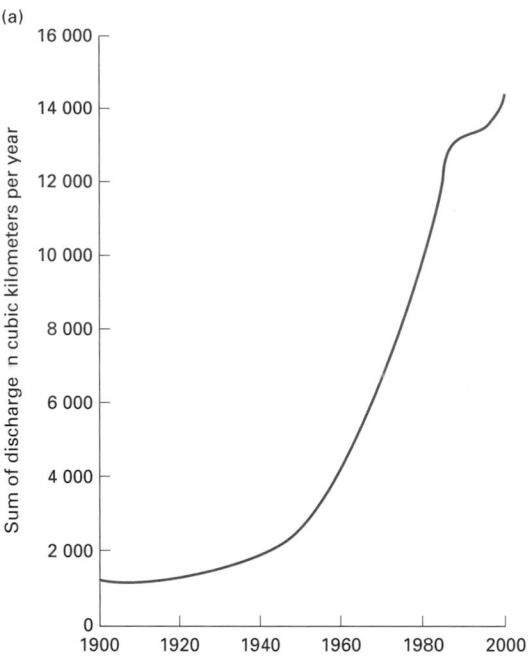

(b)

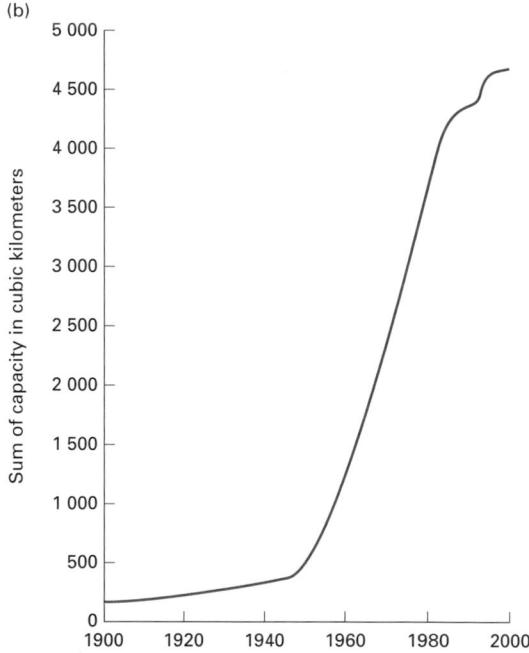

Figure 5.1 Global trends in (a) intercepted runoff and (b) reservoir capacity (modified from Millennium Ecosystem Assessment, 2005).

Table 5.2 Human impacts on river flow characteristics

Engineering
Dams, ponds, reservoirs
Swamp drainage
Land drainage
Interbasin water transfers
Groundwater abstraction
Channelization, embankments, etc.
Log jam removal
Land changes
Urbanization
Afforestation and deforestation
Removal or introduction of riparian vegetation
Anthropogenic climate change
(See Chapter 10)

types may be very different. Cumulatively, however, the presence of many small dams in an area may have a great effect on river flow and on sediment retention (Boardman and Foster, 2011).

There are some 75,000 dams in the USA. Most are small, but the bulk of the storage of water is associated with a relatively limited number of structures. Those dams creating reservoirs of more than $1.2 \times 10^9 \, m^3$ (1×10^6 acre-feet) account for only 3% of the total number of structures, but they account for 63% of the total storage. In all, the dams are capable of storing a volume of water almost equalling 1 year's runoff, and they store around $5000 \, m^3$ (4 acre-feet) of water per person. The decade of the 1960s saw the greatest spate of dam construction in American history (18,833 dams were built then). Since the 1980s, however, there have been only relatively minor increases in storage. The dam building era is over, but the environmental effects remain and the physical integrity of many rivers has been damaged (Graf, 2001).

The construction of large dams increased markedly, especially between 1945 and the early 1970s. Engineers have now built more than 45,000 large dams around the world and, as Table 5.1b shows, such large dams (i.e. more than 15 m high) are still being constructed at an appreciable rate, especially in Asia. Indeed, one of the most striking features of dams and reservoirs is that they have become increasingly large. Thus in the 1930s the Hoover or Boulder Dam in the USA (221 m high) was by far the tallest in the world and it impounded the biggest reservoir, Lake Mead, which contained just under 38 billion m^3 of water. By the

Figure 5.2 The Kariba Dam on the Zambezi River between Zambia and Zimbabwe. Such large dams can provide protection against floods and water shortages, and generate a great deal of electricity. However, they can have a whole suite of environmental consequences. (See Plate 33)

Table 5.3 Peak flow reduction downstream from selected British reservoirs. Source: After Petts and Lewin (1979: 82, table 1)

Reservoir	% of catchment inundated	% of peak flow reduction
Avon, Dartmoor	1.38	16
Fernworth, Dartmoor	2.80	28
Meldon, Dartmoor	1.30	9
Vyrnwy, mid-Wales	6.13	69
Sutton Bingham, Somerset	1.90	35
Blagdon, Mendip	6.84	51
Stocks, Forest of Bowland	3.70	70
Daer, Southern Uplands	4.33	56
Camps, Southern Uplands	3.13	41
Catcleugh, Cheviots	2.72	71
Ladybower, Peak District	1.60	42
Chew Magna, Mendips	8.33	73

2000s it was exceeded in height by at least 29 others, and four of these impounded reservoirs (Kariba, Bratsk, Nasser and Volta) with more than four times the volume of Lake Mead. The massive new Three Gorges Dam in China is a major cause of current environmental concern, as is the damming of the Narmada River in India.

Large dams are capable of causing almost total regulation of the streams they impound but, in general, the degree to which peak flows are reduced depends on the size of the dam and the impounded lake in relation to catchment characteristics. In Britain, as Table 5.3 shows, peak flow reduction downstream from selected reservoirs varies considerably, with some tendency for the greatest degree of reduction to occur in those catchments where the reservoirs cover the largest percentage of the area. In the USA, Fitzhugh and Vogel (2011) have mapped the impacts of dams on flood flows. Across the country as a whole the estimated reduction in median annual flood for large rivers averages 29%, for medium rivers 15% and for small rivers 7%. The greatest reductions have taken place to the west of the Mississippi and especially in the Great Plains, the deserts of the south-west and in northern California.

When considering the magnitude of floods of different recurrence intervals before and after dam construction, it is clear that dams have much less effect on rare events of high magnitude (Petts and Lewin,

Table 5.4 The ratios of post- to pre-dam discharges for flood magnitudes of selected frequency. Source: After Petts and Lewin (1979: 84, table 2)

Reservoir	Recurrence interval (years)			
	1.5	2.3	5.0	10.0
Avon, R. Avon	0.90	0.89	0.93	1.02
Stocks, R. Hodder	0.83	0.86	0.84	0.95
Sutton Bingham, R. Yeo	0.52	0.61	0.69	0.79

1979), and this is brought out in Table 5.4. Nonetheless, most dams achieve their aim: to regulate river discharge. They are also highly successful in fulfilling the needs of surrounding communities: millions of people depend upon them for survival, welfare and employment.

However, dams may have a whole series of environmental consequences that may or may not have been anticipated (Figure 5.3) (World Commission on Dams, 2000). Several of these processes may in turn affect the viability of the scheme for which the dam was created. Some of these are dealt with in greater detail elsewhere in this volume, such as subsidence, earthquake triggering, the transmission and expansion in the range of organisms, inhibition of fish migration, the build-up of soil salinity, changes in ground-water levels creating slope instability and water-logging. Regulation of river

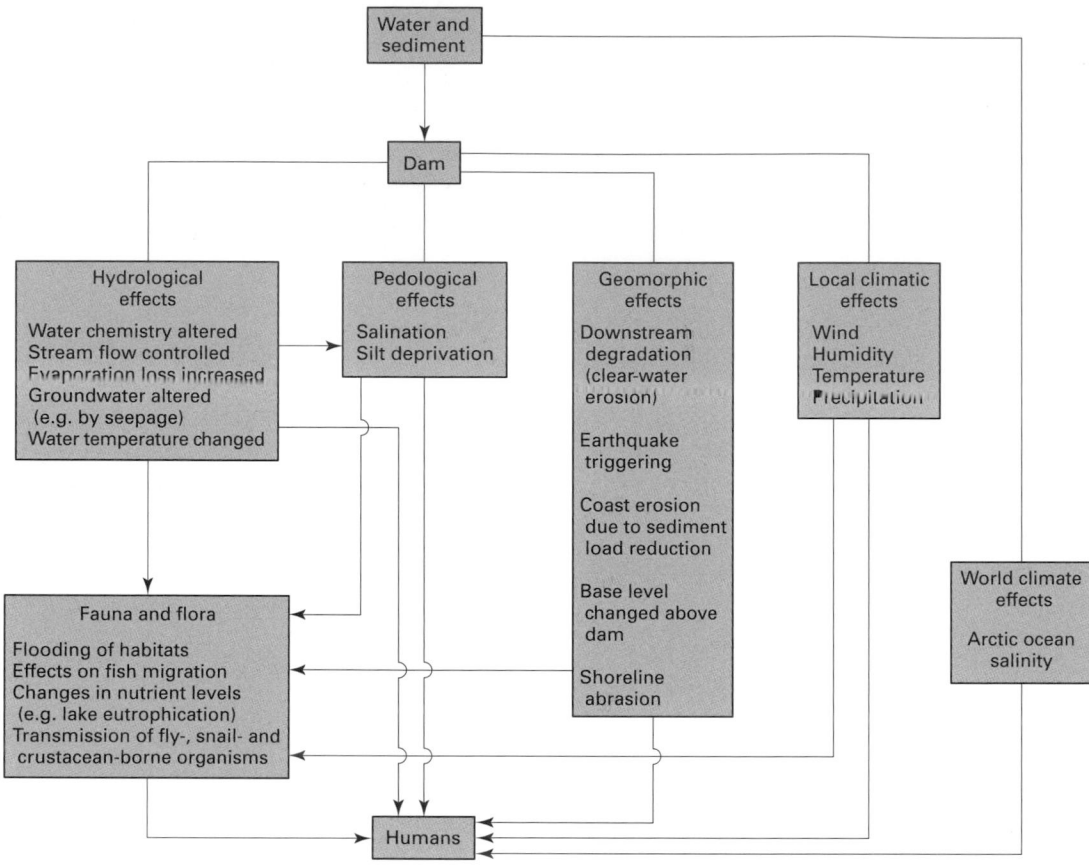

Figure 5.3 Generalized representation of the possible effects of dam construction on human life and various components of the environment.

flows may also affect aeolian landscapes by modifying sources of sediment and sand transport paths, as has happened on the Colorado River (Draut, 2012).

Sediment deposition in reservoirs in turn has various possible consequences, including a reduction in flood-deposited nutrients on fields, less nutrients for fish in the south-east Mediterranean Sea, accelerated erosion of the Nile Delta and accelerated riverbed erosion since less sediment is available to cause bed **aggradation**. The last process is called 'clear-water erosion' (see Beckinsale, 1972), and in the case of the Hoover Dam it affected the river channel of the Colorado for 150 km downstream by causing incision. In turn, such channel incision may initiate headward erosion in tributaries and may cause the lowering of groundwater tables and the undermining of bridge piers and other structures downstream of the dam. On the other hand, in regions such as northern China, where modern dams trap silt, the incision of the river channel downstream may alle-

viate the strain on **levées** and lessen the expense of levée strengthening or heightening.

However, clear-water erosion does not always follow from silt retention in reservoirs. There are examples of rivers where, before impoundment, floods carried away the sediment brought into the main stream by steep tributaries. Reduction of the peak discharge after the completion of the dam leaves some rivers unable to scour away the sediment that accumulates as large fans of sand or gravel below each tributary mouth (Dunne and Leopold, 1978). The bed of the main stream is raised and if water-intakes, or other structures lie alongside the river they can be threatened again by flooding or channel shifting across the accumulating wedge of sediment. Rates of aggradation of a metre a year have been observed, and tens of kilometres of channel have been affected by sedimentation. One of the best-documented cases of aggradation concerns the Colorado River below Glen Canyon Dam in the USA. Since dam closure the extremes of river flow have been largely eliminated so that the 10 years' recurrence interval flow has been reduced to less than one-third. The main channel flow is no longer capable of removing sediment provided by flash-flooding tributaries (Petts, 1985: 133). In the Rio Grande, vertical channel accretion of 2.75–3.0 m occurred between 1991 and 2008 (Dean and Schmidt, 2011). The reduction in peak floods below dams can also permit the expansion of riparian vegetation, such as Tamarisk, which may trap sediment (Allred and Schmidt, 1999; Birken and Cooper, 2006) and so add to the amount of accretion that occurs. Rapid aggradation has also been noted in the channel of the Yellow River in China, partly because inputs of aeolian sands are no longer so effectively flushed out (Ta et al., 2008).

Some landscapes in the world are dominated by dams, canals and reservoirs. Probably the most striking example of this is the 'tank' landscape of south-east India where myriads of little streams and areas of overland flow have been dammed by small earth structures to give what Spate (Spate and Learmonth, 1967: 778) likened to 'a surface of vast overlapping fish-scales'. In the northern part of the subcontinent, in Sind, the landscape changes wrought by hydrology are no less striking, with the mighty snow-fed Indus being controlled by large embankments (*bunds*) and interrupted by great barrages. Its waters are distributed over thousands of square kilometres by a canal network that has evolved over the past 4000 years (Figure 5.4). Another landscape where equally far-reaching changes have been wrought is the Netherlands. Coates (1976) has calculated that, before 1860, reclamation of that country from the sea, in the extension of drainage lines, involved the movement of 1000 million m^3 of material. The area is dominated by human constructions: canals, rivers, drains and lakes.

Another direct means of river manipulation is channelization. This involves the construction of embankments, dikes, levées and floodwalls to confine floodwaters; and improving the ability of channels to transmit floods by enlarging their capacity through straightening, widening, deepening or smoothing (Table 5.5).

Some of the great rivers of the world are now lined by extensive embankment systems such as those that run for more than 1000 km alongside the Nile, 700 km along the Hwang Ho, 1400 km by the Red River in Vietnam and over 4500 km in the Mississippi Valley

Figure 5.4 An irrigation canal in the plains of Haryana, India. (See Plate 34)

Table 5.5 Selected terminologies for the methods of river channelization in the USA and the UK. Source: Brookes (1985, table 1)

American term	British equivalent	Method involved
Widening	Resectioning	Increase of channel capacity
Deepening	Resectioning	by manipulating width and/or depth variable
Straightening	Realigning	Increasing velocity of flow by steepening the gradient
Diking	Embanking	Raising of channel banks to confine floodwaters
Bank stabilization	Bank protection	Methods to control bank erosion for example, gabions and concrete structures
Clearing and snagging	Pioneer tree clearance	Removal of obstructions from a watercourse, thereby
	Control of aquatic plants	decreasing the resistance and increasing the velocity
	Dredging of sediments	of flow
	Urban clearing	

(Ward, 1978). Like dams, embankments and related structures often fulfil their purpose but they may also create some environmental problems and have some disadvantages. For example, they reduce natural storage for floodwaters, both by preventing water from spilling on to much of the floodplain and by stopping bank storage in cases where impermeable floodwalls are used. Likewise the flow of water in tributaries may be constrained. In addition, embankments may occasionally exacerbate the flood problem they were designed to reduce by preventing floodwaters downstream of a breach from draining back into the channel once the peak has passed.

Channel improvement, designed to improve water flow, may also have unforeseen or undesirable effects. For example, the more rapid movement of water along improved channel sections can aggravate flood peaks further downstream and cause excessive erosion. On the other hand, the lower reaches of a channelized system can see accumulation of the eroded sediment from the upper reaches, so that plugs of sediment accumulate. These sediment plugs can block channels and cause water and sediment to be forced onto the neighbouring floodplain even at times of low flow, so creating swamp conditions along, for example the rivers of Tennessee and elsewhere in the US coastal plain (Shankman

and Smith, 2004; Hupp et al., 2009). In addition, lined channels may obstruct soil water movement (interflow) and shallow ground water flow, and so cause surface saturation. The diversity of habitats in natural meandering channels means that straitening may lead to a reduction in the diversity of organisms (Garcia et al., 2012). Brookes (1985) and Gregory (1985) provide useful reviews of the impact of channelization.

Channelization may also have miscellaneous effects on fauna through the increased velocities of water flow, reductions in the extent of shelter in the channel bed and by reduced nutrient inputs due to the destruction of overhanging bank vegetation (see Keller, 1976). In the case of large swamps, like those of the Sudd in Sudan or the Okavango in Botswana, the channelization of rivers could completely transform the whole character of the swamp environment. Figure 5.5 illus-

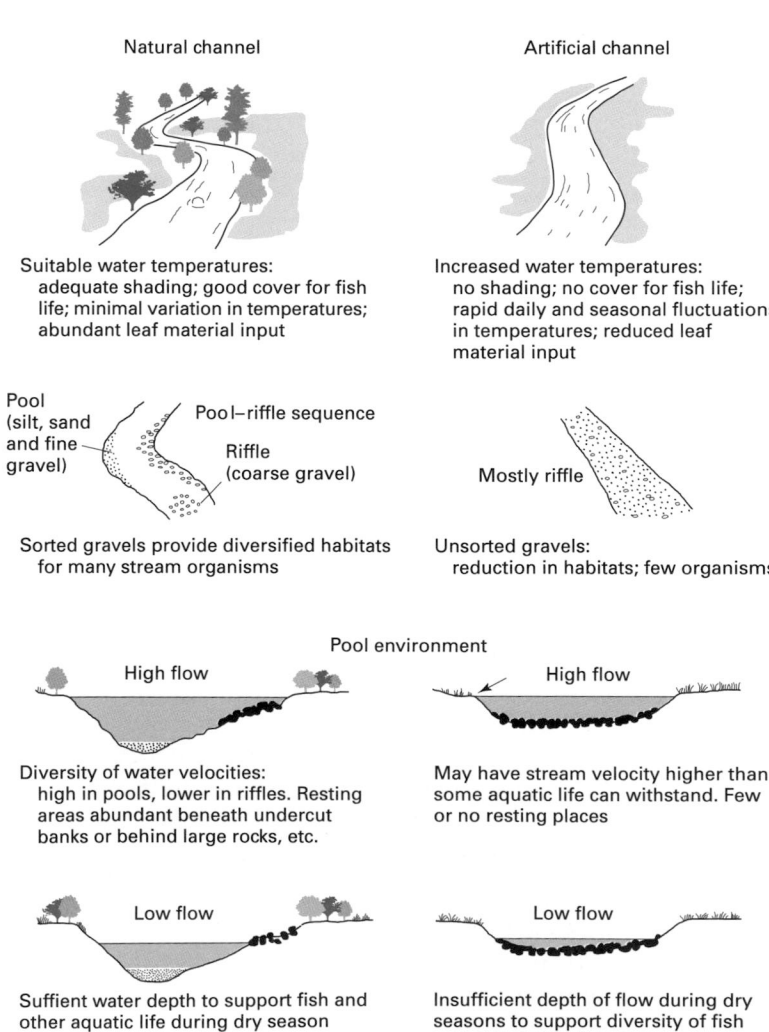

Figure 5.5 Comparison of the natural channel morphology and hydrology with that of a channelized stream, suggesting some possible ecological consequences (modified from Keller, 1976, figure 4).

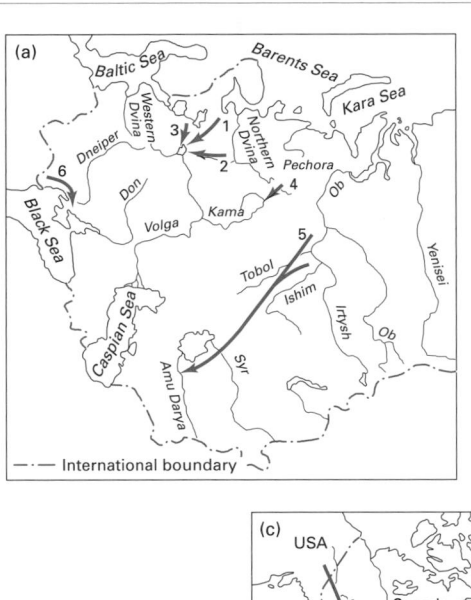

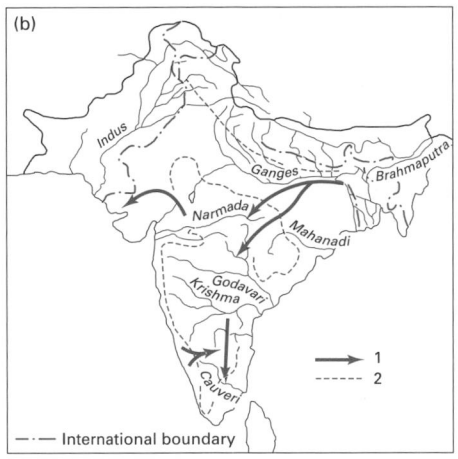

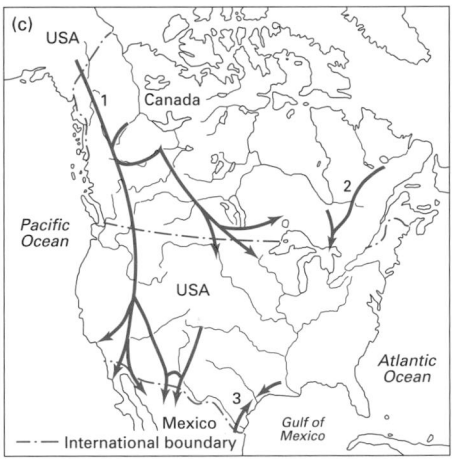

Figure 5.6 Some major schemes proposed for large-scale interbasin water transfers: (a) Projected water transfer systems in the CIS: 1. From the Onega River and in future from Onega Bay. 2. From the Sukhona and Northern Dvina Rivers. 3. From the Svir River and Lake Onega. 4. From the Pechora River. 5. From the Ob River. 6. From the Danube Delta; (b) Projected systems for water transfers in India: 1. Scheme of the national water network. 2. Scheme of the Grand Water Garland; (c) Some major projects for water transfers in North America: 1. NAWAPA. 2. Grand Canal. 3. Texas River Basins (after Shiklomanov, 1985, figures 12.6, 12.9 and 12.11, in *Facets of Hydrology II*, ed. J.C. Rodda, with permission from John Wiley and Sons Ltd).

trates some of the differences between natural and artificial channels.

Deliberate modification of a river regime can be achieved by long-distance interbasin water transfers (Shiklomanov, 1985), transfers necessitated by the unequal spatial distribution of water resources and by the increasing rates of water consumption. The total volume of water in the various transfer systems in operation and under construction on a global scale is about 300 km³ a year, with the largest countries in terms of volume of transfers being Canada, the former Soviet Union, the USA and India.

In future decades it is likely that many even greater schemes will be constructed (Figure 5.6); route lengths

of some hundreds of kilometres will be common, and the water balances of many rivers and lakes will be transformed. This has happened in the former Soviet Union (Figure 5.7), where the operation of various anthropogenic activities of this type have caused runoff in the most intensely cultivated central and southern areas to decrease by 30–50% compared to normal natural runoff. At the same time, inflows into the Caspian and Aral Seas have declined sharply. The level of the Aral has fallen and its area decreased.

When one turns to the coastal portions of rivers, to estuaries, the possible effects of another human impact, dredging, can be as complex as the effects of dams and reservoirs upstream. Dredging and filling are certainly

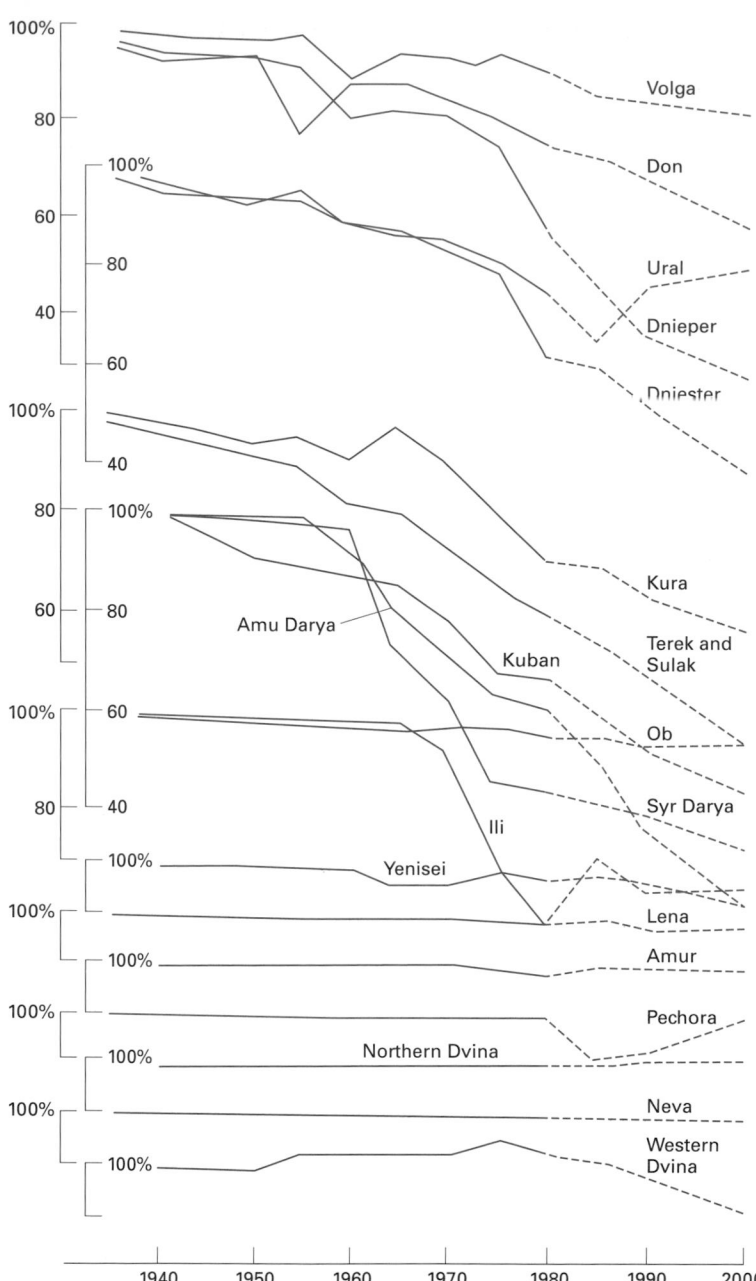

Figure 5.7 Changes in annual runoff in the Commonwealth of Independent States (former Soviet Union) due to human activity during 1936–2000 (from Shiklomanov, 1985, figure 12.7, in *Facets of Hydrology II*, ed. J.C. Rodda, with permission from John Wiley and Sons Ltd).

widespread and often desirable. Dredging may be performed to create and maintain canals, navigation channels, turning basins, harbours and marinas; to lay pipelines; and to obtain a source of material for fill or construction. Filling is the deposition of dredged materials to create new land. There are miscellaneous ecological effects of such actions. In the first place, filling directly disrupts habitats. Second, the generation of large quantities of suspended silt tends physically to smother bottom-dwelling plants and animals; tends to smother fish by clogging their gills; reduces photosynthesis through the effects of turbidity; and tends to lead to eutrophication by an increased nutrient release. Likewise, the destruction of marshes, mangroves and sea grasses by dredge and fill can result in the loss of these natural purifying systems. The removal of vegetation may also cause erosion. Moreover, as silt deposits stirred up by dredging accumulate elsewhere in the

Table 5.6 Stages of urban growth and their miscellaneous hydrological impacts. Source: Modified from Savini and Kammerer (1961)

Stage	Impact
1 Transition from pre-urban stage:	
(a) Removal of trees or vegetation	Decrease in transpiration and increase in storm flow
(b) Construction of scattered houses with limited sewerage and water facilities	
(c) Drilling of wells	Some lowering of water table
(d) Construction of septic tanks, etc.	Some increase in soil moisture and perhaps some contamination
2 Transition from early-urban to middle-urban stage:	
(a) Bulldozing of land	Accelerated land erosion
(b) Mass construction of houses, etc.	Decreased infiltration
(c) Discontinued use and abandonment of some shallower wells	Rise in water table
(d) Diversion of nearby streams for public supply	Decrease in runoff between points of diversion of disposal
(e) Untreated or inadequately treated sewerage into streams and wells	Pollution of streams and wells
3 Transition from middle-urban to late-urban stage:	
(a) Urbanization of area completed by addition of more buildings	Reduced infiltration and lowered water table, higher flood peaks and lower low flows
(b) Larger quantities of untreated waste into local streams	Increased pollution
(c) Abandonment of remaining shallow wells because of pollution	Rise in water table
(d) Increase in population requiring establishment of new water supply and distribution systems	Increase in local stream flow if supply is from outside basin
(e) Channels of streams restricted at least in part to artificial channels and tunnels	Higher stage for a given flow (therefore increased flood damage) changes in channel geometry and sediment load
(f) Construction of sanitary drainage system and treatment plant for sewerage	Removal of additional water from area
(g) Improvement of storm drainage system	
(h) Drilling of deeper, large-capacity industrial wells	Lowered water pressure, some subsidence, saltwater encroachment
(i) Increased use of water for air conditioning	Overloading of sewers and other drainage facilities
(j) Drilling of recharge wells	Raising of water pressure surface
(k) Wastewater reclamation and utilization	Recharge of groundwater aquifers: more efficient use of water resources

estuary they tend to create a 'false bottom'. Characterized by shifting, unstable sediments, the dredged bottom, fill deposits or spoil areas are slowly – if at all – recolonized by fauna and flora. Furthermore, dredging tends to change the configuration of currents, the rate of freshwater drainage and may provide avenues for saltwater intrusion.

Urbanization and its effects on river flow

The process of urbanization has a considerable hydrological impact, in terms of controlling rates of erosion and the delivery of pollutants to rivers, and in terms of influencing the nature of runoff and other hydro-logical characteristics (Hollis, 1988; Chin, 2006). An attempt to generalize some of these impacts using a historical model of urbanization is summarized usefully by Savini and Kammerer (1961) and reproduced here in Table 5.6.

Urbanization affects flood runoff. Research both in the USA and in Britain has shown that, because urbanization produces extended impermeable surfaces of bitumen, tarmac, tiles and concrete, there is a tendency for flood runoff to increase in comparison with rural sites. Sheng and Wilson (2009) provide an analysis of how urbanization in the Los Angeles metropolitan region has increased flood hazard. City drainage densities may be greater than those in natural conditions (Graf, 1977) and the installation of sewers and storm drains accelerates runoff, as illustrated in Figure 5.8a. The greater the area that is underlain by sewers, the greater is the discharge for a particular recurrence level (Figure 5.8b). Peak discharges are higher and occur sooner after runoff starts in basins that have been affected by urbanization and the installation of sewers. Some runoff may be generated in urban areas because low vegetation densities mean that evapotranspiration is limited.

However, in many cases the effect of urbanization is greater on small floods; as the size of the flood and its recurrence interval increase, so the effect of urbanization diminishes (Hollis, 1975). A probable explanation for this is that, during a severe and prolonged storm event, a non-urbanized catchment may become so saturated and its channel network so extended that it begins to behave hydrologically as if it were an impervious catchment with a dense surface-water drain network. Under these conditions, a rural catchment produces floods of a type and size similar to those of its urban counterpart. Moreover, a further mechanism probably operates in the same direction, for in an urban catchment it seems probable that some throttling of flow occurs in surface-water drains during intense storms, tending to attenuate the very highest discharges. Thus, Hollis believed, whilst the size of small frequent floods is increased many times by urbanization, large, rare floods (the ones likely to cause extreme damage) are not significantly affected by the construction of suburban areas within a catchment area (Figure 5.8c). Hollis's findings may not be universally applicable. For example, Wilson (1967) working in Jackson, Mississippi, found that the 50-year flood for an urbanized catchment was three times higher than that of a rural one, and Sheng and Wilson (2009), working in Los Angeles, found that both frequent and rare floods were sensitive to urbanization.

A whole series of techniques have been developed in an attempt to reduce and delay urban storm runoff (Table 5.7). Of particular current interest are Sustainable Urban Drainage Systems (SUDS). These tackle urban surface runoff problems at source using features such as soakaways or infiltration basins (Brander et al., 2004), permeable pavements (Brattebo and Booth, 2003), grassed swales or vegetated filter strips, infiltration trenches, ponds (detention and retention basins) and wetlands to attenuate flood peak flows.

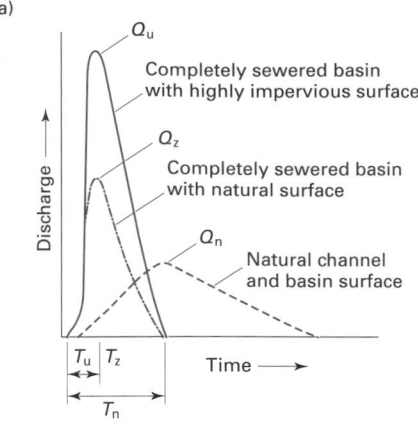

(a)

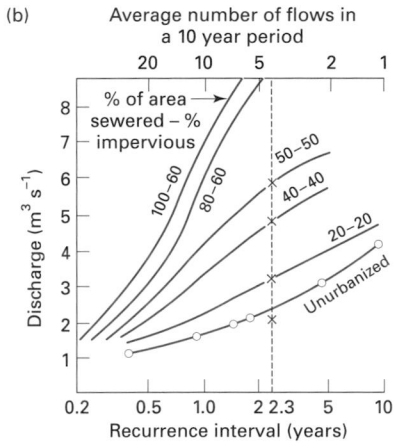

(b)

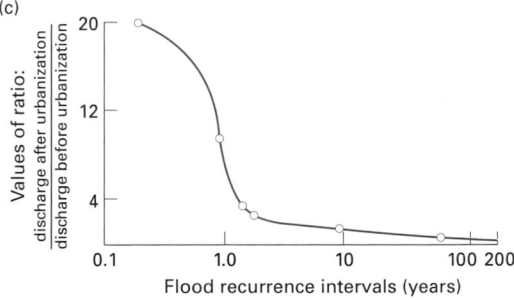

(c)

Figure 5.8 Some hydrological consequences of urbanization: (a) Effect of urban development on flood hydrographs. Peak discharges (Q) are higher and occur sooner after runoff starts (T) in basins that have been developed or sewered (modified from Fox, 1979, figure 3); (b) Flood frequency curves for a $1\,mi^2$ basin in various states of urbanization (modified from US Geological Survey in Viessman et al., 1977, figure 11.33); (c) Effects of flood magnitude of paving 20% of a basin (modified from Hollis, 1975).

Table 5.7 Measures for reducing and delaying urban storm runoff, including various types of Sustainable Urban Drainage Systems. Source: After US Department of Agriculture, Soil Conservation Service (1972) in Viessman et al. (1977: 569)

Area	Reducing runoff	Delaying runoff
Large flat roof	Cistern storage	Ponding on roof by constricted drainpipes
	Rooftop gardens	Increasing roof roughness
	Pool storage or fountain storage	(1) Rippled roof
	Sod roof cover	(2) Gravelled roof
Car parks	Porous pavement	Grassy strips on car parks
	(1) Gravel car parks	Grassed waterways draining car parks
	(2) Porous or punctured asphalt	Ponding and detention measures for impervious areas
	Concrete vaults and cisterns beneath car parks in high value areas	(1) Rippled pavement
	Vegetated ponding areas around car parks	(2) Depressions
	Gravel trenches	(3) Basins
Residential	Cisterns for individual homes or groups of homes	Reservoir or detention basin
	Gravel drives (porous)	Planting a high-delaying grass (high roughness)
	Contoured landscape	Grassy gutters or channels
	Groundwater recharge	Increased length of travel of runoff by means of gutters, diversions and so on
	(1) Perforated pipe	
	(2) Gravel (sand)	
	(3) Trench	
	(4) Porous pipe	
	(5) Dry wells	
	Vegetated depressions	
General	Gravel alleys	Gravel alleys
	Porous pavements	

Vegetation modification and its effect on river flow

As we have already noted in Chapter 1, one of the first major indications that humans could inadvertently adversely affect the environment was the observation that deforestation, produced by felling and/or fire (Scott, 1997) could create torrents and floods. Whether or not deforestation promotes serious flooding is,

however, the subject of continuing investigation and debate (Bradshaw et al., 2007).

The reasons why the removal of a forest cover and its replacement with pasture, crops or bare ground has such important effects on stream flow are many. First, forests greatly modify the accumulation and ablation of snow. Secondly, a mature forest probably has a higher rainfall interception rate, a tendency to reduce rates of overland flow, and probably generates soils with a higher infiltration capacity and better general structure. All these factors will tend to produce both a reduction in overall runoff levels and less extreme flood peaks.

The first experimental study, in which a planned land-use change was executed to enable observation of the effects of stream flow, began at Wagon Wheel Gap, Colorado, in 1910. Here stream flows from two similar watersheds of about 80 hectares each were compared for 8 years. One valley was then clear-felled and the records were continued. After the clear-felling the annual water yield was 17% above that predicted from the flows of the unchanged control valley. Peak flows are also increased. Studies on two small basins in the Australian Alps (Wallace's Creek, 41 km²; Yarrango Billy River, 224 km²), which were burned over, showed that rain storms, which from previous records would have been expected to give rise to flows of 60–80 m³ per second, produced a peak of 370 m³ per second (a five or sixfold increase). Likewise catchment experiments in Arizona have shown that when **chaparral** scrub is burned there is a tenfold increase in water yield.

Experiments with tropical catchments have shown typical maximum and mean annual stream-flow increases of 400–450 mm per year on clearance with increases in water yield of up to 6 mm per year for each percentage reduction in forest area above a 15% change in cover characteristics (Anderson and Spencer, 1991: 49). Additional data for the effects of land-use change on annual runoff levels are presented for tropical areas in Table 4.9. In central Queensland, Australia, Siriwardena et al. (2006) investigated the effects that clearance of Brigalow (*Acacia harpophylla*) forest in the 1960s and its replacement with grass and cropland had had on runoff. They found that runoff in the post-clearing period was greater by 58% than if clearing had not occurred.

As vegetation regenerates after a forest has been cut or burned, so stream flow tends to revert to normal,

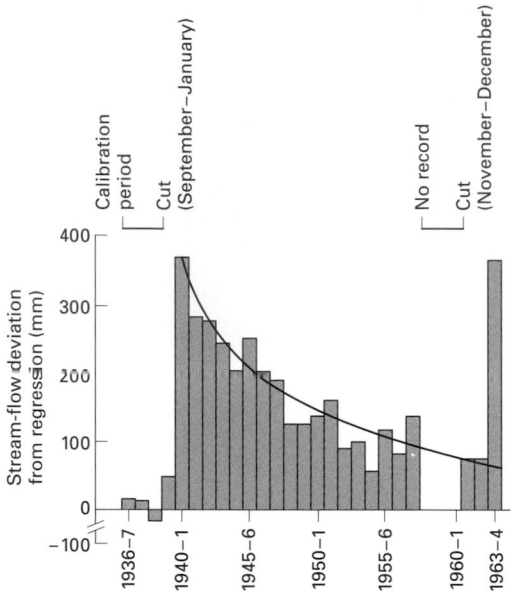

Figure 5.9 The increase of water yield after clear-felling a forest: a unique confirmation from the Coweeta catchment in North Carolina, USA (modified from Pereira, 1973, figure 8).

though the process may take some decades. This is illustrated in Figure 5.9 which shows the dramatic effects produced on the Coweeta catchments in North Carolina by two spasms of clear-felling, together with the gradual return to normality in between.

The substitution of one forest type for another may also affect stream flow. This can again be exemplified from the Coweeta catchments, where two experimental catchments were converted from a mature deciduous hardwood forest cover to a cover of white pine (*Pinus strobus*). Fifteen years after the conversion, annual stream flow was found to be reduced by about 20% (Swank and Douglass, 1974). The reason for this notable change is that the interception and subsequent evaporation of rainfall is greater for pine that it is for hardwoods during the dormant season.

Fears have also been expressed that the replacement of tall natural forests by eucalyptus will produce a decline in stream-flow. However, most current research does not support this contention, for transpiration rates from eucalyptus are similar to those from other tree species (except in situations with a shallow groundwater table) while their interception losses are if anything generally rather less than those from other

tree species of similar height and planting density (Bruijnzeel, 1990).

In many studies the runoff from clean-tilled land tends to be greater than that from areas under a dense crop cover. In the Upper Mississippi river basin in the USA, the tendency for the discharge of the river since the 1940s to increase more than can be explained by increasing precipitation alone, has been attributed to the great spread in soybean cultivation (Schilling et al., 2010).

However, tilling the soil surface does not always increase runoff (Gregory and Walling, 1973: 345). There were reports from the former Soviet Union suggesting that the reverse may be the case, and that autumn ploughing can decrease surface runoff, presumably because of its effect on surface detention and on soil structure. Indeed, with careful management the replacement of forest by other land-use types need not be detrimental in terms of either sediment loss or flood generation. In Kenya, for example, the tea plantations with shade trees, protective grass covers and carefully designed culverts were found to be 'a hydrologically effective substitute for natural forest' (Pereira, 1973: 127).

Infiltration capacities, which are so important in determining the nature of stream runoff, may be modified by various human activities that cause soil compaction, including livestock grazing and trampling (Trimble and Mendel, 1995), off-road vehicular movements (Webb, 1982) and the replacement of grasslands by shrublands (Bhark and Small, 2003). The relationships between grazing pressures and soil infiltration capacities are complex. On the one hand, moderate stocking levels may increase infiltration capacities by breaking down surface biological or rainbeat crusts, while on the other, high stocking levels may remove all vegetation cover, cause breakdown of soil aggregates, and produce severe trampling and soil compaction, thereby decreasing soil infiltration rates (du Toit et al., 2009).

The infiltration capacities of desert surfaces are highly variable, and one cause of such variability is the nature of vegetation cover, and humans have often, in areas like the western USA, caused grassland to be replaced by shrubland. The differences between grass and shrub surfaces have been the subject of an extensive literature (Ravi et al., 2009), and Parsons et al. (1996), for example, found that as a result of rainfall simulation experiments, compared to grasslands, the

Table 5.8 Impact of water use by invading alien plants on the mean annual runoff (MAR) in primary catchment areas of South Africa and Lesotho. Source: Le Maitre et al., (2000). Reproduced with permission

River system	Mean annual runoff (millions of m^3)	Condensed invaded area (hectares)	Incremental water use (millions of m^3)	Water use (% of MAR)	Reduction in rainfall equivalents (mm)
Limpopo	2381.82	122,457	190.38	7.99	155
Olifants	2904.10	217,855	290.44	10.00	133
Vaal	4567.37	64,632	190.53	4.17	295
Orange	7147.76	141,012	141.40	1.98	100
Olifants, Sout and Doring	1008.35	37,623	35.52	3.52	94
Namaqualand coast	25.01	46,618	22.76	91.00	49
W Cape and Agulhas coast	2056.75	384,636	646.50	31.43	168
Breede and Riversdale coast	2088.35	84,398	181.63	8.70	215
Gouritz	670.63	59,399	74.79	11.15	126
S Cape coast	1297.30	52,993	134.46	10.36	254
Gamtoos	494.71	34,289	96.53	19.51	282
PE Coast, Swartkops and Coega	150.04	11,358	40.18	26.78	354
Sundays	279.89	3964	8.34	2.98	210
Bushmans and Alexandria coast	172.92	22,894	73.08	42.26	319
Gt Fish	520.72	6980	21.12	4.06	303
Border Coast	578.91	12,483	55.58	9.60	445
Great Kei	1042.35	30,694	138.22	13.26	450
Former Transkei	7383.76	68,493	217.38	2.94	317
S KwaZulu-Natal	3121.20	46,442	126.37	4.05	272
Tugela	3990.88	62,151	104.67	2.62	168
N KwaZulu-Natal	4741.74	100,574	229.86	4.85	229
Komati to Nwanedzi	2871.4	124,494	283.26	9.86	228
RSA	49,495.96	1,736,438	3303.00	6.67	190

inter-rill portions of shrubland hillslopes had higher runoff rates, higher overland flow velocities and greater rates of erosion. In general, it can be argued that under a shrub canopy infiltration capacities will be relatively high because the addition of organic matter and the activity of roots increase soil porosity. On the other hand, the decreased infiltration capacity in intercanopy areas may more than offset higher infiltration capacities under the shrub canopies, with the net result that runoff from shrub-dominated hillslopes may be many times greater than those dominated by grassland. Furthermore, as Turnbull et al. (2010: 410) explain on the basis of their erosion and runoff plot studies in New Mexico:

Over the grassland to shrubland transition, the connectivity of bare areas where runoff tends to be preferentially generated increases. Therefore, from the grass, grass-shrub, shrub-grass to shrub plots, flow lines become increasingly well connected, which increases the capacity for flow to entrain and transport sediment, leading to the greater sediment yields monitored over shrubland.

In some parts of the world, such as southern Africa, the spread of invasive exotic plants (**see Chapter 2**) may cause greater loss of water then the native vegetation and so cause reduction in stream flow. Some calculations of this are presented in Table 5.8. The incremental water use of alien plants is estimated to be 3300 million m^3 per year, equivalent to a 190-mm reduction in rainfall and equivalent to almost three-quarters of the virgin mean annual runoff of the huge Vaal River (Le Maitre et al., 2000).

Changes in river-bank vegetation may have a particularly strong influence on river flow. In the south-west USA, for instance, many streams are lined by the salt cedar (*Tamarix pentandra*). With roots either in the water table or freely supplied by the capillary fringe, these shrubs have full potential transpiration opportunity. The removal of such vegetation can cause large increases in stream flow. It is interesting to note that the salt cedar itself is an alien, native to Eurasia, which was introduced by humans (Chew, 2009). It has spread explosively in the southwest USA, increasing from about 4000 hectares in 1920 to almost 400,000 hectares in the early 1960s (Harris, 1996). In the Upper Rio Grande valley in New Mexico, the salt cedar was introduced to try and combat anthropogenic soil erosion, but it spread so explosively that it came to consume approximately 45% of the area's total available water (Hay, 1973). Some recent studies, however, suggest that the hydrological impact of non-native riparian vegetation may not be as serious as has sometimes been proposed (e.g. Hultine and Bush, 2011).

A process which is often associated with afforestation is peat drainage. The hydrological effects of this are the subject of controversy, since there are cases of both increased and decreased flood peaks after drainage (Holden et al., 2004). It has been suggested that differences in peat type alone might account for the different effects. Thus it is conceivable that the drainage of a *Sphagnum* moss catchment would lead to increased flooding since *Sphagnum* compacts with drainage, reducing its storage volume and its permeability. On the other hand, in the case of non-*Sphagnum* peat there would be relatively less change in structure, but there would be a reduction in moisture content and an increase in storage capacity, thereby tending to reduce flood flows. The nature of the peat is, however, but one feature to consider (Robinson, 1979). The intensity of the drainage works (depth, spacing etc.) may also be important. In any case, there may be two (sometimes conflicting) processes operating as a result of peat drainage: the increased drainage network will facilitate rapid runoff, and the drier soil conditions will provide greater storage for rainfall. Which of these two tendencies is dominant will depend on local catchment conditions.

However, the impact of land drainage upon downstream flood incidence has long been a source of controversy. Much depends on the scale of study, the nature of land management and the character of the soil that has been drained. After a detailed review of experience in the UK, Robinson (1990) found that the drainage of heavy clay soils that are prone to prolonged surface saturation in their undrained state generally led to a reduction of large and medium flow peaks. He attributed this to the fact that their natural response is 'flashy' (with limited soil water storage available), whereas their drainage of permeable soils, which are less prone to such surface saturation, improves the speed of subsurface flow, thereby tending to increase peak flow levels.

The human impact on lake levels

One of the results of human modification of river regimes is that lake levels have suffered some change, though it is not always possible to distinguish between the part played by humans and that played by natural climatic changes.

A lake basin for which there are particularly long records of change is the Valencia Basin in Venezuela (Böckh, 1973). It was the declining level of the waters in this lake which so struck the great German geographer von Humboldt in 1800. He recorded its level as being about 422 m above sea level, and some previous observations on its level were made by Manzano in 1727, which established it as being at 426 m. The 1968 level was about 405 m, representing a fall of no less than 21 m in about 240 years (Figure 5.10). Humboldt believed that the cause of the declining level was the deforestation brought about by humans, and this has been supported by Böckh, (1973), who points also to the abstraction of water for irrigation. This remarkable fall in level meant that the lake ceased to have an overflow into the River Orinoco. It has as a consequence become subject to a build-up in salinity and is now eight times more saline than it was 250 years ago.

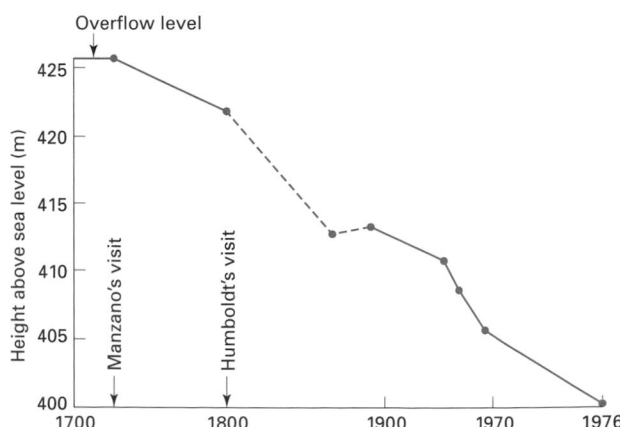

Figure 5.10 Variations in the level of Lake Valencia, Venezuela, to 1976 (modified from Böckh, 1973, figure 18.2, and other sources).

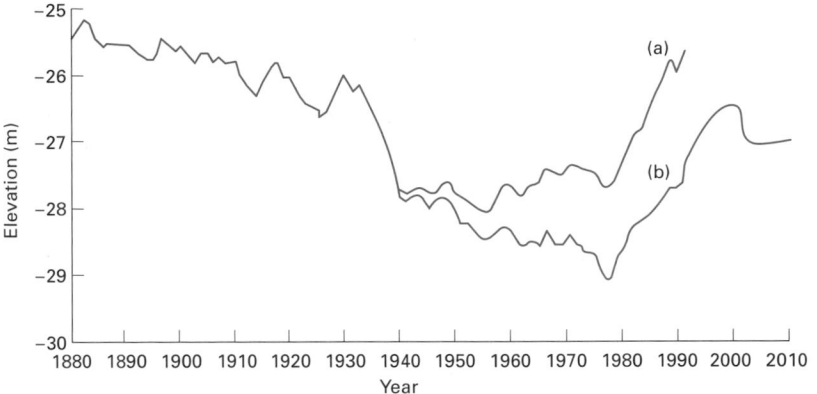

Figure 5.11 Annual fluctuations in the level of the Caspian Sea, for the period 1880–2010. Curve (a) shows the changes in level which would have occurred but for anthropogenic influences, while curve (b) shows the actual observed levels (modified from World Meteorological Organization, 1995, figure 15.3, and Sapozhnikov et al., 2010, figure 1).

Even the world's largest lake, the Caspian, has been modified by human activities. The most important change was the fall of 3 m in its level between 1929 and the late 1970s (Figure 5.11). This decline was undoubtedly partly the product of climatic change (Micklin, 1972), for winter precipitation in the northern Volga Basin, the chief flow-generating area of the Caspian, was generally below normal for that period because of a reduction in the number of moist cyclones penetrating into the Volga Basin from the Atlantic. None the less, human actions have contributed to this fall, particularly since the 1950s because of reservoir formation, irrigation, municipal and industrial withdrawals, and agricultural practices. In addition to the fall in level, salinity in the northern Caspian increased by 30% since the early 1930s. A secondary effect of the changes in level has been a decline in fish numbers due to the disappearance of the shallows. These are biologically the most productive zones of the lake, providing a food base for the more valuable types of fish and also serving as spawning grounds for some species. There are plans to divert some water from northward-flowing rivers in Siberia towards the Volga to correct the decline in Caspian levels, but the possible climatic impacts of such action have caused some concern. In any event, an amelioration of climate since the late 1970s has caused some 2 m of recovery in the level of the lake (Sapozhnikov et al., 2010).

Perhaps the most severe change to a major inland sea is that taking place in the Aral Sea (Figure 5.12) (Saiko and Zonn, 2000). Between 1960 and 1990, largely because of diversions of river flow, it had lost more than 40% of its area and about 60% of its volume, and its level had fallen by more than 14 m. By 2002 its level

had dropped another 6 m. By 2008, the area of the Aral Sea was just 15.7% of that in 1961 (Kravtsova and Tarasenko, 2010) and had become separated into four water bodies. This has lowered the artesian water table over a band 80–170 km in width, has exposed 24,000 km^2 of former lake bed to desiccation and has created salty surfaces from which salts are deflated to be transported in dust storms, to the detriment of soil quality. The mineral content of the Aral has increased by more than 20 fold (Micklin, 2010).

Water abstraction from the Jordan River has caused a decline in the level of the Dead Sea (Figure 5.13) (Maugh, 1979). During the last four decades increasing amounts of water have been diverted from surface and groundwater sources in its catchment. Under current conditions on average the annual inflow from all sources to the Dead Sea has been one-half to one-quarter that of the inflow prior to development. The water level has fallen about 20 m over that period (Figure 5.14).

From time to time there have been proposals for major augmentation of lake volumes, either by means of river diversions or by allowing the ingress of seawater through tunnels or canals. Among such plans have been those to flood the salt lakes of the Kalahari by transferring water from the rivers of central Africa such as the Zambezi; the scheme to transfer Mediterranean water to the Dead Sea and to the Quattara Depression in the eastern Sahara; and the Zaire-Chad scheme. Perhaps the most ambitious idea has been the so-called 'Atlantropa' project, whereby the Mediterranean would be empoldered, a dam built at the Straits of Gibraltar and water fed into the new lake from the Zaire River (Cathcart, 1983). There have also

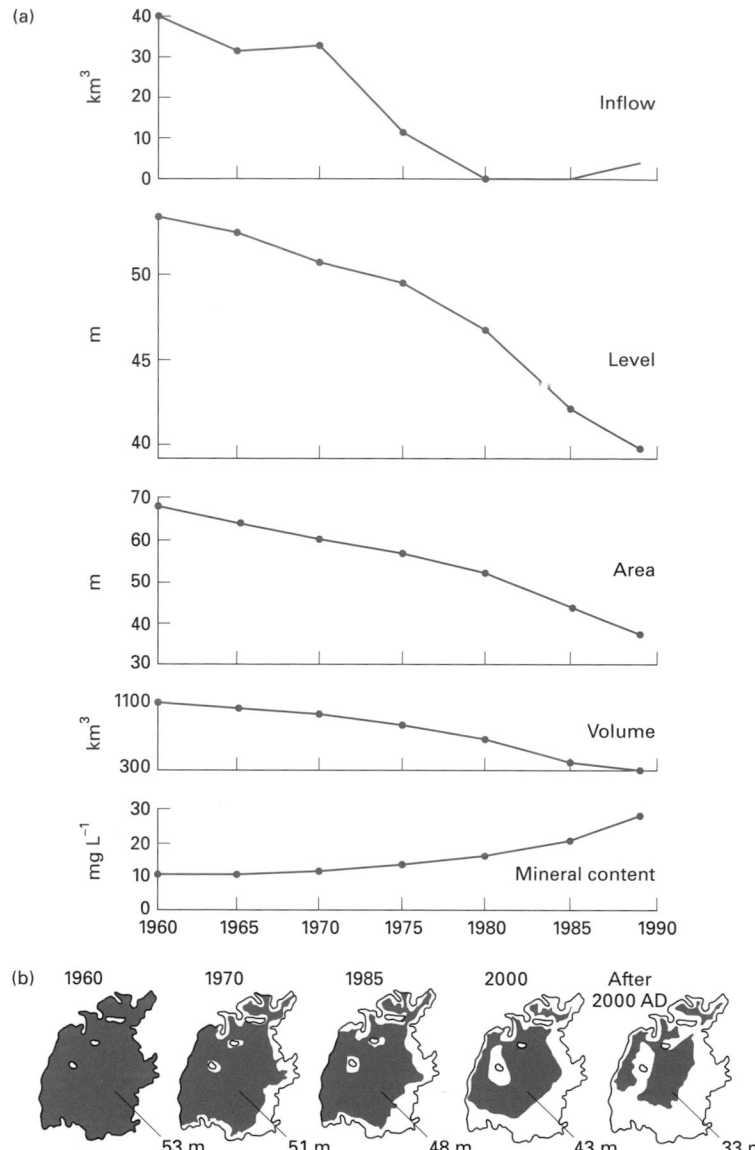

(a)

Inflow

Level

Area

Volume

Mineral content

(b)

1960   1970   1985   2000   After 2000 AD

53 m   51 m   48 m   43 m   33 m

Figure 5.12 Changes in the Aral Sea: (a) 1960–1989 (data from Kotlyakov, 1991) and (b) 1960 to after 2009 (after Micklin, 2010, figure 5). Reproduced with permission.

Figure 5.13 Recessional shorelines caused by the ongoing decline in the level of the Dead Sea, Jordan. (See Plate 35)

been proposals that the level of the Aral Sea could be restored by transferring water from the Black Sea into the Caspian and then into the Aral (Badescu and Cathcart, 2011).

Changes in groundwater conditions

In many parts of the world humans obtain water supplies by pumping from groundwater (Wada et al., 2010, 2012a, b). Now about 38% of the world's irrigated areas are groundwater-based and about 43% of irrigation water is derived from underground aquifers (Siebert et al., 2010). Groundwater is also important as a source of water for municipal consumption. Increas-

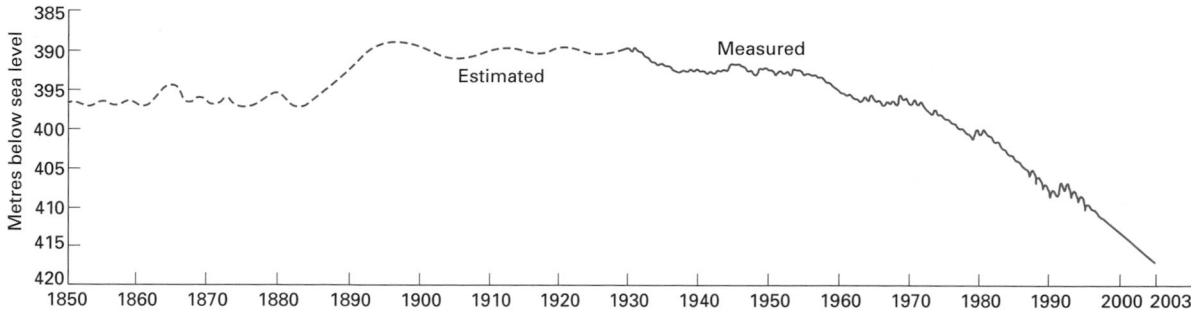

Figure 5.14 The level of the Dead Sea since 1850. Notice the 20-m fall since the 1960s.

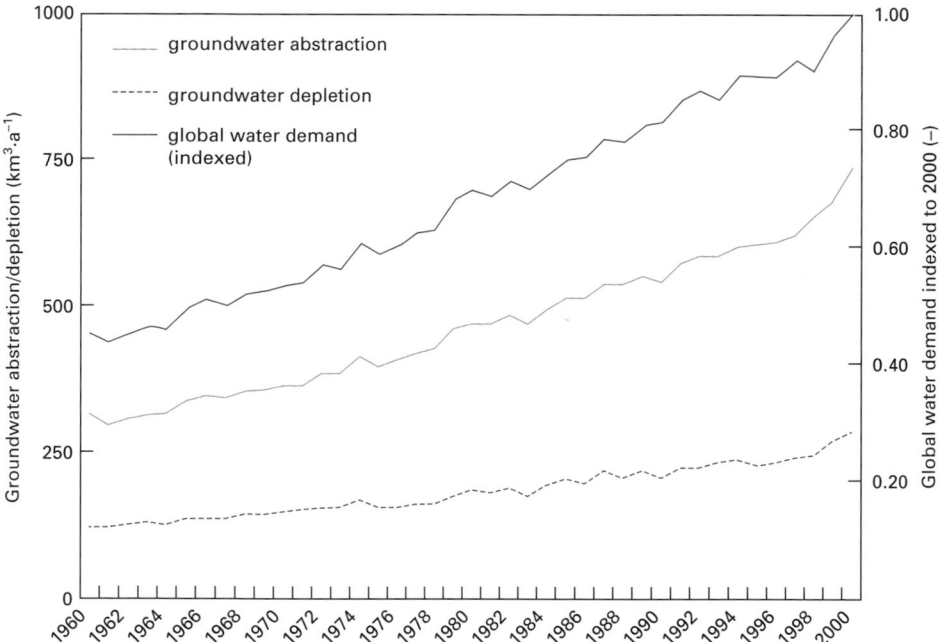

Figure 5.15 1960–2000 trends in total global water demand (right axis indexed for the year 2000), global groundwater abstraction and depletion (left axis) (modified from Wada et al., 2010, figure 3).

ing population levels and the adoption of new exploitation techniques (e.g. the replacement of irrigation methods involving animal or human power by electric and diesel pumps) has increased these problems.

Global groundwater abstraction increased from c. 312 km³ per year in 1960, to c. 734 km³ per year in 2000, and this abstraction exceeded recharge, so that over the same period groundwater reserves became depleted (Figure 5.15). This had two main effects: the reduction in the levels of water tables and the replacement, in coastal areas, of freshwater by saltwater (Barlow and Reichard, 2010). Environmental consequences of these two phenomena include ground subsidence and soil salinization.

Some of the reductions in groundwater levels that have been caused by abstraction are considerable. Figure 5.16a shows the rapid increase in the number of wells tapping groundwater in the London area from 1850 until after the Second World War, while Figures 5.16b illustrates the substantial changes in groundwater conditions that resulted. The **piezometric** surface in the confined chalk aquifer has fallen by more than 60 m over hundreds of square kilometres. Likewise, beneath Chicago, Illinois, pumping since the late nineteenth century has lowered the piezometric head by some 200 m. The drawdown that has taken place in the Great Artesian Basin of Australia locally exceeds 80–100 m (Lloyd, 1986).

(a)

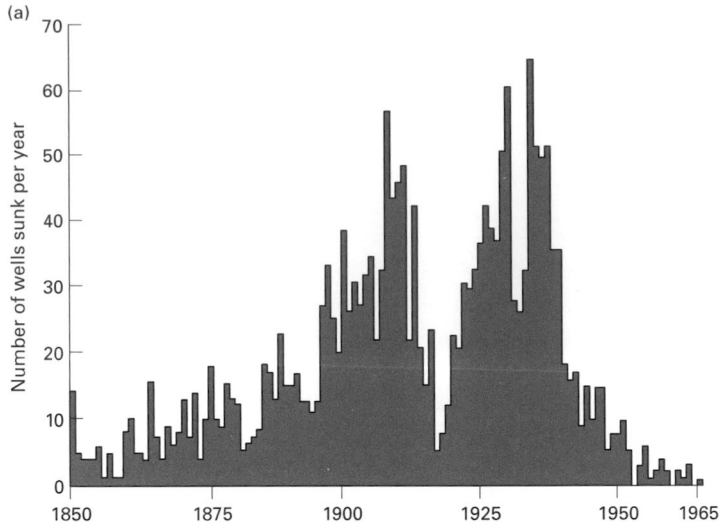

(b)

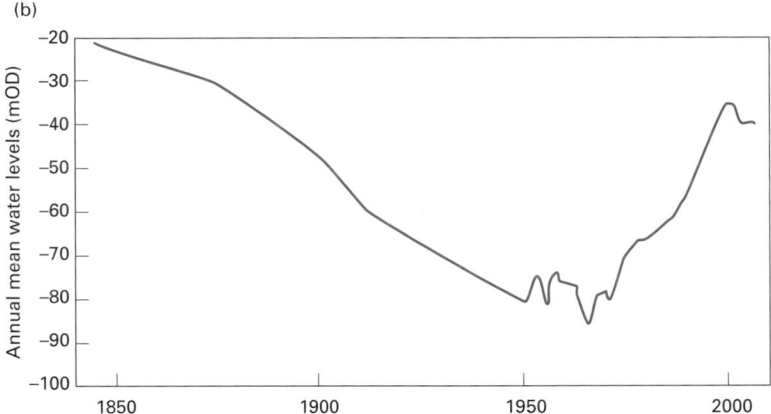

Figure 5.16 (a) Construction of wells tapping the confined aquifer below London, 1850–1965. (b) Changing groundwater levels beneath Trafalgar Square, Central London, from 1845 to 2005, showing the sharp decline until the 1950s and the 1960s, and the substantial rise since then (data from Centre for Ecology and Hydrology, Wallingford, DEFRA, 2004).

The reductions in water levels that are taking place in the High Plains from Nebraska to Texas are some of the most serious (Sophocleous, 2010; Scanlon et al., 2012) and threaten the long-term viability of irrigated agriculture in that area. Before irrigation development started in the 1930s, the High Plains groundwater system, one of the largest in the world and covering some 450,000 km² , was in a state of dynamic equilibrium, with long-term recharge equal to long-term discharge. However, the groundwater is now being mined at a rapid rate to supply centre-pivot (Figure 5.17) and other schemes. In a matter of only 50 years or less, the water level declined by 30–60 m, particularly in a large area to the north of Lubbock. This decline is ongoing (Figure 5.18). One consequence of this decline in water levels is that streams in the area now receive less groundwater input, so that their flows have declined (Kustu et al., 2010).

Figure 5.17 In the High Plains of the USA, fields are irrigated by centre-pivot irrigation schemes which use groundwater. Groundwater levels have fallen rapidly in many areas because of the adoption of this type of irrigation technology. (See Plate 36)

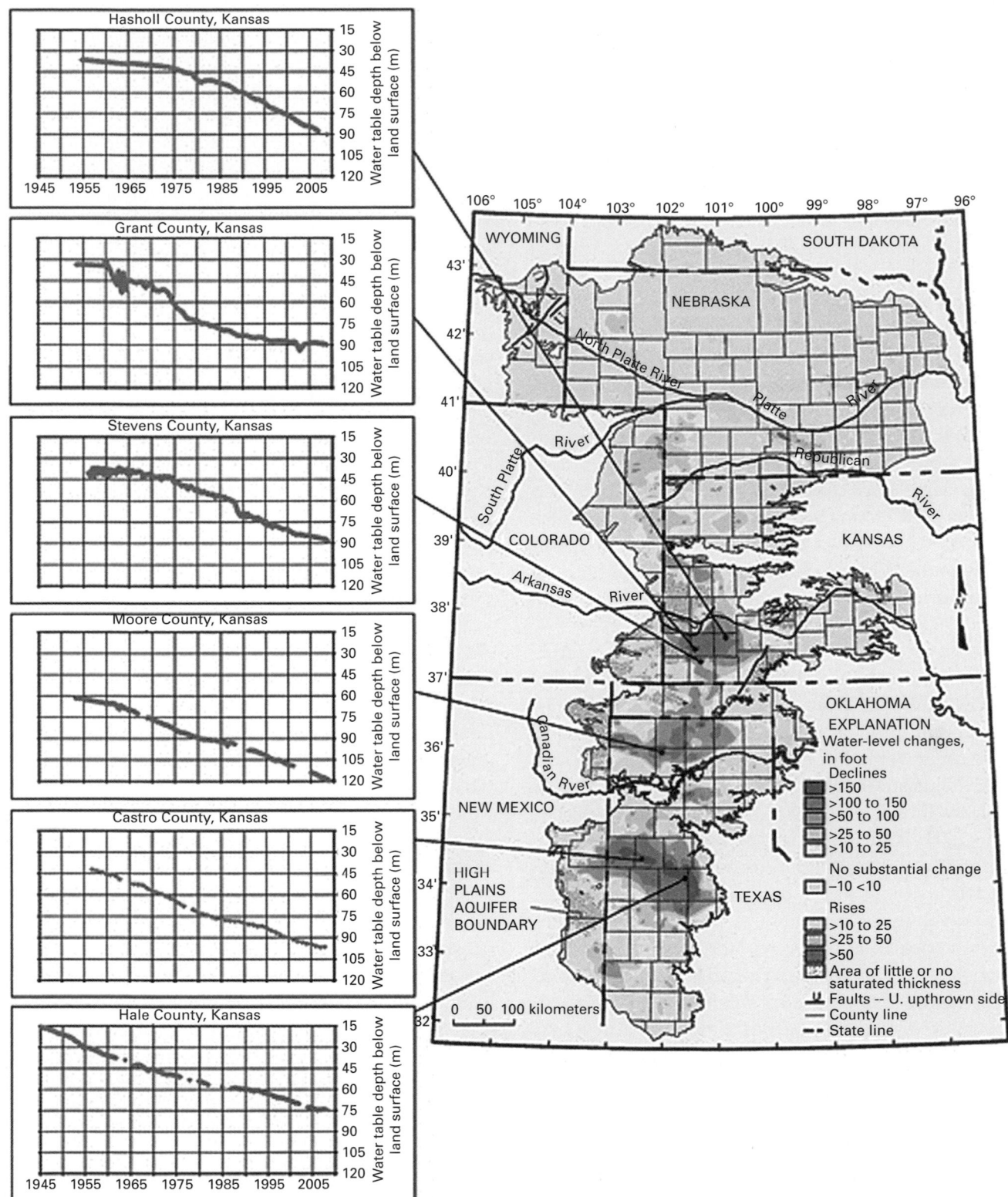

Figure 5.18 Water level change map of the High Plains aquifer from predevelopment to 2007 and selected water level hydrographs from predevelopment to 2008–2009 (modified from Sophocleous, 2010, figure 2). Reproduced with permission from Springer.

Another example of excessive 'mining' of a finite resource is the exploitation of groundwater resources in the oil-rich kingdom of Saudi Arabia. Most of Saudi Arabia is desert, so climatic conditions are not favourable for rapid large-scale recharge of aquifers. Also, much of the groundwater that lies beneath the desert is a fossil resource, created during more humid conditions – pluvials – that existed in the Late Pleistocene, between 15,000 and 30,000 years ago. In spite of these inherently unfavourable circumstances, Saudi Arabia's demand for water is growing inexorably as its economy develops. In 1980 the annual demand was 2.4 billion m^3. By 1990 it had reached 12 billion m^3 (a fivefold increase in just a decade), and by 2009 it had reached 18.7 billion m^3. Only a very small part of the demand can be met by runoff; over three-quarters of the supply is obtained from predominantly non-renewable groundwater resources. The drawdown on its aquifers is thus enormous. It was calculated that by 2010 the deep **aquifers** will contain 42% less water than in 1985. Much of the water is used ineffectively and inefficiently in the agricultural sector (Al-Ibrahim, 1991), to irrigate crops that could easily be grown in more humid regions and then imported.

The North China Plain aquifer is another example of over exploitation. A significant proportion of this shallow aquifer has seen water levels fall by 20 m since 1960, and in some parts the decline is over 40 m. Similarly severe depletion is also occurring in the Indus River Plains aquifer on the border between India and Pakistan (Rodell et al., 2009), and according to Tiwari et al. (2009: 1) this area has 'probably the largest rate of groundwater loss in any comparable-sized region on Earth'.

However, there are situations where humans deliberately endeavour to increase the natural supply of groundwater by attempting artificial recharge of groundwater basins (Peters, 1998). This has been practised, for example, in southeastern Florida (Barlow and Reichard, 2010). Where the materials containing the aquifer are permeable (as in some alluvial fans, coastal sand dunes or glacial deposits) the technique of water-spreading is much used. In relatively flat areas river water may be diverted to spread evenly over the ground so that infiltration takes place. Alternative water-spreading methods may involve releasing water into basins which are formed either by natural processes (like the playas of the High Plains in the USA),

or by excavation, or by the construction of dikes or small dams. On alluvial plains water can also be encouraged to percolate down to the water table by distributing it to a series of ditches or furrows. In some situations natural channel infiltration can be promoted by building small check dams down a stream course. In irrigated areas surplus water can be spread by irrigating with excess water during the dormant season. When artificial groundwater recharge is required in sediments with impermeable layers such water-spreading techniques are not effective and the appropriate method may then be to pump water into deep pits or into wells. This last technique is used on the coastal plain of Israel, both to replenish the groundwater reservoirs when surplus irrigation water is available and to attempt to diminish the problems associated with saltwater intrusion.

In some industrial areas, reductions in industrial activity have caused a recent reduction in groundwater abstraction, and as a consequence groundwater levels have begun to rise – a trend that is exacerbated by considerable leakage losses from ancient, deteriorating pipe and sewer systems. In cities like London, Liverpool and Birmingham an upward trend has already been identified (Brassington and Rushton, 1987; Price and Reed, 1989). In London, because of a 46% reduction in groundwater abstraction, the water table in the Chalk and Tertiary beds has risen by as much as 20 m. Such a rise has numerous implications, which are listed in Table 5.9. The trend in groundwater levels beneath Trafalgar Square are shown in Figure 5.16b.

There is, finally, a series of ways in which unintentionally changes in groundwater conditions can occur in urban areas. As we have clearly seen, in cities surface runoff is increased by the presence of impermeable surfaces. One consequence of this would be that less water went to recharge groundwater. However, there is an alternative point of view, namely that groundwater recharge can be accelerated in urban areas because of leaking water mains, sewers, septic tanks and soakaways (Figure 5.19). In cities in arid areas there is often no adequate provision for storm runoff, and the (rare) increased runoff from impermeable surfaces will infiltrate into the permeable surroundings. In some cities recharge may result from over-irrigation of parks and gardens. Indeed, where the climate is dry, or where large supplies of water are imported, or where pipes and drains are poorly maintained, groundwater recharge in urban areas is likely to exceed that in rural areas.

Water pollution

Water pollution is not new, but is frequently undesirable: it causes disease transmission through infection; it may poison humans and animals; it may create objectionable odours and unsightliness; it may be the cause of the unsatisfactory quality even of treated water; it may cause the eutrophication of water bodies, and it may affect economic activities like shellfish culture.

The causes and forms of water pollution created by humans are many and can be classified into groups as follows (after Strandberg, 1971):

1 sewage and other oxygen-demanding wastes;
2 infectious agents;
3 organic chemicals;
4 other chemical and mineral substances;
5 sediments (turbidity);
6 radioactive substances;
7 heat (thermal pollution).

Moreover, many human activities can contribute to changes in water quality, including agriculture, fire, urbanization, industry, mining, irrigation and many others. Of these, agriculture is probably the most important. Some pollutants merely have local effects, while others, such as acid rain or DDT, may have continental or even planetary implications.

It is also possible to categorize water pollutants according to whether or not they are derived from 'point' or 'non-point' (also called 'diffuse') sources.

Table 5.9 Possible effects of a rising groundwater level. Source: Modified from Wilkinson and Brassington (1991, table 4.3: 42)

Increase in spring and river flows
Re-emergence of 'dry springs'
Surface water flooding
Pollution of surface waters and spread of underground pollution
Flooding of basements
Increased leakage into tunnels
Reduction of slope and retaining wall stability
Reduction in bearing capacity of foundations and piles
Increased hydrostatic uplift and swelling pressures on foundations and structures
Swelling of clays
Chemical attack on foundations

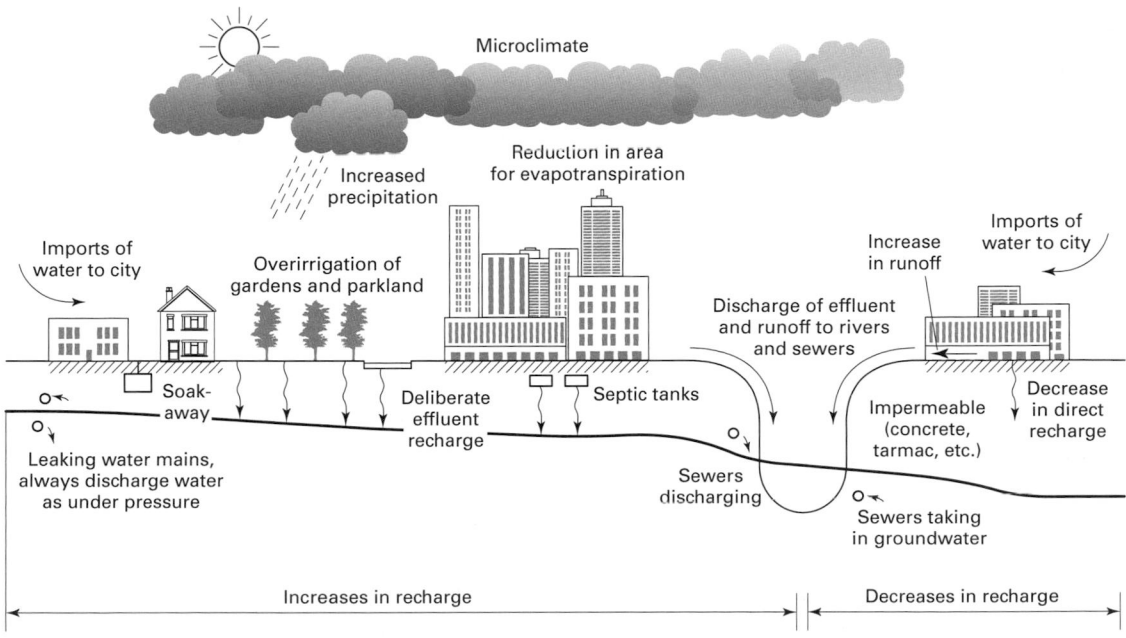

Figure 5.19 Urban effects on groundwater recharge (after Lerner, 1990, figure 2). Reproduced with permission.

Municipal and industrial wastes tend to fall into the former category because they are emitted from one specific and identifiable place (e.g. a sewage pipe or industrial outfall). Pollutants from non-point sources include agricultural wastes, many of which enter rivers in a diffuse manner as chemicals percolate into groundwater or are washed off into fields, as well as some mining pollutants, uncollected sewage and some urban storm-water runoff.

It is plainly not a simple matter to try to estimate the global figure for the extent of water pollution caused by humans. For one thing we know too little about the natural long-term levels of dissolved materials in the world's rivers. None the less, Meybeck (1979) calculated that about 500 million tonnes of dissolved salts reach the oceans each year as a result of human activity. These inputs have increased by more than 30% the natural values for sodium, chloride and sulfate, and have created an overall global augmentation of river mineralization by about 12%. Likewise, Peierls et al. (1991) demonstrated that the quantity of nitrates in the world's rivers now appears to be closely correlated to human population density. Using published data for 42 major world rivers they found a highly significant correlation between annual nitrate concentration and human population density that explained 76% of the variation in nitrate concentration for the 42 rivers. They maintain that 'human activity clearly dominates nitrate export from land.' Meybeck (2001a) argued that on a global scale, nutrient inputs to the oceans by rivers have already increased 2.2 times for nitrate and four times for ammonia.

There are three main classes of chemical pollutant that deserve particular attention: nitrates and phosphates; metals; and synthetic and industrial organic pollutants.

Nitrates and phosphates, trends in the concentration of which are reviewed by Heathwaite et al. (1996), are an important cause of a process called eutrophication. Nitrates normally occur in drainage waters and are derived from soil nitrogen, from nitrogen-rich glacial deposits and from atmospheric deposition. Anthropogenic sources include synthetic fertilizers, sewage and animal wastes from feedlots. Land-use changes (e.g. logging) can also increase nitrate inputs to streams. Phosphate levels are also rising in some parts of the world. Major sources include detergents, fertilizers and human wastes. On the other hand, in the UK, due to phosphorous stripping at sewage plants, phosphate concentrations in river waters appear to have gone down (Bowes et al., 2011), though in Europe more generally, in spite of EC regulations, nutrient concentrations have not shown any very great decline (Bouraouni and Grizzetti, 2011).

Metals, like nitrates and phosphates, occur naturally in soil and water. However, as the human use of metals has burgeoned, so has the amount of water pollution they cause. In addition, some metal ions reach river waters because they become more quickly mobilized as a result of acid rain. Aluminium is a notable example of this. From a human point of view, the metals of greatest concern are probably lead, mercury, arsenic and cadmium, all of which have adverse health effects. Other metals can be toxic to aquatic life, and these include copper, silver, selenium, zinc and chromium.

The anthropogenic sources of metal pollution include the industrial processing of ores and minerals, the use of metals, the leaching of metals from garbage and solid waste dumps, and animal and human excretions. Nriagu and Pacyna (1988) estimated the global anthropogenic inputs of trace metals into aquatic systems (including the oceans) and concluded that the sources producing the greatest quantities were, in descending order, the following (the metals produced by each source are listed in parentheses):

- domestic wastewater effluents (arsenic, chromium, copper, manganese, nickel);
- coal-burning power stations (arsenic, mercury, selenium);
- non-ferrous metal smelters (cadmium, nickel, lead, selenium);
- iron and steel plants (chromium, molybdenum, antimony, zinc);
- the dumping of sewage sludge (arsenic, manganese, lead).

However, in some parts of the world metal pollution may be derived from other sources. There is increasing evidence, for example, that in the western USA water derived from the drainage of irrigated lands may contain high concentrations of toxic or potentially toxic trace elements such as arsenic, boron, chromium, molybdenum and selenium. These can cause human health problems and poison fish and wildlife in desert wetlands (Lently, 1994).

Synthetic and industrial organic pollutants have been manufactured and released in very large quantities since the 1960s. The dispersal of these substances into watercourses has resulted in widespread environmental contamination. There are many tens of thousands of synthetic organic compounds currently in use, and many are thought to be hazardous to life, even at quite low concentrations – concentrations possibly lower than those that can routinely be measured by commonly available analytical methods. Among these pollutants are synthetic organic pesticides, including chlorinated hydrocarbon insecticides (e.g. DDT). Some of these can reach harmful concentrations as a result of biological magnification in the food chain. Other important organic pollutants include PCBs, which have been used extensively in the electrical industry as di-electrics in large transformers and capacitors; PAHs, which result from the incomplete burning of fossil fuels; various organic solvents used in industrial and domestic processes; phthalates, which are plasticizers used, for example, in the production of polyvinyl chloride resins; and DBPs, which are a range of disinfection by-products. The long-term health effects of cumulative exposure to such substances are difficult to quantify. However, some work suggests that they may be implicated in the development of birth defects and certain types of cancer.

Chemical pollution by agriculture and other activities

Agriculture may be one, if not the most, important cause of pollution, either by the production of sediments or by the generation of chemical wastes. With regard to the latter, it has been suggested that denitrification processes in the environment are incapable of keeping pace with the rate at which atmospheric nitrogen is being mobilized through industrial fixation processes and being introduced into the biosphere in the form of commercial fertilizers. Nitrogen, with phosphorus, tends to regulate the growth of aquatic plants and therefore the eutrophication of inland waters. Excess nitrates can also cause health hazards to humans and animals.

Eutrophication is the enrichment of waters by nutrients (Ryding and Rast, 1989). The process occurs naturally during, for example, the slow ageing of lakes, but it can be accelerated both by runoff from fertilized agricultural land and by the discharge of domestic sewage and industrial effluents (Lund, 1972). This process, often called 'cultural eutrophication', commonly leads to excessive growths of algae followed in some cases by a serious depletion of dissolved oxygen as the algae decay after death. Oxygen levels may become too low to support fish life, resulting in fish kills. Changes in diatom assemblages can also occur (see Battarbee, 1977). The nature of these changes is summarized in Table 5.10. Trends in lake eutrophication in the Anthropocene have been analysed by Keatley et al. (2011). Analysing data from Europe and North America since 1945 they have demonstrated that European lakes have undergone greater cultural eutrophication as indicated from diatom-inferred total phosphorus levels than North American ones. They attribute this to Europe's relatively longer history of intensive agriculture as well as to its higher current and historical population density. However, lake eutrophication is becoming evident in many parts of the world, and is even detectable in such large water

Table 5.10 Characteristics of lakes experiencing 'cultural eutrophication'. Source: Modified from Mannion (1992, table 11.4). Reproduced with permission

Biological factors
(a) *Primary productivity*: usually much higher than in unpolluted water and is manifest as extensive algal blooms.
(b) *Diversity of primary producers*: initially green algae increase, but blue-green algae rapidly become dominant and produce toxins. Similarly, macrophytes (e.g. reed maces) respond well initially, but due to increased turbidity and anoxia (see below) they decline in diversity as eutrophication proceeds.
(c) *Higher trophic level productivity*: overall decrease in response to factors given in this table.
(d) *Higher trophic level diversity*: decreases due to factors given in this table. The species of macro- and micro-invertebrates which tolerate more extreme conditions increase in numbers. Fish are also adversely affected and populations are dominated by surface dwelling coarse fish such as pike and perch.

Physical factors
(a) *Mean depth of water body*: as infill occurs the depth decreases.
(b) *Volume of hypolimnion*: varies.
(c) *Turbidity*: this increases, as sediment input increases, and restricts the depth of light penetration which can become a limiting factor for photosynthesis. It is also increased if boating is a significant activity.

bodies as Lake Malawi (Otu et al., 2011) and Lake Victoria (Sitoki et al., 2010) in Africa.

As agriculture has, in the developed world, become of an increasingly specialized and intensive nature, so the river pollution impact has increased. In the UK, a mix of factors has led to increases in river nitrate levels. The traditional mixed farm tended to be a more or less closed system which generated relatively few external impacts. This was because crop residues were fed to livestock or incorporated in the soil; and manure was returned to the land in amounts that could be absorbed and utilized. Many farms have become more specialized with the separation of crop and livestock activities, large numbers of stock may be kept on feedlots, silage may be produced in large silos, and synthetic fertilizers may be applied to fields in large quantities (Conway and Pretty, 1991). Some nitrate pollution may be derived from organic wastes. There has also been some anxiety in Britain over a possible decline in the amount of organic matter present in the soil, which could limit its ability to assimilate nitrogen. Moreover, the pattern of tillage may have affected the liberation of nitrogen via mineralization of organic matter. The increased area and depth of modern ploughing accelerates the decay of residues and may change the pattern of water movement in the soil. Finally, tile drainage has also expanded in area very greatly in recent decades in England. This has affected the movement of water through the soil and hence the degree of leaching of nitrates and other materials (Burt and Haycock, 1992).

Global trends in nitrate loadings of rivers are discussed by Green et al. (2004). Nitrate trends in most rivers in Europe and North America reveal a marked increase since the 1950s. This can be attributed to the growth in use of nitrate fertilizers (Meybeck, 2001b). Trends for the Mississippi, Danube, Rhine, Seine and Thames are shown in Figure 5.20. Trends in nitrogen and phosphorus concentrations in lowland regions of North America are not as great as those reported in lowland regions of Britain. This probably reflects the higher population density and greater intensity of land use in Britain (Heathwaite et al., 1996).

Although there are considerable fluctuations from year to year, the trend in nitrate levels in English rivers is clearly apparent (Royal Society Study Group, 1983). By 1980 they were 50–400% higher than they were 20 years before. The River Thames, which provides the

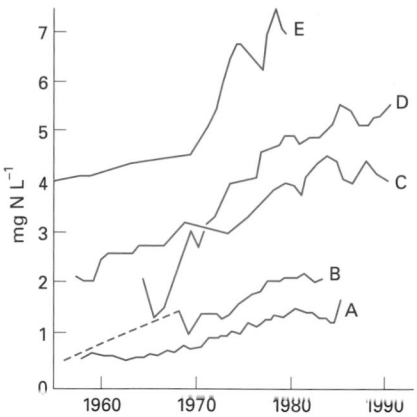

Figure 5.20 Recent trends of nitrate concentrations in some rivers. (A) Mississippi at mouth; (B) Danube at Budapest; (C) Rhine at the Dutch/German border; (D) Seine at mouth; (E) Thames at mouth (from Meybeck, 2001b, figure 17.6). Reproduced with permission from Freie Universtat Berlin.

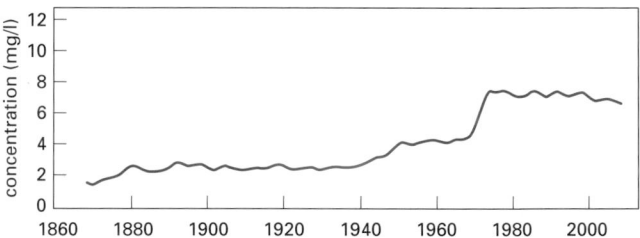

Figure 5.21 Time series of nitrate concentrations with an approximate 1-year moving average for the River Thames, England (after Howden et al., 2010, figure 1). Reproduced with permission.

major supply of London's water, increased its mean annual NO_3 concentration from around 11 mg per litre in 1928 to 35 mg per litre in the 1980s. There have been attempts in recent years to limit fertilizer application in what are called Nitrate Vulnerable Zones. Moreover, large-scale grassland to arable conversion and agricultural intensification has become less, so that less nitrate leaching consequent to ploughing takes place. Nitrate concentrations in many UK rivers now appear to have levelled off, though because of a lag effect, concentrations in groundwater-dominated watersheds may still be rising (Burt et al., 2010). Nonetheless, as Figure 5.21 shows, the Thames has yet to recover to any marked degree from the step-wise increase in nitrate concentrations that took place in the 1970s (Howden et al., 2010) and this is the case with many European rivers (Bouraouni and Grizzetti, 2011).

The trend in nitrate concentrations in British groundwaters is still a matter for concern. Investigations have revealed a large quantity of nitrate in the unsaturated zone of the principal aquifers (mainly Chalk and Triassic sandstone), and this is slowly moving down towards the main groundwater body. The slow transit time means that in many water supply wells increased nitrate concentrations will not occur for 20–30 years, but they will then be above acceptable levels for human health.

Given the uncertainties surrounding the question of nitrate pollution, and bearing in mind the major advances in agricultural productivity that they have permitted, attempts at controlling their use have not been received with complete favour. Indeed, as Viets (1971) pointed out, if fertilizer use were curtailed there would be less vegetation cover and increased erosion and sediment delivery to rivers. This, he suggested, would necessitate an increase in land area to maintain production levels which would also cause greater erosion; at the same time the increased ploughing would lead to greater nitrate loss from grasslands. Nevertheless some farmers are inefficient and wasteful in their use of fertilizers and there is scope for economy.

In addition to the influence of nitrate fertilizers, the confining of animals on feedlots results in tremendously concentrated sources of nutrients and pollutants. Animal wastes in the USA are estimated to be as much as 1.6 billion tonnes per year, with 50% of this amount originating from feedlots. U.S. animal feedlots produce 100,000 metric tonnes of manure per minute (Clay, 2004). It needs to be remembered that 1000 head of beef produce the equivalent organic load of 6000 people (Sanders, 1972) and that the amount of solid manure produced by U.S. farm animals is 130 times the amount produced by the human population. Cattle feedlot runoff is a high-strength organic waste, high in oxygen-demanding material and has caused many fish kills in rivers in states such as Kansas. In North Carolina and Virginia, industrial style hog and poultry operations have caused large waste spills, especially during severe storms, and these too have caused massive fish kills.

One very important way of reducing the input of pollutants to stream courses is to provide a buffer zone of vegetation between them and the fields from which the pollution comes. Alternatively, bans can be instituted to prevent fertilizer applications in particularly sensitive zones with respect to pollutants such as nitrates.

Pesticides are another source of chemical pollution brought about by agriculture and which can have adverse effects on fish populations (Scholz et al., 2012). There is now a tremendous range of pesticides, and they differ greatly in their mode of action, in the length of time they remain in the biosphere and in their toxicity. Much of the most adverse criticism of pesticides has been directed against the chlorinated hydrocarbon group of insecticides, which includes DDT and Dieldrin. These insecticides are toxic not only to the target organism but to other insects too (that is, they are non-specific). They are also highly persistent. Appreciable quantities of the original application may survive in the environment in unaltered form for years. This can have two rather severe effects: global dispersal and the 'biological magnification' of these substances in food chains. This last problem is well illustrated for DDT in Figure 5.22.

Changes in surface-water chemistry may be produced by the washing out of acids that pollute the air in precipitation. This is particularly clear in the heavily industrialized area of north-western Europe, where there is a marked zonation of acid rain with pH values that often fall below 4 or 5. A broadly comparable situation exists in North America (Johnson, 1979). Very

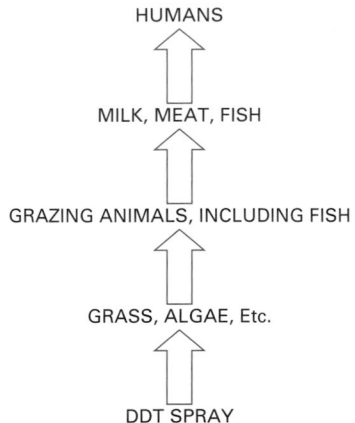

Figure 5.22 Biological concentration occurs when relatively indestructible substances (e.g. DDT) are ingested by lesser organisms at the base of the food pyramid. An estimated 1000 kg of plant plankton are needed to produce 100 kg of animal plankton. These in turn are consumed by 10 kg of fish, the amount needed by one person to gain 1 kg. The ultimate consumer (man or woman) then takes in the DDT taken in by the 1000 kg of the lesser creatures at the base of the food pyramid, when he or she ingests enough fish to gain 1 kg.

considerable fears have been expressed about the possible effects that acid deposition may have on aquatic ecosystems and particularly on fish populations. It is now generally accepted that decades of increasing lake and river acidification caused fish kills and stock depletion. Fishless lakes now occur in areas like the Adirondacks in the north-east USA. Species of fish vary in their tolerance to low pH, with rainbow trout being largely intolerant to values of pH below 6.0, while salmon, brown and brook trout are somewhat less sensitive. Values of 4.0–3.5 are lethal to salmonids, while values of 3.5 or less are lethal to most fish. Numerous lacustrine molluscs and crustaceans are not found even at weakly acid pH values of 5.8–6.0, and crayfish rapidly lose their ability to recalcify after moulting as the pH drops from 6.0 to 5.5 (Committee on the Atmosphere and the Biosphere, 1981).

Fish declines and deaths in areas afflicted by acid rain are not solely caused by water acidity *per se*. It has become evident that under conditions of high acidity certain metal ions, including aluminium, are readily mobilized, leading to their increasing concentration in freshwaters. Some of these metal ions are highly toxic and many mass mortalities of fish have been attributed to aluminium poisoning rather than high acidity alone. The aluminium adversely affects the operation of fish gills, causing mucus to collect in large quantities; this eventually inhibits the ability of the fish to take in necessary oxygen and salts. In addition, aluminium's presence in water can reduce the amount of available phosphates, an essential food for phytoplankton and other aquatic plants. This decreases the available food for fish higher in the food chain, leading to population decline (Park, 1987).

Fortunately, however, during the 1980s and 1990s rates of acid deposition decreased over large portions of North America and Europe (Stoddart et al., 1999; Cooper and Jenkins, 2003), and some recovery in streams and lakes has resulted. Reversal of acidification, while not universal, has been noted, for example, in the English Lake District (Tipping et al., 2000), in Scandinavia (Fölster and Wilander, 2002), in Finland (Vuorenmaa and Forsius, 2008), in central Europe (Vesely et al., 2002) and in the English Pennines (Evans and Jenkins, 2000). In Europe, recovery has been strongest in the Czech Republic and Slovakia, moderate in Scandinavia and the UK, and weakest in Germany (Evans et al., 2001). The period required for recovery may in some areas by considerable. In the Adirondacks of the eastern USA recovery may take many decades because, on the one hand, they have been exposed to acid deposition for long periods of time so that they have been relatively depleted of substances that can neutralize acids, and on the other, their soils are thin and so can offer less material to neutralize the acid from precipitation (US General Accounting Office, 2000).

Mining can also create serious chemical pollution by creating what is known as acid mine drainage (AMD) or acid rock drainage (ARD). In the western USA, the forest service estimates that between 20,000 and 50,000 mines are currently generating acid on forest service lands, impacting upon 8000–16,000 km of stream. In the eastern USA more than 7000 km of stream are affected by acid drainage from coal mines (US Environmental Protection Agency, 1994).

In Britain, coal seams and mudstones in the coal measures contain pyrite and marcasite (ferrous sulfide). In the course of mining the water table is lowered, air gains access to these minerals and they are oxidized. Sulfuric acid may be produced. Should mining cease, as is the situation in the coalfields of Britain, pumping may stop, groundwater may rise through the worked mine, and acid derived from it may discharge into more and more surface waters. Acidity may also be produced by the mining of other minerals that contain sulfide, including zinc and gold, as in South Africa (McCarthy, 2011).

As a consequence, mine drainage waters may be very acid and have pH values as low as 1 or even less (Romero et al., 2011; Tao et al., 2012). They have high sulfate and iron concentrations, they may contain toxic metals such as arsenic and cadmium and because of the reaction of the acid on clay and silicates in the rocks they may also contain appreciable amounts of calcium, magnesium, aluminium and manganese. Following reactions with sediments or mixing with alkaline river waters, or when chemical or bacterial oxidation of ferrous compounds occurs, the iron may precipitate as ferric hydroxide. This may discolour the water and leave unsightly deposits. Some of the reactions involved in the development of acid rock drainage or acid mine drainage have been set down by Down and Stocks (1977: 110–111).

The nature of acid drainage from mines and the remedial measures, such as lime neutralization and the use of aerobic wetlands, that can be adopted to deal with it are discussed by Robb and Robinson (1995) and Johnson and Hallberg (2005), while the adverse effects on stream organisms are discussed by Hogsden and Harding (2012). Affected soils may prevent growth of vegetation creating 'AMD barrens' (Lupton et al., 2012).

Rock salt (sodium chloride) has been used in increasing quantities since the Second World War for minimising the dangers to motorists and pedestrians from icy road and pavements. With the rise in the number of vehicles there has been a tendency throughout Europe and North America for a corresponding increase in the use of salt for de-icing purposes (see e.g. Howard and Beck, 1993). Data for the UK between 1960 and 1991 are shown in Figure 5.23 and demonstrate that while there is considerable inter-annual variability (related to weather severity), there has also been a general upward trend so that in the mid-1970s to end-1980s more than 2 million tonnes of de-icing salt were being purchased each year (Dobson, 1991). Data on de-icing salt application in North America are given by Jackson and Jobbágy (2005). The total use of de-icing salts in the USA increased at a nearly exponential rate between the 1940s and the 1970s, increasing from about 200,000 metric tonnes in 1940 to approximately 9,000,000 metric tonnes in 1970. This represented a doubling time of about 5 years. By the early 2000s the annual rate exceeded 18 million metric tonnes (Figure 5.24). Runoff from salted roads can cause ponds and lakes (Novotny et al., 2008) to become salinized, concrete to decay and susceptible trees to die. It has also caused a considerable increase in the salinity of streams in the north eastern United States. Kaushal et al. (2005) found that some streams in Maryland, New Hampshire and New York had winter chloride concentrations of up to 25% of the concentration of seawater and that summer chloride concentrations remained up to 100 times greater than unimpacted forest streams.

Similarly, storm-water runoff from urban areas may contain large amounts of contaminants, derived from litter, garbage, car-washings, horticultural treatments, vehicle drippings, industry, construction, animal droppings and the chemicals used for snow and ice clearance. Comparison of contaminant profiles for urban runoff and raw domestic sewage, based on surveys throughout the USA, indicates the importance of pollution from urban runoff sources (Table 5.11). In New York City some half million dogs leave up to 20,000 tonnes of pollutant faeces and up to 3.8 million litres of urine in the city streets each year, all of which is flushed by gutters to storm-water sewers. Taking

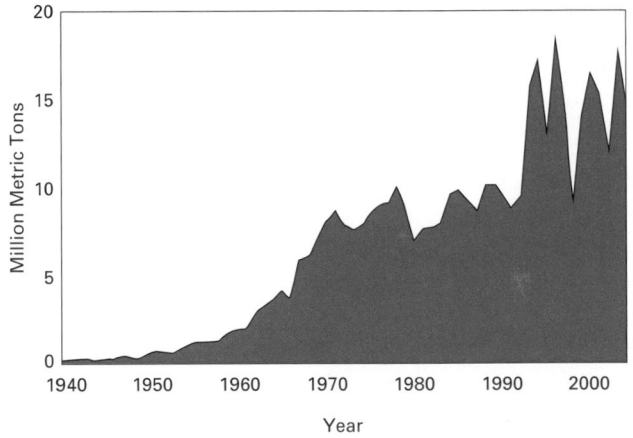

Figure 5.24 Sales of rock salt for highway use in the USA from 1940 to 2004 (modified from Jackson and Jobbágy, 2005, figure 1).

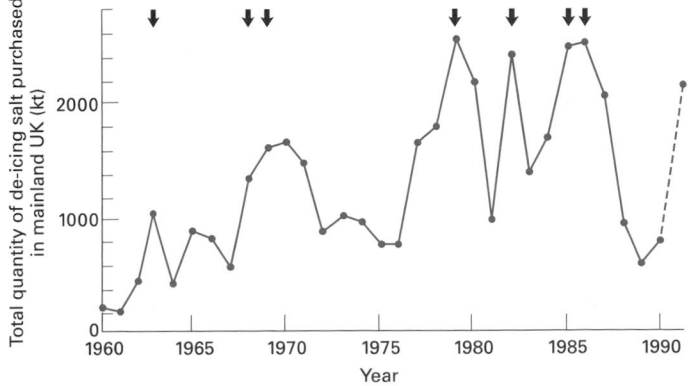

Figure 5.23 Estimates of the total quantity of de-icing salt purchased annually in mainland Britain during the period 1960–1991. Arrows highlight severe winters. Data provided by ICI in Dobson (1991, figure 1.1). Reproduced with permission.

Table 5.11 Comparison of contaminant profiles for urban surface runoff and raw domestic sewage, based on surveys throughout the USA. Source: Burke (1972, table 7.36.1). Reproduced with permission

Constituent[a]	Urban surface runoff	Raw domestic sewage
Suspended solids	250–300	150–250
BOD[b]	10–250	300–350
Nutrients		
(a) Total nitrogen	0.5–5.0	25–85
(b) Total phosphorus	0.5–5.0	2–15
Coliform bacteria (MPN/100 mL)	10^4–10^6	10^6 or greater
Chlorides	20–200	15–75
Miscellaneous substances		
(a) Oil and grease	Yes	Yes
(b) Heavy metals	(10–100) times sewage concentration	Traces
(c) Pesticides	Yes	Seldom
(d) Other toxins	Potential exists	Seldom

[a]All concentrations are expressed in mg/L unless stated otherwise.
[b]Biochemical oxygen demand.
MPN, most probable number.

Britain as a whole, dogs deposit 1000 tonnes of excrement and three million gallons of urine on the streets every day (Ponting, 1991).

None the less, it would be unfair to create the impression that water pollution is either a totally insoluble problem or that levels of pollution must inevitably rise. It is true that some components of the hydrological system – for example, lakes, because they may act as sumps – will prove relatively intractable to improvement, but in many rivers and estuaries striking developments have occurred in water quality in recent decades. This is clear in the case of the River Thames, which suffered a serious decline in its quality and in its fish as London's pollution and industries expanded. However, after about 1950, because of more stringent controls on effluent discharge, the downward trend in quality was reversed (see Figure 5.25), and many fish, long absent from the Thames, returned (Gameson and Wheeler, 1977). Stringent controls on phosphorus pollution have been highly effective in improving the quality of lake waters in parts of the UK (May et al., 2012). There also appears to have been some improvement in pollution levels in the North American Great Lakes since the mid-1970s as indicated by phosphorus loadings and DDT and PCB levels (Figure 5.26).

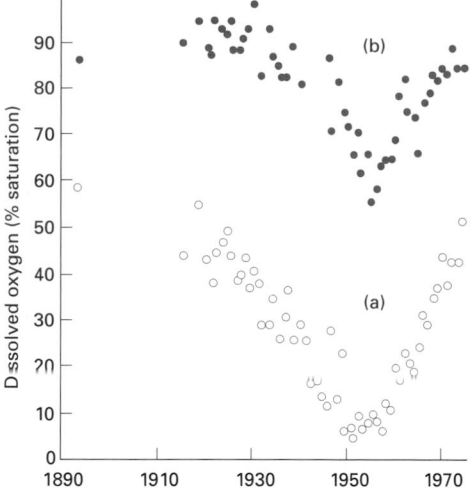

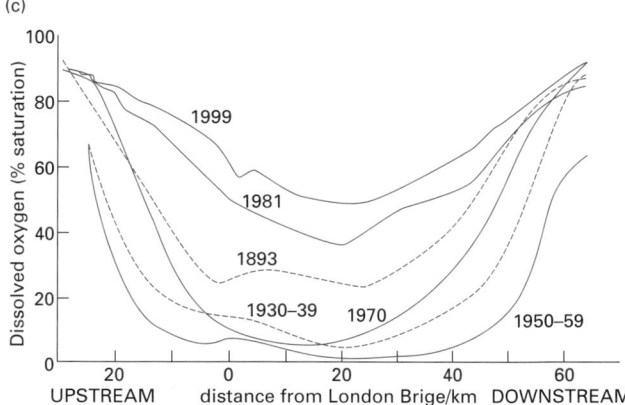

Figure 5.25 The average dissolved oxygen content of the River Thames at half-tide in the July–September quarter since 1890: (a) 79 km below Teddington Weir; (b) 95 km below Teddington Weir (modified from Gameson and Wheeler, 1977, figure 4); (c) the pattern from 1893 to 1999.

Deforestation and its effects on water quality

The removal of a tree cover not only may affect river flow and stream-water temperature, it may also cause changes in stream-water chemistry. The reason for this is that in many forests large quantities of nutrients are cycled through the vegetation, and in some cases (notably in humid tropical rain forest) the trees are a great store of the nutrients. If the trees are destroyed the cycle of nutrients is broken and a major store is disrupted. The nutrients thus released may become available for crops, and shifting cultivators, for example,

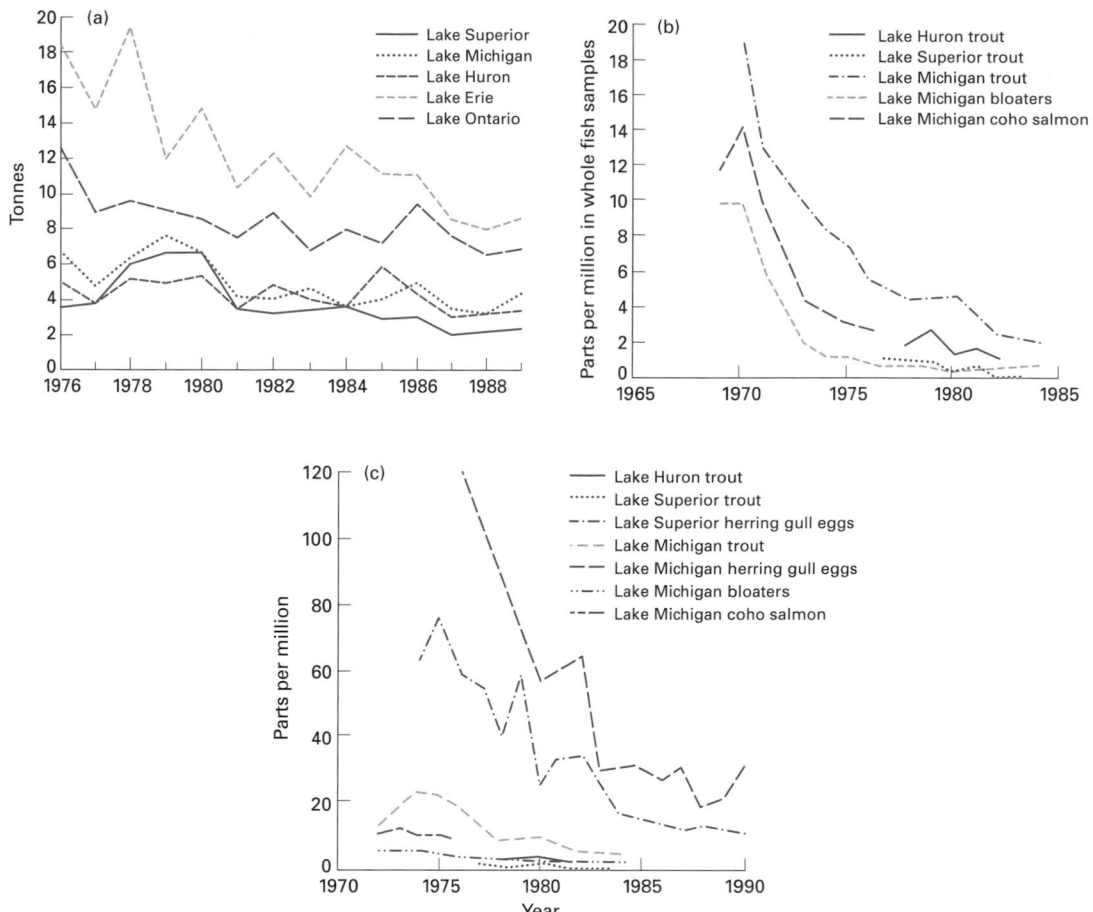

Figure 5.26 Changes in water pollution in the Great Lakes of North America: (a) Estimated phosphorus loadings; (b) DDT levels; (c) PCB levels (after Council on Environmental Quality, 17th (1986) and 22nd (1992) Annual Reports).

may utilize this fact and gain good crop yields in the first year after burning and felling forest. Some of the nutrients may, however, be leached out of the soils and appear as dissolved load in streams. Large increases in dissolved load may have undesirable effects, including eutrophication, soil salinization and a deterioration in public water supply.

A classic but, it must be added, extreme exemplification of this is the experiment that was carried out in the Hubbard Brook catchments in New Hampshire, USA (Bormann et al., 1968). These workers monitored the effects of 'savage' treatment of the forest on the chemical budgets of the streams. The treatment was to fell the trees, leave them in place, kill lesser vegetation and prevent re-growth by the application of herbicide. The effects were dramatic: the total dissolved inorganic material exported from the basin was about

fifteen times larger after treatment, and, in particular, there was a very substantial increase in stream nitrate levels. Their concentration increased by an average of fifty times.

It has been argued, however, that the Hubbard Brook results are atypical (Sopper, 1975) because of the severity of the treatment. Most other forest clear-cutting experiments in the US (Table 5.12) indicate that nitrate-nitrogen additions to stream water are not so greatly increased. Under conventional clear-cutting the trees are harvested rather than left, all saleable material is removed, and encouragement is given to the establishment of a new stand of trees. Aubertin and Patric (1974), using such conventional clear-cutting methods rather than the drastic measures employed in Hubbard Brook, found that clear-cutting in their catchments in West Virginia had a negligible effect on stream

Table 5.12 Nitrate-nitrogen losses from control and disturbed forest ecosystems. Source: Modified from Vitousek et al. (1979)

Site	Nature of disturbance	Nitrate-nitrogen loss (kg/ha/y)	
		Control	Disturbed
Hubbard Brook	Clear-cutting without vegetation removal, herbicide inhibition of re-growth	2.0	97
Gale River (New Hampshire)	Commercial clear-cutting	2.0	38
Fernow (West Virginia)	Commercial clear-cutting	0.6	3.0
Coweeta (North Carolina)	Complex	0.05	7.3[a]
H. J. Andrews Forest (Oregon)	Clear-cutting with slash-burning	0.08	0.26
Alsea River (Oregon)	Clear-cutting with slash-burning	3.9	15.4
Mean		1.44	26.83

[a]This value represents the second year of recovery after a long-term disturbance. All other results for disturbed ecosystems reflect the first year after disturbance.

temperatures (they left a forest strip along the stream), pH and the concentration of most dissolved solids.

Even if clear-cutting is practised, there is some evidence of a rapid return to steady-state nutrient cycling because of quick regeneration. As Marks and Bormann (1972) have put it:

Because terrestrial plant communities have always been subjected to various forms of natural disturbances, such as wind storms, fires, and insect outbreaks, it is only reasonable to consider recovery from disturbance as a normal part of community maintenance and repair.

Likewise Hewlett et al. (1984) found no evidence in the Piedmont region of the USA that clear-cutting would create such extreme nitrate loss that soil fertility would be impaired or water eutrophication caused. They pointed out that rapid growth of vegetation minimizes nutrient losses from the ecosystem by three main mechanisms. It channels water from runoff to evapotranspiration, thereby reducing erosion and nutrient loss; it reduces the rates of decomposition of organic matter through moderation of the microclimate so that the supply of soluble ions for loss in drainage water is reduced; and

it causes the simultaneous incorporation of nutrients into the rapidly developing biomass so that they are not lost from the system.

Thermal pollution

The pollution of water by increasing its temperature is called thermal pollution. Many organisms are affected by temperature, so this environmental impact has some significance (Langford, 1990 for metabolic rates, physiology, life history traits, nutrient cycling, productivity and habitat suitability (Poole and Berman, 2001). An increase in water temperature causes a decrease in the solubility of oxygen which is needed for the oxidation of biodegradable wastes. At the same time, the rate of oxidation is accelerated, imposing a faster oxygen demand on the smaller supply and thereby depleting the oxygen content of the water still further. Temperature also affects the lower organisms, such as plankton and crustaceans. In general, the more elevated the temperature is, the less desirable the types of algae in water. In cooler waters diatoms are the predominant phytoplankton in water that is not heavily eutrophic; with the same nutrient levels, green algae begin to become dominant at higher temperatures and diatoms decline; at the highest water temperatures, blue-green algae thrive and often develop into heavy blooms. One further ecological consequence of thermal pollution is that the spawning and migration of many fish are triggered by temperature and this behaviour can be disrupted by thermal change. On the other hand, the temporary shutdown of power plants may create severe cold-shock kills of fish in discharge-receiving waters in winter (Clark, 1977).

In industrial countries probably the main source of thermal pollution is from condenser cooling water released from electricity generating stations. Water discharged from power stations has been heated some 6–9°C, but usually has a temperature of less than 30°C. The extent to which water affects river temperature depends very much on the state of flow. For example, below the Ironbridge power station in England, the Severn River undergoes a temperature increase of only 0.5°C during floods, compared with an 8°C increase at times of low flow (Rodda et al., 1976).

Over the past few decades both the increase in the capacity of individual electricity power-generating units and the improvements in their thermal efficiency have led to a diminution in the heat rejected in relation to the amount of cooling water per unit of production. Economic optimization of the generating plant has cut down the flow of cooling water per unit of electricity, though it has raised its temperature. This increased temperature rise of the cooling water is more than offset by the reduction achieved in the volume of water utilized. None the less, the expansion of generating capacity has meant that the total quantity of heat discharged has increased, even though it is much less than would be expected on a simple proportional basis.

Thermal pollution of streams may also follow from urbanization (Kinouchi et al., 2007). This results from various sources: wastewater from municipal treatment plants; changes in the temperature regime of streams brought about by reservoirs according to their size, their depth and the season; changes produced by the urban heat island effect and the spread of paved surfaces (Herb et al., 2008); changes in the configuration of urban channels (e.g. their width-depth ratio); changes in the degree of shading of the channel, either by covering it over or by removing natural vegetation cover Hester and Doyle, 2011); changes in the volume of storm runoff; reduction in flow because of water withdrawals; and changes in the ground-water contribution. Pluhowski (1970) found that the basic effects of cities on river-water temperatures on Long Island, New York, was to raise temperatures in summer (by as much as 5–8°C) and to lower then in winter (by as much as 1.5–3.9°C).

Reservoir construction can also affect stream-water temperatures. Crisp (1977) found, for example, that as a result of the construction of a reservoir at Cow Green (Upper Teesdale, northern England) the temperature of the river downstream was modified. He noted a reduction in the amplitude of annual water-temperature fluctuations, and a delay in the spring rise of water temperature by 20–50 days, and in the autumn fall by up to 20 days. In the Ebro valley of Spain, Prats et al. (2010) found that as a result of reservoir construction and emissions of warm water from a nuclear plant, a increase in mean annual water temperature of 2.3°C occurred.

In rural areas human activities can also cause significant modifications in river-water temperature. Deforestation is especially important (Lynch et al., 1984). Swift and Messer (1971), for example, examined stream temperature measurements during six forest-cutting experiments on small basins in the southern Appalachians, USA. They found that with complete cutting, because of shade removal, the maximum stream temperatures in summer went up from 19 to 23°C. They believed that such temperature increases were detrimental to temperature-sensitive fish like trout. Likewise, in Wales Stott and Marks (2000) found that after clear felling summer temperatures in a catchment were as much as 7°C higher than they had been under plantation forest.

Pollution with suspended sediments

Probably the most important effect that humans have had on water quality is the contribution to levels of suspended sediments in streams (see Chapter 6). This is a theme which is intimately tied up with that of soil erosion, on the one hand, and with channel manipulation by activities such as reservoir construction, on the other. The clearance of forest, the introduction of ploughing and grazing by domestic animals, the construction of buildings and the introduction of spoil materials into rivers by mineral extraction industries have all led to very substantial increases in levels of stream turbidity. Frequently sediment levels are a whole order of magnitude higher than they would have been under natural conditions. However, the introduction of soil conservation measures, or a reduction in the intensity of land use, or the construction of reservoirs, can cause a relative (and sometimes absolute) reduction in sediment loads The suspended load of rivers is a theme that will be returned to in Chapter 6.

Marine pollution

It is not within the scope of this book to discuss human impacts on the oceans to any great extent. However, it is worth making a few points on this subject, which has been well reviewed by Clark (2001). At first sight, as Jickells et al. (1991: 313) point out, two contradictory thoughts may cross our minds on this issue:

The first is the observation of ocean explorers, such as Thor Heyerdahl, of lumps of tar, flotsam and jetsam, and other products of human society thousands of kilometres from inhabited land. An alternative, vaguer feeling is that given the vastness of the oceans (more than 1,000 billion billion litres of water!), how can man have significantly polluted them?

What is the answer to this conundrum? Jickells et al. (p. 330) draw a clear distinction between the open oceans and regional seas and in part come up with an answer:

The physical and chemical environment of the open oceans has not been greatly affected by events over the past 300 years, principally because of their large diluting capacity . . . Material that floats and is therefore not diluted, such as tar balls and litter, can be shown to have increased in amount and to have changed in character over the past 300 years.

In contrast to the open oceans, coastal areas (including estuaries) in close proximity to large concentrations of people show evidence of increasing concentrations of various substances that are almost certainly linked to human activities (Howarth et al., 2011). This has produced what have been termed 'dead zones' (Diaz and Rosenberg, 2008). Thus the partially enclosed North Sea and Baltic (Conley et al., 2011) show increases in phosphate concentrations as a result of discharges of sewage and agriculture. The same is true of the more enclosed Black Sea (Mee, 1992). The Mediterranean also suffers from many threats, including overfishing, species invasion, aquaculture, increased sedimentation, and pollution from growing urban centres and from intensive agriculture (Claudet and Fraschetti, 2010). Chesapeake Bay (Kemp et al., 2005) and the Gulf of Mexico are other examples (Kemp et al., 2005).

On a global basis, organic point source pollution has been identified as one of the greatest threats (Halpern et al., 2007). This can cause accelerated eutrophication – often called 'cultural eutrophication' – which can lead to excessive growths of algae and to **hypoxia** Selman et al., 2008; Zhang et al., 2010). Coastal and estuary water are sometimes affected by algal foam and scum, often called 'red tides'. Some of these blooms are so toxic that consumers of seafood that have been exposed to them can be affected by diarrhoea, sometimes fatally. A thorough analysis of eutrophication in Europe's coastal waters is provided by the European Environment Agency (2001).The nature of red tides has been discussed by Anderson (1994), who points

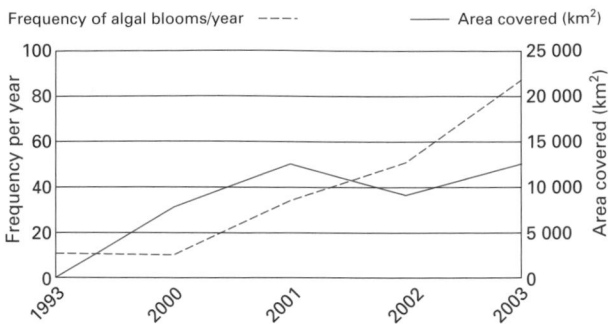

Figure 5.27 The number of harmful algal blooms in the East China Sea (modified from Geo4, UNEP, 2007, figure 4.9).

out that these blooms, produced by certain types of phytoplankton (tiny pigmented plants), can grow in such abundance that they change the colour of the seawater not only red but also brown or even green. They may be sufficiently toxic to kill marine animals such as fish and seals. Long-term studies at the local and regional level in many parts of the world suggest that these so-called red tides are increasing in extent and frequency as coastal pollution worsens and nutrient enrichment occurs more often. For example, Figure 5.27 shows the trend in the number of harmful algal blooms that have occurred in the East China Sea. There were 10 in 1993 and 86 in 2003. This is largely a consequence of a huge increase in the application of fertilizer in the catchments that drain into the sea.

Eutrophication also has adverse effects on coral reefs, as explained by Weber (1993: 49):

Initially, coral productivity increases with rising nutrient supplies. At the same time, however, corals are losing their key advantage over other organisms: their symbiotic self-sufficiency in nutrient poor seas. As eutrophication progresses, algae start to win out over corals for newly opened spaces on the reef because they grow more rapidly than corals when fertilized. The normally clear waters cloud as phytoplankton begin to multiply, reducing the intensity of the sunlight reaching the corals, further lowering their ability to compete. At a certain point, nutrients in the surrounding waters begin to overfertilize the corals' own zooxanthellae, which multiply to toxic levels inside the polyps. Eutrophication may also lead to black band and white band disease, two deadly coral disorders thought to be caused by algal infections. Through these stages of eutrophication, the health and diversity of reefs declines, potentially leading to death.

Likewise it is clear that pollution in the open ocean is, as yet, of limited biological significance. GESAMP (1990), an authoritative review of the state of the marine environment for the United Nations Environment Programme, reported (p. 1):

The open sea is still relatively clean. Low levels of lead, synthetic organic compounds and artificial radionuclides, though widely detectable, are biologically insignificant. Oil slicks and litter are common along sea lanes, but are, at present, of minor consequence to communities of organisms living in open-ocean waters.

On coastal waters it reported (p. 1):

The rate of introduction of nutrients, chiefly nitrates but sometimes also phosphates, is increasing, and areas of eutrophication are expanding, along with enhanced frequency and scale of unusual plankton blooms and excessive seaweed growth. The two major sources of nutrients to coastal waters are sewage disposal and agricultural run-off from fertilizer-treated fields and from intensive stock raising.

Attention is also drawn to the presence of synthetic organic compounds – chlorinated hydrocarbons, which build up in the fatty tissues of top predators such as seals which dwell in coastal waters. Levels of contamination are decreasing in northern temperate areas but rising in tropical and subtropical areas due to continued use of chlorinated pesticides there.

The world's oceans have been greatly contaminated by oil. The effects of this pollution on animals have been widely studied. The sources of pollution include tanker collisions with other ships, the explosion of individual tankers because of the build-up of gas levels, the wrecking of tankers on coasts through navigational or mechanical failure, seepage from offshore oil installations, the flushing of tanker holds and explosive releases from offshore drilling rigs. There is no doubt that the great bulk of the oil and related materials polluting the oceans results from human action, though humans should not be attributed with all the blame, for natural seepages are reasonably common Paradoxically, human actions mean that many natural seepages have diminished as wells have drawn down the levels of hydrocarbons in the oil-bearing rocks. Likewise, the flow of asphalt that once poured from the Trinidad Pitch Lake into the Gulf of Paria has also ceased because mining of the asphalt has lowered the lake below its outlet.

Of the various serious causes of oil pollution, a more important mechanism than the much-publicized role of tanker accidents is the discharge of the ballast water taken into empty tankers to provide stability on the return voyage to the loading terminal. It is in this field, however, that technical developments have led to most amelioration. For example, the quantity of oil in tanker ballast water has been drastically reduced by the LOT (load on top) system, in which the ballast water is allowed to settle so that the oil rises to the surface. The tank is then drained until only the surface oil in the tank remains, and this forms part of the new cargo. In the future large tankers will be fitted with separate ballast tanks not used for oil cargoes, called segregated or clean ballast systems. Given such advances it appears likely than an increasing proportion of oil pollution will be derived from accidents involving tankers of ever increasing size.. The spills from the *Atlantic Empress* (1979, off Tobago), the *ABT Summer* (1991, off Angola), the *Castilo de Bellver* (1983, off South Africa) and the *Amoco Cadiz* (1978, off Brittany) all exceeded 200,000 tonnes of oil. Even here, however, some progress has been made. Since 1980 there has been a reduction in the number of major oil spills, partly as a consequence of diminished long-distance oil transport to Western Europe (Figure 5.28).

The increasing exploitation of offshore oil reserves, some of them located in deep waters, has created a new dimension with regard to oil pollution of ocean waters. This was brought out very clearly in April, 2010, when a rig, the *Deepwater Horizon*, exploded in the Gulf of Mexico and released huge amounts of oil, some of which reached the coastlines of Louisiana, Florida and other states bordering the Gulf. This, however, was not a unique event (Jernelöv, 2010).

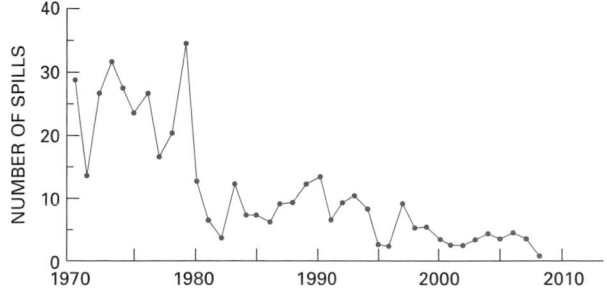

Figure 5.28 Number of large oil spills (over 700 tonnes) from 1970 to 2009 (Reproduced with permission from International Tanker Owners Pollution Federation).

A further threat that is attracting increasing attention is the pollution of the marine environment by plastic debris (Derraik, 2002; Moore, 2008; Barnes et al., 2009; Goldstein et al., 2012). This threatens marine life due to ingestion and due to entanglement (especially in discarded fishing gear). Some plastics contain polychlorinated biphenyls (PCBs), which can lead, inter alia to reproductive disorders and death. Drift plastics can also provide a platform for invasive species. Another topic of current debate is the extent to which new technologies and pressures, including ocean acidification, may begin to impact on the 'Last Great Wilderness', the deep recesses of the oceans (Ramirez-Llodra et al., 2011).

Points for review

What are the ecological effects of dams?

How do humans modify river channels?

To what extent do land cover changes lead to changes in river regimes?

Why should we be concerned about changes in groundwater levels?

Assess the causes and trends in nitrate pollution of water.

What is hypoxia and what causes it?

Guide to reading

Clark, R.B. (2001) *Marine Pollution* (5th edn). Oxford: Clarendon Press. The standard work on this important topic.

Downs, P.W. and Gregory, K.J. (2004) *River Channel Management*. London: Arnold. A comprehensive treatment of river channels, human impacts, and river system management techniques.

Gleick, P.H. (ed.) (1993) *Water in Crisis: A Guide to the World's Freshwater Resources*. New York: Oxford University Press. A compendium of information on trends in water quality and consumption.

Kabat, P. and 9 others (eds) (2004) *Vegetation, Water, Humans and Climate*. Berlin: Springer, A high level research collection that investigates in particular the effects of vegetation change on climate and hydrological processes.

Petts, G.E. (1985) *Impounded Rivers: Perspectives for Ecological Management*. Chichester: Wiley, A treatment of the many consequences of dam construction.

6 HUMAN AGENCY IN GEOMORPHOLOGY

Chapter Overview

Humans create a whole suite of landforms deliberately through excavation, construction and dumping. However, there are also the many incidental consequences of human activities, including accelerated erosion and sedimentation, ground subsidence, changes in the weathering environment, acceleration and triggering of mass movements and coastal erosion and deposition.

Introduction

The human role in creating landforms and modifying the operation of geomorphological processes such as weathering, erosion and deposition is a theme of great importance though one that, particularly in the Western world, has not received the attention it deserves.

The range of the human impact on both forms and processes is considerable (Syvitski and Kettner, 2011).

For example, in Table 6.1 there is a list of some anthropogenic landforms together with some of the causes of their creation. There are very few spheres of human activity which do not, even indirectly, create landforms (Haigh, 1978; Szabó et al., 2010). It is, however, useful to recognize that some features are produced by *direct* anthropogenic processes (Table 6.2). These tend to be more obvious in their form and origin and are frequently created deliberately and knowingly. They

The Human Impact on the Natural Environment: Past, Present and Future, Seventh Edition. Andrew S. Goudie.
© 2013 John Wiley & Sons, Ltd. Published 2013 by John Wiley & Sons, Ltd.

Table 6.1 Some anthropogenic landforms

Feature	Cause
Pits and ponds	Mining, marling, cattle pens (*sirikwa*)
Norfolk Broads	Peat extraction
Spoil heaps	Mining, waste disposal
Terracing, lynchets	Agriculture
Ridge and furrow	Agriculture
Cuttings and sunken lanes	Transport
Embankments	Transport, river and coast management
Dikes, polders	River and coast management
Mounds	Defence, memorials
Craters	War, *qanat* construction
City mounds (*tells*)	Human occupation
Canals	Transport, irrigation
Reservoirs	Water management, cooling basins
Subsidence depressions	Mineral and water extraction
Moats	Defence
Banks along roads	Noise abatement

Table 6.2 Major anthropogeomorphic processes

Direct anthropogenic processes

Constructional
  Tipping, molding, plowing, terracing, reclamation
Excavational
  Digging, cutting, mining, blasting of cohesive or noncohesive
  materials, cratering, tramping, churning, trawling of sea bed
Hydrological interference
  Flooding, damming, canal construction, dredging, channel
  modification, draining, coastal protection

Indirect anthropogenic processes

Acceleration of erosion and sedimentation
  Agricultural activity and clearance of vegetation, engineering,
  especially road construction and urbanization
  Incidental modifications of hydrological regime
Subsidence: collapse, settling
  Mining (e.g. of coal and salt)
  Hydraulic (e.g. groundwater and hydrocarbon pumping)
  Thermokarst (melting of permafrost)
Slope failure: landslides, flows, accelerated creep
  Loading
  Undercutting
  Shaking
  Lubrication
Earthquake generation
  Loading (reservoirs)
  Lubrication (fault plane)
Weathering
  Acidification of precipitation
  Accelerated salinization
  Lateritization

include landforms produced by constructional activity (such as tipping), excavation (Figure 6.1), travel, hydrological interference and farming. Hillsides have been terraced in many parts of the world for many centuries, notably in the arid and semi-arid highlands of the New World (Donkin, 1979; Denevan, 2001), but examples are also known from southern England where in Roman and Medieval times *strip lynchets* have been produced by ploughing on steep slopes.

Landforms produced by *indirect* anthropogenic processes are often less easy to recognize, not least because they tend to involve, not the operation of a new process or processes, but the acceleration of natural processes. They are the result of environmental changes brought about inadvertently by human technology. Nonetheless, it is probably this indirect and inadvertent modification of process and form which is the most crucial aspect of **anthropogeomorphology**. By removing natural vegetation cover – through the agency of cutting, burning and grazing (see Trimble and Mendel, 1995) – humans have accelerated erosion and sedimentation. Sometimes the results will be obvious, for example, when major gully systems rapidly develop; other results may have less immediate effect on landforms but are, nevertheless, of great importance. By other indirect means humans may create subsidence features and hazards, trigger off mass movements like landslides and even influence the operation of phenomena like earthquakes.

Figure 6.1 The Rössing uranium mine near Swakopmund in Namibia, southern Africa. The excavation of such mines involves the movement of prodigious amounts of material. (See Plate 37)

Finally, there are situations where, through a lack of understanding of the operation of processes and the links between different processes and phenomena, humans may deliberately and directly alter landforms and processes and thereby set in train a series of events which were not anticipated or desired. There are, for example, many records of attempts to reduce coast erosion by expensive engineering solutions, which, far from solving erosion problems, only exacerbated them.

Landforms produced by excavation

Of the landforms produced by direct anthropogenic processes, those resulting from excavation are widespread and may have some antiquity. For example, Neolithic peoples in the Breckland of East Anglia in England used antler picks and other means to dig a remarkable cluster of deep pits in the chalk. The purpose of this was to obtain good-quality non-frost-shattered flint to make stone tools. In many parts of Britain, chalk has also been excavated to provide marl for improving acidic, light, sandy soils, and Prince (1962, 1964) made a meticulous study of the 27,000 pits and ponds in Norfolk that have resulted mainly from this activity – an activity particularly prevalent in the eighteenth century. Very large numbers of marl pits, dating to the Iron Age and the Roman period, are also found in north-eastern France (Etienne et al., 2011).

It is often difficult in individual cases to decide whether the depressions are the results of human intervention, for solutional and periglacial depressions are often evident in the same area, but pits caused by human action do tend to have some distinctive features: irregular shape, a track leading into them, proximity to roads and so on (see the debate between Prince, 1979 and Sperling et al., 1979). Difficulties of identifying the true origin of excavational features were also encountered in explaining the Broads, a group of 25 freshwater lakes in the county of Norfolk in eastern England. They are of sufficient area and depth for early workers to have precluded a human origin. Later it was proposed instead that they were natural features caused by uneven alluviation and siltation of river valleys which were flooded by the rapidly rising sea level of the Holocene (Flandrian) transgression. It was postulated by Jennings (1952) that the Broads were initiated as a series of discontinu-

ous natural lakes, formed beyond the limits of a thick estuarine clay wedge laid down in Romano-British times by a transgression of the sea over earlier valley peats. The waters of the Broads were thought to have been impounded in natural peaty hollows between the flanges of the clay and the marginal valley slopes, or in tributary valleys whose mouths were blocked by the clay.

It is now clear, however, that the Broads are the result of human work (Lambert et al., 1970). Some of them have rectilinear and steep boundaries, most of them are not mentioned in early topographic books, and archival records indicate that peat-cutting (*turbary*) was widely practised in the area. On these and other grounds it is believed that peat-diggers, before AD 1300, excavated 25.5 million m^3 of peat and so created the depressions in which the lakes have formed. The flooding may have been aided by sea-level change. Comparably extensive peat excavation was also carried on in the Netherlands, notably in the fifteenth century.

In East Africa there are some large depressions that were created as cattle pens – the *sirikwa* (Sutton, 1965).

Other excavational features result from war, especially craters caused by bomb or shell impact. Regrettably, human power to create such forms is increasing. It has been calculated (Westing and Pfeiffer, 1972) that between 1965 and 1971, 26 million craters, covering an area of 171,000 hectares, were produced by bombing in Indo-China. This represents a total displacement of no less than 2.6 billion m^3 of earth, a figure much greater than calculated as being involved in the peaceable creation of the Netherlands.

Some excavation is undertaken on a large scale for purely aesthetic reasons, when Nature offends the eye (Prince, 1959), while in many countries where land is scarce, whole hills are levelled and extensive areas stripped to provide fill for harbour reclamation. One of the most spectacular examples of this kind was the deliberate removal of steep-sided hills in the centre of Brazil's Rio de Janeiro, for housing development.

An excavational activity of a rather specialized kind is the removal of limestone pavements – areas of exposed limestone in northern England which were stripped by glaciers and then moulded into bizarre shapes by solutional activity – for ornamental rock gardens. These pavements which consist of arid, bare rock surfaces (clints) bounded by deep, humid fissures (grikes) have both an aesthetic and a biological signifi-

Figure 6.2 The mining of sand from large linear sand dunes to provide material for the construction industry in Dubai, United Arab Emirates. (See Plate 38)

Table 6.3 Total excavation of material in Great Britain until 1922. Source: Sherlock (1922: 86)

Activity	Approximate volume (m^3)
Mines	15,147,000,000
Quarries and pits	11,920,000,000
Railways	2,331,000,000
Manchester Ship Canal	41,154,000
Other canals	153,800,000
Road cuttings	480,000,000
Docks and harbours	77,000,000
Foundations of buildings and street excavation	385,000,000
Total	30,534,954,000

cance (Ward, 1979). However, since the Wildlife and Countryside Act of 1981, nearly all outcrops are protected by Limestone Pavement Orders.

One of the most important causes of excavation is still mineral extraction (Figure 6.2), producing open-pit mines, strip mines, quarries for structural materials, borrow pits along roads and similar features (Doerr and Guernsely, 1956). Of these, the environmental devastation produced by strip mining is exceptional. This form of mining is a particular environmental problem in the states of Pennsylvania, Ohio, West Virginia, Kentucky and Illinois. Oil shales, a potential source of oil that at present is relatively untapped, can be exploited by open-pit mining, by traditional room and pillar mining, and by underground *in situ* pyrolysis. Vast reserves exist (as in Canada), but the amount of excavation required will probably be about three times the amount of oil produced (on a volume basis), suggesting that the extent of both the excavation and subsequent dumping of overburden and waste will be considerable. An early attempt to provide a general picture of the importance of excavation in the creation of the landscape of Britain was given by Sherlock (1922). He estimated (see Table 6.3) that up until the time in which he wrote, human society had excavated around 31 billion m^3 of material in the pursuit of its economic activities. That figure must now be a gross underestimate, partly because Sherlock himself was not in a position to appreciate the anthropogenic role

in creating features like the Norfolk Broads, and partly because, since his time, the rate of excavation has greatly accelerated. Sherlock's (1922) study covers a period when earthmoving equipment was still ill-developed. Nonetheless, on the basis of his calculations, he was able to state that 'at the present time, in a densely peopled country like England, Man is many times more powerful, as an agent of denudation, than all the atmospheric denuding forces combined' (p. 333). The most notable change since Sherlock wrote has taken place in the production of aggregates for concrete. Demand for these materials in the UK grew from 20 million tonnes per annum in 1900 to 202 million tonnes in 2001, a 10-fold increase. Douglas and Lawson (2001) estimated that in Britain the total deliberate shift of earth-surface materials is between 688 and 972 million tonnes per year, depending on whether or not the replacement of overburden in opencast mining is taken into account.

Hooke (1994) has produced some data on the significance of deliberate human earthmoving actions in the USA globally, and these are shown in Table 6.4. In all he calculates that deliberate human earthmoving causes 30 billion metric tonnes to be moved per year on a global basis. Douglas and Lawson (2001) give a rather larger figure – 57 billion tonnes per year. It has been estimated that the amount of sediment carried into the ocean by the world's rivers each year amounts to between 8.3 and 51.1 Gt per year (Walling, 2006). Thus, the amount of material moved by humans is rather greater than that moved by the world's rivers to the oceans (Price et al., 2011). As technology changes, this ability increases still further (Haff, 2010). Further-

Table 6.4 Humans as earthmovers. Source: Hooke (1994). Reproduced with permission

Omitting the indirect effects of actions such as deforestation and cultivation, the following are the magnitudes of deliberate human earthmoving actions in the USA:

Billion metric tonnes per year

Excavation for housing and other construction	0.8
Mining	3.8
Road work	3.0
Total USA	7.6
World total (roughly four times the USA total)	30.0

For comparison, the following figures represent estimated world totals due to natural earthmoving processes:

River transport	
(a) To oceans and lakes	14
(b) Short-distance transport within river basins	40
Tectonic forces lifting continents	14
Volcanic activity elevating sea floor	30
Glacial transport	4.3
Wind transport	1.0

more, land-use change, and in particular the development of agriculture, has caused a leap in the amount of erosion of the world's land surfaces. Wilkinson and McElroy (2007) calculated that 'natural' sediment fluxes to the world's rivers are about 21 Gt per year and that 'anthropogenic' losses may be around 75 Gt per year.

Landforms produced by construction and dumping

The process of constructing mounds and embankments and the creation of dry land where none previously existed is long-standing. In the Middle East and other areas of long-continued human urban settlement, the accumulated debris of life has gradually raised the level of the land surface, and occupation mounds (*tells*) are a fertile source of information to the archaeologist (Menze and Ur, 2012). In Britain the 37 m high mound at Silbury Hill dates back to the Neolithic, while the pyramids of Central America, Egypt and the Far East are even more spectacular early feats of landform creation. Likewise in the Americas, Native Indians, prior to the arrival of Europeans, created large numbers of mounds of different shapes and sizes for temples, burials, settlements and effigies (Denevan, 1992: 377). In the same way, hydrological management has involved, over many centuries, the construction of massive banks and walls – the ultimate result being the landscape of the present-day Netherlands. Transport developments have also required the creation of large constructional landmarks, but probably the most important features are those resulting from the dumping of waste materials, especially those derived from mining. It has been calculated that there are at least 2000 million tonnes of shale lying in pit heaps in the coalfields of Britain (Richardson, 1976). Today, with the technical ability to build that humans have, even estuaries may be converted from ecologically productive environments into suburban sprawl by the processes of dredging and filling. Indeed, one of the striking features of the distribution of the world's population is the tendency for large human concentrations to occur near vast expanses of water. Many of these cities have extended out on to land that has been reclaimed from the sea (e.g. Hong Kong, Figure 6.3), thereby providing valuable sites for development, but sometimes causing the loss of rich fishing grounds and ecologically valuable wetlands.

At the present time, large quantities of waste are sent to landfill sites. In 1995, member states of the European Union (EU) landfilled more than 80% of their waste, but under the EU Landfill Directive, this figure has been reduced substantially to around 37% in 2010. Germany, Austria, Sweden and Belgium send less than 1% of their waste to landfill, but Latvia, Lithuania, Bulgaria and Romania still send over 90%.

The ocean floors are also being affected because of the vast bulk of waste material that humankind is creating. Waste-solid disposal by coastal cities is now sufficiently large to modify shorelines, and it covers adjacent ocean bottoms with characteristic deposits on a scale large enough to be geologically significant. This has been brought out dramatically by Gross (1972: 3174), who undertook a quantitative comparison of the amount of solid wastes dumped into the Atlantic by humans in the New York metropolitan region with the amount of sediment brought into the ocean by rivers:

The discharge of waste solids exceeds the suspended sediment load of any single river along the U.S. Atlantic coast. Indeed, the discharge of wastes from the New York metropolitan

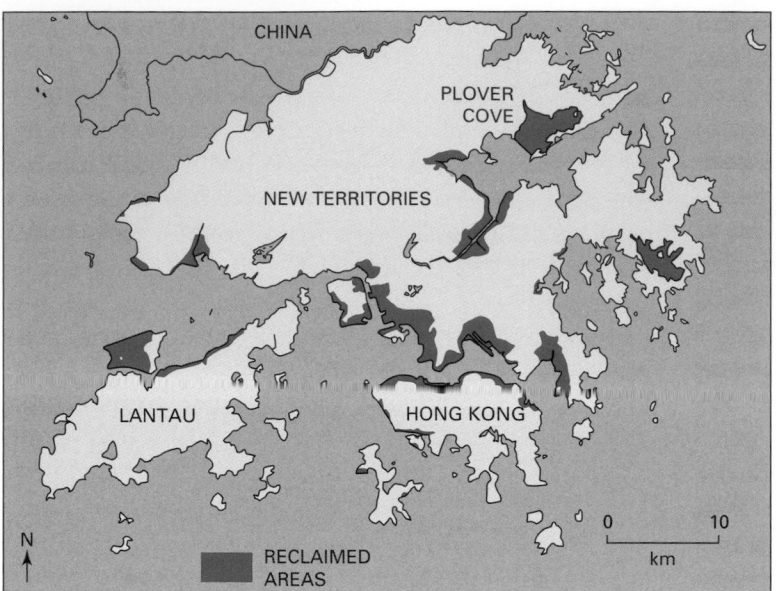

Figure 6.3 Map of Hong Kong showing main urban reclamation areas.

region is comparable to the estimated suspended-sediment yield (6.1 megatons per year) of all rivers along the Atlantic coast between Maine and Cape Hatteras, North Carolina.

Not only are the rates of sedimentation high, but the anthropogenic sediments tend to contain abnormally high contents of such substances as carbon and heavy metals (Goldberg et al., 1978). Because of this, Congress in 1988 banned the dumping of sewage sludge, and no dumping of this material took place after June, 1992.

Many of the features created by excavation in one generation are filled in by another, since phenomena like water-filled hollows produced by mineral extraction are often both wasteful of land and also suitable locations for the receipt of waste. Watson (1976), for example, has mapped the distribution of hollows, which were largely created by marl diggers in the lowlands of south-west Lancashire and north-west Cheshire, as they were represented on mid-nineteenth-century topographic maps (Figure 6.4a). When this distribution is compared with the present-day distribution for the same area (Figure 6.4b), it is evident that a very substantial proportion of the holes have been filled in and obliterated by humans, with only 2114 out of 5380 remaining. Hole densities have fallen from 121–47 per km². The loss of ponds in Great Britain was a fairly general phenomenon, with their numbers falling from c. 800,000 in the late nineteenth century to

only c. 200,000 by the mid-1980s (Jeffries, 2012). Since that time, partly because their biodiversity value is now being appreciated, the number of ponds in Great Britain has shown a modest increase (Williams et al., 2010).

Sediment transport by rivers

Sediment transport by rivers has been modified by humans (Wasson, 2012) in two main ways. On the one hand, the construction of dams (see Chapter 5) has caused much sediment to be trapped in reservoirs. On the other, sediment delivery to rivers has been increased as a result of accelerated rates of soil erosion. In the south-east USA, the results of these two tendencies have been analysed by McCarney-Castle et al. (2010). Under pre-European times (1680–1700), the mean annual suspended sediment load transfer rate was 6.2 Mt per year, under pre-dam (1905–1925) conditions the rate was 15.04 Mt per year, whereas now, following dam construction, the rate is 5.2 Mt per year.

Sediment retention is also well illustrated by the Nile (Table 6.5), both before and after the construction of the great Aswan High Dam. Until its construction the late summer and autumn period of high flow was characterized by high silt concentrations, but since it has been finished the silt load is rendered lower throughout the year, and the seasonal peak is removed.

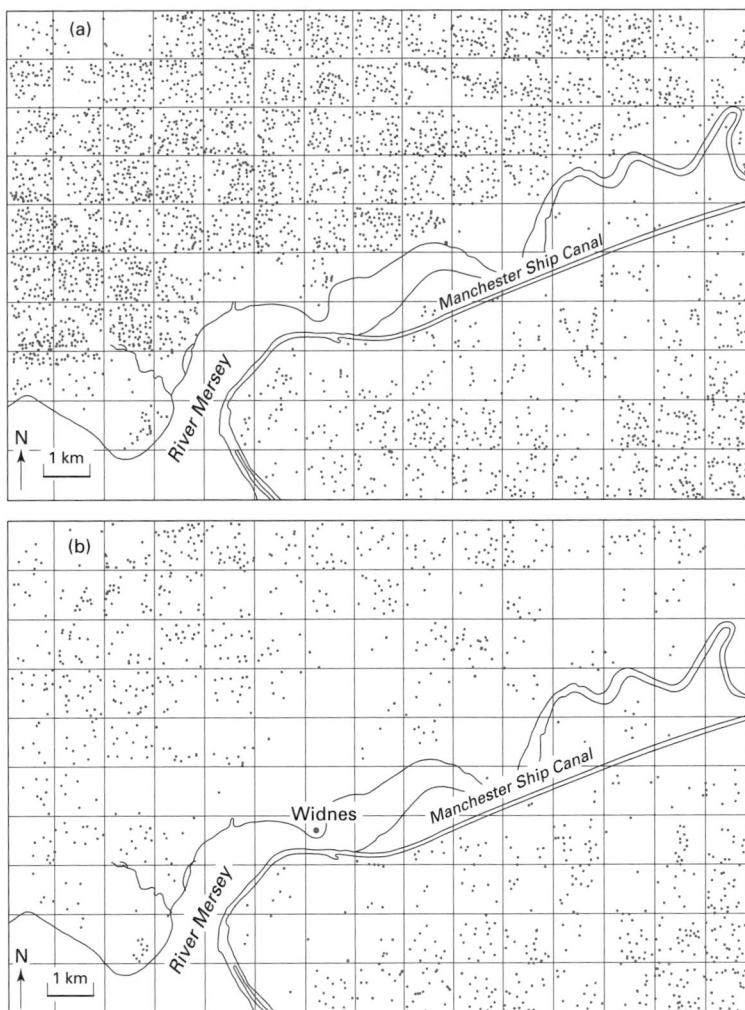

Figure 6.4 The distribution of pits and ponds in a portion of north-western England (a) in the mid-nineteenth century and (b) in the mid-twentieth century (modified from Watson, 1976).

Table 6.5 Silt concentrations (in parts per million) in the Nile at Gaafra before and after the construction of the Aswan High Dam. Source: Abul-Atta (1978: 199)

Before (averages for the period 1958–1963)											
January											*December*
64	50	45	42	43	85	674	2702	2422	925	124	71
After											
44	47	45	50	51	49	48	45	41	43	48	47
Ratio											
1.5	1.1	1.0	0.8	0.8	1.7	14.0	60.0	59.1	21.5	2.58	1.63

Petts (1985, table 18) indicates that the Nile now only transports 8% of its natural load below the Aswan High Dam, although this figure seems to be exceptionally low. Other rivers for which data are available carry between 8% and 50% of their natural suspended loads below dams.

A similarly clear demonstration of this effect has been given for the South Saskatchewan River in Canada (Table 6.6) by Rasid (1979). During the pre-dam period, typified by 1962, the total annual suspended loads at Saskatoon and Lemsford Ferry were remarkably similar. As soon as the reservoir began to fill late in 1963, however, some of the suspended sediment began to be trapped, and the transitional period was marked by a progressive reduction in the proportion of sediment which reached Saskatoon. In the 4 years after the dam was fully operational, the mean annual sediment load at Saskatoon was only 9% that at Lemsford Ferry.

Even more dramatic are the data for the Colorado River in the USA (Figure 6.5). Prior to 1930, it carried around 125–150 million tonnes of suspended sediment

Table 6.6 Total yearly suspended load (thousand imperial tonnes) of the South Saskatchewan River at Lemsford Ferry and Saskatoon, 1962–1970. Source: Rasid (1979, table 1). Reproduced with permission

Period of record	Lemsford Ferry	Saskatoon	Difference at Saskatoon (%)
Pre-dam 1962			
1962	1813	1873	+3
Transitional			
1963	4892	447	−8
1964	7711	4146	−46
1905	0732	?7?1	−72
1966	5228	1675	−68
Mean	6891	3255	−53
Post-dam			
1967	12,619	446	−96
1968	2661	101	−96
1969	10,562	2146	−80
1970	5643	118	−98
Mean	7871	703	−91

(a) Suspended-sediment discharge (10^6 tons year^{-1})

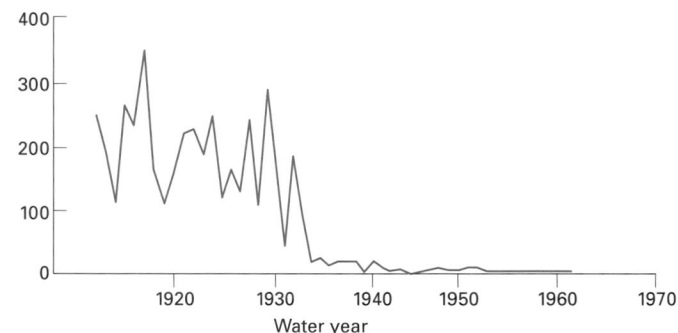

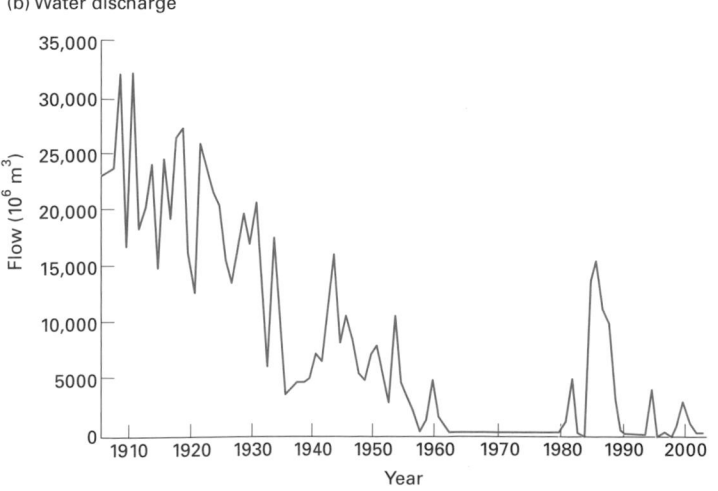

Figure 6.5 Historical (a) sediment and (b) water discharge trends for the Colorado River, USA (after the United States Geological Survey in Schwarz et al., 1991).

per year to its delta at the head of the Gulf of California. Following a series of dams the Colorado now discharges neither sediment nor water to the sea (Schwarz et al., 1991). There have also been marked changes in the amount of sediment passing along the Missouri and Mississippi rivers (Figure 6.6a) (Tweel and Turner, 2012), and Meade (1996) attempted to compare the situation in the 1980s with that which existed before humans started to interfere with those rivers (*c.* AD 1700) (Figure 6.6b). Before 1900, the Missouri–Mississippi

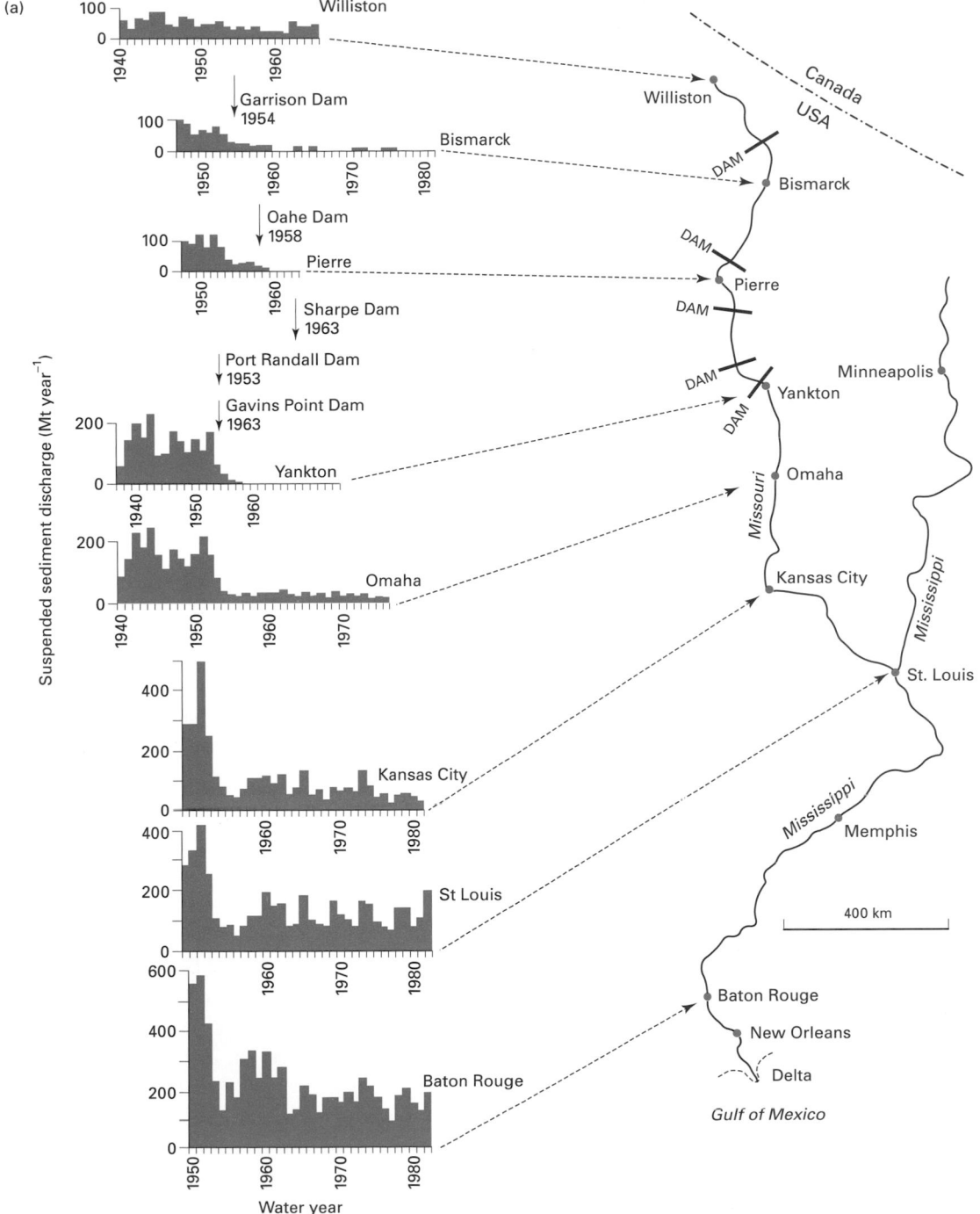

Figure 6.6 (a) Suspended sediment discharge on the Mississippi and Missouri Rivers between 1939 and 1982 (after Meade and Parker, 1985, with modifications). (b) Long-term average discharges of suspended sediment in the lower Mississippi River *c.* 1700 and *c.* 1980.

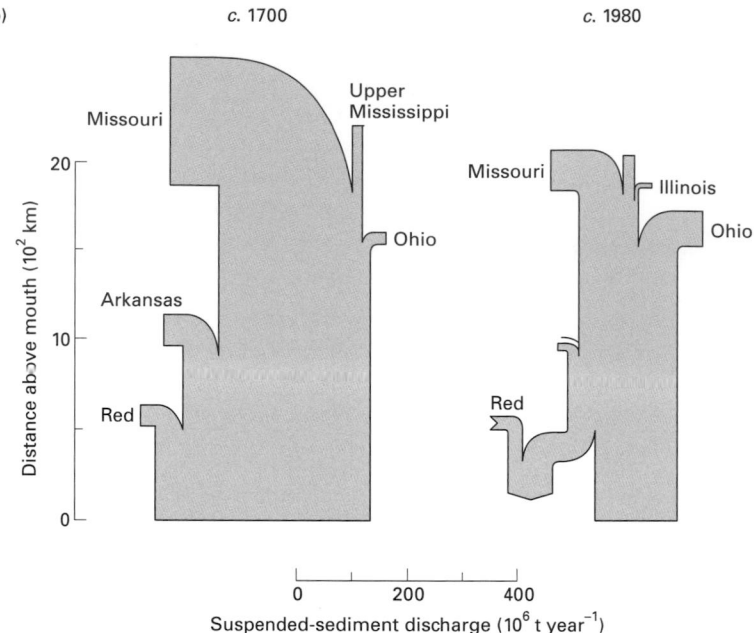

(b)

c. 1700   *c.* 1980

Missouri   Upper Mississippi

Ohio

Arkansas

Red

Missouri   Illinois

Ohio

Red

Distance above mouth (10² km)

Suspended-sediment discharge (10⁶ t year⁻¹)

0   200   400

Figure 6.6 (*Continued*)

river system carried an estimated 400 million metric tonnes of sediment per year to coastal Louisiana. Between 1987 and 2006, this figure had fallen to an average of 145 million metric tonnes per year. About half of this decline was accounted for by sediment trapping behind dams, and the rest by such actions as bank revetments and soil erosion management (Meade and Moody, 2010). Also in the USA, the Columbia River has seen a >50% reduction in total sediment transport since the mid-nineteenth century, largely because of flow regulation (Naik and Jay, 2011).

In India, dams on the Krishna and Godavari rivers have also caused major changes in discharges and sediment loads (Rao et al., 2010). The Krishna discharged 61.9 km³ of water between 1951 and 1959, but this had reduced to 11.8 km³ by 2000–2008. Its suspended sediment load was 9 million tonnes during 1966–1969 but was as low as 0.4 million tonnes by 2000–2005. This change has promoted accelerated recession of its delta.

The Indus River has also undergone a marked reduction in flow and sediment transport, especially since the completion of the Tarbela Dam in the early 1970s. Between 1931 and 1947, the river at Kotri had a flow of about 70–150 km³ per year and a sediment discharge

of 190–330 million tonnes per year. For the period from 1962 to 1986 the respective figures were 10–100 km³ per year and 10–130 million tonnes per year (Meadows and Meadows, 1999).

In the case of the Yangtze in China, in 2003–2005, sediment load in the upper river was only 17% of that in the 1950s–1960s (Xu et al., 2007). The Three Gorges Dam has had a particularly marked impact since it began to impound sediment and water in 2003 (Xu and Milliman, 2009).

On a global basis, Syvitski et al. (2005) have calculated that sediment retention behind dams has reduced the net flux of sediment reaching the world's coasts by *c.* 1.4 billion metric tonnes per year and that over 100 billion metric tonnes have been trapped within the last 50 years. Reservoirs behind dams now trap *c.* 26% of the global sediment delivery to the coastal ocean (Syvitski and Milliman, 2007).

Conversely, Syvitski et al. (2005) have calculated that humans have simultaneously increased the sediment transport by global rivers through soil erosion by *c.* 2.3 billion metric tonnes per year. On balance, therefore, global river sediment loads have gone up more as a result of soil erosion than they have been reduced by dam retention. However, plainly there are great differences between rivers and between areas. In central Japan, for example, sediment loads were reduced in the last decades of the twentieth century because land that had formerly been cultivated intensively had become urbanized (Siakeu et al., 2004). In the case of some major Chinese rivers draining into the western Pacific, recent declines in sediment load have been caused by a combination of factors (Chu et al., 2009): dam retention (56%), soil and water conservation (23%), water abstraction (15%) and in-channel sand mining (6%). The same is true of the Sacramento River in California. Its sediment load decreased by about one-half during the period between 1957 and 2001, and this has been attributed to a depletion in the amount of old hydraulic mining debris, trapping of sediment in reservoirs, riverbank protection, the construction of levees and altered land use (e.g. urbanization replacing agriculture) (Wright and Schoellhamer, 2004).

Accelerated sedimentation

Land-use changes have a range of impacts on slopes and valleys, though it is not always easy to separate the effects of such changes from those produced by climate change. Among the consequences of land-use change that have been identified in Holocene Britain are slope instability, which leads to gullying and deposition of debris cones and alluvial fans in upland catchments, and colluviation in the lowlands; valley floor sedimentation; increased runoff leading to incision (mainly in upland catchments); and high water tables and flooding (mainly in lowland catchments) (Foulds and Macklin, 2006). That said, one inevitable consequence of the accelerated erosion produced by human activities has been accelerated sedimentation (see e.g. Komar et al.'s 2004 study of sedimentation in Tillamook Bay, Oregon). Some of the eroded material accumulates downslope as colluvium (Verstraeten et al., 2009) (Figure 6.7a), some on floodplains, some in lakes and some in estuaries. This has been heightened by the deliberate addition of sediments to stream channels as a result of the need to dispose of mining and other wastes.

Before European colonization of North America, there is little or no evidence of any substantial human geomorphological impacts on erosion and sedimentation (James, 2011). However, the situation changed dramatically after European contact. Serious sedimentation of bays and estuaries caused by human activity occurred on the eastern coast of America. As Gottschalk (1945: 219) wrote:

Both historical and geological evidence indicates that the preagricultural rate of silting of eastern tidal estuaries was low. The history of sedimentation of ports in the Chesapeake Bay area is an epic of the effects of uncontrolled erosion since the beginning of the wholesale land clearing and cultivation more than three centuries ago.

He calculated that at the head of the Chesapeake Bay, 65 million m^3 of sediment were deposited between 1846 and 1938. The average depth of water over an area of $83\,km^2$ was reduced by $0.76\,m$. New land comprising 318 hectares was added to the state of Maryland and, as Gottschalk remarked, 'the Susquehanna River is repeating the history of the Tigris and Euphrates'. The trend in sedimentation rate (Pasternack et al., 2001) over the last three centuries is shown in Figure 6.7b and demonstrates the low rate in pre-European settlement times, the high rates in the mid-nineteenth century at a time of peak deforestation and of intensive agriculture, and the decline that took place in the twentieth century. Salt marshes expanded rapidly at this time as a result of rapid rates of sediment delivery (Kirwan et al., 2011).

Much of the material entrained by erosive processes on upper slopes as a result of agriculture in Maryland, however, was not evacuated as far as the coast. Costa (1975) has suggested, on the basis of the study of sedimentation, that only about one-third of the eroded material left the river valleys. The remainder accumulated on floodplains as alluvium and colluvium at rates of up to $1.6\,cm$ per year. Similarly, Happ (1944), working in Wisconsin, carried out an intensive augering survey of floodplain soils and established that, since the development of agriculture, floodplain aggradation had proceeded at a rate of approximately $0.85\,cm$ per year. He noted that channel and floodplain aggradation had caused the flooding of low alluvial terraces to be more frequent, more extensive and deeper. The rate of sedimentation has since declined (Trimble, 1976, 2008) because of less intensive land use and the institution of effective erosion control measures on farmland (see also Magilligan, 1985 and Trimble, 2011).

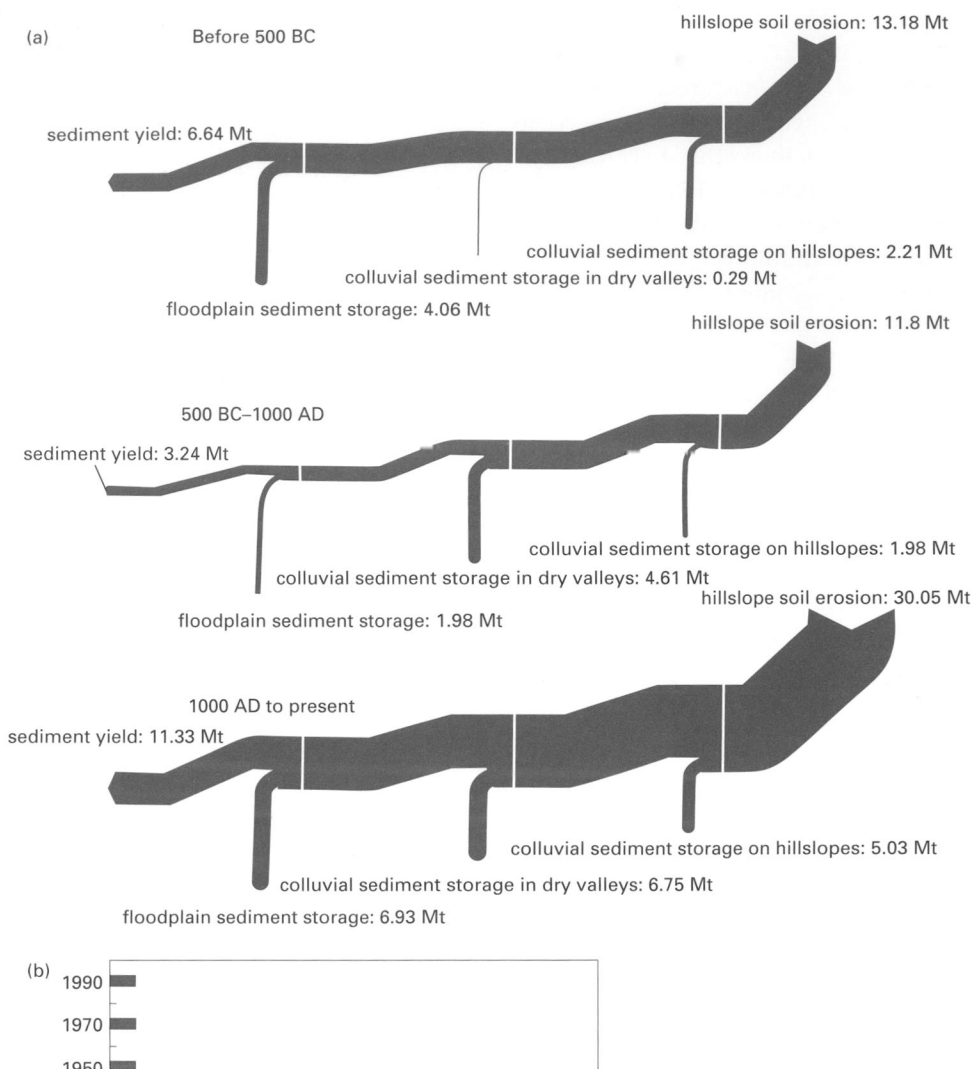

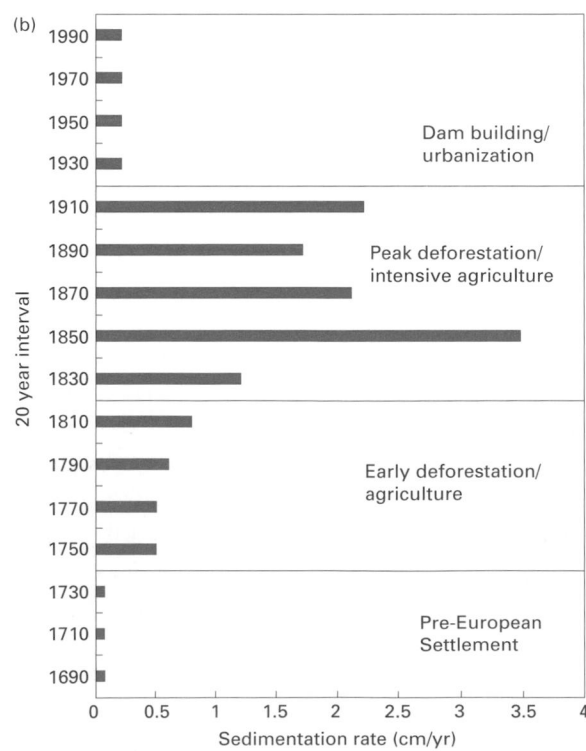

Figure 6.7 (a) Changing sediment dynamics through time in the Nethen catchment (central Belgium) as illustrated by a tentative sediment budget for three different time periods. (b) Sedimentation chronology for Chesapeake Bay, USA, for three centuries (modified from Pasternack et al., 2001). Reproduced with permission from Elsevier.

Table 6.7 Accelerated sedimentation in Britain in prehistorical and historical times

Location	Source	Evidence and date
Howgill Fells	Harvey et al. (1981)	Debris cone production following tenth century AD introduction of sheep farming
Upper Thames Basin	Robinson and Lambrick (1984)	River alluviation in late Bronze Age and early Iron Age
Lake District	Pennington (1981)	Accelerated lake sedimentation at 5000 BP as a result of Neolithic agriculture
Mid-Wales	Macklin and Lewin (1986)	Floodplain sedimentation (Capel Bangor unit) on Breidol as a result of early Iron Age sedentary agriculture
Brecon Beacons	Jones et al. (1985)	Lake sedimentation increase after 5000 BP at Llangorse due to forest clearance
Weald	Burrin (1985)	Valley alluviation from Neolithic onwards until early Iron Age
Bowland Fells	Harvey and Renwick (1987)	Valley terraces at 5000–2000 BP (Bronze or Iron Age settlement) and after 1000 BP (Viking settlement)
Southern England	Bell (1982)	Fills in dry valleys: Bronze and Iron Age
Callaly Moor (Northumberland)	Macklin et al. (1991)	Valley fill sediments of late Neolithic to Bronze Age
Ribble, north-west England	Foster et al. (2009)	Iron Age/Roman and Medieval.
Semer Water and Raydale, North Yorkshire	Chiverrell et al., 2008	Bronze Age

Such valley sedimentation is by no means restricted to the newly settled terrains of North America. There is increasing evidence to suggest that silty valley fills in Germany, France and Britain, many of them dating back to the Bronze Age and the Iron Age, are the result of accelerated slope erosion produced by the activities of early farmers (Bell, 1982). Indeed, in recent years, various studies have been undertaken with a view to assessing the importance of changes in sedimentation rate caused by humans at different times in the Holocene in Britain (e.g. Chiverrell et al., 2009; Foster et al., 2009). Among the formative events that have been identified are initial land clearance by Mesolithic and Neolithic peoples; agricultural intensification and sedentarization in the late Bronze Age; the widespread adoption of the iron plough in the early Iron Age; settlement by the Vikings and the introduction of sheep farming (Table 6.7). In British and Irish catchments, Foulds and Macklin (2006) suggest that geomorphic instability linked with Holocene land-use changes was especially intense in the Bronze Age and the Iron Age. Rates of sedimentation on British floodplains appear to have accelerated greatly in the last thousand or so years (Figure 6.8) (Macklin et al., 2010).

A core from Llangorse Lake in the Brecon Beacons of Wales (Jones et al., 1985) provides excellent long-term data on changing sedimentation rates:

Period (years BP)	Sedimentation rate (cm 100/year)
9000–7500	3.5
7500–5000	1.0
5000–2800	13.2
2800–AD 1840	14.1
c. AD 1840–present	59.0

The 13-fold increase in rates after 5000 BP seems to have occurred rapidly and is attributed to initial forest clearance. The second dramatic increase of more than fourfold took place in the last 150 years and is a result of agricultural intensification.

In the last two centuries rates of sedimentation in lake basins have changed in different ways in different basins according to the differing nature of economic activities in catchments. Some data from various sources are listed for comparison in Table 6.8. In the case of the Loe Pool in Cornwall (south-west England), rates of sedimentation were high while mining industry was active, but fell dramatically when mining was curtailed. In the case of Seeswood Pool in Warwickshire, a dominantly agricultural catchment area in central England, the highest rates have occurred since 1978 in response to various land management changes, such as larger fields, continuous cropping and increased dairy herd size. In other catchments, pre-afforestation

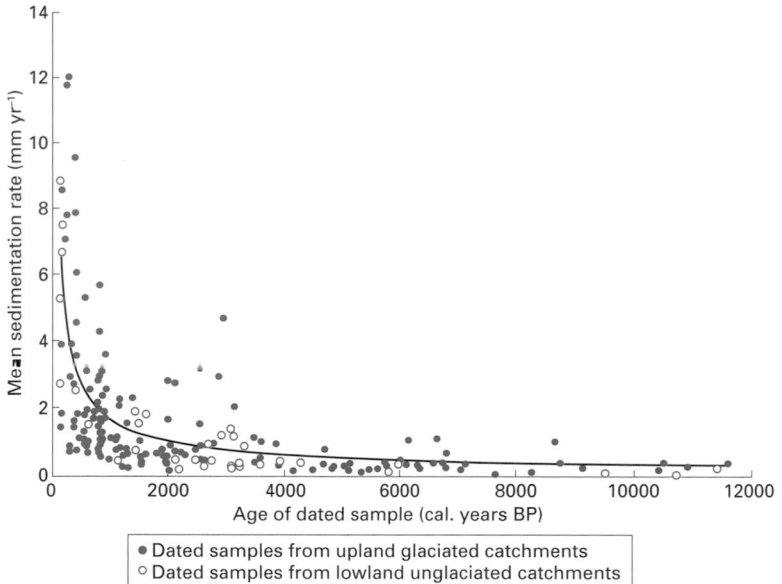

Figure 6.8 Holocene floodplain sedimentation rates in Great Britain plotted for upland (glaciated) and lowland (unglaciated) catchments (modified from Macklin et al., 2010, figure 18). Reproduced with permission from Elsevier.

Table 6.8 Data on rates of erosion and sedimentation in two English lakes in the last two centuries

Dates	Activity	Rates of erosion in catchment (R. Cober) as determined from lake sedimentation rates ($t\ km^{-2}\ y^{-1}$)
(a) Loe Pool (Cornwall) (from O'Sullivan et al., 1982)		
1860–1920	Mining and agriculture	174
1930–1936	Intensive mining and agriculture	421
1937–1938	Intensive mining and agriculture	361
1938–1981	Agriculture	12
(b) Seeswood Pool (Warwickshire) (from Foster et al., 1986)		
1765–1853		7.0
1854–1880		12.2
1881–1902		8.1
1903–1919		9.6
1920–1925		21.6
1926–1933		16.1
1934–1947		12.7
1948–1964		12.0
1965–1972		13.9
1973–1997		18.3
1978–1982		36.2

ploughing may have caused sufficient disturbance to cause accelerated sedimentation. For example, Battarbee et al. (1985b) looked at sediment cores in the Galloway area of south-west Scotland and found that in Loch Grannoch the introduction of ploughing in the catchment caused an increase in sedimentation from 0.2 cm per year to 2.2 cm per year.

The work of Binford et al. (1987) on the lakes of the Peten region of northern Guatemala (Central America), an area of tropical lowland dry forest, is also instructive with respect to early agricultural colonization. Combining studies of archaeology and lake sediment stratigraphy, they were able to reconstruct the diverse environmental consequences of the growth of Mayan civilization (Figure 6.9). This civilization showed a dramatic growth after 3000 years BP, but collapsed in the ninth century AD. The hypotheses put forward to explain this collapse include warfare, disease, earthquakes and soil degradation. The population has remained relatively low ever since, and after the first European contact (AD 1525), the region was virtually depopulated. The period of Mayan success saw a marked reduction in vegetation cover, an increase in lake sedimentation rates and in catchment soil erosion, an increased supply of inorganic silts and clays to the lakes, a pulse of phosphorus derived from human wastes and a decrease in lacustrine productivity caused by high levels of turbidity. In parts of Guatemala, a

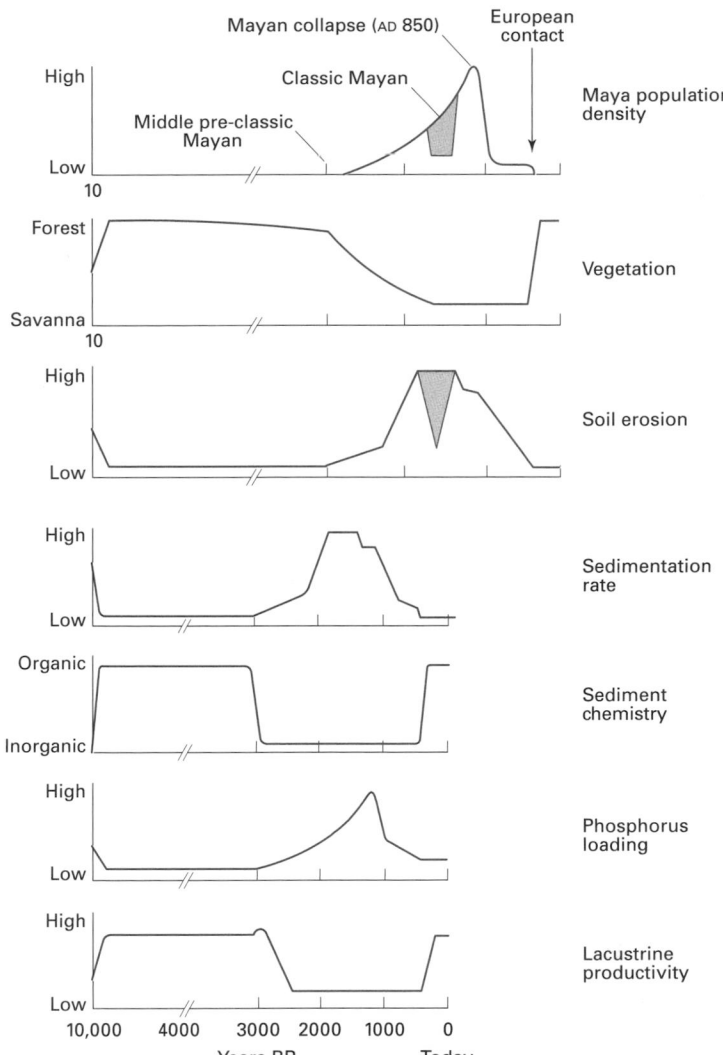

Figure 6.9 The human and environmental history of the Peten Lakes, Guatemala. The hatched areas indicate a phase of local population decline (modified from Binford et al., 1987). Reproduced with permission from Elsevier.

spasm of soil erosion occurred during the Maya Late Classic phase (AD 550–830), and here depressions record 1–3 m of aggradation in two centuries (Beach et al., 2006).

In a classic study, G.K. Gilbert (1917) demonstrated that hydraulic gold mining in the Sierra Nevada Mountains of California led to the addition of vast quantities of sediments into the river valleys draining the range. Much of this sediment remains to this day (James, 1989), and some of it may be contaminated with the mercury that was used as part of the process of collecting the gold (Lecce et al., 2008). This legacy sediment in itself raised channel bed levels, changed their channel configurations and caused the flooding of lands that had previously been immune. Of even

greater significance was the fact that the rivers transported vast quantities of debris into the estuarine bays of the San Francisco system and caused extensive shoaling which in turn diminished the tidal prism of the bay. Gilbert calculated the volume of shoaling produced by hydraulic mining since the discovery of gold to be 846 million m³. Mining waste being dumped in channels has also caused rapid aggradation in Tasmania (Knighton, 1991). Conversely, channel sediments may themselves be mined to provide aggregate for the construction industry (Rinaldi et al., 2005).

Sedimentation has severe economic implications because of the role it plays in reducing the effective lifetime of reservoirs. In the tropics capacity depletion through sedimentation is commonly around 2% per annum (Myers, 1988: 14), so that the expected useful life of the Paute hydroelectric project in Ecuador (cost US$ 600 million) is 32 years, that of the Mangla Dam in Pakistan (cost US$ 600 million) is 57 years and that of the Tarbela Dam, also in Pakistan, is just 40 years.

Ground subsidence

Not all ground subsidence is caused by humans. For example, limestone solutional processes can, in the absence of humankind, create a situation where a cavern collapses to produce a surface depression or sinkhole, and permafrost will sometimes melt to produce a thermokarst depression without human intervention. Similarly, deltaic subsidence can be caused by a whole range of natural processes. In the case of the Mississippi delta of the USA, for instance, six categories of process have been identified (Yuill et al., 2009) that can cause subsidence, four of which are natural, including tectonics (faulting and halokinesis), Holocene sediment compaction (Tornquist et al., 2008), sediment loading by the delta and glacial isostatic adjustment. Fluid removal and surface water drainage and management are the two anthropogenic processes. Indeed, ground subsidence can be caused or accelerated by humans in a variety of ways: by the transfer of subterranean fluids (such as oil, gas and water); by the removal of solids through underground mining (Bell et al., 2000) or by dissolving solids and removing them in solution (e.g. sulfur and salt); by the disruption of permafrost; and by the compaction or reduction of sediments because of drainage and irrigation (Johnson, 1991; Barends et al., 1995).

Some of the most dangerous and dramatic collapses have occurred in limestone areas because of the dewatering of limestone caused by mining activities. In the Far West Rand of South Africa, gold mining has required the abstraction of water to such a degree that the local water table has been lowered by more than 300 m. The fall of the water table caused miscellaneous clays and other materials filling the roofs of large caves to dry out and shrink so that they collapsed into the underlying void. One collapse created a depression 30 m deep and 55 m across, killing 29 people. In Alabama in the southern USA, water-level decline consequent upon pumping has had equally serious consequences in a limestone terrain; and Newton (1976) has

estimated that since 1900, about 4000 induced sinkholes or related features have been formed, while fewer than 50 natural sinkholes have been reported over the same interval. Sinkholes that may result from such human activity are also found in Georgia, Florida, Tennessee, Pennsylvania and Missouri. In some limestone areas, however, a reverse process can operate. The application of water to overburden above the limestone may render it more plastic so that the likelihood of collapse is increased. This has occurred beneath reservoirs, such as the May Reservoir in central Turkey, and as a result of the application of wastewater and sewerage to the land surface. Williams (1993) provides a survey of the diverse effects of human activities on limestone terrains.

In areas where thick halite (salt) deposits have accumulated as a result of their precipitation in saline lake basins, solution may cause sink holes to develop. An example of this is provided by the Dead Sea basin, where the recent decline in level caused by water abstraction has promoted their formation. More than a thousand potentially dangerous sinkholes have developed along its shorelines since the early 1980s as a result of the flow of undersaturated groundwater dissolving the evaporites. As Yechieli et al. (2006: 1075) explain it in the context of the Israeli side of the basin,

The abrupt appearance of the sinkholes, and their accelerated expansion thereafter, reflects a change in the groundwater regime around the shrinking lake and the extreme solubility of halite in water. The eastward retreat of the shoreline and the declining sea level cause an eastward migration of the fresh-saline water interface. As a result the salt layer, which originally was saturated with Dead Sea water over its entire spread, is gradually being invaded by freshwater at its western boundary, which mixes and displaces the original Dead Sea brine.

Similar problems have been encountered on the Jordanian side of the Dead Sea, with karstic collapse creating major problems for the chemical plants on the Lisan Peninsula (Closson et al., 2007).

Subsidence can be accelerated by the direct solution of susceptible rocks. For example, collapses have occurred in gypsum bedrock because of solution brought about by the construction of a reservoir. In 1893, the MacMillan Dam was built on the Pecos River in New Mexico, but within 12 years, the whole river flowed through caves which had developed since construction. Both the San Fernando and Rattlesnake Dams in California suffer severe leakage for similar reasons.

Subsidence produced by oil abstraction is a problem in some parts of the world. The classic area is Los Angeles, where 9.3 m of subsidence occurred as a result of exploitation of the Wilmington oilfield between 1928 and 1971. The Inglewood oilfield displayed 2.9 m of subsidence between 1917 and 1963. Some coastal flooding problems occurred at Long Beach because of this process. Similar subsidence has been recorded from the Lake Maracaibo field in Venezuela (Prokopovich, 1972) and from some Russian fields (Nikonov, 1977). Subsidence in the Mississippi Delta and elsewhere in the Gulf Coast region of the USA may be in part the result of hydrocarbon extraction (Morton et al., 2005, 2006).

A more widespread problem is posed by groundwater abstraction for industrial, domestic and agricultural purposes. Table 6.9 presents some data for such subsidence from various parts of the world. The ratios of subsidence to water level decline are strongly dependent on the nature of the sediment composing the aquifer. Ratios range from 1:7 for Mexico City, to 1:80 for the Pecos in Texas and to less than 1:400 for London, England (Rosepiler and Reilinger, 1977). The extent of subsidence that has taken place in the USA as a result of groundwater abstraction has recently been assessed by Chi and Reilinger (1984). In Japan, subsidence has also now emerged as a major problem (Nakano and Matsuda, 1976). In 1960, only 35.2 km² of the Tokyo lowland was below sea level, but continuing subsidence meant that by 1974 this had increased to 67.6 km², exposing a total of 1.5 million people to major flood hazard. Shanghai and neighbouring parts of the Yangtze Delta region in China is another low-lying area that has suffered from subsidence, with as much as 2–3 m since 1921 (Chai et al., 2004). Similar amounts of subsidence are also known from Tianjin and Xi'an (Xue et al., 2005), but as China develops the amount of subsidence that is occurring is becoming more widespread, and over 50 Chinese cities are threatened (Hu, 2006; Yin et al., 2006). In Bangkok, subsidence in the early 1980s was proceeding at a rate of as much as 120 mm per year, and by 2002, 2.05 m of subsidence had occurred (Phien-wej et al., 2006), causing flooding and building damage. Subsidence caused by groundwater and gas abstraction has also been a problem in Venice, where it has greatly increased flood risk, but

Table 6.9 Ground subsidence. Source: Data from Prokopovich (1972); Cooke and Doornkamp (1974); Rosepiler and Reilinger (1977); Holzer (1979); Nutalaya and Ran (1981); Johnson (1991); Phien-wej et al., (2006)

Location	Amount (m)	Date	Rate (mm/year)
(a) Ground subsidence produced by oil and gas abstraction			
Azerbaydzhan, USSR	2.5	1912–1962	50
Atravopol, USSR	1.5	1956–1962	125
Wilmington, USA	9.3	1928–1971	216
Inglewood, USA	2.9	1917–1963	63
Maracaibo, Venezuela	5.03	1929–1990	84
(b) Ground subsidence produced by groundwater abstraction			
London, England	0.06–0.08	1865–1931	0.91–1.21
Savannah, Georgia (USA)	0.1	1918–1955	2.7
Mexico City	7.5	–	250–300
Houston, Galveston, Texas	1.52	1943–1964	60–76
Central Valley, California	8.53	–	–
Tokyo, Japan	4	1892–1972	500
Osaka, Japan	>2.8	1935–1972	76
Niigata, Japan	>1.5	–	–
Pecos, Texas	0.2	1935–1966	6.5
South-central Arizona	2.9	1934–1977	96
Bangkok, Thailand	2.05	1933–2002	Up to 120
Shanghai, China	2.62	1921–1965	60

the rate of subsidence has fallen since c. 1970 because of controls on this abstraction (Carbognin et al., 2010).

The subsidence caused by mining (which led to court cases in England as early as the fifteenth century as a consequence of associated damage to property) is perhaps the most familiar, though its importance varies according to such factors as the thickness of seam removed, its depth, the width of working, the degree of filling with solid waste after extraction, the geological structure and the method of working adopted (Wallwork, 1974). In general terms, however, the vertical displacement by subsidence is less than the thickness of the seam being worked and decreases with an increase in the depth of mining. This is because the overlying strata collapse, fragment and fracture, so that the mass of rock fills a greater space than it did when naturally compacted. Consequently, the surface expression of deep-seated subsidence may be equal to little more than one-third of the thickness of the material removed. Studies in the Ruhr district of Germany indicate values that can be as high as 5.16 m, though the mean value is 1.6 m (Harnischmacher, 2010). Subsidence associated with coal mining may disrupt surface drainage (Sidle et al., 2000), and the resultant depressions then become permanently flooded. In the Ruhr, a lake, Lake Lanstrop, formed between 1963 and 1967 in response to up to 9 m of subsidence (Bell et al., 2000).

Coal-mining regions are not the only areas where subsidence problems are serious. In Cheshire, northwest England, rock salt is extracted from two major seams in the Triassic Mercia Mudstone Group, each about 30 m in thickness. Moreover, these seams occur at no great depth – the uppermost being at about 70 m below the surface. A further factor to be considered is that the rock salt is highly soluble in water, so the flooding of mines may cause additional collapse. These three conditions – thick seams, shallow depth and high solubility – have produced optimum conditions for subsidence, and many subsidence lakes called 'flashes' have developed (Wallwork, 1956). In the Perm district of Russia, huge sinkholes (up to 200 m deep) have developed in association with phosphate mines.

Some subsidence is created by a process called hydrocompaction, which is explained thus. Moisture-deficient, unconsolidated, low-density sediments tend to have sufficient dry strength to support considerable effective stresses without compacting. However, when such sediments, which may include alluvial fans or loess, are thoroughly wetted for the first time (e.g. by percolating irrigation water), the inter-granular strength of the deposits is diminished, rapid compaction takes place, and ground surface subsidence follows. Unequal subsidence can create problems for irrigation schemes.

Land drainage can promote subsidence of a different type, notably in areas of organic soils. The lowering of the water table makes peat susceptible to oxidation and deflation so that its volume decreases. One of the longest records of this process, and one of the clearest demonstrations of its efficacy, has been provided by the measurements at Holme Fen Post in the English Fenlands (Dawson et al., 2010). Approximately 3.8 m of subsidence occurred between 1848 and 1957 (Fillenham, 1963), with the fastest rate occurring soon after drainage had been initiated (Figure 6.10). The present rate averages about 1.4 cm per year (Richardson and

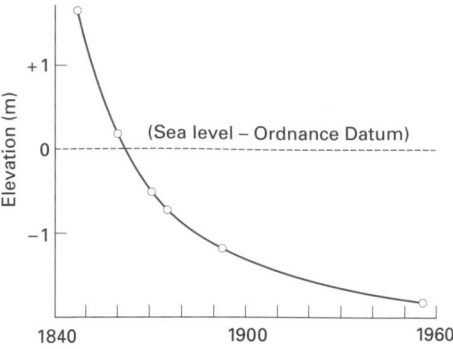

Figure 6.10 The subsidence of the English Fenlands peat in Holme Fen Post from 1842 to 1960 following drainage (data from Fillenham, 1963). Reproduced with permission from John Wiley & Sons Ltd.

Smith, 1977). At its maximum natural extent, the peat of the English Fenland covered around 1750 km². Now only about one-quarter (430 km²) remains.

A further type of subsidence, sometimes associated with earthquake activity, results from the effects on the earth's crust of large masses of water impounded behind reservoirs. As we shall see later, seismic effects can be generated in areas with susceptible fault systems, and this may account for earthquakes recorded at Koyna (India) and elsewhere. This process whereby a mass of water causes coastal depression is called hydro-isostasy.

In tundra regions ground subsidence is associated with **thermokarst** development, this being irregular, hummocky terrain produced by the melting of ground ice, permafrost (Figure 6.11). The development of thermokarst is due primarily to the disruption of the thermal equilibrium of the permafrost and an increase in the depth of the active layer. This is illustrated in Figure 6.12. Following French (1976: 106), consider an undisturbed tundra soil with an active layer of 45 cm. Assume also that the soil beneath 45 cm is supersaturated permafrost and yields on a volume basis upon thawing 50% excess water and 50% saturated soil. If the top 15 cm were removed, the equilibrium thickness of the active layer, under the bare ground conditions, might increase to 60 cm. As only 30 cm of the original active layer remains, 60 cm of the permafrost must thaw before the active layer can thicken to 60 cm, since 30 cm of supernatant water will be released. Thus, the surface subsides 30 cm because of thermal melting associated with the degrading permafrost, to give an overall depression of 45 cm.

Figure 6.11 Permafrost in Siberia. (See Plate 39)

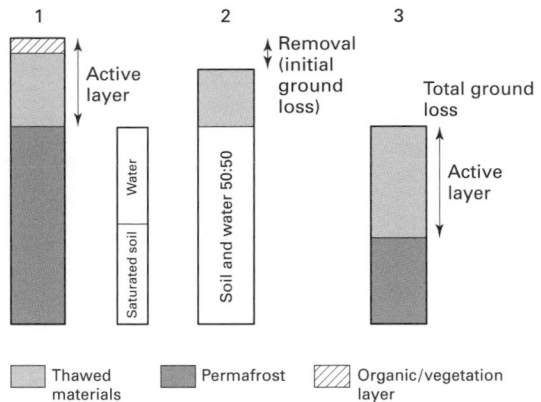

Figure 6.12 Diagram illustrating how the disturbance of high ice content terrain can lead to permanent ground subsidence. 1–3 indicate stages before, immediately after and subsequent to disturbance (after Mackay in French, 1976, figure 6.1). Reproduced with permission.

among the effects are broken dams, cracked buildings (Figure 6.13), offset roads and railways, fractured well casings, deformed canals and ditches, bridges that need re-levelling, saline encroachment and increased flood damage.

Arroyo trenching, gullies and peat haggs

In the south-western USA many broad valleys and plains became deeply incised with valley-bottom gullies (*arroyos*) over a short period between 1865 and 1915, with the 1880s being especially important (Cooke and Reeves, 1976). This cutting had a rapid and detrimental effect on the flat, fertile and easily irrigated valley floors, which are the most desirable sites for settlement and economic activity in a harsh environment. The causes of the phenomenon have been the subject of prolonged debate (Elliott et al., 1999; Gonzalez, 2001).

Many students of this phenomenon have believed that thoughtless human actions caused the entrenchment, and the apparent coincidence of white settlement and arroyo development tended to give credence to this viewpoint. The range of actions that could have been culpable is large: timber felling, overgrazing, cutting grass for hay in valley bottoms, compaction along well-travelled routes, channelling of runoff from trails and railways, disruption of valley-bottom sods

Thus the key process involved in thermokarst subsidence is the state of the active layer and its thermal relationships. When, for example, surface vegetation is cleared for agricultural or constructional purposes, the depth of thaw will tend to increase. The movement of tracked vehicles has been particularly harmful to surface vegetation, and deep channels may soon result from permafrost degradation. Similar effects may be produced by the siting of heated buildings on permafrost, and by the laying of oil, sewer and water pipes in or on the active layer (Lawson, 1986).

Thus subsidence is a diverse but significant aspect of the part humans play as geomorphological agents. The damage caused on a worldwide basis can be measured in billions of dollars each year (Coates, 1983), and

Figure 6.13 The city of Mexico has subsided by many metres as a result of groundwater abstraction. Many of the ancient buildings in the city centre have been severely damaged, and have been cracked and deformed. (See Plate 40)

by animals' feet and the invasion of grasslands by miscellaneous types of scrub.

On the other hand, study of the long-term history of the valley fills shows that there have been repeated phases of aggradation and incision and that some of these took place before the influence of humans could have been a significant factor (Waters and Haynes, 2001). This has prompted debate as to whether the arrival of white communities was in fact responsible for this particularly severe phase of environmental degradation. Huntington (1914), for example, argued

that valley filling would be a consequence of a climatic shift to more arid conditions. These, he believed, would cause a reduction in vegetation which in turn would promote rapid removal of soil from devegetated mountain slopes during storms and overload streams with sediment. With a return to humid conditions vegetation would be re-established, sediment yields would be reduced and entrenchment of valley fills would take place. Bryan (1928) put forward a contradictory climatic explanation. He argued that a slight move towards drier conditions, by depleting vegetation cover and reducing soil infiltration capacity, would produce significant increases in storm runoff which would erode valleys. Another climatic interpretation was advanced by Leopold (1951), involving a change in rainfall intensity rather than quantity. He indicated that a reduced frequency of low-intensity rains would weaken the vegetation cover, while an increased frequency of heavy rains at the same time would increase the incidence of erosion. Support for this contention comes from the work of Balling and Wells (1990) in New Mexico. They attributed early twentieth-century arroyo trenching to a run of years with intense and erosive rainfall characteristics that succeeded a phase of drought conditions in which the productive ability of the vegetation had declined. It is also possible, as Schumm et al. (1984) have pointed out, that arroyo incision could result from neither climatic change nor human influence. It could be the result of some natural geomorphological threshold being crossed. Under this argument, conditions of valley-floor stability decrease slowly over time until some triggering event initiates incision of the previously 'stable' reach.

It is therefore clear that the possible mechanisms that can lead to alternations of cut-and-fill of valley sediments are extremely complex and that any attribution of all arroyos in all areas to human activities may be a serious oversimplification of the problem (Figure 6.14). In addition, it is possible that natural environmental changes, such as changes in rainfall characteristics, have operated at the same time and in the same direction as human actions.

In the Mediterranean lands there have also been controversies surrounding the age and causes of alternating phases of aggradation and erosion in valley bottoms. Vita-Finzi (1969) suggested that at some stage during historical times many of the steams in the Mediterranean area, which had hitherto been engaged pri-

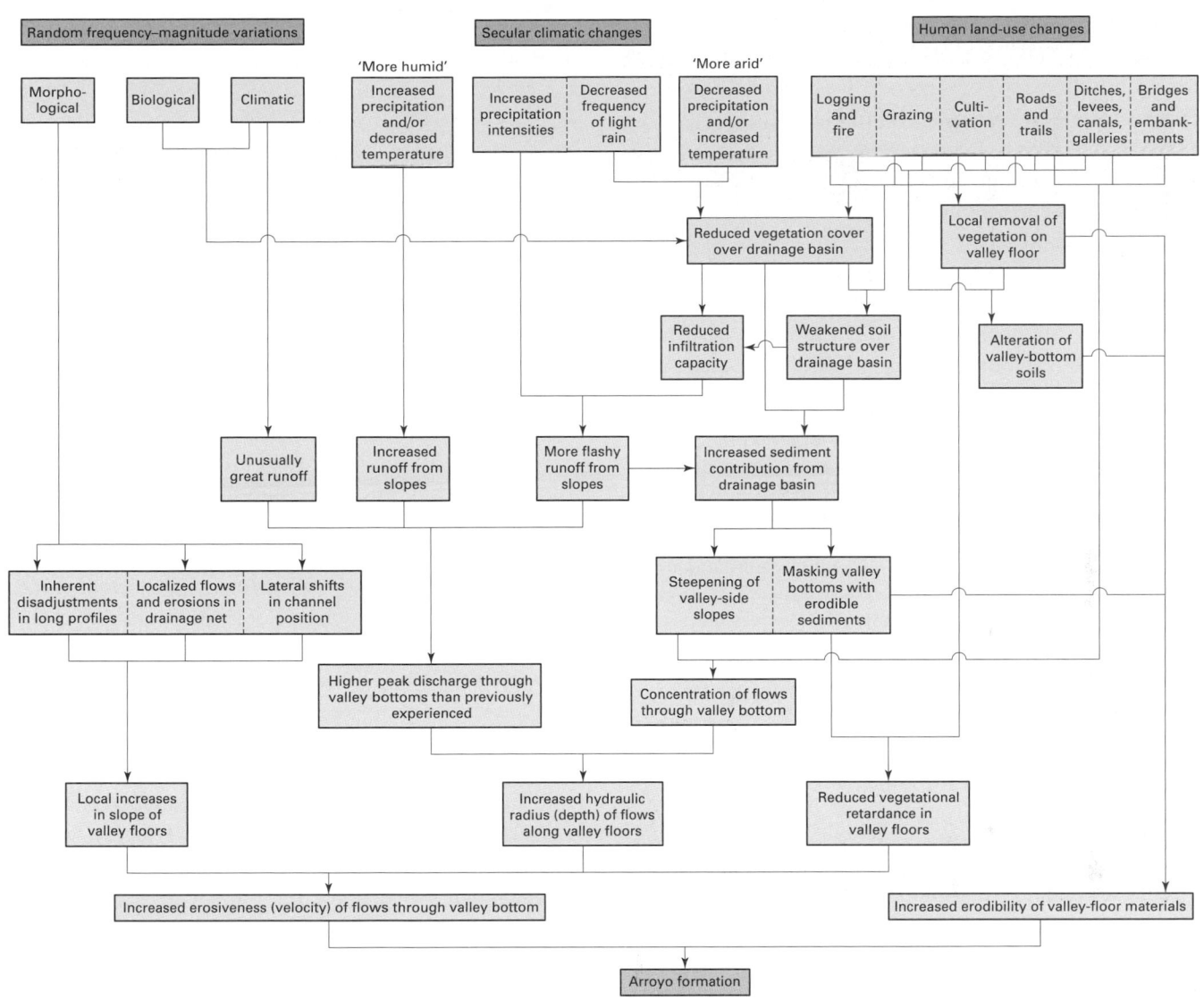

Figure 6.14 A model for the formation of arroyos (gullies) in the south-western USA (after Cooke and Reeves, 1976, figure 1.2). Reproduced with permission from Oxford University Press.

marily in downcutting, began to build up their beds. Renewed downcutting, still seemingly in operation today, has since incised the channels into the alluvial fill. He proposed that the reversal of the downcutting trend in the Middle Ages was both ubiquitous and confined in time, and that some universal and time-specific agency was required to explain it. He believed that vegetation removal by humans was not a medieval innovation and that some other mechanism was required. A solution he gave to account for the phenomenon was precipitation change during the climatic

fluctuation known as the Little Ice Age (AD 1550–1850). This was not an interpretation which found favour with Butzer (1974). He reported that his investigations showed plenty of post-Classical and pre-1500 alluviation (which could not therefore be ascribed to the Little Ice Age), and he doubted whether Vita-Finzi's dating was precise enough to warrant a 1550–1850 date. Instead, he suggested that humans were responsible for multiple phases of accelerated erosion from slopes, and accelerated sedimentation in valley bottoms, from as early as the middle of the first millennium BC.

Butzer's interpretation has found favour with van Andel et al. (1990) who have detected an intermittent and complex record of cut and fill episodes during the late Holocene in various parts of Greece. They believe that this evidence is compatible with a model of the control of timing and intensity of landscape destabilization by local economic and political conditions. This is a view shared in the context of the Algarve in Portugal by Chester and James (1991).

However, in recent years some caution has been expressed as to the extent to which humans have been responsible for landscape degradation in the Mediterranean lands. Rackham and Moody (1996: 9) have paraphrased the traditional view thus:

The persistent theory in Mediterranean studies is that of the Ruined Landscape or Lost Eden, which runs like this. Well into historic times, the land was covered with magnificent forests. Men cut down the forests to make homes and ships, or burned them to make farmland; the trees failed to grow again, and multitudes of goats devoured the remains. Trees, unlike other vegetation, have a magic power of retaining soil. The trees gone, the soil washed away into the sea or the plains, the land became 'barren', and some say that the very climate got worse.

They then go on to question it (p. 10):

Scholars play up the landscape of the past and play down the landscape of the present: they write about ancient forests when the original text mentioned only trees; they ignore modern forests or dismiss them as mere 'scrub' or 'maquis'.

This is a view that has been developed by Roberts (1998: 191):

. . . it would be wrong to assume that all Mediterranean hills are anthropogenic landscapes. A salutary lesson in this comes from the badlands of southeast Spain. Here integrated geo-archaeological studies have found in situ Bronze Age sites lying uneroded on vulnerable bare slopes. The badlands had already formed by c. 4500 years ago, and later agricultural clearances have done little to alter erosion rates. While giving the appearance of recent degradation, these and similar Mediterranean badlands . . . are ancient landscapes created by long-term geological instability rather than the ravages of the goat and the plough.

A further location with spectacular gullies, locally called *lavaka*, is Madagascar. Here too there have been debates about cultural versus natural causation. Proponents of cultural causes have argues that since humans arrived on the island in the last two thousand years, there has been excessive cattle grazing, removal of forest for charcoal and for slash-and-burn cultivation, devastating winter (dry season) burning of grasslands, and erosion along tracks and trails. However, the situation is more complex than that and the *lavaka* are polygenetic. Tectonism and natural climatic acidification may be at least as important, and given the climatic and soil types of the island many *lavaka* are a natural part of the landscape's evolution. Some of them also clearly predate primary (i.e. uncut) rain forest. The many factors, natural and cultural, involved in *lavaka* development are well reviewed by Wells and Andriamihaja (1993).

Another example of drainage incision that demonstrates the problem of disentangling the human from the natural causes of erosion is provided by the eroding peat bogs of highland Britain (Bragg and Tallis, 2001; Evans and Warburton, 2007). Over many areas, including the Pennines of northern England and the Brecon Beacons of Wales, blanket peats are being severely eroded to produce pool and hummock topography, areas of bare peat and incised gullies (*haggs*). Many rivers draining such areas are discoloured by the presence of eroded peat particles, and sediment yields of organic material are appreciable (Labadz et al., 1991).

Some of the observed peat erosion may be an essentially natural process, for the high water content and low cohesion of undrained peat masses make them inherently unstable. Moreover, the instability must normally become more pronounced as peat continues to accumulate, leading to bog slides and bursts round margins of expanded peat blankets. Conway (1954) suggested that an inevitable end-point of peat build-up on high-altitude, flat or convex surfaces is that a considerable depth of unconsolidated and poorly humidified peat overlies denser and well-humidified peat, so adding to the instability. Once a bog burst or slide occurs, this leads to the formation of drainage gullies which extend back into the peat mass, slumping-off marginal peat downslope, and leading to the drawing off of water from the pools of the hummock and hollow topography of the watershed.

Tallis (1985) believes that there have been two main phases of erosion in the Pennines. The first, initiated 1000–1200 years ago, may have been caused by natural instability of the type outlined above. However, there has been a second stage of erosion, initiated 200–300 years ago, in which miscellaneous human activities appear to have been important. Radley (1962) suggested that among the pressures that had caused erosion were heavy sheep grazing, regular burning, peat cutting, the digging of boundary ditches, the incision of packhorse tracks and military manoeuvres during the First World War. Other causes may include footpath erosion (Wishart and Warburton, 2001) and severe air pollution (Tallis, 1965), the latter causing the loss of a very important peat forming moss, *Sphagnum*. In South Wales, there is some evidence that the blanket peats have degenerated as a result of contamination by particulate pollution (soot, etc.) during the Industrial Revolution (Chambers et al., 1979). On the other hand, in Scotland lake core studies indicate that severe peat erosion was initiated between AD 1500 and 1700, prior to air pollution associated with industrial growth, and Stevenson et al. (1990) suggest that this erosion initiation may have been caused either by the adverse climatic conditions of the Little Ice Age or by an increasing intensity of burning as land-use pressures increased.

In the future it is possible that peat bog erosion may be accelerated by a combination of worsening summer droughts and increasingly erosive winter rainstorms (Evans et al., 2006).

Accelerated weathering and the tufa decline

Although fewer data are available and the effects are generally less immediately obvious, there is some evidence that human activities have produced changes in the nature and rate of weathering (Winkler, 1970). The prime cause of this is probably air pollution. It is clear that, as a result of increased emissions of sulfur dioxide through the burning of fossil fuels, there are higher levels of sulfuric acid in rain over many industrial areas. This in itself may react with stones and cause their decay. Chemical reactions involving sulfur dioxide can also generate salts such as calcium sulfate and magnesium sulfate, which may be effective in causing the physical breakdown of rock through the mechanism of salt weathering.

Similarly, atmospheric carbon dioxide levels have been rising steadily because of the burning of fossil fuels and deforestation. Carbon dioxide may combine with water, especially at lower temperatures, to produce weak carbonic acid which can dissolve limestone, marbles and dolomites. Weathering can also be accelerated by changes in groundwater levels resulting from irrigation. This can be illustrated by considering the Indus Plain in Pakistan (Goudie, 1977), where irrigation has caused the water table to be raised by about 6 m since 1922. This has produced increased evaporation and salinization. The salts that are precipitated by evaporation above the capillary fringe include sodium sulfate, a very effective cause of stone decay. Indeed buildings, such as the great archaeological site of Mohenjo-Daro or are decaying at a catastrophic rat.

In other cases accelerated weathering has been achieved by moving stone from one environment to another. Cleopatra's Needle, an Egyptian obelisk in New York City, is an example of rapid weathering of stone in an inhospitable environment. Originally erected on the Nile opposite Cairo about 1500 BC, it was toppled in about 500 BC by Persian invaders and lay partially buried in Nile sediments until, in 1880, it was moved to New York. It immediately began to suffer from scaling and the inscriptions were largely obliterated within 10 years because of the penetration of moisture which enabled frost-wedging and hydration of salts to occur.

During the 1970s and 1980s an increasing body of isotopic dates became available for deposits of tufa (secondary freshwater deposits of limestone, also known as travertine). Some of these dates suggest that over large parts of Europe, from Britain to the Mediterranean basin and from Spain to Poland, rates of tufa formation were high in the early and mid-Holocene, but declined markedly thereafter (Weisrock, 1986: 165–167). Vaudour (1986) maintains that since around 3000 BP 'the impact of man on the environment has liberated their disappearance', but he gives no clear indication of either the basis of this point of view or of the precise mechanism(s) that might be involved. If the late Holocene reduction in tufa deposition is a reality (which is contested by, e.g. Baker and Simms, 1998), then it is necessary to consider a whole range of possible mechanisms, both natural and anthropogenic (Nicod, 1986: 71–80; Table 6.10). As yet the case for an anthropogenic role is not proven (Goudie et al., 1993) and it is possible that climate changes were more a important determinant of the tufa decline (Wehrli et al., 2010).

Table 6.10 Some possible mechanisms to account for the alleged Holocene tufa decline

Climatic/natural	Anthropogenic
Discharge reduction following rainfall decline leading to less turbulence.	Discharge reduction due to overpumping, diversions and so on
Degassing leads to less deposition.	Increased flood scour and runoff variability of channels due to deforestation, urbanization, ditching and so on.
Increased rainfall cause more flood scour.	Channel shifting due to deforestation of floodplains leads to tufa erosion.
Decreasing temperature leads to less evaporation and more CO_2 solubility.	Reduced CO_2 flux in system after deforestation
Progressive Holocene peat development and soil podzol development through time leads to more acidic surface waters.	Introduction of domestic stock causes breakdown of fragile tufa structures.
	Deforestation = less fallen trees to act as foci for tufa barrages
	Increased stream turbidity following deforestation reduces algal productivity.

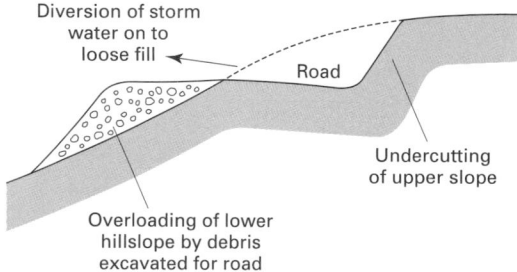

Figure 6.15 Slope instability produced by road construction.

Accelerated mass movements

There are many examples of **mass movements** being triggered by human actions (Selby, 1979). For instance, landslides can be created either by undercutting or by overloading (Figure 6.15). When a road is constructed, material derived from undercutting the upper hillside may be cast on to the lower hillslope as a relatively loose fill to widen the road bed. Storm water is then often diverted from the road on to the loose fill.

Because of the hazards presented by both natural and accelerated mass movements, humans have developed a whole series of techniques to attempt to control them. Such methods, many of which are widely used by engineers, are listed in Table 6.11. These techniques are increasingly necessary, as human capacity to change a hillside and to make it more prone to slope failure has been transformed by engineering development. Excavations are going deeper, buildings and

Table 6.11 Methods used to control mass movements on slopes. Source: After R.F. Baker and H.E. Marshall, in Dunne and Leopold (1978, table 15.16)

Type of movement	Method of control
Falls	Flattening the slope
	Benching the slope
	Drainage
	Reinforcement of rock walls by grouting with cement, anchor bolts
	Covering of walls with steel mesh
Slides and flows	Grading or benching to flatten slope
	Drainage of surface water with ditches
	Sealing surface cracks to prevent infiltration
	Sub-surface drainage
	Rock and earth buttresses at foot
	Retaining walls at foot
	Pilings through the potential slide mass

other structures are larger, and many sites which are at best marginally suitable for engineering projects are now being used because of increasing pressure on land. This applies especially to some of the expanding urban areas in the humid parts of low latitudes – Hong Kong, Kuala Lumpur, Rio de Janeiro and many others. It is very seldom that human agency deliberately accelerates mass movements; most are accidentally caused, the exception possibly being the deliberate triggering of a threatening snow avalanche (Perla, 1978).

The forces producing slope instability and landsliding can usefully be divided into disturbing forces and resisting properties (Cooke and Doornkamp, 1990: 113–114). The factors leading to an increase in shear stress (disturbing forces) can be listed as follows (modified from Cooke and Doornkamp, 1990: 113):

1 *Removal of lateral or underlying support:*
   undercutting by water (e.g. river, waves), or glacier ice; weathering of weaker strata at the toe of the slope; washing out of granular material by seepage erosion; *human cuts and excavations, drainage of lakes or reservoirs.*

2 *Increased disturbing forces:*
   natural accumulations of water, snow, talus; pressure *caused by human activity* (e.g. stockpiles of ore, tip-heaps, rubbish dumps, or buildings).

3 *Transitory earth stresses:*
   earthquakes; *continual passing of heavy traffic.*

4 *Increased internal pressure:*
   build-up of pore-water pressures (e.g. in joints and cracks, especially in the tension crack zone at the rear or the slide).

Some of the factors are natural, while others (italicized) are affected by humans. Factors leading to a decrease in the shearing resistance of materials making up a slope can also be summarized (also modified from Cooke and Doornkamp, 1990: 113):

1 *Materials:*
   beds which decrease in shear strength if water content increases (clays, shale, mica, schist, talc, serpentine) (e.g. *when local water-table is artificially increased in height by reservoir construction*), or as a result of
   stress release (vertical and/or horizontal) following slope formation;
   low internal cohesion (e.g. consolidated clays, sands, porous organic matter);
   in bedrock: faults, bedding planes, joints, foliation in schists, cleavage,
   brecciated zones and pre-existing shears.

2 *Weathering changes:*
   weathering reduces effective cohesion and to a lesser extent the angle of shearing resistance;
   absorption of water leading to changes in the fabric of clays (e.g. loss of bonds between particles or the formation of fissures).

3 *Pore-water pressure increase:*
   high groundwater table as a result of increased precipitation or *as a result of human interference* (e.g. *dam construction*) (see 1 earlier).

Once again the italics show that there are a variety of ways in which humans can play a role.

Some mass movements are created by humans piling up waste soil and rock into unstable accumulations that fail spontaneously. At Aberfan, in South Wales, a major disaster occurred in 1966 when a coal-waste tip 180 m high began to move as an earth flow. The tip had been constructed not only as a steep slope but also upon a spring line. This made an unstable configuration which eventually destroyed a school and claimed over 150 lives. In Hong Kong, where a large proportion of the population is forced to occupy steep slopes developed on deeply weathered granites and other rocks, mass movements are a severe problem, and So (1971) has shown that many of the landslides and washouts, (70% of those in the great storm of June 1966, e.g.) were associated with road sections and slopes artificially modified through construction and cultivation. In south eastern France humans have accelerated landslide activity by building excavations for roads and by loading slopes with construction material (Julian and Anthony, 1996). The undercutting and removal of the trees of slopes for the construction of roads and paths has also led to landsliding in many parts of southeast Asia (Sidle et al., 2006) and in the Himalayas (Barnard et al., 2001). This is also the case in mountainous Nepal, where the number of fatal landslides shot up during the 1990s (Petley et al., 2007) (Figure 6.16). This seems to be correlated with the rapid development of the road network after about 1990. Similarly, landslides that were triggered by a great earthquake in Kashmir in 2005 occurred preferentially in areas where road construction had taken place (Owen et al., 2008). Many landslides in China appear to have been triggered by surface excavation and mining activity (Huang and Chan, 2004) and also by road construction (Sidle et al., 2011).

One of the most serious mass movements partly caused by human activity was that which caused the Vaiont Dam disaster in Italy in 1963, in which 2600 people were killed (Kiersch, 1965). Heavy antecedent rainfall and the presence of young, highly folded sedimentary rocks provided the necessary conditions for a

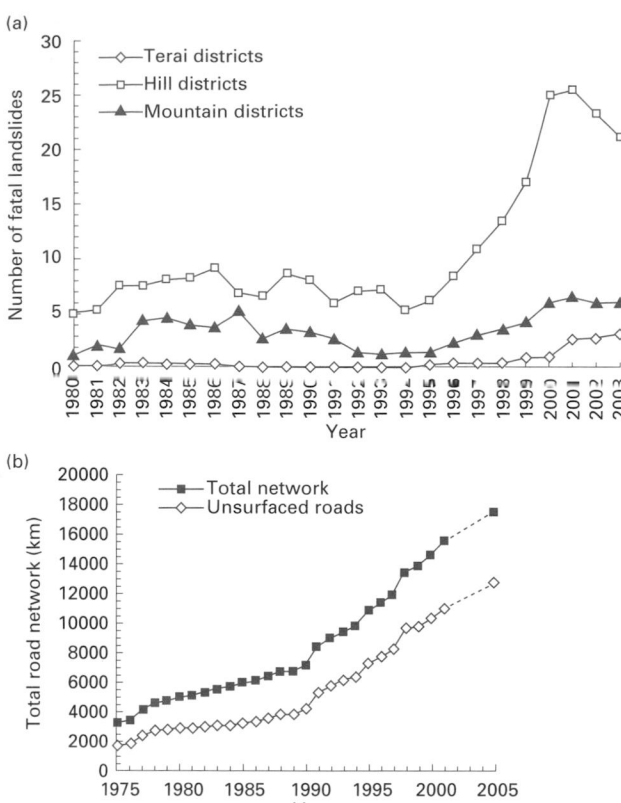

Figure 6.16 Landslides in Nepal. (a) Number of fatal landslides per year (smoothed by 5-year running mean) and (b) the growth of the road network (modified from Petley et al., 2007, figures 12b and 13b).

slip to take place, but it was the construction of the Vaiont Dam itself which changed the local groundwater conditions sufficiently to affect the stability of a rock mass on the margins of the reservoir: 240 million m³ of ground slipped, causing a rise in water level which overtopped the dam and caused flooding and loss of life downstream. Comparable slope instability resulted when the Franklin D. Roosevelt Lake was impounded by the Columbia River in the USA (Coates, 1977), but the effects were, happily, less serious.

Although the examples of accelerated mass movements that have been given here are essentially associated with the effects of modern construction projects, more long-established activities, including deforestation and agriculture, are also highly important. For example, Innes (1983) demonstrated, on the basis of lichenometric dating of debris-flow deposits in the Scottish Highlands, that most of the flows have developed in the last 250 years, and he suggests that inten-

sive burning and grazing may be responsible. The arrival of European settlers in New Zealand, particularly since the 1840s, had a profound effect on landslide activity, as they cleared the forest and converted it to pasture (Glade, 2003). Similarly, in the Italian Alps grazing and deforestation have reduced the retention of debris by vegetation and has resulted in thicker debris flow layers within lake sediments (Schneider et al., 2010). In the case of northern Spain, changes in landslide occurrence in the Upper Pleistocene and Holocene can partly be attributed to periods of intense precipitation or seismic activity, but the sharp increase in landslide numbers since the early 1980s appears to be related to increasing human pressures (Remondo et al., 2005). In the case of an alpine area in Switzerland, where the area affected by landslides increased by 92% from 1959 to 2004, the causes were a combination of an increase in the occurrence of torrential rain events and an increase in cattle stocking (Meusberger and Alewell, 2008). However, on a longer time span of 3600 years, pulses of increased landslide activity in the Swiss Alps have been linked to phases of deforestation (Dapples et al., 2002).

Fire, whether natural or man-induced, can be a major cause of slope instability and debris flow generation by removing or reducing protective vegetation, by increasing peak stream flows, and by leading to larger soil moisture contents and soil-water pore pressures (because of reduced interception of rainfall and decreased moisture loss by transpiration) (Wondzell and King, 2003). Examples of fire-related debris flow generation are known from many sites in the USA, including Colorado (Cannon et al., 2001b), New Mexico (Cannon et al., 2001a), the Rocky Mountains and the Pacific North West (Wondzell and King, 2003).

However as with so many environmental changes of that nature in the past, there are considerable difficulties in being certain about causation. This has been well expressed by Ballantyne (1991: 84):

Although there is growing evidence for Late Holocene erosion in upland Britain, the causes of this remain elusive. A few studies have presented evidence linking erosion to vegetation degradation and destruction due to human influence, but the validity of climatic deterioration as a cause of erosion remains unsubstantiated. This uncertainty stems from a tendency to link erosion with particular causes only through assumed coincidence in timing, a procedure fraught with difficulty because of imprecision in the dating of both putative causes and erosional effects. Indeed, in many reported instances, it is impossible to refute the possibility that the timing of erosional events or episodes may be linked to high magnitude storms of random occurrence, and bears little relation to either of the casual hypotheses outlined above. . . .

Deliberate modification of channels

Both for purposes of navigation and flood control humans have deliberately straightened many river channels (Figure 6.17). Indeed, the elimination of meanders contributes to flood control in two ways. First, it eliminates some overbank floods on the outside of curves, against which the swiftest current is thrown and where the water surface rises highest. Second, and more importantly, the resultant shortened course increases both the gradient and the flow velocity, and the floodwaters erode and deepen the channel, thereby increasing its flood capacity.

It was for this reason that a programme of channel cut-offs was initiated along the Mississippi in the early 1930s. By 1940 it had lowered flood stages by as much as 4 m at Arkansas City, Arkansas. By 1950 the length of the river between Memphis, Tennessee and Baton Rouge, Louisiana (600 km down the valley) had been reduced by no less than 270 km as a result of 16 cut-offs.

Some landscapes have become dominated by artificial channels, normally once again because of the need for flood alleviation and drainage. This is especially evident in an area like the English Fenlands where

straight constructed channels contrast with the sinuous courses of original rivers such as the Great Ouse.

Non-deliberate river-channel changes

There are thus many examples of the intentional modification of river-channel geometry by humans – by the construction of embankments, by channelization and by other such processes. The complexity and diversity of causes of stream channel change is brought out in the context of Australia (Table 6.12) (Rutherford, 2000) and more generally by Downs and Gregory (2004). Major changes in the configuration of channels can be achieved accidentally (Table 6.13), either because of human-induced changes in stream discharge or in

Table 6.12 Human impacts on Australian stream channel morphology

Channel incision by changes to resistance of valley flow (drains, cattle tracks)
Enlargement due to catchment clearing and grazing
Channel enlargement by sand and gravel extraction
Erosion by boats
Scour downstream from dams
Channelization and river training
Acceleration of meander migration rates by removal of riparian vegetation
Channel avulsion because of clearing of floodplain vegetation
Sedimentation resulting from mining
Channel contraction below dams
Channel invasion and narrowing by exotic vegetation

Table 6.13 Accidental channel changes

Phenomenon	Cause
Channel incision	Clearwater erosion below dams caused by sediment removal
Channel aggradation	Reduction in peak flows below dams
	Addition of sediment to streams by mining, agriculture and so on
Channel enlargement	Increase in discharge level produced by urbanization
Channel diminution	Discharge decrease following water abstraction or flood control
	Trapping and stabilizing of sediment by artificially introduced plants
Channel plan-form	Change in nature of sediment load and its composition, together with flow regime

Figure 6.17 A strait, lined channelized ephemeral river in Gran Canaria. (See Plate 41)

their sediment load: both parameters affect channel capacity (Park, 1977). The causes of observed cases of riverbed degradation are varied and complex and result from a variety of natural and human changes. A useful distinction can be drawn between degradation that proceeds downstream and that which proceeds upstream, but in both cases the complexity of causes is evident.

Deliberate channel straightening causes various types of sequential channel adjustment both within and downstream from straightened reaches, and the types of adjustment vary according to such influences as stream gradient and sediment characteristics. Brookes (1987) recognized five types of change *within* the straightened reaches (types W1 to W5) and two types of change downstream (types D1 and D2). They are illustrated in Figure 6.18.

Type W1 is degradation of the channel bed, which results from the fact that straightening increases the slope by providing a shorter channel path. This in turn increases its sediment transport capability.

Type W2 is the development of an armoured layer on the channel bed by the more efficient removal of fine

materials as a result of the increased sediment transport capability referred to above.

Type W3 is the development of a sinuous thalweg in streams which are not only straightened but which are also widened beyond the width of the natural channel.

Type W4 is the recovery of sinuosity as a result of bank erosion in channels with high slope gradients.

Type W5 is the development of a sinuous course by deposition in streams with a high sediment load and a relatively low valley gradient.

Types D1 and D2 result from deposition downstream as the stream tries to even out its gradient, the deposition occurring as a general raising of the bed level, or as a series of accentuated point bar deposits.

It is now widely recognized that the urbanization of a river basin results in an increase in the peak flood flows in a river (Bledsoe and Watson, 2001). It is also recognized that the morphology of stream channels is related to their discharge characteristics and especially to the discharge at which bank full flow occurs. As a result of urbanization, the frequency of discharges which fill the channel will increase, with the effect that the beds and banks of channels in erodible materials will be eroded so as to enlarge the channel (Trimble, 1997). Data from humid and temperate environments around the world indicate that urbanization causes channels to enlarge to 2–3 times and as much as 15 times the original size (Chin, 2006). This in turn will lead to bank caving, possible undermining of structures and increases in turbidity (Hollis and Luckett, 1976). Other changes include decreases in channel sinuosity and increases in bed material size (Grable and Harden, 2006). Trimble (2003) provides a good historical analysis of how the San Diego Creek in Orange County, California, has responded to flow and sediment yield changes related to the spread of both agriculture and urbanization.

Changes in channel morphology also result from discharge diminution and sediment load changes produced by flood-control works and diversions for irrigation (Brandt, 2000). This can be shown for the North Platte and the South Platte in America, where both peak discharge and mean annual discharge have declined to 10–30% of their pre-dam values. The North Platte, 762–1219 m wide in 1890 near the Wyoming Nebraska border, has narrowed to about 60 m at present; while the South Platte River was about 792 m wide, 89 km above its junction with the North Platte in

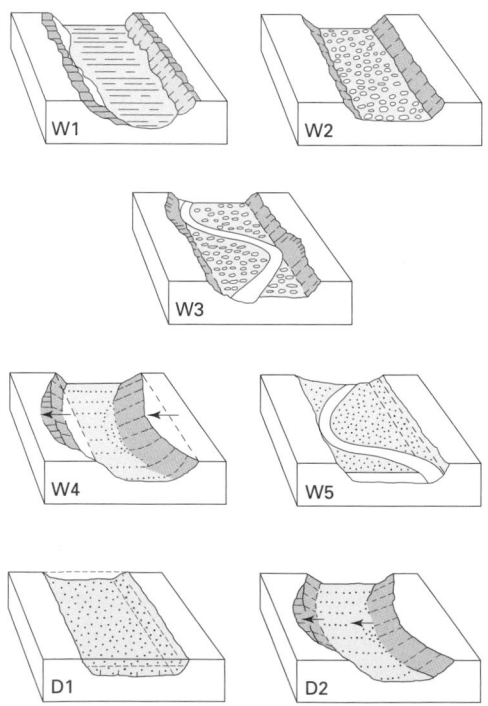

Figure 6.18 Principal types of adjustment in straightened river changes (after Brookes, 1987, figure 4). Reproduced with permission. For an explanation of the different types, see text.

Plate 1 (Figure 1.3) A view of the Olduvai Gorge in Tanzania – one of the great sites for the investigation of early man.

Plate 2 (Figure 1.4) A cluster of Palaeolithic hand axes from Olorgesailie in East Africa. Courtesy of Jean-Marc Jancovici, www.manicore.com, 2012.

The Human Impact on the Natural Environment: Past, Present and Future, Seventh Edition. Andrew S. Goudie.
© 2013 John Wiley & Sons, Ltd. Published 2013 by John Wiley & Sons, Ltd.

Plate 3 (Figure 1.5) Fire was one of the first and most powerful tools of environmental transformation employed by humans. The high grasslands of southern Africa may owe much of their character to regular burning, as shown here in Swaziland.

Plate 4 (Figure 1.9) Irrigation using animal power, as here in Rajasthan, India, is an example of the use of domesticated stock to change the environment.

Plate 5 (Figure 1.10) The development of ploughs provided humans with the ability to transform soils. This simple type is in Pakistan.

Plate 6 (Figure 1.11) In the Secondary Products Revolution, cattle, such as these in Haryana, India, were used as beasts of burden, for pulling ploughs, and as a source of products such as leather.

Plate 7 (Figure 1.12) Carthage, in northern Tunisia, was one of the great cities of the ancient world.

Plate 8 (Figure 1.13) Urbanization (and, in particular, the growth of large conurbations such as Kuwait) is an increasingly important phenomenon. Urbanization causes and accelerates a whole suite of environmental problems.

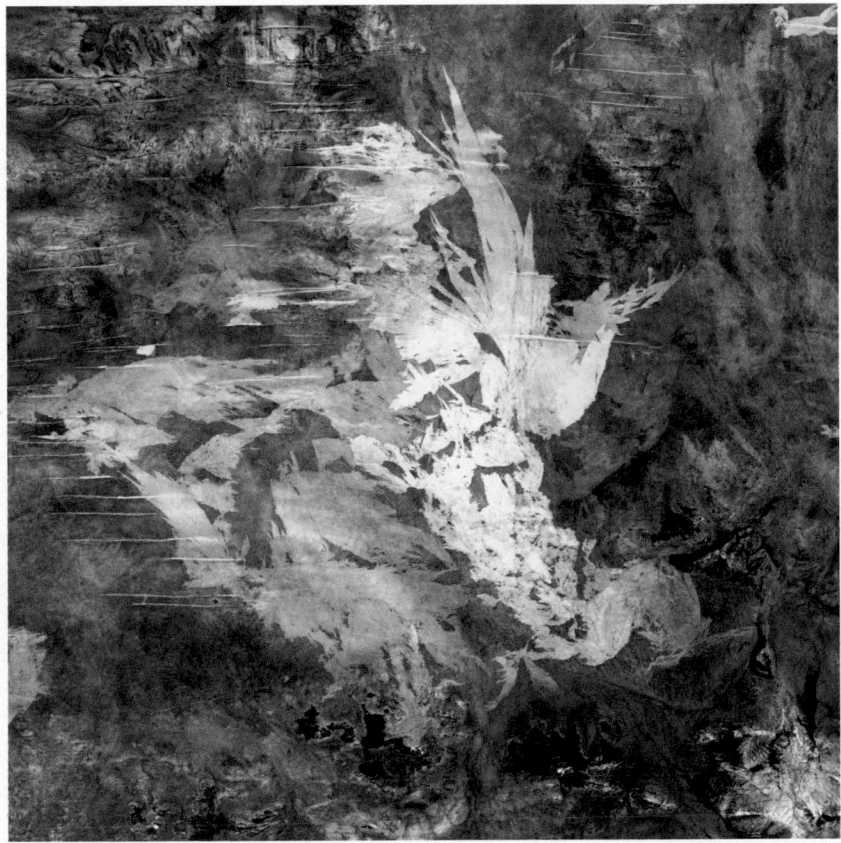

Plate 9 (Figure 2.2) The pindan bush of Australia, composed of *Eucalyptus*, *Acacia* and grasses, is frequently burned. The pattern of the burning shows up clearly on Landsat imagery.

Plate 10 (Figure 2.5) Intensive grazing by domestic stock, such as these cattle near Belo Horizonte in Brazil, has a major influence on many biomes, including grasslands and savannas.

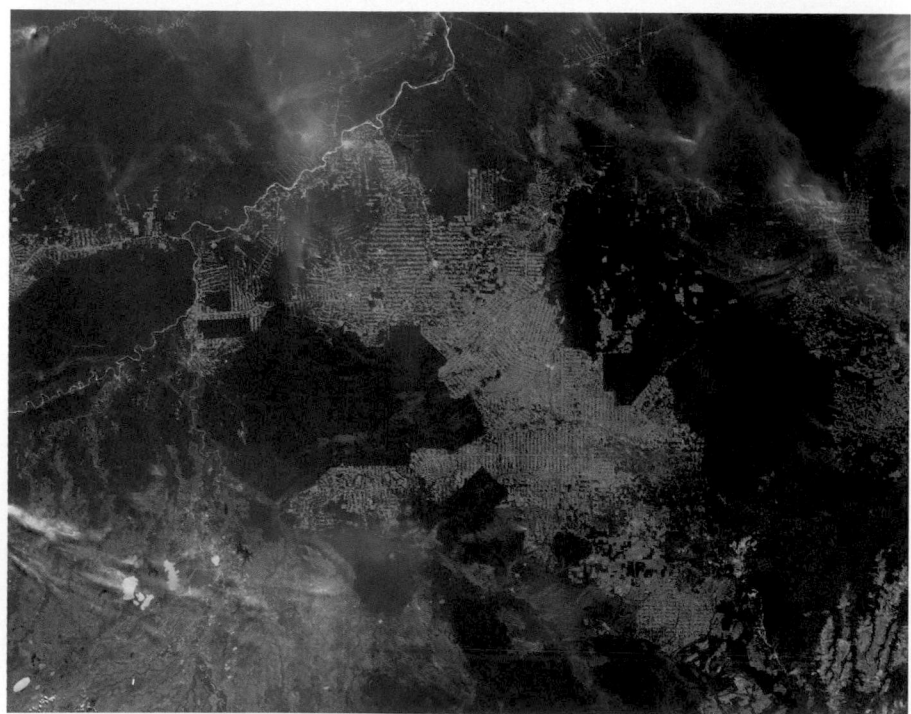

Plate 11 (Figure 2.7) A satellite image of the spread of agriculture and settlement through part of Amazonia (courtesy of NASA).

Plate 12 (Figure 2.10) Mangrove swamps are highly productive and diverse ecosystems that are being increasingly abused by human activities.

Plate 13 (Figure 2.12) A tropical savanna in the Kimberley District of north-west Australia.

Plate 14 (Figure 2.16) A lichen field, near Swakopmund in Namibia, shows the effects of vehicular traffic. Such scars can take a long time to recover.

Plate 15 (Figure 2.17) Deforestation for charcoal production and for firewood is a major cause of environmental change in the dry savanna woodlands of Swaziland.

Plate 16 (Figure 2.18) In the drylands of north-west India, the use of buffalo dung as fuel rather than as fertilizer, may have a deleterious effect on soil fertility.

Plate 17 (Figure 2.22) A monument in the botanic gardens in Mauritius, drawing attention to the value of introduced plants.

Plate 18 (Figure 2.24) The invasion of Australian acacia (background) is a major threat to the Fynbos (foreground) of the Cape region of South Africa.

Plate 19 (Figure 2.26) The great smelter at Sudbury in Canada. Pollution derived from the fumes from such sources can have many deleterious impacts on vegetation in their neighbourhood.

Plate 20 (Figure 2.28) The prickly pear (*Opuntia*) is a plant that has been introduced from the Americas to Africa and Australia. It has often spread explosively. Recently it has been controlled by the introduction of moths and beetles, an example of biological control.

Plate 21 (Figure 3.1) The consequences of domestication for the form of animals is dramatically illustrated by this large-horned but dwarf cow being displayed in a circus in Chamonix, France.

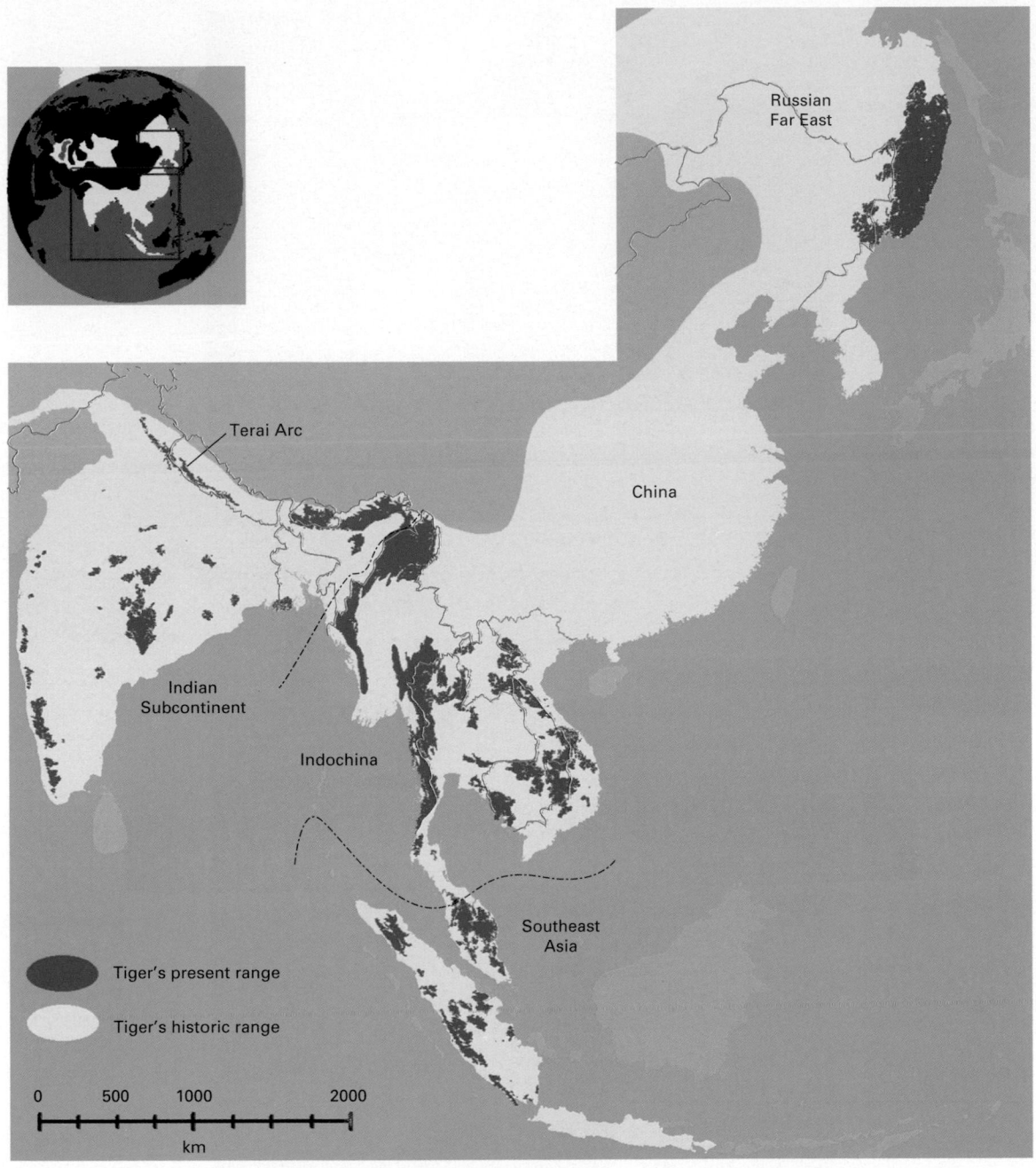

Plate 22 (Figure 3.7) The historic and present distribution of tigers. They once occupied portions of central Asia around the Caspian Sea (see inset globe) (modified from Dinerstein et al., 2007, figure 1).

Plate 23 (Figure 3.10) A major cause of animal decline is trapping. Here we see a sample of the many snares that have been found in a game reserve in Swaziland, southern Africa.

Plate 24 (Figure 3.13) A reconstruction of the demise of one member of the mega-fauna, a mastodon, at the La Brea Tar Pits, Los Angeles (ASG).

Plate 25 (Figure 4.1) The extension of irrigation in the Indus valley of Pakistan by means of large canals has caused widespread salination of the soils. Waterlogging is also prevalent. The white efflorescence of salt in the fields has been termed 'a satanic mockery of snow'.

Plate 26 (Figure 4.8) A lined irrigation canal in Haryana, India. Such lining is an important way in which to reduce waterlogging and salinization.

Plate 27 (Figure 4.10) Soil erosion near Baringo in Kenya has exposed the roots of a tree, thereby indicating the speed at which soil can be lost.

Plate 28 (Figure 4.11) The removal of vegetation in Swaziland creates spectacular gully systems, which in southern Africa are called *dongas*. The smelting of local iron ores in the early nineteenth century required the use of a great deal of firewood which may have contributed to the formation of this example.

Plate 29 (Figure 4.13) Linear erosion furrows along domestic animal tracks developed in vertisols in the highlands of Ethiopia.

Plate 30 (Figure 4.15) Surface degradation on a stone pavement caused by off-road driving in the Libyan Desert, Egypt.

Plate 31 (Figure 4.16) Soil conservation banks across a small ephemeral channel (wadi) in the Matmata area of Tunisia.

Plate 32 (Figure 4.17) Palm frond fences to reduce sand movement and dune migration in Erfoud, Morocco.

Plate 33 (Figure 5.2) The Kariba Dam on the Zambezi River between Zambia and Zimbabwe. Such large dams can provide protection against floods and water shortages, and generate a great deal of electricity. However, they can have a whole suite of environmental consequences.

Plate 34 (Figure 5.4) An irrigation canal in the plains of Haryana, India.

Plate 35 (Figure 5.13) Recessional shorelines caused by the ongoing decline in the level of the Dead Sea, Jordan.

Plate 36 (Figure 5.17) In the High Plains of the USA, fields are irrigated by centre-pivot irrigation schemes which use groundwater. Groundwater levels have fallen rapidly in many areas because of the adoption of this type of irrigation technology.

Plate 37 (Figure 6.1) The Rössing uranium mine near Swakopmund in Namibia, southern Africa. The excavation of such mines involves the movement of prodigious amounts of material.

Plate 38 (Figure 6.2) The mining of sand from large linear sand dunes to provide material for the construction industry in Dubai, United Arab Emirates.

Plate 39 (Figure 6.11) Permafrost in Siberia.

Plate 40 (Figure 6.13) The city of Mexico has subsided by many metres as a result of groundwater abstraction. Many of the ancient buildings in the city centre have been severely damaged, and have been cracked and deformed.

Plate 41 (Figure 6.17) A strait, lined channelized ephemeral river in Gran Canaria.

Plate 42 (Figure 6.22) A sand fence on a dune that threatens part of the town of Walvis Bay in Namibia.

Plate 43 (Figure 6.24) Coastal defence in Weymouth, southern England. The piecemeal emplacement of expensive sea walls and cliff protection structures is often only of short-term effectiveness and can cause accelerated erosion downdrift.

Plate 44 (Figure 6.27) In the nineteenth century a jetty was built at West Bay, Dorset, southern England to facilitate entry to the harbour. By the twenty-first century accretion has taken place in the foreground and erosion in the background.

Plate 45 (Figure 6.28) (a) Coastal protection measures at Arica, northern Chile. (b) A sea wall on Portland, Dorset, southern England.

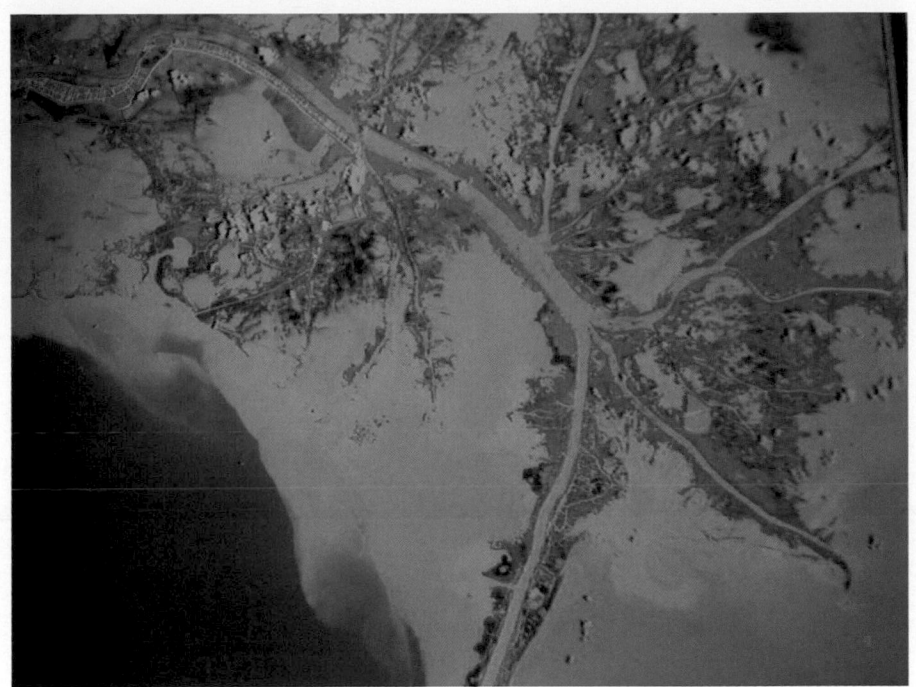

Plate 46 (Figure 6.32) A satellite image of the present day Mississippi Birdsfoot Delta (courtesy of NASA).

Plate 47 (Figure 8.4) Boreal forest in southern Sweden.

Plate 48 (Figure 9.1) An informal coastal settlement in Kota Kinabalu, Sabah, Malaysia. Such settlements will be particularly at risk as a consequence of rising sea levels (A.S.G.).

Plate 49 (Figure 9.8) Malibu Beach, California, USA, is an example of a vulnerable beach. It is very narrow and is bounded on the landward side by properties (A.S.G.).

Plate 50 (Figure 10.5) The Zambezi River at Victoria Falls.

Plate 51 (Figure 12.8) These linear dunes in the south-west Kalahari have active crests at the present day, particularly in dry, windy years, but may become still more active in coming decades.

Plate 52 (Figure 13.1) Rainforest, such as this example in Sabah, Malaysia, is a repository of great biodiversity and also controls local erosion, runoff, and hydrological conditions.

1897, but had narrowed to about 60 m by 1959 (Schumm, 1977: 161). The tendency of both rivers has been to form one narrow, well-defined channel in place of the previously wide, braided channels, and, in addition, the new channel is generally somewhat more sinuous than the old (Figure 6.19). River regulation structures have also had marked impacts on the form of Spain's Ebro River, which became much less sinuous

and much narrower during the twentieth century (Magdaleno et al., 2011).

Similarly, the building of dams can lead to channel aggradation upstream from the reservoir and channel deepening downstream because of the changes brought about in sediment loads (Figure 6.20). Some data on observed rates of degradation below dams are presented in Table 6.14. They show that the average rate

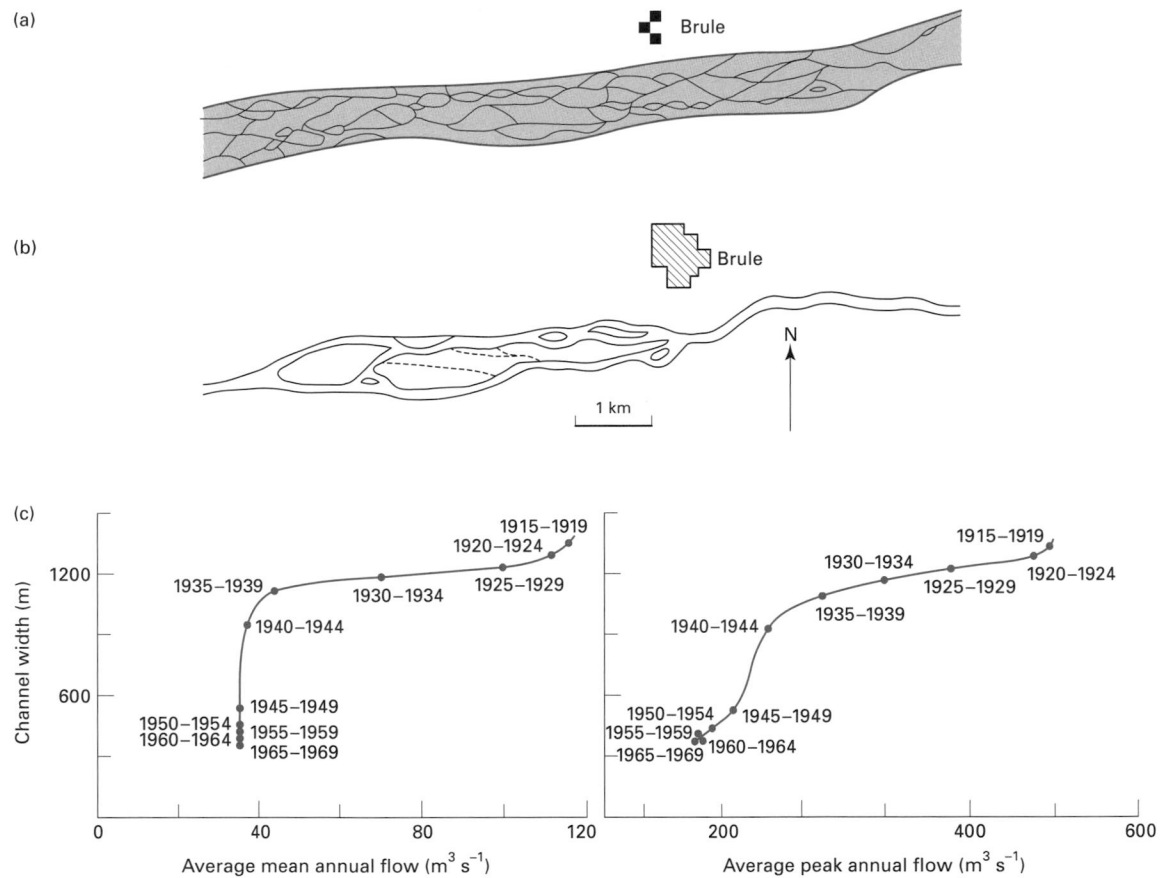

Figure 6.19 The configuration of the channel of the South Platte River at Brule, Nebraska, USA: (a) in 1897 and (b) in 1959. Such changes in channel form result from discharge diminution (c) caused by flood-control works and diversions for irrigation (modified from Schumm, 1977, figure 5.32 and Williams, 1978).

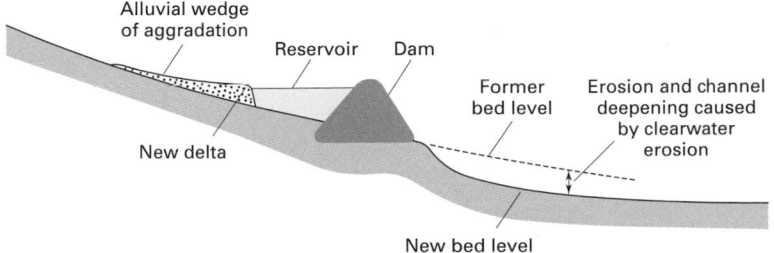

Figure 6.20 Diagrammatic long profile of a river showing the upstream aggradation and the downstream erosion caused by dam and reservoir construction.

Table 6.14 Riverbed degradation below dams. Source: Data from Galay (1983)

River	Dam	Amount (m)	Length (km)	Time (years)
South Canadian (USA)	Conchos	3.1	30	10
Middle Loup (USA)	Milburn	2.3	8	11
Colorado (USA)	Hoover	7.1	111	14
Colorado (USA)	Davis	6.1	52	30
Red (USA)	Denison	2.0	2.8	3
Cheyenne (USA)	Angostura	1.5	8	16
Saalach (Austria)	Reichenhall	3.1	9	21
South Saskatchewan (Canada)	Diefenbaker	2.4	8	12
Yellow (China)	Samenxia	4.0	68	4

of degradation has been of the order of a few metres over a few decades following closure of the dams. However, over time, the rate of degradation seems to become less or to cease altogether, and Leopold et al. (1964: 455) suggest that this can be brought about in several ways. First, because degradation results in a flattening of the channel slope in the vicinity of the dam, the slope may become so flat that the necessary force to transport the available materials is no longer provided by the flow. Second, the reduction of flood peaks by the dam reduces the competence of the transporting stream to carry some of the material on its bed. Thus if the bed contains a mixture of particle size the river may be able to transport the finer sizes but not the larger, and the gradual winnowing of the fine particles will leave an armour of coarser material that prevents further degradation.

The overall effect of the creation of reservoir by the construction of a dam is to lead to a reduction in downstream channel capacity (see Petts, 1979, for a review). This seems to amount to between about 30 m and 70% (see Table 6.15).

In recent years, partly because some old dams have become unsafe or redundant, they have been demolished. Most of the dams involved have been just a few metres high (see Table 6.16), though one example from Australia was 15 m high (Neave et al., 2009). Among the consequences of dam removal are incision into the sedimentary fill that had accumulated in the reservoir behind the dam, migration of a knickpoint upstream, deposition of liberated sediment and the formation of bars and the like downstream (Major et al., 2012). However, the precise consequences vary greatly depending on the nature and erodibility of the fill and the regime of the river. Some rivers appear to

Table 6.15 Channel capacity reduction below reservoirs. Source: Modified from Petts (1979, table 1)

River	Dam	% channel capacity loss
Republican, USA	Harlan County	66
Arkansas, USA	John Martin	50
Rio Grande, USA	Elephant Buttre	50
Tone, UK	Clatworthy	54
Meavy, UK	Burrator	73
Nidd, UK	Angram	60
Burn, UK	Burn	34
Derwent, UK	Ladybower	40

remove their fill rapidly and almost entirely, while others do not.

Equally far-reaching changes in channel form are produced by land-use changes and the introduction of soil conservation measures. Figure 6.21 is an idealized representation of how the river basins of Georgia in the USA have been modified through human agency between 1700 (the time of European settlement) and the present. Clearing of the land for cultivation (Figure 6.21b) caused massive slope erosion which resulted in the transfer of large quantities of sediment into channels and floodplains. The phase of intense erosive land use persisted and was particularly strong during the nineteenth century and the first decades of the twentieth century, but thereafter (Figure 6.21c) conservation measures, reservoir construction and a reduction in the intensity of agricultural land use led to further channel changes (Trimble, 1974). Streams ceased to carry such a heavy sediment load, they became much less turbid, and incision took place into the flood-plain sediments. By means of this active stream-bed erosion, streams

Table 6.16 Selected studies of the consequences and geomorphological effects of dam removal in the USA

Source	Location	Height (m)
Pearson et al. (2011)	Merrimack, New Hampshire	4
Rapid incision and clearance of fill		
Kibler et al. (2011)	Brownsville, Oregon	1.8–3.4
Release of gravel fill causing bars and pools and riffles downstream		
Burroughs et al. (2009)	Pine River, Michigan	5
Big headcut, but only 12% of sediment fill eroded after 10 years		
Major et al., 2008 (2012)	Marmot Dam, Oregon	14
Knickpoint migrates 1.5 km upstream, and creation of multiple thread channel over 2 km downstream		
Doyle et al. (2003)	Koshonong, Wisconsin	3.3
Little downstream aggradation as limited erosion from dam area		
Doyle et al. (2003)	Baraboo, Wisconsin	2
Channel incision upstream and temporary aggradation of point bar downstream		
Cheng and Granata (2007)	Sandusky, Ohio	2.2
Less than 1% of sediment stored in reservoir was transported downstream		
Wildman and MacBroom (2005)	Anaconda, Connecticut	3.35
Only c. one third of sediment removed after 5 years. Channel widening upstream of dam but incision limited by original channel armouring		
Rumschlag and Peck (2007)	Muroe, Ohio	3.66
Vertical incision to pre-impoundment substrate within 1 month. 1 m of aggradation downstream		

incised themselves into the modern alluvium, lowering their beds by as much as 3–4 m.

In the south-west Wisconsin a broadly comparable picture of channel change has been documented by Knox (1977). There, as in the Upper Mississippi Valley (Knox, 1987), it is possible to identify stages of channel modification associated with various stages of land use, culminating in decreased overbank sedimentation as a result of better land management in the last half century.

Other significant changes produced in channels include those prompted by accelerated sedimentation associated with changes in the vegetation communities growing along channels. The establishment or re-establishment of riparian forest has been implicated with channel narrowing in south eastern France during the twentieth century (Liébault and Piégay, 2002).

The nature of arid zone channels may be markedly affected by riparian vegetation. It is a mistake to assume that because this is generally less dense in deserts than elsewhere, it is of no consequence. In fact, phreatophytes such as tamarisk, cottonwood and willow, which draw their sustenance from groundwater at depth, can significantly influence channel geometry by increasing bank resistance to erosion, inducing deposition and increasing roughness, and by taking up so much water that discharge is reduced. Such vegetation can lead to significantly reduced channel width and, as a result, to increased overbank discharge (flooding) (Birken and Cooper, 2006). Thus, Graf (1978) showed that major river channels of the Colorado River and the plateau country in the USA had an average width reduction of 27% when tamarisk, a highly effective invasive species (Tickner et al., 2001), was established after 1930. Hadley (1961) demonstrated a similar effect on an arroyo in northern Arizona, and Graf (1981) demonstrated that in the ephemeral Gila River channel, sinuosity was increased from 1.13 to 1.23 as phreatophyte density increased in the 1950s. Tamarisk, because of its root system, is capable of resisting the hydraulic stresses of flash-floods. It is also drought tolerant and has an ability to produce denser stands than native species, which gives it a competitive advantage.

There is, however, a major question about the ways in which different vegetation types affect channel form (Trimble, 2004). Are tree-lined banks more stable than those flowing through grassland? On the one hand tree roots stabilize banks and their removal might be expected to cause channels to become wider and widening and shallower (Brooks and Brierly, 1997). On the other hand, forests produce log-jams which can cause aggradation or concentrate flow onto channel banks, thereby leading to their erosion. These issues are discussed in Trimble (1997) and Montgomery (1997).

Another organic factor which can modify channel form is the activity of grazing animals. These can break the banks down directly by trampling and can reduce bank resistance by removing protective vegetation and loosening soil (Trimble and Mendel, 1995).

The construction of water-powered mills has also been identified as a potent cause of historical channel change in the eastern USA. Walter and Merritts (2008),

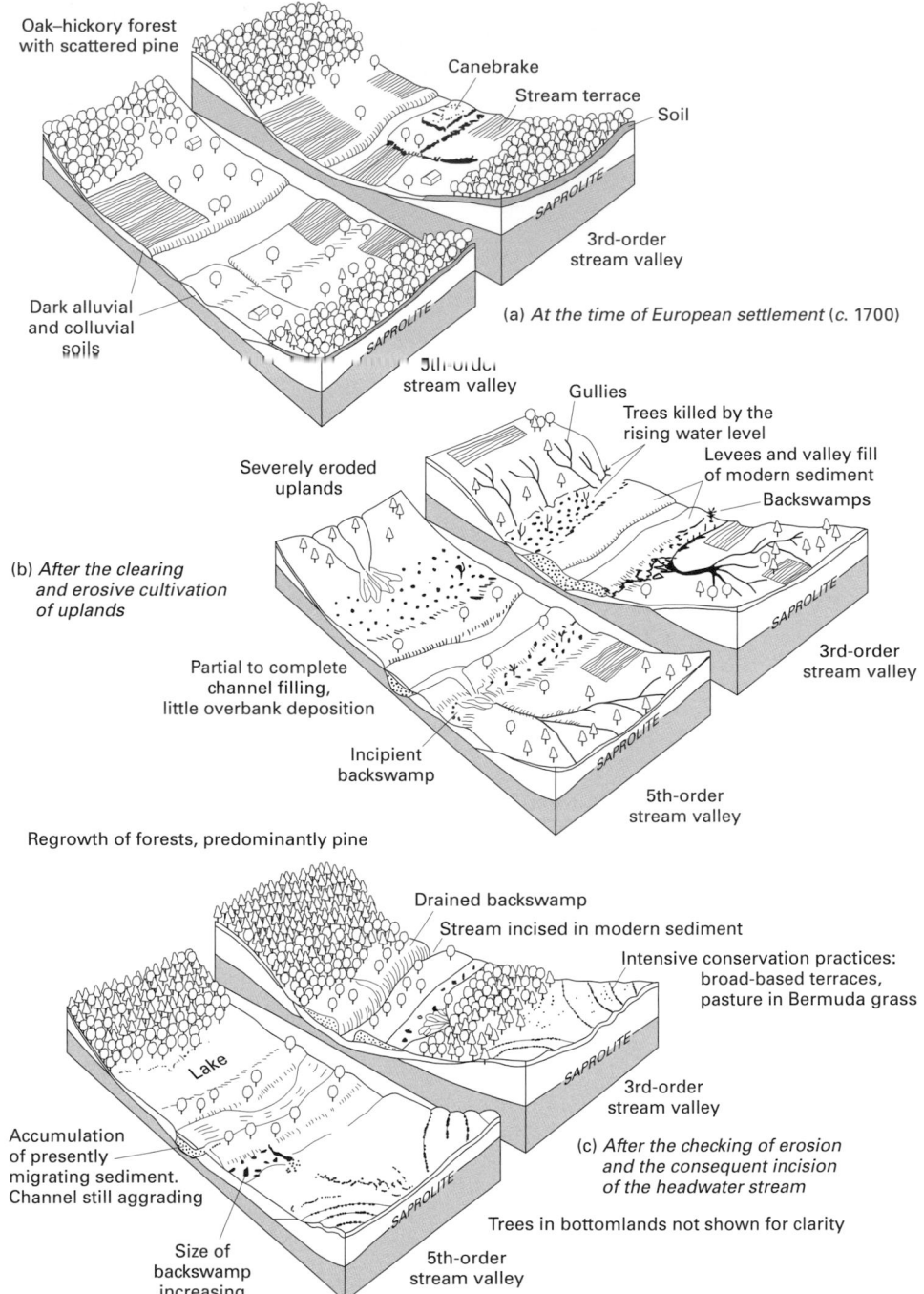

Figure 6.21 Changes in the evolution of fluvial landscapes in the Piedmont of Georgia, USA, in response to land-use change between 1970 (after Trimble, 1974: 117, in S. W. Trimble, *Man-induced soil erosion on the southern Piedmont*, Soil Conservation Society of America. © Soil Conservation Society of America). Reproduced with permission.

using old maps and archives, showed that whereas before European settlement the streams of the region were small anabranching channels within extensive vegetated wetlands, after the construction of tens of thousands of 17th- to 19th-century milldams, 1–5 m of slackwater sedimentation occurred and buried the pre-settlement wetlands with fine sediment.

Finally, the addition of sediments to stream channels by mining activity can cause channel aggradation. Mine wastes can clog channel systems (Gilbert, 1917; Lewin et al., 1983) and cause them to aggrade (Knighton, 1991). Equally, the mining of aggregates from river beds themselves can lead to channel deepening (Bravard and Petts, 1996; Rinaldi et al., 2005).

Reactivation and stabilization of sand dunes

To George Perkins Marsh the reactivation and stabilization of sand dunes, especially coastal dunes, was a theme of great importance in his analysis of human transformation of nature. He devoted 54 pages to it:

The preliminary steps, whereby wastes of loose, drifting, barren sands are transformed into wooded knolls and plains, and finally through the accumulation of vegetable mould, into arable ground, constitute a conquest over nature which proceeds agriculture – a geographical revolution – and therefore, an account of the means by which the change has been effected belongs properly to the history of man's influence on the great features of physical geography (1965: 393).

He was fascinated by 'the warfare man wages with the sand hills' and asked (1965: 410) 'in what degree the naked condition of most dunes is to be ascribed to the improvidence and indiscretion of man'.

His analysis showed quite clearly that most of the coastal dunes of Europe and North America had been rendered mobile, and hence a threat to agriculture and settlement, through human action, especially because of grazing and clearing. In Britain the cropping of dune warrens by rabbits was a severe problem, and a most significance event in their long history was the myxomatosis outbreak of the 1950s, which severely reduced the rabbit population and led to dramatic changes in stability and vegetative cover.

Appreciation of the problem of dune reactivation on mid-latitude shorelines, and attempts to overcome it,

go back a long way (Kittredge, 1948). For example, the menace of shifting sand following denudation is recognized in a decree of 1539 in Denmark which imposed a fine on those who destroyed certain species of sand plants on the coast of Jutland. The fixation of coastal sand dunes by planting vegetation was initiated in Japan in the seventeenth century, while attempts at the re-afforestation of the spectacular Landes dunes in south-west France began as early at 1717 and came to fruition in the nineteenth century through the plans of the great Bremontier: 81,000 hectares of moving sand had been fixed in the Landes by 1865. In Britain possibly the most impressive example of sand control is provided by the re-afforestation of the Culbin Sands in north-east Scotland with conifer plantations (Edlin, 1976).

Human-induced dune instability is not, however, a problem that is restricted to mid-latitude coasts. In inland areas of Europe, clearing, fire and grazing have affected some of the late Pleistocene dune fields that were created on the arid steppe margins of the great ice sheets, and in eastern England the dunes of the Breckland presented problems on many occasions. However, it is possibly on the margins of the great subtropical and tropical deserts that some of the strongest fears are being expressed about sand-dune reactivation. This is one of the facets of the process of desertification. The increasing population levels of both humans and their domestic animals, brought about by improvement in health and by the provision of boreholes, has led to an excessive pressure on the limited vegetation resources. As ground cover has been reduced, so dune instability has increased. The problem is not so much that dunes in the desert cores are relentlessly marching on to moister areas, but that the fossil dunes, laid down during the more arid phase peaking around 18,000 years ago, have been reactivated *in situ*.

A wide range of methods (Figure 6.22) is available to attempt to control drifting sand and moving dunes as follows:

1 *Drifting sand:*
  enhancement of deposition of sand by the creation of large ditches, vegetation belts and barriers and fences;
  enhancement of transport of sand by aerodynamic stream-lining of the surface, change of surface materials or panelling to direct flow;

Figure 6.22 A sand fence on a dune that threatens part of the town of Walvis Bay in Namibia. (See Plate 42)

reduction of sand supply, by surface treatment, improved vegetation cover or erection of fences;

deflection of moving sand, by fences, barriers, tree belts and so on.

2 *Moving dunes:*

removal by mechanical excavation;

destruction by reshaping, trenching through dune axis or surface stabilization of barchan arms;

immobilization, by trimming, surface treatment and fences.

In practice most solutions to the problem of dune instability and sand blowing have involved the establishment of a vegetation cover. This is not always easy. Species used to control sand dunes must be able to endure undermining of their roots, burning, abrasion and often severe deficiencies of soil moisture. Thus the species selected need to have the ability to recover after partial growth in the seedling stages, to promote rapid litter development and to add nitrogen to the soil through root nodules. During the early stages of growth they may need to be protected by fences, sand traps and surface mulches. Growth can also be stimulated by the addition of synthetic fertilizers.

In hearts of deserts sand dunes are naturally mobile because of the sparse vegetation cover. Even here, however, humans sometimes attempt to stabilize sand surfaces to protect settlements, pipelines, industrial plant and agricultural land. The use of relatively porous barriers to prevent or divert sand movement has proved comparatively successful, and palm fronds or chicken wire have made adequate stabilizers. Elsewhere surfaces have been strengthened by the application of high-gravity oil or by salt-saturated water, which promotes the development of wind-resistant surface crusts. Frequently these techniques are not particularly successful and very often the best solution is to site and design engineering structures to allow free movement of sand across them. Alternatively, by mapping different dune types and knowing their direction and rate of movement, structures should be located out of harm's way. Avoidance may be better than defence.

Various studies of the comparative effectiveness of different stabilization techniques have been undertaken in recent years. For example, Zhang et al. (2004) found that the best means of stabilizing moving dunes in Inner Mongolia, China, were wheat straw checkerboards and the planting of *Artemisia halodendron*. This finding was confirmed by a study in the Kerqin Sandy land of northern China (Li et al., 2009). Along a major highway in the Taklamakan desert, checkerboards, reed fences and nylon nets were found to be effective (Dong et al., 2004). In northwest Nigeria, Raji et al. (2004) found that shelterbelts were the most effective technique and were superior to mechanical fencing. Some success has also been claimed for chemical stabilizers (Han et al., 2007) and for geotextiles (Escalente and Pimentel, 2008). Some devices are prohibitively expensive (e.g. chemical fixers) (Dong et al., 2004), while others, such as checkerboards, are not.

In temperate areas coastal dunes have been effectively stabilized by the use of various trees and other plants (Ranwell and Boar, 1986). In Japan *Pinus thumbergii* has been successful, while in the great Culbin Sands plantations of Scotland *P. nigra* and *P. laricio* have been used initially, followed by *P. sylvestris*. Of the smaller shrubs, *Hippophae* has proved highly efficient, sometimes too efficient, at spreading. Its clearance from areas where it is not welcome is difficult precisely because of some of the properties that make is such an efficient sand stabilizer: vigorous suckering growth and the rapid regrowth of cut stems (Boorman, 1977). Different types of grass have also been employed, especially in the early stages of stabilization. These include two grasses which are moderately tolerant of salt: *Elymus farctus* (sand twitch) and *Leymus arenarius* (lime grass). Another grass which is much used, not least because of its rapid and favourable response to burial by drifting sand, is *Ammophila arenaria* (marram).

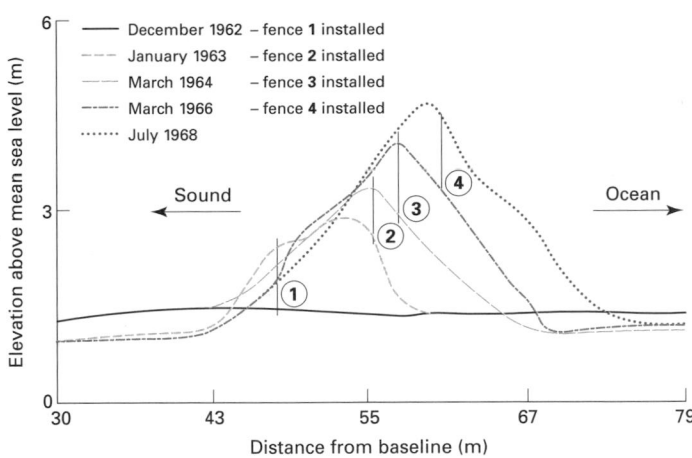

Figure 6.23 Sand accumulation using the method of multiple fences in North Carolina, USA. This raised the dune height approximately 4 m over a period of 6 years (after Savage and Woodhouse in Goldsmith, 1978, figure 36, with permission of the American Society of Civil Engineers).

Further stabilization of coastal dunes has been achieved by setting up sand fences. These generally consist of slats about 1.0–1.5 m high and have a porosity of 25–50%. They have proved to be effective in building incipient dunes in most coastal areas. By installing new fences regularly, large dunes can be created with some rapidity (see Figure 6.23). Alternative methods, such as using junk cars on the beaches at Galveston, Texas, have been attempted with little success. Larger dunes may also be created by the bulldozing of beach sediments, a process termed scraping, but such material may be easily eroded and cause downdrift accretion (Kratzmann and Hapke, 2012).

Accelerated coastal erosion

Coastal dunes are not the only parts of our seashores that are being affected by human activities. Human impacts on coasts are legion. General treatments of coastal problems and their management are provided by Viles and Spencer (1995) and by French (1997, 2001), while Reeve et al. (2012) give an account of coastal engineering from an engineering viewpoint.

Because of the high concentration of settlements, industries, transport facilities and recreational developments on coastlines, the pressures placed on coastal landforms are often acute (Nordstrom, 1994; Evans, 2008) and the consequences of excessive erosion serious. While most areas are subject to some degree of natural erosion and accretion, the balance can be upset by human activity in a whole range of different ways (Table 6.17). However, humans seldom attempt

to accelerate coastal erosion deliberately. More usually, it is an unexpected or unwelcome result of various economic projects. Frequently coast erosion has been accelerated as a result of human efforts to reduce it.

One of the best forms of coastal protection is a good beach. If material is removed from a beach, accelerated cliff retreat may take place. Removal of beach materials may be necessary to secure valuable minerals, including heavy minerals, or to provide aggregates for construction. The classic example of the latter was the mining of 660,000 tonnes of shingle from the beach at Hallsands in Devon, England, in 1887 to provide material for the construction of dockyards at Plymouth. The shingle proved to be undergoing little or no natural replenishment and in consequence the shore level was reduced by about 4 m. The loss of the protective shingle soon resulted in cliff erosion to the extent of 6 m between 1907 and 1957. The village of Hallsands was cruelly attacked by waves and is now in ruins.

Another common cause of beach and cliff erosion at one point is coast protection at another (Figure 6.24). As already stated, a broad beach serves to protect the cliffs behind, and beach formation is often encouraged by the construction of **groynes** and a range of 'hard engineering' structures is available (Figure 6.25). However, these structures sometimes merely displace the erosion (possibly in an even more marked form) further along the coast. This is illustrated in Figure 6.26.

Piers or breakwaters can have similar effects to groynes. This has occurred at various places along the British coast: erosion at Seaford resulted from the Newhaven breakwater, while erosion at Lowestoft resulted from the pier at Gorleston. Figure 6.27 shows the coast

Table 6.17 Mechanisms of human-induced erosion in coastal zones. Source: Hails (1977: 348, table 9.11). Reproduced with permission from Elsevier

Human-induced erosion zones	Effects
1 Beach mining for placer deposits (heavy minerals) such as zircon, rutile, ilmenite and monazite	Loss of sand from frontal dunes and beach ridges
2 Construction of groynes, breakwaters, jetties and other structures	Downdrift erosion
3 Construction of offshore breakwaters	Reduction in littoral drift
4 Construction of retaining walls to maintain river entrances	Interruption of littoral drift resulting in downdrift erosion
5 Construction of sea walls, revetments and so on	Wave reflection and accelerated sediment movement
6 Deforestation	Removal of sand by wind
7 Fires	Migrating dunes and sand drift after destruction of vegetation
8 Grazing of sheep and cattle	Initiation of blowouts and transgressive dunes: sand drift
9 Off-road recreational vehicles (dune buggies, trail bikes, etc.)	Triggering mechanism for sand drift attendant upon removal of vegetative cover
10 Reclamation schemes	Changes in coastal configuration and interruption of natural processes, often causing new patterns in sediment transport
11 Increased recreational needs	Accelerated deterioration, and destruction, of vegetation on dunal areas, promoting erosion by wind and wave action

Figure 6.24 Coastal defence in Weymouth, southern England. The piecemeal emplacement of expensive sea walls and cliff protection structures is often only of short-term effectiveness and can cause accelerated erosion downdrift. (See Plate 43)

at West Bay in Dorset, southern England, following the construction of a jetty. The beach in the foreground appears to have built build outwards, while the cliff behind the jetty has retreated and so needs to be protected with a sea wall and with large imported rock armour. Likewise, the construction of some sea walls (Figure 6.28a and b), erected to reduce coastal

erosion and flooding, has had the opposite effect to the one intended (see Figure 6.29). Given the extent to which artificial structures have spread along the coastlines of the world, this is a serious matter (Walker, 1988).

Problems of coastal erosion are exacerbated because there is now abundant evidence to suggest that much of the reservoir of sand and shingle that creates beaches is in some respects a relict feature. Much of it was deposited on continental shelves during the maximum of the last glaciation (around 18,000 year BP), when sea level was about 120 m below its present level. It was transported shoreward and incorporated in present-day beaches during the phase of rapidly rising post-glacial sea levels that characterized the Flandrian (Holocene) transgression until about 6000 years BP. Since that time, with the exception of minor oscillations of the order of a few metres, world sea levels have been stable and much less material is, as a consequence, being added to beaches and shingle complexes (Hails, 1977). It is because of these problems that many erosion prevention schemes now involve beach nourishment (by the artificial addition of appropriate sediments to build up the beach), or employ miscellaneous sand bypassing techniques (including pumping and dredging) whereby sediments are transferred from the accumulation side of

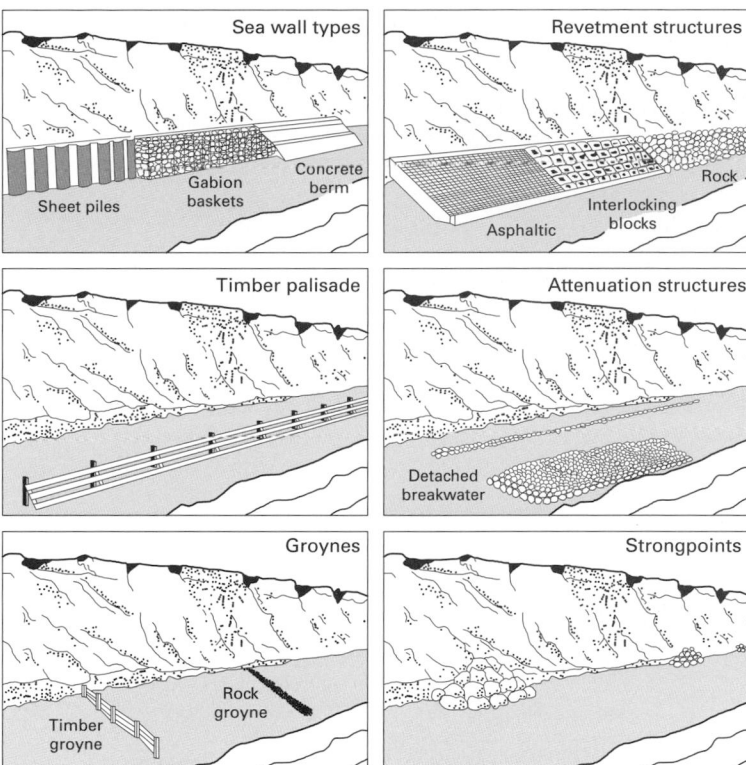

Figure 6.25 A selection of 'hard engineering' structures designed to afford coastal protection (modified from A.H. Brampton, 'Cliff conservation and protection: methods and practices to resolve conflicts', in J. Hooke, ed., *Coastal and Earth Science Conservation* (Geological Society Publishing House, 1998), figures 3.1, 3.2, 3.4, 3.5, 3.6 and 3.7)

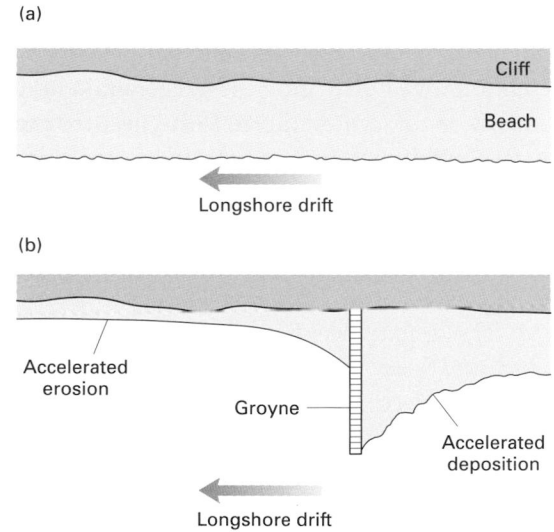

Figure 6.26 Diagrammatic illustration of the effects of groyne construction on sedimentation on a beach.

Figure 6.27 In the nineteenth century a jetty was built at West Bay, Dorset, southern England to facilitate entry to the harbour. By the twenty-first century accretion has taken place in the foreground and erosion in the background. (See Plate 44)

an artificial barrier to the erosional side. Such methods of beach nourishment are reviewed by Bird (1996). In recent years is that there has been an increasing trend towards so-called soft means of coastal protection, rather than using hard engineering structures such as sea walls or groynes. In addition to beach nourishment, the encouragement of dune formation, and promotion of salt marsh accretion are becoming recognized as being aesthetically pleasing, effective and economically advantageous.

Figure 6.28 (a) Coastal protection measures at Arica, northern Chile. (See Plate 45) (b) A sea wall on Portland, Dorset, southern England. (See Plate 45)

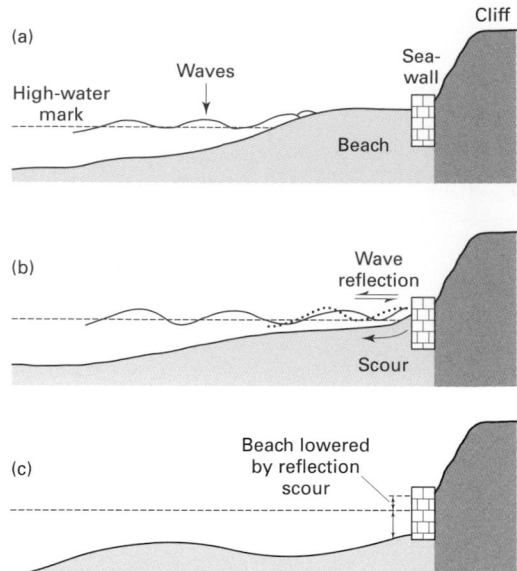

Figure 6.29 Sea walls and erosion: (a) a broad, high beach prevents storm waves breaking against a sea wall and will persist, or erode only slowly; but where the waves are reflected by the wall (b), scour is accelerated, and the beach is quickly removed and lowered (c) (modified from Bird, 1979, figure 6.3).

In some areas, however, sediment-laden rivers bring material into the coastal zone which becomes incorporated into beaches through the mechanism of longshore drift. Thus any change in the sediment load of such rivers may result in a change in the sediment budget of neighbouring beaches. When accelerated soil erosion occurs in a river basin the increased sediment load may cause coastal accretion and siltation. But where the sediment load is reduced through action such as the construction of large reservoirs, behind which sediments accumulate, coastal erosion may result. This is believed to be one of the less desirable consequences of the construction of the Aswan Dam on the Nile: parts of its delta have shown recently accelerated recession. In China the construction of around 50,000 dams in the Yangtze basin of China, has caused sediment starvation and severe erosion of its delta (Yang et al., 2011b).

In the case of the Nile, sediments, on reaching the sea, used to move eastward with the general anticlockwise direction of water movements in that part of the eastern Mediterranean, generating sand bars and dunes which contributed to delta accretion. About a century ago an inverse process was initiated and the delta began to retreat. For example, the Rosetta mouth of the Nile lost about 1.6 km of its length from 1898 to 1954. The imbalance between sedimentation and erosion appears to have started with the Delta Barrages (1861) and culminated with the High Dam itself a century later. In addition, large amounts of sediment are retained in an extremely dense network of irrigation channels and drains that has been developed in the Nile Delta itself (Stanley, 1996). Much of the Egyptian coast is now 'undernourished' with sediment and, as a result of this overall erosion of the shoreline, the sand bars bordering Lake Manzala and Lake Burullus on the seaward side are eroded and likely to collapse. If this were to happen, the lakes would be converted into marine bays, so that saline water would come into direct contact with low-lying cultivated land and freshwater aquifers.

Delta erosion is a pervasive and developing problem, not least in south and south east Asia. This is partly because they sink under their own weight of sediment, suffer from compaction of organic sediments, may be subsiding because of the removal of fluids and are subjected to ongoing sea-level rise. They also suffer from a reduced amount of sediment nourishment following dam construction upstream and from a reduction in the number of distributary channels as a consequence of a need to support navigation in a limited number of larger channels (Syvitski and Saito, 2007; Syvitski et al., 2009). Indeed, such actions have already been implicated in the changes that have taken place in the morphology of the Nile Delta and its lagoons over the last half century (El Banna and Frihy, 2009).

In Texas, where over the last century four times as much coastal land has been lost as has been gained, one of the main reasons for this change is believed to be the reduction in the suspended loads of some of the rivers discharging into the Gulf of Mexico (Table 6.18). The four rivers listed carried, in 1961–1970, on average only about one-fifth of what they carried in 1931–1940. Comparably marked falls in sediment loadings occurred elsewhere in the eastern USA (Figure 6.30). Likewise, in France the once mighty Rhône only carries about 5% of the load it did in the nineteenth century; and in Asia, the Indus discharges less than 20% of the load it did before construction of large barrages over the last half century (Milliman, 1990). On a global basis, large dams may retain 25–30% of the global flux of river sediment (Vörösmarty et al., 2003) (see also Chapter 6).

A good case study of the potential effects of dams on coastal sediment budgets is provided by California by Willis and Griggs (2003). Given that rivers provide the great bulk of beach material (75% to 90%) in the state, the reduction in sediment discharge by dammed rivers can have highly adverse effects. Almost a quarter of the beaches in California are down coast from rivers that have had sediment supplies diminished by one-third or more. Most of those threatened beaches are in southern California where much of the state's tourism and recreation activities are concentrated.

Construction of great levées on the lower Mississippi River since 1717 has also affected the Gulf of Mexico coast. The channelization of the river has

Table 6.18 Suspended loads of Texas rivers discharging into the Gulf of Mexico. Source: Modified from Hails (1977, table 9.1) after data from Stout et al. and Curtis et al.

River	Suspended load (million tonnes)		%[a]
	1931–1940	*1961–1970*	
Brazos	350	120	30
San Bernard	1	1	100
Colorado	100	11	10
Rio Grande	180	6	3
Total	631	138	20

[a]1961–1970 loads as a percentage of 1931–1940 loads.

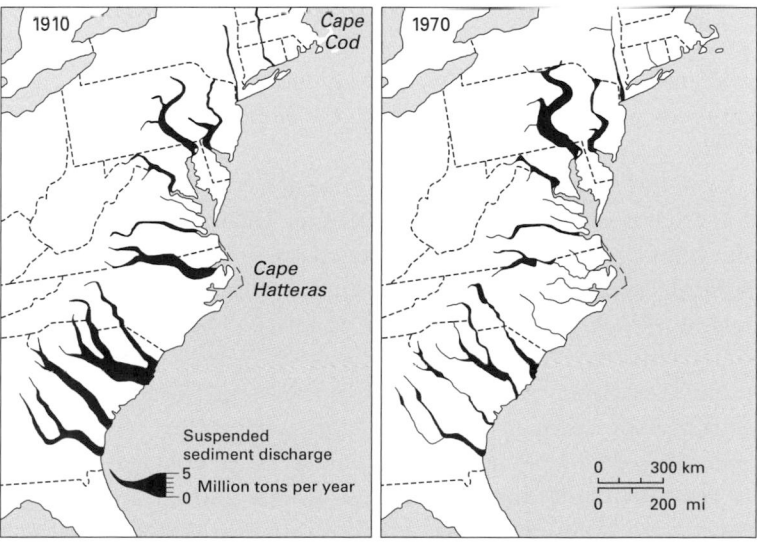

Figure 6.30 The decline in suspended sediment discharge to the eastern seaboard of the USA between 1910 and 1970 as a result of soil conservation measures, dam construction and land-use changes (after Meade and Trimble, 1974).

increased its velocity, reduced overbank deposition of silt onto swamps, marshes and estuaries, and changed the salinity conditions of marshland plants (Cronin, 1967). As a result, the coastal marshes and islands have suffered from increased erosion or a reduced rate of development. This has been vividly described by Biglane and Lafleur (1967: 691):

Like a bullet through a rifle barrel, waters of the mighty Mississippi are thrust toward the Gulf between the confines of the flood control levées. Before the day of these man-made structures, these waters poured out over tremendous reaches of the coast. . . . Freshwater marshes (salinities averaging 4-6%) were formed by deposited silts and vegetative covers of wire grass. . . . As man erected his flood protection devices, these marshes ceased to form as extensively as before.

The changes between 1956 and 1990 are shown in Figure 6.31. However, as with so many examples of environmental change, it is unlikely that just one factor, in this case channelization, is the sole cause of the observed trend. In their study of erosion loss in the Mississippi Delta and neighbouring parts of the Louisiana Coast, Walker et al. (1987) suggest that this loss is the result of a variety of complex interactions among a number of physical, chemical, biological and cultural processes. These processes include, in addition to channelization, worldwide sea-level changes, subsidence resulting from sediment loading by the delta of the underlying crust, changes in the sites of deltaic sedimentation as the delta evolves, catastrophic storm surges and subsidence resulting from subsurface fluid withdrawal. Tweel and Turner (2012) suggest that an important factor has been changes in the amount of sediment carried by the Mississippi in response to land-use changes and river engineering. Plainly, however, as seen from space (Figure 6.32) the birdfoot delta is a very fragile structure.

In some areas anthropogenic vegetation modification creates increased erosion potential. This has been illustrated for the hurricane-afflicted coast of Belize, Central America (Stoddart, 1971). He showed that natural, dense vegetation thickets on low, sand islands (*cays*) acted as a baffle against waves and served as a massive sediment trap for coral blocks, shingle and sand transported during extreme storms. However, on many islands the natural vegetation had been replaced by coconut plantations. These had an open structure easily penetrated by seawater, they tended to have

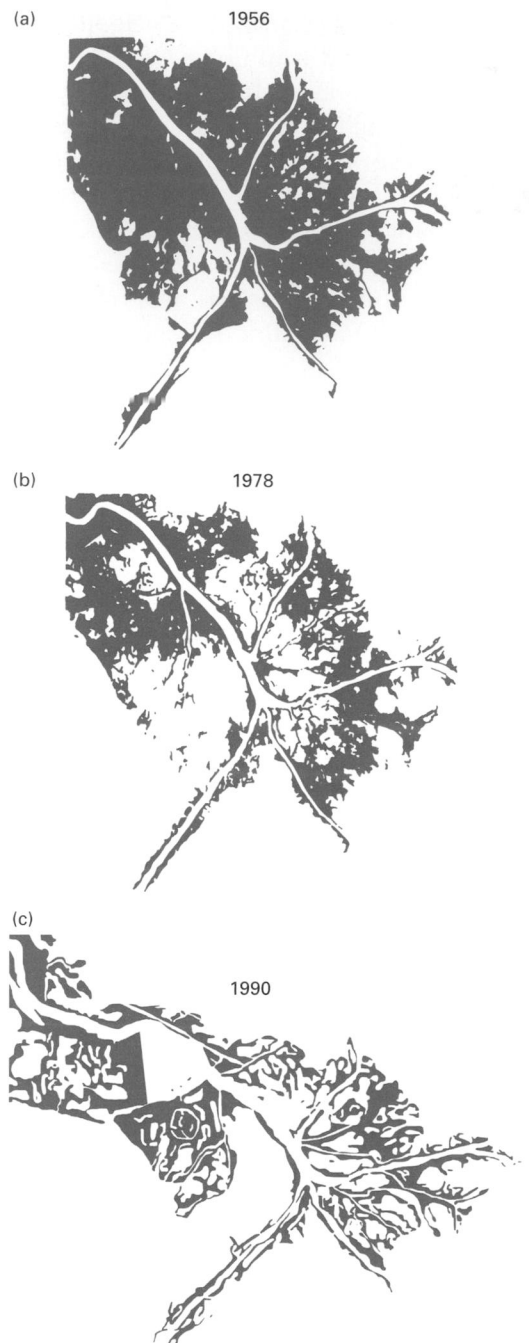

(a) 1956

(b) 1978

(c) 1990

Figure 6.31 Comparison of the outlines of the Mississippi Birdsfoot Delta from the 1930s to 1990 gives a clear indication of the transformation from marsh to open water. Artificial controls upriver have decreased the amount of sediment carried by the river; artificial levées along much of the lower course have kept flood-borne sediment from replenishing the wetlands, and in the active delta itself, rock barriers installed across breaks similarly confine the river. The Gulf of Mexico is intruding as the marshland sinks or is washed away.

Figure 6.32 A satellite image of the present day Mississippi Birdsfoot Delta (courtesy of NASA). (See Plate 46)

little or no ground vegetation (thus exposing the cay surface to stripping and channelling), and they had a dense but shallow root net easily undermined by marginal sapping. Thus Stoddart found (p.191) that 'where the natural vegetation had been replaced by coconuts before the storm (Hurricane Hattie), erosion and beach retreat led to net vertical decreases in height of 3–7 ft; whereas where natural vegetation remained, banking of storm sediment against the vegetation hedge led to a net vertical increase in height of 1–5 ft'.

Other examples of markedly accelerated coastal erosion and flooding result from anthropogenic degradation of dune ridges. Frontal dunes are a natural defence against erosion, and coastal changes may be long-lasting once they are breached. Many of those areas in eastern England which most effectively resisted the great storm and surge of 1952 were those where humans had not intervened to weaken the coastal dune belt.

Coral reefs are another coastal landform type that is undergoing profound change at the present time. Many are being heavily degraded. As Pandolfi et al. (2005: 1275) remarked, 'Large animals, like turtles, sharks and groupers, were once abundant on all coral reefs, and large long-lived corals created a complex architecture supporting diverse fish and invertebrates. Today the most degraded reefs are little more than rubble, seaweed, and slime.' This is a theme that is discussed in more detail in Chapter 3.

Changing rates of salt marsh accretion

Salt marshes are important habitats and also play a major role in coastal protection. However, in Britain, the nature of some salt marshes, and the rate at which they accrete, has been transformed by a major vegetational change, namely the introduction of a salt marsh plant, *Spartina alterniflora*. This cord-grass appears to have been introduced to Southampton Water in southern England by accident from the east coast of North America, possibly in shipping ballast. The crossing of this species with the native *Spartina maritima*, produced an invasive cord-grass of which there were two forms: *Spartina townsendii* and *Spartina anglica*, the latter of which is now the main species. It appeared first at Hythe on Southampton Water in 1870 and then spread rapidly to other salt marshes in Britain: partly because of natural spread and partly because of deliberate planning. *Spartina* now reaches as far north as the Island of Harris in the west of Scotland and to the Cromarty Firth in the east (Doody, 1984). The plant has often been effective at excluding other species and also at trapping sediment. Rates of accretion can therefore become very high. Ranwell (1964) gives rates as high as 8–10 cm per year. There is evidence that this has caused progressive silting of estuaries such as those of the Dee (Marker, 1967) and the Severn (Page, 1982). *Spartina* invasion has also occurred in other parts of the world including the Netherlands, Australia, New Zealand, California and China (Gedan et al., 2009; Strong and Ayres, 2009).

However, for reasons that are not fully understood, *Spartina* marshes have frequently suffered dieback, which has sometimes led to marsh recession (Hughes and Paramor, 2004). In the case of Poole Harbour and Beaulieu Estuary, this may date back to the 1920s, but elsewhere on the south coast it has generally been rapid and extensive since about 1960 (Tubbs, 1984).

Among the hypotheses that have been put forward to explain dieback in various parts of the world are the role of rising sea level, pathogenic fungi, increased wave attack, competition from and bioturbation by the polychaete, *Nereis diversicolor*, grazing by the crab *Sesarma reticulatum* (Bertness and Silliman, 2008), and the onset of waterlogging and anaerobic conditions on mature marsh (Hübner et al., 2010). Some support for the latter view comes from the fact that in areas where the introduction has been more recent, for example in

Wales and north-west England, the area of *Spartina* still appears to be increasing (Deadman, 1984).

However, *Spartina* invasion is not the sole cause of changes in accretion rates on marshes (Gedan et al., 2011). Other important factors to be considered include runoff and sediment erosion from the land as a result of land cover changes (Mattheus et al., 2010), the construction of dykes and levees, the digging of canals, the throwing up of spoil banks, land reclamation schemes, grid ditches to help control mosquitoes, tidal restriction by bridges and berms, and subsidence resulting from fluid abstraction. These are reviewed in the context of the USA by Kennish (2001) and globally by Gedan et al. (2009). It seems that some of the salt marshes along the coast of North America expanded rapidly during the eighteenth and nineteenth centuries due to increased rates of sediment delivery following deforestation associated with European settlement (Kirwan et al., 2011).

The human impact on seismicity and volcanoes

The seismic and tectonic forces which mould the relief of the earth and cause such hazards to human civilization are one of the fields in which efforts to control natural events have had least success and where least has been attempted. Nonetheless, the fact that humans have been able, inadvertently, to trigger off small earthquakes by nuclear blasts, as in Nevada, by injecting water into deep wells, as in Colorado, by mining, by building reservoirs and by fluid extraction, suggests that in due course it may be possible to 'defuse' earthquakes by relieving tectonic strains gradually in as series of non-destructive, low-intensity earthquakes. One problem, however, is that there is no assurance that an earthquake purposefully triggered by human action will be a small one, or that it will be restricted to a small area. The legal implications are immense.

Perhaps the most important anthropogenically induced seismicity results from the creation of large reservoirs (Talwani, 1997; Guha, 2000; Gupta, 2002; Durá-Gómez and Talwani, 2010). Reservoirs impose stresses of significant magnitude on crustal rocks at depths rarely equalled by any other human construction. With the ever-increasing number and size of reservoirs the threat rises. There are at least six cases (Koyna, Kremasta, Hsinfengkiang, Kariba, Oroville (California), Hoover and Marathon) where earthquakes of a magnitude greater than 5, accompanied by a long series of foreshocks and aftershocks, have been related to reservoir impounding. In China there is ongoing debate as to whether the magnitude 7.9 Wenchuan earthquake of 2008, which killed 80,000 people, was triggered by the reservoir behind the 156 m high Zipingpu Dam (Qiu, 2012).

As Figure 6.33 shows, there are many more locations where the filling of reservoirs behind dams has led to appreciable levels of seismic activity, though this is often minor. Detailed monitoring has shown that earthquake clusters occur in the vicinity of some dams after their reservoirs have been filled, whereas before construction activity was less clustered and less frequent. Similarly, there is evidence from Vaiont (Italy), Lake Mead (USA), Kariba (Central Africa), Koyna (India) and Kremasta (Greece) that there is a linear correlation between the storage level in the reservoir and the logarithm of the frequency of shocks.

One reason why dams induce earthquakes is the hydro-isostatic pressure exerted by the mass of the water impounded in the reservoir, together with changing water pressures across the contact surfaces of faults. Given that the deepest reservoirs provide surface loads of only 20 bars or so, direct activation by the mass of the impounded water seems an unlikely cause (Bell and Nur, 1978) and the role of changing pore pressure assumes greater importance. Paradoxically, there are some possible examples of reduced seismic activity induced by reservoirs (Milne, 1976). One possible explanation of this is the increased incidence of stable sliding (fault creep) brought about by higher porewater pressure in the vicinity of the reservoir.

However, the ability to prove an absolutely concrete cause-and-effect relationship between reservoir activity and earthquakes is severely limited by our inability to measure stress below depths of several kilometres, and some examples of induced seismicity may have been built on the false assumption that because an earthquake occurs in proximity to a reservoir it has to be induced by that reservoir (Meade, 1991).

The demonstration that increasing water pressures could initiate small-scale faulting and seismic activity was unintentionally demonstrated near Denver (Evans, 1966), where nerve-gas waste was being disposed of at a great depth in a well in the hope of avoiding contami-

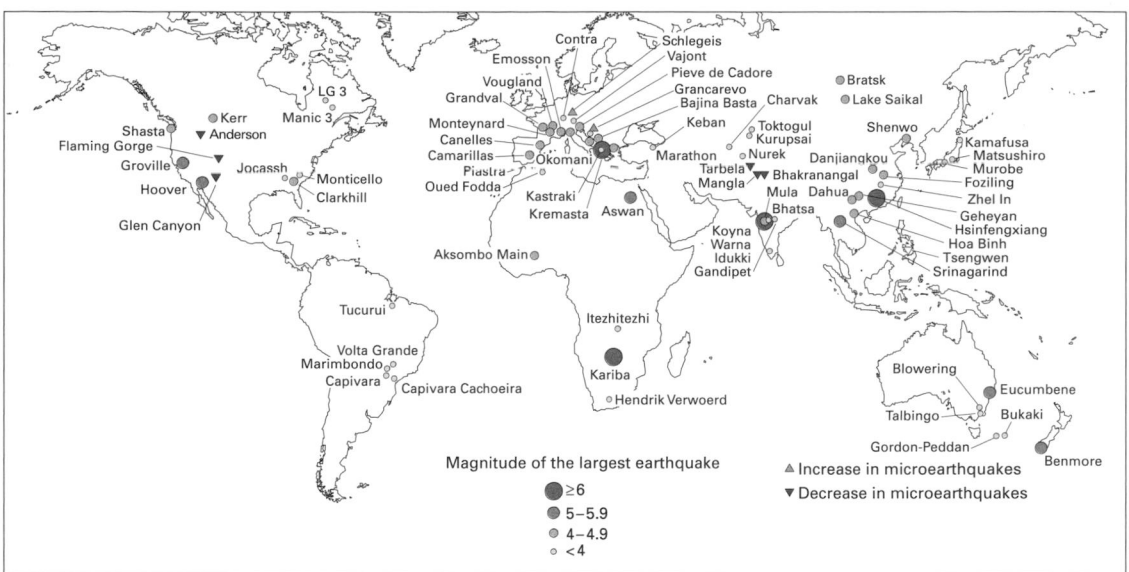

Figure 6.33 Worldwide distribution of reservoir-triggered changes in seismicity (after Gupta, 2002, figure 1). Reproduced with permission from Elsevier.

nation of groundwater supplies. The waste was pumped in at high pressures and triggered off a series of earthquakes, the timing of which corresponded very closely to the timing of waste disposal in the well. It is also now thought that the pumping of fluids into the Inglewood oil field, Los Angeles, to raise the hydrostatic pressure and increase oil recovery, may have been responsible for triggering the 1963 earthquake which fractured a wall of the Baldwin Hills Reservoir. It appears that increased fluid pressure reduces the frictional force across the contact surface of a fault, allows slippage to occur, thereby causing an earthquake. In general earthquake triggering has been related to fluid injection, but for reasons that are still obscure there are many cases where increased seismicity has resulted from fluid abstraction (Segall, 1989). Cases where seismicity and faulting can be attributed to fluid extraction come, for example, from the oilfields of Texas and California and the gas fields of the Po Valley in Italy and of Uzbekistan (Cypser and Davis, 1998; Donnely, 2009; Suckale, 2010). Some of the seismic activity may be related to a process called 'hydrofracturing' or 'fracking', whereby water is injected into the rock to create distinct fractures that increase permeability and thus facilitate the extraction of fluids and the production of geothermal energy (Majer et al., 2007). In 2011, test drilling for gas exploitation by fracking in the Blackpool area of northwest England caused some small earthquakes.

Miscellaneous other human activities appear to affect seismic levels. In Johannesburg, South Africa, for example, gold mining and associated blasting activity have produced tens of thousands of small tremors, and there is a notable reduction in the number that occurs on Sundays, a day of rest. In Staffordshire, England, coal mining has caused increased seismic activity and up to 25% of all earthquakes recorded by the British Geological Survey may be related to coal mining. Coal mining induced seismicity is also a feature of the Ruhr coalfield in Germany (Bischoff et al., 2009). In the future, seismicity might be induced by underground carbon sequestration projects.

When looking at the human impact on volcanic activity human impotence becomes apparent, though some success has been achieved in the control of lava flows. Thus in 1937 and 1947 the US Army attempted to divert lava from the city of Hilo, Hawaii, by bombing threatening flows, while elsewhere, where lava rises in the crater, breaching of the crater wall to direct lava towards uninhabited ground may be possible. In 1973 an attempt was made to halt advance of lava with cold water during the Icelandic eruption of Krikjufell. Using up to 4 million litres of pumped waste per hour, the lava was cooled sufficiently to decrease its velocity at the flow front so that the chilled front acted as a dam to divert the still fluid lava behind (Williams and Moore, 1973). There has also been some success in controlling lava flows from Mount Etna in Sicily using earthen barriers (Barberi et al., 2003).

Points for review

What are the causes and consequences of accelerated sedimentation?

How do humans cause land subsidence?

In what ways may humans accelerate mass movements?

Why are many of the world's coastlines eroding?

Guide to reading

Downs, P.W. and Gregory, K.J. (2004) *River Channel Management*. London: Arnold. A comprehensive review of river channels and their management.

James, L.A. and Marcus, W.A. (eds) (2006) 37th Binghamton Geomorphology Symposium: the human role in changing fluvial systems. *Geomorphology*, 79, 1–506. A special issue of a major journal that includes some excellent reviews of the ways in which humans have modified rivers.

Nir, D. (1983) *Man, a Geomorphological Agent: An Introduction to Anthropic Geomorphology*. Jerusalem: Keter. A general survey that was ahead of its time.

Szabó, J., Dávid, L. and Lóczy, D. (eds) (2010) *Anthropogenic Geomorphology*. Dordrecht: Springer. A largely Hungarian view of the human impact on landforms.

7 THE HUMAN IMPACT ON CLIMATE AND THE ATMOSPHERE

Chapter Overview

This chapter considers the various ways in which humans can change climate, at a global scale, concentrating on such factors as greenhouse gas loadings in the atmosphere, surface albedo changes brought about by land cover change and the presence of aerosol particles in the atmosphere. It then considers changes at the more local scale, including the climates of urban areas and the possible means and consequences of deliberate climate modification, including geoengineering. The second part of the chapter looks at issues relating to air quality, urban pollution, photochemical smogs, acidification of precipitation and stratospheric ozone depletion.

World climates

The climate of the world is now known to have fluctuated frequently and extensively in the three or so million years during which humans have inhabited the earth. The bulk of these changes have nothing to do with human intervention. Climate has changed, and is currently changing, because of a wide range of different natural factors which operate over a variety of time scales (Figure 7.1) (Anderson et al., 2013).

While humans are at present incapable of modifying some of these natural mechanisms of climate change – the output of solar radiation, the presence of fine interstellar matter, the earth's orbital variations, volcanic

The Human Impact on the Natural Environment: Past, Present and Future, Seventh Edition. Andrew S. Goudie.
© 2013 John Wiley & Sons, Ltd. Published 2013 by John Wiley & Sons, Ltd.

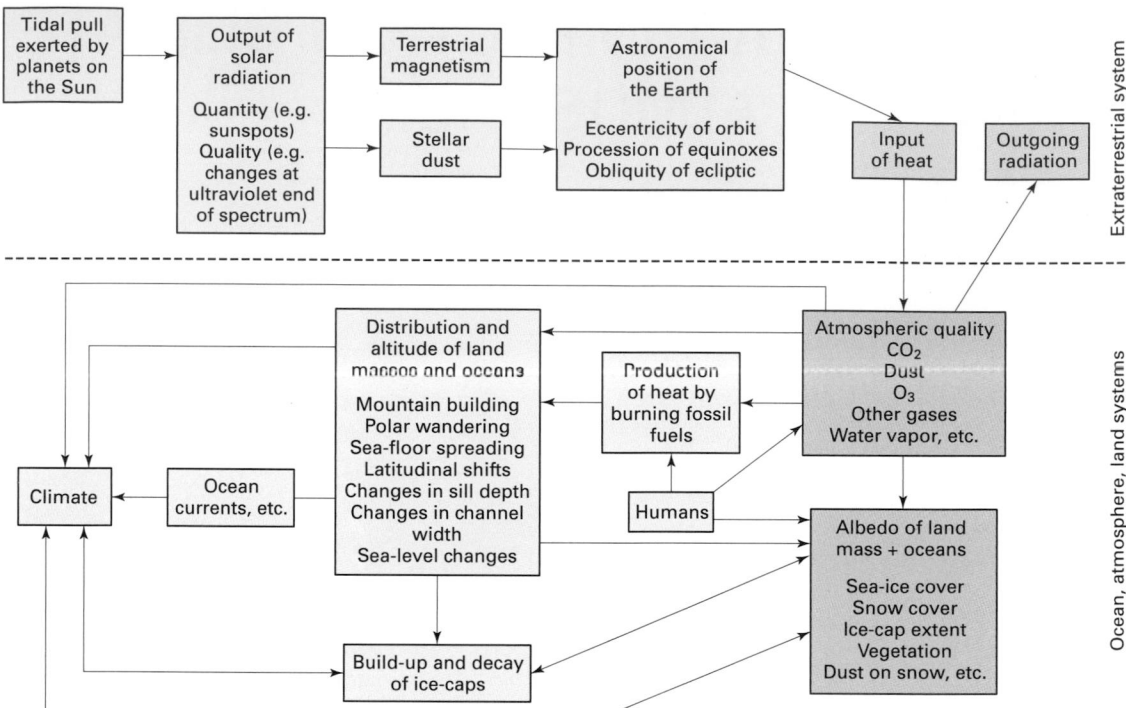

Figure 7.1 A schematic representation of some of the possible influences causing climatic change (after Goudie, 1992, figure 1). Reproduced with permission.

eruptions, mountain building and the overall pattern of land masses and oceans – there are some key areas where humans may be capable of making significant changes to global climates (Figure 7.2). The most important categories of influence are in terms of the chemical composition of the atmosphere and the albedo of Earth's surface, though, as the following list shows, there are some others that also need to be considered.

Possible mechanisms
Gas emissions
 carbon dioxide
 methane
 chlorofluorocarbons (CFCs)
 nitrous oxide
 water vapour
 miscellaneous trace gases
 stratospheric ozone depletion
Aerosol generation
 dust, smoke, black carbon, etc.
Thermal pollution
Albedo change
 dust and black carbon addition to ice caps
 deforestation and afforestation
 overgrazing

Extension of irrigation
Alteration of ocean currents by constricting straits
Diversion of fresh waters into oceans, affecting
 thermohaline circulation system
Impoundment of large reservoirs

Figure 7.2, based on the Intergovernmental Panel on Climate Change report (IPCC, 2007a) shows the relative importance of different human factors in forcing climate between 1750 (the start of the industrial revolution) and 2005. That warming of Earth's climate has taken place over the last four to five decades is unequivocal, and the IPCC (2007a: 10) asserts that 'Most of the observed increase in global average temperatures since the mid-20th century is *very likely* due to the observed increase in atmospheric greenhouse concentrations.'

The greenhouse gases – carbon dioxide

The greenhouse effect occurs in the atmosphere because of the presence of certain gases that absorb infrared radiation (Figure 7.3). Light and ultra-violet radiation from the sun are able to penetrate the atmosphere and warm Earth's surface. This energy is re-radiated

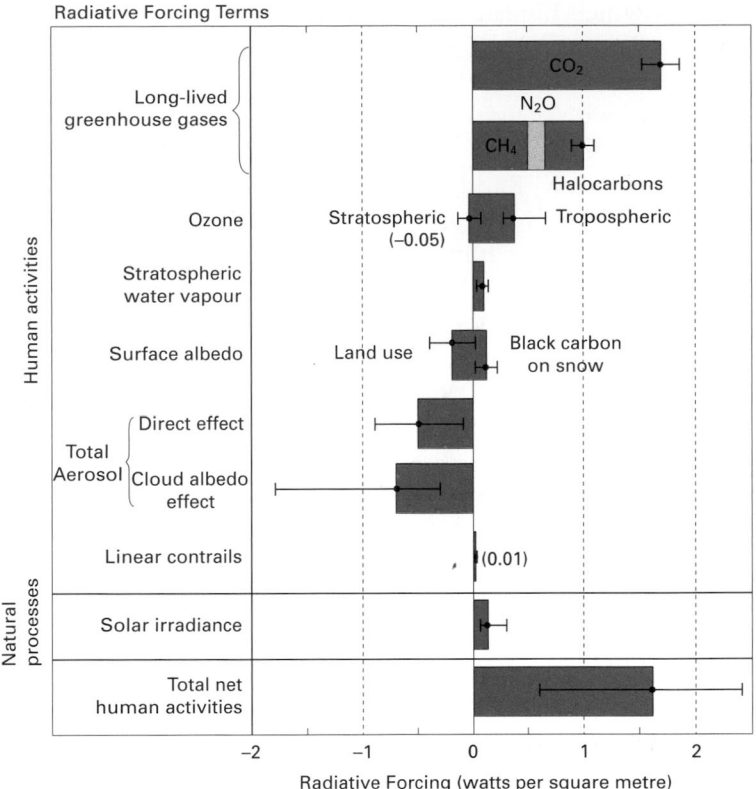

Figure 7.2 Summary diagram of the principal components of the radiative forcing of climate change between 1750 and 2005 (after IPCC, 2007a, FAQ 2.1, figure 2). Reproduced with permission.

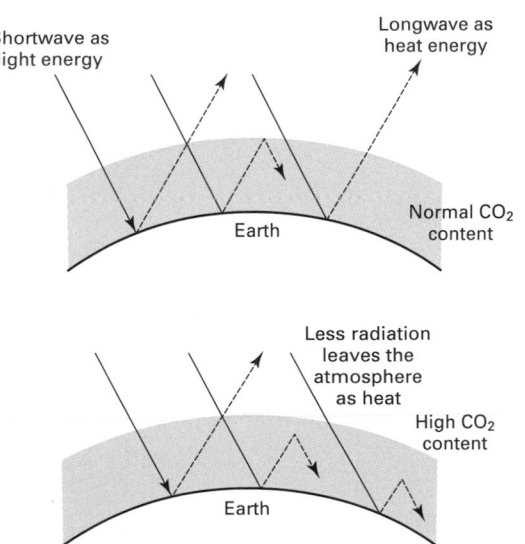

Figure 7.3 The greenhouse effect: short-wave radiation from the sun is absorbed by the earth's surface which, in turn, radiates heat at far longer wavelengths because of its temperature of around 280°K, compared with 6000°K for the sun.

as infrared radiation, which, because of its longer wavelength, is absorbed by certain substances such as water vapour, carbon dioxide and other trace gases. This causes the average temperature of the earth's surface and atmosphere to increase. Should the quantities of each substance in the atmosphere be increased, then the greenhouse effect will become enhanced (Hansen et al., 1981).

In reality the term 'greenhouse effect' is something of a misnomer. As Henderson-Sellers and Robinson (1986: 60) explain:

We know that a greenhouse maintains its higher internal temperature largely because the shelter if offers reduces the turbulent transfers of energy away from the surface rather than because of any radiative considerations. Thus while the greenhouse effect remains valid, and vital, for the atmosphere it might be better to think of the physical processes in terms of the 'leaky bucket' analogy. . . . Here an increase in the amount of gas with absorption bands in the infrared part of the spectrum is represented by a decrease in the size of the hole in the bucket. The surface temperature, represented by the depth of the water in the bucket, rises as more absorbing gases enter the atmosphere.

There are a number of ways in which humans have been enhancing the greenhouse effect, the most significant of which is to increase carbon dioxide levels in the atmosphere. This may have started with deforestation in the Holocene (Ruddiman, 2003; Boyle et al., 2011) and with the spread of rice cultivation (Ruddiman et al., 2011) but has accelerated in recent centuries. Since the beginning of the industrial revolution humans have been taking stored carbon out of the earth in the form of coal, petroleum and natural gas, and burning it to make carbon dioxide (CO_2), heat, water vapour and smaller amounts of sulfur dioxide (SO_2) and other gases. The pre-industrial level of carbon dioxide is a matter of some debate, but may have been as low as 260–70 parts per million by volume (ppmv). The 2012 level is around 396 ppmv, and the upward trend is evident in records from various parts of the world (Figure 7.4). At the present rate it would reach 500 ppmv by the end of the twenty first century. The prime cause of increased carbon dioxide emissions is fossil fuel combustion and cement production (c. 5.5 ± 0.5 GtC/year in the 1980s and now approaching 9 GtC/year), but with the release of carbon dioxide by changes in tropical land use (primarily deforestation) being a significant factor (c. 1.6 ± 1.0 GtC/year). The amount of carbon derived from deforestation has increased greatly from about 0.4 GtC/year in 1850 (Figure 7.5) (Woodwell, 1992) and now contributes between 6% and 17% of global anthropogenic CO_2 emissions (Baccini et al., 2012). The various relationships between land-use change and the build up of greenhouse gases are reviewed by Adger and Brown (1994).

Other gases

In addition to carbon dioxide, it is probable that other gases will contribute to the greenhouse effect (Table 7.1). Individually their effects may be minor, but as a group they may be major (Ramanathan, 1988). Indeed, molecule for molecule some of them may be much more effective as greenhouse gases than CO_2, as the data in Table 7.2 show.

One of the more important of the trace gases is methane (CH_4), which has a strong infrared absorption band at 7.66 μm. This gas is produced from a number of sources (Table 7.3), including cattle, the burning of fossil fuels, rice fields and landfill waste (Karakurt et al., 2012). Ice core studies and recent direct observations (Figure 7.6b) suggest that until the beginning of the industrial revolution in the eighteenth century background levels were relatively stable at around 600 parts per billion by volume (ppbv) although they may have been increased prior to that by rice farming and other agricultural activities (Ruddiman and Thomsen, 2001). They rose steadily between AD 1700 and 1900, and then increased still more rapidly, attaining levels that averaged 1300 ppbv in the early 1950s and 1600 ppbv by the mid-1980s (Khalil and Rasmussen, 1987) and over 1774 ppbv in 2005 (Figure 7.7a).

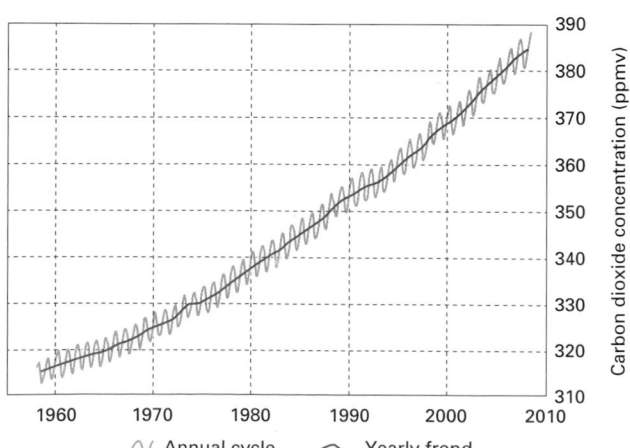

Figure 7.4 Atmospheric carbon dioxide concentrations at Mauna Loa, Hawaii (parts per million by volume).

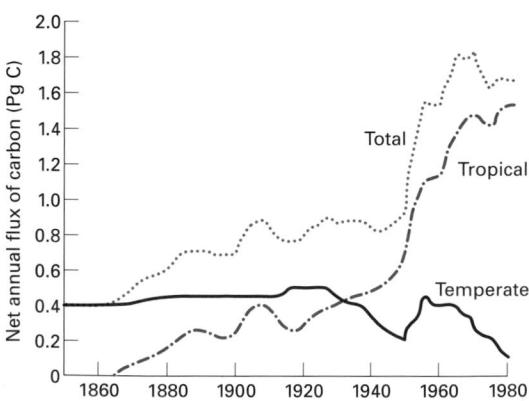

Figure 7.5 The net annual flux of carbon from deforestation in tropical and temperate zones globally, 1850–1980 (after Houghton and Skole, 1990, figure 23.2, with permission from Cambridge University Press, and Woodwell, 1992, figure 5.2).

Table 7.1 Sources of four principal greenhouse gases[a]

GAS	Natural sources	Human-derived sources
Carbon dioxide	Terrestrial biosphere Oceans	Fossil fuel combustion Cement production Land-use modification
Methane	Natural wetlands Termites Oceans and freshwater lakes	Fossil fuels (natural gas production, coal mines, petroleum industry, coal combustion) Enteric fermentation (e.g. cattle) Rice paddies Biomass burning Landfills Animal waste Domestic sewage
Nitrous oxide	Oceans Tropical soils (wet forests, dry savannas) Temperate soils (forests, grassland)	Nitrogenous fertilizers Industrial sources Land-use modification (biomass burning, forest clearing) Cattle and feed lots
Chlorofluorocarbons[b]	Nil	Rigid and flexible foam Aerosol propellants Teflon polymers Industrial solvents

[a]Sources listed in order of decreasing magnitude of emission except where otherwise indicated.
[b]Sources of chlorofluorocarbons not in order of decreasing magnitudes of emission.

Table 7.2 Radiative forcing relative to CO_2 per unit molecule change in the atmosphere. Source: Extracted from Houghton et al. (1990: 53, table 2.3). Reproduced with permission

Gas	Relative radiative forcing	Residence time in atmosphere (years)
CO_2	1	100
CH_4 (methane)	21	10
N_2O (nitrous oxide)	206	100–200
CFC-11	12,400	65
CFC-12	15,800	130

This increase of 2.5 times over background levels results primarily from increased rice cultivation in waterlogged paddy fields, the enteric fermentation produced in the growing numbers of flatulent and belching domestic cattle, and the burning of oil and natural gas (Crutzen et al., 1986). In some countries,

Table 7.3 Sources and strengths of nitrous oxide and methane

(a) Estimated sources and sinks of methane in millions of tonnes per year. Source: After Houghton (2009, table 3.2)

	Best estimate
Sources	
Natural	
Wetlands	150
Termites	20
Ocean	15
Other (including hydrates)	15
Human-generated	
Coal mining, natural gas, petroleum industry	100
Rice paddies	60
Enteric fermentation	90
Waste treatment	25
Landfills	40
Biomass burning	40
Sinks	
Atmospheric removal	545
Removal by soils	30
Atmospheric increase	22

(b) Estimates of nitrous oxide (N_2O) source strengths and sinks. Source: Data in UNEP (1991)

Sources/sinks	Range (10^6 t a^{-1})
Sources	
Oceans	1.4–2.6
Soils (tropical forests)	2.2–3.7
Soils (temperate forests)	0.7–1.5
Fossil fuel combustion	0.1–0.3
Biomass burning	0.02–0.2
Fertilizer (including groundwater)	0.01–2.2
Sinks	
Removal by soils	Unknown
Photolysis in the stratosphere	7–13
Atmospheric increase	3–4.5

including the UK, (Figure 7.7b) emissions have recently declined, and reductions in emissions from rice cultivation in Asia also occurred in the last years of the twentieth century (Kai et al., 2011) as a result of increases in fertilizer application and reductions in water use.

Chlorofluorocarbons (CFCs), despite their relatively trace amounts in the atmosphere, have increased very markedly in terms of their emissions (Figure 7.7) and their concentrations in recent decades resulting from their use as refrigerants, foam makers, fire control

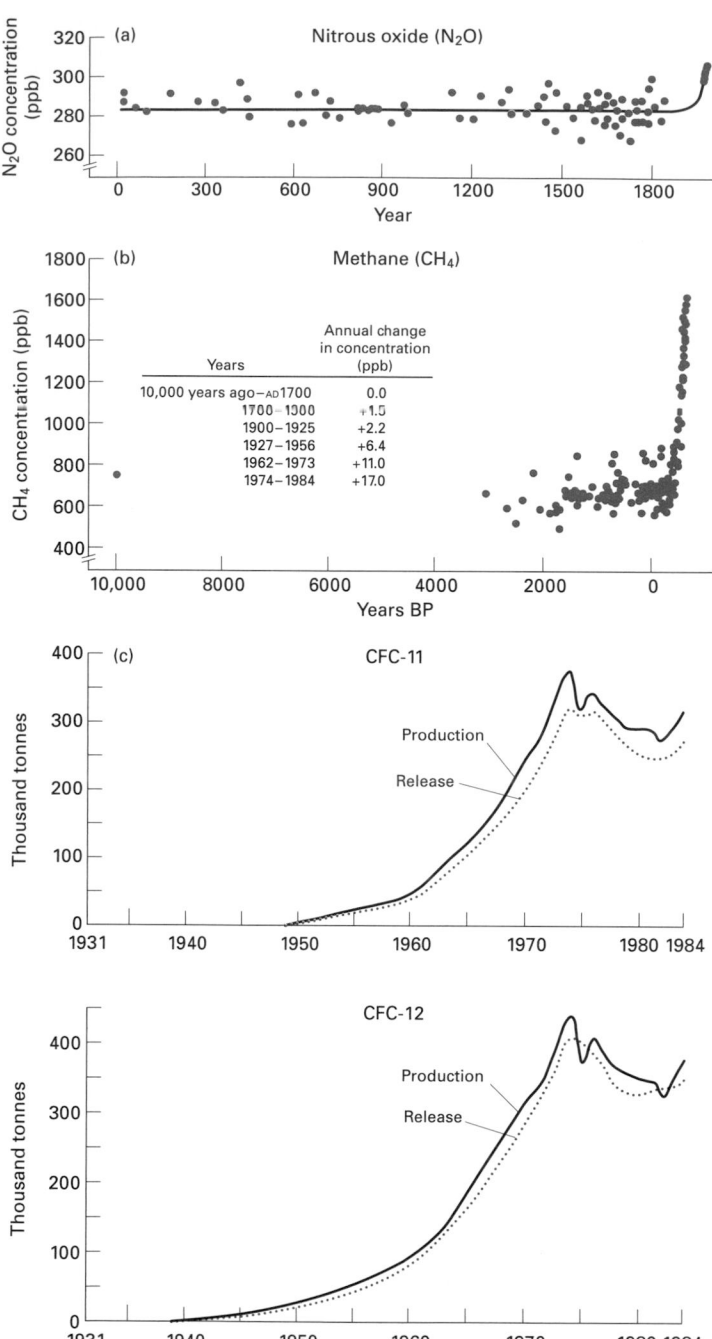

Figure 7.6 The changing concentrations of accessory greenhouse gases in the atmosphere: (a) nitrous oxide. Note these remained fairly constant between 23,000 years ago and AD 1850 at approximately 285 parts per billion; (b) methane (a and b, after Khalil and Rasmussen, 1987, reproduced with permission from Elsevier); (c) the changing production and release of two CFC gases (CFC-11 and CFC-12) between 1931 and 1992.

agents and propellants in aerosol cans. They have a very strong greenhouse effect even in relatively small amounts. On the other hand, the ozone depletion they have caused in the stratosphere may to some limited extent counteract this effect, for stratospheric ozone depletion results in a decrease in radiative forcing (Houghton et al., 1992). Conversely, the build up of lower-level **tropospheric** ozone can contribute to the greenhouse effect. As a consequence of the Montreal Protocol, CFC concentrations in the atmosphere are now starting to decline.

Nitrous oxide (N_2O) is also no laughing matter, for it can contribute to the greenhouse effect, primarily by absorption of infrared at the 7.8 and 17 μm bands.

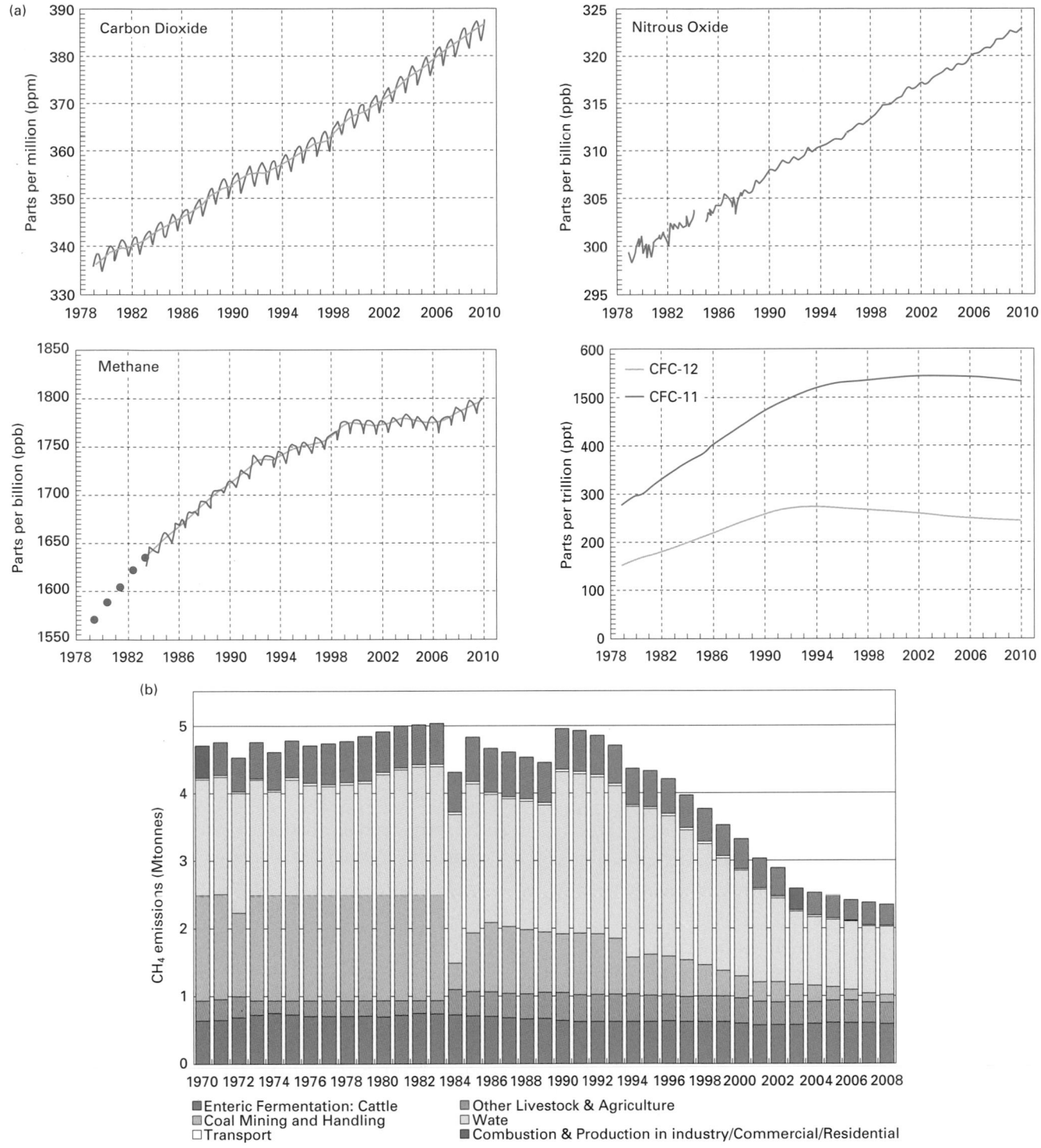

Figure 7.7 (a) Changes in concentrations of some greenhouse gases from 1978 to 2010 (source NOAA); (b) Time series and sources of UK methane emissions (data in National Atmospheric Emissions Inventory, 2010, figure 7.6).

Combustion of hydrocarbon fuels, the use of ammonia-based fertilizers, deforestation and biomass burning are among the processes that could lead to an increase in atmospheric N_2O levels (Figure 7.7 and Table 7.3b). Atmospheric N_2O concentrations increased from around 275 ppbv in pre-industrial times to 319 in 2005.

Other trace gases that could play a greenhouse role include bromide compounds, carbon tetraflouride, carbon tetrachloride and methyl chloride.

The continued role of greenhouse gases other than CO_2 in changing the climate is already not greatly less important than that of CO_2. If present trends continue, the combined concentrations of atmospheric CO_2 and other greenhouse gases would be radiatively equivalent to a doubling of CO_2 from pre-industrial levels possibly as early as the 2030s. The relative amounts of **radiative forcing** for different greenhouse gases since pre-industrial times are, according to the Intergovernmental Panel on Climate Change (1996), as follows:

CO_2	1.56 Wm^{-2}
CH_4	0.47 Wm^{-2}
N_2O	0.14 Wm^{-2}
CFCs and HCFCs	0.25 Wm^{-2}
Tropospheric ozone	0.40 Wm^{-2}

Another important feature of the various greenhouse gases is their residence time in the atmosphere. CH_4 has a residence time of about 10 years, the shortest of all the greenhouse gases. This means that if we could stop the enhanced emissions of that gas, its concentration in the atmosphere should fall to its natural level in a decade. By contrast N_2O (100–200 years) and CO_2 (c. 100 years) have much longer residence times, so that even if we could control their sources immediately it would till take a very long time for them to fall to their natural levels.

Global temperatures have been climbing since the end of the nineteenth century (Figure 7.8) and it is now regarded as highly probable that increased greenhouse gas loadings in the atmosphere have contributed to this.

Ozone depletion and climate change

The decline in stratospheric ozone concentrations and the development of an ozone hole over Antarctica (see

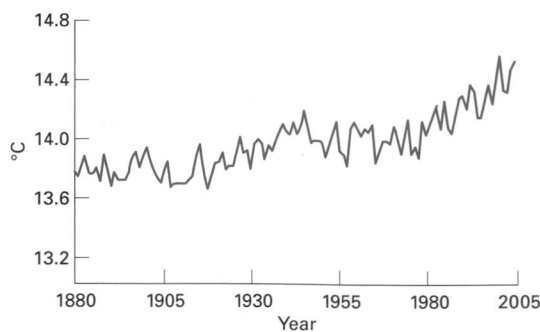

Figure 7.8 Global average temperature at Earth's surface, 1880–2002.

Chapter 7) is now recognized as having potential climatic significance, especially in the Southern Hemisphere. The ozone depletion causes cooling over Antarctica which in turn causes atmospheric the Hadley Cell to shift poleward which alters precipitation patterns in the subtropics (Kang et al., 2011) and cause a southward expansion of the subtropical dry zones (Son et al., 2009; Feldstein, 2011; Polvani et al., 2011; Purich and Son, 2012).

Aerosols

Aerosols are finely divided solid or liquid particles dispersed in the atmosphere. In general terms it is believed that they affect climate because they intercept and scatter a small portion of incoming radiation from the sun, thus reducing the energy reaching the ground. This is called the 'direct effect' of aerosols. The direct effect increases with both the number and size of aerosols in the atmosphere. Aerosols also have an 'indirect effect'. This is because they are key elements in cloud formation. The number of cloud droplets that a cloud possesses is determined by the number of aerosols available onto which water vapour can condense. Clouds with large numbers of droplets reflect more sunlight back into space and so can also contribute to cooling. Aerosols do not, however, inevitably cause cooling. So, for example, and Brazel and Idso (1979) point to the two contrasting tendencies of dust: the backscattering effect producing cooling and the thermal-blanketing effect causing warming. The second of these absorbs some of the earth's thermal radiation that would otherwise escape to space and then re-radiates a portion of this back to the land surface, raising surface temperatures. They believe that natural dust from volcanic emissions tends to enter the stratosphere (where backscattering and cooling are the prime consequences), while anthropogenic dust more frequently occurs in the lower levels of the atmosphere, causing thermal blanketing and warming.

There is a variety of ways in which human activities have increased atmospheric aerosol loadings. These include biomass burning and industrial emissions of smoke and dust particles. In addition, intensive agricultural exploitation of desert margins could create a dust pall in the atmosphere by exposing larger areas of surface materials to deflation in dust storms and there is evidence that humans have caused desert dust loadings in the atmosphere to increase (see Chapter 4) (Mahowald et al., 2010; Ginoux et al., 2012). Dust palls could so change atmospheric temperature that convection, and thus rainfall, would be reduced. There is thus the possibility that human-induced desertification generates dust which could in turn increase the degree of desertification by its effect on rainfall levels. Studies of the links between dust in the atmosphere and temperatures, precipitation and clouds is a major area of current research (see e.g. Tegen et al., 1996; Miller and Tegen, 1998). Much of the current interest in dust storms relates to their possible role in the Earth System (see Goudie and Middleton, 2006, table 1:1). Dust loadings may affect air temperatures through the absorption and scattering of solar radiation (Durant et al., 2009), modify cloud formation (Toon, 2003) and convectional activity, influence sulfur dioxide levels in the atmosphere, either by physical absorption or by heterogeneous reactions (Adams et al., 2005), and influence marine primary productivity and thus atmospheric carbon dioxide levels (Ridgwell, 2003). Wong et al. (2008) has linked the strength of African dust outbreaks in the Saharan air layer to changes in hurricane intensity in the North Atlantic Region (see also Evan et al., 2006). There are still huge uncertainties about the relationship between atmospheric mineral dust aerosol levels and radiative forcing, and an increasing number of observational and modelling studies are investigating this issue (see, e.g. Zhu et al., 2007). It remains a major research priority, not least because of the possibility that dust aerosols could be an accelerant of aridity trends.

There is also some evidence that other types of aerosol, caused by air pollution, may affect precipitation levels. For example, midweek increases in summer rainfall and storm heights in the south-east USA have been detected and are correlated with higher particulate concentrations compared to that which occurs at weekends, when industrial emissions are less (Bell et al., 2008).

In recent years, there has been a growing interest in the role of black carbon (soot) in the atmosphere (Allen et al., 2012). This is derived from fossil fuel combustion, biomass burning (associated with deforestation and crop residue burning) and cooking with biofuels (Ramanathan and Carmichael, 2008). Making up a significant part of what has been called an 'atmospheric brown cloud' over Asia, it absorbs and scatters solar radiation and may contribute to global warming, not least in the Arctic (Shindell and Faluvegi, 2009). It can also contribute to a weakening of the South Asian summer monsoon and a reduction in precipitation, especially in the tropics (Meehl et al., 2008; Bollasina et al., 2011; Ganguly et al., 2012). When deposited on ice or snow it darkens the surface, significantly enhances heat absorption and so accelerates melting of glaciers (Yasunari et al., 2012) and sea ice.

The most catastrophic effects of anthropogenic aerosols in the atmosphere could be those resulting from a nuclear exchange between the great powers (Dörries, 2011). Explosion, fire and wind might generate a great pall of smoke and dust in the atmosphere which would make the world dark and cold (Westing, 2013). It has been estimated that if the exchange reached a level of several thousand megatons, a 'nuclear winter' would be as low as −15°C to −25°C (Turco et al., 1983), though more recent simulations by Schneider and Thompson (1988) suggested that some previous estimates may have been exaggerated. They suggested that in the Northern Hemisphere maximum average land surface summertime temperature depressions might be of the order of 5–15°C. The concept is discussed by Cotton and Pielke (2007, Chapter 10).

Fears were also expressed that as a result of the severe smoke palls generated by the Gulf War in 1991 there might be severe climate impacts. Studies have suggested that because most of the smoke generated by the oil well fires stayed in the lower troposphere and only had a short residence time in the air, the effects were local (some cooling) rather than global, and that the operation of the monsoon was not affected to any significant degree (Bakan et al., 1991; Browning et al., 1991). Furthermore, in the event the emissions of smoke particles were less than some forecasters had predicted, and they were also rather less black (Hobbs and Radke, 1992).

Aircraft, both civil and military, discharge some water vapour into the atmosphere as contrails. At present, the water content of the stratosphere is very low, as is the exchange of air between the lower stratosphere and other regions. Consequently, comparatively modest amounts of water vapour discharge by aircraft could have a significant effect on the natural balance. It is possible that contrails and the development of thin cirrus clouds could lead to warming of the Earth's surface. Observations of cirrus cloud cover over the last 50 years near to major USA flight corridors show a clear upward trend that seems to correlate with US domestic jet fuel trends over the same period (Yang et al., 2010b). However, the Intergovernmental Panel on Climate Change (IPCC, 2007a: 30) concludes 'No best estimates are available for the net forcing from spreading contrails. Their effects on cirrus cloudiness and the global effect of aviation aerosol on background cloudiness remain unknown'. The whole role of aircraft impacts on the atmosphere and climate is discussed by Lee et al. (2010).

Over the world's oceans a major source of aerosols is dimethylsulfide (DMS). This is produced by planktonic algae in seawater and then oxidized in the atmosphere to form sulfate aerosols. Because the albedo of clouds (and thus the Earth's radiation budget) is sensitive to cloud-condensation nuclei density, any factor that controls planktonic algae may have an important impact on climate. The production of such plankton could be affected by water pollution in coastal areas or by global warming (Charlson et al., 1987). However, an even more important source of sulfate aerosols is the burning of fossil fuels and the subsequent emission of sulfur dioxide (SO_2) (Charlson et al., 1992), and these types of sulfate aerosol are concentrated over and downwind of major industrial regions. They have probably served to reduce the rate of global warming that has taken place in this century and may help to explain the cessation in global warming that took place in some regions between the 1940s and 1970s. Indeed, climate models that have predicted the amount of increase in global average temperature as a result of the rising concentrations of greenhouse gases have given a greater amount of temperature rise since the last century than has actually occurred. The newer climate models, which include the effect of these aerosols, produce predicted changes that have considerable similarity to the observed patterns of change (Taylor and Penner, 1994).

Global dimming and global brightening

There is some evidence that the amount of solar radiation that reached the Earth's surface declined by *c.* 0.2–0.3% per year between *c.* 1960 and *c.* 1990 – a phenomenon called 'Global Dimming' (Stanhill and Cohen, 2001; Wild, 2009). Possible causes could include changes in clouds, increasing amounts of anthropogenic aerosols and reduced atmospheric transparency after explosive volcanic eruptions (Pinker et al., 2005). Air pollution may have worked in two ways – by reflecting sunlight back into space and by making clouds more reflective. Global dimming may to a certain degree have masked the greenhouse effect. Since about 1990, the trend has been reversed and is referred to as 'Global Brightening' (Wild et al., 2005; Wild, 2012). This may have enabled the effects of rising greenhouse gas loadings to have become more evident during the 1990s (Wild et al., 2007). The reason for this change may be that in some respects levels of atmospheric pollutants, particularly sulfur dioxide and black carbon, have declined in many regions (e.g. over eastern Europe) (Streets et al., 2006). Elsewhere, however, as in China, sunshine hours have decreased appreciably in many areas, in response to increased atmospheric pollution (Li et al., 2012; Wang et al., 2012).

Vegetation and albedo change

Incoming radiation of all wavelengths is partly absorbed and partly reflected. Albedo is the term used to describe the proportion of energy reflected and hence is a measure of the ability of the surface to reflect radiation.

Land-use changes create differences in albedo which have important effects on the energy balance, and hence on the water balance, of an area. Tall rainforest

Table 7.4 Albedo values for different land-use types. Source: From miscellaneous data from Pereira (1973) collated by author

Surface type	Location	Albedo (%)
Tall rainforest	Kenya	9
Lake	Israel	11.3
Peat and moss	England and Wales	12
Pine forest	Israel	12.3
Heather moorland	England and Wales	15
Evergreen scrub (maquis)	Israel	15.9
Bamboo forest	Kenya	16
Conifer plantation	England and Wales	16
Citrus orchard	Israel	16.8
Towns	England and Wales	17
Open oak forest	Israel	17.6
Deciduous woodland	England and Wales	18
Tea bushes	Kenya	20
Rough grass hillside	Israel	20.3
Agricultural grassland	England and Wales	24
Desert	Israel	37.3

may have an albedo as low as 9%, while the albedo of a desert may be as high as 37% (Table 7.4).

There has been growing interest recently in the possible consequences of deforestation on climate through the effect of albedo change. Ground deprived of a vegetation cover as a result of deforestation and overgrazing (as in parts of the Sahel) has a very much higher albedo than ground covered in plants. This could affect temperature levels. Satellite imagery of the Sinai-Negev region of the Middle East shows an enormous difference in image between the relatively dark Negev and the very bright Sinai-Gaza strip area. This line coincides with the 1948–1949 armistice line between Israel and Egypt and results from different land-use and population pressures. Otterman (1974) suggested that this albedo change has produced temperature changes of the order of 5°C.

Charney et al. (1975) argued that the increase in surface albedo, resulting from a decrease in plant cover, would lead to a decrease in the net incoming radiation and an increase in the radiative cooling of the air. Consequently, they argue, the air would sink to maintain thermal equilibrium by adiabatic compression, and cumulus convection and its associated rainfall would be suppressed. A positive feedback

mechanism would appear at this stage, for the lower rainfall would in turn adversely affect plants and lead to a further decrease in plant cover. However, this view was disputed by Ripley (1976) who suggested that Charney and his colleagues, while considering the impact of vegetation changes on albedo, had completely ignored the effect of vegetation on evapotranspiration. He pointed out that vegetated surfaces are usually cooler than bare ground since much of the absorbed solar energy is used to evaporate water and concludes from this that protection from overgrazing and deforestation might, in contrast to Charney's views, be expected to lower surface temperatures and thereby reduce, rather than increase, convection and precipitation.

Removal of humid tropical rainforests has also been seen as a possible mechanism of anthropogenic climatic change through its effect on albedo. Potter et al. (1975) proposed the following model for such change:

Deforestation
↓
Increased surface albedo
↓
Reduced surface absorption of solar energy
↓
Surface cooling
↓
Reduced evaporation and sensible heat flux from the surface
↓
Reduced convective activity and rainfall
↓
Reduced release of latent heat, weakened Hadley circulation and cooling in the mid and upper troposphere
↓
Increased tropical lapse rates
↓
Increased precipitation in the latitude bands 5–25°N and 5–25°S, and a decrease in the Equator-pole temperature gradient
↓
Reduced meridional transport of heat and moisture out of equatorial regions
↓
Global cooling and decrease in precipitation between 45–85°N and 40–60°S

However, some studies (e.g. Potter et al., 1981) suggested that globally over the past few thousand years the climatic effects of albedo changes wrought by humans have been small and probably undetectable. Similarly, Henderson-Sellers and Gornitz (1984) sought to model the possible future effects of albedo changes produced by humans and also predicted that there would be but little alteration brought about by current levels of tropical deforestation. On the other hand, Lean and Warrilow (1989) used a general circulation model (GCM) which showed greater changes than previous models and suggested that Amazon basin deforestation would, through the effects of changes in surface roughness and albedo, lead to reductions in both precipitation and evaporation. Likewise a UK Meteorological Office GCM indicated that the deforestation of both Amazonia and Zaire would by changing surface albedo cause a decrease in precipitation levels (Mylne and Rowntree, 1992). There are now an increasing number of modelling experiments that suggest vegetation removal can have important regional and even global effects (Table 7.5), though there are considerable divergences between different models (Nobre et al., 2004). Nonetheless, most show *decreases* in mean evapotranspiration of from 25.5 to 985.0 mm per year, *increases* in mean surface temperatures of 0.1–3.8°C and *reductions* in regional precipitation (Sampaio et al., 2007). It is even possible that the effects of Amazonian deforestation on precipitation could extend some distance away from there, to the Dakotas and the Midwest Triangle in the USA, as modelling by Werth and Avissar (2002) has shown, while deforestation and afforestation in monsoon Asia could influence precipitation in the Sahara and Arabia (Dallmeyer and Claussen, 2011).

Albedo effects may be especially sensitive in higher latitudes as well. As Betts (2000) has pointed out, in a snowy environment, forests are generally darker than open land, because trees generally remain exposed when cultivated land can become entirely snow-covered. Snow-free foliage is darker than snow. This means that forest has a smaller surface albedo and so may exert a warming influence. Thus land cover changes in the boreal forest zone can have substantial climatic implications. Betts (2001) suggested that mid-latitude agricultural regions would be approximately 1 2K cooler in winter and spring in comparison with their previously forested state, due to deforestation increasing the surface albedo during periods of snow cover. Conversely, mid-latitude afforestation, which has taken place in recent decades, by increasing surface temperatures, may have a widespread influence on the general atmospheric circulation, even in the tropics, and on amounts of precipitation (Swann et al., 2012).

The Intergovernmental Panel on Climate Change (IPCC, 2007a: 30) concluded that 'The impacts of land-use change on climate are expected to be locally significant in some regions, but are small at the global scale in comparison with greenhouse gas warming'.

Forests, irrigation and climate

The replacement of forest with crops, in addition to leading to a change in surface albedo of the type just discussed, also changes some other factors that may have climatic significance, including surface aerodynamic roughness, leaf and stem areas and amount of evapotranspiration (Betts, 2001). In particular large amounts of moisture may be transpired by deep-

Table 7.5 Some recent studies of climatic effects of vegetation removal

Source	Location	Subject
Fuller and Ottka (2002)	West Africa	Albedo and desertification
Fu (2003)	East Asia	Reduced atmospheric and soil moisture in East Asian monsoon region
Chase et al. (2000)	Global	Effects on main circulation features
Werth and Avissar (2002)	Amazonia	Reduction of local precipitation, evapotranspiration and cloudiness and also global effects
Berbet and Costa (2003)	Amazonia	Precipitation variability
Reale and Dirmeyer (2000)	Mediterranean Basin	Increased precipitation prior to deforestation in Roman times
Taylor et al. (2002)	Sahel	Rainfall decrease
Sampaio et al. (2007)	Amazonia	Rainfall decrease
Doughty et al. (2012)	South America	Rainfall decrease

rooting plants, which means that moisture is pumped back into the atmosphere, leading to increased levels of precipitation. Conversely, removal of such deep-rooting vegetation could exacerbate drought.

The belief that forests can increase precipitation levels has a long history (Thornthwaite, 1956; Grove, 1997), and it has been the basis of action programmes in many lands. For example, the American Timber Culture Act of 1873 was passed in the belief that if settlers were induced to plant trees on the Great Plains and prairies, precipitation would be increased sufficiently to eliminate the climatic hazards to agriculture. On the other hand, at much the same time, the view was expressed that 'rain follows the plough'. Aughey, working in Nebraska, for example, believed that, after the soil is 'broken', rain as it falls is absorbed by the soil 'like a high sponge', and that the soil gives this absorbed moisture slowly back to the atmosphere by evaporation (cited by Thornthwaite, 1956: 569), and so increases the rainfall.

These two early and contradictory views illustrate the confusion that still surrounds this question today. Forests undoubtedly influence rates of evapotranspiration, the flow of streams, the level of groundwater and microclimates, but there is little reliable evidence to suggest that regional rainfall is either significantly increased by forest that attempts to augment rainfall levels on desert margins by widespread planting of forest belts are likely to achieve relatively much, for the aridity of deserts and their margins is controlled dominantly by the gross features of the general circulation, especially the subsiding air associated with the big high-pressure cells of the subtropics.

Although forests may not necessarily have a proven effect on regional or continental rainfall levels, they are far more effective than other vegetation types at trapping other kinds of precipitation, especially cloud, fog and mist. Hence deforestation or afforestation can affect water budgets through the degree to which they intercept non-rainfall precipitation.

There is one other land-use change that may result in measurable changes in precipitation; namely large-scale crop irrigation in arid and semi-arid regions. The High Plains of the USA are normally covered with sparse grasses and have dry soils throughout the summer; evapotranspiration is then very low. In the last five decades irrigation has been developed throughout large parts of the area, greatly increasing summer evapotranspiration levels. Barnston and Schickedanz (1984) produced strong statistical evidence of warm-season rainfall enhancement through irrigation in two parts of this area: one extending through Kansas, Nebraska and Colorado, and a second in the Texas Panhandle. The largest absolute increase was in the latter area and, significantly, occurred in June, the wettest of the three heavily irrigated months. The effect appears to be especially important when stationary weather fronts occur, for this is a situation which allows for maximum interaction between the damp irrigated surface and the atmosphere. Hailstorms and tornados are also significantly more prevalent than over non-irrigated regions (Nicholson, 1988). However, Moore and Rojstaczer (2001) thought that overall the irrigation effect is both difficult to quantify unambiguously and probably of minor significance. Irrigation may also lead to lower air surface temperatures compared to that found above natural desert surfaces, and this has been found in the northern Great Plains of the USA (Mahmood et al., 2006). Finally, Lee et al. (2011) have suggested that there is now so much irrigated land over Asia that not only does it lead to a decrease in surface temperatures, but may also affect upper level pressure and wind conditions.

Bonan (1997) tried to model the climatic consequences of replacing the natural forests of the USA with crops and argued that it would cause cooling of up to 2°C in the summer months over a wide region of the central USA. He suggested that 'land use practices that resulted in extensive deforestation in the Eastern USA, replacing forests with crop, have resulted in a significant climate change that is comparable to other well known anthropogenic climate forcings' (p. 484). Similarly, the clearing of native vegetation in eastern Australia seems to have led to an increase in the number of dry and hot days, thereby intensifying drought conditions (Deo et al., 2009).

The possible effects of water diversion schemes

The levels of the Aral and Caspian Seas in Central Asia have fallen (see Chapter 5), as have water tables all over the wide continental region. There have been proposals to divert some major rivers to help overcome these problems. However, this raises difficult questions 'because it appears to touch a peculiarly sensitive spot in the existing climatic regime of the northern hemisphere' (Lamb, 1977: 671). The low-salinity water which forms a 100–200 m upper layer to the Arctic Ocean is in part caused by the input of fresh water from the large Russian and Siberian rivers. This low-salinity water is the medium in which the pack ice at present covering the polar ocean is formed. The tapping of any large proportion of this river flow might augment the area of salt water in the Arctic Ocean and thereby reduce the area of pack ice correspondingly. Temperatures over large areas might rise, which in turn might change the position and alignment of the main thermal gradients in the northern hemisphere and, with them, the jet stream and the development and steering of cyclonic activity. However, assessment of this particular climatic impact is still very largely speculative, and some numerical models indicate that the climate of the Arctic will not be drastically affected by river diversions (Semtner, 1984).

Lakes

It has often been implied that the presence of a large body of inland water must modify the climate around its shores and therefore that artificial lakes have a significant effect on local or regional climates. Climatic changes produced by the construction of a reservoir are the result of a variety of factors (Vendrov, 1965): the creation of a body of water with a large heat capacity that reduces the continentality of the climate; the substitution of a water surface for a land surface and the rise of the groundwater level in the littoral zone supplying moisture to the evaporating surface (leading to a rise in wind velocity above the lake and in the littoral zone). Schemes have been put forward for augmenting desert rainfall by flooding desert basins in the Sahara, Kalahari and Middle East (see e.g. Schwarz, 1923). However, whether evaporation from lake surfaces can raise local precipitation levels is open to question, for precipitation depends more on atmospheric instability than upon the humidity content of the air. Moreover, most lakes are too small to affect the atmosphere materially in depth, so that their influence falls heavily under the sway of the regional circulation. In addition, one needs to remember that some of the world's driest deserts occur along coastlines. Thus a relatively small artificial lake would be even more impotent in creating rainfall. However, in the USA, particularly in areas with Mediterranean and semi-arid climates, there is some evidence that rainfall is augmented in the vicinity of large dames (Degu et al., 2011).

However, the climatic effect of artificial lakes is evident in other ways, notably in terms of a local reduction in frost hazard. In the case of the Rybinsk reservoir (c. 4500 km^2) in the CIS, it has been calculated that the climatic influence extends 10 km from the lake and that the frost-free season has been extended by 5–15 days on average (D'Yakanov and Reteyum, 1965).

Urban climates

With the increasing number of people living in big cities, there is an ever increasing interest in how urban areas modify their climates (Souch and Grimmond, 2006). There is particular interest in the production of heat by the burning of fossil fuels, which at the local scale produces the 'urban heat island' (Jones and Lister, 2009). Following an analysis of urban and rural temperature trends in 50 of the most populous cities in the USA between 1951 and 2000, Stone (2007) found that there had been an amplification of background warming rates of 0.5°C in the cities.

It has been said that 'the city is the quintessence of man's capacity to inaugurate and control changes in his habitat' (Detwyler and Marcus, 1972). One way in which such control becomes evident is in a study of urban climates (Landsberg, 1981). Individual urban areas can at times, with respect to their weather, 'have similar impacts as a volcano, a desert, and as an irregular forest' (Changnon, 1973: 146). Some of the changes that can result are listed in Table 7.6.

Compared with rural surfaces, city surfaces (Table 7.6b) absorb significantly more solar radiation, because a higher proportion of the reflected radiation is retained by the high walls and dark-coloured roofs of the city streets. The concreted city surfaces have both great thermal capacity and conductivity, so that heat is stored during the day and released by night. By contrast the plant cover of the countryside acts like an insulating blanket, so that rural areas tend to experience relatively lower temperatures by day and night, an effect enhanced by the evaporation and transpiration taking place. In addition, the energy partitioned

Table 7.6 Some urban climate characteristics (from Griffiths 1976, reproduced with permission)

(a) Average changes in climatic elements caused by cities

Element	Parameter	Urban compared with rural (–, less; +, more)
Radiation	On horizontal surface	–15%
	Ultraviolet	–30% (winter); –5% (summer)
Temperature	Annual mean	+0.7°C
	Winter maximum	+1.5°C
	Length of freeze-free season	+2 to 3 weeks (possible)
Wind speed	Annual mean	–20% to –30%
	Extreme gusts	–10% to –20%
	Frequency of calms	+5% to 20%
Relative humidity	Annual mean	–6%
	Seasonal mean	–2% (winter); –8% (summer)
Cloudiness	Cloud frequency + amount	+5% to 10%
	Fogs	+100% (winter); –30% (summer)
Precipitation	Amounts	+5% to 10%
	Days	+10%
	Snow days	–14%

(b) Effect of city surfaces. Source: H. Landsberg in Griffiths (1976: 108)

Phenomenon	Consequence
Heat production (the heat island)	Rainfall +
	Temperature +
Retention of reflected radiation by high walls and dark-coloured roofs	Temperature +
Surface roughness increase	Wind –
	Eddying +
Dust increase (the dust dome)	Fog +
	Rainfall + (?)

for evapotranspiration is less in urban areas, leading to greater surface heating. Another thermal change in cities, contributing to the development of the 'urban heat island', is the large amount of artificial heat produced by industrial, commercial and domestic users.

In general the highest temperature anomalies are associated with the densely built-up area near the city centre and decrease markedly at the city perimeter. Observations in Hamilton, Ontario and Montreal, Quebec, suggested temperature changes of 3.8 and 4.0°C, respectively, per kilometre (Oke, 1978). Tem-

perature differences also tend to be highest during the night. The form of the urban temperature effect has often been likened to an 'island' protruding distinctly out of the cool 'sea' of the surrounding landscape. The rural-urban boundary exhibits a steep temperature gradient or 'cliff' to the urban heat island. Much of the rest of the urban area appears as a 'plateau' of warm air with a steady but weaker horizontal gradient of increasing temperature towards the city centre. The urban core may be a 'peak' where the urban maximum temperature is found. The difference between this value and the background rural temperature defines the *urban heat island intensity* (T_{u-r} (max)) (Oke, 1978: 225).

The heat island of New York City is of impressive size and impacts upon a large area. The difference in temperature between the urban core and the surrounding rural areas averages about 4°C on summer nights. Figure 7.9 shows the urban heat island that was evident on a warm, summer night on 14th August, 2002. Surface air temperature readings showed that the city was several degrees warmer than the suburbs and up to 8°C warmer than the rural areas within 100 km of the city (Rosenzweig et al., 2009).

Table 7.7 lists the average annual urban-rural temperature differences for several large cities. Values range from 0.6 to 1.8°C. In their analysis of 419 cities with a population of over one million, Peng et al. (2012) found that the mean vales were 1.5°C by day and 1.1°C by night. The relationship between city size and urban-rural difference, however, is not necessarily linear; sizeable nocturnal temperature contrasts have been measured even in relatively small cities. Factors such as building density are at least as important as city size, and high wind velocities will tend to reduce the heat island effect (Schlünzen et al., 2010). The presence or absence of urban vegetation is also significant (Peng et al., 2012).

Nonetheless, Oke (1978: 257) has found that there is some relation between heat-island intensity and city size. Using population as a surrogate of city size T_{u-r} (max) is found to be proportional to the log of the population. Other interesting results of this study include the tendency for quite small centres to have a heat island, the observation that the maximum thermal modification is about 12°C, and the recognition of a difference in slope between the North American and the European relationships (Figure 7.10a). The

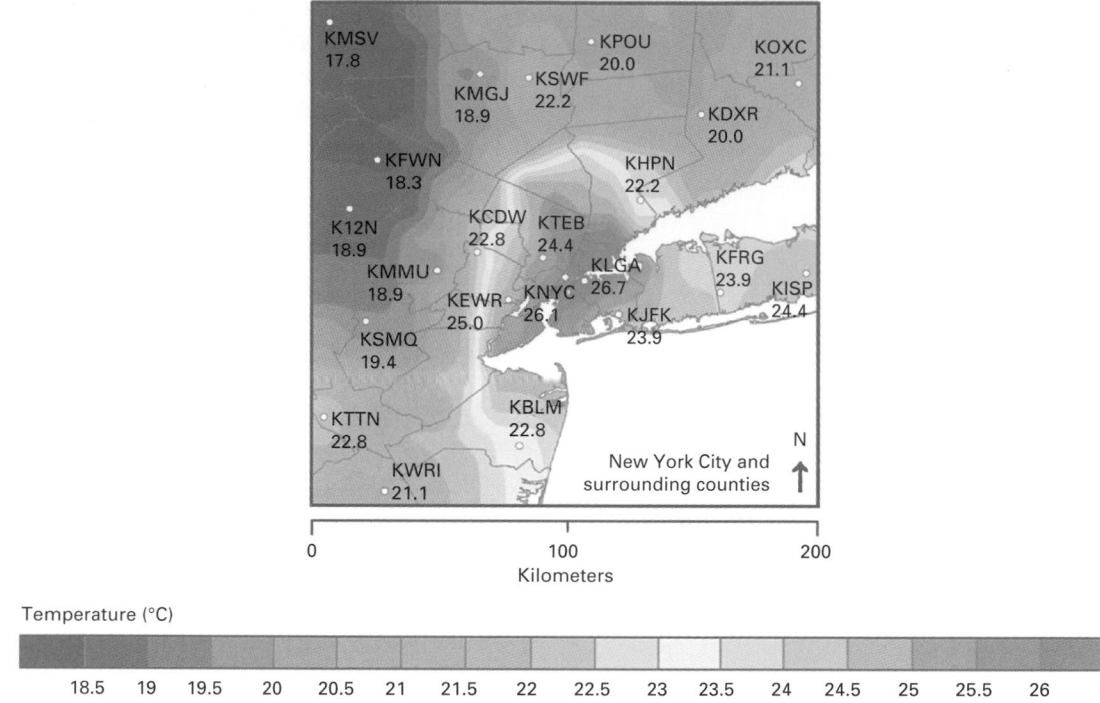

Figure 7.9 New York City's urban heat island (°C) at 0600 EST on 14 August 2002 (modified from Rosenzweig et al., 2009, figure 1).

Table 7.7 Annual mean urban-rural temperature differences of cities. Source: Data from Detwyler (1971) and Wilby (2003). Reproduced with permission

City	Temperature differences (°C)
Chicago, USA	0.6
Washington, DC, USA	0.6
Los Angeles, USA	0.7
Paris, France	0.7
Moscow, Russia	0.7
Philadelphia, USA	0.8
Berlin, Germany	1.0
New York, USA	1.1
London, UK	1.8

explanation for this last result is not clear, but it may be related to the fact that population is a surrogate index of the central building density. Imhoff et al. (2010), working on a large number of American cities also found a strong correlation between city size and the size of the urban heat island.

The relationship between maximum heat island intensity and urban population, the sky-view factor (a measure of building height and density) and the impermeable surface coverage of cities, are shown in Figure 7.10c. The heat island intensity increases with all three indices (Nakagawa, 1996).

In many older towns and cities in Western Europe and North America, the process of 'counter-urbanization' has in recent years led to a decline in population, and it is worth considering whether this is reflected in a decline in the intensity of urban heat islands. One attempt to do this, in the context of London (Lee, 1992), revealed the perplexing finding that the heat-island intensity has decreased by day, but increased by night. The explanation that has been tentatively advanced to explain this is that there has been a decrease in the receipt of daytime solar radiation as a result of vehicular atmosphere pollution, whereas at night the presence of such pollution absorbs and re-emits significant amounts of out-going terrestrial radiation, maintaining higher urban nocturnal minimum temperatures. The urban heat island may also be more marked at night because reduced nocturnal turbulent mixing keeps the warmer air near the surface. In mid-latitude cities like London, urban heat island effects are gener-

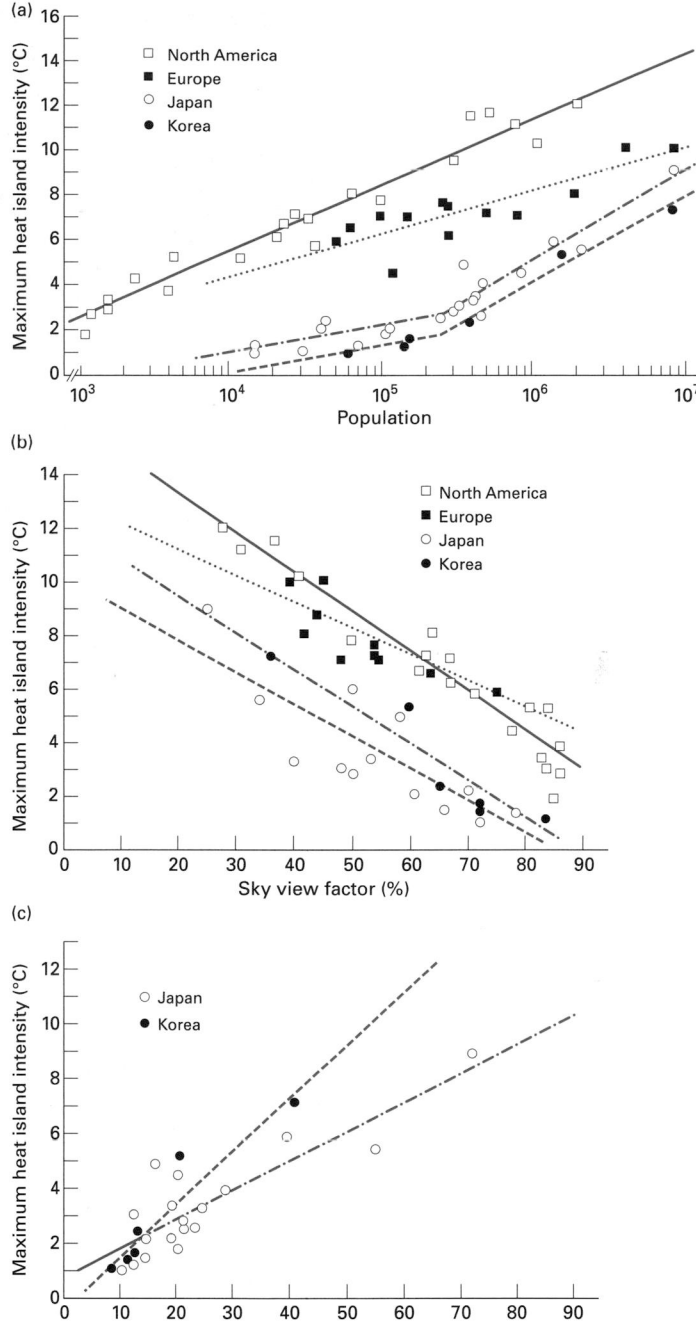

Figure 7.10 Relationships between the maximum heat island intensity and (a) urban population for Japanese, Korean, North American and European cities; (b) the sky view factor for Japanese, Korean, North American and European cities; (c) the ratio of impermeable surface coverage for Japanese and Korean cities (after Nakagawa, 1996, figures 2, 3 and 4). Reproduced with permission from Association of Japanese Geographers.

ally stronger in summer than in winter because of higher levels of solar radiation being absorbed by building materials during the day. In winter the urban-rural contrast is weaker because solar energy absorption is lower and hence there is less energy to radiate, despite higher levels of urban space heating (Wilby, 2003).

The existence of the urban heat island has a number of implications; city plants bud and bloom earlier, some birds are attracted to the thermally more favourable urban habitat, humans find added warmth stressful if the city is already situated in a warm area, during summer heat waves exacerbated temperatures may cause mortality among sensitive members of the population as has been proved in Shanghai, China (Tan et al., 2010), less winter space-heating is required, but, conversely, more summer air-conditioning is necessary.

The urban-industrial effects on clouds, rain, snow-fall and associated weather hazards such as hail and thunder are harder to measure and explain than the temperature changes (Darungo et al., 1978). The changes can be related to various influences (Chang-non, 1973: 143):

thermally induced upward movement of air;
increased vertical motions from mechanically
    induced turbulence;
increased cloud and raindrop nuclei;
industrial increases in water vapour.

Table 7.8 illustrates the differences in summer rainfall, thunderstorms and hailstorms between various rural and urban areas in the USA. These data indicate that in cities rainfall increased from 9% to 27%, the incidence of thunderstorms by 10% to 42%, and hailstorms by 67% to 430%. Such changes have been identified in Houston, Texas, where the average warm-season rain-fall amount in the urban area increased by 25% from the pre- to the post-urban period, while the amount in the upwind control region declined by 8% (Burian and Shepherd, 2005). Urban effects on summer thunder-storm rainfall have also been identified in Atlanta, Georgia (Rose et al., 2008; Shem and Shepherd, 2009) and more generally in the south-east USA (Ashley

et al., 2011). Convective storms have also been intensi-fied over Sydney, Australia (Gero et al., 2006), and lightning activity seems to be affected by the metro-politan area of São Paulo, Brazil (Farias et al., 2009). Urban aerosols appear to play an important role both in terms of enhancing precipitation and enhancing lightning (Heever and Cotton, 2007; Kar et al., 2009).

An interesting example of the effects of major con-urbations on precipitation levels is provided by the London area. In this case it seems that the mechanical effect of the city was dominant in creating localized maxima of precipitation both by being a mechanical obstacle to air flow, on the one hand, and by causing frictional convergence of flow, on the other (Atkinson, 1975). A long-term analysis of thunderstorm records for south-east England is highly suggestive – indicat-ing the higher frequencies of thunderstorms over the conurbation compared to elsewhere (Atkinson, 1968). The similarity in the morphology of the thunderstorm isopleth and the urban area is striking (Figure 7.11a and b). Moreover, Brimblecombe (1977) shows a stead-ily increasing thunderstorm frequency as the city has grown (Figure 7.11c).

Similarly the detailed Metromex investigation of St Louis in the USA (Changnon, 1978) showed that in the summer the city affects precipitation and other varia-bles within a distance of 40 km. Increases were found in various thunderstorm characteristics (about +10% to +115%), hailstorm condition (+3% to +330%), various heavy rainfall characteristics (+35% to +100%) and strong gusts (+90% to +100%).

Table 7.8 Areas of maximum increases (urban-rural difference) in summer rainfall and severe weather events for eight American cities (from Changnon, 2003)

City	Rainfall		Thunderstorms		Hailstorms	
	%	Location[a]	%	Location[a]	%	Location[a]
St Louis	+15	B	+25	B	+276	C
Chicago	+17	C	+38	A, B, C	+246	C
Cleveland	+27	C	+42	A, B	+90	C
Indianapolis	0	–	0	–	0	–
Washington, DC	+9	C	+36	A	+67	B
Houston	+9	A	+10	A, B	+430	B
New Orleans	+10	A	+27	A	+350	A, B
Tulsa	0	–	0	–	0	–

[a]A = within city perimeter, B = 8–24 km downwind, C = 24–64 km downwind.

Figure 7.11 Thunder in south-east England: (a) total thunder rain in south-east England, 1951–1960, expressed in inches (after Atkinson, 1968, figure 6). Reproduced with permission; (b) number of days with thunder overhead in south-east England, 1951–1960 (after Atkinson, 1968, figure 5). Reproduced with permission; (c) thunderstorms per year in London (decadal means for whole year) (after Brimblecombe, 1977, figure 2). Reproduced with permission.

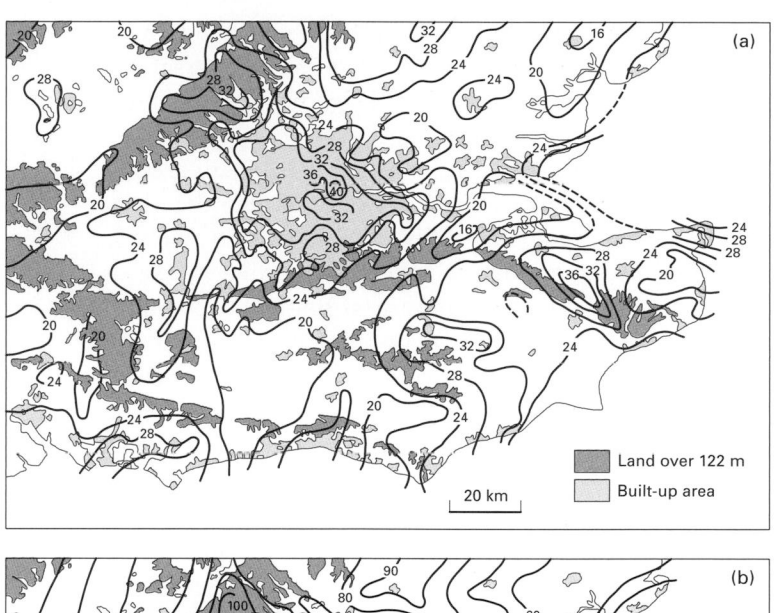

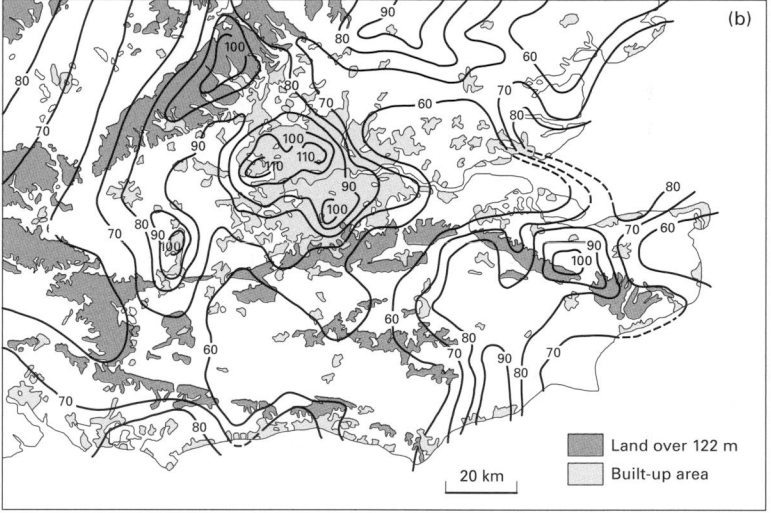

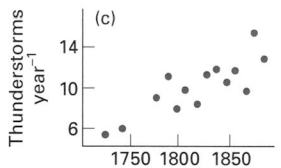

However, increases in precipitation are not universal over urban areas, and some studies show the reverse. Work in the Pearl River delta cities of southeast China (Kaufmann et al., 2007) produced the concept of an 'urban precipitation deficit'. The reasons for this may have been a reduced flow of moisture into the air because of the replacement of rural vegetation with concrete surfaces, and also the effects of urban aerosol pollution on clouds and the precipitation from them.

Precipitation suppression as a result of aerosol build up in cities has also been identified in Brisbane and the Gold Coast of Australia (Bigg, 2008) and also in parts of Israel and California (Givati and Rosenfeld, 2004).

Cities also affect winds in their vicinity. Two main factors are involved in this: the rougher surface they present in comparison with rural areas; and the frequently higher temperatures of the city fabric. Buildings, especially those in cities with a highly differentiated skyline, exert a powerful frictional drag on air moving over and around them (Chandler, 1976). This creates turbulence, with characteristically rapid spatial and temporal changes in both direction and speed. The average speed of the winds is lower in built-up areas than over rural areas, but Chandler found that in London, when winds are light, speeds are greater in the inner city than outside, whereas the reverse relationship exists when winds are strong. The overall annual reduction of wind speed in central London is about 6%, but for the higher velocity winds (more than 1.5 m per second) the reduction is more than doubled.

Studies in both Leicester and London (Chandler, 1976), England, have shown that on calm, clear nights, when the urban heat-island effect is at its maximum, there is a surface inflow of cool air towards the zones of highest temperatures. These so-called 'country breezes' have low velocities and become quickly decelerated by intense surface friction in the suburban areas. A practical implication of these breezes is that they transport pollution from the outer parts of an urban area into the city centre, accentuating the pollution problem during smogs.

Deliberate climatic modification

It has long been a human desire to modify weather and climate, but it is only since the Second World War and the development of high-altitude observations of clouds that serious attempts have been made to modify such phenomena as rainfall, hailstorms and hurricanes (Cotton and Pielke, 2007). It was in 1946 that Irving Langmuir discovered that tiny particles of dry ice and silver iodide could stimulate the precipitation process by providing a nucleus to which cloud-borne moisture could cling. Both civilian and military interest arose and there was considerable USA government support for work on deliberate climate modification, support which is now miniscule compared to that of the 1970s (Harper, 2008). Research continues in some other countries, however, including China (Guo and Zheng, 2009).

Following on from Langmuir's work, the most fruitful human attempts to augment natural precipitation have been through cloud seeding. Rainmaking experiments of this type are based on three main assumptions (Chorley and More, 1967: 159).

1 Either the presence of ice crystals in a supercooled cloud is necessary to release snow and rain, or the presence of comparatively large water droplets is necessary to initiate the coalescence process.
2 Some clouds precipitate inefficiently or not at all, because these components are naturally deficient.
3 The deficiency can be remedied by seeding the clouds artificially, either with solid carbon dioxide (dry ice) or silver iodide, to produce crystals, or by introducing water droplets or large hygroscopic nuclei (e.g. salt).

These methods of seeding are not universally productive and in many countries, expenditure on such procedures has been reduced. In the USA the glory days of weather modification were in the 1970s, and after that federal expenditure has declined (Cotton and Pielke, 2007). Under conditions of orographic lift and in thunderstorm cells, when nuclei are insufficient to generate rain by natural means, some augmentation may be attained, especially if cloud temperatures are of the order of −10 to −15°C. The increase of precipitation gained under favourable conditions may be of the order of 10–20% in any one storm. In lower latitudes, where cloud-top temperatures frequently remain above 0°C, silver-iodide or dry-ice seeding is not applicable. Therefore alternative methods have been introduced whereby small water droplets of 50 mm diameter are sprayed into the lower layers of deep clouds, so that the growth of cloud particles will be stimulated by coalescence. Other techniques of warm cloud seeding include the feeding of hygroscopic particles into the lower air layers near the updraught of a growing cumulus cloud (Breuer, 1980).

The results of the many experiments now carried out on cloud seeding are still controversial, very largely because we have an imperfect understanding of the physical processes involved. This means that the evidence has to be evaluated on a statistical rather than a scientific basis so, although precipitation may occur after many seeding trials, it is difficult to decide to what extent artificial stimulation and augmentation is responsible. It also needs to be remembered that this form of planned weather modification applies to small areas for short periods. As yet, no means exist to change precipitation appreciably over large areas on a sustained basis.

Other types of deliberate climatic modification have also been attempted (Hess, 1974). For example, it has been thought that the production of many more hailstone embryos by silver-iodide seeding will yield smaller hailstones which would both be less damaging and more likely to melt before reaching the ground. Some results in Russia have been encouraging, but one cannot exclude the possibility that seeding may sometimes even increase hail damage (Atlas, 1977). Similar experiments have been conducted in lightning suppression. The concept here is to produce in a thundercloud, again by silver-iodide seeding, an abnormal abundance of ice crystals that would act as added corona points and thus relieve the electrical potential gradient by corona discharge before a lightning strike could develop (Panel on Weather and Climate Modification, 1966: 4–8).

Hurricane modification is perhaps the most desirable aim of those seeking to suppress severe storms, because of the extremely favourable benefit-to-cost ratio of the work (Cotton and Pielke, 2007, section 2.6). The principle once again is that of introducing freezing nuclei into the ring of clouds around the hurricane centre to trigger the release of the latent heat of fusion in the eye-wall cloud system which, in turn, diminishes the maximum horizontal temperature gradients in the storm, causing a hydrostatic lowering of the surface temperature. This eventually should lead to a weakening of the damaging winds (Smith, 1975: 212). A 15% reduction in maximum winds is theoretically possible, but as yet work is largely at an experimental stage. A 30% reduction in maximum winds was claimed following seeding of Hurricane Debbie in 1969 but US attempts to seed hurricanes were discontinued in the 1970s following the development of new computer models of the effects of seeding on hurricanes. The models suggested that although maximum winds might be reduced by 10–15% on average, seeding may increase the winds just outside the region of maximum winds by 10–15%, may either increase or decrease the maximum storm surge and may or may not affect the direction of the storm (Sorkin, 1982). Seeding of hurricanes is unlikely to recommence until such uncertainties can be resolved.

Fog dispersal, vital for airport operation, is another aim of weather modification. Seeding experiments have shown that fog consisting of supercooled drop-lets can be cleared by using liquid propane or dry ice. In very cold fogs this seeding method causes rapid transformation of water droplets into ice particles. Warm fogs with temperatures above freezing point occur more frequently than supercooled fogs in mid-latitudes and are more difficult to disperse. Some success has been achieved using sodium chloride and other hygroscopic particles as seeding agents but the most effective method is to evaporate the fog. The French have developed the 'turboclair' system in which jet engines are installed alongside the runway at major airports and the engines produce short bursts of heat to evaporate the fog and improve visibility as an aircraft approaches (Hess, 1974).

In regions of high temperature, dark soils then become overheated, and the resultant high evapo-transpiration rates lead to moisture deficiencies. Applications of white powders (India and Israel) or of aluminium foils (Hungary) increase the reflection from the soil surface and reduce the rate at which insulation is absorbed. Temperatures of the soil surface and sub-surface are lowered (by as much as 10°C), and soil moisture is conserved (by as much as 50%).

The planting of windbreaks is an even more important attempt by humans to modify local climate deliberately. Shelterbelts have been in use for centuries in many windswept areas of the globe, both to protect soils from blowing and to protect the plants from the direct effects of high velocity winds. The size and effectiveness of the protection depends on their height, density, shape and frequency. However, the belts may have consequences for microclimate beyond those for which they were planted. Evaporation rates are curtailed; snow is arrested, and its melting waters are available for the fields; but the temperatures may become more extreme in the stagnant space in the lee of the belt, creating an increase in frost danger.

Traditional farmers in many societies have been aware of the virtues of microclimatic management (Wilken, 1972). They manage shade by employing layered cropping systems or by covering the plant and soil with mulches; they may deliberately try to modify albedo conditions. Tibetan farmers, for example, reportedly throw dark rocks on to snow-covered fields to promote late spring melting; and in the Paris area of France some very dense stonewalls were constructed to absorb and radiate heat.

Geoengineering

In coming decades it is possible that more ambitious attempts at global climatic modification may be attempted, to which the term **geoengineering** is often applied (Lovelock, 2008). In particular such techniques may be developed to combat global warming. Geoengineering methods can be divided into two basic 'classes' (Royal Society, 2009a). The first of these is a group of carbon dioxide removal (CDR) techniques which aim to address the root cause of climate change by removing greenhouse gases from the atmosphere: land-use management to protect or enhance land carbon sinks; the use of biomass for carbon sequestration as well as a carbon neutral energy source; enhancement of natural rock weathering processes to remove CO_2 from the atmosphere; direct engineered capture of CO_2 from ambient air; and the enhancement of oceanic uptake of CO_2 by fertilization of the oceans with naturally scarce nutrients, or by increasing upwelling of ocean water. The second class of techniques are called solar radiation management (SRM) techniques. These aim to offset the effects of increased greenhouse gas concentrations by causing the Earth to absorb less solar radiation. Among the methods put forward are: placing shields or reflectors in space to reduce the amount of solar energy reaching the Earth; brightening the surface of the Earth (e.g. by covering deserts with mirrors or other reflective material) so that more incoming radiation is reflected back to space; and injecting sulfate aerosols into the low stratosphere to mimic the cooling effect achieved by volcanic eruptions.

Urban air pollution

The concentration of large numbers of people, factories, power stations and cars mean that large amounts of pollutants may be emitted into urban atmospheres. If weather conditions permit, the level of pollution may build up. The nature of the pollutants (Table 7.9) has changed as technologies have changed. For example, in the early phases of the industrial revolution in Britain the prime cause of air pollution in cities may have been the burning of coal, whereas now it may be vehicular emissions. Different cities may have very different levels of pollution, depending on factors such as the level of technology, size, wealth and anti-pollution legislation. Differences may also arise because of local topographic and climatic conditions. Photochemical smogs, for example, are a more serious threat in areas subjected to intense sunlight.

Fenger (1999) has argued that the development of urban air pollution shows certain general historical trends. At the earlier stages air pollution increases to high levels. There then follows various abatement measures that cause a stabilization of air quality to occur. Levels of pollution then fall as high technology solutions are applied, though this may be countered to a certain extent by growth in vehicular traffic.

In some developed cities concentrations of pollutants have indeed tended to fall over recent decades. This can result from changes in industrial technology or from legislative changes (e.g. clean air legislation, restriction on car use, etc.). In many British cities, for example, legislation since the 1950s has reduced the

Table 7.9 Major urban pollutants

Type	Some consequences
Suspended particulate matter (characteristically 0.1–25 μm in diameter)	Fog, respiratory problems, carcinogens, soiling of buildings
Sulfur dioxide (SO_2)	Respiratory problems, can cause asthma attacks. Damage to plants and lichens, corrosion of buildings and materials, production of haze and acid rain
Photochemical oxidants: ozone and peroxyacetyl nitrate (PAN)	Headaches, eye irritation, coughs, chest discomfort, damage to materials (e.g. rubber), damage to crops and natural vegetation, smog
Oxides of nitrogen (NO_x)	Photochemical reactions, accelerated weathering of buildings, respiratory problems, production of acid rain and haze
Carbon monoxide (CO)	Heart problems, headaches, fatigue, and so on
Toxic metals: lead	Poisoning, reduced educational attainments and increased behavioural difficulties in children
Toxic chemicals: dioxins and so on	Poisoning, cancers and so on

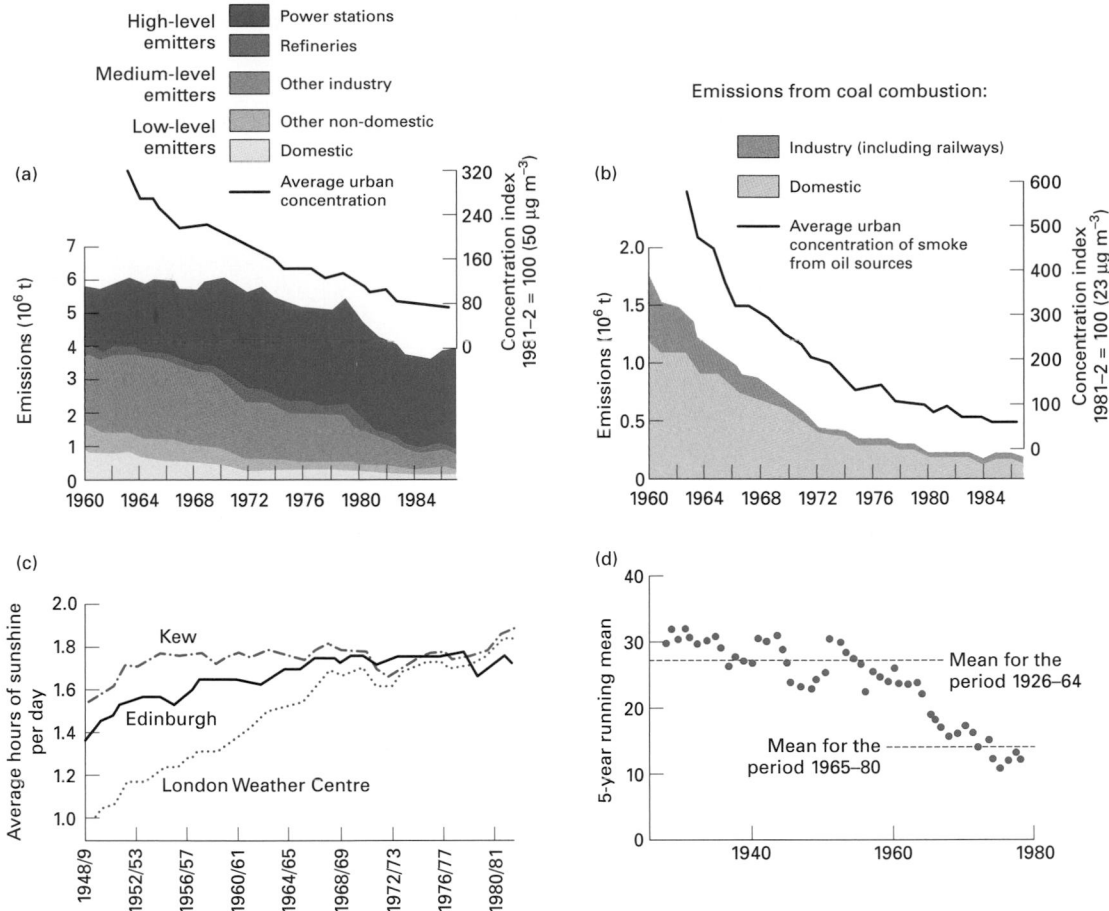

Figure 7.12 Trends in atmospheric quality in the UK: (a) sulfur dioxide emissions from coal combustion and average urban concentrations; (b) smoke emissions from coal combustion and average urban concentrations of oil smoke; (c) increase in winter sunshine (10-year moving average) for London and Edinburgh city centres and for Kew, outer London; (d) annual fog frequency at 0900 GMT in Oxford, central England, 1926–1980 (after Department of the Environment data and Gomez and Smith, 1984, figure 3).

burning of coal. As a consequence, fogs have become less frequent and the amount of sunshine has increased. Figure 7.12 shows the overall trends for the UK, and highlights the decreasing fog frequency and increasing sunshine levels (Musk, 1991). Similarly lead concentrations in the air in British cities have declined sharply following the introduction of unleaded petrol (Figure 7.12b) (Kirby, 1995), as they have in Copenhagen (Figure 7.13c) (Fenger, 1999). The concentrations of various pollutants have also been reduced in the Los Angeles area of California (Figure 7.14). Here, carbon monoxide, non-methane hydrocarbon, nitrogen oxide and ozone concentrations have all fallen steadily over the period since the late 1960s (Lents and Kelly, 1993). Urban air quality has also improved in New York State

(Buckley and Mitchell, 2010) since the 1980s. Carbon monoxide, sulfur dioxide, nitrogen dioxide, ozone, lead and particulate matter all showed statistically significant downward trends. A further clear example of a recent decline in urban pollution is provided by New Jersey (Lioy and Georgopoulos, 2011) (Figure 7.15).

However, these examples of improving trends come from developed countries. In many cities in poorer countries, some types of pollution are increasing at present. In certain countries, heavy reliance on coal, oil and even wood for domestic cooking and heating means that their levels of sulfur dioxide and suspended particulate matter (SPM) are high and climbing. In addition, rapid economic development is bringing increased emissions from industry and motor vehicles,

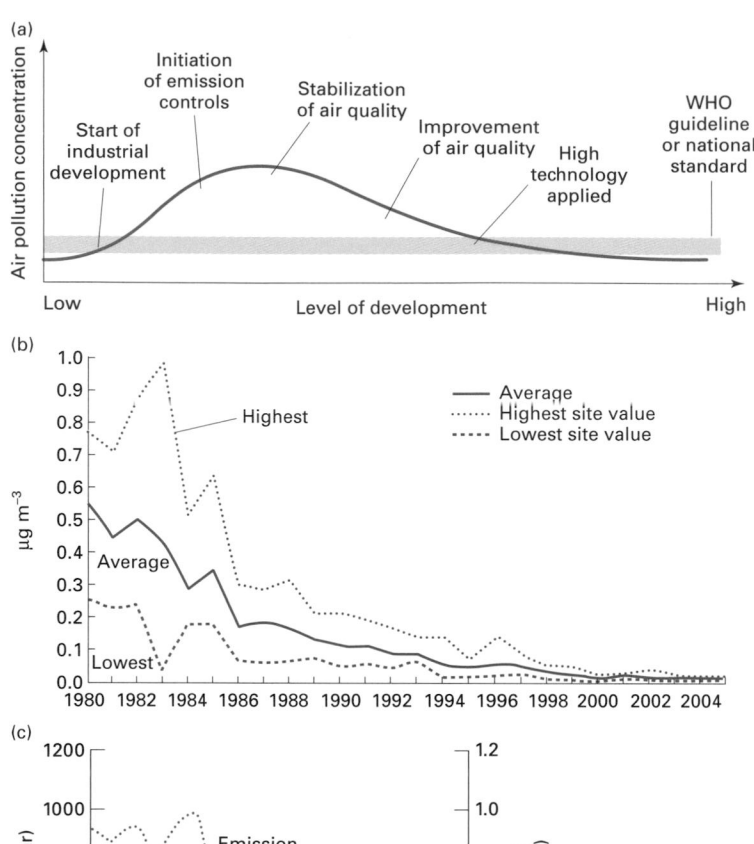

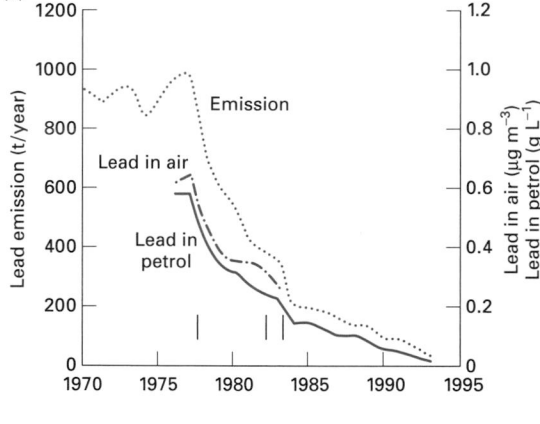

Figure 7.13 (a) Schematic presentation of a typical development of urban air pollution levels (after Fenger, 1999, figure 3). Reproduced with permission. (b) Lead concentration (annual means) in UK urban sites from 1980–2005 (DEFRA data). (c) Annual average values for the total Danish lead emissions 1969–1993, the lead pollution in Copenhagen since 1976, and the average lead content in petrol sold in Denmark. The dates of tightening of restrictions on lead content are indicated with bars. Lead concentrations for the recent years can be found in Kemp et al. (1998) (after Fenger, 1999, figure 12).

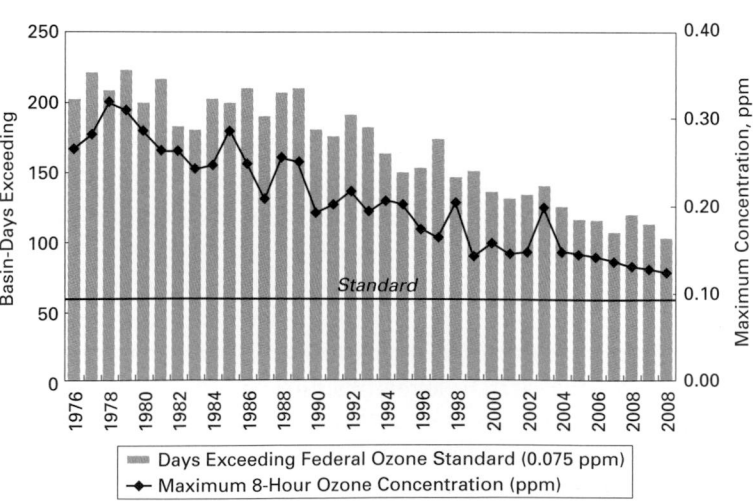

Figure 7.14 The decline in urban ozone pollution in the Los Angeles area, from 1976 to 2010. Reproduced courtesy of South Coast Air Quality Management District (SCAQMD).

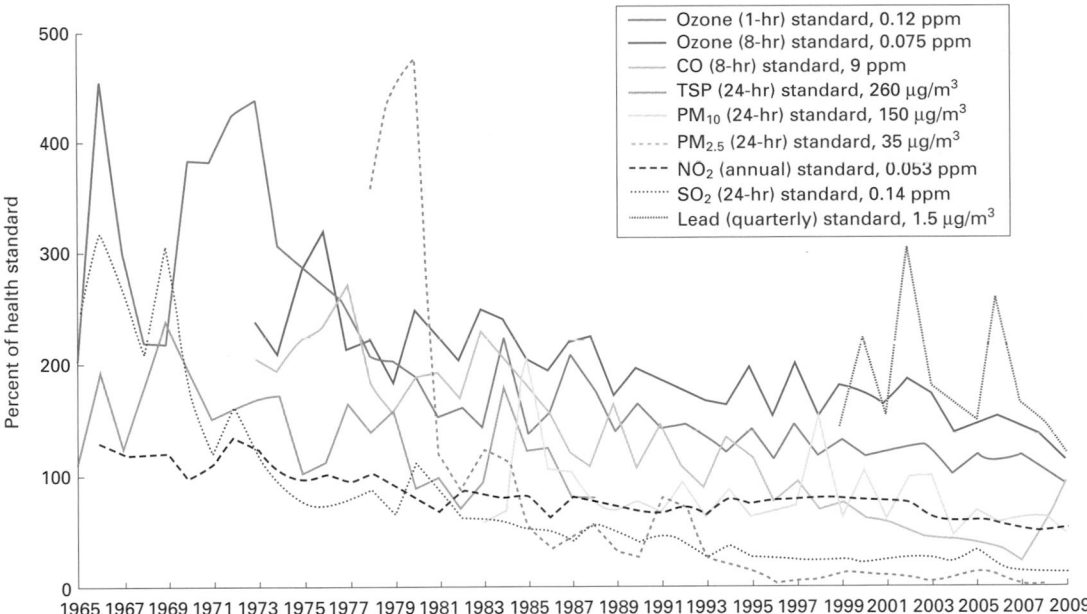

Figure 7.15 The decline in various atmospheric pollutants in New Jersey, USA from 1965–2009 (from Lioy and Georgopoulos, 2011). Reproduced with permission from Environmental Health Perspectives.

which are generating progressively more serious air-quality problems. NO_x emissions from vehicles have, for example, climbed dramatically in China (UNEP, 2007).

Particular attention is being paid at the present time to the chemical composition of SPMs and particularly to those particles that are small enough to be breathed in (i.e. smaller than 10 μm, and so often known as PM10s). Also of great concern in terms of human health are elemental carbon (e.g. from diesel vehicles), polynuclear aromatic hydrocarbons (PAHs) and toxic base metals (e.g. arsenic, lead, cadmium and mercury), in part because of their possible role as carcinogens.

A major cause of urban air pollution is the development of photochemical smog. The name originates from the fact that most of the less desirable properties of such fog result from the products of chemical reactions induced by sunlight. Unburned hydrocarbons play a major role in this type of smog formation and result from evaporation of solvents and fuels, as well as incomplete combustion of fossil fuels. In the presence of oxides of nitrogen, strong sunlight and stable meteorological conditions, complex chemical reactions occur, forming a family of peroxyacyl nitrates (sometimes collectively abbreviated to PANs).

Photochemical smog appears 'cleaner' than other kinds of fog in the sense that it does not contain the very large particles of soot that are so characteristic of smog derived from coal burning. However, the eye irritation and damage to plant leaves it causes make it unpleasant. Its unpleasant properties also include a high lead content. Photochemical smog occurs particularly where there is large-scale combustion of petroleum products, as in car-dominated cities like Los Angeles and is triggered by a series of chemical reactions involving sunlight. For example, a photochemical decomposition of nitrogen dioxide into nitric oxide and atomic oxygen occurs, and the atomic oxygen can react with molecular oxygen to form ozone (O_3). Further ozone may be produced by the reaction of atomic oxygen with various hydrocarbons. Photochemical smogs are not universal. Because sunlight is a crucial factor in their development they are most common in the tropics or during seasons of strong sunshine. Their especial notoriety in Los Angeles is due to a meteorological setting dominated at times by subtropical anticyclones with weak winds, clear skies and a subsidence inversion, combined with the general topographic situation and the high vehicle density (>1500 vehicles per square kilometre). However, photochemical ozone pollution can, on certain summer days, with anti-cyclonic conditions bringing air in from Europe reach appreciable levels even in the UK (Jenkins et al., 2002), especially in large cities such as London.

Rigorous controls on vehicle emissions can greatly reduce the problem of high urban ozone concentrations and this has been a major cause of the reduction in ozone levels in Los Angeles since the 1970s, in spite of a growth in that city's population and vehicle numbers. A full discussion of tropospheric ozone trends on a global basis is provided by Guicherit and Roemer (2000) Oltmans et al. (2006) and by The Royal Society (2008). The latter report suggests that under hot and sunny conditions, O_3 concentrations in polluted locations may peak at over 200 parts per billion by volume (ppb), but background concentrations of ground-level ozone have also climbed. Between the late 19th century and the present they have more than doubled to 35–40 ppb.

Air pollution: some further effects

This chapter has already made much reference to the ways in which humans have changed the turbidity of the atmosphere and the gases within it. However, the consequences of air pollution go further than either their direct impact on human health or their impact on local, regional and global climates.

First of all, the atmosphere acts as a major channel for the transfer of pollutants from one place to another, so that some harmful substances have been transferred long distances from their sources of emission. DDT is one example; lead is another. Thus, from the start of the industrial revolution, the lead content of the Greenland ice cap, although far removed from the source of the pollutant (which is largely derived from either industrial or auto-mobile emissions) rose very substantially (Figure 7.15a). The same applies to its sulfate content (Figure 7.15b). An analysis of pond sediments from a remote part of North America (Yosemite) indicates that lead levels were raised as a result of human activities, being more than 20 times the natural levels. The lead, which came in from atmospheric sources, showed a fivefold elevation in the plants of the area and a fiftyfold elevation in the animals compared with natural levels (Shirahata et al., 1980). A study of lead pollution history from a varved lake in Finland (Mer- iläinen et al., 2011) provides data back to Roman times on rates of lead deposition. During Roman times they were 0.2–0.3 mg m^2 per year, they then fell to less than this after the collapse of Rome, climbed to 2.6 mg m^2 per year between 1420 and 1895, and then attained values of 11–22 mg m^2 per year between 1926 and 1985.

Some workers have also compared the total quantities of heavy metals that humans are releasing into the atmosphere with emissions from natural sources (Nriagu, 1979). The increase was eighteenfold for lead, ninefold for cadmium, sevenfold for zinc and threefold for copper.

There are signs that as a result of pollution control regulations some of these trends are now being reversed. Boutron et al. (1991), for example, analysed ice and snow that has accumulated over Greenland in the previous two decades and found that lead concentrations had decreased by a factor of 7.5 since 1970 (Figure 7.16c). They attribute this to a curbing of the use of lead additives in petrol. Likewise, in Sweden, lead deposition rates have fallen by more than 90% since 1970 (Bindler, 2011). Over the same period cadmium and zinc concentrations have decreased by a factor of 2.5. It is very clear than since 1970, the emissions of cadmium, lead and mercury have plummeted (Table 7.10). Again, the precipitous decline in lead emissions is due primarily to the prohibition of lead additives in vehicle fuel. This downward trend in lead in the UK is shown in Figure 7.16d.

A second example of the possible widespread and ramifying ecological consequences of atmospheric pollution is provided by 'acid rain' or other types of acid precipitation, including fog and cloud water (Likens and Bormann, 1974; Likens, 2010). Acid rain is rain which has a pH of less than 5.65, this being the pH which is produced by carbonic acid in equilibrium with atmospheric CO_2. In many parts of the world, rain may be markedly more acid than this normal, natural background level. Snow and rain in the north-east USA have been known to have pH values as low as 2.1, while in Scotland in one storm the rain was the acidic equivalent of vinegar (pH 2.4). In the eastern USA the average annual precipitation acidity values tend to be between pH 4 and 4.5 (see Figure 7.17), and the degree of acidulation appears to have increased between the 1950s and 1970s.

It needs to be remembered that not all environmental acidification is caused by acid rain in the narrow

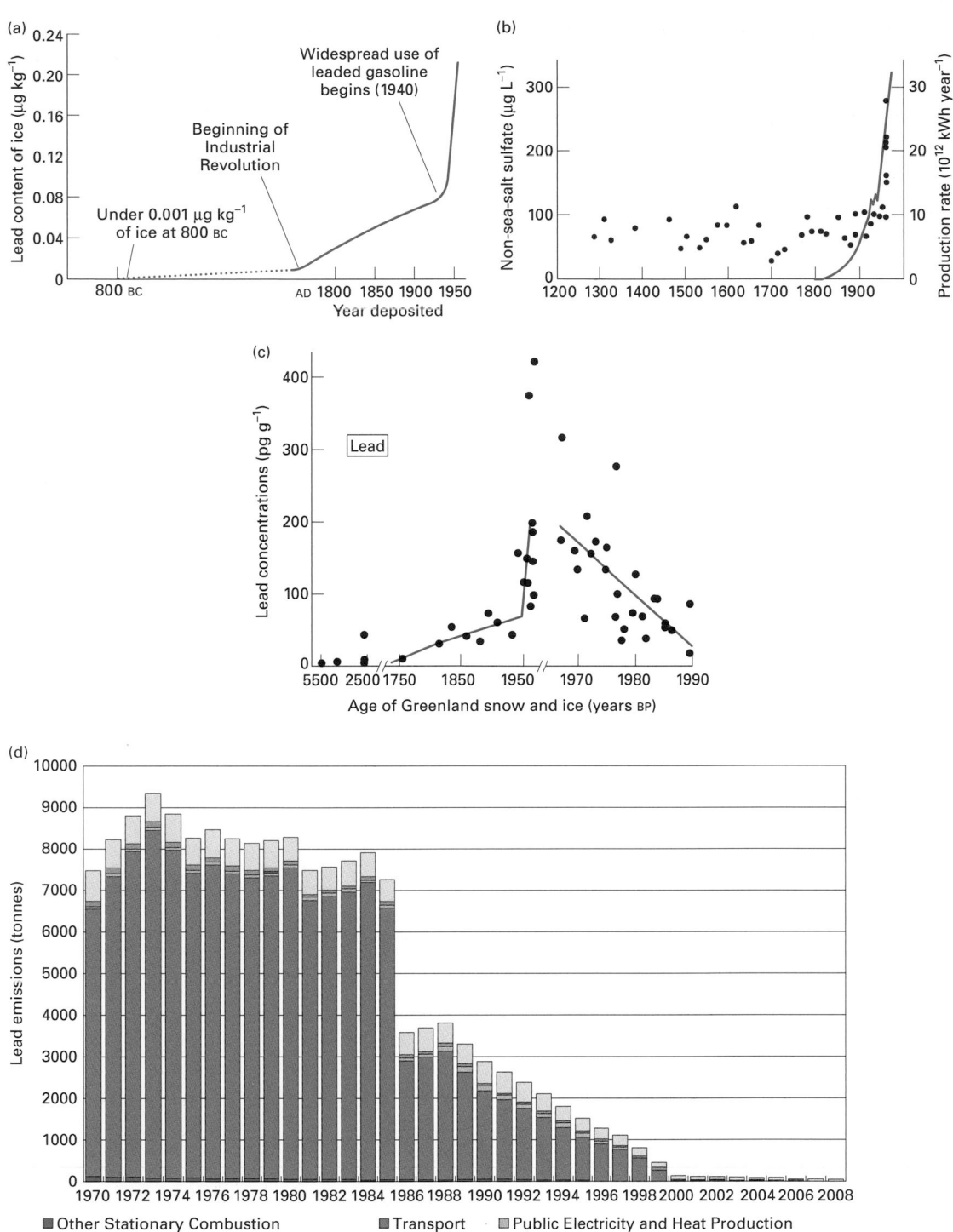

Figure 7.16 Trends in atmospheric quality: (a) lead content of the Greenland ice cap due to atmospheric fallout of the mineral on the snow surface. A dramatic upturn in worldwide atmospheric levels of lead occurred at the beginning of the industrial revolution in the nineteenth century and again after the more recent spread of the automobile (modified from Murozumi et al., 1969: 1247); (b) the sulfate concentration on a sea-salt-free basis in north-west Greenland glacier ice samples as a function of year. The curve represents the world production of thermal energy from coal, lignite and crude oil (modified from Koide and Goldberg, 1971, figure 1); (c) lead concentrations in Greenland snow (after Boutron et al., 1991. Reprinted with permission from *Nature*. Copyright 1991. Macmillan Magazines Limited); (d) Time series of lead emissions in the UK from 1970 to 2008, showing major sources (from MacCarthy et al., 2012, figure 3.5, based on data from AEA National Atmospheric Emissions Inventory, DEFRA, 2010).

sense. Acidity can reach the ground surface without the assistance of water droplets. This is as particulate matter and is termed 'dry deposition'. Furthermore, there are various types of 'wet precipitation' by mist, hail, sleet or snow, in addition to rain itself. Thus some people prefer the term 'acid deposition' to 'acid rain'. The acidity of the precipitation in turn leads to greater acidity in rivers and lakes.

The causes of acid deposition are the quantities of sulfur oxides and nitrogen oxides emitted from fossil-fuel combustion. Figure 7.18 shows the increase in the global emissions of sulfur dioxide from anthropogenic

Table 7.10 UK emissions of selected heavy metals (in tonnes per year) from 1970 to 2000 (from data provided by National Atmospheric Emissions Inventory)

	1970	1980	1990	2000	Decrease (x)
Cadmium	26.8	20.6	20.3	7.2	3.7
Lead	7339	8151	2780	193	38.0
Mercury	44.9	35.2	31.6	8.8	5.1

sources. Sulfate levels increased in European precipitation between the 1950s and 1970s. Two main factors contributed to the increasing seriousness of the problem at the time. One was the replacement of coal by oil and natural gas. The second, paradoxically, was a result of the implementation of air pollution control measures (particularly increasing the height of smoke

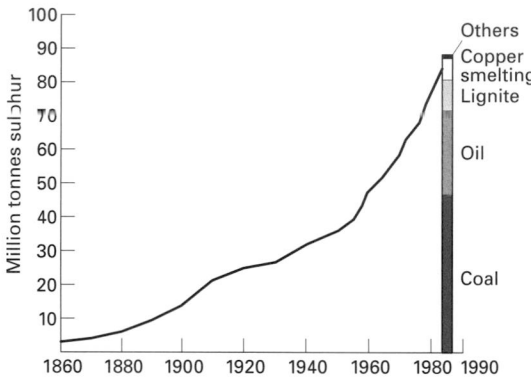

Figure 7.18 Global SO$_2$ emissions from anthropogenic sources, including the burning of coal, lignite and oil, and copper smelting.

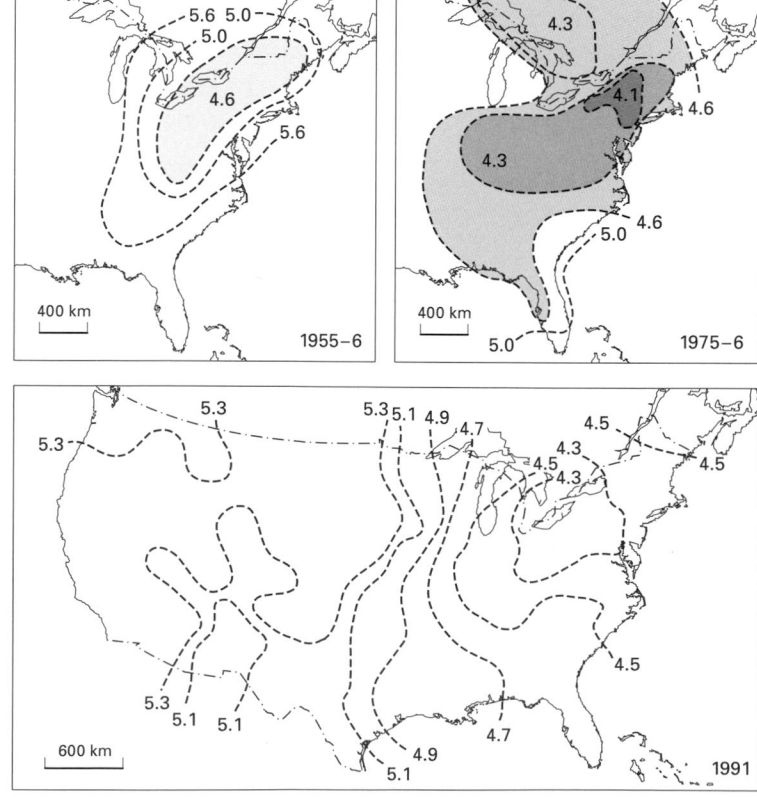

Figure 7.17 Isopleths showing average pH for precipitation in North America. Note the low values for eastern North America and the relatively higher values to the west of the Mississippi. (Combined from Likens et al., 1979 and Graedel and Crutzen, 1995, with modifications).

stacks and installing particle precipitators). These appear to have transformed a local 'soot problem' into a regional 'acid rain problem'. Coal burning produced a great deal of sulfate, but was largely neutralized by high calcium contents in the relatively unfiltered coal smoke emissions. Natural gas burning creates less sulfate but that which is produced is not neutralized. The new higher chimneys pump the smoke so high that it is dispersed over wide areas, whereas previously it returned to earth nearer the source.

Very long-term records of lake acidification provided dramatic evidence for the recent magnification of the acid deposition problem. These were obtained by extracting cores from lake floors and analysing their diatom assemblages at different levels. The diatom assemblages reflect water acidity levels at the time they were living. Two studies serve to illustrate the trend. In south-west Sweden (Renberg and Hellberg, 1982) for most of post-glacial time (i.e. the last 12,500 years) the pH of the lakes appears to have decreased gradually from around 7.0 to about 6.0 as a result of natural ageing processes. However, especially since the 1950s, a marked decrease occurred to present-day values of about 4.5. In Britain, the work of Battarbee et al. (1985a, b) showed that at sensitive sites pH values before around 1850 were close to 6.0 and that since then pH declines have varied between 0.5 and 1.5 units. More monitoring of lake acidity also demonstrated that significant changes were taking place. Studies by Beamish et al. (1975) demonstrated that between 1961 and 1975 pH had declined by 0.13 pH units per year in George Lake, Canada, and a comparable picture emerges from Sweden, where Almer et al. (1974) found that the pH in some lakes had decreased by as much as 1.8 pH units since the 1930s.

The effects of acid rain (Figure 7.19) are especially serious in areas underlain by highly siliceous types of bedrock (e.g. granite, some gneisses, quartzite and quartz sandstone), such as the old shield areas of the Fenno-Scandian shield in Scandinavia and the Laurentide shield in Canada (Likens et al., 1979). This is because of the lack of buffering by cations. Mobile anions are able to cause the leaching of basic cations (nutrients).

The ecological consequences of acid rain have been the subject of some debate. Krug and Frink (1983) argued that acid rain only accelerates natural processes and point out that the results of natural soil formation in humid climates include the leaching of nutrients, the release of aluminium ions, and the acidification of soil and water. They also note that acidification by acid rain may be superimposed on longer-term acidification induced by changes in land use. Thus the re-growth of coniferous forests in what are now marginal agriculture areas, such as New England and highland Western Europe, can increase acidification of soils and water. Similarly, it is possible, though in general unproven (see Battarbee et al., 1985b), that in areas like western Scotland, a decline in upland agriculture and the regeneration of heathland could play a role in increasing soil and water acidification. Likewise Johnston et al. (1982) suggested that acid rain can cause either a decrease or an increase in forest productivity, depending on local factors. For example in soils where cation nutrients are abundant and sulfur or nitrogen are deficient, moderate inputs of acid rain are very likely to stimulate forest growth.

In general, however, it is the negative consequences of acid rain that have been stressed. One harmful effect is a change in soil character. The high concentration of hydrogen ions in acid rain causes accelerated leaching of essential nutrients, making them less available for plant use. Furthermore, the solubility of aluminium and heavy metal ions increases and instead of being fixed in the soil's sorption complex, these toxic substances become available for plants or are transferred into lakes, where they become a major physiological stress for some aquatic organisms.

Fresh-water bodies with limited natural cations are poorly buffered and thus vulnerable to acid inputs. The acidification of thousands of lakes and rivers in southern Norway and Sweden during the late twentieth century has been attributed to acid rain, and this increased acidity resulted in the decline of various species of fish, particularly trout and salmon. But fish are not the only aquatic organisms that may be affected. Fungi and moss may proliferate, organic matter may start to decompose less rapidly, and the number of green algae may be reduced. Forest growth can also be affected by acid rain, though the evidence is not necessarily proven. Acid rain can damage foliage, increase susceptibility to pathogens, affect germination and reduce nutrient availability. However, since acid precipitation is only one of many environmental stresses, its impact may enhance, be enhanced by, or be swamped by other factors. For example, Blank (1985),

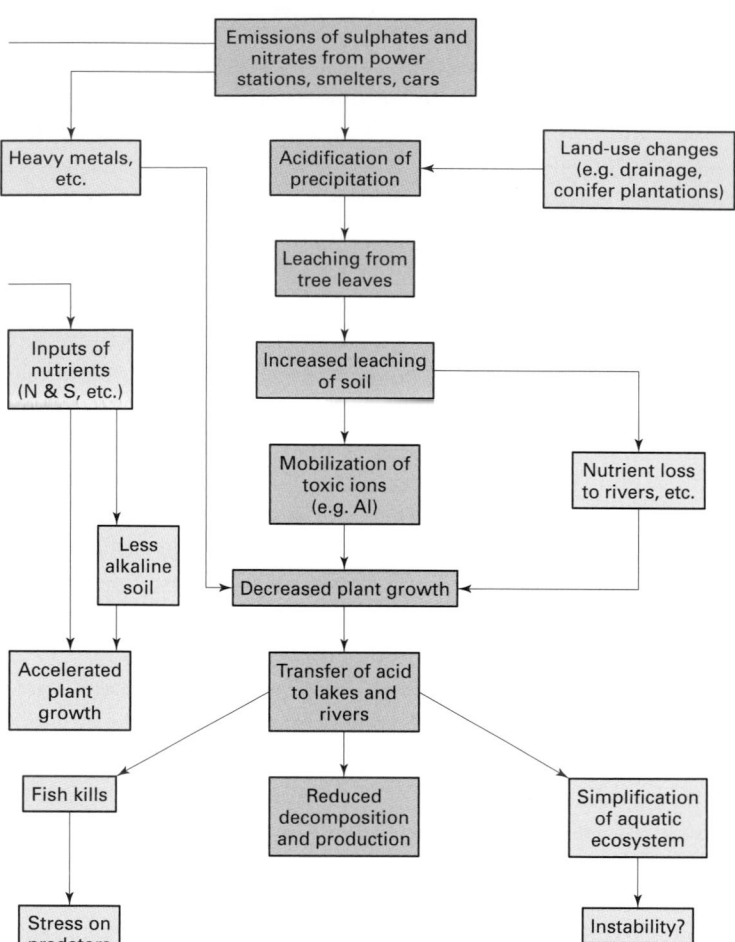

Figure 7.19 Pathways and effects of acid precipitation through different components of the ecosystem, showing some of the adverse and beneficial consequences.

in considering the fact that an estimated one-half of the total forest area of the former West Germany was showing signs of damage, referred to the possible role of ozone or of a run of hot, dry summers on tree health and growth. The whole question of forest decline is addressed in Chapter 4.

The seriousness of acid rain caused by sulfur dioxide emissions in the western industrialized nations peaked in the mid-1970s or early 1980s (Figure 7.20). Changes in industrial technology, in the nature of economic activity, and in legislation caused the output of SO_2 in Britain to decrease by 35% between 1974 and 1990. In 2009, SO_2 emissions were 89% lower than 1990 levels (DEFRA, 2011; see http://archive.defra.gov.uk/environment/business/reporting/pdf/110819-guidelines-ghg-conversion-factors.pdf). Nitrogen oxide emissions also fell in the UK (Figure 7.21a and b). There is now clear evidence that over the last two decades the rainfall and surface waters in the UK have become less

acid (Kernan et al., 2010). This was also the case in many other industrialized countries, including the USA (Malm et al., 2002). Figure 7.22 shows how power station emissions of sulfates and nitrates have declined in New England. However, there has been a shift in the geographical sources of sulfate emissions, so that whereas in 1980 60% of global emissions were from the US, Canada and Europe, by 1995 only 38% of world emissions originated from this region (Smith et al., 2001b). There are also increasing controls on the emission of NO_x in vehicle exhaust emissions, although emissions of nitrogen oxides in Europe have not declined (Table 7.11). A similar picture emerges from Japan (Seto et al., 2002) where sulfate emissions have fallen because of emission controls, whereas nitrate emissions have increased with an increase in vehicular traffic. This means that the geography of acid rain may change, with it becoming less serious in the developed world, but with it increasing in locations like China,

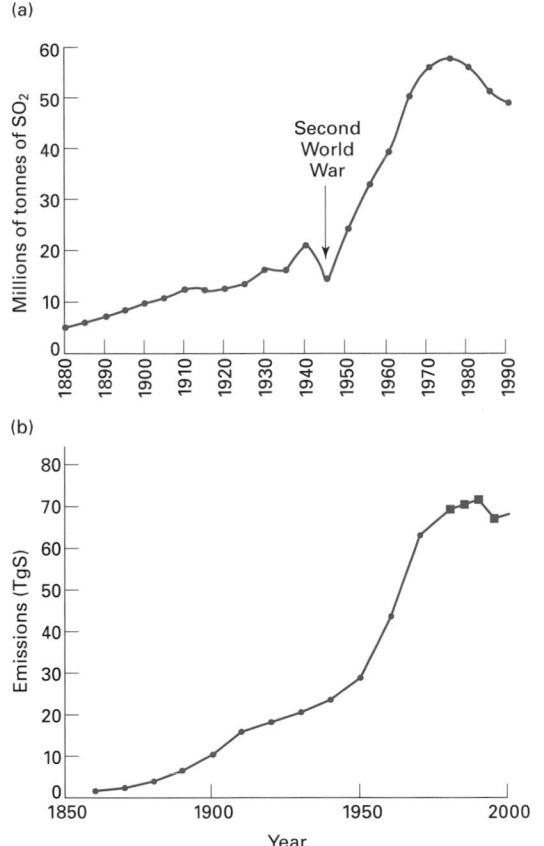

Figure 7.20 (a) Trends in sulfate emissions in Europe 1880–1990, based on data in Mylona (1996). (b) Historical global sulfur dioxide emissions (TgS) estimates from anthropogenic sources. From Smith et al. (2001b, figure 1) with modifications.

or sulfate gases (e.g. hydropower or nuclear power); and by removing the pollutants before they reach the atmosphere. For example, after combustion at a power station, sulfur can be removed ('scrubbed') from flue gases by a process known as flue gas desulfurization (FGD), in which a mixture of limestone and water is sprayed into the flue gas which converts the sulfur dioxide (SO_2) into gypsum (calcium sulfate). NO_x in flue gas can be reduced by adding ammonia and passing it over a catalyst to produce nitrogen and water (a process called selective catalytic reduction or SCR). NO_x produced by cars can be reduced by fitting a catalytic converter.

The improving situation with regard to acid precipitation in such areas as the eastern USA and western Europe does not mean that ecosystems will recover immediately. Likens (2010) suggests that stream macro-invertebrates may recover relatively rapidly (i.e. within 3 years), while lake zooplankton may need a decade or more to fully re-establish. Fish populations in lakes and streams should recover in 5–10 years following the recovery of the macro-invertebrates and zooplankton that act as their food sources. With regard to terrestrial ecosystems, given the relatively long life span of trees and the long delay in the response of soils to decreases in acid deposition, decades may be required for affected trees on sensitive sites to recover.

Stratospheric ozone depletion

A very major area of concern in pollution studies has been the current status of stratospheric ozone levels. The atmosphere has a layer of relatively high concentration of ozone (O_3) at a height of about 16 and 18 km in the polar latitudes and of about 25 km in equatorial regions. This ozone layer is important because it absorbs incoming solar ultraviolet radiation, thus warming the stratosphere and creating a steep inversion of temperature at heights between about 15 and 50 km. This in turn affects convective processes and atmospheric circulation, thereby influencing global weather and climate. However, the role of ozone in controlling receipts of ultraviolet radiation at the earth's surface has great ecological significance, because it modifies rates of photosynthesis. One class of organism that has been identified as being especially prone to the effects of increased ultraviolet radiation

where economic development will continue to be fuelled by the burning of low quality sulfur-rich coal in enormous quantities. From 2000 to 2006 sulfate emissions in China increased by 53%, an annual growth rate of 7.3% (Lu et al., 2010) and precipitation became increasingly acid (Tang et al., 2010).

In those countries where acid rain remains or is increasing as a problem there are various methods available to reduce its damaging effects. One of these is to add powdered limestone to lakes to increase their pH values. However, the only really effective and practical long-term treatment is to curb the emission of the offending gases. This can be achieved in a variety of ways: by reducing the amount of fossil fuel combustion; by using less sulfur-rich fossil fuels; by using alternative energy sources that do not produce nitrate

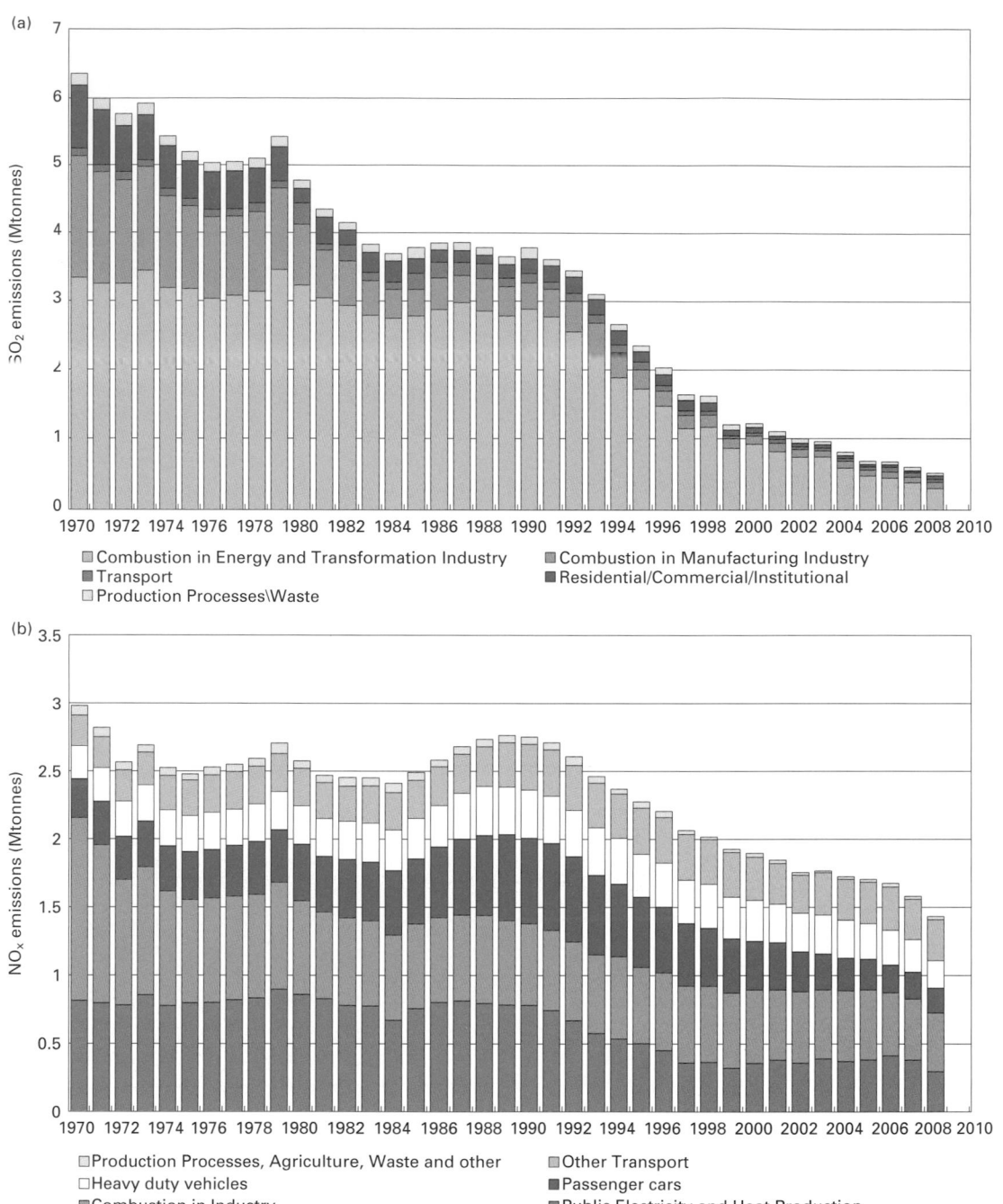

Figure 7.21 (a) Sulfur dioxide emissions and sources in the UK from 1970 to 2010 (b) NO_x emissions and sources in the UK from 1970 to 2010 (data from MacCarthy et al., 2012, figures 2.24 and 2.12, based on data from AEA National Atmospheric Emissions Inventory, DEFRA, 2010).

consequent upon ozone depletion are phytoplankton – aquatic plants that spend much of their time near the sea surface and are therefore exposed to such radiation. A reduction in their productivities would have potentially ramifying consequences, because these plants directly and indirectly provide the food for almost all fish. Ozone also protects humans from adverse effects of ultraviolet radiation, which include damage to the eyes, suppression of the immune system and higher rates of skin cancer.

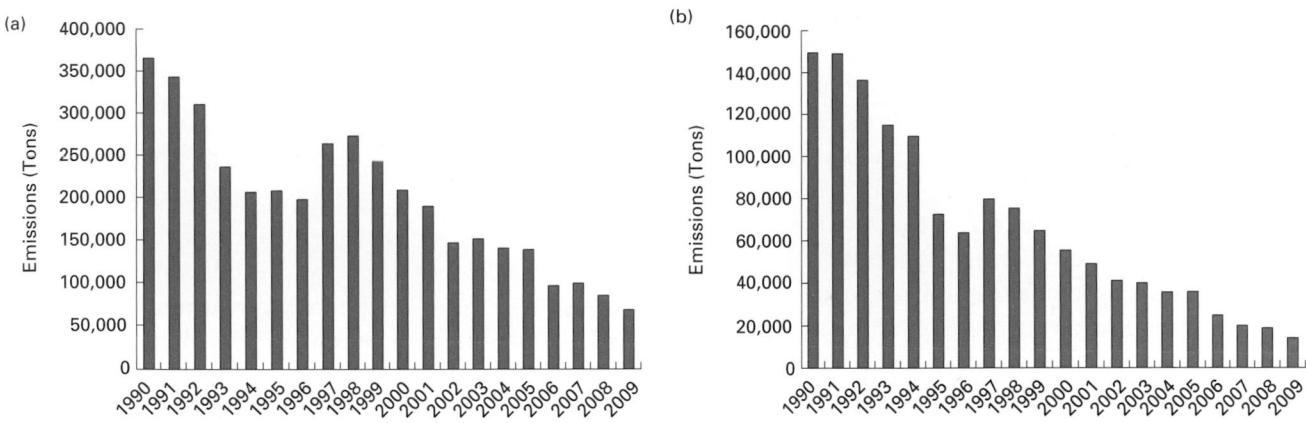

Figure 7.22 Trends in emissions of (a) SO_2 and (b) NO_x from power plants in New England since 1990 (data from Environmental Protection Agency).

Table 7.11 Nitrogen oxide and sulfur dioxide emissions. Source: Smith et al. (2001b). Reproduced with permission from Cambridge University Press

(a) Emissions of sulfur and nitrogen oxides (as NO_2) in 1000 tonnes a year for selected European countries in 1980 and 1993

Country	Sulfur			Nitrogen oxides		
	1980	*1993*	*1993 as % of 1980*	*1980*	*1993*	*1993 as % of 1980*
Czech Republic	1128	710	62.9	937	574	61.26
Denmark	226	78	34.51	274	264	96.35
Finland	292	60	20.55	264	253	95.83
France	1669	568	34.03	1823	1519	83.32
Germany	3743	1948	52.04	2440	2904	84.42
Ireland	111	78	70.27	73	122	167.12
Italy	1900	1126	59.26	1480	2053	138.72
Netherlands	244	84	34.43	582	561	96.39
Norway	70	18	25.71	185	225	120.96
Poland	2050	1362	66.44	1500	1140	76.00
Spain	1660	1158	69.76	950	1257	132.32
Sweden	254	50	19.69	424	399	94.10
UK	2454	1597	65.08	2395	2355	98.32
			47.28			103.47

(b) Percentage contributions for four regions to global sulfur dioxide emissions. Source: Smith et al. (2001b). Reproduced with permission from Cambridge University Press

	Sulfur dioxide emissions by geographic region (%)				
	1980	*1985*	*1990*	*1995*	*2000*
US/Canada	21	18	18	16	15
Europe	39	36	31	22	19
Asia	26	31	35	43	46
Rest of world	14	15	16	18	20

Human activities appear to have caused ozone depletion in the stratosphere, most notably over the south polar regions, where an 'ozone hole' has been identified (Staehelin et al., 2001). Possible causes of ozone depletion are legion and include various combustion products emitted from high-flying military and civil supersonic aircraft; nitrous oxide released from nitrogenous chemical fertilizers; and chlorofluorocarbons (CFCs) used in aerosol spray cans, refrigerant systems and in the manufacture of foam fast-food containers. However, the greatest attention has been focused on the role of CFCs, the production of which climbed greatly in the decades after the Second World War (Figure 7.23). These CFCs may diffuse upwards into the stratosphere where solar radiation causes them to become dissociated to yield chlorine atoms which react with and destroy the ozone. The process has been described thus by Titus and Seidel (1986: 4):

Because CFCs are very stable compounds, they do not break up in the lower atmosphere (known as the troposphere). Instead, they slowly migrate to the stratosphere, where ultraviolet radiation breaks them down, releasing chlorine.

Chlorine acts as a catalyst to destroy ozone; it promotes reactions that destroy ozone without being consumed. A chlorine (Cl) atom reacts with ozone (O_3) to form ClO and O_2. The ClO later reacts with another O_3 to form two molecules of O_2, which releases the Cl atoms. Thus two molecules of ozone are converted to three molecules of ordinary oxygen, and the chlorine is once again free to start the process. A single chlorine atom can destroy thousands of ozone molecules. Eventually, it returns to the troposphere, where it is rained out as hydrochloric acid.

Global production of CFC gases increased from around 180 million kg per year in 1960 to nearly 1100 million kg per year in 1990. However, in response to the thinning of the ozone layer, many governments signed an international agreement called the Montreal Protocol in 1987. This pledged them to a rapid phasing out of CFCs and halons. Production has since dropped substantially. However, because of their stability, these gases will persist in the atmosphere for decades or even centuries to come. Even with the most stringent controls that are now being considered, it will be the middle of the twenty-first century before the chlorine content of the stratosphere falls below the level that triggered the formation of the Antarctic 'ozone hole' in the first place. In spite of controls on CFC production, it is likely that the Antarctic ozone hole 'will be present for the better part of the 21st century' (Salby et al., 2011).

The Antarctic ozone hole has been identified through satellite monitoring and by monitoring of atmospheric chemistry on the ground. The decrease in ozone levels at Halley Bay is shown in Figure 7.24. The area of the Antarctic Ozone hole has now stabilized (Figure 7.25). The reasons why this zone of ozone depletion is so well developed over Antarctica include the very low temperatures of the polar winter, which seem to play

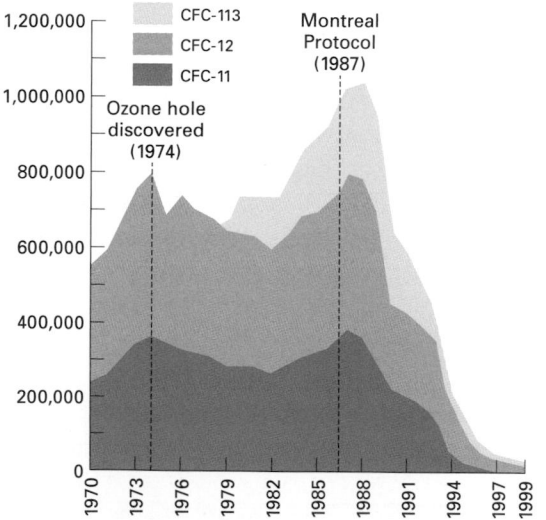

Figure 7.23 World production of major chlorofluorocarbons (tonnes/year). World production of the three major CFCs peaked in about 1988 and has since declined to very low values.

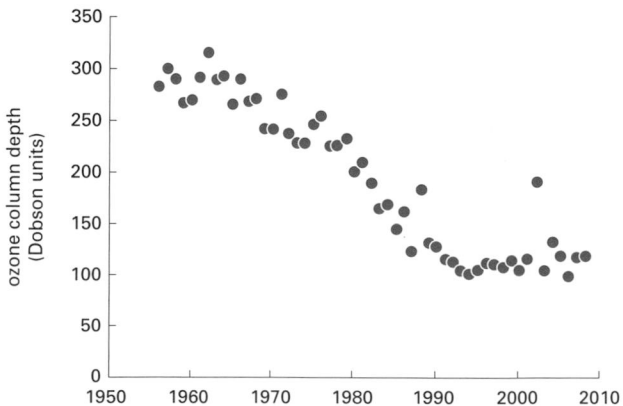

Figure 7.24 Ozone concentrations in DOBSON Units (DUs) over Halley Bay, Antarctica (based on data provided by the British Antarctic Survey).

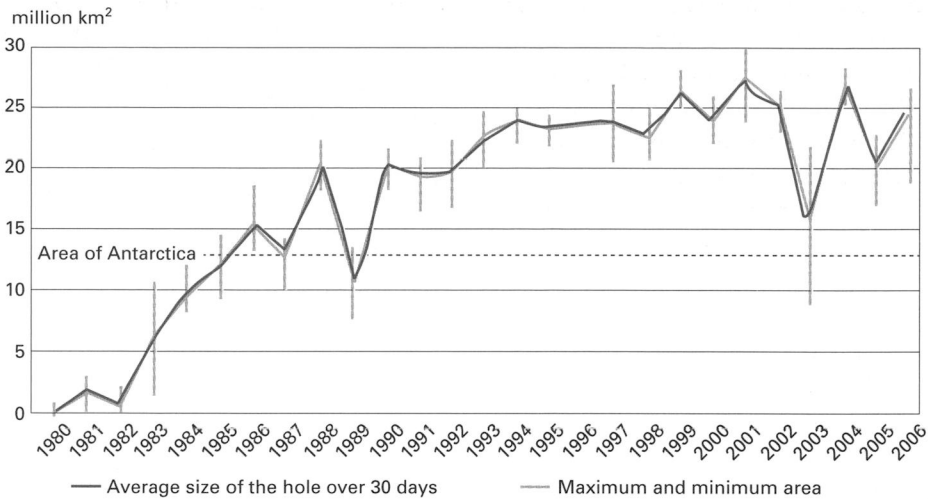

Figure 7.25 Size of the Antarctic ozone hole over time (modified from Geo4, UNEP, 2007, figure 2.23).

a role in releasing chlorine atoms; the long sunlight hours of the polar summer, which promote photochemical processes; and the existence of a well-defined circulation vortex. This vortex is a region of very cold air surrounded by strong westerly winds, and air within this vortex is isolated from that at lower latitudes, permitting chemical reactions to be contained rather than more widely diffused. No such clearly defined vortex exists in the Northern Hemisphere, though ozone depletion does seem to have occurred in the Arctic as well (Proffitt et al., 1990), with a particularly severe ozone loss occurring in 2011 (Manney et al., 2011). Furthermore, observations in the last few years indicate that the Antarctic ozone hole is spreading over wider areas and persisting longer into the Antarctic summer (Solomon, 1999). Decreases in stratospheric ozone levels on a global basis were analysed by Harris et al. (2003). The most negative trends occurred at mid- to high latitudes. Northern Hemisphere mid-latitude trends since 1978 have been at −2% to −3% per decade, while average losses in the Southern Hemisphere high latitude have been at up to 8% per decade.

Conclusions

Changes in the composition of Earth's atmosphere as a result of human emissions of trace gases and changes in the nature of land cover have caused great concern in recent years. Global warming, ozone depletion and acid rain have become central issues in the study of environmental change. Although most attention is often paid to climatic change resulting from greenhouse gases, there is a whole series of other mechanisms which have the potential to cause climatic change. Most notably, this chapter has pointed to the importance of other changes in atmospheric composition and properties, whether these are caused by aerosol generation or albedo change.

However, the greenhouse effect and global warming may prove to have great significance for the environment and for human activities. Huge uncertainties remain about the speed, degree, direction and spatial patterning of potential change. Nonetheless, if the earth warms up by a couple of degrees over the next hundred or so years, the impacts, some negative and some positive, are unlikely to be trivial. This is something that will form a focus of the following chapters.

For many people, especially in cities, the immediate climatic environment has already been changed. Urban climates are different in many ways from those of their rural surroundings. The quality of the air in many cities has been transformed by a range of pollutants, but under certain circumstances clean air legislation and other measure can cause rapid and often remarkable improvements in this area.

The same is true of two major pollution issues – stratospheric ozone depletion and acid deposition. Both processes have serious environmental consequences and their effects may remain with us for many years, but both can be slowed down or even reversed by regulating the production and output of the offending gases.

Points for review

What do you understand by the term 'the greenhouse effect'?

Explain the major processes involved in global warming

What role may changes in aerosols and albedo play in climate change?

How does the climate of cities differ from that of the surrounding countryside?

Give examples of some locations where levels of air pollution are falling. Why are they?

What is acid rain and how does it affect the environment?

What is 'the ozone hole' and how did it form?

Guide to reading

Colls, J. and Tiwary, A. (2009) *Air Pollution: Measurement, Modelling and Mitigation* (3rd edn.) London: Routledge. This much used text covers the three major issues connected with the study of air pollution.

Cotton, W.R. and Pielke, R.A. (2007) *Human Impacts on Weather and Climate* (2nd edn.) Cambridge: Cambridge University Press. A discussion of deliberate and non-deliberate attempts to modify weather and climate.

Harvey, L.D.D. (2001) *Global Warming: The Hard Science*. Harlow: Prentice Hall. A summary of the science behind global warming.

Houghton, J.T. (2009) *Global Warming: The Complete Briefing* (4th edn.) Cambridge: Cambridge University Press. An introduction by a leading scientist.

Kemp, D.D. (2004) *Exploring Environmental Issues: An Integrated Approach*. London: Routledge. An introductory survey.

Metcalfe, S. and Derwent, D. (2005) *Atmospheric Pollution and Environmental Change*. London: Hodder Arnold. An introductory review that is comprehensive in scope.

Ruddiman, W.F. (2005) *Plows, Plagues and Petroleum. How Humans Took Control of Climate*. Princeton and Oxford: Princeton University Press. A discussion of the ways in which humans may have altered global climates through time.

Weart, S.R. (2003) *The Discovery of Global Warming*. Cambridge, MA: Harvard University Press. A review of the development of the global warming theory.

Williams, M.A.J. and Balling, R.C. (1996) *Interactions of Desertification and Climate*. London: Arnold. A study of land cover changes and their effects.

Part II

The Future

8 THE FUTURE: INTRODUCTION

Chapter Overview

The main focus of this chapter is on future climate change. Current views on the main types of change that may occur are outlined, and then the point is stressed that some environments will be especially sensitive to change. There is also discussion of the question of 'tipping points' that cause positive feedbacks to magnify changes in the environment. Finally, general consideration is given of likely changes in the biosphere and to landforms and geomorphological processes.

Introduction

According to a range of modelling studies, it is highly likely that human-induced global warming, or, to use, Lovelock's more frightening term, 'global heating' (Lovelock, 2008), produced by the emissions of a cocktail of greenhouse gases into the atmosphere, is now occurring and will continue to affect the global climate for decades and centuries. The Royal Society (2010: 13) believes that 'There is strong evidence that changes in greenhouse gas concentrations due to human activity are the dominant cause of the global warming that has taken place over the last half century.' Concentrations of carbon dioxide, methane, nitrous oxide and the chlorofluorocarbons, all greenhouse gases, have increased. The international body charged with such matters, the Intergovernmental Panel on Climate Change (IPCC), set up in the 1980s, has considered

The Human Impact on the Natural Environment: Past, Present and Future, Seventh Edition. Andrew S. Goudie.
© 2013 John Wiley & Sons, Ltd. Published 2013 by John Wiley & Sons, Ltd.

these changes and their impacts in great detail (e.g. IPCC, 2007a, b).

The atmospheric concentration of CO_2 has increased by 31% since 1750 and the present CO_2 concentration, 392 ppm in 2011, has not been exceeded during the past 420,000 years and possibly during the past 20 million years (IPCC, 2001: 7). The atmospheric concentration of methane has more than doubled in the past 150 years, rising from c. 715 ppb in pre-industrial times to 1800 ppb in 2009 (IPCC, 2007a). Compared to 1750, nitrous oxide concentrations have increased by 18%. CFCs, which are entirely artificial, have been emitted, and although the Montreal Protocol and its amendments have effectively curtailed emissions, they have a long residence time in the atmosphere.

During the twentieth century the global average surface temperature increased by around 0.6°C. The 1990s were the warmest decade in the instrumental record and in the Northern Hemisphere the increase in temperature is likely to have been the largest of any century during the last 1000 years. The decade 2000–2009 was, globally, around 0.15°C warmer than the decade 1990–1999 and was characterized by heatwaves and precipitation extremes that may not bode well for the future (Coumou and Rahmstorf, 2012).

The IPCC (2007a) suggests that global average surface temperatures will rise by 1.8–7.1°C compared to pre-industrial levels by 2100, with a best estimate of 2.5–4.7°C. This represents a much larger rate of warming than that observed over the twentieth century and, based on palaeoclimatic data, is very likely to surpass anything experienced over the last 10,000 years. In its turn, increased temperatures will transform the behaviour of the atmospheric heat engine that drives the general circulation of the air and the oceans, leading to changes in the nature, pattern and amount of precipitation. Table 8.1, adapted from the IPCC (2007a, table SPM?), indicates some of the projections for future climate. In particular it is likely that northern high latitudes will warm more rapidly than the global average, probably by more than 40%.

There may also be marked changes in precipitation, with a tendency for increases to occur globally, but particularly over northern mid- to high latitudes and Antarctica in winter. Figure 8.1 shows the predicted pattern of future precipitation change in North America and shows that in general it will get wetter in the north and drier in the south. In lower latitudes some areas will see reductions in precipitation (which will exacerbate soil moisture losses caused by increased evapotranspiration), while others may become wetter. In general climate models predict that precipitation will generally increase in areas with already high amounts of precipitation and generally decrease in areas with

Table 8.1 Recent trends, assessment of human influence on the trend and projections for extreme weather events for which there is an observed late twentieth century trend. Source: Modified from IPCC, 2007a, table SPM 2

Phenomena and direction of trend	Likelihood that trend occurred in late twentieth century (typically post 1960)	Likelihood of a human contribution to observed trend	Likelihood of future trends based on projections for twenty-first century
Warmer and fewer cold days and nights over most land areas	Very likely	Likely	Virtually
Warmer and more frequent hot days and nights over most land areas	Very likely	Likely (nights)	Virtually certain
Warm spells/heat waves. Frequently increases over most land areas	Likely	More likely than not	Very likely
Heavy precipitation events. Frequently (or proportion of total rainfall from heavy falls) increases over most areas	Likely	More likely than not	Very Likely
Areas affected by droughts increases	Likely in many regions since 1970s	More likely than not	Likely
Intense tropical cyclone activity increases	Likely in some regions since 1970	More likely than not	Likely
Increased incidence of extreme high sea level (excludes tsunamis)	Likely	More likely than not	Likely

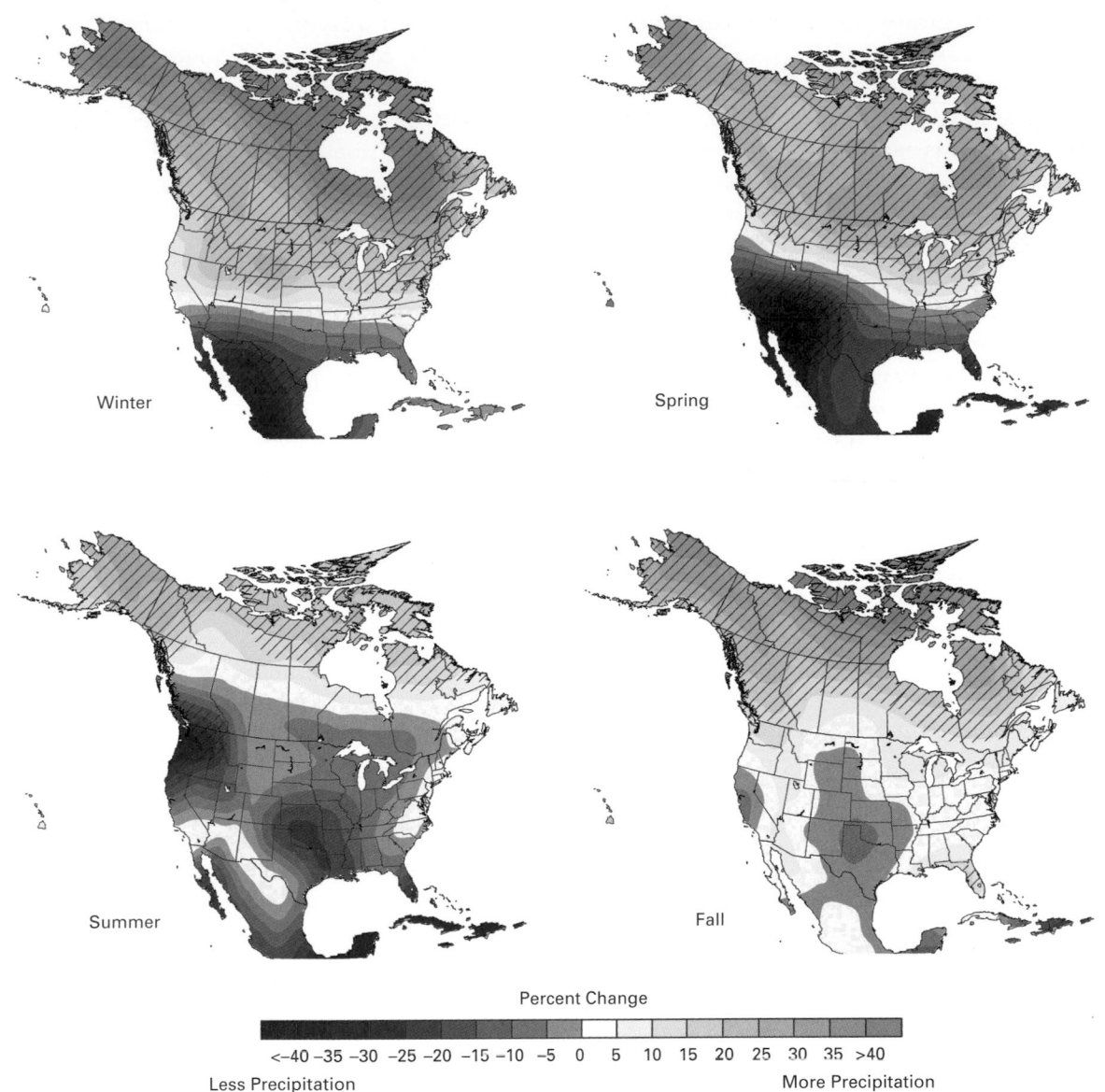

Winter

Spring

Summer

Fall

Percent Change

<-40 -35 -30 -25 -20 -15 -10 -5 0 5 10 15 20 25 30 35 >40

Less Precipitation   More Precipitation

Figure 8.1 Projected changes in precipitation in North America by 2080–2099 compared to the recent past. Confidence in the projected changes is greatest in the hatched areas (modified from Karl et al., 2009: 31).

low amounts of precipitation (New et al., 2011). Monsoonal areas may become greater, be more intense and have higher amounts of precipitation (Hsu et al., 2012).

If global warming and associated changes in rainfall and soil moisture amounts occur, then the effects on natural environments may be substantial and rapid.

Some environments will change more than others – 'geomorphological hotspots', especially when crucial thresholds are crossed. There are four types of reasons why some geomorphological processes, hazards and

landform assemblages will show particularly substantial modification as climate changes.

The first of these is threshold reliance. Some landforms and land-forming processes change across crucial climatic thresholds. For example, aeolian activity in drylands is dependent on wind energy, sand supply and the nature and extent of vegetation cover. If the last falls below a certain level due to a reduction in moisture availability, wind action is sharply intensified. Another example of threshold dependence is the coral reef. These features of tropical coastlines are highly sensitive to any changes in cyclone activity but also to coral bleaching (Teneva et al., 2012) caused by elevated sea-surface temperatures (as has happened during recent El Niño years in the Caribbean and elsewhere) and to sea-level rise itself. Terminal lake basins, such as those of the African Rift valleys or Lake Chad, are other landforms that have shown very rapid and substantial variations in response to Holocene and twentieth century changes in temperature, precipitation and evapotranspiration.

Secondly, there are many examples of landform change being promoted by a combination of climate changes and other human pressures – the compound effect. Indeed, desertification is often most intense when climatic (drought) and human pressures (e.g. overgrazing, deforestation or over-cultivation) coincide. In coastal regions, the effects of rising global sea levels are compounded by local geophysical changes (Sallenger et al., 2012) and by local subsidence caused by fluid (water, hydrocarbons) abstraction while beaches, marshes and deltas starved of sediment by the damming of rivers and the construction of coastal defences will be especially prone to erosion and inundation as sea levels rise (Blum and Roberts, 2009).

Thirdly, some landforms are robust, while others are less so. For example, *dongas* (erosion scars) in colluvial aprons in southern Africa are gully systems that have a history of cut and fill in response to climate and land cover changes. This is because they are developed in materials that have high exchangeable sodium percentages and thus are prone to severe erosion if the vegetation cover is changed. Likewise, once their protective vegetation cover is removed, sand dunes in semi-arid areas are easily reactivated. Similarly, muddy and sandy coastlines will plainly be more prone to erosion than hard-rock coastlines, and heavily lithified reefs will be more resistant that less lithified ones (Woodroffe, 2008).

Fourthly, the severity of climate change, and thus its likely impact, will vary spatially. Reductions in soil moisture and stream flows may be especially great in some areas that are currently relatively dry, so that stream networks will shrink (De Wit and Stanciewicz, 2006; Goudie, 2006). Features such as the dambos of central Africa are among those fluvial systems that may be considerably altered. Warming will have an especially strong impact on river behaviour in areas

where winter precipitation currently falls as snow, such as the mountains of South Asia or the Andes. Giorgi (2006) developed a Regional Climate Change Index (RCCI) based on four variables: changes in mean regional precipitation and inter-annual precipitation variability, and changes in regional surface air temperatures and their inter-annual variability. On this basis certain climate change hot-spots were identified, including the Mediterranean and north-eastern European regions, Central America, Southern Equatorial Africa and the Sahara, and eastern North America. Similarly, Sheffield and Wood (2008) recognized some potential future long-term drought hot spots, including the Mediterranean, West Africa, Central Asia and Central America. Markedly increased aridity in the Mediterranean lands has also been proposed by Gao and Giorgi (2008), with the central and southern portions of the Iberian, Italian, Greek and Turkish peninsulas among the areas that will be particularly vulnerable to water stress and desertification under global warming conditions.

There are, of course, considerable uncertainties built into any consideration of the future (Schiermeier, 2007). Our models are crude and relatively simple and it is difficult to build in all the complexities and feedbacks in the atmosphere, pedosphere, lithosphere, oceans, biosphere and cryosphere. Different models often provide very different future scenarios, particularly with respect to future rainfall amounts. Other difficulties are presented by the existence of non-linearities and thresholds in natural systems. There may be complex responses to change and sometimes indeterminable time lags. There are also great uncertainties about how emissions of greenhouse gases will change in coming decades as a result of changes in the global economy, technological changes and the adoption of mitigation strategies. Among the uncertainties identified by the Royal Society (2010) are (1) those produced by the inadequacies of some of our observations of change through time, (2) limited knowledge about the extent to which clouds and water vapour can be incorporated into global warming models, (3) a poor understanding of the future uptake of carbon dioxide by the land and oceans, (4) an inability to speak with confidence about changes in the Atlantic's ocean circulation, (5) a poor understanding of the future behaviour of the Greenland and West Antarctic ice sheets, and (6) our limited ability to make confident predictions at the regional scale. This is because many of our models have a coarse spatial resolution and are difficult to use at more local scales. It is also problematic to use past warm phases as analogues for the future, for the driving mechanisms may be different. Finally, changes brought about by enhanced greenhouse loadings cannot be seen in isolation. They will occur concurrently with other natural and anthropogenic climatic changes. In some cases other human activities could compound (or reverse) the effects of global warming.

There is also an increasing concern that there are some key **tipping points** in the climate system and that we may be crossing various thresholds or 'planetary boundaries' (Rockström et al., 2009). As Walker (2006) has written (p 802), 'The idea that passing some hidden threshold will drastically worsen man-made change has been around for decades, normally couched in technical terms such as "nonlinearity," "positive feedback" and "hysteresis".' Recently, the term tipping point has come into vogue to describe the moment at which internal dynamics start to propel change previously driven by external forces. It has become recognized that there are some specific locations or environments – 'tipping elements' (Lenton et al., 2008) – where a small change may make a big difference. Among the questions are being asked are 'Is climatic change getting out of control?' and 'Are we at a point of no return?' For example does the melting of Arctic ice lead to such a change in albedo (surface reflectivity) that it is self-reinforcing? Studies by Flanner et al. (2011) and Lenton (2012) suggest that it is. Could increasing amounts of melting of Arctic sea ice and the Greenland ice cap cause the thermohaline circulation of the oceans to flip from one state to another? Could melting of permafrost release great amounts of carbon dioxide and methane, so adding to the greenhouse effect? (Schuur and Abbott, 2011; Kort et al., 2012)? Could a rise in temperature cause the Amazon rainforest to be replaced by scrub or desert, thereby causing the Earth to lose another cooling mechanism? (Lovelock, 2006: 51). These and other examples of potential nonlinear climate changes are discussed by McNeall et al. (2011).

As Flannery (2005: 83) has argued, global warming may change climate in jerks, during which patterns jump from one state to another. As he put it, 'The best analogy is perhaps that of a finger on a light switch.

Nothing happens for a while, but if you slowly increase the pressure a certain point is reached, a sudden change occurs, and conditions swiftly alter from one state to another.' Flannery recognized three potential climatic tipping points: a slowing or collapse of the Gulf Stream, the demise of the Amazon rainforest and the release of gas hydrates (methane) from the floor of the Arctic. However, there are doubtless more tipping points than these, including disintegration of the West Antarctic Ice sheet, or a shift to a more persistent El Niño regime (Kriegler et al., 2009) and a map of such locations is provided by Kemp (2005).

The causes, consequences and controversies associated with global warming have been extensively treated elsewhere (e.g. Harvey, 2000). The purpose of the following chapters of this book is not to review and revisit this literature, but to examine the implications that global warming has for landscapes and habitats.

Changes in the biosphere

Climate change and changes in the concentrations of carbon dioxide in the atmosphere are likely to have a whole suite of biological consequences (Gates, 1993; Joyce et al., 2001). The ranges and the productivity of organisms will change.

Altitudinal changes in vegetation zones will be of considerable significance. In general, Peters (1988) believes that with a 3°C temperature change, vegetation belts will move about 500 m in altitude. One consequence of this would be the probable elimination of Douglas fir (*Pseudotsuga taxifolia*) from the lowlands of California and Oregon, because rising temperatures would preclude the seasonal chilling this species requires for seed germination and shoot growth. In the twentieth century there is some evidence that warming did indeed affect the position of the tree line, and Kullman (2001), for example, found that in Sweden a 0.8°C warming in the last 100 years caused the tree limit to migrate upwards by more than 100 m. In addition, changes in precipitation may have an influence on vegetation zonation in mountains (Engler et al., 2011). Particularly sensitive areas because of what can be called the altitudinal squeeze include the European Alps (Dirnböck et al., 2011; Dullinger et al., 2012), and tropical alpine regions and cloud forests (Buytaert et al., 2011; Rojas-Soto et al., 2012), where important habitats for endemic species may be lost.

Vegetation will also change latitudinally and some models suggest that wholesale change will occur in the distribution of biomes (Hickler et al., 2012). Theoretically a rise of 1°C in mean temperature could cause a pole-ward shift of vegetation zones of about 200 km (Ozenda and Borel, 1990). However, uncertainties surround the question of how fast plant species would be able to move to and to settle new habitats suitable to the changed climatic conditions. Post-glacial vegetation migration rates appear to have been in the range of a few tens of kilometres per century. For a warming of 2–3°C forest bioclimates could shift northwards about 4–6° of latitude in a century, indicating the need for a migration rate of some tens of kilometres per decade. Furthermore, migration could be hampered because of natural barriers, ecological fragmentation, zones of cultivation and so on.

Changes in forest composition or location could be slow, for mature trees tend to be long-lived and resilient. This means that they can survive long periods of marginal climate. However, it is during the stage of tree regeneration, or seedling establishment that they are most vulnerable to climate change. Seedlings are highly sensitive to temperature and may not be able to grow under altered climatic conditions. This means that if climate zones do indeed shift at a faster rate than trees migrate, established adults of appropriate species and genotypes will be separated from the place where seedling establishment is needed in the future (Snover, 1997). It is probable that future climate change will not cause catastrophic dieback of forests, but that faster growing tree species will enter existing forests over extended time periods (Hanson and Weltzin, 2000).

An early attempt to model changes on a global basis was made by Emmanuel et al. (1985). Using the GCM developed by Manabe and Stouffer (1980) for a doubling of CO_2 levels, and mapping the present distribution of ecosystem types in relation to contemporary temperature conditions, they found *inter alia* that the following changes would take place: boreal forests would contract from their present position of comprising 23% of total world forest cover to less than 15%; grasslands would increase from 17.7% of all world vegetation types to 28.9%; deserts would increase from 20.6% to 23.8%; and forests would decline from 58.4% to 47.4%.

Smith et al. (1992b) attempted to model the response of Holdridge's Life Zones to global warming and asso-

Table 8.2 Changes in areal coverage ($km^2 \times 10^3$) of major biomes as a result of climatic conditions predicted for a $2 \times CO_2$ would by various General Circulation Models. Source: From Smith et al. (1992b, table 3)

Biome	Current	OSU	GFDL	GISS	UKMO
Tundra	939	−302	−5.5	−314	−573
Desert	3699	−619	−630	−962	−980
Grassland	1923	380	969	694	810
Dry forest	1816	4	608	487	1296
Mesic forest	5172	561	−402	120	−519

OSU, Oregon State University; GFDL, Geophysical Fluid Dynamic Laboratory; GISS, Goddard Institute for Space Studies; UKMO, UK Meteorological Office.

ciated precipitation changes as predicted by a range of different GCMs. These are summarized for some major biomes in Table 8.2. All the GCMs predict conditions that would lead to a very marked contraction in the tundra and desert biomes, and an increase in the areas of grassland and dry forests. There is, however, some disagreement as to what will happen to mesic forests as a whole, though within this broad class the humid tropical rainforest element will show an expansion.

Zabinski and Davis (1989) modelled the potential changes in the range of certain tree species in eastern North America, using a GISS GCM and a doubling of atmospheric carbon dioxide levels. The difference between their present ranges and their predicted ranges is very large (Figure 8.2). Melillo et al. (2001) have modelled shifts in major vegetation types for the whole of the USA, using the Hadley (HadCM2) and Canadian Centre for Climate Modelling and Analysis simulations and various biogeography models. The results of their simulations for 2099 are summarized in Table 8.3.

Vegetation will also probably be changed by variations in the role of certain extreme events that cause habitat disturbance, including fire (O'Connor et al., 2011), drought (Hanson and Weltzin, 2000), heat stress (Allen et al., 2010), windstorms, hurricanes (Lugo, 2000) and coastal flooding (Overpeck et al., 1990).

Other possible non-climatic effects of elevated CO_2 levels on vegetation include changes in photosynthesis, stomatal closure and carbon fertilization (Idso, 1983), though these are still matters of controversy (Karnosky et al., 2001; Wullschleger et al., 2002; Karnosky, 2003). However, especially at high altitudes, it is possible that elevated CO_2 levels would have poten-

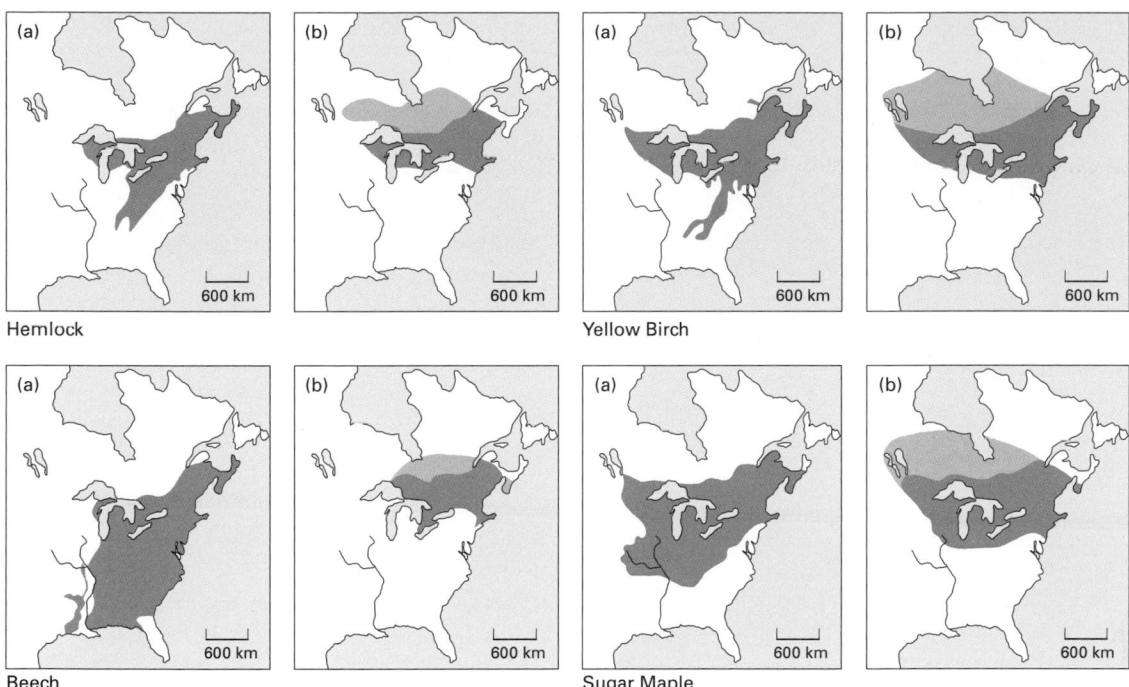

Figure 8.2 Present and future range of four tree species in eastern North America: (a) present range; (b) range in AD 2090 under the GISS GCM $2 \times CO_2$ scenario. The black area is the projected occupied range considering a rate of migration of 100 km per century. The grey area is the potential projected range with climate change (after Zabinski and Davis, 1989).

tially significant effects on tree growth, causing growth enhancement. Likewise, warm, drought stressed ecosystems such as the chaparral in the south-west USA might be very responsive to elevated CO_2 levels (Oechel et al., 1995). Increased carbon dioxide levels have also been implicated in the increasing density of trees that has been observed in some savanna areas (Bond and Midgley, 2012; Buitenwerf et al., 2012).

Peterson (2000) has investigated the potential impact of climate change on the role of catastrophic winds from tornadoes and downbursts in causing disturbance to forests in North America. It is feasible that with warmer air masses over middle latitude areas, the temperature contrast with polar air masses will be greater, providing more energy and thus more violent storms, but much more work needs to be undertaken before this can be said with any certainty. It is still far from clear whether tropical hurricane frequencies and intensities will increase in a warmer world. However, whether they increase or decrease, there are likely to be a number of possible ecosystem responses (Table 8.4).

The frequency and intensity of fires is an important factor in many biomes (Nitschke and Innes, 2012), and fires are highly dependent on weather and climate.

Fires, for example, are more likely to occur in drought years or when there are severe lightning strikes. Wind conditions are also a significant control of fire severity. Fire frequencies have increased substantially in the boreal forests of Russia, Canada and the USA in recent warming decades (Soja et al., 2007) (Figure 8.3). An analysis by Flannigan et al. (2000) suggests that future fire severity could increase over much of North America. They anticipate increases in the area burned in the USA of 25–50% by the middle of the twenty-first century, with most of the increases occurring in Alaska and the southeast USA. For Canada they forecast that the area burned could increase by 74–118% by the end of the century (Flannigan et al., 2005).

Other disturbances to vegetation communities could be brought about by changing patterns of insects and pathogens (Ayres and Lombardero, 2000; Pautasso et al., 2012). For example, in temperate and boreal forests, increases in summer temperatures could accelerate the development rate and reproductive potential of insects, whereas warmer winter temperatures could increase over-winter survival. In addition, it is possible that the ranges of introduced alien invasive species could change (Simberloff, 2000).

Table 8.3 Projected changes in major vegetation types in the USA (after Melillo et al., 2001)

North-east

Under both simulated climates, forests remain the dominant natural vegetation, but the mix of forest types changes. For example, winter-deciduous forests expand at the expense of mixed conifer-broadleaf forests.

Under the climate simulated by the Canadian model, there is a modest increase in savannas and woodlands

South-east

Under the climate simulated by the Hadley model, forest remains the dominant natural vegetation, but once again the mix of forest types changes.

Under the climate simulated by the Canadian model, all three biogeography models show an expansion of savannas and grasslands at the expense of forests. For two of biogeography models, LPJ and MAPSS, the expansion of these non-forest ecosystems is dramatic by the end of the twenty-first century. Both drought and fire play an important role in the forest break-up.

Midwest

Under both simulated climates, forests remain the dominant natural vegetation, but the mix of forest types changes.

One biogeography model, LBJ, simulates a modest expansion of savannas and grasslands.

Great Plains

Under the climate simulated by the Hadley model, two biogeography models project an increase in woodiness in this region, while the third projects no change in woodiness.

Under the climate simulated by the Canadian model, the biogeography models project either no change in woodiness or a slight decrease

West

Under the climate simulated by both the Hadley and Canadian models, the area of desert ecosystems shrinks and the area of forest ecosystem grows.

North-west

Under both simulated climates, the forest area grows slightly.

Table 8.4 Potential response of ecosystems to changes in hurricane characteristics (modified from Lugo, 2000)

Response associated with increased hurricane frequency and intensity:

A larger fraction of the natural landscape will be set back in successional stage; that is, there will be more secondary forests.

Forest aboveground biomass and height will decrease because vegetation growth will be interrupted more frequently or with greater intensity.

Familiar species combinations will change as species capable of thriving under disturbance conditions will increase in frequency at the expense of species that require long periods of disturbance-free conditions to mature.

Response associated with decreased hurricane frequency and intensity:

A larger fraction of the natural landscape will advance in successional stage; that is, there will be more mature forests and fewer secondary forests.

Forest aboveground biomass and height will increase because the longer disturbance-free periods allow greater biomass accumulation and tree height.

Species combinations will change as species capable of thriving under disturbance conditions will decrease in frequency, and species typical of disturbance-free conditions will increase.

Let us now turn to a consideration of some particular biomes. First, the boreal forest (Figure 8.4; Table 8.5), which is dominated by such species as spruce (*Picea*), pine (*Pinus*) and larch (*Larix*), is the northernmost and coldest forested biome in the Northern Hemisphere (Chapin et al., 2010). Potential changes in the geographical extent of boreal forests in the Northern Hemisphere are especially striking (Kauppi and Posch, 1988). The southern boundary will move from the southern tip of Scandinavia to the northernmost portion. Basically, boreal forest will be lost to other biomes (such as temperate forest) on its southern margins but will expand into tundra on its northern margins (Figure 8.5) (Chapin and Danell, 2001). This biome will also be subject to increasing fire frequencies and insect infestations (Calef, 2010), for there is evidence that these have been rising in recent decades (Soja et al., 2007). Within the forest species composition

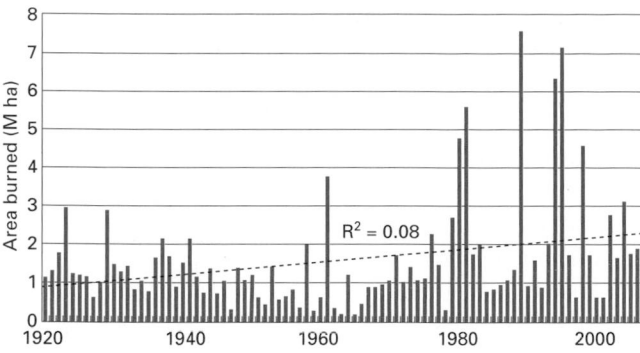

Figure 8.3 Area burned annually in Canada since 1920. Note the increase in the number of extreme fire years over the last five decades (after Soja et al., 2007, figure 6). Reproduced with permission from Elsevier.

Figure 8.4 Boreal forest in southern Sweden. (See Plate 47)

Table 8.5 Potential drivers of future boreal forest composition and extent

Primary climatic changes with effects on plant growth
  Temperature}
    Evapotranspiration
  Precipitation}
  Winds

Secondary climatic changes
  Droughts
  Floods
  Frost frequency and intensity
  Snowpack extent and duration

Environmental responses to climate change
  Invasive species
  Pests
  Disease
  Fire
  State of permafrost

Human responses to climate change
  Ease of access for logging and so on.
  Changes in land use and spread of agriculture with longer growing season
  Tree planting for carbon sequestration

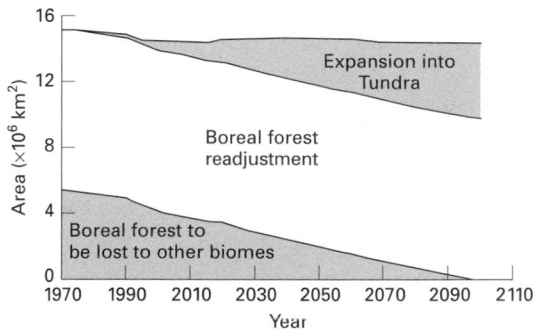

Figure 8.5 Changes in areas of boreal forest projected by the IMAGE model in response to scenarios of future change in climate and land use. Although the total area of boreal forest is projected to remain relatively constant, about one-third of the present boreal forest is projected to be converted to other biomes, and an additional third will be added as trees advance into tundra (Source: Chapin and Danell, 2001, figure 6.1). Reproduced with permission from Springer.

may change, because some species, such as white spruce, appear to be highly influenced by temperature conditions (Beck et al., 2011). In Siberia deciduous larch (*Larix*) may be replaced by evergreen conifers (Shuman et al., 2011).

One biome that may suffer a particularly severe loss in area as a result of warming is arctic and alpine tundra, an environment where climatic impacts have already been detected (Grabherr et al., 2010) and where changes in the cryosphere will be considerable (see Chapter 11). As the tree line moves up mountains and migrates northwards in the Northern Hemisphere, the extent of tundra will be greatly reduced, perhaps by as much as 55% in total (Walker et al., 2001). The alpine tundra zone may disappear completely from some mountain tops. On the other hand, recent Arctic warming has led to increases in tundra shrub cover (alder and willow) in the Eurasian Arctic (Macias-Fauria et al., 2012).

A somewhat precarious biome, located as it is at the southernmost tip of Africa and so with little scope for displacement, is the very diverse Fynbos Biome. It is a predominantly sclerophyll shrubland, characterized by the pre-eminence of hard-leaved shrubs, many of which are proteas. In addition to being extraordinarily diverse, this biome contains many endemic species. Modelling, using the Hadley Centre CM2 GCM, indicates a likely contraction of the extent of Fynbos by *c.* 2050 (Figure 8.6) (Midgley et al., 2003).

In the UK, the MONARCH programme (UK Climate Impacts Programme, 2001) has investigated potential changes in the distribution of organisms. Among the consequences of global warming they identify as being

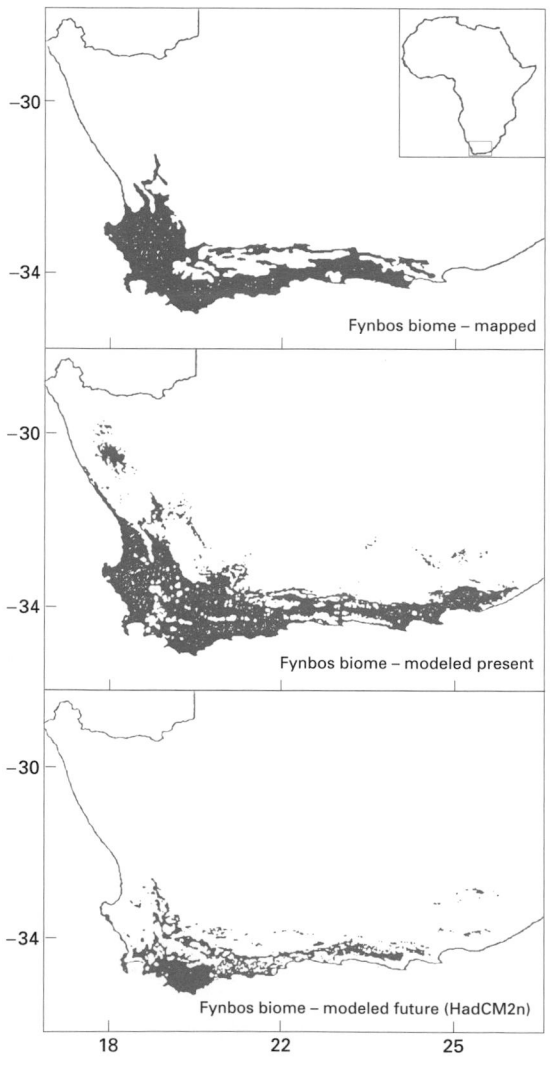

Figure 8.6 Current mapped Fynbos Biome (upper panel, after Rutherford and Westfall, 1994) and the modelled extent of the biome under current (middle panel) and future (*c.* 2050) climate conditions (lower panel), the latter based on climate change projections for the region generated by the GCM HadCM2 (Source: Midgley et al., 2003, figure 1). Reproduced with permission from Elsevier.

especially serious are the loss of montane heaths in the mountains of Scotland and the dieback of beach (*Fagus*) woodlands in southern Britain as summer droughts become more intense. However, it needs to be remembered that climate is not the only control on the distribution of species and that other factors such as biotic interactions and species' dispersal are also likely to be significant (Pearson and Dawson, 2003).

Mangroves, important components of coastlines in the warmer parts of the world, are also sensitive to temperature. The best developed mangrove swamps are found near the Equator where the temperature in the coldest month does not fall below 20°C in Indonesia, New Guinea and the Philippines, and the number of species decreases with increasing latitude. Walter (1984) indicates that the last outposts of mangrove are to be found at 30°N and 33°S in Eastern Africa, 37–38°S in Australia and New Zealand, 20°S in Brazil and 32°N in the Bermudas. Mangroves in such areas might expand their pole-ward range under warmer conditions.

Another type of marine habitat at risk is seagrass communities (Short et al., 2011). These are composed of flowering plants that include the widely distributed genera *Zostera*, *Thalassia* and *Posidonia*. They are highly productive ecosystems that support marine food webs and provide essential habitat for many coastal species. They are currently declining fast (Waycott et al., 2009) but in areas like the Mediterranean may decline further as seawater temperatures rise (Jordà et al., 2012).

Finally, great interest is emerging in the future of Amazonia. As we saw in Chapter 7, the replacement of rainforest with other forms of land use is likely in itself to cause a reduction in annual rainfall amounts over the area. However, in addition, the concern is that global warming will also modify climatic conditions in the area so that rainfall totals are reduced, droughts become more common, the forest starts to die back, and, as a consequence, more carbon is released into the atmosphere thereby accentuating the greenhouse effect, while increased temperatures may themselves have deleterious effects on some plants (Corlett, 2011). In other words, there is a worry that Amazonia could be at a tipping point (Nepstad et al., 2008; Nobre and Borma, 2009). The 2005 drought in Amazonia raised awareness of the potential threat of a more drought-prone Amazon (Betts et al., 2008) and the potential role that changes in sea surface conditions could have on rainfall amounts. One particularly grim modelling simulation by Cook and Vizy (2008) suggested that there could be a 70% reduction in the extent of the Amazon rainforest by the end of the century and a large eastward expansion of the drought resistant *caatinga* that is prominent in parts of north-east Brazil today. They concluded (p. 558):

The coupled regional model simulation projects a 70% reduction in the extent of the Amazon rain forest. . . . Rain forest vegetation disappears entirely from Bolivia, Paraguay, and Argentina, and most of Brazil and Peru. Rain forest in Columbia is largely maintained, as all of the surviving rain forest is concentrated close to the equator. Venezuela and French Guiana experience small reductions. . . . North of about 15°S the rain forest is primarily replaced by savanna vegetation.

Salazar et al. (2007) came up with a much lower figure for the reduction in tropical forest area in South America: *c.* 18% by the end of the century. What is predicted for South America does not necessarily apply to other parts of the tropical world, and forest expansion might, for example, occur in the Congo Basin (Zelazowski et al., 2011).

The wholesale changes in the biomes discussed so far in this chapter, together with the changes that will take place in, for example, the Arctic and in coral reefs will impact upon individual organisms and species and may mean that protected areas are in the wrong place. In addition, climate changes will affect a whole raft of factors that may be important: for individual species: disease incidence and distribution; the occurrence of fires, floods, droughts or windstorms; the success of ecological invaders; habitat loss or gain due to sea-level change; changes in the timing and success of breeding, hibernation or migration; the extinction of habitats at the tops of mountains; the sex determination of embryos; and changes in food availability. Some organisms may prove to be relatively resilient in the face of such changes, but those that are highly specialized, have restricted geographical distributions, have long gestation times and only migrate slowly may be more prone to contraction in their numbers and range. The potential effects of climate change on marine mammals is reviewed by Learmonth et al. (2006), while the vulnerability of one particular types of marine mammal – the marine turtle – is covered by Poloczanska et al. (2010). The vulnerability of land mammals in Africa is assessed by Thuiller et al. (2006), that of birds by Crick (2004) and that of fish by (Rijnsdorp et al., 2009).

Climate and geomorphology

Attempts to relate landforms and land-forming processes to climatic conditions have been long continued and climatic geomorphology has a long and distinguished history (Gutierrez-Elorza, 2001). Certainly, any change in climatic variability and extreme events would have considerable geomorphological significance (Viles and Goudie, 2003) (Table 8.6).

It is self evident that some landforms and land-forming processes are intimately related to temperature. In general terms, for example, permafrost today only occurs in areas where the mean annual temperature is less than −2°C and it is virtually ubiquitous north of the −6 to −8°C isotherm in the Northern Hemisphere. Likewise, temperature is one of the major controls of ablation of glacier ice. At the other extreme, coral reefs have a restricted distribution in the world's oceans which reflects their levels of temperature tolerance. They do not thrive in cool waters (i.e. <20°C mean annual sea surface temperature), but equally many coral species cannot tolerate temperatures greater than about 30°C and have been stressed by warm El Niño–Southern Oscillation (ENSO) years.

Some relatively clear relationships have also been established between precipitation levels and geomorphic

Table 8.6 Examples of geomorphological effects of decadal- to century-scale oscillations

Environment affected	Impacts upon
Terrestrial hydrology	Glacier mass balance
	Lake levels
	River flows
	Snow cover
	Permafrost
Terrestrial geomorphology	Soil erosion
	Floodplain sedimentation and erosion
	Slope instability/mass movements
	Dune movements
	Geochemical sediment growth
	Effects of fire frequency with knock-on effects on weathering
	Runoff and slope instability
Coastal/marine ecology and geomorphology	Coastal erosion
	Mangrove defoliation/land loss
	Coral bleaching
	Coastal dune activation

processes and forms. Large sand dunes only occur under dry conditions where rainfall levels are less than 150–200 mm per year, and dust storms appear to be a feature of hyper-arid areas where annual rainfall totals are around 0–150 mm per year. Desert crusts vary according to rainfall, according to the solubility of their components, with, for example, nitrate and halite crusts only occurring in the driest deserts and calcretes in areas where rainfall is less than around 500 mm per year.

However, for other phenomena the climatic controls may be rather more complex. Take, for example, the case of lake basins. These, and particularly those with no outlets, can respond in a dramatic manner to changes in their hydrological balance. During times of positive water budgets (due to high rates of water input and/or low rates of evapotranspirational loss), lakes may develop and expand over large areas, only to recede and desiccate during times of negative water balance. Shoreline deposits and sediment cores extracted from lake floors can provide a detailed picture of lake fluctuations (e.g. Street and Grove, 1979). However, given the range of factors affecting rates of evaporation and runoff over a lake basin (Table 8.7), their precise climatological interpretation needs to be determined with care. A full review is provided by Mason et al. (1994).

Many of the studies of the links between climate and geomorphology have employed rather crude climatic parameters such as mean annual temperature or mean annual rainfall. It is, however, true that climatic extremes and climatic variability may be of even greater significance. Unfortunately, they are not very effectively dealt with in General Circulation Models. We have imperfect knowledge of how extreme events and climate variability will change in a warmer world. It is, however, possible that climatic variability such as the **ENSO** phenomenon could be influenced by global warming (Latif and Keenlyside, 2009). The situation here is far from clear, for as An et al. (2008: 3) have put it 'dynamical understanding of ENSO responses to global warming is still in a toddling stage!' Some models suggest a weakening of ENSO activity under a global warming scenario, while others do not.

The impacts of climatic variability on ecological and geomorphic systems may be non-linear, as recently found for rainfall erosivity and ENSO in the southwest USA (D'Odorico et al., 2001). Indeed, Brunsden (2001)

Table 8.7 Factors affecting rates of evaporation and runoff. Source: Bradley (1985)

Evaporation	Runoff
Temperature (daily means and seasonal range)	Ground temperature
Cloudiness and solar radiation receipts	Vegetation cover and type
Wind speed	Soil type (infiltration capacity)
Humidity (vapour pressure gradient)	Precipitation type (rain, snow and so on)
Depth of water in lake and basin morphology (water volume)	Precipitation intensity (event magnitude and duration)
Duration of ice cover	Precipitation frequency and seasonal distribution
Salinity of lake water	Slope gradients (stream size and number)

has drawn attention to the complex responses of landforms to environmental changes and events, while Knox (1993) has shown how modest changes in climate can cause very large changes in the flood response of rivers. The complexity of future changes in the environment makes effective prediction and modelling extremely difficult (Blum and Törnqvist, 2000). As Bogaart et al. (2003) have pointed out landscape response to climate change is highly non-linear and characterized by numerous feedbacks between different variables and by lead-lag phenomena. The example they give is of an increase in precipitation in an initially semi-arid area. This would by itself increase hillslope erosion and sediment transport capacity. However, over time, soil and vegetation conditions would adjust to the new environmental conditions, resulting in a better soil structure and more vegetation cover. As a result after a lag of time hillslope erosion and sediment yield might decrease.

Some ideas on the clustering of events and complex responses to them can be presented as conceptual diagrams. Figure 8.7a indicates in a general way how clusters of climatic events can produce variable overtopping of geomorphic thresholds, although it represents a simple linear view of geomorphic response which may be an oversimplification for many geomorphic systems. Figure 8.7b illustrates how complex chains of linkages between climatic, vegetation and geomorphic processes produce a complex geomorphic response, and Figure 8.7c provides a simple conceptual

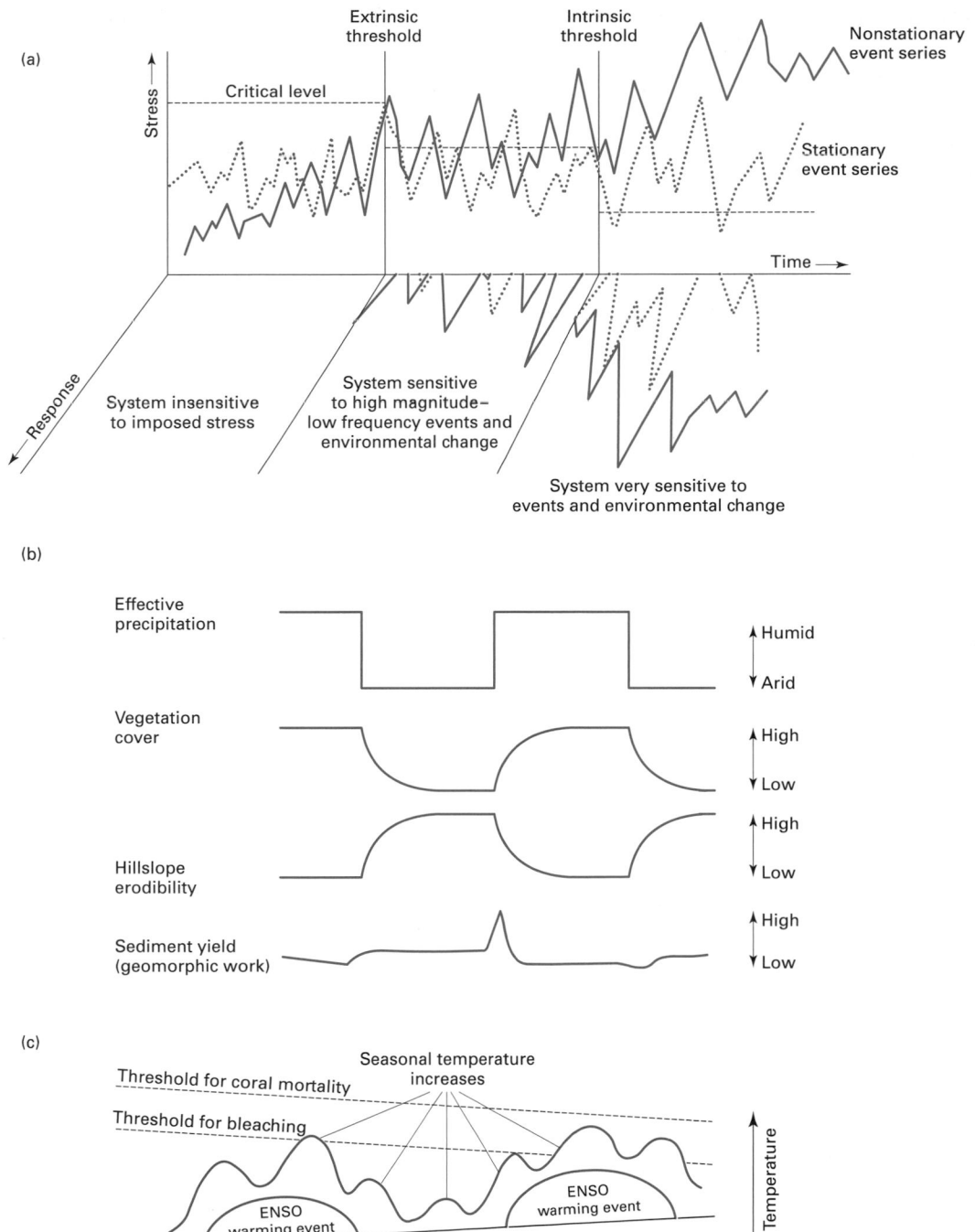

Figure 8.7 Representation of the impacts of climatic variability on geomorphic systems. (a) Stress–response sequences including thresholds under stable and changing climate conditions (adapted from Brunsden, 2001, figure 1). (b) A simplified view of the biogeomorphic response model (adapted from Knox, 1972 and Roberts and Barker, 1993, figure 6.1). (c) A model of the possible interactions of different time scales of warming and their impact on coral bleaching (adapted from Williams and Bunkley-Williams, 1990 and Viles and Spencer, 1995, figure 6.11).

model of the synergistic associations between different scales of warming producing coral bleaching. Such conceptual diagrams provide a useful starting point for analysing the relationships between climatic variability and geomorphology as a prelude to more detailed empirical and computational studies.

The term 'climate change' involves changes in a whole package of climatic parameters, such as rainfall amounts, rainfall intensity, temperature amounts and variability, wind intensity and so on. This in itself means that the response of land forming processes to climate change will be complex. This is brought out in Table 8.8, which is based on the work of Crozier (2010) and provides a framework for understanding slope stability and how it might be impacted upon by diverse climatic parameters.

In the following chapters we shall consider some of the challenges that face some major types of environment in the face of global warming and other trends that have been discussed in this brief introduction.

Table 8.8 Direct changes and potential slope stability responses to climate change (modified from Crozier, 2010, table 1)

Climate change	Condition/process affected	Slope stability response
Increase in precipitation totals	Wetter antecedent conditions	Less rainfall in an event required to achieve critical water content
		Reduction in soil capillarity – reduction in cohesion. Softened layers can act as lubricants.
		Higher water tables – reduction in shear strength
	Increased weight (surcharge)	Increased bulk density, leading to decrease in shear strength/stress ratio in cohesive material
	Higher water tables for longer periods	More frequent attainment of critical water content during rainfall events
	Increased lubrication of contact surfaces between certain minerals	Reduction in friction (only occurs with certain platy minerals, e.g. micas)
	Increase in river discharge	Increased bank scour and removal of lateral and basal support from slopes
		Higher lake levels, increase in bordering slope water tables
		Increase in rapid draw down events and higher drag forces, removal of lateral confining pressure plus perched groundwater levels on flood recession, increasing shear stress
Increase in rainfall intensity	Infiltration more likely to exceed subsurface drainage rates.	Landslides triggering by reduction in effective normal stress leading to reduction in shear strength
	Rapid build-up of perched water tables	Increase in cleft water pressures
	Increased throughflow	Increase in seepage and drag forces, particle detachment and piping. Piping removes underlying structural support. Enhanced drainage unless blockage occurs
Shift in cyclone tracks and other rain bearing weather systems	Areas previously unaffected, subject to high rainfall	Rapid adjustment of slopes to new climate regime
Increased variability in precipitation and temperature	More frequent wetting and drying cycles	Increase fissuring, widening of joint systems
		Reduction in cohesion and rock mass joint friction
	Reduction in antecedent water conditions through evapotranspiration	
Increased temperature	Reduction in interstitial ice and permafrost	Build-up of porewater pressure and strength reduction
	Rapid snow melt-runoff and infiltration	Removal of lateral support to valley side slopes
	Reduction in glacier volume	Enhanced basal erosion on coasts, increase in groundwater levels on costal slopes
	Increased sea level	
Increased wind speed and duration	Enhanced evapotranspiration	Reduction of soil moisture
	Enhanced root levering by trees	Enhanced drying and cracking
	Increased wave action on shorelines (enhanced by higher sea levels)	Loosening and dislodging of joint blocks
		Removal of slope lateral support

Points for review

How may vegetation belts respond to future climate changes?

Which landscape types may be especially sensitive to global warming?

Guide to reading

Archer, D. (2007) *Global Warming: Understanding the Forecast.* Oxford: Blackwell. This book provides the scientific background required to understand global warming.

Arnell, N. (1996) *Global Warming, River Flows and Water Resources.* Chichester: Wiley. A study of potential future hydrological changes.

Chapin, F.S., Sala, O.E. and Huber-Sannwald, E. (eds) (2001) *Global Biodiversity in a Changing Environment.* Berlin: Springer Verlag. A wide ranging assessment of how biodiversity may change in response to climate and land use changes.

Flannery, T. (2005) *The Weather Makers.* London: Allen Lane. A hard-hitting 'popular' treatment of the consequences of future climate change.

Gates, D.M. (1993) *Climate Change and Its Biological Consequences.* Sunderland, Mass: Sinauer Associates. A textbook that considers some of the effects of climate change on the biosphere.

Intergovernmental Panel on Climate Change (IPCC) (2007a) *Climate Change 2007: The Physical Science Basis.* Cambridge: Cambridge University Press. A report by the international body that co-ordinates research on climate change which deals with the causes and nature of climate change.

Intergovernmental Panel on Climate Change (IPCC) (2007b) *Climate Change 2007; Impacts, Adaptation and Vulnerability.* Cambridge: Cambridge University Press. A full discussion of the possible consequences of climate change for humans and their environment.

Karl, T.R., Melillo, J.M. and Peterson, T.C. (2009) *Global Climate Change Impacts in the United States.* Cambridge: Cambridge University Press. An accessible and well-illustrated discussion of future climate change impacts in the USA.

9 THE FUTURE: COASTAL ENVIRONMENTS

Chapter Overview

One consequence of global warming is that sea levels will rise, partly in response to the thermal expansion of the oceans (the steric effect) and partly because of the melting of glacier ice. Sea level is presently rising, and the rate will accelerate in the future. It poses a particular threat to areas that are also subsiding. Among the coastal environments that may be sensitive to sea-level rise are coral reefs, mangrove swamps, salt marshes, deltas, estuaries, cliffed coasts and sandy beaches. Some coastal environments will also be affected by changes in storm intensity, and coral reefs may be affected by increases in ocean acidity caused by the build-up of carbonic acid.

Introduction

There is a strong likelihood that if temperatures climb in coming decades, so will sea levels. Indeed, on a global basis, sea levels are rising at the present time, and always have in warmer periods in the past. The reasons why sea levels will rise include the thermal expansion of seawater (the **steric effect**), the melting of the cryosphere (glaciers, ice sheets and permafrost) and miscellaneous anthropogenic impacts on the hydrological cycle (which will modify how much water is stored on land). Rising sea levels will have substantial geomorphological consequences for the world's coastlines. Moreover, as Viles and Spencer

The Human Impact on the Natural Environment: Past, Present and Future, Seventh Edition. Andrew S. Goudie.
© 2013 John Wiley & Sons, Ltd. Published 2013 by John Wiley & Sons, Ltd.

(1995) have pointed out, about 50% of the population in the industrialized world lives within 1 km of a coast, about 60% of the world's population live in the coastal zone (Figure 9.1), and two-thirds of the world's cities with populations of over 2.5 million people are near estuaries. Rising sea levels will thus have an impact on a large proportion of the Earth's human population (Leatherman, 2001; FitzGerald et al., 2008). Thirteen of the world's twenty largest cities are located on coasts. Areas at less than 10 m in altitude (the so-called Low Elevation Coastal Zone – the LECZ) account for c. 2% of the world's land area but contain 10% of the population. While only 13% of urban settlements with populations of less than 100,000 occur in the LECZ, the figure rises to over 65% among cities of 5 million or more. Many of the most severely impacted countries with respect to the effects of sea-level rise and higher storm surges are in developing countries (Dasgupta et al., 2009; Brecht et al., 2012), including Vietnam, Pakistan, the Philippines, Guyana, Suriname, Belize, Benin, Egypt, Jamaica and Tunisia. Jakarta, in Indonesia, already has about 40% of its area below sea level, and because of subsidence caused by groundwater abstraction, the threat of inundation is rising (Ward et al., 2011). Nicholls et al. (2011) suggest that were global temperatures to rise by 4°C, then sea levels would rise by 0.5–2.0 m, causing the displacement of between 72 and 187 million people.

Florida provides an excellent example of the threats posed to low-lying coastal areas. Approximately 10% of the state's land area is less than 1 m above present sea level. However, parts of it are especially vulnerable. A 0.6 m rise in sea level would inundate about 70% of the total land surface of the Florida Keys, 71% of the population and 68% of property, while a 1.5 m rise would inundate 91% of the land surface, 71% of the population and 68% of the property (Noss, 2011). Severe consequences have also been proposed as likely for California (Heberger et al., 2011; Revell et al., 2011).

However, it is not only the rise of sea levels that one needs to consider (Nicholls et al., 2007). Also important are the effects of high carbon dioxide levels on ocean acidification and the health of corals, the effects of higher seawater temperatures on various marine organisms, changes in the freshwater inputs to estuaries and lagoons, and changes in wave energy and storm surges. In addition, there are many other human influences on coastlines, including changes in the sediment budgets of rivers that nourish beaches and accelerated ground subsidence brought about by fluid removal. The range of climate drivers for coastal systems listed by the IPCC (2007a) is shown in Table 9.1.

Figure 9.1 An informal coastal settlement in Kota Kinabalu, Sabah, Malaysia. Such settlements will be particularly at risk as a consequence of rising sea levels (A.S.G.). (See Plate 48)

Table 9.1 Main climate drivers for coastal systems

Climate drivers	Effects
CO_2 concentration increase	Ocean acidification
Sea surface temperature increase	Change in ocean circulation, reduced sea ice in high latitudes, coral bleaching, stimulation of coral growth in cooler seas, poleward species migration, more algal blooms and red tides
Sea level	Inundation, flood and storm surge damage, beach erosion, saltwater intrusion up estuaries, etc., rising water tables, impeded drainage, wetland loss
Storm intensity	Increased extreme water levels and wave heights
Storm frequency	Altered surges and storm waves
Wave climate	Changes in wave conditions, leading to altered patterns of erosion and accretion; reorientation of beach plan form
Runoff	Altered flood risk in coastal lowlands, water quality/salinity, fluvial sediment supply, and circulation and nutrient supply

The steric effect

As the oceans warm up, their density decreases and their volume increases. This thermal expansion, the steric effect, causes sea levels to rise. Uncertainties arise as to the rate at which different parts of the oceans will warm up in response to increased atmospheric temperatures (Gregory et al., 2001). Long-term records of ocean warming and sea-level change are also sparse.

Church et al. (2001, table 11.2) presented data on global average sea-level rise due to thermal expansion during the twentieth century. Rates between 1910 and 1990 ranged between 0.25 and 0.75 mm per year while those between 1960 and 1990 ranged between 0.60 and 1.09 mm per year. The steric effect over the period 1910–1990 accounts for at least one-third of the observed eustatic change over that period.

With regard to the future, sea-level rise caused by thermal expansion between 1990 and 2100 is thought likely to be around 0.28 m, out of a total predicted sea-level rise due to climate change of 0.49 m (Church et al., 2001, table 11.14). The IPCC (2007a, b) suggested that thermal expansion is likely to account for 70–75% of sea-level rise by the end of the century. In other words, it will comfortably exceed in importance the role of ice cap and glacier melting. However, this proportion might change if the contribution of the melting of the Greenland Ice Cap becomes a more important component of the total (Vermeer and Rahmstorf, 2009).

Anthropogenic contribution to sea-level change

Human activities have an impact upon the hydrological cycle in a wide range of ways (see Chapter 5), so that changes in sea level could result from such processes as groundwater exploitation, drainage of lakes and wetlands, the construction of reservoirs behind dams, and modification of runoff and evapotranspiration rates from different types of land cover. Some of these processes could accelerate sea-level rise by adding more water to the oceans, whereas others would decelerate the rate of sea-level rise by impounding water on land. Unfortunately, quantitative data are sparse and sometimes contradictory, and so it is difficult to assess the importance of anthropogenic contri-butions to sea-level change. The IPCC (Church et al., 2001: 658) came to three conclusions:

1 The effect of changes in terrestrial water storage on sea level might be considerable.
2 The net effect on sea level could be of either sign.
3 The rate of hydrological interference has increased over the last few decades.

Reduction in lake-water volumes

Increased use of irrigation and interbasin water transfers have contributed to reduced volumes of water being stored in some lake basins (see Chapter 5). Classic examples of this include the Aral and Caspian Seas of Central Asia and Owens Lake in the south-west USA. Unfortunately, as not all lake level declines are due to human actions, it is not always easy to disentangle these from the effects of natural decadal scale climate fluctuations. Equally it is difficult to calculate the proportion of the extracted water that reaches the world ocean by runoff and evapotranspiration. Some of it may enter groundwater stores, causing ground-water levels to rise and waterlogging of the land to occur. Controversy has thus arisen on this topic. On the one hand, Sahagian (2000) has argued that this could be a process of some significance, perhaps contributing around 0.2 mm per year to sea-level rise, whereas Gornitz et al. (1997) believe that the net effect of the drying up of interior lakes is probably only small and indirect.

Water impoundment in reservoirs

Recent decades have seen the construction of large numbers of major dams and reservoirs. Today, nearly 500,000 km^2 of land worldwide are inundated by reservoirs that are capable of storing 6000 km^3 of water (Glieck, 2002).

Newman and Fairbridge (1986) calculated that between 1957 and 1982, as much as 0.75 mm per year of sea-level rise potential was stored in reservoirs and irrigation projects. More recently, Gornitz et al. (1997) have estimated that 13.6 mm of sea-level rise potential has been impounded in reservoirs, equivalent since the 1950s to a potential average reduction in sea-level rise of 0.34 mm per year. The figure given more recently by

Pokhrel et al. (2012) is 0.39. However, as Sahagian (2000: 43) reported, 'the total amount of impounded water is not known because there have been no comprehensive inventories of the millions of small reservoirs such as farm ponds and rice paddies. As a result, the total contribution of impounded water to global hydrological balance has remained unclear and most likely severely underestimated.'

Groundwater mining

Groundwater mining is the withdrawal of groundwater in excess of natural recharge. Many large groundwater aquifers are currently being heavily mined, leading to massive falls in aquifer levels and volumes. Examples include the Ogallala aquifer of the High Plains in the USA, the Nubian Sandstone aquifers of Libya and elsewhere in North Africa, and the great aquifers of the Arabian Peninsula. Estimates of the volumes of groundwater that are being removed from storage on a global basis are around $1000-1300\,km^3$, but not all of it is transferred to the oceans. Nonetheless, Church et al. (2001: 657) believed that it is probably the largest positive anthropogenic contributor to sea-level rise (apart from anthropogenic climatic change), amounting to perhaps $0.2-1.00\,mm$ per year. This is a view that is largely confirmed by Wada et al. (2012b) and Pokhrel et al. (2012). This source may not continue indefinitely, however, because groundwater contributions may become exhausted in some regions.

Urbanization and runoff

Urbanization leads to a net increase in total runoff from the land surface (see Chapter 5) due to the spread of areas of impermeable ground (e.g. concrete, tile and tarmac covered surfaces) which impede groundwater replenishment. In addition, water may be evacuated efficiently from urban areas in storm water drains and sewers. Plausible rates of sea-level rise due to this mechanism are in the range of $0.35-0.41\,mm$ per year.

Deforestation and runoff

As discussed in Chapter 5, deforestation can lead to increases in runoff from land surfaces. For example, experiments with tropical catchments have shown typical increases in stream flow of $400-450\,mm$ per year (Anderson and Spencer, 1991). The reasons for this include changes in rainfall interception, transpiration and soil structure. Combining data on rates of tropical deforestation with average values for increases in runoff, Gornitz et al. (1997: 153) came up with an estimate that this could cause a rate of sea-level rise of $0.13\,mm$ per year.

Wetland losses

The reclamation or drainage of wetlands may release stored water that can then enter the oceans. Gornitz et al. (1997: 153) do not see this as a significant mechanism of sea-level change.

Irrigation

The area of irrigated land on earth has gone up dramatically, amounting to $c.\ 45 \times 10^6$ hectares in 1900 and 240×10^6 hectares in 1990 (see Chapter 5). During the 1950s, the irrigated area increased by over 4% annually, though the figure has now dropped to only about 1%. Gornitz et al. (1997) estimated that evapotranspiration of water from irrigated surfaces would lead to an increase in the water content of the atmosphere and so to a fall in sea level of $0.14-0.15\,mm$ per year. Irrigation water might also infiltrate into groundwater aquifers, removing $0.40-0.48\,mm$ per year of sea-level equivalent.

Synthesis

As we have already noted, individual mechanisms can in some cases be significant in amount, but some serve to augment sea-level rise and some to reduce it. Gornitz et al. (1997: 158) summarized their analysis thus:

Increased runoff from groundwater mining and impermeable urbanized surfaces are potentially important anthropogenic sources contributing to sea level rise. Runoff from tropical deforestation and water released by oxidation of fossil fuel and biomass, including wetlands clearance, provide a smaller share of the total. Taken together, these processes could augment sea level by some $0.6-1.0\,mm/year$.

On the other hand, storage of water behind dams, and losses of water due to infiltration beneath reservoirs and irrigated fields, along with evaporation from these surfaces could prevent the equivalent of 1.5–.8mm/year from reaching the ocean. The net effect of all of these anthropogenic processes is to withhold the equivalent of 0.8 ± 0.4mm/year from the sea. This rate represents a significant fraction of the observed recent sea level rise of 1–2mm/year, but opposite in sign.

Permafrost degradation

Increases in temperatures in cold regions will lead to substantial reductions in the area of permanently frozen subsoil (permafrost). In effect, ground ice will be converted to liquid water which could in principle contribute to a rise in sea level. However, a proportion, unknown, of this water could be captured in ponds, thermokarst lakes and marshes, rather than running off into the oceans. Another uncertainty relates to the volume of ground ice that will melt. Bearing these caveats in mind, Church et al. (2001: 658) suggested that the contribution of permafrost to sea-level rise between 1990 and 2100 will be somewhere between 0 and 25mm (0–0.23mm per year).

Melting of glaciers and sea-level rise

Many glaciers are expected to be reduced in area and volume as global warming occurs and their meltwater will flow into the oceans, causing sea level to rise (Gregory and Oerlemans, 1998; Radić and Hock, 2011). In some situations, on the other hand, positive changes in precipitation may nourish glaciers, causing them to maintain their selves or to grow in spite of a warming tendency. Other uncertainties are produced by a lack of knowledge about how the **mass balance of glaciers** will respond to differing degrees of warming. The IPCC (2001, table 11.14) suggested that between 1990 and 2100, the melting of glaciers will contribute c. 0.01–0.23m of sea-level change, with a best estimate of 0.16m, second, therefore, only to steric effects. The problems of mass balance estimations have been well summarized by Arendt et al. (2002):

Conventional mass balance programs are too costly and difficult to sample adequately the >160,000 glaciers on Earth. At present, there are only but 40 glaciers worldwide with continuous balance measurements spanning more than 20 years. High-latitude glaciers, which are particularly important because predicted climate warming may be greatest there, receive even less attention because of their remote locations. Glaciers that are monitored routinely are often chosen more for their ease of access and manageable size than for how well they represent a given region or how large a contribution they might make to changing sea level. As a result, global mass balance data are biased toward small glaciers (<20km^2) rather than those that contain the most ice (>100km^2). Also, large cumulative errors can result from using only a few point measurements to estimate glacier-wide mass balances on an individual glacier.

They used laser altimetry to estimate volume changes of Alaskan glaciers from the mid-1950s to the mid-1990s and suggested that they may have contributed 0.27 ± 0.10mm per year of sea-level change. This is considerably more than had previously been appreciated.

Ice sheets and sea-level rise

The ice sheets of Greenland and Antarctica have proved pivotal in ideas about future sea-level changes. In the 1980s, there were fears that ice sheets could decay at near catastrophic rates, causing sea levels to rise rapidly and substantially, perhaps by 3m or more by 2100. Since that time this has become thought to be less likely and that far from suffering major decay, the ice sheets (particularly of Antarctica) might show some accumulation of mass as a result of increasing levels of nourishment by snow. However, considerable debates still exist on this, and there remain major uncertainties about how the two great polar ice masses may respond in coming decades and whether the West Antarctic Ice Sheet (WAIS) is inherently and dangerously unstable (see e.g. Oppenheimer, 1998; Sabadini, 2002; van der Ween, 2002). These are issues that will be addressed in Chapter 11.

Were the WAIS to collapse into the ocean, the impacts of the resulting 5m increase in sea level would be catastrophic for many coastal lowlands. The IPCC (2001) thought that it was very unlikely that this would occur in the twenty-first century. They believed that the contribution that the Antarctica and Greenland ice caps would make to sea-level change would be modest and less than those of glaciers or thermal expansion (the steric effect). The Greenland contribution between

1990 and 2100 was thought to be in the range from −0.02 to 0.09 m and for the Antarctic from −0.17 to 0.02 m.

How fast are sea levels rising?

Observational estimates of sea-level change are based on the short (only a decade or so) satellite record of sea-level height (Cabanes et al., 2001) and on the larger, but geographically sparse and uneven tide-gauge network. Tide gauges are rare in mid-ocean locations and very inadequate for the ocean-dominated Southern Hemisphere. In addition, difficulties in determining global (eustatic) sea-level changes are bedevilled by the need to make allowances for land motion caused by post-glacial isostatic rebound and tectonic movements (Church, 2001). In general it has been estimated that rates of sea-level rise from 1910 to 1990 average 1.0–2.0 mm per year (Church et al., 2001, table 11.10). On the other hand, over the period 1993–1998, the Topex/Poseidon satellite suggested a global mean rate of sea-level rise of 3.2 ± 0.2 mm per year (Cabanes et al., 2001). The IPCC (2007a, b) curve of global mean sea-level rise since 1870 is shown in Figure 9.2.

The amount of sea-level rise by 2100

Over the years, there has been a considerable diversity of views about how much sea-level rise is likely to occur by 2100. In general, however, estimates have

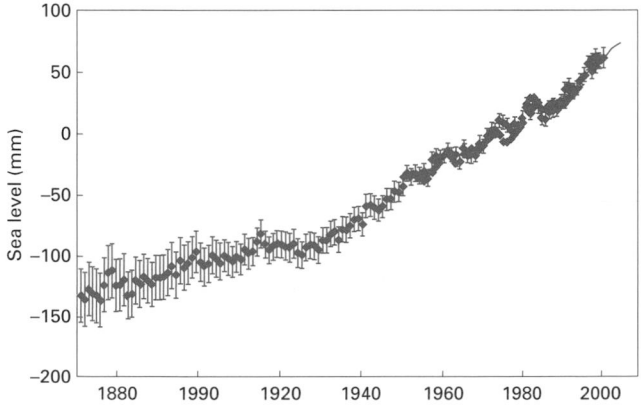

Figure 9.2 Sea-level rise since 1870 (after IPCC, 2007a, b, figure TS.18). Reproduced with permission. Error bars are 90% confidence intervals.

tended to be revised downward through time (French et al., 1995; Pirazzoli, 1996) (Figure 9.3). They have now settled at best estimates of just under 50 cm by 2100. This implies rates of sea-level rise of around 5 mm per year, which compares with a rate of about 1.5–2.0 mm during the twentieth century (Miller and Douglas, 2004). However, with increasing evidence that polar ice sheets are melting rapidly (Rignot et al., 2011), there are those who believe that the rates should be revised upwards and that by the end of the century sea level might be 0.5–1.4 m above the 1990 level (Rahmstorf, 2007), or even 75–190 cm (Vermeer and Rahmstorf, 2009). Nicholls et al. (2011) believe that with a 4°C or more rise in temperature that a credible upper bound for twenty-first-century sea-level rise is 2 m.

Land subsidence

The effects of global sea-level rise will be compounded in those areas that suffer from local subsidence as a result of local tectonic movements, isostatic adjustments and fluid abstraction (see Chapter 6). Areas where land is rising because of **isostasy** (e.g. Fennoscandia or the Canadian Shield) or because of tectonic uplift (e.g. much of the Pacific coast of the Americas) will be less at risk than subsiding regions (e.g. the deltas of the Mississippi and Nile Rivers) (Figure 9.4). In general, those sectors of the world's coastline that have been subsiding in recent decades include a large tract of the eastern seaboard of the USA, south-eastern England and some of the world's great river deltas (e.g. India, Ganges, Mekong, Tigris-Euphrates and Zambezi).

Areas of appreciable subsidence include some ocean islands. Indeed, crucial to Darwin's model of atoll evolution is the idea that subsidence has occurred, and the presence of guyots and seamounts in the Pacific Ocean attest to the fact that such subsidence has been a reality over wide areas. The causes of this subsidence are a matter of some debate (see Lambeck, 1988: 506–509). Some of it may be caused by the loading of volcanic material onto the crust, but some may be due to a gradual contraction of the seafloor as the ocean lithosphere moves away from either the ridge or the hotspot that led to the initial formation of the island volcanoes.

Another type of situation prone to subsidence is the river delta. Loading of sediment onto the crust by the

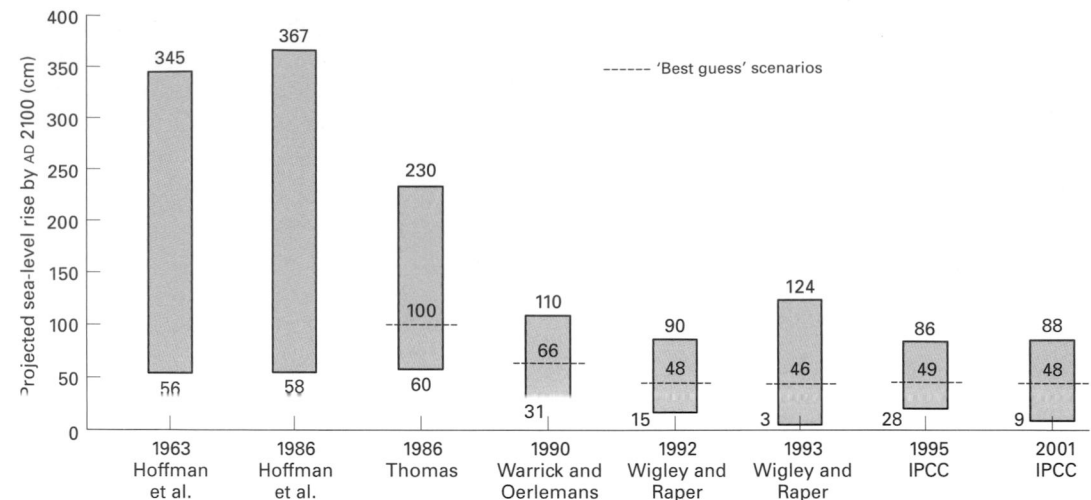

Figure 9.3 Revisions of anticipated sea-level rise by 2100 (after French et al., 1995, with modifications).

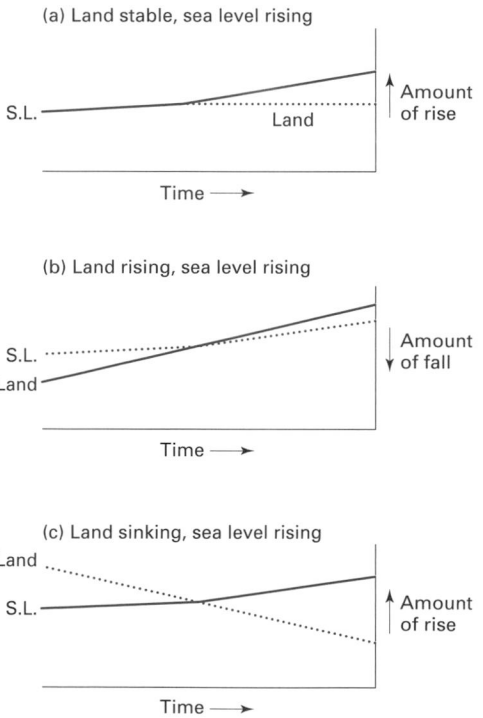

Figure 9.4 The effects of global sea-level rise will be either compounded or mitigated according to whether the local environment is one that is stable, rising or sinking.

river causes subsidence to occur. So, for example, Fairbridge (1983) calculated subsidence of the Mississippi Delta at a rate of c. 15 mm per year during the Holocene, while Stanley and Chen (1993) calculated Holocene subsidence rates for the Yangtze Delta in China at

1.6–4.4 mm per year. The Rhone Delta has subsided at between 0.5 and 4.5 mm per year (L'Homer, 1992), while the Nile Delta is subsiding at c. 4.7 mm per year (Sherif and Singh, 1999).

Some areas are prone to subsidence because of ongoing adjustment to the application and removal of ice loadings to the crust in the Pleistocene. During glacials, areas directly under the weight of ice caps were depressed, whereas areas adjacent to them bobbed up by way of compensation (the so-called peripheral bulge). Conversely, during the Holocene, following removal of the ice load, the formerly glaciated areas have rebounded, whereas the marginal areas have foundered. A good example of this is the Atlantic coast of North America (Engelhart et al., 2009).

Elsewhere, human actions can promote subsidence (see Chapter 6): the withdrawal of groundwater, oil and gas (Table 6.6); the extraction of coal, salt, sulfur and other solids through mining; the hydro-compaction of sediments; the oxidation and shrinkage of organic deposits such as peats and humus-rich soils; the melting of permafrost; and the catastrophic development of sinkholes in karstic terrain.

Coral reefs

Future changes in sea level and in other environmental factors will have impacts on a range of sensitive coastal environments. Among these are coral reefs, which are landforms that are more or less restricted to the tropics and that have fascinated scientists for over 150 years. Coral reefs are extremely important habitats. Not only do they have an estimated area of 600,000 km² globally, they are also the hosts to a great diversity of species, especially in the warm waters of the Indian Ocean and the western Pacific. They have been called the marine version of the tropical rainforests, rivalling their terrestrial counterparts in both richness of species and biological productivity. They are also of aesthetic importance. As Norris (2001: 37) wrote:

Clear blue sea, brightly coloured fish and amazing underwater structures built up over generations from the calcified remains of the tiny animals called coral polyps. It's a spectacular sight. No one who has ever visited one of the world's great coral reefs is ever likely to forget it. But it looks like we are one of the last generations who will have the opportunity to experience this spectacle. Reefs as we know them are on their way out.

The reason for Norris's pessimism is that coral reefs are under a whole series of direct anthropogenic threats (pollution, sedimentation, dynamiting, overfishing, etc.) and that they face a suite of potential threats from climate change and sea-level rise (Hoegh-Guldberg, 2011).

One potential change is that of hurricane frequency, intensity and distribution. They might build some coral islands up, erase others, and through high levels of runoff and sediment delivery they could change the turbidity and salinity of the water in which corals grow. If hurricane frequency, intensity and geographical spread were to change, there would be significant implications. This is a matter that is discussed further in Chapter 8.

Increased sea surface temperature could have deleterious consequences for corals which are near their thermal maximum (Hoegh-Guldberg, 2001). Most coral species cannot tolerate temperature greater than about 30°C, and even a rise in seawater temperature of 1–2°C could adversely affect many shallow water coral species.

Increased temperatures in recent years have been identified as a cause of widespread **coral bleaching** (loss of symbiotic zooxanthellae) (Baker et al., 2008). Those corals stressed by temperature or pollution might well find it more difficult to cope with rapidly rising sea levels than would healthy corals. Moreover,

it is possible that increased ultraviolet (UV) radiation because of ozone layer depletion could aggravate bleaching and mortality caused by global warming. Various studies suggest that coral bleaching was a widespread feature in the warm years of the 1980s and 1990s. The frequency of thermal stress events is likely to increase with ocean warming, leading to declines in coral cover, shifts in the composition of corals and other reef-dwelling organisms, and stress on the human populations who rely upon these ecosystems for food, income and shoreline protection (Donner, 2009).

Coral bleaching, which can produce mass mortality of corals in extreme cases, has been found to be strongly correlated with elevated water temperatures and high UV solar irradiance (e.g. Brown, 1997; Spencer et al., 2000). Although bleaching itself is a complex phenomenon to which corals can respond in a variety of ways (Brown et al., 2000; Fitt et al., 2001; Loya et al., 2001), El Niño–Southern Oscillation (ENSO)-related heating, cooling and migrations of ocean water masses have been found to be important controls of mass bleaching episodes (Spencer et al., 2000). For example, in 1998, sea surface temperatures in the tropical Indian Ocean were as much as 3–5°C above normal, and this led to up to 90% coral mortality in shallow areas (Wilkinson et al., 1999; Edwards et al., 2001). McClanahan (2000) notes that warm conditions of between 25 and 29°C favour coral growth, survival and species richness, and that somewhere about 30°C there are species-, environment-, or regionally specific thresholds above which many of the dominant coral species are lost. As Hoegh-Guldberg (1999), Souter and Linden (2000), Sheppard (2003) and Baker et al. (2008) have suggested, continued warming trends superimposed on interannual and decadal patterns of variability are likely to increase the incidence of bleaching and coral mortality unless significant adaptation to increased temperatures occurs. Donner (2009) has argued that the majority of the world's reefs may experience harmfully frequent thermal stress events before 2050, and Teneva et al. (2012) suggest that the most threatened reefs may be in the Central and Western Equatorial Pacific.

Indeed, Goreau and Hayes (1994) have produced maps of global coral bleaching episodes between 1983 and 1991 and have related them to maps of sea surface temperatures over that period. They find that areas of severe bleaching are related to what they describe as ocean 'hotspots' where marked positive temperature anomalies exist. They argue that coral reefs are ecosystems that may be uniquely prone to the effects of global warming:

If global warming continues, almost all ecosystems can be replaced by migration of species from lower latitudes, except for the warmest ecosystems. These have no source of immigrants already adapted to warmer conditions. Their species must evolve new environmental tolerances if their descendants are to survive, a much slower process than migrations. (pp. 179–180)

However, Kinsey and Hopley (1991) believed that few of the reefs in the world are so close to the limits of temperature tolerance that they are likely to fail to adapt satisfactorily to an increase in ocean temperature of 1–2°C, provided that there are not very many more short-term temperature deviations. Indeed, in general, they believed that reef growth will be stimulated by the rising sea levels of a warmer world, and they predict that reef productivity could double in the next 100 years from around 900 to 1800 million tonnes per year. They did, however, point to a range of subsidiary factors that could serve to diminish the increase in productivity: increased cloud cover in a warmer world could reduce calcification because of reduced rates of photosynthesis; increased rainfall levels and hurricane activity could cause storm damage and freshwater kills; and a drop in seawater pH might adversely affect calcification. In cooler ocean regions, however, warming may lead to a rapid poleward expansion in the range of tropical reef corals, as appears to have taken place since the 1930s (Yamano et al., 2011).

In the 1980s, there were widespread fears that if rates of sea-level rise were high (perhaps 2–3 m or more by 2100), then coral reefs would be unable to keep up, and submergence of whole atolls might occur. Particular concern was expressed about the potential fate of Tokelau, the Marshall Islands, Tuvalu, the Line Islands, and Kiribati in the Pacific Ocean, and of the Maldives in the Indian Ocean. However, with the reduced expectations for the degree of sea-level rise that may occur, there has arisen a belief that coral reefs may survive and even prosper with moderate rates of sea-level rise. As is the case with coastal marshes and other wetlands, reefs are dynamic features that may be able to respond adequately to rises in sea level (Spencer, 1995). It is also important to realize that their condition depends on factors other than the rate of submergence.

An example of the pessimistic tone of opinion in the 1980s is provided by Buddemeier and Smith (1988). Employing 15 mm per year as the probable rise of sea level over the next century, they suggest that this would be 'five times the present modal rate of vertical accretion on coral reef flats and 50% greater than the maximum vertical accretion rates apparently attained by coral reefs' (p. 51). Using a variety of techniques they believed 'the best overall estimate of the sustained maximum of reef growth to be 10 mm/year . . ..' (p. 53). They predicted that

Inundated reef flats in areas of heavy seas will be subjected to progressively more destructive wave activity as larger waves move across the deepening flats . . . .. Reef growth on the seaward portions of inundated, wave swept reef flats may therefore be negligible compared to sea level rise over the next century, and such reef flats may become submerged by almost 1.5 m. (p. 54)

Reefs have a range of topographies, and low-lying reef islands on the rims of atolls may be especially vulnerable to the effects of sea-level rise (Woodroffe, 2008). In addition to the potential effects of submergence, there is the possibility that higher sea levels could promote accelerated erosion of reefs (Dickinson, 1999). Sea-level rise could also cause enhanced sedimentation and turbidity as tides ebb and flow over shallow, fringing reefs (Field et al., 2011).

Reefs may also suffer from increasing ocean acidification (Royal Society, 2005). The oceans are likely to become more acidic in the future, for a proportion of the extra carbon dioxide being released into the atmosphere by the burning of fossil fuels and biomass is absorbed by seawater. As carbon dioxide combines with water it produces carbonic acid. An increase in carbonic acid in seawater will cause that water to become more acidic (i.e. it will have a lower pH than now). The absorption of carbon dioxide has already caused the pH of modern surface waters to be about 0.1 lower (i.e. more acidic and less alkaline) than in pre-industrial times. Ocean pH may fall an additional 0.3 by 2100 (Gattuso et al., 1998; Caldeira and Wickett, 2003), which means that the oceans may be more acidic than they have been for 25 million years. Several centuries from now, if we continue to add carbon dioxide to the atmosphere, ocean pH will be lower than at any time in the past 300 million years (Doney, 2006). It is also notable that the rate of change in acidity will be very rapid. This will be particularly harmful to those organisms (corals, molluscs and plankton) that depend on the presence of carbonate ions to build their shells (or other hard parts) out of calcium bicarbonate (Orr et al., 2005a, b; Pelejero et al., 2010). A coral reef, as Kleypas et al. (1999) have pointed out, represents the net accumulation of calcium carbonate produced by corals and other calcifying organisms. Equally, Leclercq et al. (2000) have calculated that the calcification rate of scleractinian-dominated communities could decrease by 21% between the pre-industrial period (1880) and the year (2065) at which atmospheric carbon dioxide concentrations will double. Thus were calcification to decline, then reef-building capacity would also decline. It is also possible that ocean acidification will accelerate coral bleaching (Anthony et al., 2008).

To conclude this section, reefs play an important role in coastal protection against storms so that 'we can anticipate that decreasing rates of reef accretion, increasing rates of bio-erosion, rising sea levels, and intensifying storms may combine to jeopardize a wide range of coastal barriers' (Hoegh-Guldberg et al. 2007: 1742).

Salt marshes and mangrove swamps

Salt marshes, including the mangrove swamps of the tropics, are extremely valuable ecosystems that are potentially highly vulnerable in the face of sea-level rise (Gedan et al., 2011), particularly in those circumstances where sea defences and other barriers prevent the landward migration of marshes as sea level rises. In such locations what is termed 'coastal squeeze' occurs (Doody, 2012). Sediment supply, organic and inorganic, is a crucial issue (Reed, 2002). On coasts with limited sediment supply, a rise in sea level will impede the normal process of marsh progradation, and increasing wave attack will start or accelerate erosion along their seaward margins. The tidal creeks that flow across the marsh will tend to become wider, deeper and more extended headwards as the marsh is submerged. The marsh will attempt to move landwards, and where the hinterland is low lying, the salt marsh vegetation will tend to take over from freshwater or terrigenous communities. Such landward movement is impossible where sea walls or embankments have been built at the inner margins of a marsh (Figure 9.5). Equally, if there is very limited availability of sediment, the marsh may not build up and inwards, so that in such circumstances the salt marsh will cease to exist (Bird, 1993). Marshes would be further threatened if climate change caused increased incidence of severe droughts (Thomson et al., 2002). Table 9.2 demonstrates the differences between marshes in terms of their sensitivity. However, salt marshes are highly dynamic features and in some situations may well be able to cope, even with quite rapid rises of sea level (Reed, 1995).

Reed (1990) suggests that salt marshes in riverine settings may receive sufficient inputs of sediment that

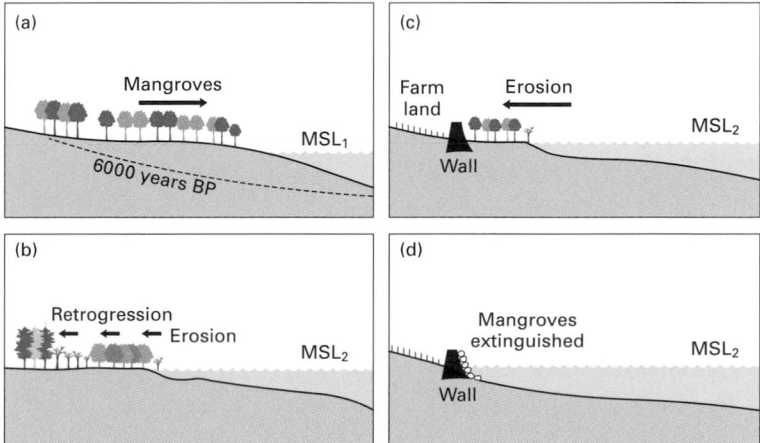

Figure 9.5 Changes on mangrove-fringed coasts as sea level rises (a and b) will be modified where they are backed by a wall built to protect farmland (c). The mangrove fringe will then be narrowed by erosion, and may eventually disappear (d), unless there is a sufficient sediment supply to maintain the substrate and enable the mangroves to persist. MSL is mean sea level (modified from Bird, 1993, figure 45).

Table 9.2 Salt marsh vulnerability

Less sensitive

Areas of high sediment input

Areas of high tidal range (high sediment transport potential)

Areas with effective organic accumulation

More sensitive

Areas of subsidence

Areas of low sediment input (e.g. cyclically abandoned delta areas)

Mangroves (longer life cycle, therefore slower response)

Constraint by sea walls, etc. (nowhere to go)

Microtidal areas (rise in sea level represents a larger proportion of total tidal range)

Reef settings (lack of allogenic sediment)

they are able to accrete rapidly enough to keep pace with projected rises of sea level. Areas of high-tidal range, such as the marshes of the Severn Estuary in England/Wales, or the Tagus Estuary of Portugal, are also areas of high sediment-transport potential and may thus be less vulnerable to sea-level rise (Simas et al., 2001). Likewise, some vegetation associations, for example, *Spartina* swards, may be relatively more effective than others at encouraging accretion, and organic matter accumulation may itself be significant in promoting vertical build-up of some marsh surfaces. For marshes that are dependent upon inorganic-sediment accretion, increased storm activity and beach erosion which might be associated with the greenhouse effect could conceivably mobilize sufficient sediments in coastal areas to increase their sediment supply.

Marsh areas that may be highly prone to sea-level rise include areas of deltaic sedimentation where, because of sediment movement controls (e.g. reservoir construction) or because of cyclic changes in the location of centres of deposition, rates of sediment supply are low. Such areas may also be areas with high rates of subsidence. A classic example of this is portions of the Mississippi Delta. Park et al. (1986) undertook a survey of coastal wetlands in the USA and suggested that sea-level change could, by 2100, lead to a loss of between 22 and 56% of the 1975 wetland area, according to the degree of sea-level rise that takes place.

Ray et al. (1992) have examined the past and future response of salt marshes in Virginia in the USA to changes in sea level and have been particularly interested in the response of mid-lagoon marshes. Between 6000 and 2000 years ago, when the rate of sea-level rise was about 3 mm per year, the lagoons were primarily open water environments with little evidence of mid-lagoon marshes. After AD 1200, land subsidence in the area slowed down, and sea-level rise was only about 1.3 mm per year. This permitted mid-lagoon marshes to expand and flourish. Such marshes in general lack an inorganic sediment supply so that their upwards growth rate approximates only about 1.5 mm per year. Present rates of sea-level rise exceed that figure (they are about 2 mm per year), and cartographic analysis shows a 16% loss of marsh between 1852 and 1968. As sea-level rise accelerates as a result of global warming, almost all mid-lagoon marsh will be lost.

Rises in sea level will increase near shore water depths and thereby modify wave refraction patterns. This means that wave energy amounts will also change at different points along a particular shoreline. Pethick (1993) maintained that this could be significant for the classic Scolt Head Island salt marshes of Norfolk, eastern England, which are at present within a low to medium wave energy zone. After a 1.0 m rise in sea level these marshes will experience high wave energy because of the migration of wave foci. Pethick remarked:

The result will be to force the long shore migration of salt marsh and mudflat systems over distances of up to ten kilometres in 50 years – a yearly migration rate of 200 meters. It is doubtful whether salt marsh vegetation could survive in such a transitory environment – although intertidal mudflat organisms may be competent to do so – and a reduction of total loss of these open coast wetlands may result. (p. 166)

One particular type of marsh that may be affected by anthropogenically accelerated sea-level rise is the mangrove swamp. As with other types of marsh the exact response will depend on the local setting, sources and rates of sediment supply, and the rate of sea-level rise itself. However, mangroves may respond rather differently from other marshes in that they are composed of relatively long-lived trees and shrubs, which means that the speed of zonation change will be less (Woodroffe, 1990). With increasing temperatures and fewer coastal freezes, mangroves may expand their latitudinal range (Comeaux et al., 2012).

Like salt marshes, however, mangroves trap sediment to construct a depositional terrace in the upper intertidal zone. Where there is only a modest sediment supply, submergence by rising sea level may cause dieback of vegetation and erosion of their seaward margins. As their seaward margin erodes backwards, the mangroves will attempt to spread landward, displacing existing freshwater swamps or forests. As with normal salt marshes this would not be possible if they were backed by walls or embankments.

The degree of disruption is likely to be greatest in microtidal areas, where any rise in sea level represents a larger proportion of the total tidal range than in macrotidal areas. The setting of mangrove swamps will be very important in determining how they respond. River-dominated systems with large **allochthonous** sediment supply will have faster rates of shoreline progradation and deltaic plain accretion and so may be able to keep pace with relatively rapid rates of sea-level rise. By contrast, in reef settings, in which sedimentation is primarily autochthonous, mangrove surfaces are less likely to be able to keep up with sea-level rises. This is the view of Ellison and Stoddart (1990: 161) who argued that low island mangrove ecosystems (mangals) have in the past been able to keep up with a sea-level rise of up to 8–9 cm per hundred years, but that at rates over 12 cm per hundred years, they had not been able to persist.

Snedaker (1995) found it difficult, however, to reconcile the Ellison and Stoddard view with what has happened in South Florida, where relative sea level rose by about 30 cm over 147 years (equivalent to 23 cm per hundred years). The mangrove swamps of the area did not for the most part appear to have been adversely stressed by this. Snedaker argued that precipitation and catchment runoff changes also need to be considered, as for any given sea-level elevation, reduced rainfall and precipitation would result in higher salinity and greater seawater-sulfate exposure. These in turn would be associated with decreased production and increased organic matter decomposition that would lead to subsidence. On the other hand, under conditions with higher rainfall and runoff, the reverse would occur, so that mangrove production would increase and sediment elevations would be maintained.

The ability of mangrove propagules to take root and become established in intertidal areas subjected to a higher mean sea level is in part dependant on species (Ellison and Farnsworth, 1997). In general, the large propagule species (e.g. *Rhizophora* spp.) can become established in rather deeper water than can the smaller propagule species (e.g. *Avicennia* spp.). The latter has aerial roots which project only vertically above tidal muds for short distances (Snedaker, 1993).

In arid areas, such as the Arabian Gulf in the Middle East, extensive tracts of coastline are fringed by low-level salt-plains called *sabkhas*. These features are generally regarded as equilibrium forms that are produced by depositional processes (e.g. wind erosion and storm surge effects). They tend to occur at or about high tide level. Because of the range of depositional processes involved in their development they might be able to adjust to a rising sea level, but quantitative data on present and past rates of accretion are sparse. A large proportion of the industrial and urban infrastructure of the United Arab Emirates is located on or in close proximity to sabkhas.

Salt marsh, swamp and sabkha regression caused by climatic and sea-level changes will compound the problems of wetland loss and degradation caused by other human activities (Wells, 1996; Kennish, 2001), including reclamation, ditching, diking, dredging, pollution and sediment starvation. More than half of the original salt marsh habitat in the USA has already been lost, and Shriner and Street (1998: 298) suggest that a 50-cm rise in sea level would inundate approximately 50% of North American coastal wetlands in the twenty-first century. At a global scale, Nicholls et al. (1999) have suggested that by the 2080s, sea-level rise could cause the loss of up to 22% of the world's coastal wetlands. When combined with other losses due to direct human action, up to 70% of the world's coastal wetlands could be lost by that date.

River deltas

Deltaic coasts and their environs are home to large numbers of people. Fifty-one of the world's deltas have a combined population of over 325 million, and the Nile Delta alone has a population of around 50 million (Syvitski, 2008). They are likely to be threatened by submergence as sea levels rise, especially where prospects of compensating sediment accretion are not evident. Many deltas are currently zones of subsidence because of the isostatic effects of the sedimentation that caused them to form. This will compound the effects of eustatic sea-level rise (Milliman and Haq, 1996). The Niger Delta has high rates of subsidence because of oil and gas abstraction. Even without accelerating sea-level rise, the Nile Delta has been suffering accelerated recession because of sediment retention by dams.

It needs to be remembered, however, that deltas will not solely be affected by sea-level changes. The delta lands of Bangladesh (Warrick and Ahmad, 1996), for example, receive very heavy sediment loads from the rivers that feed them so that it is the relative rates of accretion and inundation that will be crucial (Milliman et al., 1989). Land-use changes upstream, such as deforestation, could increase rates of sediment accumulation. Conversely, many deltas, such as the Mississippi, suffer because they have an insufficient sediment supply consequent upon the trapping of sediment by upstream dams (Ericson et al., 2006; Blum and Roberts, 2009). Deltas could also be affected by changing tropical cyclone activity, which might cause storm surge effects to be magnified (Karim and Mimura, 2008).

Broadus et al. (1986) calculated that were the sea level to rise by just 1 m in 100 years, 12–15% of Egypt's arable land would be lost, and 16% of the population would have to be relocated. With a 3-m rise the figures would be a 20% loss of arable land and a need to relocate 21% of the population. The cities of Alexandria, Rosetta and Port Said are at particular risk, and even a sea-level rise of 50 cm could mean that 2 million people would have to abandon their homes (El-Raey, 1997). In Bangladesh (Figure 9.6), a 1-m rise would inundate 11.5% of the total land area of the state and affect 9% of the population directly, while a 3-m rise in sea level would inundate 29% of the land area and affect 21% of the population. It is sobering to remember that at the present time, approximately one-half of Bangladesh's rice production is in the area that is less than 1 m above sea level. Many of the world's major conurbations might be flooded in whole or in part, sewers and drains rendered inoperative (Kuo, 1986) and peri-urban agricultural productivity reduced by saltwater incursion (Chen and Zong, 1999).

Bangkok is an example of a city that is being threatened by a combination of accelerated subsidence and accelerated sea-level rise (Nutalaya et al., 1996). More than 4550 km² of the city was affected by land subsidence between 1960 and 1988, and 20–160 cm of depression of the land surface occurred. The situation is

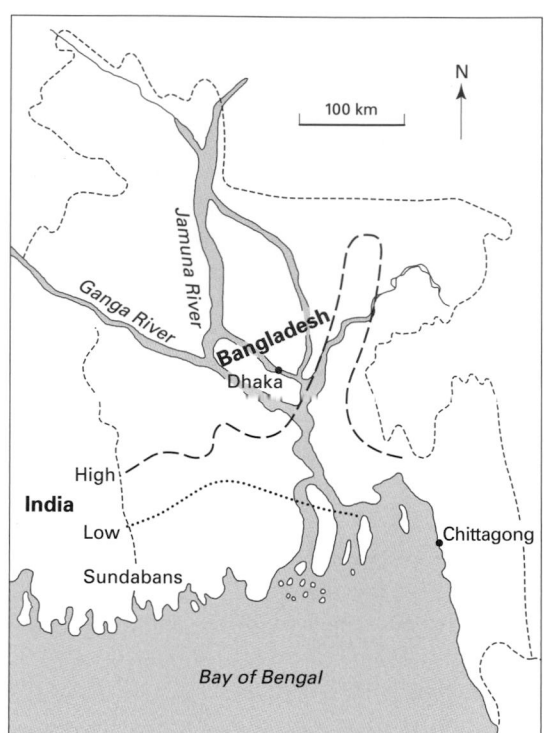

Figure 9.6 Projected areas of flooding as a result of sea-level change in Bangladesh, for two scenarios (low = 1 m and high = 3 m) (modified from Broadus et al., 1986, figure 7).

critical because Bangkok is situated on a very flat, low-lying area, where the ground level elevations range from only 0 to 1.5 m above mean sea level (Figure 9.7).

Estuaries

Estuaries, located between the land and the sea, are unique and sensitive ecosystems that are locations of many major ports and industrial concerns (Dyer, 1995). They could be impacted upon by a range of climate-related variables, including the amount of freshwater and sediment coming from the land, the temperature and salinity of the estuarine water, and tidal range (Day et al., 2008).

The overall effect of sea-level rise has been summarized by Chappell et al. (1996: 224):

In the earliest stages of sea level rise, the extent of tidal flooding on high tide flats will increase and networks of small tidal creeks will expand into estuarine floodplains. Levels of tidal rivers will be breached and brackish water will invade

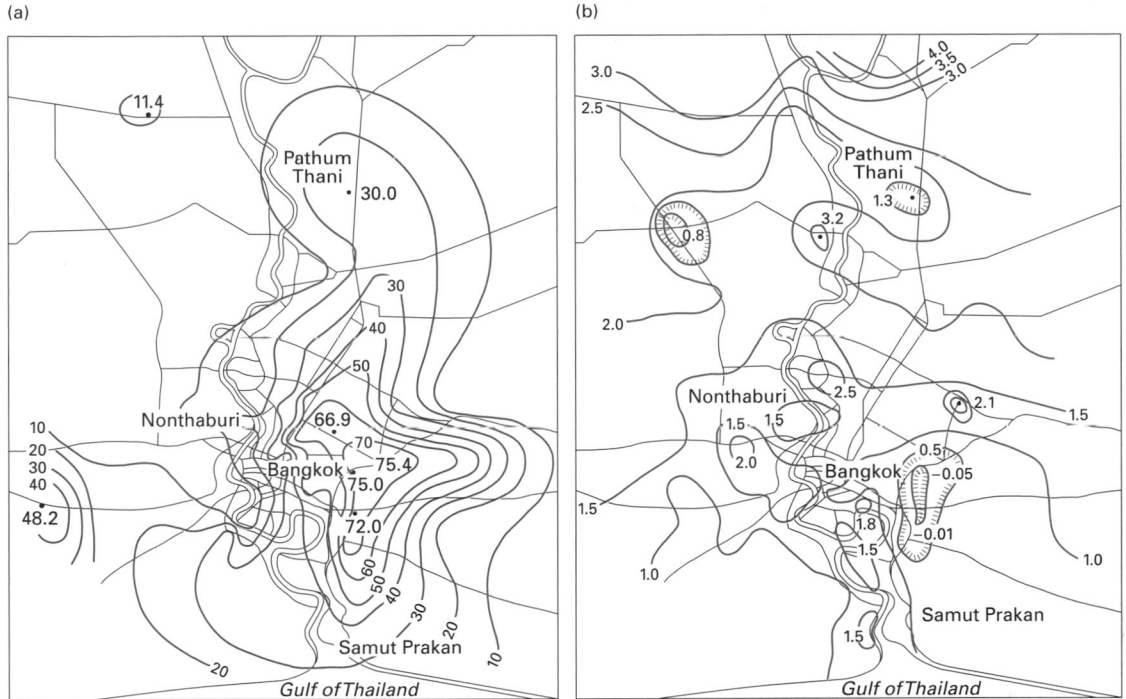

Figure 9.7 Total subsidence (in centimetres), 1978–1987 (a) and ground elevation of Bangkok, 1987 (b) (modified from Nutalaya et al., 1996, figures 3 and 9).

freshwater floodplains. Eventually, tidal flow in main channels will increase to carry the enlarged tidal prism, and channels are likely to widen. In systems with extensive tidal floodplains, tidal regimes will shift towards ebb dominance and net sediment flux will tend to be offshore, except in macrotidal systems. Salt-water intrusion could be a major change, with saline water extending far inland as sea level rose. This in turn could affect swamp and marsh vegetation. (Mulrennan and Woodroffe, 1998)

Attempts have been made to model the transgression rate for landward extending estuaries. For the Humber estuary in north-east England, Pethick (2001: 34) reports that the mean migration rate of the estuary as a whole would be 1.3 m per mm rise in sea level (or 8 m per year assuming a 6 mm per year rise in sea level).

Cliffed coasts

When sea level rises, nearshore waters deepen, shore platforms become submerged, deeper water allows larger waves to reach the bases of cliffs, and the cliffs may suffer from accelerated rates of retreat, especially if they are made of susceptible materials (Ashton et al., 2011; Revell et al., 2011; Brooks and Spencer, 2012). In addition, sea-level rise is likely to cause increased frequencies of coastal landslides. Accelerated cliff retreat and land sliding may cause an augmented supply of sediment to adjacent and downdrift beaches. This could be beneficial. However, as Bray et al. (1992: 86) remarked:

The location of the benefit will depend upon sediment transport conditions and their relation to sediment supply. Where littoral transport is poorly developed, beaches will accrete in front of cliffs as sea level rises, and tend to offset any trend for increased retreat. Where littoral transport is efficient, cliff retreat is likely to increase significantly as sea level rises because extra sediments yielded are rapidly removed from the eroding cliffs. Coasts of this type are likely to become increasingly valuable in the future because they are sensitive to sea level rise and can supply large quantities of sediment to downdrift beaches. A further complicating factor is that cliff erosion products differ with respect to their mobility and coarse durable materials may be retained on the upper shoreface and provide natural armouring. Soft-rock cliffs yielding a high proportion of coarse durable products are therefore likely to be less sensitive to sea level rise.

Cliffs on high latitude coasts (e.g. in Siberia, Alaska and Canada) might be especially seriously affected by global change. On the one hand, coasts formed of weak sediments that are currently cemented by permafrost would loose strength if warming caused the permafrost to melt. On the other hand, melting of sea ice would expose them to greater wave affects from open water.

Sandy beaches

As sea level rises, many beaches, especially those in closed bays, often called pocket beaches, may suffer from accelerated erosion (Brunel and Sabatier, 2009) (Figure 9.8). Bruun (1962) developed a widely cited and elegantly simple model of the response of a sandy beach to sea-level rise in a situation where the beach was initially in equilibrium, neither gaining nor losing sediment (Figure 9.9). As Bird (1993: 56) has explained:

Erosion of the upper beach would then occur, with removal of sand to the nearshore zone in such a way as to restore the previous transverse profile. In effect, there would be an upward and landward migration of the transverse profile, so that the coastline would recede beyond the limits of submergence. This restoration would be completed when the sea became stable at a higher level, and coastline recession would come to an end after a new equilibrium was achieved. The extent of recession was predicted by using a formula that translates into a 'rule of thumb' whereby the coastline retreats 50–100 times the dimensions of the rise in sea level: a 1 m rise would cause the beach to retreat by 50–100 m. Since many seaside resort beaches are no more than 30 m wide, the implication is that these beaches will have disappeared by the time the sea has risen 15–30 cm (i.e. by the year 2030), unless they are artificially replaced.

This influential model, often called the Bruun Rule, has been widely used. It is important to recognize, however, that there are some constraints on its applicability (Wells, 1995; Healy, 1996). The rule assumes that no sand is lost to longshore transport and that an offshore 'closure depth' exists beyond which there is no sediment exchange. Moreover, the rule does not allow for shoreward transport of sediment as overwash or for those situations where the slope of the coastal plain is too gentle for sufficient sand to be available as a source for supplying the offshore. In addition, the rule was originally proposed for beaches that were initially in equilibrium. However, as Bird (1993: 58) has pointed out, only a small proportion of the world's sandy beaches can in fact be considered to be in equilibrium. Beach erosion is widespread. Furthermore, there is a lack of established criteria to ascertain wherever the original shoreline really is in dynamic equilibrium (Healy, 1991).

Figure 9.8 Malibu Beach, California, USA, is an example of a vulnerable beach. It is very narrow and is bounded on the landward side by properties (A.S.G.). (See Plate 49)

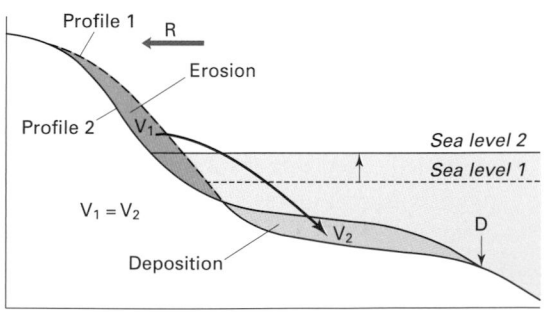

Figure 9.9 Sea-level rise and coastline changes. The Bruun Rule states that a sea-level rise will lead to erosion of the beach and removal of a volume of sand (V_1) seaward to be deposited (V_2) in such a way as to restore the initial transverse profile landward of D, the outer boundary of nearshore sand deposits. The coastline will retreat (R) until stability is restored after the sea-level rise comes to an end. The coastline thus recedes further than it would if submergence were not accompanied by erosion.

In spite of these limitations the Bruun Rule remains widely used and can be stated mathematically as follows (Gornitz et al., 2002: 68):

$$S = (A \times B)/d,$$

where S = shoreline movement, A = sea-level rise, d = maximum depth of beach profile, measured from the berm elevation for each project location to the estimated depth of closure, B − horizontal length of the profile, measured from the beginning of the berm to the intersection with the estimated depth of closure. The depth of closure is taken as the minimum water depth at which no significantly measurable change occurs in bottom depth. It is often erroneously interpreted to mean the depth at which no sediment moves in deeper water. 'Closure' is a somewhat ambiguous term in that it can vary, depending on waves and other hydrodynamic forces.

An example of the testing of the validity of the Bruun Rule is the study by List et al. (1997) of the barrier islands of Louisiana in the USA. Using bathymetric surveys over about a century, they found that only a portion of their studied profiles met the equilibrium criterion of the Bruun Rule. Furthermore, using those shore profiles that did meet the equilibrium criterion, they determined measured rates of relative sea-level rise so that they could hindcast shoreline retreat rates using the Bruun Rule formula. They found that the modelled and observed shoreline retreat rates showed no significant correlation. They suggested that if the Bruun Rule is inadequate for hindcasting it

would also be inadequate for forecasting future rates of beach retreat. There are those who now suggest that use of the Bruun Rule should be abandoned (Cooper and Pilkey, 2004; Aagaard and Sorensen, 2012).

There are various other techniques that can be used to predict rates of shoreline retreat. One of these is historical trend analysis, which is based upon extrapolating the trend of shoreline change with respect to recorded sea-level rise over a given historical period:

$$R_2 = (R_1/S_1) \times S_2,$$

where S_1 = historical sea-level rise, S_2 = future sea-level rise, R_1 = historical retreat rate, and R_2 = future retreat rate.

Barrier islands, such as those that line the southern North Sea and the eastern seaboard of the USA, are dynamic landforms that will tend to migrate inland with rising sea levels and increased intensity of overtopping by waves (Eitner, 1996; Ashton et al., 2008). If sea-level rise is not too rapid, and if they are not constrained by human activities (e.g. engineering structures and erosion control measures), they are moved inland by washover – a process similar to rolling up a rug (Titus, 1990); as the island rolls landward, it builds landwards and remains above sea level (Figure 9.10). As sea level rises, they will be exposed to higher storm surges and greater flooding. Extreme storm surges may become a more serious threat to the low-lying coasts of the North Sea (Woth et al., 2006).

Locations in the north-eastern USA, such as most of Nantucket Island, the eastern portions of Martha's

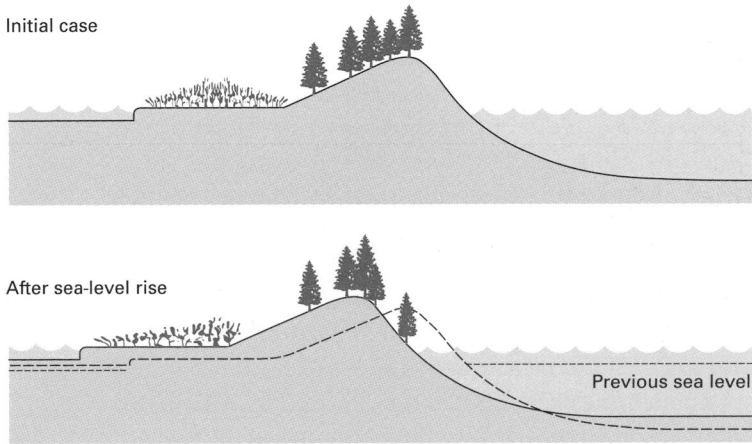

Figure 9.10 Overwash: the natural response of undeveloped barrier islands to sea level rise (Source: Titus, 1990, figure 2). Reproduced with permission from Taylor & Francis Publishers.

Vineyard, parts of Cape Cod, and areas of Long Island and New Jersey are also regarded as very vulnerable (Ashton et al., 2008). In the New York area, Gornitz et al. (2002) have calculated that by the 2080s, the return period of the 100-year storm flood could be reduced to between 4 and 60 years (depending on location). Similar predictions for the New York area have also been made by Lin et al. (2012). In the case of Long Island, Shepard et al. (2012) argued that even a modest and probable sea-level rise of 0.5 m by 2080 would vastly increase the numbers of people (47% increase) and property loss (73%) impacted by storm surges. Equally, if hurricanes become more widespread and frequent along the eastern and Gulf coasts of the USA, then wave heights will also tend to rise, as has happened during recent warming decades (Komar and Allan, 2008), and flooding will become a greater hazard (Frazier et al., 2010). For the city of Corpus Christi in Texas, hurricane flooding is projected to increase between 20% and 70% by the 2030s (Irish et al., 2010; Mousavi et al., 2011). The effects of increased surge levels will be amplified along the Atlantic coast of North America as this is an area where sea-level appears to have been rising faster than the global average (Sallenger et al., 2012).

The role of sediment starvation

Many beaches are currently eroding because they are no longer being replenished with sediment by rivers. This is because of the construction of dams, which trap much of the sediment that would otherwise go to the sea (see Chapter 5).

The River Nile now only transports about 8% of its natural load below the Aswan Dam. Even more dramatic is the picture for the Colorado River in the USA. Prior to 1930, it carried around 125–150 million tonnes of suspended sediment to its delta at the head of the Gulf of California. Following the construction of a series of dams, the Colorado now discharges neither sediment nor water into the ocean (Schwarz et al., 1991). Similarly, rivers on the eastern seaboard of the USA draining into the Gulf of Mexico or the Atlantic have shown marked falls in sediment loadings, and four major Texan rivers carried in 1961–1970 an average only about one-fifth of what they carried in 1931–1940. Likewise, in France, the Rhône, because of hydroelec-

tric dam construction, reafforestation and a decline in exceptional flood events (Brunel and Sabatier, 2009), only carries about 5% of the load that it did in the nineteenth century, while in Asia the Indus discharges less than 20% of the load it did before the construction of large barrages over the last half century (Milliman, 1990).

Conclusions

Coastal environments are already suffering from many human impacts, including sediment starvation. However, these dynamic regions, at the interface between land and sea and the home to many people, will also be subjected in the future to the combined effects of climate change and sea-level rise (Hoegh-Guldberg and Bruno, 2010). Plainly coastlines will be directly impacted by climatic change through such mechanisms as changes in storm surges and hurricanes, but they will also be impacted by sea-level changes resulting from global climatic change, local subsidence and miscellaneous modifications of the hydrological cycle. If, on average, sea level rises by around half a meter during this century, it will greatly modify susceptible and sensitive coastlines, including marshes and swamps, estuaries, soft cliffs, barrier islands and sandy beaches.

Points for review

In what ways may human activities influence global sea levels?

How may coral reefs respond to global warming?

Which coastal environments may be particularly susceptible to sea-level rise?

What do you understand by the Bruun Rule? Is it a rule that has wide applicability?

Guide to reading

Bird, E.C.F. (1993) *Submerging Coasts*. Chichester: Wiley. A geomorphological overview of how coasts will respond to sea-level rise.

Douglas, B.C., Kearney, M.S. and Leatherman, S.P. (eds) (2001) *Sea Level Rise: History and Consequences*. San Diego: Academic Press. An edited volume, with a long time perspective, which considers some of the consequences of sea-level change.

Eisma, D. (ed.) (1995) *Climate Change. Impact on Coastal Habitation*. Boca Raton, FL: Lewis Publishers. A study of the effects of climate change and sea-level rise.

Gattuso, J.-P. and Hansson, L. (eds) (2011) *Ocean Acidification*. Oxford: Oxford University Press. An edited guide to the nature, causes and impacts of ocean acidification.

Milliman, J.D. and Haq, B.V. (eds) (1996) *Sea-level Rise and Coastal Subsidence*. Dordrecht: Kluwer. An edited discussion of the ways in which sea-level rise and land subsidence may impact upon coastal environments.

10 THE FUTURE: HYDROLOGICAL AND GEOMORPHOLOGICAL IMPACTS

Chapter Overview

In a warmer world, changes in the type, amount and intensity of moisture received as precipitation, together with increased rates of evapotranspiration, will have major, though sometimes opposite, impacts on the hydrological environment. In addition, the presence of increased levels of carbon dioxide in the atmosphere may affect moisture transpiration by plants. There is evidence that some parts of the world will have more available moisture for stream flow and that other areas will have less. The chapter will consider the possible future runoff characteristics at a global scale, in cold regions, and also consider regional examples from the UK, Europe, the USA and elsewhere. It will also consider how such geomorphological processes as mass movements, soil erosion and stone weathering may respond to changed environmental conditions.

Introduction

Global warming will affect hydrological systems and fluvial geomorphology in a whole range of ways (Arnell, 1996; Jones et al., 1996). Increasing temperatures will tend to melt snow and ice and promote greater evapotranspirational losses. There will be changes in the amount, intensity, duration and timing of precipitation, which will affect river flows and groundwater recharge. Vegetation cover will respond to tempera-

The Human Impact on the Natural Environment: Past, Present and Future, Seventh Edition. Andrew S. Goudie.
© 2013 John Wiley & Sons, Ltd. Published 2013 by John Wiley & Sons, Ltd.

ture and precipitation changes, as will land use. Higher CO_2 levels may lead to changes in plant growth and physiology, which can lead to changes in transpiration. Higher atmospheric CO_2 levels may stimulate plant growth and lead to changes in plant water use efficiency and thus to transpiration and runoff (Eckhardt and Ulbrich, 2003; Morgan et al., 2004; Gedney et al., 2006). Global warming may also affect soil properties (such as organic matter content) which could alter runoff generation processes. Climate change will also lead to human interventions in the hydrological system with, for example, greater use of irrigation in areas subject to increased drought risk, and the continued spread of engineering controls on flooding and erosion.

The nature of the changes that will take place in fluvial systems depends to a certain extent on scale. As Ashmore and Church (2001: 5) explained:

The nature of the changes caused by climatic change depends to some extent on the size of the drainage basin in question. Drainage basin size also affects the relative impact of widespread climate change compared to local land-use change. The effects of land-use change are expected to dominate in smaller basins where a large proportion of the basin area may be affected, leading to substantial changes in runoff and erosion throughout the basin. In larger drainage basins, land use change is seldom sufficiently widespread to affect the entire basin and climatic effects will dominate. The impacts of climatic change will also vary with basin size. Thus, small basins will be affected by changes in local, high-intensity storms, whereas larger basins will show a greater response to cyclonic events or basin-wide snowmelt effects.

Equally, as Ashmore and Church (2001: 5) maintain, there are many ways in which river systems may change:

Potential consequences of climatic change for river processes include changes to the magnitude of flood flows; modification of river channel dimensions and form; changes to bank stability, bank erosion rates, and channel migration; modification of in-channel erosion and deposition; on-set of long term aggradation or degradation of river channels; changes to intensity and frequency of overbank flooding and ice-jams; and changes to the stability of valley sides. These changes in channel processes present a significant risk to structures both in and near streams including dams, bridges, water intakes, and outfalls. Structures on or near floodplains or valley margins are also at risk. Other environmental, scenic and economic attributes of streams will also be affected, especially in stream and riparian habitat.

Rainfall Intensity

Rainfall intensity is a major factor in controlling such phenomena as flooding, rates of soil erosion and mass movements (Sidle and Dhakal, 2002). Under increased greenhouse gas concentrations some general circulation models (GCMs) exhibit enhanced mid-latitude and global precipitation intensity and shortened return periods of extreme events (Jones and Reid, 2001; New et al., 2001). The IPCC (2007b: 91) suggests that 'Extremes of daily precipitation are *likely* to increase in many regions. The increase is considered to be *very likely* in northern Europe, south Asia, East Asia, Australia and New Zealand'. There are various reasons to expect increases in extreme precipitation if and when significant warming occurs. There will be more moisture in the atmosphere, and there is likely to be greater thermodynamic instability (Kunkel, 2003).

There is some evidence of increased rainfall events in various countries over recent warming decades, which lends some support to this notion. Examples are known from the USA (Kunkel, 2003), Canada (Francis and Hengeveld, 1998), Australia (Suppiah and Hennessy, 1998), Japan (Iwashima and Yamamoto, 1993), South Africa (Mason et al., 1999), and Europe (Forland et al., 1998). In the UK there has been an upward trend in the heaviest winter rainfall events (Osborn et al., 2000). Heavy precipitation events appear to have been intensified over very large parts of Northern Hemisphere land areas (Min et al., 2011).

Probabilistic analysis of GCMs by Palmer and Räisänen (2002), applied to western Europe and the Asian monsoon region, shows under global warming a clear increase in extreme winter precipitation for the former and for extreme summer precipitation for the latter. Increases in extreme winter precipitation are also likely in the western USA (Dominguez et al., 2012). Increased monsoonal rainfall events are widely predicted (Hsu et al., 2012) and would have potentially grave implications for flooding in places like Bangladesh.

In their analysis of flood records for 29 river basins from high and low latitudes with areas greater than $200,000 \, km^2$, Milly et al. (2002) found that the frequency of great floods had increased substantially during the twentieth century, particularly during its warmer later decades. Their model suggested that this trend would continue.

Changes in tropical cyclones

Tropical cyclones (hurricanes) are highly important geomorphological agents, in addition to being notable natural hazards, and are closely related in their places of origin to sea temperature conditions. They only develop where sea surface temperatures (SSTs) are in excess of 26.5°C. Moreover, their frequency over the last century has changed in response to changes in temperature, and there is even evidence that their frequency was reduced during the Little Ice Age (see Spencer and Douglas, 1985). It is, therefore, possible that as the oceans warm up, so the geographical spread and frequency of hurricanes will increase (Figure 10.1). Furthermore, it is also likely that the intensity of these storms will be magnified (Figure 10.2). Emanuel (1987) used a GCM which predicted that with a doubling of present atmospheric concentrations of CO_2 there will be an increase of 40–50% in the destructive potential of hurricanes. Subsequently, Knutson et al. (1998) and Knutson and Tuleya (1999) simulated hurricane activity for an SST warming of 2.2°C and found that this yielded hurricanes that were more intense by 3–7 m per second for wind speed, an increase of 5–12%.

An increase in hurricane intensity and frequency would have numerous geomorphological consequences in low latitudes, including accentuated river flooding and coastal surges (Mousavi et al., 2011), severe coast erosion, accelerated land erosion and siltation, and the killing of corals (because of freshwater and siltation effects) (De Sylva, 1986).

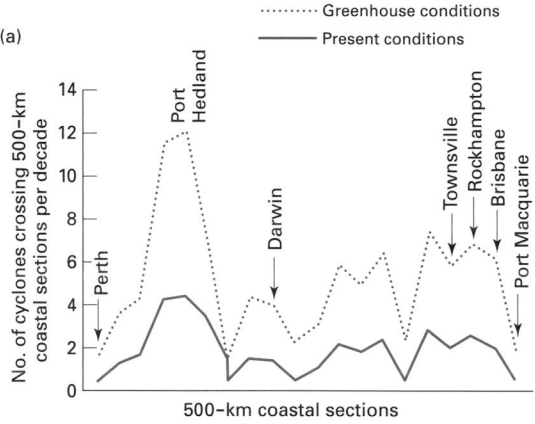

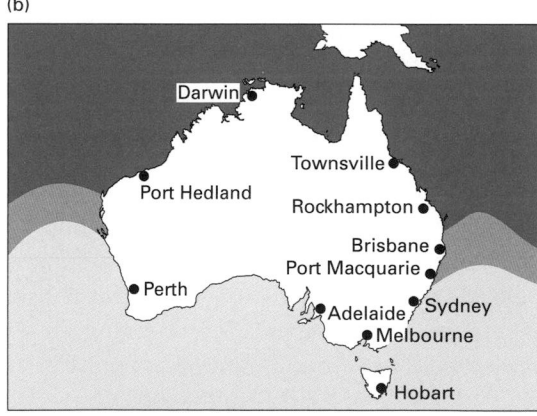

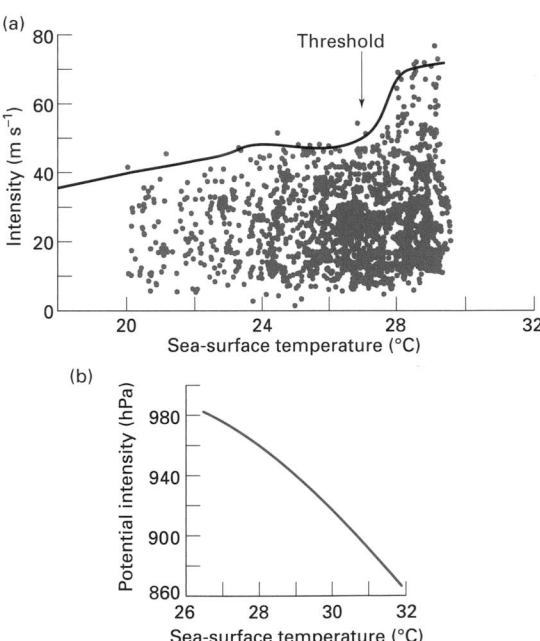

Figure 10.1 (a) The present frequency of cyclones crossing 500-km long sections of the Australian coast, and an estimate of the frequency under conditions with a 2°C rise in temperature; (b) The area where February sea surface temperatures around Australia are currently greater than 27°C (stippled) and the additional area with such temperatures with a 2°C rise in temperature (hatched) (modified from Henderson-Sellers and Blong, 1989, figures 4.12, 4.14). Reproduced with permission from New South Wales University Press.

Figure 10.2 (a) Scatter diagram of monthly mean sea surface temperature and best-track maximum wind speeds (after removing storm motion) for a sample of North Atlantic tropical cyclones. The line indicates the 99th percentile and provides an empirical upper bound on intensity as a function of ocean temperature; (b) The derived relationship between sea surface temperature and potential intensity of tropical cyclones (a and b, modified from Holland et al., 1988, figures 5 and 6. Reproduced with permission).

Figure 10.1 indicates one scenario of the likely latitudinal change in the extent of warm, cyclone-generating seawater in the Australian region, using as a working threshold for cyclone genesis a summer (February) SST of 27°C. Although cyclones do occur to the south of this line, they are considerably more frequent to the north of it. Under greenhouse conditions it is probable that on the margins of the Great Sandy Desert near Port Hedland the number of cyclones crossing the coast will approximately triple from around 4 per decade to 12 per decade (Henderson-Sellers and Blong, 1989).

Although some caution is necessary with regard to the extent to which warming will stimulate tropical cyclone activity (Gray, 1993; Walsh and Pittock, 1998), the relationship between SST increase and increasing global hurricane activity has been confirmed by Hoyos et al. (2006) and Saunders and Lea (2008). Santer et al. (2006) have demonstrated that human factors have caused the increase in SSTs and cyclone development in both the Atlantic and Pacific regions, while Webster et al. (2005) and Elsner et al. (2008) have shown that the strongest tropical cyclones have become more intense in recent decades. There is also the possibility that tropical cyclones, at present relatively unimportant in South America, could become more significant, as presaged by Hurricane Catarina, which hit southern Brazil in March, 2004 (Pezza and Simmonds, 2005). On the other hand, Knutson et al. (2008) argued that there might not be any great increase in hurricane frequency driven by increases in atmospheric greenhouse gas concentrations. Hurricane activity is linked to the El Niño–Southern Oscillation (ENSO) phenomenon, so that if this was to be changed by global warming then so might the frequency, distribution and intensity of hurricanes.

Severe El Niños, like that of 1997/1998, can have a remarkable effect on rainfall amounts. This was shown with particular clarity in the context of Peru (Bendix et al., 2000), where normally dry locations suffered huge storms. At Paita (mean annual rainfall 15 mm), there was 1845 mm of rainfall while at Chulucanas (mean annual rainfall 310 mm), there was 3803 mm. Major floods resulted (Magilligan and Goldstein, 2001). ENSO also affects tropical cyclone activity. As Landsea (2000: 149) remarked, 'Perhaps the most dramatic effect that El Niño has upon the climate system is changing tropical cyclone characteristics around the world.' In some regions, an El Niño phase brings increases in tropical cyclone formation (e.g. the South Pacific and the North Pacific between 140°W and 160°E), while others tend to see decreases (e.g. the North Atlantic, the Northwest Pacific and the Australian region). La Niña phases typically bring opposite conditions. Landsea envisages a variety of reasons why ENSO should relate to cyclone activity: modulation of the intensity of the local monsoon trough, repositioning of the location of the monsoon trough and alteration of the tropospheric vertical shear.

The differences in cyclone frequency between El Niño and La Niña years are considerable (Bove et al., 1998). For example, the probability of at least two hurricanes striking the USA is 28% during El Niño years, 48% during neutral years and 66% during La Niña years. There can be very large differences in hurricane landfalls from decade to decade. In Florida, for instance, over the period 1851–1996, the number of hurricane landfalls ranged from 3 per decade (1860s and 1980s) to 17 per decade (1940s) (Elsner and Kara, 1999). Given the importance of hurricanes for slope, channel and coastal processes, changes of this type of magnitude have considerable geomorphological significance. Mangroves, for example, are highly susceptible to hurricanes, being damaged by high winds and surges (Doyle and Girod, 1997).

The IPCC report (2007b: 15) concludes thus on tropical cyclones:

Based on a range of models, it is *likely* that future tropical cyclones (typhoon and hurricanes) will become more intense, with larger peak wind speeds and more heavy precipitation associated with ongoing increases in tropical sea surface temperatures.

Runoff response

Studies of the sensitivity of runoff to climate changes have tended to indicate that annual runoff volume is more sensitive to changes in precipitation than to changes in potential evapotranspiration and that a given percentage change in precipitation results in a greater percentage change in runoff (Arnell, 1996: 99; Najjar, 1999) with arid catchments showing a greater sensitivity than humid climates. As Table 10.1 shows, an increase in annual precipitation of 10% is enough

Table 10.1 Percentage change in annual runoff with different changes in annual precipitation and an increase in temperature of 2°C. Source: From various sources in Arnell (1996, table 5.1)

River	Location	Percent change in precipitation				
		−20	−10	0	10	20
White	Colorado basin, USA	−23	−14	−4	7	19
East	Colorado basin, USA	−28	−19	−9	1	12
Animas	Colorado basin, USA	−26	−17	−7	3	14
Upper Colorado	Colorado basin, USA		−23	−12	1	
Lower Delaware	USA	−23		−5		−12
Saskatchewan	Canada	−51	−28	−15	11	40
Pease	Texas, USA	−76	−47	−12	40	100
Leaf	Mississippi, USA	−50	−30	−8	−16	40
Nzoia	Kenya	−65	−44	−13	17	70
Mesohora	Greece	−32	−18	−2	11	25
Pyli	Greece	−25	−13	−1	13	25
Jardine	North-East Australia	−32	−17	−1	11	28
Corang	South-East Australia	−38	−21	−3	16	33

to offset the higher evaporation associated with a 2°C rise in temperature. The effects of increasing or decreasing precipitation are greatly amplified in those catchments with the lowest runoff coefficients (Pease, Nzoia and Saskatchewan).

Another key aspect of the runoff response to climate change is that, as historical records show, higher average annual precipitation leads not only to higher stream flow but also to higher flood discharges. In Canada, for example, Ashmore and Church (2001) found that in the Southern Prairies and the Atlantic coast the magnitude of large floods (with a 10-year recurrence interval) increases by up to 50–100% for only 5–15% increases in annual precipitation. Flood discharges increase proportionately much more than mean flows.

An attempt to map future runoff trends on a global basis has been made by Arnell (2002). What is striking in this work is the large range there is in responses. Some areas will become very markedly prone to greatly reduced annual runoff, while others will see an enhancement of flows. The degree of change will vary substantially according to the levels of CO_2 in the atmosphere and the consequent amount of temperature change. However, the patterning at a global scale indicates that by the 2080s, high latitudes in the Northern Hemisphere, together with parts of Central Africa

and Central Asia, will have higher annual runoff levels, whereas Australia, southern Africa, north-west India, the Middle East and the Mediterranean basin will show reduced runoff levels. There is some tendency, to which the Taklamakan of Central Asia appears to be a major exception, for some of the major deserts (e.g. Namib, Kalahari, Australian, Thar, Arabian, Patagonian and North Sahara) to become even drier.

More recently, Sperna Weiland et al. (2011) have analysed available GCMs and suggest that there will be consistent decreases in discharge for southern Europe, southern Australia, parts of Africa and south-western South America, while discharge will increase slightly in some monsoonal areas and in the Arctic. In the USA, models tend to suggest substantial reductions in runoff the south-west and central USA, but increases in the north-east (Karl et al., 2009).

On a global basis it is possible that runoff will increase in a warmer world because of a global increase in precipitation (Douville et al., 2002), and historical discharge records indicate that global runoff increases by *c.* 4% for each 1°C rise in temperature (Labat et al., 2004). Liu et al. (2011) have suggested that in the Yellow River Basin of China, the effects of increased evapotranspiration will be more than offset by increases in precipitation so that by 2080 that river's annual discharge could increase by 35–43%.

Table 10.2 Potential hydrological changes in permafrost regions. Source: A.S.G., from information in Woo (1996)

Increased precipitation

Reduced proportion of snow, increased proportion of rain

Earlier melting of snowpacks

More groundwater storage in deeper active layers

Increased summer evaporative losses due to higher temperatures or replacement of lichens and mosses by transpiring plants

Less overland flow across impermeable permafrost

Degradation of permafrost will add water during the transient phase of climatic change

Changes in timing, extent and duration of ice jams

Increase in surface ponding due to thermokarst development

Subterranean flow conduits may reopen

Cold regions

There are many factors to be considered in an analysis of the response of cold region hydrological systems to climate change (Table 10.2), and good reviews are provided by Woo (1996) and Rouse et al. (1997). River channel forms are also likely to be transformed (Anisimov et al., 2008).

In the Arctic, overall, the amount of moisture in the atmosphere is expected to increase as the atmosphere warms, leading to a general increase in precipitation and in river discharges. Peterson et al. (2002) analysed discharge records for the six largest Eurasian rivers that flow into the Arctic Ocean and have shown that as surface air temperatures have increased, so has the average annual discharge of freshwater. It grew by 7% from 1936 to 1999. They suggested that with increased levels of warming (1.5–5.8° by 2100) there would be an 18–70% augmentation in Eurasian Arctic river discharge over present conditions. Wetherald and Manabe (2002) suggested that by the middle of this century the runoff from such major rivers as the Mackenzie and Ob' could be increased by more than 20%. This analysis was largely confirmed by McClelland et al. (2004) and Wu et al. (2005), who believed that the main driver of the larger river discharge in the area is the increasing northward transport of moisture as a result off global warming. Prowse and Beltaos (2002) and Beltaos and Burrell (2003) explored the effects that climate change could have on ice jams and thus on the size and seasonality of floods and low flows, with the occurrence of ice break-up occurring earlier in the year, as appears

to have been the case in the last several decades (Yoo and D'Odorico, 2002).

A particularly interesting analysis of recent temperature and stream flow trends in the permafrost region of north-east China was provided by Liu et al. (2003). Warming has changed the depth of the active layer, enabling enhanced transport of subsurface water through unfrozen soils and a greater contribution of this flow to winter base flow. Stream flow data for the period 1958 1998 indicated significantly greater runoff, with flow during February and March in the 1990s increasing by 80–100% over prior values.

The amount of snowpack accumulation is another major control on hydrological conditions in cold regions (Nijssen et al., 2001; Barnett et al., 2005), and winter snow accumulation in alpine watersheds provides most of the stream flow runoff in western North America and similar regions of the world. Snow cover in the Northern Hemisphere has indeed shown a decline during recent warming decades (Stewart, 2009). Following a global analysis of future hydrological trends, Nijssen et al. (2001) suggested that the largest changes in the hydrological cycle are predicted for the snow-dominated basins of mid to higher latitudes, and in particular there are likely to be marked changes in the amplitude and phase of the annual water cycle (Arora and Boer, 2001). Lapp et al. (2005) have modelled the likely response of snowpack accumulation to global warming in the Canadian Rockies, and have suggested that (1) there will be a substantial decline in the over-winter snow accumulation in most years, and (2) that spring runoff volumes may substantially decrease with the decline in snowpacks. This is confirmed by the work of Mote et al. (2005), who provide a historical analysis of changes in snowpack condition in recent warming decades, and who indicate that the losses in snowpack to date will continue and even accelerate, with faster losses in milder climates like the Cascades and the slowest losses in the high peaks of the Rockies and southern Sierra. Tree-ring chronologies for the North American cordillera have shown that the late twentieth century snowpack reductions have been greater than at any time in the last millennium (Pederson et al., 2011). However, it needs to be remembered that snowpack amounts will not only be affected by rising temperatures. Changes in precipitation totals may also be significant and in some areas may work in the same or opposite direction as warming (Adam et al., 2009; Stewart, 2009).

The other major control on future river flow in cold regions will be the melting of glaciers (see Chapter 11). If this occurs, discharges may initially increase, but as the glacier mass declines through time, so will stream flows (Bradley et al., 2006; Rees and Collins, 2006; Immer Zeel et al., 2010). Barnett et al. (2005: 306) have remarked with respect to the Himalayas and Hindu Kush that 'It appears that some areas of the most populated region on Earth are likely to "run out of water" during the dry season if the current warming and glacial melting trends continue for several more decades'.

Changes in runoff in the UK

Arnell (1996) has studied the likely changes that will occur in the UK as a result of global warming. With higher winter rainfalls and lower summer rainfalls (particularly in the south-east) he forecasts for the 2050s:

1 An increase in average annual runoff in the north of Britain of between 5 and 15%
2 A decrease in the south of between 5 and 15%, but up to 25% in the south-east
3 An increased seasonal variation in flow, with proportionately more of the total runoff occurring during winter
4 High flows increased in northern catchments and decreased in the south.

Figure 10.3 shows the change in monthly stream flow under two climatic scenarios for six British catchments. As can be seen, in southern England, lower summer rainfall, coupled with increased evaporation, means that stream flows decrease during summer, whereas in the catchments in northern Britain, stream flow increases in winter, but changes little in summer. A more recent analysis by Prudhomme et al. (2012) confirms that there are likely to be decreases in summer flows, especially in the south and east, but indicates that there remains uncertainty as to whether winter flows will increase or decrease.

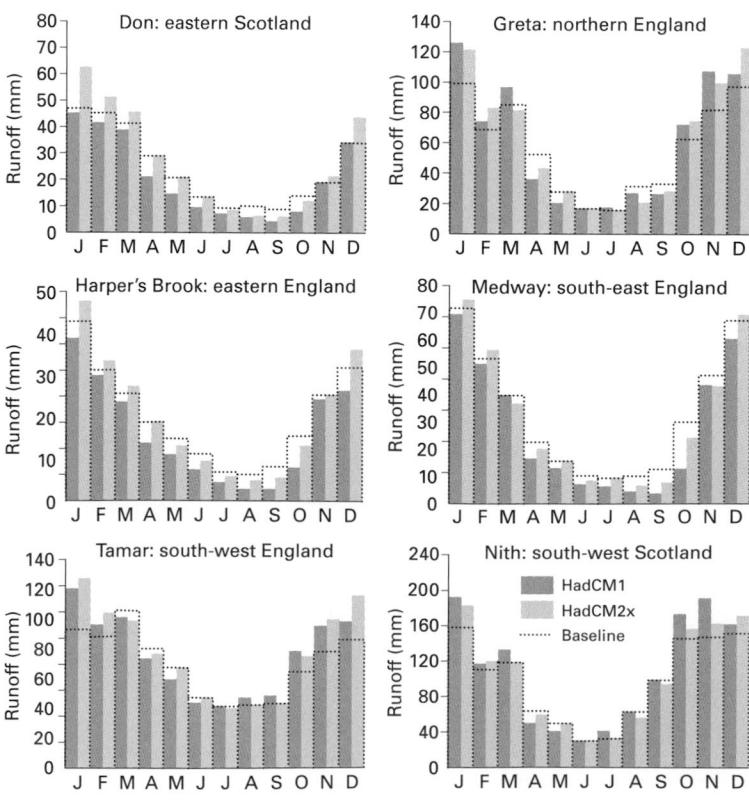

Figure 10.3 Monthly runoff by the 2050s under two scenarios for six British catchments (modified from Arnell and Reynard, 2000, figure 7.19).

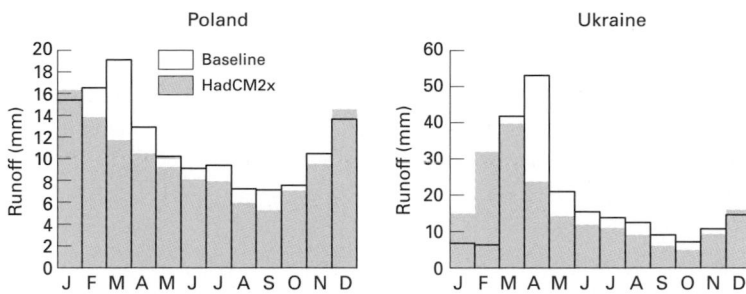

Figure 10.4 Effect of a climate change scenario on stream flow in two European snow-affected catchments by 2050s (modified from Arnell, 2002, figure 7.20).

Werritty (2002) looked specifically at Scottish catchments and noted a particularly marked increase in precipitation in the north and west in the winter half of the year. He predicted that by the 2050s Scotland as a whole will become wetter than at present and that average river flows will increase, notably in the autumn and winter months. He believed that high flows could become more frequent, increasing the likelihood of valley floor inundation. However, a study of the history of valley floors in upland Scotland suggests that they are relatively robust, so that although they may become subject to extensive reworking, they are unlikely to undergo large-scale destabilization (Werritty and Lees, 2001).

Europe

Arnell (1999b) modelled potential changes in hydrological regimes for Europe, using four different GCM-based climate scenarios. While there are differences between the four scenarios, each indicates a general reduction in annual runoff in Europe south of around 50°N and an increase polewards of that. The decreases could be as great as 50% and the increases up to 25%. The proposed decrease in annual runoff in southern Europe is also confirmed by the work of Menzel and Burger (2002) in Germany, who also suggest that peak flows will be very substantially reduced.

Rising temperatures will affect snowfall and when it melts (Seidel et al., 1998). Under relatively mild conditions, even a modest temperature rise might mean that snow becomes virtually unknown, so that the spring snowmelt peak would be eliminated. It would be replaced by higher flows during the winter. Under more extreme conditions, all winter precipitation would still fall as snow, even with a rise in tempera-

ture. As a consequence, the snowmelt peak would still occur, although it might occur earlier in the year. In Figure 10.4, for instance, in the Polish example the snowmelt peak is eliminated by the 2050s, whereas in the Ukraine example it is brought forward.

One of the most important rivers in Europe is the Rhine. It stretches from the Swiss Alps to the Dutch coast and its catchment covers 185,000 km^2 (Shabalova et al., 2003). Models suggest that the Rhine's discharge will become markedly more seasonal with mean discharge decreases of about 30% in summer and increases by about 30% in winter, by the end of the century. The increase in winter discharge will be caused by a combination of increased precipitation, reduced snow storage and increased early melt. The decrease in the summer discharge is related mainly to a predicted decrease in precipitation combined with increases in evapotranspiration. Glacier melting in the Alps also contributes to the flow of the Rhine. Once these glaciers have wasted away, this contribution will diminish sharply.

In the Mediterranean region a number of processes operating in tandem may lead to a severe reduction in available water resources (García-Ruiz et al., 2011). A general decrease in precipitation and a rise in evapotranspiration rates will lead to flow reductions, as will the presence of dams and reservoirs, overexploitation of groundwater, and an expansion of the area that is forested following on from farm abandonment in upland regions.

Other examples

1 *Zambezi*

A study of the Zambezi River in central Africa, using different GCMs, and projecting conditions for 2080

Figure 10.5 The Zambezi River at Victoria Falls. (See Plate 50)

Table 10.3 Predicted change in climate variables for different GCM scenarios

(a) Climate change scenarios for California for the 2080s (relative to 1961–1990 mean)

	HADCM2	HADCM2-S	ECHAM4
Precipitation change (%)	−12.5	−17.6	−1.6
Temperature change (°C)	+5.3	+4.4	+5.0

(b) Seasonal changes in runoff for California under GCM scenarios

GCM scenario	Runoff change (%)		
	Annual	Wet season (January–July)	Dry season (August–December)
ECHAM4	−10.0	−9.5	−12.1
HadCM2	−28.3	−28.2	−28.9
HadCM2-S	−35.5	−36.1	−32.6

(c) Estimated change in runoff in California (from Smith et al. 2001a, table 1) using Canadian and Hadley GCMs

Historical runoff (mm/year), 1961–1990	Change in annual runoff (mm/year), 2025–2034		Change in annual runoff (mm/year), 2090–2099	
	Canadian	Hadley	Canadian	Hadley
232	60	63	320	273

NB: Hadley model for California proposes for 2090s a 4.9°C increase in winter temperatures and a 4.6°C increase in summer temperatures. Annual precipitation will increase by 30%. The Canadian model equivalents are 7.1°C and 4.3°C and c. 70%.

indicates that river flow may decline substantially. Simulations indicate that for three scenarios annual flow levels at Victoria Falls (Figure 10.5) reduce between 10 and 35.5% (Harrison and Whittington, 2002).

2 *Susquehanna, eastern USA, and New England*
The Susquehanna which flows into Chesapeake Bay, the largest estuary in USA, is an example of a river system which will enjoy larger annual flows because of increased precipitation over its catchment (Najjar, 1999). For a 17.5% increase in annual precipitation, and a temperature increase of 2.5°C, the total predicated increase in annual stream flow is 24%. This contrasts with the prediction of future flows for rivers in New England, where Huntington (2003) has suggested that annual stream flow would be reduced by 11–3%.

3 *California and the Colorado River*
One of the most pronounced features of some recent GCMs (Hadley Centre and Canadian) is that they show a projected increase in precipitation for California and the south-west. This would be the result of a warmer Pacific Ocean causing an increase mainly in wintertime precipitation (MacCracken et al., 2001). Smith et al. (2001a), using the Hadley and Canadian models, estimated that California runoff will increase by the 2030s by about three-fifths and double by the 2090s (Table 10.3). On the other hand, Maurer and Duffy (2005) believed that there will be decreases in summer flows and increases in winter flows, and a shift to flow earlier in the year. Barnett

and Pierce (2009) have suggested that annual runoff in the Colorado River will be reduced by 10–30%.

4 *Pacific Northwest of USA*
Wigmosta and Leung (2002) modelled the response of the American River in the Pacific Northwest of the USA. More winter precipitation falling as rain rather than snow, and also leading to more rain-on-snow events, produced a future with greater winter flooding. However, the reduced snowpack caused less flows in the spring and summer.

5 *Bangladesh*
In a warmer world, it is probable that there will be a general increase in precipitation, caused by enhancement in summer monsoon activity. This could be a cause of increased flood risk in areas like

Table 10.4 Changes in precipitation under various warming scenarios and corresponding mean peak discharge for three south Asian rivers. Source: Mirza (2002), with modifications

GCM	ΔT (°C)	Ganges			Brahmaputra			Meghna		
		ΔP (%)	Mean peak discharge (m³/s)	[a]	ΔP (%)	Mean peak discharge (m³/s)	[a]	ΔP (%)	Mean peak discharge (m³/s)	[a]
CSIRO9	2	8.5	57,790	107	−0.5	64,853	97	4.5	15,171	108
	4	17.0	62,900	116	−1.0	64,840	97	9.0	16,267	116
	6	25.5	68,010	126	−1.5	64,827	97	13.5	17,378	124
HADCM2	2	−2.8	50,963	94	7.2	70,308	105	9.7	16,017	114
	4	−5.6	49,240	91	14.4	75,757	113	19.4	17,971	128
	6	−8.4	47,523	88	21.6	81,199	121	29.1	19,927	142
GFDL	2	4.6	55,419	103	7.2	67,487	101	11.3	16,844	120
	4	9.2	58,159	108	14.4	70,107	105	22.6	19,614	140
	6	13.8	60,898	113	21.6	72,728	109	33.9	22,412	160
LLNL	2	0.5	52,996	98	1.4	65,385	98	7.7	15,958	114
	4	1.0	53,312	99	2.8	65,904	98	15.4	17,842	127
	6	1.5	53,628	99	4.2	66,423	99	23.1	19,740	141

[a] = Percentage of present peak discharge value.

Bangladesh. In Table 10.4, from the work of Mirza (2002), four GCM scenarios are presented for changes in temperature and precipitation for three major catchments that create floods in Bangladesh. In addition, predicted mean peak discharges for those scenarios are presented. The current peak discharges (cubic metres per second) are 54,000 for the Ganges, 67,000 for the Brahmaputra and 14,000 for the Meghna. The four GCMs display some differences, as do predicted peak discharge values. However, overall, the discharge values show a range from a modest decline to a substantial increase.

Geomorphological consequences of hydrological and other changes

It is likely that changes in river flow will cause changes in river morphology, particularly in sensitive systems, which include fine-grained alluvial streams. Bedrock streams will probably be less sensitive. Ashmore and Church (2001: 41) summarized some of the potential effects:

The potential impacts of increased discharge include channel enlargement and incision, a tendency toward either higher sinuosity single channels or braided patterns, increased bank erosion, and more rapid channel migration. Increased magnitude of large floods will result in sudden changes to channel characteristics that may trigger greater long-term instability of rivers. Increased frequency of large floods will tend to keep rivers in the modified and unstable state. Decreased discharge often results in channel shrinkage, vegetation encroachment into the channel, sedimentation in side channels, and channel pattern change toward more stable, single-channel patterns. In entrenched or confined valleys there may be reductions in the stability of the valley walls and, hence, increases in the rate of erosion caused by a greater tendency for streams to erode the valley walls. Increased valley-side erosion will increase sediment delivery to the streams with consequences for stream morphology.

One possible response of river basins to changes in runoff is that the channel length per unit area (i.e. drainage density) will change. After a consideration of the relationship between mean annual rainfall and drainage density in Africa, De Wit and Stankiewicz (2006) proposed that in areas with 500 mm per year, a 10% decrease in precipitation could reduce drainage density by as much as 50%.

Changes in river flow and direct inputs of water from precipitation may affect future lake levels (see Chapter 12), including the Great Lakes of North

America, which are likely to show modest falls (MacKay and Seglenieks, 2011).

Slope stability is another aspect of geomorphology which will be impacted upon by climate change, and Crozier (2010) provides a good general analysis of the factors that may alter future landslide activity (see also Table 8.8). Mass movements on slopes will also be impacted upon by changes in hydrological conditions. In particular, slope stability and landslide activity are greatly influenced by groundwater levels and pore-water pressure fluctuation in slopes.

In the south-east of England, Collison et al. (2000) have also sought to model the impact of climate change on the Lower Greensand escarpment of Kent. Using the Hadley Centre GCM (HADCMZ), combined with downscaling and a GIS-based combined slope hydrology/stability model, they found that because increases in rainfall would largely be matched by increases in evapotranspiration, the frequency of large landslides would be unchanged over the next 80 years. They argued that other factors, such as land-use change and human activity would be likely to have a greater impact than climate change.

Various other attempts have been made to model potential slope responses. For example, in the Italian Dolomites, Dehn et al. (2000) suggested that future landslide activity would be reduced because there would be less storage of precipitation as snow. Therefore, the release of meltwater, which under present conditions contributes to high groundwater levels and strong landslide displacement in early spring, would be significantly diminished. However, because of the differences between GCMs and problems of downscaling, there are still great problems in modelling future landslide activity, and Dehn and Buma (1999) found that the use of three different GCM experiments for the assessment of the activity of a small landslide in south-east France did not show a consistent picture of future landslide frequencies. In the Swiss Alps, there is some evidence that debris flow activity has declined since the Little Ice Age, and if summer precipitation events occur less frequently in the future, their incidence could decline still further (Stoffel and Beniston, 2006). Comparable predictions have been made for the French Alps (Jomelli et al., 2009), where it is proposed that debris flow frequency could be reduced by 30% in the late twenty-first century. On the other hand, analysis of mass movement failures in high mountains during

recent warm decades has provided no statistically rigorous evidence of a trend in activity in recent decades (Huggel et al., 2012).

The peatlands of cold temperate regions are another landscape type that may be impacted by global change, though uncertainties exist about how they will respond (Moore, 2002). On the one hand, higher temperatures will tend to enhance plant productivity and organic matter accumulation but on the other will also lead to faster decomposition rates. Equally any decrease in water availability in the summer could cause peat loss and bog contraction.

Geomorphologists are starting to model changes in soil erosion that may occur as a consequence of changes in rainfall amounts and intensity (Yang et al., 2003; Mullan et al., 2012), though it is difficult to determine the likely effects of climate change compared to future land-use management practices (Wilby et al., 1997). For example, Sun et al. (2002) calculated runoff erosivity changes for China and Nearing (2001) for the USA using the Revised Soil Loss Equation (RSLE) of Renard and Freid (1994). They adopted the UKMO Hadley climate scenario for their China study and produced a map of rainfall erosivity for 2061–2099 and suggested that for China as a whole, assuming current land cover and land management conditions, soil erosion rates will increase by 37–93% across China. For Brazil, Favis-Mortlock and Guerra (1999) used the Hadley Centre HADCM2 GCM and erosion model (Water Erosion Predictions Project or WEPP). They found that by 2050 the increase in mean annual sediment yield in their area in the Mato Grosso would be 27%. For the southeast of the UK, where winter rainfall is predicted to increase, albeit modestly, Favis-Mortlock and Boardman (1995) recognized that changes in rainfall not only impacted upon erosion rates directly, but also through their effects on rates of crop growth and on soil properties. Nonetheless, they showed that erosion rates were likely to rise, particularly in wet years.

In the USA, after analyzing a range of different models, Nearing et al. (2005: 151) remarked:

If rainfall amounts during the erosive times of the year were to increase roughly as they did during the last century in the United States, the increase in rainfall would be on the order of 10%, with greater than 50% of that increase due to increase in storm intensity. If these numbers are correct, and if no changes in land cover occurred, erosion could increase by something on the order of 25–55% over the next century . . . .

Both storm water runoff and soil erosion are likely to increase significantly under climate change unless offsetting amelioration measures are taken.

Sediment delivery by rivers may also be impacted by climate change. It has been suggested for example, that fire activity may increase in a warmer and drier environment, and that this could cause an increase in suspended loads as slopes are subjected to greater erosion and debris flow generation (Goode et al., 2012).

It is, of course, likely, that climate change will cause farmers to change the ways in which they manage their crops and, in particular, to change the sorts of crops they plant in any particular area. This in turn may alter runoff and erosion. O'Neal et al. (2005) suggested, for example, that in the Midwest of the USA there might be a shift, because of price and yield advantages, from maize and wheat to soybean cultivation. Rates of runoff and erosion under soybeans would tend to be substantially greater than under the crops they replace.

There has also been interest recently in the role that global warming may play in accentuating or triggering various geohazards (McGuire, 2010). For example, the accelerated melting of submarine permafrost and the release of gas hydrates might promote submarine slope failure with consequent tsunami formation (Day and Maslin, 2010). Equally, changes in the extent of ice sheets would alter the amount of loading on the Earth's crust and so might affect earthquake and volcanic activity.

Weathering

Remarkably little work has been done on how weathering rates of rock will change with changes in climate. Various attempts, however, deserve mention. The first of these is the study by Viles (2002) on stone deterioration in the UK. There she contrasted the likely impacts of climate change on building stones between the north-west and the south-east of the country (Table 10.5). She argued that in the north-west, which would have warmer and wetter winters, chemical weathering's importance might be increased, whereas in the south-east, which would have warmer, drier summers, processes like salt weathering might become more important. Viles (2003) has also tried to model the impacts of climate change on limestone weathering and karstic processes in the UK (Figure 10.6), but the model has wider significance and could be applied to karst areas worldwide. Smith et al. (2010) suggested that higher winter rainfall amounts in the UK could cause the 'greening' of buildings by algal films, while deeper penetration of moisture might affect the operation of processes like salt weathering. Finally, as the climate warms, the number of freeze–thaw cycles to produce frost weathering in areas like Central England

Table 10.5 Possible consequences of future climate change for building stone decay in the UK. Source: Viles (2002, table 3).

	Dominant building stone type	Process responses	Other threats	Overall response
North-west – warmer, wetter winters	Siliceous sandstones and granites	Increased chemical weathering. Less freeze–thaw weathering. More organic growths contributing to soiling.	Increased storm activity may cause episodic damage. Increased wave heights may encourage faster weathering in coastal areas. Increased flooding may encourage decay.	Enhanced chemical decay processes and biological soiling, reduced physical decay processes.
South-east – warmer, drier summers	Limestones	Less freeze–thaw weathering. Reduced chemical weathering as a result of less available water. Increased salt crystallization in summer. More deteriorating organic growths.	Increased drying of soils (especially clay-rich soils) will encourage subsidence and building damage. Low-lying coastal areas will be particularly prone to marine encroachment. Increased drought frequency may encourage decay.	Enhanced physical and biological weathering, more dust for soiling, reduced chemical weathering.

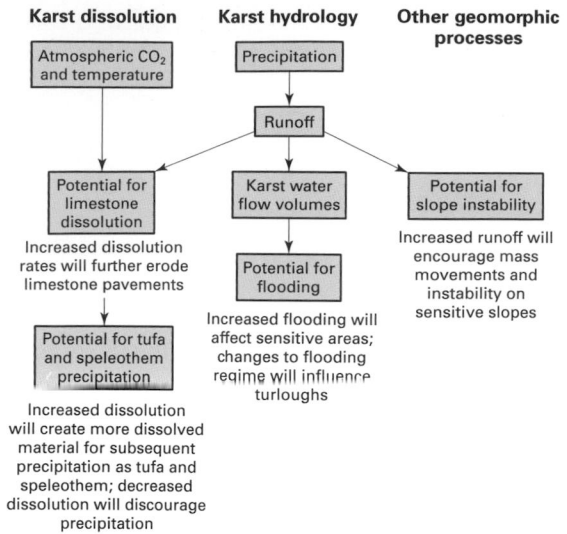

Figure 10.6 Conceptual model of the impacts of effective precipitation, air temperature and atmospheric CO_2 concentration on karst dissolution, hydrology and other geomorphic processes (Source: Viles, 2003, figure 1). Reproduced with permission from Elsevier.

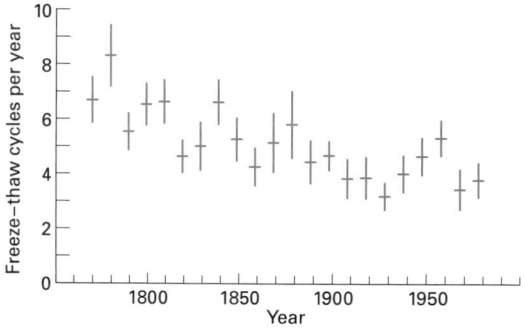

Figure 10.7 Decadal means of the freeze–thaw cycles in Central England (Source: Brimblecombe and Camuffo, 2003, figure 3). Reproduced with permission from Imperial College.

are likely to decrease, as they did during the warming of the nineteenth and twentieth centuries (Figure 10.7) (Brimblecombe and Camuffo, 2003). The possibility that salt weathering activity will change in drylands is a topic that is discussed in Chapter 12.

Points for review

Will rainfall amounts and intensities change in a warming world? Where may these be greatest?

What are the likely ways in which rivers in (1) periglacial and (2) arid environments respond to global warming?

To what extent may rates of soil erosion, mass movements and rock weathering change in coming decades?

Guide to reading

Arnell, N.W. (2002) *Hydrology and Global Environmental Change*. Harlow: Prentice Hall. A discussion of the responses of river systems to climate change.

Collier, M. and Webb, R.H. (2002) *Floods, Droughts and Climate Change*. Tucson: University of Arizona Press. With a strong USA focus, this accessible book discusses the relationship between climate change and the incidence of floods and droughts.

Sidle, R.C. (ed.) (2002) *Environmental Change and Geomorphic Hazards in Forests*. Wallingford: CABI. A consideration of the ways in which mass movements and other hazards may change in the future.

11 THE FUTURE: THE CRYOSPHERE

Chapter Overview

The cryosphere consists of ice sheets, ice caps, glaciers, sea, river and lake ice, and frozen ground (permafrost). Because of its very nature, ice in its various forms will be susceptible to the effects of future warming. This chapter examines what may happen to the polar ice sheets and their associated ice shelves, the glaciers of mountainous regions, the sea ice of the Arctic and Antarctic, and the great spreads of permafrost that occupy around a quarter of Earth's land surface.

The nature of the cryosphere

The cryosphere is that part of the Earth's surface or subsurface environment that is composed of water in the solid state. It includes snow, sea ice, the polar ice sheets, mountain glaciers, river and lake ice, and permafrost (permanently frozen subsoil).

The cryosphere contains nearly 80% of all Earth's freshwater. Perennial ice covers about 11% of the earth's land surface and 7% of the world's oceans, while permafrost underlies about 25% of it. Seasonal snow has the largest area of any component of the global land surface; at its maximum in late winter it covers almost 50% of the land surface of the Northern Hemisphere.

Good background studies on the cryosphere include Knight (1999), Benn and Evans (2010) on glaciation and French (1996) on permafrost and periglacial environments.

The Human Impact on the Natural Environment: Past, Present and Future, Seventh Edition. Andrew S. Goudie.
© 2013 John Wiley & Sons, Ltd. Published 2013 by John Wiley & Sons, Ltd.

Because of the obvious role of temperature change in controlling the change of state of water to and from the liquid and solid sates, global warming has the potential to cause very major changes in the state of the cryosphere.

The polar ice sheets and ice caps

Ice sheets are ice masses that cover more than 50,000 km². The Antarctic ice sheet covers a continent that is a third bigger than Europe or Canada and twice as big as Australia. It attains a thickness that can be greater than 4000 m, thereby inundating entire mountain ranges. The Greenland ice sheet only contains 8% of the world's freshwater ice (Antarctica has 91%), but

nevertheless covers an area 10 times that of the British Isles. The Greenland ice fills a huge basin that is rimmed by ranges of mountains and has depressed the Earth's crust beneath.

The Antarctic ice sheet (Figure 11.1) is bounded over almost half of its extent by ice shelves. These are floating ice sheets nourished by the seaward extensions of the land-based glaciers or ice streams and by the accumulation of snow on their upper surfaces. Ice-shelf thicknesses vary, and the seaward edge may be in the form of an ice cliff up to 50 m above sea level with 100–600 m below. At its landward edge the Ross Ice Shelf is 1000 m thick. It covers an area greater than that of California.

Ice caps have areas that are less than 50,000 km² but still bury the landscape. The world's alpine or moun-

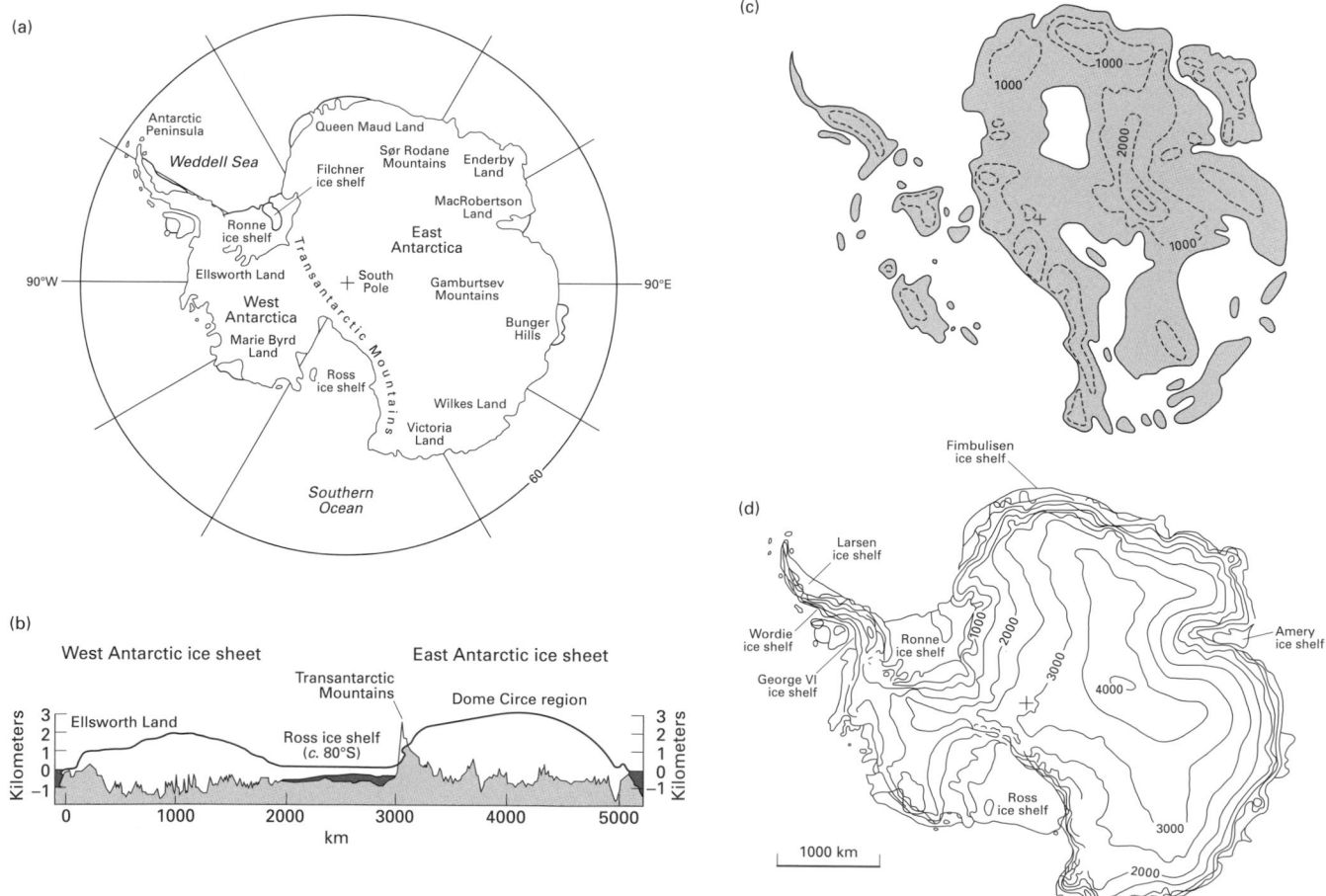

Figure 11.1 The Antarctic ice sheet and shelves. (a) Location map of Antarctica. (b) Cross-section through the East and West Antarctic ice sheets, showing the irregular nature of the bedrock surface, ice thickness, and the floating ice shelves. (c) Subglacial relief and sea level. The white areas are below sea level. (d) Surface elevations on the ice sheet in metres.

tain glaciers are numerous, and in all there are probably over 160,000 on the face of the earth. Their total surface area is around 530,000 km².

Three main consequences of warming may be discerned for ice sheets (Drewry, 1991): ice temperature rise and attendant ice flow changes; enhanced basal melting beneath ice shelves and related dynamical response; and changes in mass balance.

Temperatures of the ice sheets will rise due to the transfer of heat from the atmosphere above. However, because of the slow vertical conduction of heat through the thick ice column, the timescale for this process is relatively long (10^2–10^3 years). As the ice warms, it softens and can undergo enhanced deformation, which could increase the discharge of ice into the oceans. However, Drewry argues that it can be discounted as a major factor on a timescale of decades to a century.

Ice shelves would have enhanced basal melt rates if sea surface temperatures were to rise. This could lead to thinning and weakening of ice shelves (Warner and Budd, 1990). Combined with reduced underpinning from grounding points as sea level rose, this would result in a reduction of backpressure on ice flowing from inland. Ice discharge through ice streams, the arteries of the ice sheets (Bennett, 2003), might therefore increase. There are some studies (e.g. De Angelis and Skvarca, 2003) that have indeed found evidence of ice streams increasing their velocities when ice shelves have collapsed, but this is not invariably the case (Vaughan and Doake, 1996). Nonetheless, following the disintegration of the Larsen B ice shelf in 2002, the glaciers that had previously been sheltered behind the ice shelf became tidewater glaciers and then flow rates increased greatly (Rott et al., 2010). Indeed, Cook and Vaughan (2010) have demonstrated the loss or substantial retreat of ice shelves around the Antarctic Peninsula in recent decades and the profound acceleration and thinning of ice streams that has occurred. The West Antarctic Ice Sheet (WAIS), because it is a marine ice sheet that rests on a bed well below sea level, is likely to be much more unstable than the East Antarctic Ice Sheet, from which it is separated by the Transantarctic Mountains. If the entire WAIS were discharged into the ocean, the sea level would rise by 5 or 6 m. Fears have been expressed over many years that it could be subject to rapid collapse, with disastrous consequences for coastal regions all over the world.

One extreme view was postulated by Mercer (1978), who believed that predicted increases in temperature at 80°S would start a 'catastrophic' deglaciation of the area, leading to a sudden 5-m rise in sea level. A less extreme view was put forward by Thomas et al. (1979). They recognized that higher temperatures will weaken the ice sheets by thinning them, enhancing lines of weakness, and promoting calving, but they contended that deglaciation would be rapid rather than catastrophic, the whole process taking 400 years or so.

The WAIS has part of its ice grounded on land below sea level, and part in the form of floating extensions called ice shelves that move seaward but are confined horizontally by the rocky coast. The boundary between grounded and floating ice is called the grounding line (Oppenheimer, 1998; Ross et al., 2012). In 1974, Weertman suggested that very rapid grounding line retreat could take place in the absence of ice shelf backpressure. He demonstrated (see Figure 11.2) that if the bedrock slopes downward in the inland direction away from the grounding line, as is the case with some of the WAIS ice streams, then if an ice shelf providing buttressing becomes unpinned owing to either melting or global sea-level rise, the grounded ice would accelerate its flow, become thinner and rapidly float off its bed.

Whether the WAIS will collapse catastrophically is still being debated (Oppenheimer, 1998; Joughlin and Alley, 2011), but there is some consensus that the WAIS will most probably not collapse in the next few centuries (Vaughan and Spouge, 2002). Bentley (1997) is amongst those who have expressed doubts about the imminence of collapse, believing 'that a rapid rise in sea level in the next century or two from a West Antarctic cause could only occur if a natural (not induced) collapse of the WAIS is imminent, the chances of which, based on the concept of a randomly timed collapse on the average of once every 100,000 years, are on the order of 0.1%' (see also Bentley, 1998).

Greenland and Antarctica will react very differently to global warming. This is in part because Antarctica, particularly in its interior, is very cold indeed. This means that the moisture content of the atmosphere over it is very low, ultimately suppressing the amount of snow and ice accumulation that can occur. An increase in temperature may therefore lead to more precipitation and greater accumulation of ice. With Greenland, the situation is different. Whereas in Figure

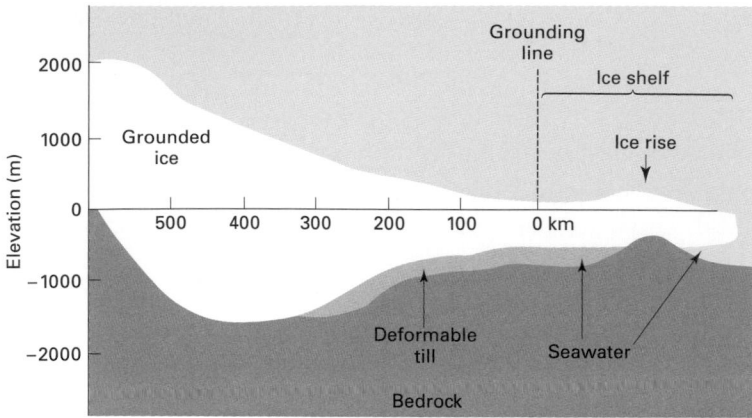

Figure 11.2 Cross-section of an ice stream and ice shelf of a marine ice sheet, indicating location of grounding line, bedrock rise on the ocean floor, and possible extent of deformable till. The thickness of the till layer, actually a few metres, is exaggerated for clarity. Sea-level rise due to collapse of WAIS was estimated by T. Hughes (after Oppenheimer, 1998, figure 2) reproduced with permission, p. 327 in *Nature*, volume 393, 28 May 1998 © Macmillan Publishers Ltd 1998.

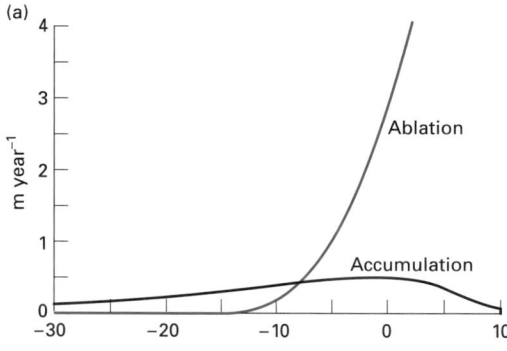

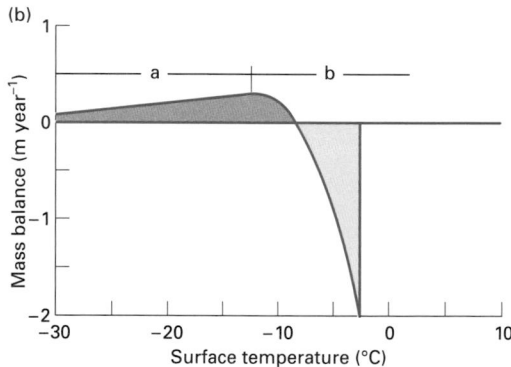

Figure 11.3 (a and b) Generalized curves of ablation and accumulation in dependence of (surface) temperature. Depending on which part of the *M-curve* applies, a climatic warming may lead to increasing (Antarctic) or decreasing (Greenland) mass balance. (after Oerlemans, 1993, figure 9.4, reproduced with permission from Springer).

11.3 Antarctica lies to the left of the mass balance maximum, much of Greenland lies to the right of the maximum so that the total surface mass balance will decrease in the event of global warming (Oerlemans, 1993: 153). Indeed, Gregory et al. (2004a) have argued

that if the average temperature in Greenland increases by more than about 3°C, the ice sheet would be eliminated, causing the global average sea level to rise by 7 m over a period of 1000 years or more.

However, the state of the Greenland ice sheet does not solely depend on the rate of surface ablation. Flow of the ice is enhanced by rapid migration of surface meltwater to the ice bedrock interface, and studies have shown that ice accelerates in summer (when melt occurs), that there is a near coincidence of ice acceleration with the duration of surface melting, and that inter-annual variations in ice acceleration are correlated with variations in the intensity of surface melting. This coupling between surface melting and ice sheet flow provides a mechanism for rapid response of ice sheets to warming (Zwally et al., 2002; Parizek and Alley, 2004). Some confirmation of this is provided by laser altimeter surveys, which show rapid ice thinning below 1500 m (Krabill et al., 1999; Rignot and Thomas, 2002; Rignot and Kanagaratnam, 2006). They argue that the glaciers are thinning as a result of increased creep rates brought about by decreased basal friction consequent upon water penetrating to the bed of the glacier. They see this as a mechanism for transfer of ice sheet mass to the oceans that is potentially larger than could be achieved by surface melting alone. The overall loss of mass of the Greenland ice sheet has accelerated in recent years (Thomas et al., 2006; van den Broeke et al., 2009).

Valley glaciers and small ice caps

Together the present Greenland and Antarctic ice sheets contain enough water to raise sea level by almost 70 m.

By contrast, valley glaciers and small ice caps contain only a small amount – *c*. 0.5 m of sea-level equivalent. However, because of their relatively short response times, associated with large mass turnover, they contribute significantly to sea-level fluctuations on a century timescale. Indeed, there is a clear observational record that valley glaciers are sensitive to climate change, and we have many examples of changes in glacier length of several kilometres since the end of the **Little Ice Age** (LIA) (Oerlemans et al., 1998).

Since the nineteenth century, many of the world's alpine glaciers have retreated up their valleys as a consequence of the climatic changes, especially warming, that have occurred in the last hundred or so years since the ending of 'The Little Ice Age' (Oerlemans, 1994). Globally, the World Glacier Monitoring Service (WGMS) (2008) suggests that the average annual mass loss of glaciers during the decade from 1996 to 2005 was twice that of the previous decade (1986–1995) and over four times that of the decade from 1976 to 1985. Studies of the changes of snout positions obtained from cartographic, photogrammetric and other data therefore permit estimates to be made of the rate at which retreat can occur. The rate has not been constant, or the process uninterrupted. Indeed, some glaciers have shown a tendency to advance for some of the period. However, if one takes those glaciers that have shown a tendency for a fairly general retreat (Table 11.1) it becomes evident that as with most geomorphological phenomena, there is a wide range of values, the variability of which is probably related to such variables as topography, slope, size, altitude, accumulation rate and ablation rate. It is also evident, however, that rates of retreat can often be very high, being of the order of 20–70 m per year over extended periods of some decades in the case of the more active examples. It is therefore not unusual to find that over the last hundred or so years, alpine glaciers in many areas have managed to retreat by some kilometres.

Fitzharris (1996: 246) suggested that since the end of the LIA, the glaciers of the European Alps have lost about 30–40% of their surface area and about 50% of their ice volume. In Alaska (Arendt et al., 2002), glaciers appear to be thinning at an accelerating rate, which in the late 1990s amounted to 1.8 m per year. In North America, glaciers have shown a general retreat after the LIA maximum, particularly at lower elevations and southern latitudes. In the Rocky Mountains,

Table 11.1 Retreat of glaciers in metres per year in the twentieth century

(a) Individual glaciers. Source: Tables, maps and text in Grove (1988)

Location	Period	Rate
Breidamerkurjökull, Iceland	1903–1948	30–40
	1945–1965	53–62
	1965–1980	48–70
Lemon Creek, Alaska	1902–1919	4.4
	1919–1929	7.5
	1929–1948	32.9
	1948–1958	37.5
Humo Glacier, Argentina	1914–1982	60.4
Franz Josef, New Zealand	1909–1965	40.2
Nigardsbreen, Norway	1900–1970	26.1
Austersdalbreen, Norway	–	21
Abrekkbreen, Norway	–	17.7
Brikdalbreen, Norway	–	11.4
Tunsbergdalsbreen, Norway	–	11.4
Argentière, Mont Blanc, France	1900–1970	12.1
Bossons, Mont Blanc France	1900–1970	6.4
Oztal Group, Switzerland	1910–1980	3.6–12.9
Grosser Aletsch, Switzerland	1900–1980	52.5
Carstenz, New Guinea	1936–1974	26.2

(b) Region/country. Source: Oerlemans (1994)

Region	Period	Mean rate
Rocky Mountains	1890–1974	15.2
Spitzbergen	1906–1990	51.7
Iceland	1850–1965	12.2
Norway	1850–1990	28.7
Alps	1850–1988	15.6
Central Asia	1874–1980	9.9
Irian Jaya	1936–1990	25.9
Kenya	1893–1987	4.8
New Zealand	1894–1990	25.9

the area lost since the LIA is about 25%. In the Sierra Nevadas, ice volume loss from 1903 to 2004 averaged no less than 55% and showed a rapid acceleration in the early 2000s. In China, monsoonal temperate glaciers have lost an amount equivalent to 30% of their modern glacier area since the maximum of the LIA. New Zealand glaciers lost between one quarter and almost half of their volume between the LIA maximum and the 1970s. A further net ice volume loss between

1977 and 2005 has also been reported. In the Pyrenees, since the first half of the nineteenth century, about two-thirds of the ice cover has been lost, while in South America, except for a few cases in Patagonia and Tierra del Fuego, there has been a general retreat, with enhanced rates in recent decades.

Not all glaciers have retreated in recent decades (Figure 11.4) and current glacier tendencies for selected regions are shown. These are expressed in terms of their mass balance, which is defined as the difference between gains and losses (expressed in terms of water equivalent). In the European Alps (Figure 11.4a) a general trend toward mass loss, with some interruptions in the mid-1960s, late 1970s, and early 1980s, is observed. In Scandinavia (Figure 11.4b), glaciers close

to the sea have seen a very strong mass gain since the 1970s, but mass losses have occurred with the more continental glaciers. The mass gain in western Scandinavia could be explained by an increase in precipitation, which more than compensates for an increase in ablation caused by rising temperatures. Western North America (Figure 11.4c) shows a general mass loss near the coast and in the Cascade Mountains (Hoelzle and Trindler, 1998).

The positive mass balance (and advance) of some Scandinavian glaciers in recent decades, notwithstanding rising temperatures, has been attributed to increased storm activity and precipitation inputs coincident with a high index of the North Atlantic Oscillation (NAO) in winter months since 1980 (Nesje et al.,

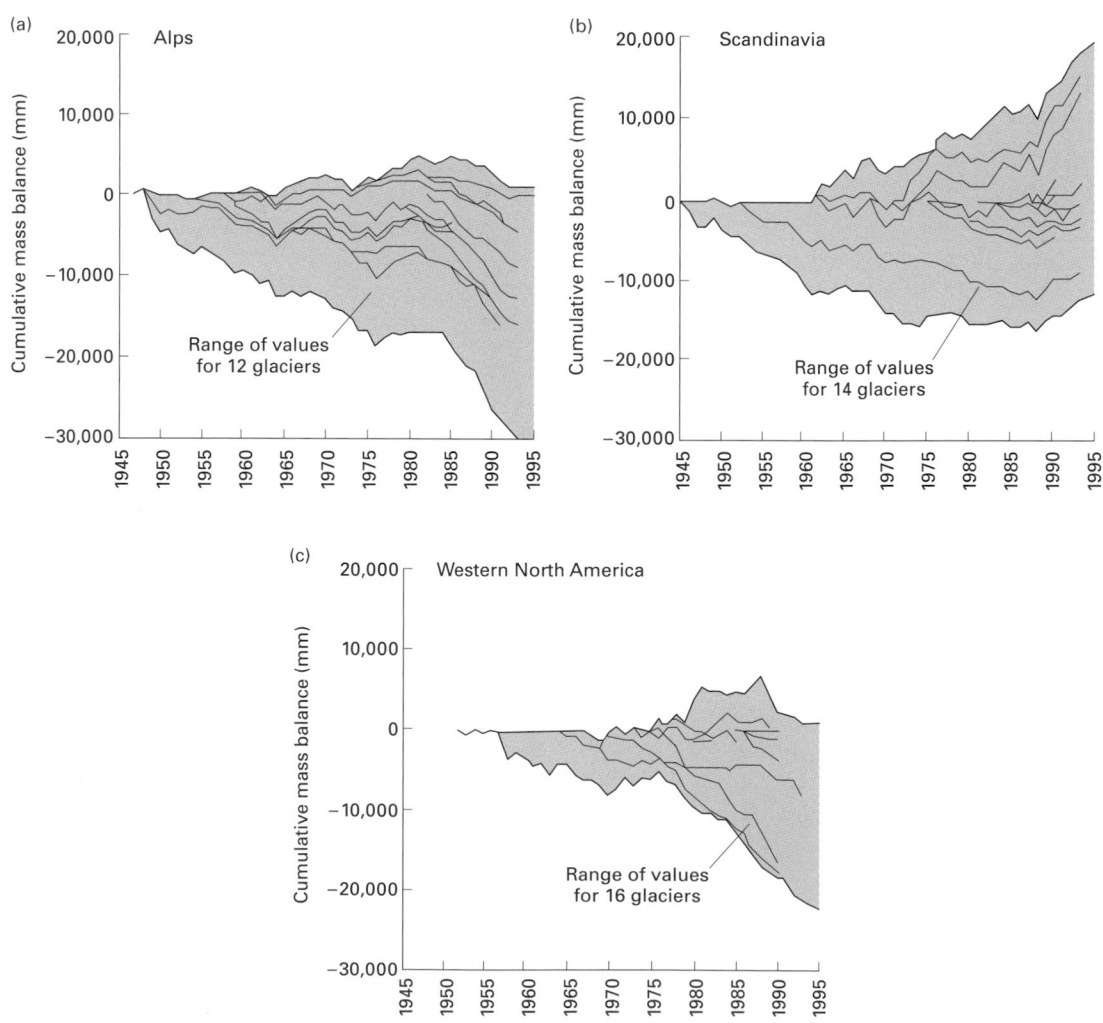

Figure 11.4 Cumulative mass balance for the Alps (a), Scandinavia (b) and western North America (c).

Table 11.2 Area of tropical glaciers. Source: Data from Kaser (1999, table 2).

Location	Date	Area (km²)	Date	Area (km²)	%
Irian Jaya, Indonesia	c. 1850	19.3	c. 1990	3.0	15.5
Mount Kenya, Kenya	1899	1.563	1993	0.413	26.4
Lewis Glacier, Kenya	1899	0.63	1993	0.20	31.7
Kibo (Kilimanjaro), Tanzania	c. 1850	20.0	1989	3.3	16.5
Ruwenzori, Uganda	1906	6.509	c. 1990	1.674	25.7

2000; Zeeberg and Forman, 2001). In the case of Nigardsbreen (Norway), there is a strong correlation between mass balance and the NAO index (Reichert et al., 2001). A positive mass balance phase in the Austrian Alps between 1965 and 1981 has been correlated with a negative NAO index (Schoner et al., 2000). Indeed, the mass balances of glaciers in the north and south of Europe are inversely correlated (Six et al., 2001). However, since 2000, the Norwegian glaciers, which were still advancing in the 1990s, have started to retreat at a rapid rate. If this continues, then by 2100, about 98% of Norwegian glaciers are likely to disappear and the glacier area may be reduced by around a third (Nesje et al., 2008).

Glaciers that calve into water can show especially fast rates of retreat, which could be more than 1 km per year (Venteris, 1999). In Patagonia, in the 1990s, retreat rates of up to 500 m per year were observed. This rapid retreat is accomplished by iceberg calving with icebergs detaching from glacier termini when the ice connection is no longer able to resist the upward force of flotation and/or the downward force of gravity. The rapid retreat is favoured by the thinning of ice near the termini, its flotation and its weakening by bottom crevasses. Tidewater glaciers on the Antarctic Peninsula have also been retreating and thinning fast (Pritchard and Vaughan, 2007).

The Columbia tidewater glacier in Alaska retreated around 13 km between 1982 and 2000. Equally the Mendenhall Glacier, which calves into a proglacial lake, has displayed rapid rates of retreat, with 3 km of terminus retreat in the twentieth century (Motyka et al., 2002). Certainly, calving permits much larger volumes of ice to be lost to the glacier than would be possible through surface ablation (van der Ween, 2002) and great phases of calving in the past help to explain the Heinrich events of the Pleistocene and the rapid

demise of the Northern Hemisphere ice sheets at the end of the last glacial period.

Some glaciers at high-altitude locations in the tropics have also displayed fast rates of decay over the last hundred or so years, to the extent that some of them now only have around one-sixth to one-third of the area they had at the end of the nineteenth century (Table 11.2) (Kaser, 1999). In the mountains of New Guinea, nearly all the ice caps have disappeared since the late nineteenth century and the cessation of the LIA. In East Africa, the ice bodies of Mount Kilimanjaro have shrunk from c. 20 km² in 1880 to c. 2.5 km² in 2003. The rate of glacier loss on Kilimanjaro continues unabated (Thompson et al., 2009). At present rates of retreat, the glaciers and ice caps of Mount Kilimanjaro in East Africa will have disappeared by 2010–2020. The reasons for the retreat of the East African glaciers include not only an increase in temperature. Also important may have been a relatively dry phase since the end of the nineteenth century (which led to less accumulation of snow) and a reduction in cloud cover (Mölg et al., 2003). In the tropical Andes, the Quelccaya ice cap in Peru also has an accelerating and drastic loss over the last 40 years (Thompson, 2000). In neighbouring Bolivia, the Glacier Chacaltya lost no less than 40% of its average thickness, two-thirds of its volume and more than 40% of its surface area between 1992 and 1998. Complete extinction is expected within 10–15 years (Ramirez et al., 2001). The recent response of Himalayan and Tibetan Plateau glaciers shows great regional variations depending on such factors as the amount of debris cover that occurs on glacier surfaces and also on differing precipitation trends (Yao et al., 2012). In the west of the region, around half of the westerlies-influenced Karakoram glaciers have been advancing or have been stable, showing a slight positive mass balance in recent years (Gardelle et al., 2012),

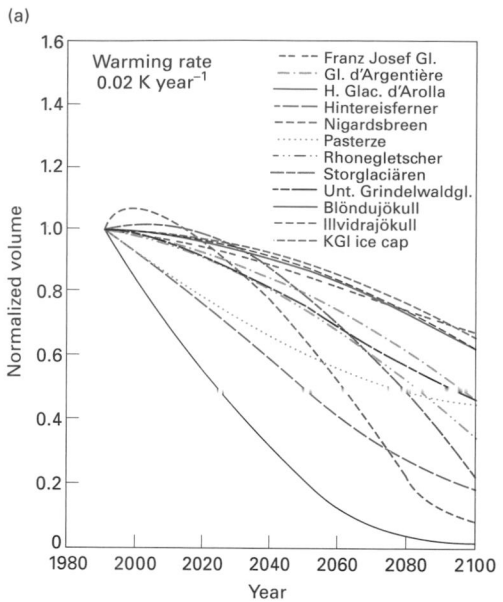
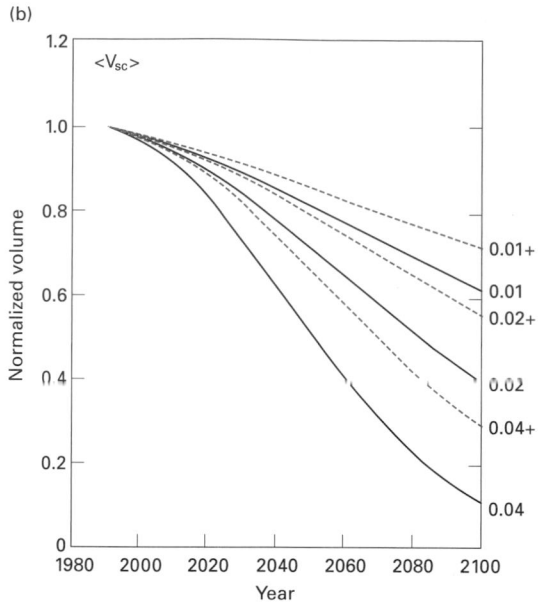

Figure 11.5 Changes in glacier volume 1980–2100 (as modelled by Oerlemans et al., 1998 reproduced with permission from Springer) for 12 glaciers: Franz Josef Glacier (New Zealand), Glacier d'Argentière (France), Haut Glacier d'Arolla (Switzerland), Hintereisferner (Austria), Nigardsbreen (Norway), Pasterze (Austria), Rhonegletscher (Switzerland), Storglaciären (Sweden), Unt. Grindelwaldgl. (Switzerland), Blondujökull (Iceland), Illvidrajökull (Iceland), ice cap (Antarctica). (a) Ice volume change with a warming rate of 0.02 K/a without a change in precipitation. Volume is normalized with the 1990 volume. (b) Scaled ice volume change for six climate change scenarios. + refers to an increase in precipitation of 10% degree warming.

while the great bulk of the monsoon-influenced glaciers in the central Himalaya have been retreating, especially those with a limited debris cover (Scherler et al., 2011).

Predicted rates of glacier retreat and some environmental consequences

The monsoonal temperate glaciers of China, which occur in the south-eastern part of the Tibetan Plateau, are an example of glaciers that will suffer great shrinkage if temperatures rise. Indeed, Su and Shi (2002) have, using modelling studies, calculated that if temperatures rise by 2.1°C, these glaciers will lose about 9900 km^2 of their total area of 13,203 km^2 (c. 75%) by the year 2100. The subpolar and polar glaciers of China will shrink by c. 20% over the same period (Shi and Liu, 2000).

Oerlemans et al. (1998) attempted to model the responses of 12 glaciers (from Europe, New Zealand and Antarctica) to three warming rates (0.01, 0.02 and 0.04 K per year), with and without concurrent changes

in precipitation (Figure 11.5). For a warming rate of 0.04 K per year, without an increase in precipitation, they believed that few glaciers would survive until 2100. On the other hand, with a low rate of warming (0.01 K per year) and an increase in precipitation of 10% per degree of warming, they predicted that overall loss of volume by 2100 would be 10–20% of the 1990 volume. On a global basis, mass balance modelling by Braithwaite and Zhang (1999) suggested that the most sensitive glaciers in terms of temperature change would be maritime and tropical glaciers. The latter, as in East Africa, are literally hanging on to high peaks, and so are very sensitive to quite modest changes in the height of the snowline. Radić and Hock (2011) have tried to assess the glacier volume loss that will occur in various parts of the world by 2100. They suggest that for the Earth as a whole, the total glacier volume will be reduced by about a fifth, but that in some locations, notably New Zealand and the European Alps, the loss will be about three quarters.

Although an increase in glacier melting initially increases runoff, the eventual disappearance of glaciers will cause abrupt changes in stream flow because of the lack of a glacial buffer or reservoir during the dry season (Bradley et al., 2006; Vuille et al., 2008). The disappearance of glaciers can cause slope instability (Vilímek et al., 2005) by reducing their buttressing effect and by exposing rock slopes to severe weathering (Fischer et al., 2006) that produces rock falls. Such slopes (Holm et al., 2004) can become unstable, generating a risk of increased landsliding and debris avalanches (Kirkbride and Warren, 1999; Haeberli and Burn, 2002). It is also possible that melting glaciers will produce large proglacial lakes which may subsequently fail, so producing down valley floods (Vilímek et al., 2005; Harrison et al., 2006). Glacial lakes in areas such as the Himalayas of Nepal and Bhutan are rapidly expanding as they are fed by increasing amounts of meltwater. Outburst floods from such lakes are extremely hazardous. Increased rates of glacier melting may indeed for a period of years cause an increased incidence of summer meltwater floods (see e.g. the study by Liu et al., 2009 on the Yangtze), but when the glaciers have disappeared, river flow volumes may be drastically reduced (Braun et al., 2000), leading to severe water shortages.

Sea ice in the Arctic and Antarctic

Since the 1970s, sea ice in the Arctic has declined in area (Vinnikov et al., 1999; Stocker, 2001: 446). Its thickness may also have become markedly less, with a near 40% decrease in its summer minimum thickness, though the data are imperfect (Holloway and Sou, 2002; Maslowski et al., 2012). Figure 11.6 shows the decrease in area of Northern Hemisphere summer (as recorded in September each year) sea ice between 1979 and 2012. The average extent from 1979 to 2000 was 6.70 million km², whereas in 2012 it was just 3.41, which is a reduction by almost a half.

By contrast, changes in Antarctic sea ice have been described by the Intergovernmental Panel on Climate Change (IPCC) as 'insignificant' (Stocker, 2001: 446). It has proved difficult to identify long-term trends because of the limited length of observations and the inherent interannual variability of Antarctic sea ice extent, but it is possible that there has been some decline since the 1950s (Curran et al., 2003). Figure 11.7 shows the extent of Antarctic sea ice since 1979, and shows little evidence for any long-term change of the

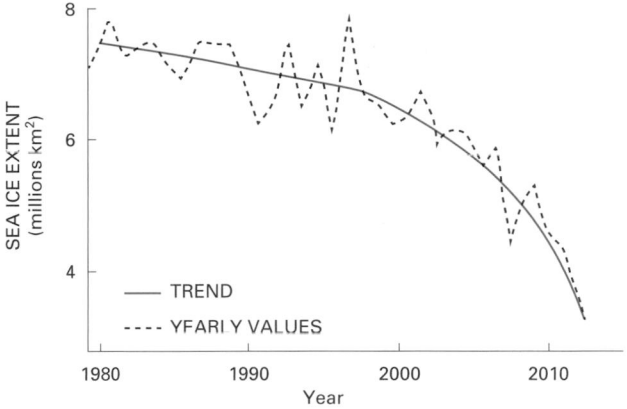

Figure 11.6 Decrease of Arctic sea ice extent since the late 1970s (Reproduced with permission from Fetterer, et al., 2002, updated 2009, from National Snow and Ice Data Center).

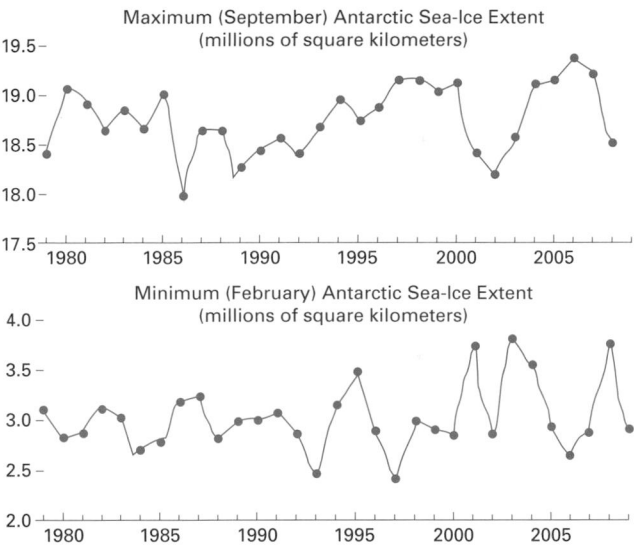

Figure 11.7 Antarctic sea ice extent since the late 1970s. No decrease is evident. Courtesy of NASA.

sort seen in the Arctic. The IPCC has also recognized that with a doubling of atmospheric greenhouse loadings, Antarctic sea ice will undergo a volume reduction of about 25–45%.

Model projections tend to show a substantial future decrease of Arctic sea ice cover. By 2050, there will be a roughly 20% reduction in annual mean Arctic sea ice extent (Stocker, 2001: 446). The extent and volume of summer ice may be especially sensitive, and it is possible that the Arctic could become free of summer sea ice at some point after 2050 (Zhang et al., 2010c). Such a reduction in sea ice area would allow increased absorption of solar radiation so that a further increase in temperature could occur, though this might to some extent be mitigated by increasing cloud cover (Miller and Russell, 2002). However, the IPCC (Anisimov and Fitzharris, 2001: 819) believes that a sea ice albedo feedback, with amplified warming, mean that 'At some point, with prolonged warming a transition to an Arctic Ocean that is ice free in summer – and perhaps even in the winter – could take place.'

Perversely, it is possible that one feedback of a decreasing Arctic sea ice cover will be an increase in winter snowfall in various parts of the Northern Hemisphere due to changes in the pattern of blocking anticyclones and an increase in available atmospheric moisture (Overland et al., 2011; Liu et al., 2012).

Fears have been expressed that the melting of Arctic sea ice will have an adverse impact on polar bears (*Ursus maritimus*). These are almost completely dependent on sea ice for access to sustenance, require it to be sufficiently thick to take their weight, and use it as the platform from which they hunt seals. Because of their

specialized nature and long generation times, 'it is unlikely that polar bears will survive as a species if the sea ice disappears completely as has been predicted by some' (Derocher et al., 2004).

Permafrost regions

Permafrost underlies large expanses of Siberia, Canada, Alaska, Greenland, Spitzbergen and north-western Scandinavia (Figure 11.8). Permafrost also exists offshore, particularly in the Beaufort Sea of the western Arctic and in the Laptev and East Siberian Seas, and at high elevation in mid-latitudes, such as the Rocky Mountains of North America and the interior plateaux of central Asia. It occurs not only in the tundra and polar desert environments poleward of the tree line, but also in extensive areas of the boreal forest and forest-tundra environments.

Above the layer of permanently frozen ground, there is usually a layer of soil in which temperature

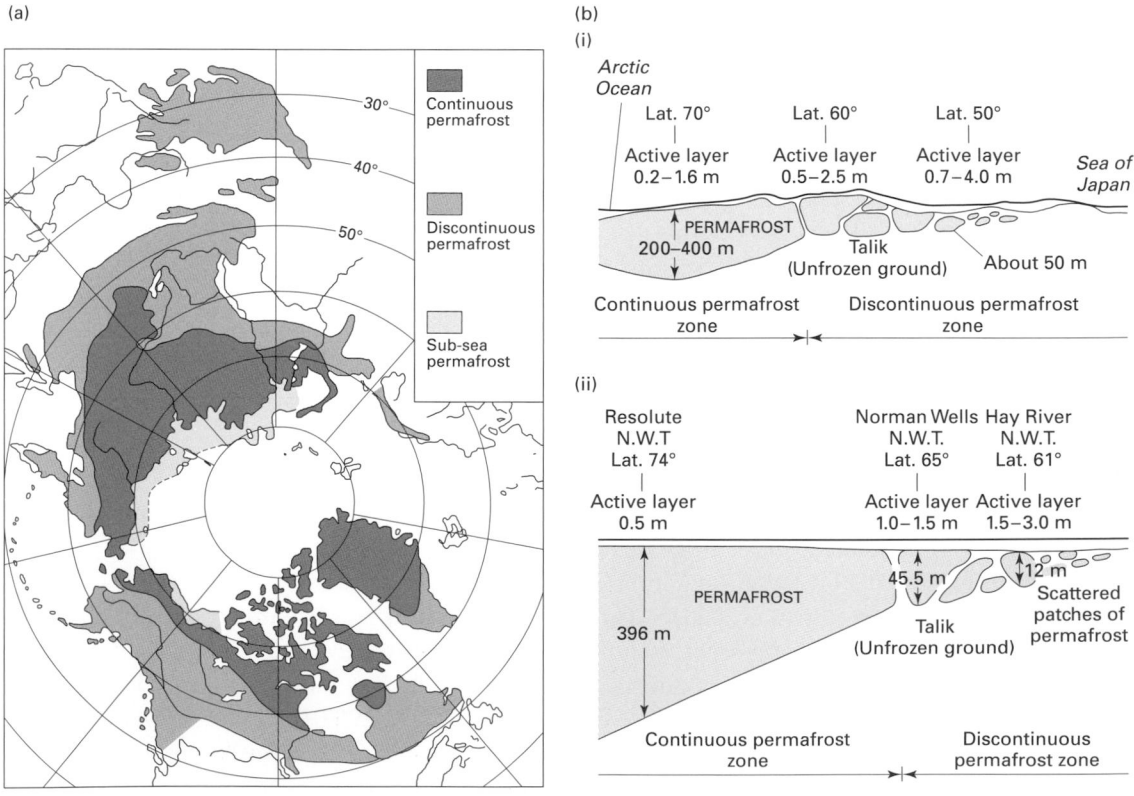

Figure 11.8 (a) The distribution of the main permafrost types in the Northern Hemisphere. (b) Vertical distribution of permafrost and active zones in longitudinal transects through (i) Eurasia and (ii) Northern America.

conditions vary seasonally, so that thawing occurs when temperatures rise sufficiently in summer but freeze in winter or on cold nights. This zone of freeze–thaw processes is called the *active layer*. It varies in thickness, ranging from 5 m where unprotected by vegetation to typical values of 15 cm in peat.

Conventionally, two main belts of permafrost are identified. The first is the zone of continuous permafrost; in this area, permafrost is present at all localities except for localized thawed zones, or *taliks*, existing beneath lakes, river channels and other large water bodies which do not freeze to their bottoms in winter. In the discontinuous permafrost zone, small-scattered unfrozen areas appear.

Maximum known depths of permafrost reach 1400–1450 m in northern Russia and 700 m in the north of Canada, regions of intense winter cold, short cool summers, minimal vegetation and limited snowfall. In general, the thickness decreases equatorwards. Sporadic permafrost tends to occur between the −1°C and −4°C mean annual air temperature isotherms, while continuous permafrost tends to occur to the north of the −7°C to −8°C isotherm.

Regions underlain by permafrost may be especially prone to the effects of global climate change (Bowden, 2010). First of all, the permafrost itself is by its very nature and definition susceptible to the effects of warming. Secondly, the amount of temperature increase predicted for high latitude environments is greater than the global mean, and the loss of permafrost could have a direct effect on atmospheric conditions in the lower atmosphere. Thirdly, permafrost itself is an especially important control of a wide range of geomorphological processes and phenomena, including slope stability, rates of erosion, ground subsidence and surface runoff. Fourthly, the nature (e.g. rain rather than snow) and amount of precipitation may also change substantially. Fifthly, the northern limits of some very important vegetation zones, including boreal forest, shrub-tundra and tundra may shift latitudinally by some hundreds of kilometres. Changes in snow cover and vegetation type may themselves have a considerable impact on the state of permafrost because of their role in insulating the ground surface. For example, if spring temperatures were to increase and early spring snowfall events were to become rain events, the duration of snow cover would decrease, surface albedo would also decrease, leading to an increase in air temperatures, and the snow

might provide less insulation. This could cause relatively rapid permafrost degradation (Ling and Zhang, 2003). Conversely, if the warming were to occur in rather more severe periglacial climates, where temperatures are predicted to remain below or at freezing, the winters could become warmer, and wetter, producing an increase in the longevity and depth of the snowpack. This could lead to more insulation (retarding the penetration of the winter cold wave) and a reduction of warming because of an increase in surface albedo. A good general review of the likely consequences of climate change in the tundra of Canada is provided by Smith (1993).

There is some evidence that permafrost has been degraded by the warming of recent decades. For example, Kwong and Gan (1994) reported a northward migration of permafrost along the Mackenzie Highway in Canada. Between 1962 and 1988, the mean annual temperature in the area rose by 1°C. Over the same period, the southern fringe of the discontinuous permafrost zone moved northwards by about 120 km. In Alaska, Osterkamp and Romanovsky (1999) found that in the late 1980s to mid-1990s, some areas experienced warming of the permafrost table of 0.5–1.5°C and that associated thawing rates were about 0.1 m per year. However, thawing rates at the base of the permafrost were an order of magnitude slower, indicating 'that time scales of the order of a century are required to thaw the top 10 m of ice rich permafrost, which would be primarily responsible for environmental and engineering problems' (p. 35).

Recent permafrost degradation has also been identified in the Qinghai–Tibet Plateau (Jin et al., 2000), particularly in marginal areas. Both upward and downward degradation have occurred.

Various attempts have been made to assess future extents of permafrost (Figure 11.9). On a Northern Hemisphere basis, Nelson and Anisimov (1993) have calculated the areas of continuous, discontinuous and sporadic permafrost for the year 2050 (Table 11.3). They indicate an overall reduction of 16% by that date. In Canada, Woo et al. (1992: 297) suggest that if temperatures rise by 4–5°C, as they may in this high latitude situation, 'Permafrost in over half of what is now the discontinuous zone could be eliminated'. They add,

The boundary between continuous and discontinuous permafrost may shift northwards by hundreds of kilometres but

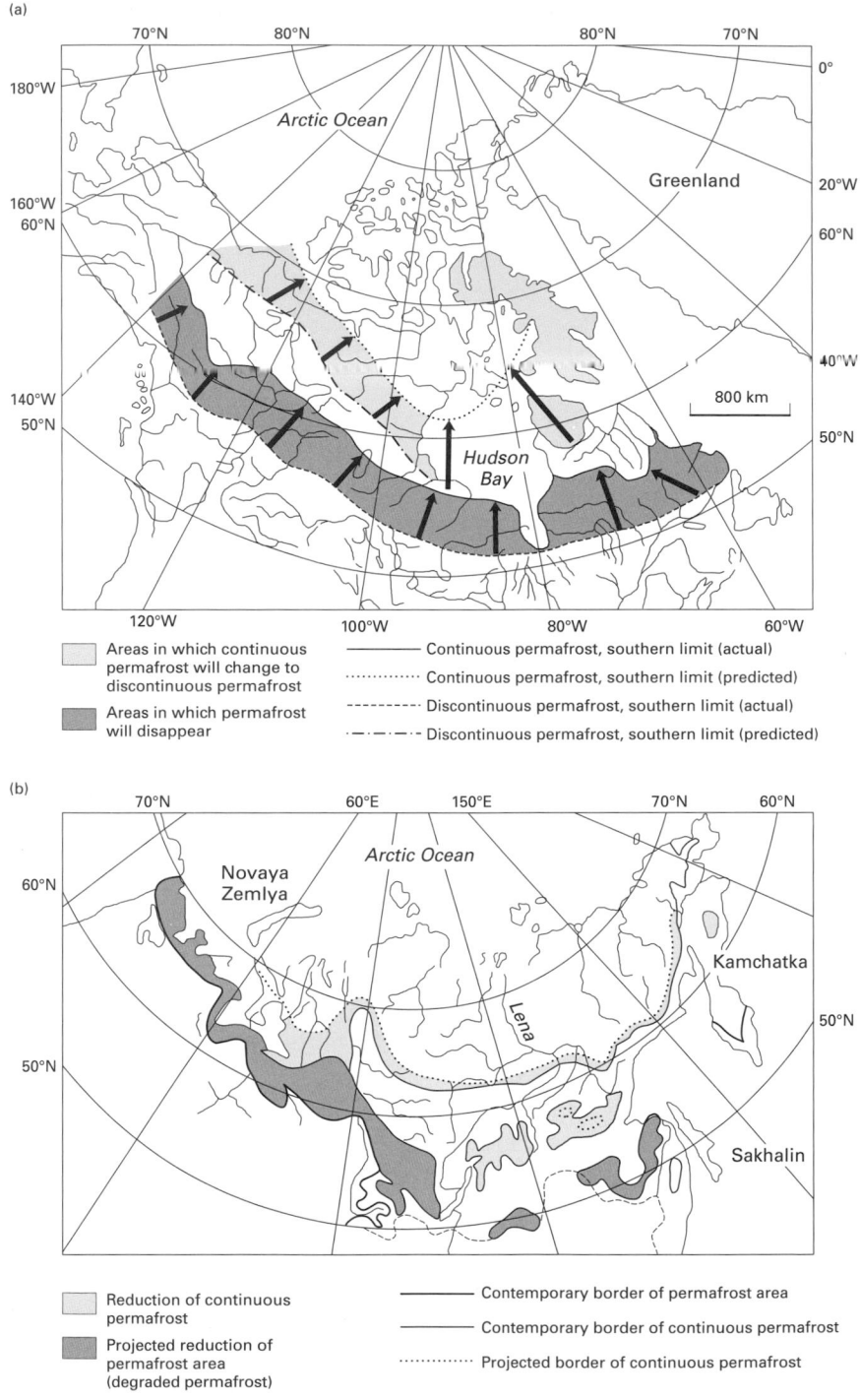

Figure 11.9 Projection of changes in permafrost with global warming for (a) North America and (b) Siberia (modified from French, 1996, figure 17.5 and Anisimov, 1989).

Table 11.3 Calculated contemporary and future area of permafrost in the Northern Hemisphere by 2050. Source: From Nelson and Anisimov (1993).

Zone	Contemporary	2050	% change
Continuous	11.7	8.5	−27
Discontinuous	5.6	5.0	−11
Sporadic	8.1	7.9	−2
Total	25.4	21.4	−16

Table 11.4 Forecast permafrost degradation on the Qinghai–Tibet Plateau during the next 100 years. Source: Jin et al. (2000, table 11).

Temperature increase (°C) (year)	Permafrost area ($10^6 \, km^2$)	Percentage decrease
0.51 (2009)	1.190	8
1.10 (2049)	1.055	18
2.91 (2099)	0.541	58

because of its links to the tree line, its ultimate position and the time taken to reach it are more speculative. It's possible that a warmer climate could ultimately eliminate continuous permafrost from the whole of the mainland of North America, restricting its presence only to the Arctic Archipelago.

Barry (1985) estimated that an average northward displacement of the southern permafrost boundary by 150 ± 50 km would be expected for each 1°C warming so that a total *maximum* displacement of between 1000 and 2000 km is possible.

Jin et al. (2000: 397) have attempted to model the response of permafrost to different degrees of temperature rise in the Tibetan Plateau region (Table 11.4) and suggest that with a temperature rise of 2.91°C by the end of the twenty-first century, the permafrost area will decrease by 58%. Similar estimates have been provided by Li et al. (2008).

Degradation of permafrost may lead to an increasing scale and frequency of slope failures (Gruber and Haeberli, 2007). Thawing reduces the strength of both ice-rich sediments and frozen jointed bedrock (Davis et al., 2001). Ice-rich soils undergo thaw consolidation during melting, with resulting elevated pore-water pressures, so that formerly sediment-mantled slopes may become unstable. Equally, bedrock slopes may be destabilized if warming reduces the strength of ice-

bonded open joints or leads to groundwater movements that cause pore pressures to rise (Harris et al., 2001). Increases in the thickness of the active layer may make more material available for debris flows.

Likewise, coastal bluffs may be subject to increased rates of erosion if they suffer from thermal erosion caused by permafrost decay. This would be accelerated still further if sea ice were to be less prevalent, for sea ice can protect coasts from wave erosion and debris removal (Carter, 1987). Local coastal losses to erosion of as much as 40 m per year have been observed in some locations in both Siberia and Canada in recent years, while erosive losses of up to 600 m over the past few decades have occurred in Alaska (Parson et al., 2001). The number and total area of thaw slumps along the Yukon coastline increased by 125% and 160%, respectively, between 1952 and 2000 (Lantuit and Pollard, 2008).

Slope failures in cold environments could be exacerbated by glacier recession (Haeberli and Burn, 2002). The association between glacier retreat since the LIA and slope movement processes such as gravitational rock slope deformation, rock avalanches, debris flows and debris slides is evident. This is the case where recent ice retreat has removed buttress support to glacially undercut and oversteepened slopes (Holm et al., 2004).

One of the severest consequences of permafrost degradation is ground subsidence and the formation of thermokarst phenomena (Nelson, 2002; Nelson et al., 2001), including multitudes of thaw lakes. This is likely to be a particular problem in the zones of relatively warm permafrost (the discontinuous and sporadic zones), and where the permafrost is rich in ice (Woo et al., 1992).

The excess ice content determines how much surface settlement will occur as a result of the thawing of permafrost. It is the volume of water that exceeds the space available within the pores when the soil or sediment is thawed (Woo et al., 1992: 297):

$$EX = 100 \times (V_W / V_T),$$

where EX is the excess ice content (%), V_W is the volume of water that is released on thawing the sample (cubic metre) and V_T is the total volume of the thawed sample (cubic metre). Thus, to deepen an active layer by 0.5 m, 1.0 m of permafrost with an EX of 50% must thaw, and

Surface settlement (cm)	25	12.5	6.25	0.05
Cumulative settlement (cm)	25	37.5	43.75	50

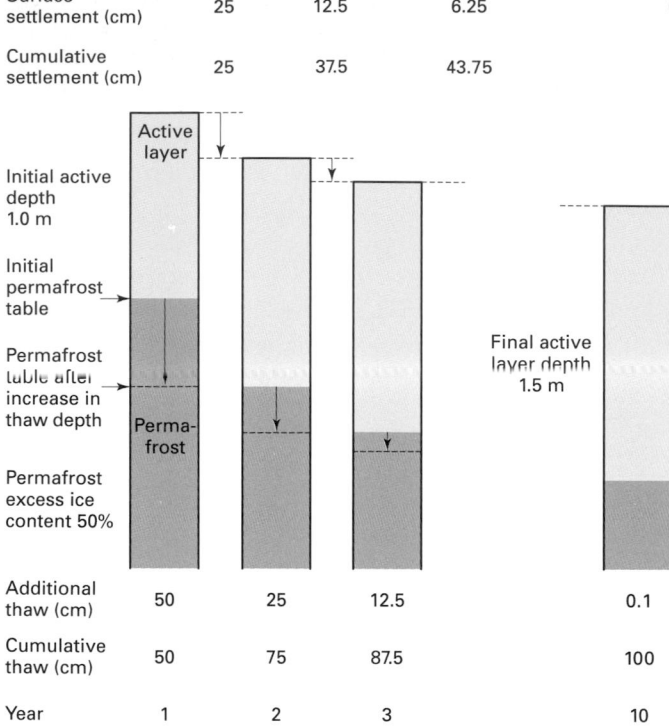

Additional thaw (cm)	50	25	12.5	0.1
Cumulative thaw (cm)	50	75	87.5	100
Year	1	2	3	10

Figure 11.10 Ground settlement in response to a thickening of the active layer in permafrost with an excess ice content of 50%. The active layer increases from 1.0 to 1.5 m and in so doing, 1.0 m of permafrost is thawed and the surface settles by 0.5 (after Woo et al., 1992, figure 8). Reproduced with permission from Bellwether.

this ultimately will result in surface settlement of 0.5 m (see Figure 11.10).

Thawing ice-rich areas, such as ice-cored mounds (*pingos*), will settle more than those with lesser ice contents, thereby producing irregular hummocks and depressions. Water released from ice melt may accumulate in such depressions. The depressions may then enlarge into thaw lakes as a result of thermal erosion (Harris, 2002), which occurs because summer heat is transmitted efficiently through the water body into surrounding ice-rich material (Yoshikawa and Hinzman, 2003). Once they have started, thaw lakes can continue to enlarge for decades to centuries because of wave action and continued thermal erosion of the banks. However, if thawing eventually penetrates the permafrost, drainage occurs that leads to ponds drying up.

Increasing temperatures are likely to cause the thickness of the seasonally thawed active layer to become deeper. A simulation by Stendel and Christensen (2002) indicated a 30–40% increase by the end of the twenty-first century for most of the permafrost area in the Northern Hemisphere.

Points for review

What are the main components of the cryosphere?
How are valley glaciers likely to respond to global warming?
Will the polar ice sheets respond catastrophically to global warming?
What are the consequences of permafrost melting?

Guide to reading

Harris, C. and Stonehouse, B. (eds) (1991) *Antarctica and Climatic Change*. London: Belhaven Press. A study of the response of Antarctica to future warming and a consideration of the global implications.

Hassol, S.J. (2004) *Impacts of a Warming Arctic*. Cambridge: Cambridge University Press. A report of the Arctic Climate Impact Assessment.

Nesje, A. and Dahl, S.O. (2000) *Glaciers and Environmental Change*. London: Arnold. A general text which is very strong on the changing mass balance of glaciers.

Slaymaker, O. and Kelly, R.E.J. (2007) *The Cryosphere and Global Environmental Change*. Oxford: Blackwell. This text considers all aspects of the cryosphere, including glaciers and permafrost, and their response to environmental changes.

Wadhams, P., Dowdeswell, J.A. and Schofield, A.N. (eds) (1996) *The Arctic And Environmental Change*. Amsterdam: Gordon and Breach. A collection of papers presented to the Royal Society and covering many aspects of cryospheric change in northern regions.

12 THE FUTURE: DRYLANDS

Chapter Overview

Some components of desert landscapes seem to be highly susceptible to environmental changes. They have demonstrated this in the ways they have responded to past climatic changes. Rainfall amounts and temperatures in coming decades will have an impact on a whole range of desert phenomena: wind erosion, dust storm emissions, dunes, runoff, lake levels, desert coastlines and salt weathering.

Introduction

In the past, arid environments often appear to have been prone to rapid geomorphological and hydrological changes in response to apparently modest climatic stimuli, switching speedily from one state to another when a particular threshold was reached (Goudie, 1994).

1 Desert valley bottoms appear to have been subject to dramatic alternations of cut-and-fill during the course of the Holocene and a large literature has developed on the arroyos (see Chapter 6) of the American south-west (see e.g. Cooke and Reeves, 1976; Balling and Wells, 1990). Schumm (1977) has proposed that valley bottom trenching may occur when a critical valley slope gradient is crossed for a

The Human Impact on the Natural Environment: Past, Present and Future, Seventh Edition. Andrew S. Goudie.
© 2013 John Wiley & Sons, Ltd. Published 2013 by John Wiley & Sons, Ltd.

drainage area of a particular areal extent, but other examples result from changes in rainfall intensity and grazing pressure.

2 The heads of alluvial fans often display fan head trenching, suggesting recent channel instability. There has been considerable debate in the literature about the trigger for such trenching, be it tectonic, extreme flood events, climatic change, or an inherent consequence of fluctuating sediment–discharge relationships during the course of a depositional cycle (e.g. Harvey, 1989).

3 Fluvial systems, as, for example, in the drier lands around the Mediterranean Basin, display a suite of terraces which demonstrate a complex record of cut-and-fill during the late Holocene (van Andel et al., 1990). There has been prolonged debate as to whether the driving force has been climatic change or anthropogenic activities.

4 Colluvial sections in subhumid landscapes (including southern Africa) show complex consequences of deposition, stability and incision (Watson et al., 1984). Ongoing luminescence dating is beginning to give an indication of the complexity of chronology.

5 Terminal lake basins (pans, playas, etc.) can respond dramatically, in terms of both their extent and the rapidity of change that can occur, to episodic rainfall events in their catchments. The history of Lake Eyre in Australia in the twentieth century bears witness to this fact (Figure 12.1).

6 West African dust storm activity has shown very marked shifts in the past few decades in response to runs of dry years and increasing land-use pressures

(a)

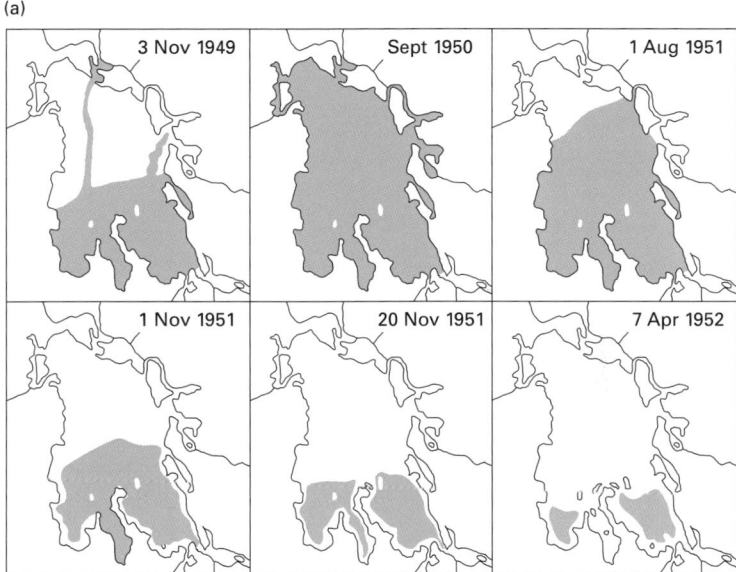

(b)

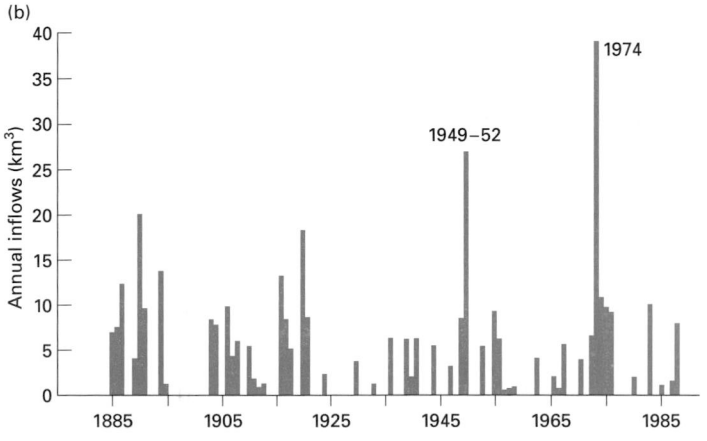

Figure 12.1 The flooding of Lake Eyre. (a) Extent of flooding in 1949–1952 (after Bonython and Mason in Mabbutt, 1977, figure 54). (b) Estimated annual inflows to Lake Eyre North for the period 1885–1989 (based on Kotwicki and Isdale, 1991, figure 2). Both 1949–1952 and 1974 were strong La Niña events (Source: Viles and Goudie, 2003, reproduced with permission from Elsevier).

(Goudie and Middleton, 1992). The data for Nouak-chott (Mauritania) are especially instructive, revealing a sudden acceleration in dust storm events since the 1960s. A broadly similar picture could be presented for the High Plains of the USA during the 'Dust Bowl years' of the 1930s.

7 Some dune fields have also proved to be prone to repeated fluctuations in deposition and stabilization in the Holocene, and as more ^{14}C and luminescence dates become available, the situation is likely to prove to be even more complex than hitherto believed. A good illustration of this is provided by Gaylord's (1990) work in the Clear Creek area of south-central Wyoming where over the last 7500 years at least four episodes of enhanced aeolian activity and aridity are recorded.

8 Various studies have shown how rates of denudation (e.g. Langbein and Schumm, 1958) and drainage density can change very rapidly in semi-arid areas either side of a critical rainfall or precipitation/evapotranspiration (P/E) value related to vegetation cover that constitutes a particular threshold between equilibrium states (Figure 12.2 and Figure 12.3).

This apparent instability and threshold dependence of a range of arid zone landforms, rates and processes lead one to believe that such areas may be especially susceptible to the effects of global warming, should this occur in the coming decades (Figure 12.4).

Climate changes in the past

Past climatic changes have had dramatic effects on deserts (Goudie, 2002; Anderson et al., 2013), and this makes it likely that future climatic changes will also have dramatic effects. For example, climatic changes in the Holocene have led to the expansion and contraction of lakes. In the Sahara there are huge numbers of Holocene pluvial lakes both in the Chotts of Tunisia and Algeria, in Mali (Petit-Maire et al., 1999) and in the south (e.g. Mega-Chad). In the Western Desert of Egypt and the Sudan, there are many closed depression or playas, relict river systems and abundant evidence of prehistoric human activity (Hoelzmann et al., 2001). Playa sediments indicate that they once contained substantial bodies of water, which attracted Neolithic settlers. Many of these sediments indicate the ubiquity of an early to mid-Holocene wet phase, which has often been termed the 'Neolithic Pluvial'. A large lake, 'The West Nubian Palaeolake', formed in the far north west of Sudan (Hoelzmann et al., 2001). It was especially extensive between 9500 and 4000 years BP, and may have covered as much as 7000 km². If it was indeed that

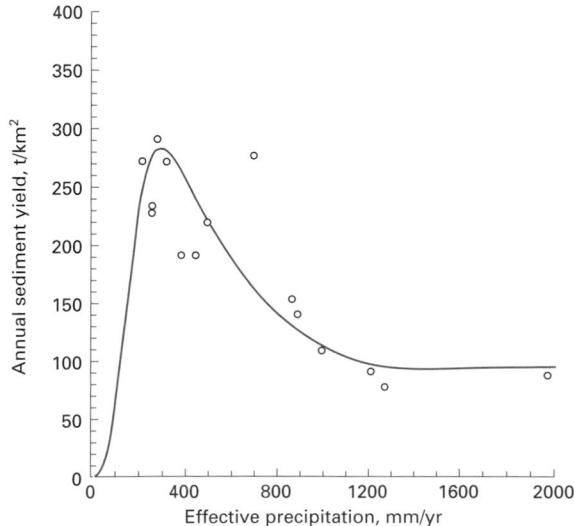

Figure 12.2 Variation of sediment yield with climate as based on data from small watersheds in the USA (after Langbein and Schumm, 1958).

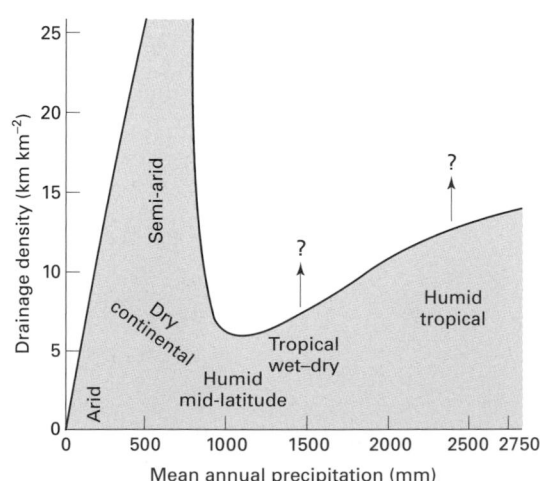

Figure 12.3 Relation between drainage density and mean annual precipitation (after Gregory, 1976. Copyright © 1976. Reprinted with permission from John Wiley and Sons, Ltd).

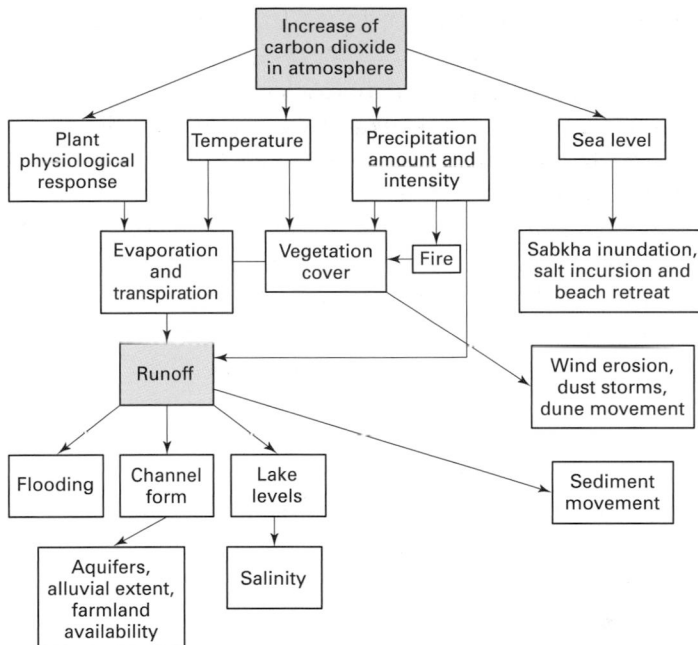

Figure 12.4 The potential responses of desert environments to increases of carbon dioxide in the atmosphere.

big, then a large amount of precipitation would have been needed to maintain it – possibly as much as 900 mm compared to the less than 15 mm it receives today. The Sahara may have largely disappeared during what has been called 'The Greening of Africa'.

Conversely, some deserts experienced abrupt and relatively brief drought episodes which caused extensive dune mobilization in areas like the American High Plains (Forman et al., 2001, 2006)

At the present day, deserts continue to change, sometimes as a result of climatic fluctuations like the run of droughts that have afflicted the West African Sahel over recent decades. The changing vegetation conditions in the Sahel, which have shown marked fluctuations since the mid-1960s, have been mapped by Tucker et al. (1991). Interannual variability in the position of the southern boundary of the Sahara, as represented by the 200 mm isohyet, can be explained in large measure by changes in the North Atlantic Oscillation and the El Niño Southern Oscillation (ENSO) (Oba et al., 2001). The drought also led to reductions in the flow of rivers like the Senegal and Niger, and to an increase in dust storm activity (Goudie and Middleton, 2006). The extent of Lake Chad's water surface has also fluctuated dramatically. It was particularly badly affected by the drought, which caused the lake surface area to fall from c. 25,000 km² in 1960 to just a tenth of that figure in the mid-1980s (Birkett, 2000).

We also know from recent centuries that ENSO fluctuations have had important consequences for such phenomena as droughts and dune reactivation (Forman et al., 2001), and through their effect on vegetation cover have had major impacts on slope stability and soil erosion (Viles and Goudie, 2003). Lake levels have responded to El Niño influences (Arpe et al., 2000), large floods have occurred, valleys have been trenched (Bull, 1997), erosivity patterns have altered (D'Odorico et al., 2001), and landslides and debris flows have been generated (Grosjean et al., 1997).

Future changes in climate in drylands

The IPCC (2007b) suggested that in drylands temperatures could increase between 1 and 7°C by 2017–2100 compared to 1961–1990, and that precipitation levels could decrease by as much as 10–20% in the case of the Sahara but increase by as much as 10–15% in the Chinese deserts. Droughts may become more prevalent (Dai, 2011). The models of Sylla et al. (2010) suggest that this will be the case in the West African Sahel, partly because of a weaker monsoon flow and reduced easterly wave activity. While is likely that many areas that are currently dry may see enhanced aridity because of reductions in precipitation, even within an area such as the Middle East, some parts may become wetter and some may become drier. The Saudi Arabian desert may, for instance, become wetter because of a more northerly intrusion by the Intertropical Convergence Zone (ITCZ) (Evans, 2010). There is a wide consensus that the south-western USA will become more arid (Seager et al., 2007; Seager and Vecchi, 2010). A review of global climate models for Australia suggests that over many areas, droughts will become more intense and more frequent (Kirono et al., 2011). Zeng and Yoon (2009) have suggested that as conditions become drier and

vegetation cover is reduced, there may be vegetation-albedo feedbacks which will serve to enhance any aridity trend. By 2099, their model indicates that globally, the warm desert area may expand by 8.5 million km² or by 34%. However, some areas may experience more frequent hurricane activity, and there may also be changes in ENSO frequency and intensity, though this latter aspect of climate change is still characterized by highly divergent model results (Latif and Keenlyside, 2009). Figure 12.5 shows those areas in the

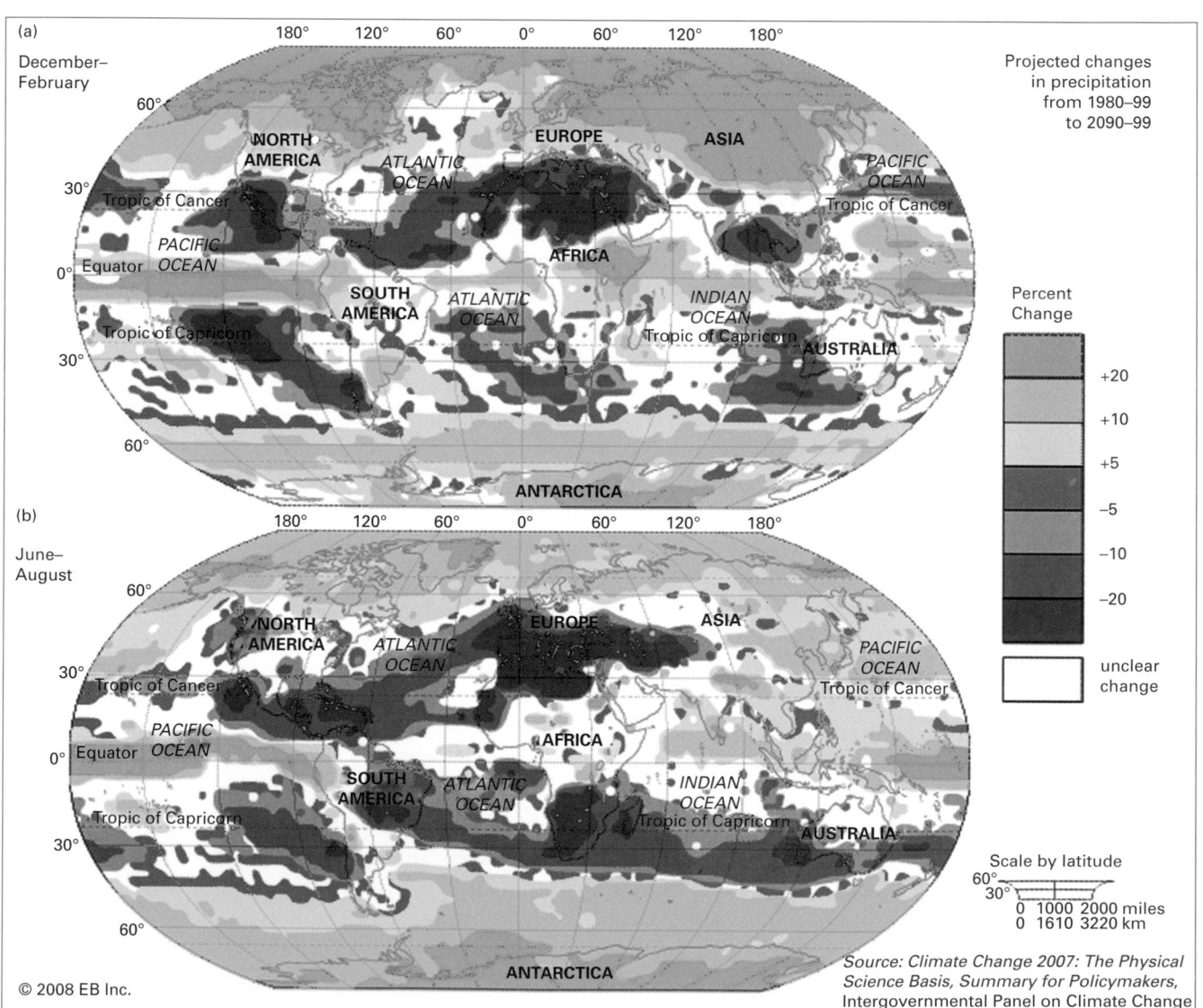

Figure 12.5 The global pattern of future precipitation change for this century as predicted by the IPCC (2007b, reproduced with permission) for (a) December, January, February and (b) June, July, August. The shaded areas are those where precipitation amounts will fall by 10% or greater.

world where the IPCC (2007b) proposes that precipitation levels may fall by over 10%, and the increasing dryness forecast for the Sahara is clear.

Wind erosivity and erodibility

Changes in climate could affect wind erosion either through their impact on erosivity or through their effect on erodibility. The former is controlled by a range of wind variables including velocity, frequency, duration, magnitude, shear and turbulence. For example, Bullard et al. (1996) have shown how dune activity varies in the south-west Kalahari in response to decadal scale variability in wind velocity, while over a longer timescale there is evidence that trade-wind velocities may have been elevated during the Pleistocene glacials. Unfortunately, general circulation models (GCMs) as yet give little indication of how these characteristics might be modified in a warmer world, so that prediction of future changes in wind erosivity is problematic. Erodibility is largely controlled by vegetation cover and surface type, both of which can be influenced markedly by climate. In general, vegetation cover, which protects the ground surface and modifies the wind regime, decreases as conditions become more arid. Likewise, climate affects surface materials and their erodibility by controlling their moisture content, the nature and amount of clay mineral content (cohesiveness) and organic levels. However, modelling the response of wind erosion to climatic variables on agricultural land is vastly complex, not least because of the variability of soil characteristics, topographic variations, the state of plant growth and residue decomposition, and the existence of windbreaks. To this needs to be added the temporal variability of aeolian processes and moisture conditions and the effects of different land management practices (Leys, 1999), which may themselves change with climate change.

Future dust storm activity

The changes in temperature and precipitation conditions that occurred in the twentieth century (combined with land cover changes) had an influence on the development of dust storms (Goudie and Middleton, 1992). These are events in which visibility is reduced to less than 1 km as a result of particulate matter, such as valuable topsoil, being entrained by wind. This is a process that is most likely to happen when there are high winds and large soil-moisture deficits. Probably the greatest incidence of dust storms occurs when climatic conditions and human pressures combine to make surfaces susceptible to wind attack as with the Dust Bowl of the 1930s in the USA (see Chapter 4). Attempts to relate past dust storm frequencies to simple climatic parameters or antecedent moisture conditions have frequently demonstrated rather weak relationships (Bach et al., 1996), confirming the view that complex combinations of processes control dust emissions. Nevertheless, evidence is now emerging that relates dust emissions from Africa to changes in the North Atlantic Oscillation (Moulin et al., 1997).

If, however, soil moisture levels decline as a result of changes in precipitation and/or temperature, there is the possibility that dust storm activity could increase in a warmer world (Wheaton, 1990). A comparison between the Dust Bowl years of the 1930s and model predictions of precipitation and temperature for the Great Plains of Kansas and Nebraska (Figure 12.6) indicates that mean conditions could be similar to those of the 1930s under enhanced greenhouse conditions (Smith and Tirpak, 1990), or even worse (Rosenzweig and Hillel, 1993). Munson et al. (2011) also argue that with predicted increases in droughts in the western USA, perennial vegetation cover will be reduced and aeolian activity will be increased.

If dust storm activity were to increase as a response to global warming, it is possible that this could have a feedback effect on precipitation that would lead to further decreases in soil moisture (Tegen et al., 1996; Miller and Tegen, 1998). It is also possible that increased dust delivery to the oceans could affect biogeochemical cycling, by providing nutrients to plankton, which in turn could draw down carbon dioxide from the atmosphere (Ridgwell, 2002). However, the impact and occurrence of dust storms will depend a great deal on land management practices, and recent decreases in dust storm activity in North Dakota have resulted from conservation measures (Todhunter and Chihacek, 1999).

The nature of future dust activity will depend on three main factors: anthropogenic modification of desert surfaces (Mahowald and Luo, 2003; Mahowald et al.,

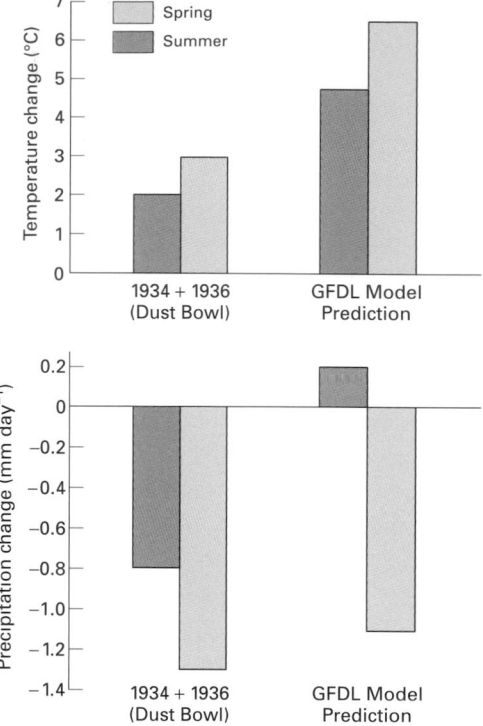

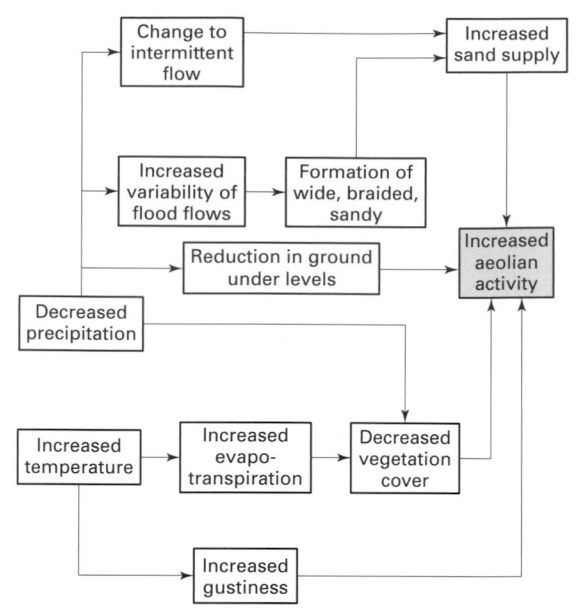

Figure 12.7 The influence of decreased precipitation and increased temperatures on aeolian activity.

Figure 12.6 Comparison between two 'Dust Bowl years' (1934 and 1936) and the Geophysical Fluid Dynamic Laboratory (GFDL) model prediction for a × 2 CO_2 situation for the Great Plains (Kansas and Nebraska) (modified from Smith and Tirpak, 1990, figure 7.3).

2006); natural climatic variability (e.g. in the El Niño Southern Oscillation or the North Atlantic Oscillation); and changes in climate brought about by global warming. In northern China, there is some evidence that dust storm activity has decreased in recent warming decades, partially in response to changes in the atmospheric circulation and associated wind conditions (Jiang et al., 2009) and might decrease therefore still further in a warming world (Zhu et al., 2008).

Sand dunes

Sand dunes, because of the crucial relationships between vegetation cover and sand movement, are highly susceptible to changes of climate, though there are huge challenges in predicting just how much dune systems may change (Knight et al., 2004; Thomas and Wiggs, 2008). One problem is that active and stabilized

dunes often coexist under the same climatic conditions (Yizhaq et al., 2007). There may also be lags in terms of response and these may be related to physical–biological interactions. So for example, a dune may become vegetated when the wind power is sufficiently low, and once vegetated, a much higher wind stress is needed to destroy the vegetation and re-activate the dunes. Vegetation dynamics are a major consideration (Hugenholtz and Wolfe, 2005). In addition, there are multiple environmental factors involved in determining dune activity: wind energy, rainfall, snow cover, soil moisture, groundwater conditions, vegetation cover, fire frequency, and so on (Figure 12.7).

Some areas, such as the south-west Kalahari (Stokes et al., 1997) or portions of the High Plains of the USA (Gaylord, 1990) may have been especially prone to changes in precipitation and/or wind velocity because of their location in climatic zones that are close to a climatic threshold between dune stability and activity (Figure 12.8). Development in the use of luminescence dating of sand grains and studies of explorers' accounts (Muhs and Holliday, 1995), have led to the realization that such marginal dune fields have undergone episodic and repeated phases of change at decadal and century timescales in response to extended drought

Figure 12.8 These linear dunes in the south-west Kalahari have active crests at the present day, particularly in dry, windy years, but may become still more active in coming decades. (See Plate 51)

events during the Holocene (Forman et al., 2006, 2008). In the Canadian Prairies, changes in precipitation amounts and decreases in wind velocity have led to increasing dune stability since the mid 1900s (Hugenholtz and Wolfe, 2005), while over the last few hundreds of years relatively stable parabolic dunes have tended to replace active barchans (Wolfe and Hugenholtz, 2009).

The mobility of desert dunes (M) is directly proportional to the sand-moving power of the wind, but indirectly proportional to vegetation cover (Lancaster, 1995: 238). An index of the wind's sand-moving power is given by the percentage of the time (W) the wind blows above the threshold velocity (4.5 m per second) for sand transport. Vegetation cover is a function of the ratio between annual rainfall (P) and potential evapotranspiration (PE). Thus, $M = W/(P/PE)$. Empirical observations in the USA and southern Africa indicate that dunes are completely stabilized by vegetation when M is <50, and are fully active when M is >200.

Muhs and Maat (1993) used the output from GCMs combined with this dune mobility index to show that sand dunes and sand sheets on the Great Plains are likely to become reactivated over wide areas, particularly if the frequencies of wind speeds above the threshold velocity were to increase by even a moderate amount. For another part of the USA, Washington State, Stetler and Gaylord (1996) suggested that with a

4°C warming vegetation would be greatly reduced and that as a consequence dune mobility would increase by over 400%.

Detailed scenarios for dune remobilization by global warming have been developed for the mega-Kalahari in southern Africa (Thomas et al., 2005) (Figure 12.9). Much of this vast thirstland is currently vegetated and stable, but GCMs suggest that by 2099 all dune fields, from South Africa and Botswana in the south to Zambia and Angola in the north will be reactivated. This could disrupt pastoral and agricultural systems and might lead to loss of agricultural land, the overwhelming of buildings, roads, canals, runways and the like, abrasion of structures and equipment, damage to crops, and the impoverishment of soil structure. However, the methods used to estimate future dune field mobility are still problematic, and much more research is needed before we can have confidence in them (Knight et al., 2004).

Once again, however, it is important to stress that the activity of dunes and of dust storms may be dependent on drought interludes, but that the effects of such interludes, both in the past (see, e.g. Cook et al., 2009; Seifan, 2009) and in the future may be greatly affected by miscellaneous anthropogenic activities (Yizhaq et al., 2009): deforestation, overgrazing, dune removal, extension of irrigation, surface disturbance, mining, fire and so on.

Recent research in China has demonstrated that over the last five decades or so there have been changes in wind activity that have been related to warming trends. The general message is that as warming has occurred wind velocities have fallen (e.g. Xu et al., 2006; Wang et al., 2007) (Figure 12.10). It has also been suggested that as a consequence dunes in northern China have become more stable (Wang et al., 2007), though this is not invariably true, partly because of human pressures. Thus it is possible that as warming continues and wind velocities decrease dunes may become still more stable in the region, but a great deal will also depend upon future soil moisture conditions which in turn will depend on precipitation amounts and levels of evapotranspiration. Various modelling studies have suggested that more generally in low latitudes extreme wind events will become less frequent with global warming, and this has been confirmed for the USA (Breslow and Sailor, 2002). Another possible influence on future dune activity, as it has been in the past, may

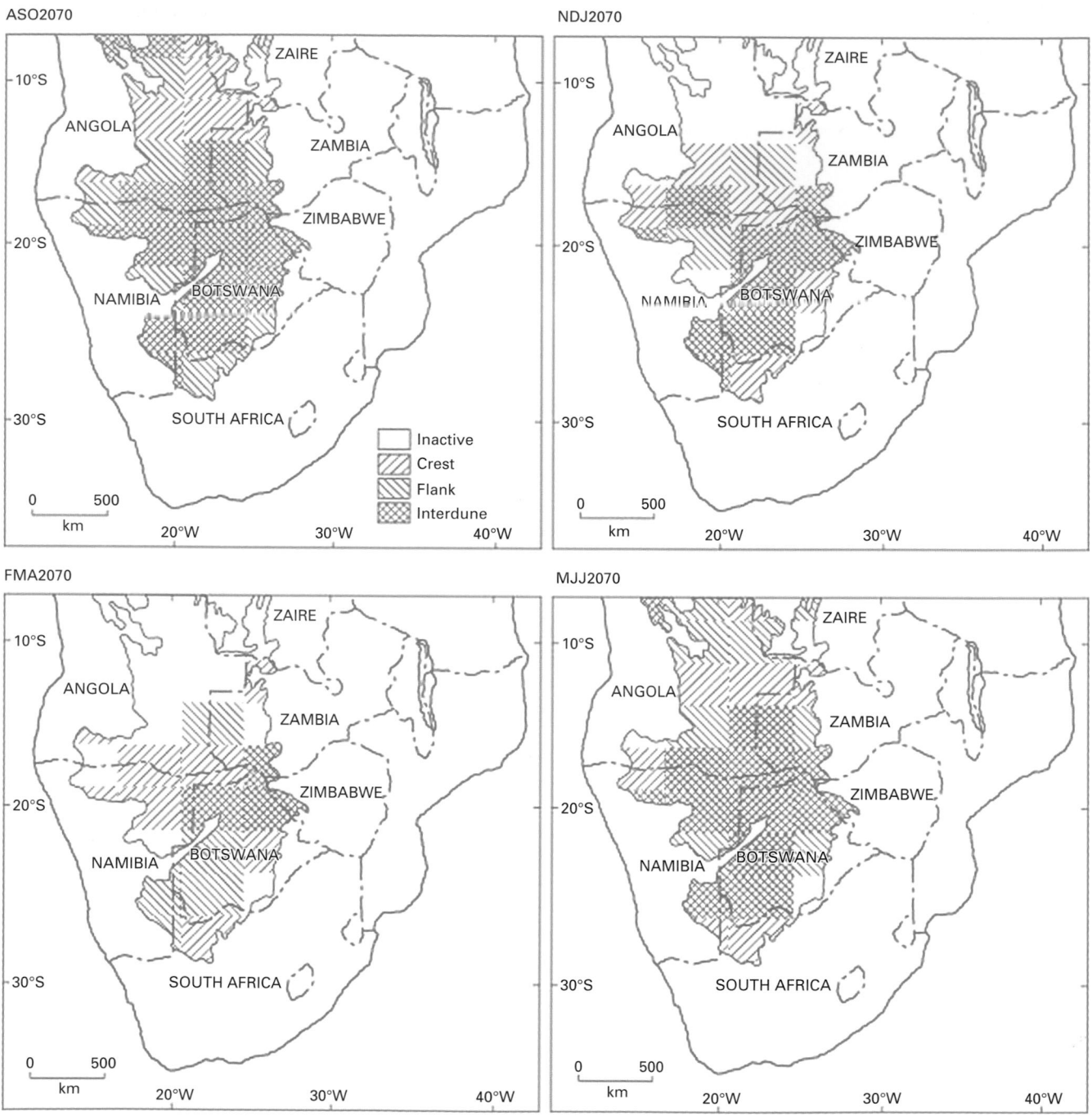

Figure 12.9 Modelled dunefield activity by the end of the twenty-first century (modified from Thomas et al., 2005, figure 8). Reproduced with permission from McMillan.

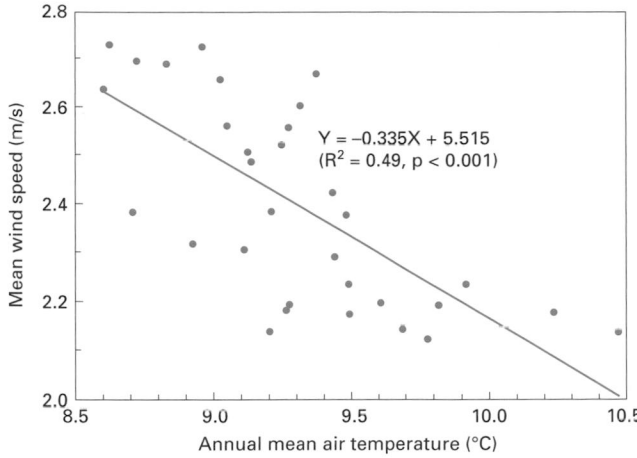

Figure 12.10 The relationship between annual mean wind speed and air temperature over China, 1969–2000 (modified from Xu et al., 2006, figure 4).

be changes in the frequency and severity of wildfires (Sankey et al., 2009).

Rainfall and runoff

Some dry regions will suffer particularly large diminutions in soil moisture levels (Wetherald and Manabe, 2002) and annual runoff (sometimes 60% or more). Drylands appear to be more vulnerable than humid regions (Chiew et al., 1996; Guo et al., 2002). The sensitivity of runoff to changes in precipitation is complex, but in some environments quite small changes in rainfall can cause proportionally larger changes in runoff. As rainfall amounts decrease, the proportion that is lost to stream flow through evapotranspiration increases. Our modelling capability in this area is still imperfect, and different types of model indicate differing degrees of sensitivity to climatic change (Nash and Gleick, 1991).

However, highly significant runoff changes and rates of aquifer recharge (Rosenberg et al., 1999) may also be anticipated for the semi-arid environments of the south-west USA (Thomson et al., 2005). Revelle and Waggoner (1983) argued that the effects of increased evapotranspiration losses as a result of a 2°C rise in temperature would be particularly serious in those regions where the mean annual precipitation is less than about 400 mm (Table 12.1). Projected summer

Table 12.1 Approximate percentage decrease in runoff for a 2°C increase in temperatures. Source: Data from Revelle and Waggoner (1983)

Initial mean temperature (°C)	Precipitation (mm/year)					
	200	300	400	500	600	700
−2	26	20	19	17	17	14
0	30	23	23	19	17	16
2	39	30	24	19	17	16
4	47	35	25	20	17	16
6	100	35	30	21	17	16
8		53	31	22	20	16
10		100	34	22	22	16
12			47	32	22	19
14			100	38	23	19

Table 12.2 Results of assessments of impacts of climate change on annual river runoff (basins and areas of water deficit). Source: After various sources in Shiklomanov (1999, table 2.1)

Region and basin	Temperature	Precipitation	Change in annual runoff (%)
Mean for seven large basins in the western USA	+2	−10	−40 to −76
Hypothetical arid basins	+1	−10	
Colorado River, USA	+1	−10	−50
Peace River, USA	+1	−10	−50
River basins in Utah and Nevada, USA	+2	−10	−60
River basins in the steppe zones of European Russia	+1	−10	−60

dryness in such areas may be accentuated by a positive feedback process involving decreases in cloud cover and associated increases in radiation absorption on the ground consequent upon a reduction in soil moisture levels (Manabe and Wetherald, 1986). Shiklomanov (1999) has suggested that in arid and semi-arid areas as a whole, an increase in mean annual temperature by 1–2° and a 10% decrease in precipitation could reduce annual river runoff by up to 40–70% (Table 12.2).

One factor that makes estimates of rainfall–runoff relationships complicated is the possible effect of higher CO_2 concentrations on plant physiology and transpiration capacity. At higher CO_2 concentrations transpiration rates are less, and this could lead to increases in runoff (e.g. Idso and Brazel, 1984). It is also important to remember that future runoff will be conditioned by non-climatic factors, such as land use and land cover change, the construction of reservoirs, groundwater storage and water demand (Conway et al., 1996).

A recent attempt to estimate the effects of global warming on runoff is provided by the UK Meteorological Office (Arnell, 1999a). What is clear from this work is that there will be clear differences at a global scale, with some areas generating more runoff and some generating less.

River channels in arid regions are particularly sensitive to changes in precipitation characteristics and runoff (Nanson and Tooth, 1999). They can display rapid switches between incision and aggradation over short time periods in response to quite modest changes in climate. This is particularly true in the case of the arroyos of the American south-west (Cooke and Reeves, 1976; Balling and Wells, 1990), which have displayed major changes in form since the 1880s (see Chapter 6). There has been considerable debate as to the causes of phases of trenching, and it is far from easy to disentangle anthropogenic from climatic causes, but in many cases it is fluctuations in either rainfall amount or intensity that have been the controlling factor (Hereford, 1984; Graf et al., 1991). Thus, the sorts of changes in runoff discussed in the last section could have a profound influence on channel characteristics. These in turn impact on humans because they can lead to changes in the agricultural suitability of bottomlands, modify local aquifers, modify sediment inputs into reservoirs and cut into engineering structures. Arroyo incision can also lead to the draining of riverbed marshes (*cienegas*). It may even have produced settlement abandonment (Hereford et al., 1995).

Lake levels

Closed depressions are widespread in arid lands, and their water levels and salinity characteristics respond rapidly and profoundly to climatic changes (Grimm et al., 1997). This generalization applies both to large and small lakes. In the twentieth century, for example, some of the largest arid zone lakes (e.g. Chad, the Aral Sea, the Caspian and the Great Salt Lake of Utah) have shown large variations in their extents, partly in some cases because of human activities, but also because of climatic fluctuations within their catchments. They are sensitive to climate variability that would be of minor influence in systems with outflows.

In the early 1960s, prior to the development of the Sahel Drought, Lake Chad had an area of 23,500 km², but by the 1980s had split into two separate basins and had an area of only 1500 km². The Caspian Sea was −29.10 m in 1977, but in 1995 had risen to −26.65 m, an increase of 2.45 m in just 17 years. Similarly impressive changes have occurred in recent decades in the level of the Great Salt Lake in Utah, with a particularly rapid rise taking place between 1964 and 1985 of nearly 20 ft (6 m).

Changes in lake level of these sorts of magnitude have an impact on a diverse range of human activities, ranging from fisheries and irrigation to recreation. Moreover, the drying-up of lake beds can have adverse effects on air quality and human health through the liberation of dust, as has been found as a result of the humanly induced desiccation of the Aral Sea (Indoitu et al., 2012) and Owens Lake (Reheis, 1997).

Lakes will respond to the temperature and precipitation changes that may result from global warming. Falls of up to 9 m by the end of the present century have been predicted for the Caspian (Elguindi and Giorgi, 2006).

Sea-level rise and arid zone coastlines

One consequence of global warming will be a rise in sea level at around 5 mm per year (50 cm per century). Arid coastlines will be affected by this process, which could potentially be especially serious for low-lying coastlines such as the Sabkhas of the United Arab Emirates and elsewhere in the Middle East. Sabkhas result from the interaction of various depositional and erosional processes that create a low-angle surface in the zone of tidal influence. This means that they are subject to periodic inundation and might be vulnerable to modest sea-level rise and to any increase in storm-surge events. Given the degree of infrastructural development that has taken place in their proximity, this is a serious issue for cities like Abu Dhabi. However, it is likely that many sabkhas will be able to cope with modestly rising sea levels, for a range of processes contribute to their accretion. These include algal stromatolite growth, faecal pellet deposition, aeolian inputs and evaporite precipitation. Some of these can cause markedly rapid accretion (Schrieber, 1986), even in the absence of a very well-developed plant cover. Moreover, as sea level (and groundwater) rises, surface lowering by deflational processes will be reduced. An example of a sabkha that may well maintain itself, or even continue to aggrade in spite of sea-level rise, is provided by the Umm Said Sabkha of Qatar, where aeolian dune inputs from inland cause the sabkha to build out into the Arabian Gulf.

Among arid-zone coastal environments that may be particularly susceptible to sea-level rise are deltaic areas subject to subsidence and sediment starvation (e.g. the Nile) and areas where ground subsidence is occurring as a result of fluid abstraction (e.g. California). Whereas the IPCC prediction of sea-level rise is 30–100 cm per century, rates of deltaic subsidence in the Nile Valley are 35–50 cm per century, and in other parts of the world, rates of land subsidence produced by oil, gas or groundwater abstraction can be up to 500 cm per century.

Rising sea levels can be expected to cause increased flooding, accelerated erosion and accelerated incursion of saline water up estuaries and into aquifers. Coastal lagoon, spit and barrier systems (such as those of Ras Al Khaimah) may be especially sensitive (Goudie et al., 2000), as will coastlines that have been deprived of sediment nourishment by dam construction across rivers.

Salt weathering

Salt weathering is a major geomorphological process in many drylands, and it poses a threat to many man-made structures, both ancient and modern (Goudie and Viles, 1997).

Global warming could influence salt weathering in a variety of ways (Table 12.3). In low-lying coastal areas, sea-level rise will impact upon groundwater conditions. The balance between fresh aquifers and seawater incursion is a very delicate one. Recent modelling, using the Ghyben–Herzberg relationship, suggests that in the Nile Delta, where subsidence is occurring at a rate of 4.7 mm per year, overpumping is

Table 12.3 Global warming and salinization

Negative effects	Positive effects
Increased moisture stress will lead to a greater need for irrigation.	Groundwater levels may fall.
Higher evaporation losses will lead to increased salt concentrations.	
Lesser freshwater inputs will reduce estuarine flushing and decrease replenishment in coastal aquifers.	
Higher sea levels will cause change of salinity in estuaries and in coastal aquifers.	

taking place, and less freshwater flushing is being achieved by the dammed Nile, a 50-cm rise in Mediterranean level will cause an additional intrusion of salt water by 9.0 km into the Nile Delta aquifer (Sherif and Singh, 1999).

In addition, many arid areas may become still more arid. Moisture deficits will increase, and stream flows will decrease, partly because of increased evaporative loss promoted by higher temperature, and partly because of decreased precipitation inputs. Lakes and reservoirs will suffer more evaporative losses and so could become more saline. In many areas stream flow will decline by 60–70%. Greater moisture stress may lead to an expansion in the quest for irrigation water, there may be less freshwater inputs to flush out estuarine zones, and rates of evaporative concentration of salt may increase (Utset and Borroto, 2001). Such tendencies could lead to an increase in salinization (Imeson and Emmer, 1992; Szabolcs, 1994). Conversely, in some areas, falling groundwater levels may follow on from decreased recharge, and this could lead to improvements in the incidence of groundwater-induced salinization.

Points for review

Are desert areas prone to large and rapid geomorphological and hydrological changes in response to apparently modest climatic stimuli?
To what extent may wind erosion of soils, dust storm activity and dune mobility change in a warmer world?

Guide to reading

Goudie, A.S. (2002) *Great Warm Deserts of the World: Landscapes and Evolution*. Oxford: Oxford University Press. A survey of the response of the world's deserts to past climatic changes.

Williams, M.A.J. and Balling, R.C. Jr. (1996) *Interactions of Desertification and Climate*. London: Arnold. A consideration of human impacts on desert environments, including a discussion of the effects of future climate changes.

13 CONCLUSION

Chapter Overview

This concluding chapter provides some general reflections on some of the issues that have been discussed elsewhere in the book. It stresses the role of non-industrial and pre-industrial societies in modifying the Earth's environment, discusses how human impacts have proliferated through time, considers to what extent the changes we have wrought are reversible, assesses to what extent some environments are particularly sensitive to change, discusses environmental problems in China, explores the problems of deciding whether certain environmental changes are natural or anthropogenic, tries to place the significance of potential future global warming in the context of other anthropogenic pressures, and concludes by highlighting the uncertainties that remain with regard to predicting the future.

The power of non-industrial and pre-industrial civilizations

It has become apparent that Marsh (1864) was correct nearly a century and a half ago to express his cogently argued views on the importance of human agency in environmental change. Since his time the impact that humans have had on the environment has increased, as has our awareness of this impact. There has been 'a screeching acceleration of so many processes that bring ecological change' (McNeill, 2000: 4). However, it is worth making the point here that, although much of

The Human Impact on the Natural Environment: Past, Present and Future, Seventh Edition. Andrew S. Goudie.
© 2013 John Wiley & Sons, Ltd. Published 2013 by John Wiley & Sons, Ltd.

the concern expressed about the undesirable effects humans have tends to focus on the role played by sophisticated industrial societies, this should not blind us to the fact that many highly significant environmental changes were and are being achieved by non-industrial societies.

In recent years it has become apparent that fire, in particular, enabled early societies to alter vegetation substantially, so that plant assemblages that were once thought to be natural climatic climaxes may in reality be in part anthropogenic fire climaxes. This applies to many areas of both savanna and mid-latitude grassland (see Chapter 2). Such alteration of natural vegetation has been shown to re-date the arrival of European settlers in the Americas (Denevan, 1992), New Zealand and elsewhere. The effects of fire may have been compounded by the use of the stone axe and by the grazing effects of domestic animals. In turn the removal and modification of vegetation would have led to adjustment in fauna. It is also apparent that soil erosion resulting from vegetation removal has a long history and that it was regarded as a threat by the classical authors.

Recent studies (see Chapter 2) tend to suggest that some of the major environmental changes in highland Britain and similar parts of western Europe that were one once explained by climatic changes can be more effectively explained by the activities of Mesolithic and Neolithic peoples. This applies, for example, to the decline in the numbers of certain plants in the pollen record and to the development of peat bogs and podzolization (see Chapter 4). Even soil salinization started at an early date because of the adoption of irrigation practices in arid areas, and its effects on crop yields were noted in Iraq more than 4000 years ago (see Chapter 4). Similarly (see Chapter 3) there is an increasing body of evidence that the hunting practices of early civilization may have caused great changes in the world's mega-fauna as early as 11,000 years ago.

In spite of the increasing pace of world industrialization and urbanization, it is ploughing and pastoralism which are responsible for many of our most serious environmental problems and which are still causing some of our most widespread changes in the landscape. Thus, soil erosion brought about by agriculture is, it can be argued, a more serious pollutant of the world's waters than is industry: many of the habitat changes which so effect wild animals are brought

Figure 13.1 Rainforest, such as this example in Sabah, Malaysia, is a repository of great biodiversity and also controls local erosion, runoff, and hydrological conditions. (See Plate 52)

about through agricultural expansion (see Chapter 3); and soil salinization and desertification can be regarded as two of the most serious problems facing the human race. Land-use changes, such as the conversion of tropical forests (Figure 13.1) to fields, may be as effective in causing anthropogenic changes in climate as the more celebrated burning of fossil fuels and emission of industrial aerosols into the atmosphere. The liberation of CO_2 in the atmosphere through agricultural expansion, changes in surface albedo values and the production of dust, are all major ways in which agriculture may modify world climates. Perhaps most remarkably of all, humans, who only represent roughly 0.5% of the total heterotroph biomass on Earth, appropriate for their use something around one-third of the total amount of net primary production on land (Imhoff et al., 2004).

The proliferation of impacts

A further point we can make is that, with developments in technology, the number of ways in which humans are affecting the environment is proliferating. It is these recent changes, because of the uncertainty which surrounds them and the limited amount of experience we have of their potential effects, which have caused greatest concern. Thus it is only since the Second World War, for example, that humans have had nuclear reactors for electricity generation, that they have used powerful pesticides such as dichlorodiphenyltrichloroethane (DDT), and that they have sent supersonic aircraft into the stratosphere. Likewise, it is only since around the turn of the century that the world's oil resources have been extensively exploited, that chemical fertilizers have become widely used, that CFCs have been manufactured, and that the internal combustion engine has revolutionized the scale and speed of transport and communications. Mass production of plastic products has only occurred since the 1950s and discarded plastic has a whole range of environmental consequences (Barnes et al., 2009), not least for the oceans (Moore, 2008). New techniques to obtain gas from shale by hydraulic fracturing (fracking) are likely to be widely adopted in coming decades, but have implications for water quality and seismic activity (Rahm, 2011). The wholesale adoption of e-technology produces what is called e-waste (Tripathy, 2010), which can contain many hazardous and polluting components (Wagner, 2009), including lithium from discarded batteries (Wanger, 2011). The world has been seeing a quadrupling in the production and consumption of rare earth elements (scandium, yttrium and 15 lathanides) since the 1970s, for they have many uses in such areas as energy-efficient lighting systems, hybrid cars, wind turbines and various low-carbon technologies. This creates environmental risks, including pollution by the acids used in the refining process, and the release of radioactivity from mine tailings and slurries.

Above all, however, the complexity, frequency and magnitude of impacts are increasing, partly because of steeply rising population levels and partly because of a general increase in *per capita* consumption (Figure 13.2). Thus some traditional methods of land use, such as shifting agriculture and nomadism, which have been thought to sustain some sort of environmental equilibrium, seem to break down and to cause environmental deterioration when population pressures exceed a particular threshold. For shifting agriculture systems fertility can be maintained under the long fallow cycles characteristic of low-density populations. However, as population levels increase, the fallow cycles necessarily become shorter, and soil fertility levels are not maintained, thereby imposing greater stresses on the land.

At the other end of the spectrum, increasing incomes, leisure and ease of communication, have generated a stronger demand for recreation and tourism in the developed nations. Environmental degradation has, for example, been caused by the proliferation of off-road vehicle use in sensitive environments like deserts. The number of these vehicles has increased more than sevenfold in the USA in the last three decades. Tourism development has created environmental problems, especially in coastal and mountain areas, and can have adverse effects on biodiversity (Pickering and Hill, 2007). Some of the environmental consequences of recreation, which are reviewed at length by Liddle (1997), can be listed as follows:

1 desecration of cave formations by speleologists;
2 trampling by human feet leading to soil compaction;
3 nutrient additions at campsites by people and their pets;
4 decrease in soil temperatures because of show compaction by snowmobiles;
5 footpath erosion and off-road vehicle erosion;
6 dune reactivation by trampling;
7 vegetation change due to trampling and collecting;
8 creating of new habitats by cutting trails and clearing campsites;
9 pollution of lakes and inland waterways by gasoline discharge from outboard motors and by human waste;
10 creation of game reserves and protection of ancient domestic breeds;
11 disturbance of wildlife by proximity of persons and by hunting, fishing and shooting;
12 conservation of woodland for pheasant shooting.

Table 13.1 indicates the remarkable growth in international tourism that has taken place since the middle of the twentieth century.

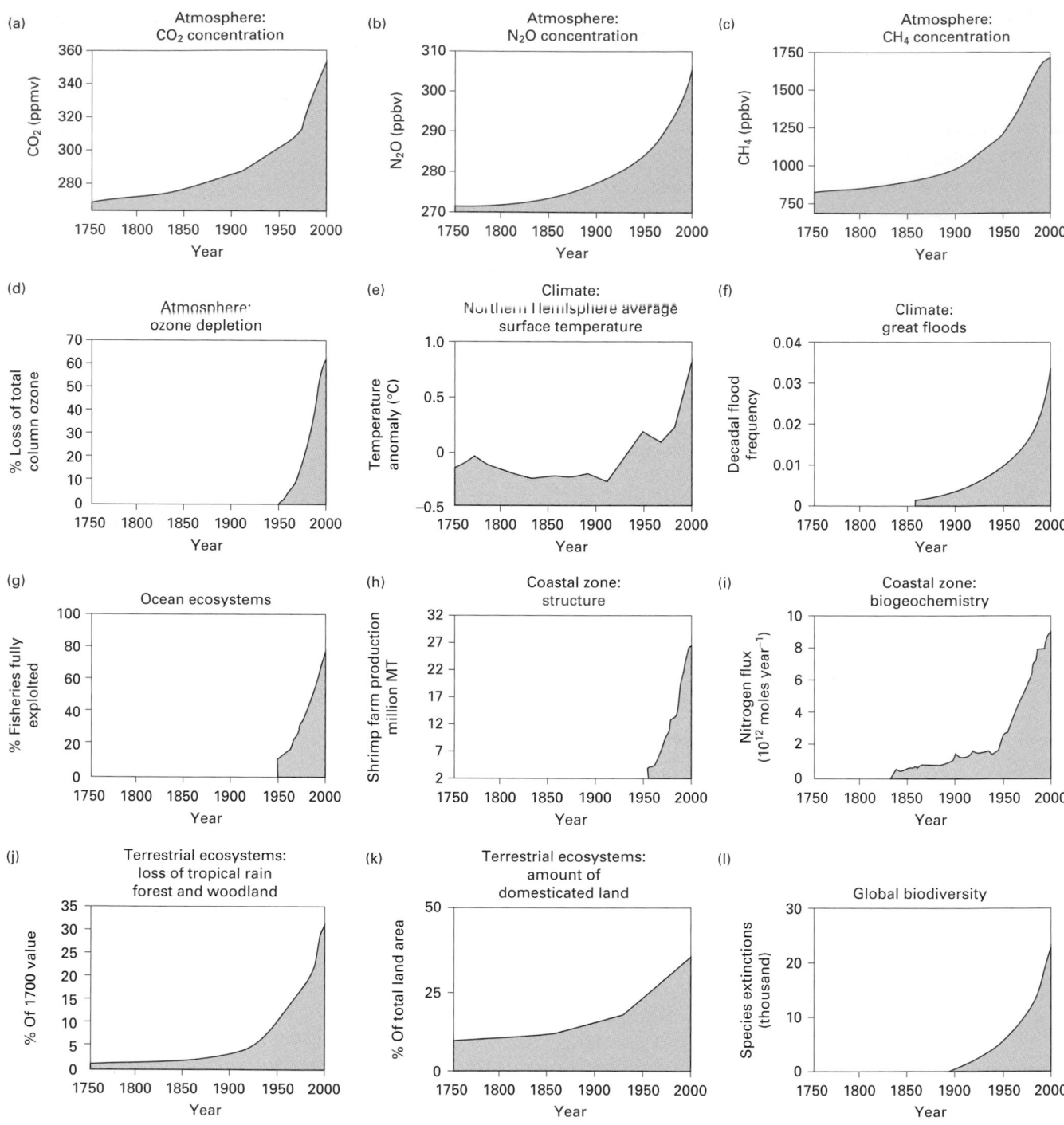

Figure 13.2 Global scale changes in the Earth System since 1750 caused by human activity (after Steffen, 2010, figure 16). Reproduced with permission.

Likewise, it is apparent when considering the range of possible impacts of one major type of industrial development that they are significant. As Table 13.2 indicates, the exploitation of an oilfield and all the activities that it involves (e.g., pipelines, new roads, refineries and drilling) have a wide range of likely effects on land, air, water and organisms.

Conversely, if one takes one ecosystem type as an example – the coral reef – once can see the diversity of

stresses to which they are now being exposed (Figure 13.3) as a result of a whole range of different human activities, which include global warming, increased sedimentation and pollution from river runoff, and overharvesting of fish and other organisms (Bellwood et al., 2004).

A very substantial amount of change has been achieved in recent decades. Table 13.3, based on the work of Kates et al. (1990), attempts to make quantitative comparisons of the human impact on 10 'component indicators of the biosphere'. For each component they defined total net change clearly induced by humans to be 0% for 10,000 years ago and 100% for 1985. They estimated dates by which each component had reached successive quartiles (i.e. 5, 50 and 75%) of its 1985 total change. They believe that about half of the components have changed more in the single generation since 1950 than in the whole of human history before that date.

It is clear that many environmental problems are now interrelated and transboundary in scope so that integrated approaches and international cooperation

Table 13.1 World international tourist arrivals (in millions). Source: UN World Tourism Organisation

Year	Number
1950	25.3
1960	69.3
1970	165.8
1980	278.1
1990	439.5
2000	687.0
2005	806.8
2020 (forecast)	1600.0

Table 13.2 Qualitative environmental impacts of mineral industries with particular reference to an oilfield. Source: Denisova (1977: 650, table 2).

Facility	Direction of the impact and reaction to the environment			
	Land	Air	Water	Biocenosis
Well	Alienation of land surface Extraction of oil associated gas, groundwater Pollution by crude oil, refined products, drilling mud Disturbance of internal structure of soil and subsoil Destruction of soil	Pollution by associated gas and volatile hydrocarbons, productions of combustion	Withdrawal of surface water and groundwater Pollution by crude oil and refined products, salination of freshwater Disturbance of water balance of both subsurface and surface waters	Pollution by crude oil and refined products Disturbance and destruction over a limited surface area
Pipeline	Alienation of land Accidental oil spills Disturbance of landforms and internal structure of soil and subsoil	Pollution by volatile hydrocarbons	Disturbance and destruction over limited surface area	
Motor roads	Alienation of land Pollution by oil products Disturbance of landforms and internal structure of soil and subsoil	Pollution by combustion products, volatile hydrocarbons, sulfur dioxide, nitrogen oxides	Pollution by combustion products Disturbances and destruction over limited surface area	
Collection point	Alienation of land Pollution by crude oil and refined products (spills) Disturbance of internal structure of soil and subsoil	Pollution by volatile hydrocarbons	Disturbance and destruction over limited surface area	

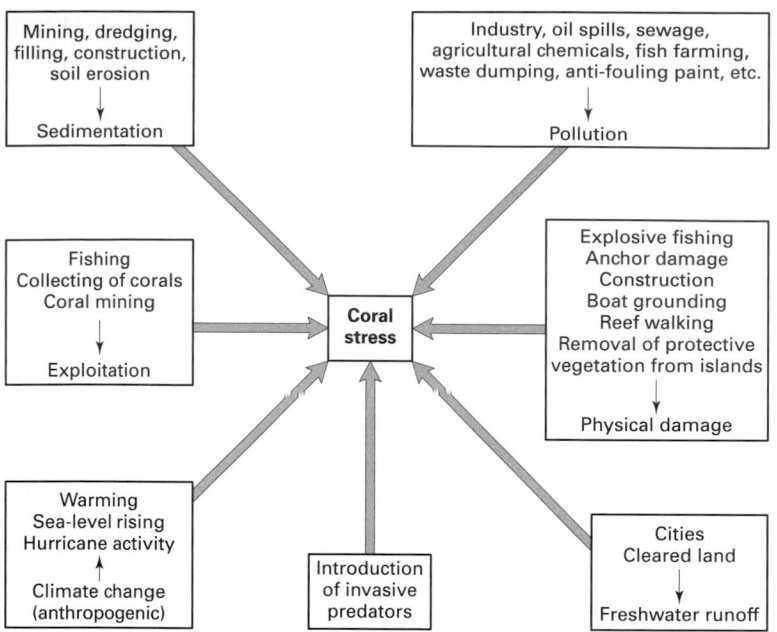

Figure 13.3 Some causes of anthropogenic stress on coral reef ecosystems.

are required. Indeed, environmental issues and environmental solutions have become globalized (Steffen et al., 2004: 290).

Human impacts on the environment in China

Some major countries have shown remarkable growth in recent decades (Table 13.4), notably China, India and Brazil. One of the most interesting facets of the study of the human impact in recent years has been to study the speed and range of human impacts that have occurred as a result of this growth. This can be illustrated by China. This huge country has seen a dramatic increase in population, urban growth, energy consumption, industrial production, water abstraction, the construction of great dams (e.g. the Three Gorges Project), car use and the like. Figure 13.4 shows the recent car and coal production. The latter trebled between 2000 and 2010. Internal tourism has also grown and has set in train whole suite of environmental concerns (Zhong et al., 2011). As a consequence of all these trends there has been a proliferation and intensification of many environmental problems. These are listed in Table 13.5. As we have seen for other parts of the world elsewhere in this book, there have been plant

invasions, loss of habitats (including forests, wetlands and mangrove swamps), extinction of mammals (most tragically the Yangtze dolphin), overexploitation of groundwater, saltwater intrusion, decreases in river flows because of dam construction and water abstraction, accelerated wind and water erosion of soils, soil acidification, desertification, land subsidence, permafrost degradation, landslide generation, glacier retreat in the face of global warming, accelerated coastal retreat because of sediment starvation, eutrophication of water bodies, the growth of algal blooms offshore, the build-up of heat in cities as a result of the heat island effect, and a wide range of types of air, water and soil pollution. Air pollution probably causes in excess of 400,000 premature deaths a year in China, a country which has been termed the 'air pollution capital of the world' (Watts, 2005). Steps are being taken to mitigate some of these problems (Yi et al., 2007), and not all environmental problems are getting worse, either because of legislation or because of more efficient use of energy sources. China has instituted a massive re-afforestation policy, though it is not always soundly based in ecological terms (Cao et al., 2011). In addition, some problems, such as increased fogs (Niu et al., 2010) and dust storms, may at least in part be the result of natural climate changes. Nevertheless, through the burning of fossil fuels, China is becoming an

Table 13.3 Chronologies of human-induced transformations

(a) Quartiles of change from 10,000 BC to mid-1980s

Form of transformation	Dates of quartiles		
	25%	50%	75%
Deforested area	1700	1850	1915
Terrestrial vertebrate diversity	1790	1880	1910
Water withdrawals	1925	1955	1975
Population size	1850	1950	1970
Carbon releases	1815	1920	1960
Sulfur releases	1940	1960	1970
Phosphorus releases	1955	1975	1980
Nitrogen releases	1970	1975	1980
Lead releases	1920	1950	1965
Carbon tetrachloride production	1950	1960	1970

(b) % change by time of Marsh and Princeton symposium.
Source: From Kates et al. (1990, table 1.3).

Form of transformation	% change	
	1860	1950
Deforested area	50	90
Terrestrial vertebrate diversity	25–50	75–100
Water withdrawals	15	40
Population size	30	50
Carbon releases	30	65
Sulfur releases	5	40
Phosphorus releases	<1	20
Nitrogen releases	<1	5
Lead releases	5	50
Carbon tetrachloride production	0	25

increasingly major source of global greenhouse gas emissions, its desire for wood drives deforestation in other countries, it is responsible for a substantial portion of world fish consumption, and it exports sulfur dioxide pollution (of which it is the world's largest culprit) to other parts of the Pacific rim (Liu and Diamond, 2005). It may also be the world's biggest emitter of nitrogen oxides (Vennemo et al., 2009).

Are changes reversible?

It is evident that while humans have imposed many undesirable and often unexpected changes on the environment, they often have the capacity to modify

Table 13.4 Some recent developments in Brazil, China and India. Source: World Bank

(a) Air transport. Passengers carried (in millions)

	1981	1991	2001	2009
Brazil	12.6	19.2	34.3	67.9
China	3.2	19.5	72.7	229.0
India	7.6	10.7	16.9	54.4

(b) Energy production (kt of oil equivalent)

	1981	1991	2001	2008
Brazil	66,541	104,779	152,152	228,127
China	612,319	894,968	1,092,157	1,993,306
India	200,049	301,567	374,522	468,307

(c) Carbon dioxide emissions (kt)

	1981	1991	2001	2007
Brazil	171,666	219,151	339,616	368,016
China	1,450,314	2,582,424	3,484,512	6,533,018
India	374,516	737,248	1,202,858	1,611,043

the rate of such changes or to reverse them. There are cases where this is not possible: once soil has been eroded from an area it cannot be restored; once a plant or animal has become extinct it cannot be brought back; and once a laterite iron pan has become established it is difficult to destroy.

However, through the work of Marsh and others, people became aware that many of the changes that had been set in train needed to be reversed or reduced in degree. Sometimes this has simply involved discontinuing a practice which has proved undesirable (such as the cavalier use of DDT or CFCs), or replacing it with another which is less detrimental in its effects. Often, however, specific measures have been taken which have involved deliberate decisions of management and conservation. Denson (1970), for example, outlines a sophisticated six-stage model for wildlife conservation:

1 Immediate physical protection from humans and from changes in the environment;
2 Educational efforts to awaken the public to the need for protection and to gain acceptance of protective measures;

(a)

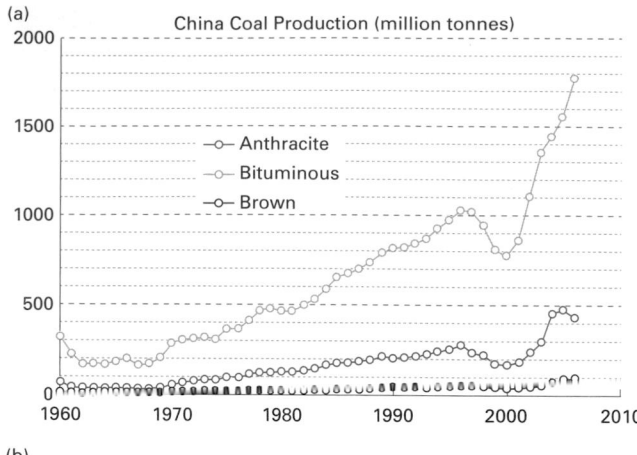

(b)

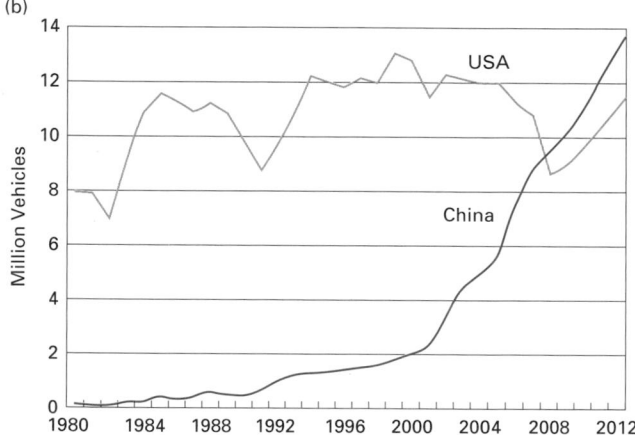

Figure 13.4 Trends in coal and car production in China.

3 Life-history studies of the species to determine their habitat requirements and the causes of their population decline;

4 Dispersion of the stock to prevent loss of the species by disease or by a chance event such as fire;

5 Captive breeding of the species to assure higher survival of young, to aid research and to reduce the chances of catastrophic loss;

6 Habitat restoration or rehabilitation when this is necessary before introducing the species.

Many conservation measures have been successful, while others have created as many problems as they were intended to solve. This applies, for example, to certain schemes for the reduction of coast erosion. On balance, however, there has been notable progress in dealing with such problems as acid rain in Europe, and the depletion of stratospheric ozone.

Table 13.5 Examples of human impacts on the environment in China

Acid rain	Tang et al. (2010)
Acidification of soils	Guo et al. (2010)
Acid mine drainage	Tao et al. (2012)
Aeolian desertification	Hu et al. (2012)
Aerosol pollution	Tie and Cao (2009)
Algal blooms	Tang et al. (2006)
Alien plant invasions	Weber et al. (2008)
Coast erosion	Xu et al. (2009)
Decreased sediment flux to coast	Chu et al. (2009)
Delta erosion	Yang et al. (2011b)
Deforestation	Zhang and Song (2006)
Groundwater depletion	Foster et al. (2004)
Heavy metal pollution	Cheng (2003)
Karst rock desertification	Huang and Cai (2007)
Lake eutrophication	Le et al. (2010)
Land degradation	Zhang et al. (2007)
Land subsidence	Xue et al. (2005)
Landslide generation	Huang and Chan (2004)
Lead pollution	Gottesfeld and Cherry (2011)
Mammal extinction	Turvey et al. (2007)
Mangrove loss	Chen et al. (2009)
Mercury pollution	Wu et al. (2006); Li et al. (2012)
Municipal waste accumulation	Zhang et al. (2010a)
Nitrate pollution	Chen et al. (2007)
Nitrogen dioxide pollution	Richter et al. (2005)
Ozone pollution	Zhao et al. (2009)
PCB pollution	Xing et al. (2005)
Permafrost degradation	Yang et al. (2010a)
Persistent organic pollutants	Fu et al. (2003)
Rocky desertification	Yang et al. (2011a)
Saltwater intrusion	An et al. (2009)
Shrinking glaciers	Li et al. (2008)
Soil erosion	Liu et al. (2010)
Soil nutrient loss	Zheng (2005)
Stream flow reduction	Xu and Milliman (2009)
Sulfur dioxide pollution	Lu et al. (2010)
Sunshine hours depletion	Wang et al. (2012)
Urban heat island build up	Tan et al. (2010)
Vehicular emissions	Wang et al. (2010)
Wetland loss	Zhang et al. (2010b)
Wind erosion	Shi et al. (2004)

The concern with preservation and conservation has been long-standing, with many important landmarks. Interest has grown dramatically in recent years. Lowe (1983) identified four stages in the history of British Nature Conservation:

1 the natural history/humanitarium (1830–1890)
  period
2 the preservation period (1870–1940)
3 the scientific period (1910–1970)
4 the popular/political period (1960–present).

The first of these stages was rooted in a strong enthusiasm for natural history and the crusade against cruelty to animals. Although many Victorian naturalists were avid collectors, numerous clubs were established to study nature and some of them sought to preserve species to make them available for observation. As we shall see, certain acts were introduced at this time to protect birds. During the preservationist period, there was the formation of a spate of societies devoted to preserving open land and its associated wildlife (e.g. the National Trust, 1894; and the Council for the Preservation of Rural England, 1926). There was a growing sense of vulnerability of wildlife and landscapes to urban and industrial expansion and geographers like Vaughan Cornish (see Goudie, 1972) campaigned for the creation of national parks and the preservation of scenery, made possible through the National Parks and Access to Countryside Act of 1949. From the First World War onwards ecological research developed, and there arose an increasing understanding of ecological relationships. Scientists pressed for the regulation of habitats and species, and the Nature Conservancy was established in 1949. Popular interest in conservation and widespread media attention first developed in the 1960s. This was partly generated by pollution incidents (such as the wrecks of the *Torrey Canyon* and *Amoco Cadiz*), and a gathering sense of impending environmental doom, generated by such persuasive books as Rachel Carson's *Silent Spring* of 1962. Ecology became a political issue in various European nations, including the UK. In many countries major developments in land use, construction and industrialization now have to be preceded by the production of an Environmental Impact Assessment, and the European Union has introduced measures such as the Landfill Directive (1999) and the Water Framework Directive (2000) to improve the disposal of waste and the ecological status of water resources.

Thus in some countries, and in connection with particular species, conservation and protection have had a long and sometimes successful impact. In Britain, for example, the Wild Birds Protection Act dates back to 1880, and the Sea Birds Protection Act even further to 1869. The various acts have been modified and augmented over the years to outlaw egg-collecting, pole-trapping, plumage importation and the capture or possession of a range of species.

One further ground which gives some basis for hope that humans may be soon reconciled with the environment is that there are some signs of a widespread shift in public attitudes to nature and the environment. These changing social values, combined with scientific facts, influence political action. This point of view, which acts as an antidote for some of the more pessimistic views of the world's future, was elegantly presented by Ashby (1978). He contended that the rudiments of a healthy environmental ethic are developing, and explained

Its premise is that respect for nature is more moral than lack of respect for nature. Its logic is to put the Teesdale Sandwort . . . into the same category of value as a piece of Ming porcelain, the Yosemite Valley in the same category as Chartres Cathedral: a Suffolk landscape in the same category as a painting of the landscape by Constable. Its justification for preserving these and similar things is that they are unique, or irreplaceable, or simply part of the fabric of nature, just as Chartres and the painting by Constable are part of the fabric of civilisation; also that we do not understand how they have acquired their durability and what all the consequences would be if we destroy them. (pp. 84–55)

Although there may be considerable controversy surrounding the precise criteria that can be used to select and manage sites that are particularly worthy of conservation (Goldsmith, 1983), there are nonetheless many motives behind the increasing desire to protect species and landscapes. These can be listed under the following general headings:

1 *The ethical:* It is asserted that wild species have a right to coexist with us on our planet, and that we have no right to exterminate them. Nature, it is maintained, is not simply there for humans to transform and modify as they please for their own utilitarian ends.
2 *The scientific:* We know very little about our surrounding environments, including, for example, the rich insect faunas of the tropical rainforest;

therefore such environments should be preserved for future scientific study.

3 *The aesthetic:* Plants and animals, together with landscapes, may be beautiful and so enrich the life of humans.

4 *The need to maintain genetic diversity:* By protecting species we maintain the species diversity upon which future plant- and animal-breeding work will depend. Once genes have been lost (see Chapter 2, section on 'The Change in Genetic and Species Diversity') they cannot be replaced.

5 *Environmental stability:* It is argued that in general, the more diverse an ecosystem is, the more checks and balances there are to maintain stability. Thus environments that have been greatly simplified by humans may be inherently unstable, and prone to disease, and so on.

6 *Recreational:* Preserved habitats and landscapes have enormous recreational value, and in the case of some game, reserves and natural parks may have economic value as well (e.g. the safari industry of East Africa).

7 *Economic:* Many of the species in the world are still little known, and there is the possibility that we have great storehouses of plants and animals, which, when knowledge improves, may become useful economic resources.

8 *Future generations:* One of the prime arguments for conservation, whether of beautiful countryside, rare species, soil or mineral resources, is that future generations (and possibly ourselves later in life) will require them, and may think badly of a generation that has squandered them.

9 *Unintended impacts:* As we have seen so often in this book, profligate or unwise actions can lead to side effects and consequences that may be disadvantageous to humans.

10 *Spiritual imperatives:* This includes a belief in the need for environmental stewardship.

Some of these arguments are more utilitarian than others (e.g. 4, 5, 6, 7 and 9), and some may be subject to doubt – it could, for example, be argued that future generations will have technology to use new resources and may not need some of those we regard as essential – but overall they provide a broadly based platform for the conservation ideal (Myers, 1979).

The world's network of protected areas has grown steadily since the latter part of the nineteenth century (Figure 13.5), and particularly since the early 1960s. Indeed, there are now over 120,000 terrestrial sites worldwide, and these cover about 12.2% of the land area, and have a combined area almost the size of South America.

The susceptibility to change

Ecosystems respond in different ways to the human impact, and some are more vulnerable to human perturbation than others (Kasperson et al., 1995). It has often been thought, for example, that complex ecosys-

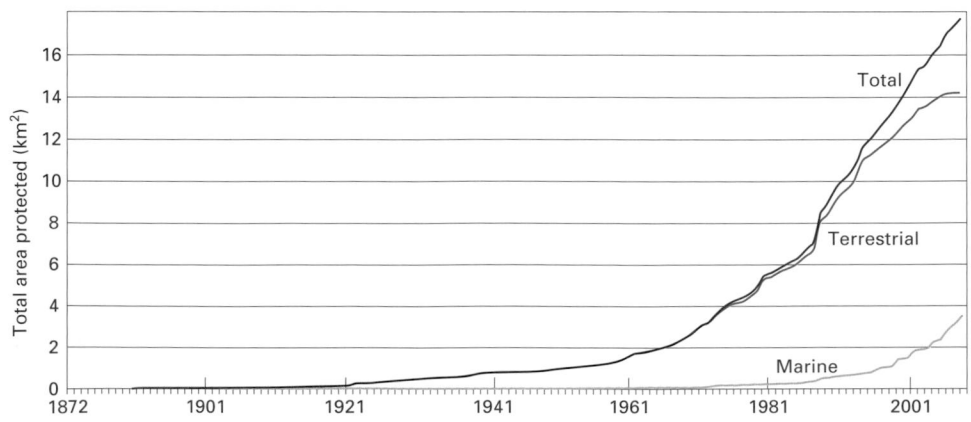

Figure 13.5 Global growth in nationally protected areas from 1872 to 2008 (based on data from UNEP and World Conservation Monitoring Centre).

tems are more stable than simple ones. Thus in Clements' Theory of **Succession** the tendency towards community stabilization was ascribed in part to an increasing level of integration of community functions. As Goodman (1975: 238) expressed it:

In general the predisposition to expect greater stability of complex systems was probably a combined legacy of eighteenth century theories of political economics, aesthetically and perhaps religiously motivated attraction to the belief that the wondrous variety of nature must have some purpose in an orderly work, and ageless folkwisdom regarding eggs and baskets.

Indeed, as Murdoch (1975) pointed out, it makes good intuitive sense that a system with many links, or 'multiple fail-safes', is more stable than one with few links or feedback loops. As an example, if a type of herbivore is attacked by several predatory species, the loss of any one of these species will be less likely to allow the herbivore to erupt or explode in numbers than if only one predator species were present and that single predator type disappeared. The basic idea therefore is that diverse groups of species are more stable because complementary species compensate for one another if one species suffers severe declines (Doak and Marvier, 2003). A diverse ecosystem will have a variety of species that help to insure it against a range of environmental upsets (Naeem, 2002).

Various other arguments have been marshalled to support the idea that great diversity and complexity affords greater ability to minimize the magnitude, duration and irreversibility of changes brought about by some external perturbation such as humans (Noy-Meir, 1974). It has been stated that natural systems, which are generally more diverse than artificial systems such as crops or laboratory populations, are also more stable. Likewise, the tropical rainforest has been thought of as more diverse and more stable than less complex temperate communities, while simple Arctic ecosystems of oceanic islands have always appeared highly vulnerable to disturbance brought about by anthropogenic plant and animal introductions (see Chapter 3).

However, considerable doubt has been expressed as to whether the classic concept of the causal linkage between diversity/complexity and stability is entirely valid (see e.g. Hurd et al., 1971). Murdoch (1975) indicated that there is not convincing field evidence that diverse natural communities are generally more

stable than simple ones. He cited various papers which show that fluctuations of microtine rodents (lemmings, field voles, etc.) are as violent in relatively complex temperate zone ecosystems as they are in the less complex Arctic zone ecosystems. This was supported by Goodman (1975: 239) who wrote:

As for the apparent stability of tropical biota, that could well be an illusion attributable to insufficient study of bewilderingly complex assemblages in which many species are so poorly represented in samples of feasible size that even considerable fluctuations might go undetected. Indeed, there are countervailing anecdotes regarding ecological instability in the tropics, such as recent reports on an insect virtually defoliating the wild Brazil-nut trees in Bolivia and of monkeys succumbing in large numbers to epidemics.

He went on to add: 'There is growing awareness of the surprising susceptibility of the rain forest ecosystems to man-made perturbation.' This is a point of view supported by May (1979) and discussed by Hill (1975). Hill pointed out that a very high species diversity is frequently associated with areas which have relatively constant physical environmental conditions over the course of a year and a series of years. The rainforest may be construed to be such an environment and one where this constancy has allowed the presence of many specialized species, each pursuing a narrow range of activities. It has been argued that because of the high degree of specialization, the indigenous species have a limited ability to recover from major stresses caused by human intervention.

Goodman (1975) also queried the sufficiency of the argument in its reference to the apparent instability of island ecosystems, suggesting that islands, being evolutionary backwaters and dead ends, may accumulate species that are especially susceptible to competitive or exploitative displacement. In this case, lack of diversity may not necessarily be the sole or prime cause of instability.

The apparent instability of agricultural compared to natural communities is also often attributed to lack of diversity, and indeed modern agriculture does involve significant ecosystem simplification. However, such instability as there is may not, once again, necessarily result from simplification. Other factors could promote instability: agricultural communities are disrupted, even destroyed more frequently and more massively as part of the cultivation process than those natural systems we tend to think of as stable; the component species of natural systems are co-evolved (co-adapted), and this is not usually true of agricultural communities. As Murdoch (1975: 799) suggests, it may be that:

Natural systems are more stable than crop systems because their interacting species have had a long shared evolutionary history. In contrast with these natural communities the dominant plant species of a crop system is thrust into an often alien landscape . . . the crops have undergone radical selection in breeding programmes, often losing their genetic defence mechanisms.

Thus the idea that complex natural ecosystems will be less susceptible to human interference and that simple artificial ecosystems will inevitably be unstable are not necessarily tenable. Nonetheless, it is apparent that there are differences in susceptibility between different ecosystem types, and these differences may result from factors other than the degree of diversity and complexity (Cairns and Dickson, 1977).

Some systems tend to be *vulnerable*. Lakes, for example, are natural traps and sinks and are thus more vulnerable to the effect of disadvantageous inputs than are rivers (which are continually receiving new inputs) or oceans (which are so much larger). Other systems display the property of *elasticity* – the ability to recover from damage. This may be because nearby epicentres exist to provide organisms to reinvade a damaged system. Small, isolated systems will often tend to possess low elasticity. Two of the most important properties, however, are *resilience* Folke et al., 2004) (being a measure of the number of times a system can recover after stress), and *inertia* (the ability to resist displacement of structural and functional characteristics).

Two systems which display resilience and inertia are deserts and estuaries. In both cases their indigenous organisms are highly accustomed to variable environmental conditions. Thus most desert fauna and flora evolved in an environment where the normal pattern is one of more or less random alternations of short favourable periods and long stress periods. They have pre-adapted resilience (Noy-Meir, 1974) so that they can tolerate extreme conditions, have the ability for rapid recovery, have various delay and trigger mechanisms (in the case of plants) and have flexible and opportunistic eating habits (in the case beasts). Estuaries, on the other hand, through the subject of increasing human pressures, also display some resilience. The

vigour of their water circulation continuously and endogenously renews the supply of water, food, larvae, etc.; this aids recovery. Also, many species have biological characteristics which provide special advantages in estuarine survival. These characteristics usually protect the species against the natural violence of estuaries and are often helpful in resisting external forces such as humans.

The relationship between biodiversity and ecosystem stability continues to be a hot topic in ecology (Loreau et al., 2002; Kareiva and Lovin, 2003; Tilman et al., 2006; Ives and Carpenter, 2007; Downing and Leibold, 2010). Some studies continue to throw doubt upon any simple relationship between biodiversity and stability (e.g. Pfisterer and Schmid, 2002), but there is perhaps an emerging consensus that diversity is crucial to ecosystem operation (McCann, 2000). As Loreau et al. (2001: 807) wrote,

There is consensus that at least some minimum number of species is essential for ecosystem functioning under constant conditions and that a larger number of species is probably essential for maintaining the stability of ecosystem processes in changing environments.

Human influence or nature?

From many of the examples given in this book it is apparent that in many cases of environmental change, it is impossible to state, without risk of contradiction, that people rather than nature are responsible. Most systems are complex and human agency is but one component of them, so that many human actions can lead to end-products which are intrinsically similar to those that may be produced by natural forces. How to distinguish between human-induced perturbations and ill-defined natural oscillations is a crucial question when considering issues like coral reef degradation (Sapp, 1999). It is a case of equifinality, whereby different processes can lead to basically similar results. Humans are not always responsible for some of the changes with which they are credited. This book has given many examples of this problem and a selection is presented in Table 13.6. Deciphering the cause is often a ticklish problem, given the intricate interdependence of different components of ecosystems, the frequency and complexity of environmental changes,

Table 13.6 Human influence or nature? Some examples, with chapter references to this book where applicable

Change	Causes		Chapter reference
	Natural	*Anthropogenic*	
Late Pleistocene animal extinction	Climate	Hunting	3
Death of savanna trees	Soil salinization through climatically induced groundwater rise	Overgrazing	4
Desertification of semi-arid areas	Climatic change	Overgrazing, etc.	2
Holocene peat-bog development in highland Britain	Climatic change and progressive soil deterioration	Deforestation and ploughing	2
Holocene elm and linden decline	Climatic change	Feeding and stalling of animals	2
Tree encroachment into alpine pastures in USA	Temperature amelioration	Cessation of burning	2
Gully development	Climatic change	Land-use change	6
Late twentieth-century climatic warming	Changes in solar emission and volcanic activity	CO_2-generated greenhouse effect	8
Increasing coast recession	Rising sea level	Disruption of sediment supply	9
Increasing coastal flood risk	Rising sea level, natural subsidence	Pumping of aquifers creating subsidence	9
Increasing river flood intensity	Higher intensity rainfall	Creation of drainage ditches	10
Ground collapse	Karstic process	Dewatering by overpumping	6
Delta subsidence	Tectonics, sediment compaction, sediment loading	Fluid abstraction, drainage, water management	6
Forest decline	Drought	Air, soil and water pollution	2

and the varying relaxation times that different ecosystem components may have when subject to a new impulse. This problem plainly does not apply to the same extent to changes which have been brought about deliberately and knowingly by humans, but it does apply to the many cases where humans may have initiated change inadvertently and unintentionally.

This fundamental difficulty means that environmental impact statements of any kind are extremely difficult to make. As we have seen, humans have been living on the earth and modifying it in different degrees for several millions of years, so that it is problematical to reconstruct any picture of the environment before human intervention. We seldom have any clear baseline against which to measure changes brought about by human society. Moreover, even without human interference, the environment would be in a perpetual state of flux on a great many different timescales. In addition, there are spatial and temporal discontinuities between cause and effect. For example, erosion in one locality may lead to deposition in another, while destruction of key elements of an animal's habitat may lead to population declines throughout its range. Likewise, in a time context, a considerable interval may elapse before the full implications of an activity are apparent. Also, because of the complex interaction between different components of different environmental systems and subsystems, it is almost impossible to measure total environmental impact. For example, changes in soil may lead to changes in vegetation, which in turn may trigger changes in water quality and in animal populations. Primary impacts give rise to myriad of successive repercussions throughout ecosystems which may be impracticable to trace and monitor. Quantitative cause-and-effect relationships can seldom be established.

Global warming and other pressures

As has been made clear in Chapter 8, Chapter 9, Chapter 10, Chapter 11, and Chapter 12, future climate changes, and in particular global warming, will have many impacts on the environment. However, global warming is not the only big global issue that we face. Also of immense and in many cases more immediate importance are other aspects of global change, particularly those brought about my modifications of land use and land cover. Biodiversity loss, problems of water shortage and quality, the undermining of the value of our soil resources because of such processes as accelerated erosion and accelerated salinization, and the impacts of many types of pollution, remain of huge significance. Deforestation and land cover change may cause climate changes at the regional scale which may be of comparable dimensions to those predicted to arise from global warming (e.g. Werth and Avissar, 2002; Sampaio et al., 2007; Deo et al., 2009; Avila et al., 2012), changes in stream runoff and sediment loads caused by land cover change or dam construction may far exceed those caused by future changes in rainfall amounts (e.g. Ericson et al., 2006; Xu et al., 2007), loss of coastal wetlands due to direct human action may exceed those caused by sea-level rise (Nicholls et al., 1999), and landslide activity may owe more to changes in human activity than to changes in climate (Crozier, 2010). Overexploitation of groundwater may be a more potent cause of saltwater intrusion in coastal area than sea-level rise caused by global warming (Ferguson and Gleeson, 2012).

Secondly, it is important to recognize that global warming is going to act on an Earth that has already been transformed by other human activities. If it works in the same direction as these other processes, then the compound effect will be greater than individual effects in isolation. Global warming may already be working on 'a groggy Earth'. For example, many of the world's coastlines are eroding quickly at the present day because of sediment starvation, but this rate of erosion may almost certainly be accelerated further as a result of sea-level rise. Equally, sea-level rise will be especially threatening in areas that are subsiding because of groundwater or hydrocarbon extraction. Deserts are suffering today from land degradation and desertification, and this could be exacerbated if droughts become more prevalent and severe in a warmer world. The state of the Amazonian rainforest is being threatened by deforestation, which though serious in itself may also cause regional rainfall declines that could be rendered still greater if global warming is associated with greater drought incidence.

Thirdly, as discussed in Chapter 8, some environments may be especially sensitive to future climate change, either because they will be subjected to particularly severe impacts, or because they are inherently weak (e.g. sandy coastlines), or because they are close to some

environmental threshold (e.g. at the edge of their temperature tolerance like some permafrost areas), or because, as discussed above, they have already been adversely impacted by other environmental changes.

Fourthly, some attempts at reducing greenhouse gas emissions may, perversely, have adverse environmental effects. This is, for example, the case with biofuel development. The increase in the production of ethanol derived from such crops as sugar cane and maize has been remarkable: global production in 1980 was around 1 billion gallons, and by 2011 had jumped to 23 billion gallons. Biomass production that places additional demands on land and water resources, and increases nutrient, pesticide and water use can have negative implications for environmental quality. Also, conversion of forests into biofuel production could lead to an increase rather than a decrease in greenhouse gas emissions (Searchinger et al., 2008). Production of biofuels may mean that grasslands, wetlands or forests are transformed. There are also the effects of indirect land-use changes. For instance, if increased US maize production for ethanol were to reduce soybean production and exports, changes in world prices might provide an incentive to increase soybean production on land previously used as pasture in Brazil which in turn might push cattle farming into Amazonia (Marshall et al., 2011). Similarly, wind energy has various adverse environmental consequences, including noise, visual intrusion, local climatic change and the killing of birds (Leung and Yang, 2012). Tidal barrage schemes have also been opposed because of their influence on sediment budgets, mud flat exposure and migration routes for fish (Kadiri et al., 2012), while wave energy schemes can create barriers, electromagnetic waves, noise and other problems for organisms (Gill, 2005; Frid et al., 2012). Photovoltaic energy systems may add to lead pollution risks (Gottesfeld and Cherry, 2011), while the use of geothermal energy might increase seismic activity (Majer et al., 2007). Geoengineering to reduce global warming (see Chapter 7) would also have an array of environmental implications (Vaughan and Lenton, 2011).

Into the unknown

During the last three decades the full significance of possible future environmental changes has become apparent, and national governments and international institutions have begun to ponder whether the world is entering a spasm of unparalleled humanly induced modification. For example, Steffen et al. (2004) have suggested that the Earth is currently operating in a no-analogue state. They remark (p. 262):

In terms of key environmental parameters, the Earth System has recently moved well outside the range of the natural variability exhibited over at least the last half million years. The nature of changes now occurring simultaneously in the Earth System, their magnitudes and rates of change are unprecedented.

Likewise, the Amsterdam Declaration of 2001 pointed to the role of thresholds and surprises (see Steffen et al., 2004: 298):

Global change cannot be understood in terms of a simple cause-effect paradigm. Human-driven changes cause multiple effects that cascade through the earth System in complex ways. These effects interact with each other and with local- and regional-scale changes in multidimensional patterns that are difficult to understand and even more difficult to predict. Surprises abound.

Earth System dynamics are characterised by critical thresholds and abrupt changes. Human activities could inadvertently trigger such changes with severe consequences for Earth's environment and inhabitants.

Our models and predictions are still highly inadequate (see Chapter 8), and there are great ranges in some of the values we give for such crucial changes as sea-level rise and global climatic warming, but the balance of scientific argument favours the view that change will occur and that change will be substantial. Some of the changes may be advantageous for humans or for particular ecosystems; others will be extremely disadvantageous.

Some environments will change very substantially during the twenty-first century in response to a rise of land-use changes and climatic changes, with some predictions suggesting that the world's grasslands and Mediterranean biomes being particularly impacted (Sala et al., 2000). Marine ecosystems will also be impacted and Jenkins (2003: 1176) suggests that by 2050: 'If present trends . . . continue, the world's marine ecosystems in 2050 will look very different from today's, large species, and particularly top predators, will be by and large extremely scarce and some will

have disappeared entirely . . .' Human populations will increase, and will probably be greater by 2–4 billion people by 2050 (Cohen, 2003).

But all change, if it is rapid and of a great magnitude, is likely to create uncertainties and instabilities, and there is always the risk that critical thresholds or boundaries will be crossed (Rockström et al., 2009; Lynas, 2011). The study of future events will not only become a major concern for the environmental sciences but will also become a major concern for economists, sociologists, lawyers and political scientists. George Perkins Marsh was a lawyer and politician, but it is only now, over a century since he wrote *Man and Nature*, that the wisdom, perspicacity and prescience of his ideas have begun to be given the praise and attention they deserve.

Points for review

Why has there been a 'screeching acceleration' in the twentieth century of so many processes that bring ecological change?

How may adverse environmental changes be reversed?

Why should one conserve nature?

In the context of ecosystems, what do you understand by such terms as 'stability', 'resilience', 'elasticity' and 'inertia'?

Why is it often difficult to disentangle natural and anthropogenic causes of environmental changes?

Is global warming the most serious environmental threat?

Guide to reading

Liddle, M. (1997) *Recreation Ecology*. London: Chapman and Hall. An excellent review of the multiple ways in which recreation and tourism have an impact on the environment.

Lynas, M. (2011) *The God Species*. London: Fourth Estate. A discussion of how humans may be transgressing various environmental thresholds or boundaries.

O'Riordan, T. (ed.) (1995) *Environmental Science for Environmental Management*. Harlow: Longman Scientific. A multi-author guide to managing environmental change.

Roberts, N. (ed.) (1994) *The Changing Global Environment*. Oxford: Blackwell. A good collection of case studies with a wide perspective.

Simon, J.L. (ed.) (1995) *The State of Humanity*. Oxford: Blackwell. A multi-author work with an optimistic message.

Simon, J.L. (1996) *The Ultimate Resource 2*. Princeton: Princeton University Press. A view that the state of the world is improving.

Steffen, W. and 10 others. (2004) *Global Change and the Earth System*. Berlin: Springer. Chapter 6 of this multi-author volume addresses ways of global management for global sustainability.

Turner, B.L. (ed.) (1990) *The Earth as Transformed by Human Action*. Cambridge: Cambridge University Press. A magnificent study of change in the last 300 years.

GLOSSARY

Note: Places in the text where glossary terms are first mentioned are marked in bold.

Aerosols Aggregations of minute particles (solid or liquid) suspended in the atmosphere. The term is often used to describe smoke, dust, condensation nuclei, freezing nuclei, fog or pollutants such as droplets containing sulfur dioxide or nitrogen dioxide.

Aggradation The building upwards or outwards of the land surface by the deposition of sediment.

Albedo A measure for the reflectivity of a body or surface, defined as the total radiation reflected by the body divided by the total radiation falling on it. Values are expressed on a scale of either 0–1 or 1–100%.

Allochthonous Formed at a distance from its present position (see Autochthonous).

Anthropocene Epoch which started about three centuries ago in which the power of humans has become a global geophysical force.

Aquifer An underground water-bearing layer of porous rock through which water can flow.

Autochthonous Formed in its present position, rather than by transport processes.

Biodegradation The breakdown or rendering harmless of a substance by natural processes.

Biodiversity The variety of species, both floral and faunal, contained within an ecosystem.

Biomass The total mass of biological material contained in a given area of the Earth's surface (expressed as dry weight or per unit area).

Biosphere The interlinked communities of animals, plants and micro-organisms that live on the Earth.

Boreal Of northern regions. A term applied both to a climatic zone characterized by cold, snowy winters and short summers, and to the coniferous forests of the high mid-latitudes in the Northern Hemisphere, also known as taiga.

Channelization The modification of river channels for the purpose of flood control, land drainage,

The Human Impact on the Natural Environment: Past, Present and Future, Seventh Edition. Andrew S. Goudie.
© 2013 John Wiley & Sons, Ltd. Published 2013 by John Wiley & Sons, Ltd.

navigation and the reduction or prevention of erosion.

Chaparral A type of stunted (scrub) woodland found in temperate regions with dry summers (Mediterranean regions). It is dominated by drought-resistant evergreen shrubs.

Chlorofluorocarbons (CFC) A range of synthetically manufactured, chemically inert compounds containing atoms of carbon, fluorine and chloride. They have been developed and widely used as solvents, refrigerant and aerosol propellants and in the manufacture of foam plastics.

Climax The final stage of plant succession, when the plant community is relatively stable and in equilibrium with the existing environmental conditions. It is normally determined by climate (climatic climax) or by soil (edaphic climax).

Coral bleaching Corals are bleached when the colourful symbiotic algae they house are lost. When the algae are absent for any length of time, the coral dies. This absence can be caused by excessively warm water temperatures.

Critical loads A concept in pollution studies which involves the idea that there is a certain pollution load level (e.g. of acid rain) above which harmful effects on biological systems will occur.

Cryosphere The portion of the Earth's surface where water is in the solid form (i.e. ice). It includes sea ice, glaciers and ice sheets and permafrost.

Deflation The removal of dry, unconsolidated material, such as dust or sand, from a surface by wind.

Deforestation The permanent removal of trees from an area of forest or woodland.

Desertification The spread of desert-like conditions in arid or semi-arid areas, due to human interference or climatic change, or both.

DNA (deoxyribonucleic acid). The substance that is the carrier of genetic information, found in the chromosomes of the nucleus of a cell.

Domestication The taming and breeding of previously wild animals and plants for human use.

Dust storm A storm, particularly in dry areas, which carries dense clouds of dust, sometimes to a great height, often obscuring visibility to below 1000 m.

Ecological footprint The area that is impacted by pollution, resource extraction, development and transport from a particular location (e.g. a city).

Ecology The science which studies the relations between living organisms and their environment.

ENSO (El Niño Southern Oscillation). Periodical disturbances of Pacific Ocean and atmosphere, with El Niño conditions being abnormally warm off the coast of South America and La Niña conditions being abnormally cool.

Eutrophication The process by which an aquatic ecosystem increases in productivity as a result of increased nutrient input. Often this is due to human induced additions of elements such as nitrogen and phosphorous (cultural eutrophication). However, the process may also be a natural phenomenon.

Forest decline The decline of forest vitality characterized by decreased and abnormal growth, leading eventually to death. The causes include poor management practices; climatic change; fungal, viral and pest attack; nutrient deficiency; and atmospheric pollution.

Forest transition The shift from net deforestation to net reforestation in response to such factors as the movement of people from the land to cities.

Fracking Also known as Induced Hydraulic Fracturing, this is a technique to release hydrocarbons, and especially natural gas, from rocks such as shale. It involves drilling into the rock and creating fractures by pumping fracturing fluids under high pressure.

Genetic engineering The technology involved in manipulating the genes (molecular building blocks) of organisms. Organisms treated in this way are called genetically modified organisms (GMOs).

Geoengineering A branch of engineering concerned with the design and construction of major structures within and on Earth. Recently the term has been taken to mean the manipulation of Earth's climate to counteract the effects of climate change.

Gleying Soil characteristics (including mottling) developed as a result of poor drainage and intermittent waterlogging reducing oxidation or causing the deoxidation of ferric compounds.

Global change Largely synonymous with 'global environmental change', it refers to changes in the global environment (including climate change) that may alter the capacity of the Earth to sustain life.

Green revolution An agricultural revolution, especially in the less developed countries of Asia in the 1960s and 1970s, which gave rise to increased food production through the introduction of new high-yielding varieties of crops and the adoption of techniques (e.g. synthetic fertilizers) necessary to grow them.

Groyne   A construction, usually at right angles to the coast and jutting into the sea, to combat longshore drifting of sediment and beach erosion.

Habitat   The place in which an organism lives, characterized by its physical features or the dominant plant types.

Heinrich event   Deposition of ice berg rafted debris in ocean core sediments because of rapid ice sheet decay during the Pleistocene. They are periods of rapid climate change.

Hominid   Primates of a family (Hominidae) which includes humans and their fossil ancestors.

Hominim   Human species.

Hybridization   The process that results from a cross between parents of differing genotypes. A good example is the mule, produced by cross-breeding an ass and a horse. Hybrids may be fertile or sterile depending on differences in the genomes of the two parents.

Hypoxia   Reduced oxygen content of water caused by pollution and associated with eutrophication and the development of dead zones.

Infiltration capacity   The capacity of the soil surface to absorb water. If the capacity is exceeded, ponding will occur and surface runoff may result.

Isostasy   A process that causes the Earth's crust to rise or sink according to whether a weight is removed or added to it. Such a weight could be, for example, an ice cap (Glacio-isostasy).

Karstic   Relating to a limestone region (or another type of soluble rock) with underground drainage and many cavities and passages caused by the solution of the rock.

Keystone species   A species whose removal from the ecosystem of which it forms a part leads to a series of adverse effects (including extinctions) in that system.

Land cover   The physical state of the land, embracing, for example, the quantity and type of surface vegetation, water and earth materials. The state may change as a result of land use changes.

Laterite   A residual deposit formed by the chemical weathering of rock, composed primarily of hydrated iron and aluminium oxides. It is extensively developed in the humid and subtropical regions.

Levée   A natural or man-made embankment along a river.

Little Ice Age   A period of glacial advance and cold weather (neoglaciation) that took place between *c.* AD 1550 and AD 1850.

Maquis   Scrub vegetation of evergreen shrubs, characteristic of the western Mediterranean; broadly equivalent to chaparral.

Mass balance of glaciers   The sum of all processes which add mass to a glacier (e.g. snowfall, avalanches) and remove mass from it (e.g. melting, ice berg calving).

Mass movement   The downward movement of material under the influence of gravity on a slope (e.g. landslips, mudflows, etc.)

Permafrost   The thermal conditions in soil and rock where temperatures are below 0° for at least two consecutive years.

Photochemical   Relating to a chemical reaction which is speeded up by particular wavelengths of electromagnetic radiation (e.g. sunlight).

Piezometric   Relating to a subterranean surface marking the level to which water will rise within an aquifer.

Podzolized   Relating to a soil that has been characterized by the acidification of the A horizon, the downward leaching of cations, metals and humic substances and their deposition in the B horizon, often precipitating to form a pan.

Radiative forcing   A change in average net radiation at the top of the troposphere resulting from a change in either solar or infrared radiation due to a change in atmospheric greenhouse gas concentrations.

Savanna   A grassland, often with scattered trees, of the tropics and subtropics.

Secondary forest   Woodland which has regenerated and colonized an area after the original (primary) forest has been removed.

Steric effect   In the context of sea-level change, the change in sea level caused by changes in the volume of water in the oceans in response to temperature changes.

Stratosphere   A layer of Earth's atmosphere between the troposphere and the mesosphere. It generally occurs at an altitude of 8–50 km.

Succession   The sequence of changes in a plant community as it develops over time and eventually leading to climax.

Sunspot   A dark area on the visible surface of the sun. Their number usually reaches a maximum every 11 years.

Synanthrope An organism that benefits from association with humans.

Thermokarst Topographical depressions resulting from the thawing of ground ice (permafrost).

Tipping point The idea that passing some hidden threshold will drastically worsen man-made climate change. Recently, the term has come into vogue to describe the moment at which internal dynamics start to propel change previously driven by external forces. Positive feedbacks mean that small changes can have big and abrupt consequences.

Trophic Relating to the positions that organisms occupy in a food chain.

Tropospheric Relating to the lowest level of the atmosphere, in which most of our weather occurs. The troposphere lies beneath the stratosphere and its thickness ranges from about 7 km at the poles to about 28 km at the equator.

Tundra The zone between the latitudinal limits of tree growth and polar ice, characterized by severe winters and a short growing season.

Turbidity A measure of the lack of clearness in a liquid caused by the presence of suspended material.

Wallace's Line A line, developed initially by A.R. Wallace, that separates the distinct flora and fauna of south east Asia from that of Australasia.

Wetland An ecosystem whose formation has been dominated by water (e.g. a marsh or swamp), and whose processes and characteristics are largely controlled by water.

REFERENCES

Aagaard, T. and Sorensen, P. (2012) Coastal profile response to sea level rise: a process-based approach. *Earth Surface Processes and Landforms*, 37, 354–362.

Abbo, S., Lev-Yadun, S. and Gopher, A. (2010) Agricultural origins: centers and noncenters; a Near Eastern appraisal. *Critical Reviews in Plant Sciences*, 29, 317–328.

Abul-Atta, A.A. (1978) *Egypt and the Nile after the Construction of the High Aswan Dam*. Cairo: Ministry of Irrigation and Land Reclamation.

Achard, F., Eva, H.D., Stibig, H.-J., Mayaux, P., Gallego, J., Richards, T. and Malingreau, J.-P. (2002) Determination of deforestation rates of the world's humid tropical forests. *Science*, 297, 999–1002.

Adam, J.C., Hamlet, A.F. and Lettenmaier, D.P. (2009) Implications of global climate change for snowmelt hydrology in the twenty-first century. *Hydrological Processes*, 23, 962–972.

Adams, J.W., Rodriguez, D. and Cox, R.A. (2005) The uptake of SO_2 on Saharan dust: a flow tube study. *Atmospheric Chemistry and Physics*, 5, 2679–2689.

Adger, W.N. and Brown, K. (1994) *Land Use and the Causes of Global Warming*. Chichester: Wiley.

Ågren, C. and Elvingson, P. (1996) *Still with Us: The Acidification of the Environment is Still Going on*. Göteborg: Swedish NGO Secretariat on Acid Rain.

Alabaster, J.S. (1972) Suspended solids and fisheries. *Proceedings of the Royal Society of London. Series B. Biological Sciences*, 180, 395–406.

Al-Ibrahim, A.A. (1991) Excessive use of ground-water resources in Saudi Arabia: impacts and policy options. *Ambio*, 20, 34–37.

Allchin, B., Goudie, A.S. and Hegde, K.T.M. (1977) *The Prehistory and Palaeogeography of the Great Indian Desert*. London: Academic Press.

Allen, C.D. et al. (2010) A global overview of drgought and heat-induced tree mortality reveals emerging climate change risks for forests. *Forest Ecology and Management*, 259, 660–684.

Allen, J.C. and Barnes, D.F. (1985) The causes of deforestation in developing countries. *Annals of the Association of American Geographers*, 75, 163–184.

Allen, R.J., Sherwood, S.C., Norris, J.R. and Zender, C.S. (2012) Recent Northern hemisphere tropical expansion primarily driven by black carbon and tropospheric ozone. *Nature*, 485, 350–354.

Allred, T.M. and Schmidt, J.C. (1999) Channel narrowing by vertical accretion along the Green River near Green River, Utah. *Geological Society of America Bulletin*, 111, 1757–1772.

Almer, B., Dickson, W., Ekstrom, C., Hornstrom, E. and Miller, U. (1974) Effects of acidification on Finnish lakes. *Ambio*, 3, 30–36.

Amills, M. (2011) Biodiversity and origin of pig breeds. *Bulletin UASVM Animal Science and Biotechnologies*, 68, 1–5.

An, Q., Wu, Y., Taylor, S. and Zhao, B. (2009) Influence of the Three Gorges Project on saltwater intrusion in the Yangtze River estuary. *Environmental Geology*, 56, 1679–1686.

An, S.-I., Kug, J.-S., Ham, Y.-G. and Kang, I.-S. (2008) Successive modulation of ENSO to the future greenhouse warming. *Journal of Climate*, 21, 3–21.

van Andel, T.H., Zangger, E. and Demitrack, A. (1990) Land use and soil erosion in prehistoric and historical Greece. *Journal of Field Archaeology*, 17, 379–396.

Anderson, A. (2009) The rat and the octopus: initial human colonization and the prehistoric introduction of domestic animals to remote Oceania. *Biological Invasions*, 11, 1503–1519.

Anderson, D.E., Goudie, A.S. and Parker, A.G. (2013) *Global Environments through the Quaternary* (2nd edn). Oxford: Oxford University Press.

Anderson, D.M. (1994) Red tides. *Scientific American*, 271 (2), 52–58.

Anderson, J.M. and Spencer, T. (1991) Carbon, nutrient and water balances of tropical rainforest ecosystems subject to disturbance: management implications and research proposals. *MAB Digest*, 7, UNESCO, Paris, France.

Anderson, R.C. (2006) Evolution and origin of the Central Grassland of North America: climate, fire, and mammalian grazers. *The Journal of the Torrey Botanical Society*, 133, 626–647.

Anisimov, O., Vandenberghe, J., Lobanov, V. and Kondratiev, A. (2008) Predicting changes in alluvial channel patterns in North-European Russia under conditions of global warming. *Geomorphology*, 98, 262–274.

Anisimov, O.A. (1989) Changing climate and permafrost distribution in the Soviet Arctic. *Physical Geography*, 10, 282–293.

Anisimov, O.A. and Fitzharris, B. (2001) Polar regions (Arctic and Antarctic. In J.J. McCarthy (ed.), *Climate Change 2001: Impacts, Adaptation, and Vulnerability*. Cambridge: Cambridge University Press, 801–841.

Anthony, K.R.N., Kline, D.I., Diaz-Pulido, G., Dove, S. and Hoegh-Guldberg, O. (2008) Ocean acidification causes bleaching and productivity loss in coral reef builders. *Proceedings of the National Academy of Sciences of the United States of America*, 105, 17442–17446.

Archer, S. et al. (1999) Arid and semi-arid land community dynamics in a management context. In T.W. Hoekstra and M. Schachak (eds), *Arid Lands Management*. Urbana and Chicago: University of Illinois Press, 48–74.

Archer, D. (2007) *Global Warming: Understanding the Forecast*. Oxford: Blackwell.

Arendt, A.A., Echelmeyer, K.A., Harrison, W.D., Lingle, C.S. and Valentine, V.B. (2002) Rapid wastage of Alaska glaciers and their contribution to rising sea level. *Science*, 297, 382–385.

Arnell, N.W. (1996) *Global Warming, River Flows and Water Resources*. Chichester: Wiley.

Arnell, N.W. (1999a) The impacts of climate change on water resources. In Meteorological Office (ed.), *Meteorological Office, Climate Change and Its Impacts*. Bracknell: Hadley Centre, 14–18.

Arnell, N.W. (1999b) The effect of climate change on hydrological regimes in Europe: a continental prospective. *Global Environmental Change*, 9, 5–23.

Arnell, N.W. (2002) *Hydrology and Global Environmental Change*. Harlow: Prentice Hall.

Arnell, N.W. and Reynard, N.S. (2000) Climate change and UK hydrology. In M.C. Acreman (ed.), *The Hydrology of the UK: A Study of Change*. London: Routledge, 3–29.

Aron, W.I. and Smith, S.H. (1971) Ship canals and aquatic ecosystems. *Science*, 174, 13–20.

Arora, V.K. and Boer, G.J. (2001) Effects of simulated climate change on the hydrology of major river basins. *Journal of Geophysical Research*, 106 (D4), 3335–3348.

Arpe, K., Bengtsson, L., Golitsyn, G.S., Mokhov, I.I. and Ettahir, E.A.B. (2000) Connection between Caspian Sea level variability and ENSO. *Geophysical Research Letters*, 27, 2693–2696.

Arzarello, M., Marcolini, F., Pavia, G., Pavia, M., Petronio, C., Petrucci, M., Rook, L. and Sardella, R. (2007) Evidence of earliest human occurrence in Europe: the site of Pirro Nord (Southern Italy). *Die Naturwissenschaften*, 94, 107–112.

Ashby, E. (1978) *Reconciling Man with the Environment*. London: Oxford University Press.

Ashley, W.S., Bentley, M.L. and Stallins, J.A. (2011) Urban-induced thunderstorm modification in the Southeast United States. *Climatic Change*, doi:10.1007/s10584-011-0324-1.

Ashmore, P. and Church, M. (2001) The impact of climate change on rivers and river processes in Canada. Geological Survey of Canada Bulletin, 555.

Ashton, A.D., Donnelly, J.P. and Evans, R.L. (2008) A discussion of the potential impacts of climate change on the shorelines of the northeastern USA. *Mitigation and Adaptation Strategies for Global Change*, 13, 719–743.

Ashton, A.D., Walkden, M.J.A. and Dickson, M.E. (2011) Equilibrium responses of cliffed coasts to changes in the rate of sea level rise. *Marine Geology*, 284, 217–229.

Atkinson, B.W. (1968) A preliminary examination of the possible effect of London's urban area on the distribution of thunder rainfall, 1951–60. *Transactions of the Institute of British Geographers*, 44, 97–118.

Atkinson, B.W. (1975) *The Mechanical Effect of an Urban Area on Convective Precipitation*. Occasional paper 3. London: Department of Geography, Queen Mary College, University of London.

Atlas, P. (1977) The paradox of hail suppression. *Science*, 195, 139–145.

Aubertin, G.M. and Patric, J.H. (1974) Water quality after clearcutting a small watershed in West Virginia. *Journal of Environmental Quality*, 3, 243–249.

Aubréville, A. (1949) *Climats, forêts et desertification de L'Afrique tropicale*. Paris: Societé d'Edition Géographiques Maritimes et Coloniales.

Avila, F.B., Pitman, A.J., Donat, M.G., Alexander, L.V. and Abramowitz, G. (2012) Climate model simulated changes in temperature extremes due to land cover change. *Journal of Geophysical Research*, 117, D04108.

Ayres, M.C. and Lombardero, M.J. (2000) Assessing the consequences of global change for forest disturbance from herbivores and pathogens. *The Science of the Total Environment*, 262, 263–286.

Baccari, N., Boussema, M.R., Lamachère, J.-M. and Nasri, S. (2008) Efficiency of contour benches, filling-in and silting-up of a hillside reservoir in a semi-arid climate in Tunisia. *Comptes Rendus Geoscience*, 340, 38–48.

Baccini, A. et al. (2012) Estimated carbon dioxide emissions from tropical deforestation improved by carbon-density maps. *Nature Climate Change*, 2, 182–185.

Bach, A.J., Brazel, A.J. and Lancaster, N. (1996) Temporal and spatial aspects of blowing dust in the Mojave and Colorado Deserts of southern California, 1973–1994. *Physical Geography*, 17, 329–353.

Badescu, V. and Cathcart, R.B. (2011) Aral Sea partial restoration.1. A Caspian importation macroproject. *International Journal of Environment and Waste Management*, 7, 161–182.

Baillie, J.E.M., Griffiths, J., Turvey, S.T., Loh, J. and Collen, B. (2010) *Evolution Lost. Status and Trends of the World's Vertebrates*. London: Zoological Society of London.

Bakan, S., Chlono, A., Cubasch, U., Feichter, J., Graf, H., Grassl, H., Hasselman, K., Kirchner, L., Latif, M., Roeckner, E., Samsen, R., Schlese, U., Schrivener, D., Schult, I., Sielman, F. and Welks, W. (1991) Climate response to smoke from the burning oil wells in Kuwait. *Nature*, 351, 367–371.

Baker, A. and Simms, M.J. (1998) Active deposition of calcareous tufa in Wessex, UK, and its implications for the 'late-Holocene tufa decline'. *The Holocene*, 8, 359–365.

Baker, A.C., Glynn, P.W. and Riegl, B. (2008) Climate change and coral reef bleaching: an ecological assessment of long-term impacts, recovery trends and future outlook. *Estuarine, Coastal and Shelf Science*, 80, 435–471.

Ball, D.F. (1975) Discussion in J. G. Evans, S. Limbrey and H. Cleere (eds), The effect of man on the landscape: the Highland zone. *Council for British Archaeology research report*, 11, 26.

Ballantyne, C.K. (1991) Late Holocene erosion in upland Britain: climatic deterioration or human influence? *The Holocene*, 1, 81–85.

Balling, R.C. and Wells, S.G. (1990) Historical rainfall patterns and arroyo activity within the Zuni river drainage basin, New Mexico. *Annals of the Association of American Geographers*, 80, 603–617.

Balme, J. (2011) Of boats and string: the maritime colonisation of Australia. *Quaternary International* Online, 1–8.

Balter, M. (2010) The tangled roots of agriculture. *Science*, 327, 404–406.

Barberi, F., Brondi, F., Carapezza, M.L., Cavarra, L. and Murgia, C. (2003) Earthen barriers to control lava flows in the 2001 eruption of Mt. Etna. *Journal of Volcanology and Geothermal Research*, 123, 231–243.

Barends, F.B.J., Brouwer, F.J.J. and Schröder, F.H. (1995) *Land subsidence*. IAHS publication 234. Wallingford, Oxon: International Association of Hydrological Sciences.

Bari, M.A. and Scholfield, N.J. (1992) Lowering of a shallow, saline water table by extensive eucalypt reforestation. *Journal of Hydrology*, 133, 273–291.

Barker, G. (2006) *The Agricultural Revolution in Prehistory. Why did Foragers Become Farmers?* Cambridge: Cambridge University Press.

Barlow, P.M. and Reichard, E.G. (2010) Saltwater intrusion in coastal regions of North America. *Hydrogeology Journal*, 18, 246–270.

Barnard, P.L., Owen, L.A., Sharma, M.C. and Finkel, R.C. (2001) Natural and human-induced landsliding in the Garhwal Himalaya of northern India. *Geomorphology*, 40, 21–35.

Barnes, D.K.A., Galgani, F., Thompson, R.C. and Barlaz, M. (2009) Accumulation and fragmentation of plastic debris in global environments. *Philosophical Transactions of the Royal Society of London. Series B, Biological Sciences*, 364, 1985–1998.

Barnett, T.P. and Pierce, D.W. (2009) Sustainable water deliveries from the Colorado River in a changing climate. *Proceedings of the National Academy of Sciences of the United States of America*, 106, 7334–7338.

Barnett, T.P., Adam, J.C. and Lettenmaier, D.P. (2005) Potential impacts of a warming climate on water availability in snow-dominated regions. *Nature*, 438, 303–309.

Barnston, A.G. and Schickedanz, P.T. (1984) The effect of irrigation on warm season precipitation in the southern Great Plains. *Journal of Climate and Applied Meteorology*, 23, 865–888.

Barry, R.G. (1985) The cryosphere and climate change. In M.C. MacCracken and F.M. Luther (eds), *Detecting the Climatic Effects of Increasing Carbon Dioxide*. Washington DC: US Dept of Energy, 111–148.

Bar-Yosef, O. (1998) On the nature of transitions: the Middle to Upper Palaeolithic and the Neolithic Revolution. *Cambridge Archaeological Journal*, 8, 141–163.

Batey, T. (2009) Soil compaction and soil management – a review. *Soil Use and Management*, 25, 335–345.

Battarbee, R.W. (1977) Observations in the recent history of Lough Neagh and its drainage basin. *Philosophical Transactions of the Royal Society of London. Series B, Biological Sciences*, 281, 303–345.

Battarbee, R.W., Flower, R.J., Stevenson, A.C. and Rippey, B. (1985a) Lake acidification in Galloway: a palaeoecological test of competing hypotheses. *Nature*, 314, 350–352.

Battarbee, R.W., Appleby, P.G., Odel, K. and Flower, R.J. (1985b) 210Pb dating of Scottish Lake sediments, afforestation and accelerated soil erosion. *Earth Surface Processes and Landforms*, 10, 137–142.

Bautista, S., Bellot, J. and Ramón Vallejo, V. (1996) Mulching treatment for postfire soil conservation in a semi-arid ecosystem. *Arid soil Research and Rehabilitation*, 10, 235–242.

Baxter, R.M. (1977) Environmental effects of dams and impoundments. *Annual Review of Ecology and Systematics*, 8, 255–283.

Beach, T., Dunning, N., Luzzadder-Beach, S., Cook, D.E. and Lohse, J. (2006) Impacts of the ancient Maya on soils and soil erosion in the central Maya lowlands. *Catena*, 65, 166–178.

Beamish, R.J., Lockhart, W.L., Van Loon, J.C. and Harvey, H.H. (1975) Long-term acidification of a lake and resulting effects on fishes. *Ambio*, 4, 98–102.

Beaumont, P.B. (2011) The edge: more on fire-making by about 1.7 million years ago at Wonderwerk Cave in South Africa. *Current Anthropology*, 52, 585–594.

Beck, P.S.A. et al. (2011) Changes in forest productivity across Alaska consistent with biome shift. *Ecology Letters*, 14 (4), 373–379.

Beckinsale, R.P. (1972) The effect upon river channels of sudden changes in sediment load. *Acta Geographica Debrecina*, 10, 181–186.

Beerung, D.J. and Osborne, C.P. (2006) The origin of the savanna biome. *Global Change Biology*, 12, 2023–2031.

Behre, K.-E. (ed.) (1986) *Anthropogenic Indicators in Pollen Diagrams*. Rotterdam: Balkema.

Beja-Pereira, A. et al. (2004) African origins of the domestic donkey. *Science*, 304, 1781.

Bell, F.G., Stacey, T.R. and Genske, D.D. (2000) Mining subsidence and its effect on the environment: some differing examples. *Environmental Geology*, 40, 13552.

Bell, M.L. (1982) The effect of land-use and climate on valley sedimentation. In A.F. Harding (ed.), *Climatic Change in Later Prehistory*. Edinburgh: Edinburgh University Press, 127–142.

Bell, M.L. and Nur, A. (1978) Strength changes due to reservoir-induced pore pressure and stresses and application to Lake Oroville. *Journal of Geophysical Research*, 83, 4469–4483.

Bell, M.L. and Walker, M.J.C. (1992) *Late Quaternary Environmental Change*. Harlow: Longman Scientific and Technical.

Bell, T.L., Rosenfeld, D., Kim, K.-M., Yoo, J.-M., Lee, M.-I. and Hahnenberger, M. (2008) Midweek increase in U.S. summer rain and storm heights suggests air pollution invigorates rainstorms. *Journal of Geophysical Research*, 113, D02209.

Bellwood, D.R., Hughes, T.P., Folke, C. and Nyström, M. (2004) Confronting the coral reef crisis. *Nature*, 429, 827–833.

Belski, A.J. (1996) Viewpoint: western juniper expansion: is it a threat to arid northwest ecosystems? *Journal of Range Management*, 49, 53–59.

Beltaos, S. and Burrell, B.C. (2003) Climatic change and river ice breakup. *Canadian Journal of Civil Engineering*, 30, 145–155.

Bendell, J.F. (1974) Effects of fire on birds and mammals. In T.T. Kozlowski and C.C. Ahlgren (eds), *Fire and Ecosystems*. New York: Academic Press, 73–138.

Bendix, J., Bendix, A. and Richter, M. (2000) El Niño 1997/1998 in Nordperu: anzeichen eines Ökosystem – Wandels? *Petermanns Geographische Mitteilungen*, 144, 20–31.

Benn, D.I. and Evans, D.J.A. (2010) *Glaciers and Glaciations* (2nd edn). London: Hodder.

Bennett, C.F. (1968) Human influences in the zoogeography of Panama. *Ibero-Americana*, 51, 1–112.

Bennett, H.H. (1938) *Soil Conservation*. New York: McGraw-Hill.

Bennett, M.R. (2003) Ice streams as the arteries of an ice sheet: their mechanics, stability and significance. *Earth-Science Reviews*, 61, 309–339.

Bentley, C.R. (1997) Rapid sea-level rise soon from West Antarctic ice sheet collapse? *Science*, 275, 1077–1078.

Bentley, C.R. (1998) Rapid sea-level rise from a West Antarctic ice-sheet collapse: a short-term perspective. *Journal of Glaciology*, 44, 157–163.

Berbet, M.L.C. and Costa, M.H. (2003) Climate change after tropical deforestation: seasonal variability of surface albedo and its effects on precipitation change. *Journal of Climate*, 16, 2099–2104.

Beringer, J.E. (2000) Releasing genetically modified organisms: will any harm outweigh any advantage? *Journal of Applied Ecology*, 37, 207–214.

Berna, F., Goldberg, P., Horwitz, L.K., Brink, J., Holt, S., Bamford, M. and Chazan, M. (2012) Microstratigraphic evidence of in situ fire in the Acheulean strata of Wonderwerk Cave, Northern Cape province, South Africa. *Proceedings of the National Academy of Sciences of the United States of America*, 109 (20), E1215–1220.

Bernabo, J.C. and Webb, T., III (1977) Changing patterns in the Holocene pollen record of northeastern North America: mapped summary. *Quaternary Research*, 8, 64–96.

Bertness, M.D. and Silliman, B.R. (2008) Consumer control of salt marshes driven by human disturbance. *Conservation Biology*, 22, 618–623.

Betts, R.A. (2000) Offset of the potential carbon sink from boreal forestation by decreases in surface albedo. *Nature*, 408, 187–190.

Betts, R.A. (2001) Biogeophysical impacts of land use on present-day climate: near surface temperature change and radiative forcing. *Atmospheric Science Letters*, 2 (1–4), 39–51.

Betts, R.A., Malhi, Y. and Roberts, J.T. (2008) The future of the Amazon: new perspectives from climate, ecosystem and social sciences. *Philosophical Transactions of the Royal Society of London. Series B, Biological Sciences*, 363, 1729–1735.

Bhagwat, S.A., Breman, E., Thekaekara, T., Thornton, T.F. and Willis, K.J. (2012) A battle lost? Report on two centuries of invasion management of *Lantana camara* L. in Australia, India and South Africa. *PLoS ONE*, 7 (3), e32407.

Bhark, E.W. and Small, E.E. 2002. Association between plant canopies and the spatial patterns of infiltration in shrubland and grassland of the Chihuahuan Desert, New Mexico. *Ecosystems*, 6, 185–196.

Bidwell, O.W. and Hole, F.D. (1965) Man as a factor of soil formation. *Soil Science*, 99, 65–72.

Bigg, E.K. (2008) Trends in rainfall associated with sources of air pollution. *Environmental Chemistry*, 5, 184–193.

Biglane, K.E. and Lafleur, R.A. (1967) Notes on estuarine pollution with emphasis on the Louisiana Gulf Coast. In Lauff, G.H. (ed.), *Estuaries. American Association for the Advancement of Science Publication*, Vol. 83. Washington, DC: AAAS, 690–692.

Bindler, R. (2011) Contaminated lead environments of man: reviewing the lead isotopic evidence in sediments, peat and soils for the temporal and spatial patterns of atmospheric lead pollution in Sweden. *Environmental Geochemistry and Health*, 33, 311–329.

Binford, M.W., Brenner, M., Whitmore, T.J., Higuera-Grundy, A., Deevey, E.S. and Leyden, B. (1987) Ecosystems, palaeoecology, and human disturbance in subtropical and tropical America. *Quaternary Science Reviews*, 6, 115–128.

Bird, E.C.F. (1979) Coastal processes. In K.J. Gregory and D.E. Walling (eds), *Man and Environmental Processes*. Folkestone: Dawson, 82–101.

Bird, E.C.F. (1993) *Submerging Coasts*. Chichester: Wiley.

Bird, E.C.F. (1996) *Beach Management*. Chichester: Wiley.

Birken, A.S. and Cooper, D.J. (2006) Processes of *Tamarix* invasion and floodplain development along the lower Green River, Utah. *Ecological Applications*, 16, 1103–1120.

Birkett, C.M. (2000) Synergist remote sensing of Lake Chad: variability of basin inundation. *Remote Sensing of Environment*, 72, 218–236.

Birks, H.J.B. (1986) Late-Quaternary biotic changes in terrestrial and lacustrine environments, with particular reference to north-west Europe. In B.E. Berglund (ed.), *Handbook of Holocene Palaeoecology and Palaeohydrology*. Chichester: Wiley, 3–65.

Birks, H.J.B. (1988) Long-term ecological change in the British uplands. *British Ecological Society special publication*, 7, 37–56.

Bischoff, M., Cete, A., Fritschen, R. and Meier, T. (2009) Coal mining induced seismicity in the Ruhr area, Germany. *Pure and Applied Geophysics*, 167, 63–75.

Blainey, G. (1975) *Triumph of the Nomads. A History of Ancient Australia*. Melbourne: Macmillan.

Blank, I.W. (1985) A new type of forest decline in Germany. *Nature*, 314, 311–314.

Bledsoe, B.P. and Watson, C.C. (2001) Effects of urbanization on channel instability. *Journal of the American Water Resources Association*, 37, 255–270.

Blum, M.D. and Roberts, H.H. (2009) Drowning of the Mississippi delta due to insufficient sediment supply and global sea-level rise. *Nature Geoscience*, 2, 488–491.

Blum, M.D. and Törnqvist, T.E. (2000) Fluvial responses to climate and sea-level change: a review and look forward. *Sedimentology*, 47 (Suppl. 1), 2–48.

Blume, H.-P. and Leinweber, P. (2004) Plaggen soils: landscape history, properties, and classification. *Journal of Plant Nutrition and Soil Science*, 167, 319–327.

Boakes, E.H., Mace, G.M., McGowan, P.J.K. and Fuller, R.A. (2010) Extreme contagion in global habitat clearance. *Proceedings. Biological Sciences/The Royal Society*, 277, 1081–1085.

Boardman, J. (1998) An average soil erosion rate for Europe: myth or reality. *Journal of Soil and Water Conservation*, 53, 46–50.

Boardman, J. and Foster, I.D.L. (2011) The potential significance of the breaching of small farm dams in the Sneeuberg region. South Africa. *Journal of Soils and Sediments*, 11, 1456–1465.

Bochet, E., Rubio, J.L. and Poesen, J. (1998) Relative efficiency of three representative matorral species in reducing water erosion at the microscale in a semi-arid climate (Valencia, Spain). *Geomorphology*, 23, 139–150.

Böckh, A. (1973) Consequences of uncontrolled human activities in the Valencia lake basin. In M.T. Farvar and J.P. Milton (eds), *The Careless Technology*. London: Tom Stacey, 301–317.

Bockman, O.C., Kaarstad, O., Lie, O.H. and Richards, I. (1990) *Agriculture and Fertilizers*. Oslo: Norsk Hydro.

Bogaart, P.W., Van Balen, R.T., Kasse, C. and Vandenberghe, J. (2003) Process-based modeling of fluvial system response to rapid climate change – I: model formulation and generic applications. *Quaternary Science Reviews*, 22, 2077–2095.

Bogucki, P. (1999) *The Origins of Human Society*. Oxford: Blackwell.

Bollasina, M.O., Yi, M. and Ramaswamy, V. (2011) Anthropogenic aerosols and the weakening of the South Asian Summer Monsoon. *Science*, 334, 502–505.

Bomford, M. and Hart, Q. (2002) Non-indigenous vertebrates in Australia. In D. Pimentel (ed.), *Biological Invasions*. Boca Raton: CRC Press, 25–44.

Bonan, G.B. (1997) Effects of land use on the climate of the United States. *Climatic Change*, 37, 449–486.

Bond, W.J. and Midgley, G.F. (2012) Carbon dioxide and the uneasy interactions of trees and savannah grasses. *Philosophical Transactions of the Royal Society of London. Series B, Biological Sciences*, 367, 601–612.

Bonnell, T.R., Reyna-Hurtado, R. and Chapman, C.A. (2011) Post-logging recovery time is longer than expected in an East African tropical forest. *Forest Ecology and Management*, 261, 855–864.

Bonnichsen, T.M., Bonnichsen, M., Anderson, K., Lewis, H.T., Kay, C.E. and Knudson, R. (1999) Native American influences on the development of forest ecosystems. In R.C. Szaro, N.C. Johnson, W.T. Sexton and A.J. Malle (eds),

Ecological Stewardship (Vol. 2). Amsterdam: Elsevier Science, 439–470.

Boorman, L.A. (1977) Sand-dunes. In R.S.K. Barnes (ed.), *The Coastline*. London: Wiley, 161–197.

Booysen, P. de V. and Tainton, N.M. (eds) (1984) *Ecological Effects of Fire in South African Ecosystems*. Berlin: Springer-Verlag.

Bormann, F.H., Likens, G.E., Fisher, D.W. and Pierce, R.S. (1968) Nutrient loss accelerated by clear cutting of a forest ecosystem. *Science*, 159, 882–884.

Bouraouni, F. and Grizzetti, B. (2011) Long term change of nutrient concentrations of rivers discharging in European seas. *The Science of the Total Environment*, 409, 4899–4916.

Boussingault, J.B. (1845) *Rural Economy* (2nd edn). London: Baillière.

Boutron, C.F., Görlach, U., Candelone, J.-P., Bolshov, M.A. and Delmas, R.J. (1991) Decrease in anthropogenic lead, cadmium and zinc in Greenland snows since the late 1960s. *Nature*, 353, 153–156.

Bouwman, A.F., van Vuuren, D.P., Derwent, R.G. and Posch, M. (2002) A global analysis of acidification and eutrophication of terrestrial ecosystems. *Water, Air, and Soil Pollution*, 141, 349–382.

Bove, M.C., Elsner, J.B., Landsea, C.W., Niu, X. and O'Brien, J.J. (1998) El Niño on US land falling hurricanes, revisited. *Bulletin of the American Meteorological Society*, 79, 2477–2482.

Bowden, W.B. (2010) Climate change in the Arctic – permafrost, thermokarst, and why they matter to the non-Arctic world. *Geography Compass*, 4, 1553–1566.

Bowes, M.J. et al. (2011) Changes in water quality of the River Frome (UK) from 1965 to 2009: is phosphorus mitigation finally working? *The Science of the Total Environment*, 409, 3418–3430.

Bowler, J., Johnsston, H., Olley, J.M., Prescott, J.R., Roberts, R.G., Shawcross, W. and Spooner, N. (2003) New ages for human occupation and climatic change at Lake Mungo, Australia. *Nature*, 421, 837–840.

Bowman, D.M.J.S. et al. (2011) The human dimension of fire regimes on Earth. *Journal of Biogeography*, 38, 2223–2226.

Boyd, M. (2002) Identification of anthropogenic burning in the paleoecological record of the Northern Prairies: a new approach. *Annals of the Association of American Geographers*, 92, 471–487.

Boyle, J.F., Gaillard, M.-J., Kaplan, J.O. and Dearing, J.A. (2011) Modelling prehistoric land use and carbon budgets: a critical review. *The Holocene*, 21, 715–722.

Bradley, R. (1985) *Quaternary Palaeoclimatology*. London: Chapman and Hall.

Bradley, R.S., Vuille, M., Diaz, H.F. and Vergara, W. (2006) Threats to water supplies in the tropical Andes. *Science*, 312, 1755–1756.

Bradshaw, C.J.A., Sodhi, N.T., Peh, K.S.H. and Brook, B.W. (2007) Global evidence that deforestation amplifies flood risk and severity in the developing world. *Global Change Biology*, 13, 2379–2395.

Bragg, O.M. and Tallis, J.H. (2001) The sensitivity of peat-covered upland landscapes. *Catena*, 42, 345–360.

Braithwaite, R.J. and Zhang, Y. (1999) Modelling changes in glacier mass balance that may occur as a result of climate changes. *Geografiska Annaler. Series A, Physical Geography*, 81, 489–496.

Brander, K.E., Owen, K.E. and Potter, K.W. (2004) Modelled impacts of development type on runoff volume and infiltration performance. *Journal of the American Water Resources Association*, 40, 961–969.

Brandt, S.A. (2000) Classification of geomorphological effects downstream of dams. *Catena*, 40, 375–401.

Brassington, F.C. and Rushton, K.R. (1987) A rising water table in central Liverpool. *Quarterly Journal of Engineering Geology and Hydrogeology*, 20, 151–158.

Brattebo, B.O. and Booth, D.B. (2003) Long-term stormwater quantity and quality performance of permeable pavement systems. *Water Research*, 37, 4369–4376.

Braun, L.N., Weber, M. and Schulz, M. (2000) Consequences of climate change for runoff from alpine regions. *Annals of Glaciology*, 31, 19–25.

Bravard, J.-P. and Petts, G.E. (1996) Human impacts on fluvial hydrosystems. In G.E. Petts and C. Amoros (eds), *Fluvial Hydrosystems*. London: Chapman & Hall, 242–262.

Bray, H.J., Carter, D.J. and Hooke, J.M. (1992) *Sea level rise and global warming: scenarios, physical impacts and policies*. Report to SCOPAC, University of Portsmouth.

Brazel, A.J. and Idso, S.B. (1979) Thermal effects of dust on climate. *Annals of the Association of American Geographers*, 69, 432–437.

Brecht, H., Dasgupta, S., Laplante, B., Murray, S. and Wheeler, D. (2012) Sea-level rise and storm surges: high stakes for a small number of developing countries. *The Journal of Environment and Development*, 21, 120–138.

Bredenkamp, G.J., Spada, F. and Kazmierczak, E. (2002) On the origin of northern and southern hemisphere grasslands. *Plant Ecology*, 163, 209–229.

Breslow, P.B. and Sailor, D.J. (2002) Vulnerability of wind power resources to climate change in the continental United States. *Renewable Energy*, 27, 585–598.

Breuer, G. (1980) *Weather Modification: Prospects and Problems*. Cambridge: Cambridge University Press.

Brewer, S. (2008) Declines in plant species richness and endemic plant species in longleaf pine savannas invaded by *Imperata cylindrica*. *Biological Invasions*, 10, 1257–1264.

Brimblecombe, P. (1977) London air pollution 1500–1900. *Atmospheric Environment*, 11, 1157–1162.

Brimblecombe, P. and Camuffo, D. (2003) Long term damage to the built environment. In P. Brimblecombe (ed.), *The Effects of Air Pollution on the Built Environment*. London: Imperial College Press, 1–30.

Broadus, J., Milliman, J., Edwards, S., Aubrey, D. and Gable, F. (1986) Rising sea level and damming of rivers: possible effects in Egypt and Bangladesh. In J.G. Titus (ed.), *Effects of Changes in Stratospheric Ozone and Global Climate*. Washington DC: UNEP/USEPA, 165–189.

Broadus, J.G. (1990) Greenhouse effects, sea level rise and land use. *Land Use Policy*, 7, 138–153.

Broodbank, C. (2006) The origins and early development of Mediterranean maritime history. *Journal of Mediterranean Archaeology*, 19, 199–230.

Brook, B.W. and Bowman, D.M.J.S. (2002) Explaining the Pleistocene megafaunal extinction: models, chronologies, and assumptions. *Proceedings of the National Academy of Sciences of the United States of America*, 99, 14624–14627.

Brookes, A. (1985) River channelization: traditional engineering methods, physical consequences, and alternative practices. *Progress in Physical Geography*, 9, 44–73.

Brookes, A. (1987) The distribution and management of channelized streams in Denmark. *Regulated Rivers: Research & Management*, 1, 3–16.

Brooks, A.P. and Brierly, G.J. (1997) Geomorphic responses of lower Bega River to catchment disturbance, 1851–1926. *Geomorphology*, 18, 291–304.

Brooks, M.L. and Matchett, J.R. (2006) Spatial and temporal patterns of wildfires in the Mojave Desert, 1980–2004. *Journal of Arid Environments*, 67, 148–164.

Brooks, S.M. and Spencer, T. (2012) Shoreline retreat and sediment release in response to accelerating sea level rise: measuring and modelling cliffloine dynamics on the Suffolk coast, UK. *Global and Planetary Change*, 80–81, 165–179.

Brown, B.E. (1997) Coral bleaching: causes and consequences. *Coral Reefs*, 16, S129–S138.

Brown, B.E., Dunne, R.P., Goodson, H.S. and Douglas, A.G. (2000) Bleaching patterns in coral reefs. *Nature*, 404, 142–143.

Brown, J.H. (1989) Patterns, modes and extents of invasions by vertebrates. In J.A. Drake (ed.), *Biological Invasions: A Global Perspective*. Chichester: Wiley, 85–109.

Brown, S. and Lugo, A.E. (1990) Tropical secondary forests. *Journal of Tropical Ecology*, 6, 1–32.

Browning, K.A., Allah, R.J., Ballard, B.P., Barnes, R.T.H., Bennetts, D.A., Maryon, R.H., Mason, P.J., McKenna, D., Mitchell, J.F.B., Senior, C.A., Slingo, A. and Smith, F. (1991) Environmental effects from burning oil wells in Kuwait. *Nature*, 351, 363–367.

Bruijnzeel, L.A. (1990) *Hydrology of Moist Tropical Forests and Effects of Conversion: A State of Knowledge Review*. Amsterdam: Free University for UNESCO International Hydrological Programme.

Brunel, C. and Sabatier, F. (2009) Potential influence of sea-level rise in controlling shoreline position on the French Mediterranean Coast. *Geomorphology*, 107, 47–57.

Brunet, M., Guy, F., Pilbeam, D., Mackay, H.T., Likius, A. and Djimboumalbaye, A. (2002) A new hominid from the upper Miocene of Chad, central Africa. *Nature*, 418, 145–151.

Brunhes, J. (1920) *Human Geography*. London: Harrap.

Bruno, J.F. and Selig, E.R. (2007) Regional decline of coral cover in the Indo-Pacific: timing, extent and subregional comparisons. *PLoS ONE*, 8, e711.

Brunsden, D. (2001) A critical assessment of the sensitivity concept in geomorphology. *Catena*, 42, 83–98.

Bruun, P. (1962) Sea level rise as a cause of shore erosion. *American Society of Civil Engineers Proceedings: Journal of the Waterways and Harbors Division*, 88, 117–130.

Bryan, G.W. (1979) Bioaccumulation of marine pollutants. *Philosophical Transactions of the Royal Society of London. Series B, Biological Sciences*, 286, 483–505.

Bryan, K. (1928) Historic evidence of changes in the channel of Rio Puerco, a tributary of the Rio Grande in New Mexico. *The Journal of Geology*, 36, 265–282.

Buckley, S.M. and Mitchell, M.J. (2010) Improvements in urban air quality: case studies from new York State, USA. *Water, Air, and Soil Pollution*, 214, 93–106.

Buddemeier, R.W. and Smith, S.V. (1988) Coral reef growth in an era of rapidly rising sea level: predictions and suggestions for long-term research. *Coral Reefs*, 7, 51–56.

Buitenwerf, R., Bond, W.J., Stevens, N. and Trollope, W.S.W. (2012) Increased tree densities in South African savannas: >50 years of data suggests CO_2 as a driver. *Global Change Biology*, 18, 675–684.

Bull, W.B. (1997) Discontinuous ephemeral streams. *Geomorphology*, 19, 227–276.

Bullard, J.E., Thomas, D.S.G., Livingstone, I. and Wiggs, G. (1996) Wind energy variations in the south-western Kalahari Desert and their implications for linear dunefield activity. *Earth Surface Processes and Landforms*, 21, 263–278.

Bulmer, S. (1982) Human ecology and cultural variation in prehistoric New Guinea. *Monographiae Biologicae* (New Guinea), 42, 169–206.

Burian, S.J. and Shepherd, J.M. (2005) Effect of urbanization on the diurnal rainfall pattern in Houston. *Hydrological Processes*, 19, 1089–1103.

Burke, R. (1972) Stormwater runoff. In R.T. Oglesby, C.A. Carlson and J.A. McCann (eds), *River Ecology and Man*. New York: Academic Press, 727–733.

Burney, A.D. (1993) Recent animal extinctions: recipes for disaster. *American Scientist*, 81, 530–541.

Burrin, P.J. (1985) Holocene alluviation in southeast England and some implications for palaeohydrological studies. *Earth Surface Processes and Landforms*, 10, 257–271.

Burroughs, B.A., Hayes, D.B., Klomp, K.D., Hansen, J.F. and Mistak, J. (2009) Effects of Stronach Dam removal on fluvial geomorphology in the Pine River, Michigan, United States. *Geomorphology*, 110, 96–107.

Burt, T.P., Donohoe, M.A. and Vann, A.R. (1983) The effect of forestry drainage operations on upland sediment yields: the results of a stormbased study. *Earth Surface Processes and Landforms*, 8, 339–346.

Burt, T.P. and Haycock, N.E. (1992) Catchment planning and the nitrate issue: a UK perspective. *Progress in Physical Geography*, 16, 379–404.

Burt, T.P., Howden, N.J.K., Worrall, F., Whelan, M.J. and Bieroza, M. (2010) Nitrate in United Kingdom rivers:

policy and its outcomes since 1970. *Environmental Science & Technology*, 24, 2657–2662.

Bush, M.B. (1988) Early Mesolithic disturbance: a force on the landscape. *Journal of Archaeological Science*, 15, 453–462.

Butzer, K.W. (1972) *Environment and Archaeology – An Ecological Approach to Prehistory*. London: Methuen.

Butzer, K.W. (1974) Accelerated soil erosion: a problem of man-land relationships. In I.R. Manners and M.W. Mikesell (eds), *Perspectives on Environments*. Washington, DC: Association of American Geographers, 55–77.

Butzer, K.W. (1976) *Early Hydraulic Civilization in Egypt*. Chicago: University of Chicago Press.

Buyanovsky, G.A. and Wagner, G.H. (1998) Carbon cycling in cultivated land and its global significance. *Global Change Biology*, 4, 131–141.

Buytaert, W., Cuesta-Camacho, F. and Tobón, C. (2011) Potential impacts of climate change on the environmental services of humid tropical alpine regions. *Global Ecology and Biogeography*, 20, 19–33.

Cabanes, C., Cazanave, A. and Le Provost, C. (2001) Sea level rise during past 40 years determined from satellite and in situ observations. *Science*, 294, 840–842.

Cairns, J. and Dickson, D.K. (1977) Recovery of streams from spills of hazardous materials. In J. Cairns, K.L. Dickson and E.E. Herricks (eds), *Recovery and Restoration of Damaged Ecosystems*. Charlottesville: University Press of Virginia, 24–42.

Caldararo, N. (2002) Human ecological intervention and the role of forest fires in human ecology. *The Science of the Total Environment*, 292, 141–165.

Caldeira, K. and Wickett, M.G. (2003) Anthropogenic carbon and ocean pH. *Nature*, 425, 365.

Calef, M.P. (2010) Recent climate change impacts on the boreal forest of Alaska. *Geography Compass*, 4, 67–80.

Callaway, R.M. and Aschehoug, E.T. (2000) Invasive plants versus their new and old neighbors: a mechanism for exotic invasion. *Science*, 290, 521–523.

Cannon, S.H., Bigio, E.R. and Mine, E. (2001a) A process for fire-related debris flow initiation, Cerro Grande fire, New Mexico. *Hydrological Processes*, 15, 3011–3023.

Cannon, S.H., Kirkham, R.M. and Parise, M. (2001b) Wildfire-related debris-flow initiation processes, Storm King Mountains, Colorado. *Geomorphology*, 39, 171–188.

Cao, S., Tian, T., Chen, L., Dong, X., Yu, X. and Wang, G. (2010) Damage caused to the environment by reforestation policies in arid and semi-arid areas of China. *Ambio*, 39, 279–283.

Cao, S., Chen, L., Shankman, D., Wang, C., Wang, X. and Zhang, H. (2011) Excessive reliance on afforestation in China's arid and semi-arid regions: lessons in ecological restoration. *Earth-Science Reviews*, 104, 240–245.

Carbognin, L., Teatini, P., Tomasin, A. and Tosi, L. (2010) Global change and relative sea level rise at Venice: what impact in term of flooding. *Climate Dynamics*, 35, 1039–1047.

Carbonell, E. et al. (2008) The first hominim of Europe. *Nature*, 452, 465–469.

Carlson, C.W. (1978) Research in ARS related to soil structure. In W.W. Emerson, R.D. Bond and A.R. Dexter (eds), *Modification of Soil Structure*. Chichester: Wiley, 279–284.

Carpenter, S.R., Stanley, E.H. and Zanden, M.J.V. (2011) State of the world's freshwater ecosystems: physical, chemical, and biological changes. *Annual Review of Environment and Resources*, 36, 75–99.

Carpenter, T.G. (ed.) (2001) *Sustainable Civil Engineering* (2nd edn). Chichester: Wiley.

Carrara, P.E. and Carroll, T.R. (1979) The determination of erosion rates from exposed tree roots in the Piceance Basin, Colorado. *Earth Surface Processes*, 4, 407–417.

Carter, L.D. (1987) Arctic lowlands: introduction. In W.L. Graf (ed.), *Geomorphic Systems of North America* (centennial special Vol. 2). Boulder: Geological Society of America, 583–615.

Carter, L.J. (1977) Soil erosion: the problem persists despite the billions spent on it. *Science*, 196, 409–411.

Catford, J.A., Vesk, P.A., Richardson, D.M. and Pyšek, P. (2012) Quantifying levels of biological invasion: towards the objective classification of invaded and invasible ecosystems. *Global Change Biology*, 18, 44–62.

Cathcart, R.B. (1983) Mediterranean Basin – Sahara reclamation. *Speculations in Science and Technology*, 6, 150–152.

Ceballos, G. and Ehrlich, P.R. (2002) Mammal population losses and the extinction crisis. *Science*, 296, 904–907.

Cerda, A. (1998) Post-fire dynamics of erosional processes under Mediterranean climatic conditions. *Zeitschrift für Geomorphologie*, 42, 373–398.

Cerdan, O. et al. (2010) Rates and spatial variations of soil erosion in Europe: a study based on erosion plot data. *Geomorphology*, 122, 167–177.

Chai, J.-C., Shen, S.-L., Zhu, H.H. and Zhang, X.L. (2004) Land subsidence due to groundwater drawdown in Shanghai. *Géotechnique*, 54, 143–147.

Chai, S.-L. and Tanner, E.V.J. (2011) 150-year legacy of land use on tree species composition in old-secondary forests of Jamaica. *The Journal of Ecology*, 99, 113–121.

Challinor, D. (1968) Alteration of surface soil characteristics by four tree species. *Ecology*, 49, 286–290.

Chambers, F.M., Dresser, P.Q. and Smith, A.G. (1979) Radiocarbon dating evidence on the impact of atmospheric pollution on upland peats. *Nature*, 282, 829–831.

Changnon, S.A., Kunkel, K.E. and Winstanley, D. (2003) Climate factors that caused the unique tall grass prairie in the Central United States. *Physical Geography*, 23, 259–280.

Chandler, T.J. (1976) The climate of towns. In T.J. Chandler and S. Gregory (eds), *The Climate of the British Isles*. London: Longman, 307–329.

Changnon, S.A. (1973) Atmospheric alterations from manmade biospheric changes. In W.R.D. Sewell (ed.), *Modifying the Weather: A Social Assessment*. Adelaide: University of Victoria, 135–184.

Changnon, S.A. (1978) Urban effects on severe local storms at St Louis. *Journal of Applied Meteorology*, 17, 578–586.

Chapin, F.S. and Danell, K. (2001) Boreal forest. In F.S. Chapin, O.E. Sala and E. Huber-Sannwald (eds), *Global Biodiversity in a Changing Environment*. New York: Springer, 101–120.

Chapin, F.S., Sala, O.E. and Huber-Sannwald, E. (2001) *Global Biodiversity in a Changing Environment*. Berlin: Springer Verlag.

Chapin, F.S. et al. (2010) Resilience of Alaska's boreal forest to climatic change. *Canadian Journal of Forest Research*, 40, 1360–1370.

Chappell, J., Cowell, P.J., Woodroffe, C.D. and Eliot, I.G. (1996) Coastal impacts of enhanced greenhouse climate change in Australia: implications for coal use. In W.J. Bouma, G.I. Pearman and M.R. Manning (eds), *Greenhouse*. Melbourne: Commonwealth Scientific and Industrial Research organisation (CSIRO), 220–234.

Charlson, R.J., Lovelock, J.E., Andreae, M.O. and Warren, S.G. (1987) Oceanic phytoplankton, atmospheric sulphur, cloud albedo and climate. *Nature*, 326, 655–661.

Charlson, R.J., Schwartz, S.E., Hales, J.M., Cess, R.D., Coakley, J.A., Hansen, J.E. and Hoffmann, D.J. (1992) Climate forcing by anthropogenic aerosols. *Science*, 255, 423–430.

Charney, J., Stone, P.H. and Quirk, W.J. (1975) Drought in the Sahara: a bio-geophysical feedback mechanism. *Science*, 187, 434–435.

Chase, T.N., Pielke, R.A., Kittle, T.G.F., Nemani, R.R. and Running, S.W. (2000) Simulated impacts of historical land cover changes on global climate in northern winter. *Climate Dynamics*, 16, 93–105.

Chazdon, R.L., Letcher, S.G., van Breugel, M., Martinez-Ramos, M., Bongers, F. and Finegan, B. (2007) Rates of change in tree communities of secondary Neotropical forests following major disturbances. *Philosophical Transactions of the Royal Society of London. Series B, Biological Sciences*, 362, 273–289.

Chen, J., Taniguchi, M., Liu, G., Miyaoka, K., Onodera, S.-I., Tokunaga, T. and Fukushima, Y. (2007) Nitrate pollution of groundwater in the Yellow River delta, China. *Hydrogeology Journal*, 15, 1605–1614.

Chen, L., Wang, W., Zhang, Y. and Lin, G. (2009) Recent progresses in mangrove conservation, restoration and research in China. *Journal of Plant Ecology*, 2, 45–54.

Chen, X. and Zong, Y. (1999) Major impacts of sea-level rise on agriculture in the Yangtze Delta area around Shanghai. *Applied Geography*, 19, 69–84.

Cheng, F. and Granata, T. (2007) Sediment transport and channel adjustments associated with dam removal: field observations. *Water Resources Research*, 43, W03444.

Cheng, S. (2003) Heavy metal pollution in China: origin, pattern and control. *Environmental Science and Pollution Research International*, 10, 192–198.

Chester, D.K. and James, P.A. (1991) Holocene alluviation in the Algarve, southern Portugal: the case for an anthropogenic cause. *Journal of Archaeological Science*, 18, 73–87.

Chew, M.K. (2009) The monstering of tamarisk: how scientists made a plant into a problem. *Journal of the History Of Biology*, 42, 231–266.

Chi, S.C. and Reilinger, R.E. (1984) Geodetic evidence for subsidence due to groundwater withdrawal in many parts of the United States of America. *Journal of Hydrology*, 67, 155–182.

Chiew, F.H.S., Wang, Q.J., McMahon, T.A., Bates, B.C. and Whetton, P.H. (1996) Potential hydrological responses to climate change in Australia. In J.A.A. Jones, E. Liu, M.-K. Woo and H.-T. Kung (eds), *Regional hydrological Response to Climate Change*. Dordrecht: Kluwer, 337–350.

Childe, V.G. (1936) *Man Makes Himself*. London: Watts.

Chin, A. (2006) Urban transformation of river landscapes in a global context. *Geomorphology*, 79, 460–487.

Chiverrell, R.C., Oldfield, F., Appleby, P.G., Barlow, D., Fisger, E., Thompson, R. and Wolff, G. (2008) Evidence for changes in Holocene sediment flux in Semer Water and Raydale, North Yorkshire, UK. *Geomorphology*, 100, 70–82.

Chiverrell, R.C., Foster, G.C., Marshall, P., Harvey, A.M. and Thomas, G.S.P. (2009) Coupling relationships: hillslope-fluvial linkages in the Hodder catchment, NW England. *Geomorphology*, 109, 222–235.

Chorley, R.J. and More, R.J. (1967) The interaction of precipitation and man. In R.J. Chorley (ed.), *Water, Earth and Man*. London: Methuen, 157–166.

Chu, Z.X., Zhai, S.K., Lu, X.X., Liu, J.P., Xu, J.X. and Xu, K.H. (2009) A quantitative assessment of human impacts on decrease in sediment flux from major Chinese rivers entering the western Pacific Ocean. *Geophysical Research Letters*, 36, L19603.

Church, J.A. (2001) How fast are sea levels rising? *Science*, 294, 802–803.

Church, J.A. et al. (2001) Changes in sea level. In J.T. Houghton (ed.), *Climate Change 2001: The Scientific Basis*. Cambridge: Cambridge University Press, 639–693.

Clark, G. (1962) *World Prehistory*. Cambridge: Cambridge University Press.

Clark, J.R. (1977) *Coastal Ecosystem Management*. New York: Wiley.

Clark, J.S., Cachier, H., Goldammer, J.G. and Stocks, B. (eds) (1997) *Sediment Records of Biomass Burning and Global Change*. Berlin: Springer Verlag.

Clark, R.B. (1997) *Marine Pollution* (4th edn). Oxford: Clarendon Press.

Clark, R.B., 2001, *Marine Pollution* (5th edn). Oxford: Clarendon Press.

Clarke, R.T. and McCulloch, J.S.G. (1979) The effect of land use on the hydrology of small upland catchments. In G.E. Hollis (ed.), *Man's Impact on the Hydrological Cycle in the United Kingdom*. Norwich: Geo Abstracts, 71–78.

Claudet, J. and Fraschetti, S. (2010) Human-driven impacts on marine habitats: a regional meta-analysis in the Mediterranean Sea. *Biological Conservation*, 143, 2195–2206.

Clay, J. (2004) *World Agriculture and the Environment*. Washington: Island Press.

Closson, D., LaMoreaux, P.E., Karaki, N.A. and Al-Fugha, H. (2007) Karst system developed in salt layers of the Lisan Peninsula, Dead Sea, Jordan. *Environmental Geology*, 52, 155–172.

Clout, M.N. and Williams, P.A. (eds) (2009) *Invasive Species Management*. New York: Oxford University Press.

Coates, D.R. (ed.) (1976) *Geomorphology and Engineering*. Stroudsburg: Dowden, Hutchinson and Ross.

Coates, D.R. (1977) Landslide perspective. *Reviews in Engineering Geology*, 3, 3–28.

Coates, D.R. (1983) Large-scale land subsidence. In R. Gardner and H. Scoging (eds), *Mega-Geomorphology*. Oxford: Oxford University Press, 21234.

Cochrane, R. (1977) The impact of man on the natural biota. In A.G. Anderson (ed.), *New Zealand in Maps*. Section 14. London: Hodder & Stoughton, 14–15.

Coe, M. (1981) Body size and the extinction of the Pleistocene megafauna. *Palaeoecology of Africa*, 13, 139–145.

Coe, M. (1982) The bigger they are. *Oryx*, 16, 225–228.

Coffey, M. (1978) The dust storms. *Natural History* (New York), 87, 72–83.

Coffin, A.W. (2007) From roadkill to road ecology: a review of the ecological effects of roads. *Journal of Transport Geography*, 15, 396–406.

Cohen, J.E. (2003) Human population: the next half century. *Science*, 302, 1172–1175.

Cole, M.M. and Smith, R.F. (1984) Vegetation as an indicator of environmental pollution. *Transactions of the Institute of British Geographers*, 9, 477–493.

Cole, S. (1970) *The Neolithic Revolution* (5th edn). London: British Museum (Natural History).

Collier, M. and Webb, R.H. (2002) *Floods, Droughts and Climate Change*. Tucson: University of Arizona Press.

Collins, P.M., Davis, B.A.S. and Kaplan, J.O. (2012) The mid-Holocene vegetation of the Mediterranean region and southern Europe, and a comparison with the present day. *Journal of Biogeography*, 39 (10), 1848–1861.

Collison, A., Wade, S., Griffiths, J. and Dehn, M. (2000) Modelling the impact of predicted climate change on landslide frequency and magnitude in S. E. England. *Engineering Geology*, 55, 205–218.

Colls, J. and Tiwary, A. (2009) *Air Pollution: Measurement, Modelling and Mitigation* (3rd edn). London: Routledge.

Comeaux, R.S., Allison, M.A. and Bianchi, T.S. (2012) Mangrove expansion in the Gulf of Mexico with climate change: implications for wetland health and resistance to rising sea levels. *Estuarine, Coastal and Shelf Science*, 96, 81–95.

Committee on the Atmosphere and the Biosphere (1981) *Atmosphere–Biosphere Interactions: Towards a Better Understanding of the Ecological Consequences of Fossil Fuel Combustion*. Washington DC: National Academy Press.

Conacher, A.J. and Conacher, J. (1995) *Rural Land Degradation in Australia*. Melbourne: Oxford University Press.

Conacher, A.J. and Sala, M. (eds) (1998) *Land Degradation in Mediterranean Environments of the World*. Chichester: Wiley.

Conley, D.J. et al. (2011) Hypoxia is increasing in the coastal zone of the Baltic Sea. *Environmental Science & Technology*, 45, 6777–6783.

Conway, D., Kroi, M., Alcamo, J. and Hulme, M. (1996) Future availability of water in Egypt: the interaction of global, regional and basin scale driving forces in the Nile Basin. *Ambio*, 25, 336–342.

Conway, G.R. and Pretty, J.N. (1991) *Unwelcome Harvest: Agriculture and Pollution*. London: Earthscan.

Conway, V.M. (1954) Stratigraphy and pollen analysis of southern Pennine blanket peats. *The Journal of Ecology*, 42, 117–147.

Cook, A.J. and Vaughan, D.G. (2010) Overview of areal changes of the ice shelves on the Antarctic Peninsula over the past 50 years. *The Cryosphere*, 4, 77–98.

Cook, B.I., Miller, R.L. and Seager, R. (2009) Amplification of the North American 'Dust Bowl' drought through human-induced land degradation. *Proceedings of the National Academy of Sciences of the United States of America*, 106, 4997–5001.

Cook, K.H. and Vizy, E.K. (2008) Effects of twenty-first-century climate change on the Amazon rain forest. *Journal of Climate*, 21, 542–560.

Cooke, R.U. and Doornkamp, J.C. (1974) *Geomorphology in Environmental Management*. Oxford: Clarendon Press.

Cooke, R.U. and Doornkamp, J.C. (1990) *Geomorphology in Environmental Management* (2nd edn). Oxford: Clarendon Press.

Cooke, R.U. and Reeves, R.W. (1976) *Arroyos and Environmental Change in the American South-West*. Oxford: Clarendon Press.

Cooke, R.U., Brunsden, D., Doornkamp, J.C. and Jones, D.K.C. (1982) *Urban Geomorphology in Drylands*. Oxford: Oxford University Press.

Coones, P. and Patten, J.H.C. (1986) *The Landscape of England and Wales*. Harmondsworth: Penguin Books.

Cooper, C.F. (1961) The ecology of fire. *Scientific American*, 204 (4), 150–160.

Cooper, D.M. and Jenkins, A. (2003) Response of acid lakes in the UK to reductions in atmospheric deposition of sulphur. *The Science of the Total Environment*, 313, 91–100.

Cooper, J.A.G. and Pilkey, O.H. (2004) Sea-level rise and shoreline retreat: time to abandon the Bruun Rule. *Global and Planetary Change*, 43, 157–171.

Corlett, R.T. (1995) Tropical secondary forests. *Progress in Physical Geography*, 19, 159–172.

Corlett, R.T. (2011) Impacts of warming on tropical lowland rainforests. *Trends in Ecology & Evolution*, 26, 606–613.

Costa, J.E. (1975) Effects of agriculture on erosion and sedimentation in the Piedmont province, Maryland. *Geological Society of America Bulletin*, 86, 1281–1286.

Cotton, W.R. and Pielke, R.A. (2007) *Human Impacts on Weather and Climate* (2nd edn). Cambridge: Cambridge University Press.

Coumou, D. and Rahmstorf, S. (2012) A decade of weather exteremes. *Nature Climate Change*, 2, 491–496.

Cowell, E.B. (1976) Oil pollution of the sea. In R. Johnson (ed.), *Marine Pollution*. London: Academic Press, 353–401.

Craddock, P.T. (2000) From hearth to furnace: evidences for the earliest metal smelting technologies in the eastern Mediterranean. *Paléorient*, 26, 151–165.

Crawford, S.S. and Muir, A.M. (2008) Global introductions of salmon and trout in the genus Oncorhynchus: 1870–2007. *Reviews in Fish Biology and Fisheries*, 18, 313–344.

Crick, H.Q.P. (2004) The impact of climate change on birds. *The Ibis*, 146 (Suppl. 1), 48–56.

Crisp, D.T. (1977) Some physical and chemical effects of the Cow Green (Upper Teesdale) impoundment. *Freshwater Biology*, 7, 109–120.

Critchley, W.R.S., Reij, C. and Willcocks, T.J. (1994) Indigenous soil and water conservation: a review of the state of knowledge and prospects for building on traditions. *Land Degradation and Rehabilitation*, 5, 293–314.

Cronin, L.E. (1967) The role of man in estuarine processes. *Publication 83, American Association for the Advancement of Science*, 667–689.

Cronk, Q.C.B. and Fuller, J.L. (1995) *Plant Invaders*. London: Chapman & Hall.

Crowley, B.E. (2010) A refined chronology of prehistoric Madagascar and the demise of the megafauna. *Quaternary Science Reviews*, 29, 2591–2603.

Croxall, J.P., Butchart, S.H.M., Lascelles, B., Stattersfield, A.J., Sullivan, B., Symes, A. and Taylor, P. (2012) Seabird conservation status, threats and priority actions: a global assessment. *Bird Conservation International*, 22, 1–34.

Crozier, M.J. (2010) Deciphering the effect of climate change on landslide activity: a review. *Geomorphology*, 124, 260–267.

Crozier, M.J., Marx, S.L. and Grant, I.J. (1978) Impact of off-road recreational vehicles on soil and vegetation. *Proceedings of the 9th New Zealand Geography Conference*, Dunedin, 76–79.

Crutzen, P.J. (2002) Geology of mankind. *Nature*, 415, 23.

Crutzen, P.J. and Goldammer, J.G., (1993) *Fire in the Environment*. Chichester: Wiley.

Crutzen, P.J., Aselmann, I. and Sepler, W. (1986) Methane production by domestic animals, wild ruminants, other herbivores, fauna and humans. *Tellus*, 38B, 271–284.

Csiki, S. and Rhoads, B.L. (2010) Hydraulic and geomorphological effects of run-of-river dams. *Progress in Physical Geography*, 34, 755–780.

Cuadra, P.E. and Vidon, P. (2011) Storm nitrogen dynamics in tile-drain flow in the US Midwest. *Biogeochemistry*, 104, 293–308.

Cuff, D.J. and Goudie, A.S. (eds) (2009) *The Oxford Companion to Global Change*. New York: Oxford University Press.

Cumberland, K.B. (1961) Man in nature in New Zealand. *New Zealand Geographer*, 17, 137–154.

Cunningham, D.A., Collins, J.F. and Cummins, T. (2001) Anthropogenically-triggered iron pan formation in some Irish soils over various time spans. *Catena*, 43, 167–176.

Curran, M.A.J., van Ommen, T.D., Morgan, V.I., Phillips, K.L. and Palmer, A.S. (2003) Ice core evidence for Antarctic sea ice decline since the 1950s. *Science*, 302, 1203–1206.

Cushman, G.T. (2011) Humboldtian science, creole meteorology, and the discovery of human-caused climate change in South America. *Osiris*, 26, 19–44.

Custodio, E., Iribar, V., Manzano, B.A. and Galofre, A. (1986) Evolution of sea water chemistry in the Llobyegat Delta, Barcelona, Spain. In *Proceedings of the 9th salt water chemistry meeting*, Delft.

Cypser, D.A. and Davis, S.D. (1998) Induced seismicity and the potential for liability under U.S. law. *Tectonophysics*, 289, 239–255.

Dai, A. (2011) Drought under global warming: a review. *Wiley Interdisciplinary Reviews: Climate Change*, 2, 45–65.

Dallmeyer, A. and Claussen, M. (2011) The influence of land cover change in the Asian monsoon region on present-day and mid-Holocene climate. *Biogeosciences*, 8, 1499–1519.

Dalzell, B.J., King, J.Y., Mulla, D.J., Finlay, J.C. and Sands, G.R. (2011) Infleunce of subsurface drainage on quantity and quality of dissolved organic matter export from agricultural landscapes. *Journal of Geophysical Research*, 116, G02023.

Daniel, T.C., McGuire, P.E., Stoffel, D. and Millfe, B. (1979) Sediment and nutrient yield from residential construction sites. *Journal of Environmental Quality*, 8, 304–308.

Dapples, F., Lotter, A.F., van Leeuwen, J.F.N., van der Knaap, W.O., Dimitriadis, S. and Oswald, D. (2002) Paleolimnological evidence for increased landslide activity due to forest clearing and land-use since 3600 cal BP in the western Swiss Alps. *Journal of Paleolimnology*, 27, 239–248.

Darby, H.C. (1956) The clearing of the woodland in Europe. In W.L. Thomas (ed.), *Man's Role in Changing the Face of the Earth*. Chicago: University of Chicago Press, 183–216.

Darling, F.F. (1956) Man's ecological dominance through domesticated animals on wild lands. In W.L. Thomas (ed.), *Man's Role in Changing the Face of the Earth*. Chicago: University of Chicago Press, 778–787.

Darungo, F.P., Allee, P.H. and Weickmann, H.K. (1978) Snowfall induced by a power plant plume. *Geophysical Research Letters*, 5, 515–517.

Dasgupta, S., Laplante, B., Meisner, C., Wheeler, D. and Yan, J. (2009) The impact of sea level rise on developing countries: a comparative analysis. *Climatic Change*, 93, 379–388.

Davidson, A.D. et al. (2012) Drivers and hotspots of extinction risk in marine mammals. *Proceedings of the National Academy of Sciences of the United States of America*, 109, 3395–3400.

Davis, M.A. (2009) *Invasion Biology*. Oxford: Oxford University Press.

Davis, M.C.R., Hamza, O. and Harris, C. (2001) The effect of rise in mean annual temperature on the stability of rock slopes containing ice-filled discontinuities. *Permafrost and Periglacial Processes*, 12, 137–144.

Dawson, Q., Kechavarzi, C., Leeds-Harrison, P.B. and Burton, R.G.O. (2010) Subsidence and degradation of agricultural peatlands in the Fenlands of Norfolk, U.K. *Geoderma*, 154, 181–187.

Day, J.W. et al. (2008) Consequences of climate change on the ecogeomorphology of coastal wetlands. *Estuaries and Coasts*, 31, 477–491.

Day, S.J. and Maslin, M. (2010) Gas hydrates: a hazard for the twenty-first century? *Philosophical Transactions. Series A, Mathematical, Physical, and Engineering Sciences*, 368, 2579–2583.

Deadman, A. (1984) Recent history of Spartina in northwest England and in North Wales and its possible future development. In P. Doody (ed.), *Spartina Anglica in Great Britain*. Shrewsbury: Nature Conservancy Council, 22–24.

Dean, D.J. and Schmidt, J.C. (2011) The role of feedback mechanisms in historic channel changes of the lower Rio Grande in the Big Bend region. *Geomorphology*, 126, 333–349.

De Angelis, H. and Skvarca, P. (2003) Glacier surge after Ice Shelf collapse. *Science*, 299, 1560–1562.

DeBano, L.F. (2000) The role of fire and soil heating on water repellency in wildland environments: a review. *Journal of Hydrology*, 231/2, 195–206.

DEFRA (2004) http://archive.defra.gov.uk/evidence/statistics/environment/inlwater/iwlevels.htm (accessed 06 February 2013).

DEFRA (2010) http://archive.defra.gov.uk/environment/biodiversity/documents/indicator/200904h1b.pdf (Accessed December 7th, 2012)

DEFRA (2011) www.defra.gov.uk/statistics/files/Wild-bird-populations-in-the-UK-1970-2010/ (Accessed December 7th, 2012)

Degu, A.M. et al. (2011) The influence of large dams on surrounding climate and precipitation patterns. *Geophysical Research Letters*, 38, L044405.

Dehn, M. and Buma, J. (1999) Modelling future landslide activity based on general circulation models. *Geomorphology*, 30, 175–187.

Dehn, M., Gurger, G., Buma, J. and Gasparetto, P. (2000) Impact of climate change on slope stability using expanded downscaling. *Engineering Geology*, 55, 193–204.

De Meyer, A., Poesen, J., Isabirye, M., Deckers, J. and Raes, D. (2011) Soil erosion rates in tropical villages: a case study from Lake Victoria Basin, Uganda. *Catena*, 84, 89–98.

Denevan, M.W. (1992) The pristine myth: the landscapes of the Americas in 1492. *Annals of the Association of American Geographers*, 82, 369–385.

Denevan, W.M. (2001) *Cultivated Landscapes of Native Amazonia and the Andes*. Oxford: Oxford University Press.

Denisova, T.B. (1977) The environmental impact of mineral industries. *Soviet Geography*, 18, 646–659.

Denson, E.P. (1970) The trumpeter swan, *Olor Buccinator*; a conservation success and its lessons. *Biological Conservation*, 2, 251–256.

Dent, D.L. and Pons, L.J. (1995) A world perspective on acid sulphate soils. *Geoderma*, 67, 263–276.

Deo, R.C., Syktus, J.I., McAlpine, C.A., Lawrence, P.J., McGowan, H.A. and Phinn, S.R. (2009) Impact of historical land cover change on daily indices of climate extremes including droughts in eastern Australia. *Geophysical Research Letters*, 36, L08705.

Department of Environment (1984) *Digest of Environmental Pollution and Water Statistics for 1983*. London: HMSO.

Derocher, A.E., Lunn, N.J. and Stirling, I. (2004) Polar bears in a warming climate. *Integrative and Comparative Biology*, 44, 163–176.

Derraik, J.G.B. (2002) The pollution of the marine environment by plastic debris: a review. *Marine Pollution Bulletin*, 44, 842–852.

De Sylva, D. (1986) Increased storms and estuarine salinity and other ecological impacts of the greenhouse effect. In J.G. Titus (ed.), *Effects of Changes in Stratospheric Ozone and Global Climate* (Vol. 4, *Sea level rise*). Washington DC: UNEP/USEPA, 153–164.

Detwyler, T.R. (ed.) (1971) *Man's Impact on Environment*. New York: McGraw-Hill.

Detwyler, T.R. and Marcus, M.G. (1972) *Urbanisation and Environment: The Physical Geography of the City*. Belmont: Duxbury Press.

De Wit, M. and Stankiewicz, J. (2006) Changes in surface water supply across Africa with predicted climate change. *Science*, 311, 1917–1921.

Diamond, J. (1997) *Guns, Germs and Steel*. London: Chatto and Windus.

Diamond, J. (2002) Evolution, consequences and future of plant and animal domestication. *Nature*, 418, 700–707.

Diaz, R.J. and Rosenberg, R. (2008) Spreading dead zones and consequences for marine ecosystems. *Science*, 321, 926–929.

Di Castri, F. (1989) History of biological invasions with special emphasis on the old world. In J.A. Drake (ed.), *Biological Invasions: A Global Perspective*. Chichester: Wiley, 1–30.

Dickinson, W.R. (1999) Holocene sea-level record on Funafuti and potential impact of global warming on Central Pacific atolls. *Quaternary Research*, 51, 124–132.

Diem, J. (2003) Potential impact of ozone on coniferous forests of the interior southwestern United States. *Annals of the Association of American Geographers*, 93, 265–280.

Dillehay, T.D. (2003) Tracking the first Americans. *Nature*, 425, 23–24.

Dillehay, T.D., Rossen, J., Andres, T.C. and Williams, D.E. (2007) Preceramic adoption of peanut, squash and cotton in Northern Peru. *Science*, 316, 1890–1893.

Dimbleby, G.W. (1974) The legacy of prehistoric man. In A. Warren and F.B. Goldsmith (eds), *Conservation in Practice*. London: Wiley, 179–189.

Dinerstein, E. et al. (2007) The fate of wild tigers. *Bioscience*, 57, 508–514.

Dirnböck, T., Essl, F. and Rabitsch, W. (2011) Disproportional risk for habitat loss of high-altitude endemic species under climate change. *Global Change Biology*, 17, 990–996.

Dirzo, R. and Raven, P.H. (2003) Global state of biodiversity and loss. *Annual Review of Environment and Resources*, 28, 137–167.

Doak, T. and Marvier, M. (2003) Predicting the effects of species loss on community stability. In P. Kareiva and S.A.

Levin (eds), *The Importance of Species*. Princeton: Princeton University Press, 140–160.

Dobson, A., Borner, M. and Sinclair, T. (2010) Road will ruin Serengeti. *Nature*, 467, 272–273.

Dobson, M.C. (1991) *De-icing Salt Damage to Trees and Shrubs*. Forestry Commission Bulletin 101. London: HMSO.

Dodd, A.P. (1959) The biological control of the prickly pear in Australia. In A. Keast, R.L. Crocker and C.S. Christian (eds), *Biogeography and Ecology in Australia*. The Hague: Junk, 565–577.

D'Odorico, P., Yoo, J.C. and Over, T.M. (2001) An assessment of ENSO-induced patterns of rainfall erosivity in the southwestern United States. *Journal of Climate*, 14, 4230–4242.

Doerr, A. and Guernsely, L. (1956) Man as a geomorphological agent: the example of coal mining. *Annals of the Association of American Geographers*, 46, 197–210.

Dominguez, F., Rivera, E., Lettenmaier, D.P. and Castro, C.L. (2012) Changes in winter precipitation extremes for the western United States under a warmer climate as simulated by regional climate models. *Geophysical Research Letters*, 39, L05803.

Doney, S.C. (2006) The dangers of ocean acidification. *Scientific American*, 294 (3), 38–45.

Dong, Z., Chen, G., He, X., Han, Z. and Wang, X. (2004) Controlling blown sand along the highway crossing the Taklimakan Desert. *Journal of Arid Environments*, 57, 329–344.

Donkin, R.A. (1979) Agricultural terracing in the Aboriginal New World. *Viking Fund Publications in Anthropology*, 561.

Donnely, L.J. (2009) A review of international cases of fault reactivation during mining subsidence and fluid abstraction. *Quarterly Journal of Engineering Geology and Hydrogeology*, 42, 73–94.

Donner, S.D. (2009) Coping with commitment: projected thermal stress on coral reefs under different future scenarios. *PLoS ONE*, 4, e5712, 1–9.

Doody, J.P. (2012) Coastal squeeze and managed realignment in southeast England, does it tell us anything about the future? *Ocean and Coastal Management* doi 10.1016j.ocecoaman.2012.05.0008.

Doody, P. (ed.) (1984) *Spartina Anglica in Great Britain*. Shrewsbury: Nature Conservancy Council.

Doolittle, W. (2000) *Cultivated landscapes of native North America*. Oxford: Oxford University Press.

Dörries, M. (2011) The politics of atmospheric sciences: 'Nuclear Winter' and global climate change. *Osiris*, 26, 198–223.

Dotterweich, M. (2008) The history of soil erosion and fluvial deposits in small catchments of central Europe: deciphering the long-term interaction between humans and the environment – a review. *Geomorphology*, 101, 192–208.

Doughty, C.E., Loarie, S.R. and Field, C.B. (2012) Theoretical impact of changing albedo on precipitation at the southernmost boundary of the ITCZ in South America. *Earth Interactions*, 16, 1–14.

Doughty, R.W. (1974) The human predator: a survey. In I.R. Manners and M.V. Mikesell (eds), *Perspectives on Environment*. Washington DC: Association of American Geographers, 152–180.

Doughty, R.W. (1978) *The English Sparrow in the American Landscape: A Paradox in Nineteenth Century Wildlife Conservation*. Research paper. Oxford: School of Geography, University of Oxford, 19.

Douglas, I. (1983) *The Urban Environment*. London: Arnold.

Douglas, I. and Lawson, N. (2001) The human dimensions of geomorphological work in Britain. *Journal of Industrial Ecology*, 4, 9–33.

Douglas, I., Goode, D., Houck, C. and Wang, R. (eds) (2011) *The Routledge Handbook of Urban Ecology*. New York: Routledge.

Douville, H., Chauvin, F., Planton, S., Royer, J.-F., Salas-Mélia, D. and Tyteca, S. (2002) Sensitivity of the hydrological cycle to increasing amounts of greenhouse gases and aerosols. *Climate Dynamics*, 20, 45–68.

Down, C.G. and Stocks, J. (1977) *Environmental Impact of Mining*. London: Applied Science Publishers.

Downing, A. and Leibold, M.A. (2010) Species richness facilitates ecosystem resilience in aquatic food webs. *Freshwater Biology*, 55, 2123–2137.

Downs, P.W. and Gregory, K.J. (2004) *River Channel Management*. London: Arnold.

Doyle, M.W., Stanley, E.H. and Habor, J.M. (2003) Channel adjustments following two dam removals in Wisconsin. *Water Resources Research*, 39, W1011.

Doyle, T.W. and Girod, G.F. (1997) The frequency and structure of Atlantic hurricanes and their influence on the structure of the South Florida mangrove communities. In H.F. Diaz and R.S. Pulwarty (eds), *Hurricanes*. Berlin: Springer, 109–120.

Dragovich, D. and Morris, R. (2002) Fire intensity, slopewash and bio-transfer of sediment in eucalypt forest, Australia. *Earth Surface Processes and Landforms*, 27, 1309–1319.

Draut, A.E. (2012) Effects of river regulation on aeolian landscapes, Colorado River, southwestern USA. *Journal of Geophysical Research*, 117, F02022.

Dregne, H.E. (1986) Desertification of arid lands. In F. El-Baz and M.H.A. Hassan (eds), *Physics of Desertification*. Dordrecht: Nijhoff, 4–34.

Dregne, H.E. and Tucker, C.J. (1988) Desert encroachment. *Desertification Control Bulletin*, 16, 16–19.

Dreibrodt, S., Lubos, C., Terhorst, B., Damm, B. and Bork, H.-R. (2010) Historical soil erosion by water in Germany: scales and archives, chronology, research perspectives. *Quaternary International*, 222, 80–95.

Drewry, D.J. (1991) The response of the Antarctic ice sheet to climate change. In C.M. Harris and B. Stonehouse (eds), *Antarctica and Global Climatic Change*. London: Belhaven Press, 90–106.

Drexler, J.Z., de Fontaine, C.S. and Deverel, S.J. (2009) The legacy of wetland drainage on the remaining peat in the Sacramento-San Joaquin Delta, California, USA. *Wetlands*, 29, 372–386.

Driscoll, C.A., Macdonald, D.W. and O'Brien, S.J. (2009) From wild animals to domestic pets, an evolutionary view of domestication. *Proceedings of the National Academy of Sciences of the United States of America*, 106, 9971–9978.

Driscoll, R. (1983) The influence of vegetation on the swelling and shrinkage of clay soils in Britain. *Géotechnique*, 33, 293–326.

Dullinger, S. et al. (2012) Extinction debt of hig-mountain plants under twenty-first-century climate change. *Nature Climate Change*, 2, 619–622.

Duncan, R.P., Blackburn, T.M. and Sol, D. (2003) The ecology of bird introductions. *Annual Review of Environment and Resources*, 28, 359–399.

Dunne, T. and Leopold, L.B. (1978) *Water in Environmental Planning*. San Francisco: Freeman.

Durá-Gómez, I. and Talwani, P. (2010) Reservoir-induced seismicity associated with the Itoiz Reservoir, Spain: a case study. *Geophysical Journal International*, 181, 343–356.

Durant, A.J., Harrison, S.P., Watson, I.M. and Balkanski, Y. (2009) Sensitivity of direct radiative forcing by mineral dust to particle characteristics. *Progress in Physical Geography*, 33, 80–102.

Durán Zuazo, V.H., Francia Martínez, J.R. and Martínez Raya, A. (2004) Impact of vegetative cover on runoff and soil erosion at hillslope scale in Lanjaron, Spain. *The environmentalist*, 24, 39–48.

Dusar, B., Verstraeten, G., Notebaert, B. and Bakker, J. (2011) Holocene environmental change and its impact on sediment dynamics in the Eastern Mediterranean. *Earth-Science Reviews*, 108, 137–157.

Du Toit, G. van N., Snyman, H.A. and Malan, P.J. (2009) Physical impact of grazing by sheep on soil parameters in the Nama Karoo subshrub/grass rangeland of South Africa. *Journal of Arid Environments*, 73, 804–810.

Duvall, C.S. (2011) Biocomplexity from the ground up: vegetation patterns in a West African savanna landscape. *Annals of the Association of American Geographers*, 101, 497–522.

D'Yakanov, K.N. and Reteyum, A.Y. (1965) The local climate of the Rybinsk reservoir. *Soviet Geography*, 6, 40–53.

Dyer, K.R. (1995) Responses of estuaries to climate change. In D. Eisma (ed.), *Climate Change on Coastal Habitation*. Boca Raton: Lewis, 85–110.

Eckhardt, K. and Ulbrich, U. (2003) Potential impacts of climate change on groundwater recharge and streamflow in a central European low mountain range. *Journal of Hydrology*, 284, 244–252.

Eden, M.J. (1974) Palaeoclimatic influences and the development of savanna in southern Venezuela. *Journal of Biogeography*, 1, 95–109.

Edington, J.M. and Edington, M.A. (1977) *Ecology and Environmental Planning*. London: Chapman & Hall.

Edlin, H.L. (1976) The Culbin sands. In J. Leniham and W.W. Fletcher (eds), *Reclamation*. Glasgow: Blackie, 1–31.

Edwards, A.J., Clark, S., Zahir, H., Rajasuriya, A., Naseer, A. and Rubens, J. (2001) Coral bleaching and mortality on artificial and natural reefs in Maldives in 1998, sea surface temperature anomalies and initial recovery. *Marine Pollution Bulletin*, 42, 7–15.

Edwards, G.P., Zeng, B., Saalfield, W.K. and Vaarzon-Morel, P. (2010) Evaluation of the impacts of feral camels. *The Rangeland Journal*, 32, 43–54.

Edwards, K.J. (1985) The anthropogenic factor in vegetational history. In K.J. Edwards and W.P. Warren (eds), *The Quaternary History of Ireland*. London: Academic Press, 187–200.

Edwards, P.J., Fletcher, M.R. and Berny, P. (2000) Review of the factors affecting the decline ofm the European brown hare, *Lepus europaeus* (Pallas, 1770) and the use of wildlife incident data to evaluate the significance of paraquat. *Agriculture, Ecosystems & Environment*, 79, 95–103.

Ehrenfeld, D.W. (1972) *Conserving Life on Earth*. New York: Oxford University Press.

Ehrlich, P.R. and Ehrlich, A.H. (1970) *Population, Resources, Environment: Issues in Human Ecology*. San Francisco: Freeman.

Ehrlich, P.R. and Ehrlich, A.H. (1982) *Extinction*. London: Gollancz.

Ehrlich, P.R., Ehrlich, A.H. and Holdren, J.P. (1977) *Ecoscience: Population, Resources, Environment*. San Francisco: Freeman.

Eitner, V. (1996) Geomorphological response to the East Frisian barrier islands to sea level rise: an investigation of past and future evolution. *Geomorphology*, 15, 57–65.

El Banna, M.M. and Frihy, O.E. (2009) Human-induced changes in the geomorphology of the northeastern coast of the Nile delta, Egypt. *Geomorphology*, 107, 72–78.

Eldridge, D.J. (1998) Trampling of microphytic crusts on calcareous soils, and its impact on erosion under rain-impacted flow. *Catena*, 33, 221–239.

Eldridge, D.J. and Robson, A.D. (1997) Bladeploughing and exclosure influence soil properties in a semi-arid Australian woodland. *Journal of Range Management*, 50, 191–198.

Eldridge, D.J., Bowker, M.A., Maester, F.T., Roger, E., Reynolds, J.F. and Whitford, W.G. (2011) Impacts of shrub encroachment on ecosystem structure and functioning: towards a global synthesis. *Ecology Letters*, 14, 709–722.

Elguindi, N. and Giorgi, F. (2006) Projected changes in the Caspian Sea level for the 21st century based on the latest AOGCM simulations. *Geophysical Research Letters*, 33, L08706.

Elling, W., Dittmar, C., Pfaffelmoser, K. and Rötzer, T. (2009) Dendroecological assessment of the complex causes of decline and recovery of the growth of silver fir (*Abies alba* Mill) in Southern Germany. *Forest Ecology and Management*, 257, 1175–1187.

Elliott, J.G., Gillis, A.C. and Aby, S.B. (1999) Evolution of arroyos: incised channels of the southwestern United States. In S.E. Darby and A. Simon (eds), *Incised River Channels*. Chichester: Wiley, 153–185.

Elliott, J.M. (1989) Wild brown trout *Salmon trutta*: an important national and international resource. *Freshwater Biology*, 21, 1–5.

Ellis, E.C. (2011) Anthropogenic transformation of the terrestrial biosphere. *Philosophical Transactions. Series A, Mathematical, Physical, and Engineering Sciences*, 369, 1010–1035.

Ellison, A.M. and Farnsworth, E.J. (1997) Simulated sea level change alters anatomy, physiology, growth and reproduction of red mangrove (*Rhizophora mangle* L.). *Oecologia*, 112, 435–446.

Ellison, J.C. and Stoddart, D.R. (1990) Mangrove ecosystem collapse during predicted sea level rise: holocene analogues and implications. *Journal of Coastal Research*, 7, 151–165.

El-Raey, M. (1997) Vulnerability assessment of the coastal zone of the Nile delta of Egypt to the impact of sea level rise. *Ocean and Coastal Management*, 37, 29–40.

Elsner, J.B. and Kara, A.B. (1999) *Hurricanes of the North Atlantic*. New York: Oxford University Press.

Elsner, J.B., Kossin, J.P. and Jagger, T.H. (2008) The increasing intensity of the strongest tropical cyclones. *Nature*, 455, 92–95.

Elsom, D. (1992) *Atmospheric Pollution* (2nd edn). Oxford: Blackwell.

Elton, C.S. (1958) *The Ecology of Invasions by Plants and Animals*. London: Methuen.

Emanuel, K.A. (1987) The dependence of hurricane intensity on climate. *Nature*, 326, 483–485.

Emmanuel, W.R., Shugart, H.H. and Stevenson, M.P. (1985) Climatic change and the broad-scale distribution of terrestrial ecosystem complexes. *Climatic Change*, 7, 29–43.

Engelhardt, F.R. (ed.) (1985) *Petroleum Effects in the Arctic Environment*. London: Elsevier Applied Science.

Engelhart, S.E., Horton, B.P., Douglas, B.C., Peltier, W.R. and Törnquist, T.E. (2009) Spatial variability of late Holocene and 20th century sea-level rise along the Atlantic coast of the United States. *Geology*, 37, 1115–1118.

Engler, R. et al. (2011) 21st century climate change threatens mountain flora unequally across Europe. *Global Change Biology*, 17, 2330–2341.

Environment Agency (2010) *Fifth otter survey of England 2009–2010. Summary report*. Bristol: Environment Agency.

Ericson, J.P., Vorosmarty, C.J., Dingham, L., Ward, L.G. and Meybeck, M. (2006) Effective sea-level rise and deltas: causes of change and human dimension implications. *Global and Planetary Change*, 50, 63–82.

Erlandson, J.M. (2010) A deep history for the Pacific: where past, present, and future meet. *Journal of Pacific Archaeology*, 1, 111–114.

Erlandson, J.M. et al. (2011) Paleoindian seafaring, maritime technologies, and coastal foraging on California's Channel Islands. *Science*, 331, 1181–1185.

Escalente, S.A. and Pimentel, A.S. (2008) Coastal dune stabilization using geotextile tubes at Las Colorados. *Geosynthetics*, 26, 16–24.

Etienne, D., Ruffaldi, P., Goepp, S., Ritz, F., Georges-Leroy, M., Pollier, B. and Dambrine, E. (2011) The origin of closed depressions in Northeastern France: a new assessment. *Geomorphology*, 126, 121–131.

European Environment Agency (2001) *Eutrophication in Europe's Coastal Waters*. Copenhagen: European Environment Agency.

Evan, A.T., Dunion, J., Foley, J.A., Heidinger, A.K. and Velden, C.S. (2006) New evidence for a relationship between Atlantic tropical cyclone activity and African dust outbreaks. *Geophysical Research Letters*, 33, L19813.

Evans, C.D. and Jenkins, A. (2000) Surface water acidification in the South Pennines II. Temporal trends. *Environmental Pollution*, 109, 21–34.

Evans, C.D. et al. (2001) Recovery from acidification in European surface waters. *Hydrology and Earth System Sciences*, 5, 283–297.

Evans, D.M. (1966) Man-made earthquakes in Denver. *Geotimes*, 10, 11–18.

Evans, G. (2008) Man's impact on the coastline. *Journal of Iberian Geology*, 34, 167–190.

Evans, J.G., Limbrey, S. and Cleere, H. (eds) (1975) *The effect of man on the landscape: the Highland zone*. Council for British Archaeology research report 11.

Evans, J.P. (2010) Global warming impact on the dominant precipitation processes in the Middle east. *Theoretical and Applied Climatology*, 99, 389–402.

Evans, M. and Warburton, J. (2007) *Geomorphology of Upland Peat*. Oxford: Blackwell.

Evans, M., Warburton, J. and Yang, J. (2006) Eroding blanket peat catchments: global and local implications of upland organic sediment budgets. *Geomorphology*, 79, 45–57.

Eve, M.D., Sperow, M., Paustian, K. and Follett, R.F. (2002) National-scale estimation of changes in soil carbon stocks on agricultural lands. *Environmental Pollution*, 116, 431–438.

Evenari, M., Shanan, L. and Tadmor, N.H. (1971) Runoff agriculture in the Negev Desert of Israel. In W.G. McGinnies, R.J. Goldman and P. Paylore (eds), *Food, Fiber and the Arid Lands*. Tucson: University of Arizona Press, 312–322.

Ewing, B., Oursler, A., Reed, A., Moore, D., Goldfinger, S. and Wackernagle, M. (2009) *The Ecolpogical Footprint Atlas*. Oakland: Global Footprint Network.

Fabricius, K.E. (2005) Effects of terrestrial runoff on the ecology of corals and coral reefs: review and synthesis. *Marine Pollution Bulletin*, 50, 125–146.

Faeth, S.H., Bang, C. and Saari, S. (2011) Urban biodiversity: patterns and mechanisms. *Annals of the New York Academy of Sciences*, 1223, 69–81.

Fahrig, L. (2003) Effects of habitat fragmentation on biodiversity. *Annual Review of Ecology, Evolution, and Systematics*, 34, 487–515.

Fairbridge, R.W. (1983) Isostasy and eustasy. In D.E. Smith and A.G. Dawson (eds), *Shorelines and Isostasy*. London: Academic Press, 3–25.

Fairhead, J. and Leach, M. (1996) *Misreading the African Landscape: Society and Ecology in a Forest-Savanna Mosaic*. Cambridge: Cambridge University Press.

Faith, J.T. and Surovell, T.A. (2009) Synchronous extinction of North America's Pleistocene mammals. *Proceedings of the National Academy of Sciences of the United States of America*, 106, 20641–20645.

FAO (2009) *The State of World fisheries and Aquaculture 2008.* Rome: FAO.

Farias, W.R.G., Pinto, O., Naccarato, K.P. and Pinto, I.R.C.A. (2009) Anomalous lightning activity over the metropolitan region of São Paulo due to urban effects. *Atmospheric Research*, 91, 485–490.

Favis-Mortlock, D. and Boardman, J. (1995) Non-linear responses of soil erosion to climate change: modelling study on the UK South Downs, *Catena*, 25, 365–387

Favis-Mortlock, D.T. and Guerra, A.J.T. (1999) The implications of general circulation model estimates of rainfall for future erosion: a case study from Brazil. *Catena*, 37, 329–354.

Fearnside, P.M. and Laurance, W.F. (2003) Comment on 'Determination rates of the world's humid tropical forests'. *Science*, 299, 1015.

Federal Research Centre for Forestry and Forest Products (2000) *Forest condition in Europe.* Geneva and Brussels: UN/ECE and EC.

Feldstein, S.B. (2011) Subtropical rainfall and the Antarctic ozone hole. *Science*, 332, 925–926.

Fenger, J. (1999) Urban air quality. *Atmospheric Environment*, 33, 4877–4900.

Fenn, M.E. et al. (2003) Ecological effects of nitrogen deposition in the western United States. *Bioscience*, 53, 404–420.

Ferguson, G. and Gleeson, T. (2012) Vulnerability of coastal aquifers to groundwater use and climate change. *Nature Climate Change*, 2, 342–345.

Fiedel, S.J. (2005) Man's best friend – Mammoth's worst enemy? A speculative essay on the role of dogs in Paleoindian colonization and magafaunal extinction. *World Archaeology*, 37, 11–25.

Field, J. and Wroe, S. (2012) Aridity, faunal adaptations and Australian Late Pleistocene extinctions. *World Archaeology*, 44, 56–74.

Field, M.E., Ogston, A.S. and Storlazzi, C.D. (2011) Rising sea level may cause decline of fringing coral reefs. *Eos*, 92, 273–280.

Fillenham, L.F. (1963) Holme Fen Post. *The Geographical Journal*, 129, 502–503.

Finegan, B. (1996) Pattern and process in neotropical secondary rain forests: the first 100 years of succession. *Trends in Ecology & Evolution*, 11, 119–124.

Fischer, L., Kääb, A., Huggel, C. and Noetzli, J. (2006) Geology, glacier retreat and permafrost degradation as controlling factors of slope instabilities in a high-mountain rock wall: the Monte Rosa east face. *Natural Hazards and Earth System Sciences*, 6, 761–772.

Fisher, I., Pain, D.J. and Thomas, V.G. (2006) A review of lead poisoning from ammunition sources in terrestrial birds. *Biological Conservation*, 131, 421–432.

Fisher, J., Simon, N. and Vincent, J. (1969) *The Red Book – Wildlife in Danger.* London: Collins.

Fitt, W.K., Brown, B.E., Warner, M.E. and Dunne, R.P. (2001) Coral bleaching: interpretation of thermal tolerance limits and thermal thresholds in tropical corals. *Coral Reefs*, 20, 51–65.

FitzGerald, D.M., Fenster, M.S., Argow, B.A. and Buynevich, I.V. (2008) Coastal impacts due to sea-level rise. *Annual review of Earth and Planetary Sciences*, 36, 601–647.

Fitzharris, B. (1996) The cryosphere: changes and their impacts. In R.T. Watson, M.C. Zinyowera, R.H. Moss and D.J. Dokken (eds), *Climate Change 1995. Impacts, Adaptation and Mitigation of Climate Change: Scientific and Technical Analyses.* Cambridge: Cambridge University Press, 880 pp.

Fitzhugh, T.W. and Vogel, R.M. (2011) The impact of dams on flood flows in the United States. *River Research and Applications*, 27, 1192–1115.

Fitzpatrick, J. (1994) *A Continent Transformed: Human Impact on the Natural Vegetation of Australia.* Melbourne: Oxford University Press.

Flanner, M.G., Shell, K.M., Barlage, M., Perovich, D.K. and Tschudi, M.A. (2011) Radiative forcing and albedo feedback from the Northern Hemisphere cryosphere between 1979 and 2008. *Nature Geoscience*, 4, 151–155.

Flannery, J. (2005) *The Weather Makers.* London: Allen Lane.

Flannigan, M.D., Stocks, B.J. and Wotton, B.M. (2000) Climate change and forest fires. *The Science of the Total Environment*, 262, 221–230.

Flannigan, M.D., Logan, K.A., Amiro, B.D., Skinner, W.R. and Stocks, B.J. (2005) Future area burned in Canada. *Climatic Change*, 72, 1–16.

Flenley, J.R. (1979) *The Equatorial Rain Forest: A Geological History.* London: Butterworth.

Flenley, J.R., King, A.S.M., Jackson, J., Chew, C., Teller, J. and Prentice, M.E. (1991) The late Quaternary vegetational and climatic history of Easter Island. *Journal of Quaternary Science*, 6, 85–115.

Folke, C., Carpenter, S., Walker, B., Scheffer, M., Elmquist, T., Gunderson, L. and Holling, C.S. (2004) Regime shifts, resilience, and biodiversity in ecosystem management. *Annual Review of Ecology, Evolution, and Systematics*, 35, 557–581.

Fölster, J. and Wilander, A. (2002) Recovery from acidification in Swedish forest streams. *Environmental Pollution*, 117, 389–399.

Food and Agriculture Organisation (FAO) (2001) *The State of the World's Forests.* Rome: FAO.

Forland, E.J., Alexandersson, H., Drebs, A., Hamssen-Bauer, I., Vedin, H. and Tveito, O.E. (1998) Trends in maximum 1-day precipitation in the Nordic region. *DNMI report 14/98, Klima.* Oslo: Norwegian Meteorological Institute, 1–55.

Forman, R.T.T. et al. (2003) *Road Ecology. Science and Solutions.* Washington DC: Island Press.

Forman, S.L., Oglesby, R. and Webb, R.S. (2001) Temporal and spatial patterns of Holocene dune activity on the Great Plains of North America: megadroughts and climate links. *Global and Planetary Change*, 29, 1–29.

Forman, S.L., Spaeth, M., Márin, L.O., Pierson, J., Gómez, J., Bunch, F. and Valdez, A. (2006) Episodic Late Holocene dune movements on the sand-sheet area, Great Sand Dunes National Park and Preserve, San Luis Valley, Colorado, USA. *Quaternary Research*, 66, 97–108.

Forman, S.L., Sagintayev, Z., Sultan, M., Smith, S., Becker, R., Kendall, M. and Márín, L. (2008) The twentieth-century migration of parabolic dunes and wetland formation at Cape Cod National Sea Shore, Massachusetts, USA: landscape response to a legacy of environmental disturbance. *The Holocene*, 18, 765–774.

Foster, G.C., Chiverrell, R.C., Thomas, G.S.P., Marshall, P. and Hamilton, D. (2009) Fluvial development and the sediment regime of the lower Calder, Ribble catchment, northwest England. *Catena*, 77, 81–95.

Foster, I.D.L., Dearing, J.A. and Appleby, R.G. (1986) Historical trends in catchment sediment yields: a case study in reconstruction from lake-sediment records in Warwickshire, UK. *Hydrological Sciences Journal*, 31, 427–443.

Foster, S., Garduno, H., Evans, R., Olson, D., Tian, Y., Zhang, W. and Han, Z. (2004) Quaternary aquifer of the North China Plain – assessing and achieving groundwater resource sustainability. *Hydrogeology Journal*, 12, 81–93.

Foulds, S.A. and Macklin, M.G. (2006) Holocene land-use change and its impact on river basin dynamics in Great Britain and Ireland. *Progress in Physical Geography*, 30, 589–604.

Fox, H.L. (1976) The urbanizing river: a case study in the Maryland piedmont. In D.R. Coates (ed.), *Geomorphology and Engineering*. Stroudsburg: Dowden, Hutchinson and Ross, 245–271.

Foxcroft, L.C., Richardson, D.M., Rejmánek, M. and Pyšek, P. (2010) Alien plant invasions in tropical and sub-tropical savannas: patterns, processes and prospects. *Biological Invasions*, 12, 3913–3933.

Francis, D. and Hengeveld, H. (1998) *Extreme Weather and Climate Change*. Downsview, Ontario: Environment Canada.

Frankel, O.H. (1984) Genetic diversity, ecosystem conservation and evolutionary responsibility. In F.D. Castri, F.W.G. Baker and M. Hadley (eds), *Ecology in Practice* (Vol. I). Dublin: Tycooly, 4315–4327.

Frazier, T.G., Wood, N., Yarnal, B. and Bauer, D.H. (2010) Influence of potential sea level rise on societal vulnerability to hurricane storm-surge hazards, Sarasota County, Florida. *Applied Geography*, 30, 490–505.

Freedman, B. (1995) *Environmental Ecology* (2nd edn). San Diego: Academic Press.

Freeland, W.J. (1990) Large herbivorous mammals: exotic species in northern Australia. *Journal of Biogeography*, 17, 445–449.

French, H.M. (1976) *The Periglacial Environment*. London: Longman.

French, H.M. (1996) *The Periglacial Environment* (2nd edn). Harlow: Longman.

French, J.R., Spencer, T. and Reed, D.J. (eds) (1995) Geomorphic responses to sea level rise: existing evidence and future impacts. *Earth Surface Processes and Landforms*, 20, 1–6.

French, P.W. (1997) *Coastal and Estuarine Management*. London: Routledge.

French, P.W. (2001) *Coastal Defences. Processes, Problems and Solutions*. London: Routledge.

Frenkel, R.E. (1970) Ruderal vegetation along some California roadsides. *University of California publications in geography*, 20.

Frid, C. et al. (2012) The environmental interactions of tidal and wave energy generation devices. *Environmental Impact Assessment Review*, 32, 133–139.

Fu, C. (2003) Potential impacts of human-induced land cover change on East Asia monsoon. *Global and Planetary Change*, 37, 219–229.

Fu, J. et al. (2003) Persistent organic pollutants in environment of the Pearl River delta, China, an overview. *Chemosphere*, 52, 1411–1422.

Fuller, D.O. and Ottka, C. (2002) Land cover, rainfall and land-surface albedo in West Africa. *Climate Change*, 54, 181–204.

Fuller, D.Q. et al. (2009) The domestication process and domestication rate in rice: spikelet bases from the lower Yangtze. *Science*, 323, 1607–1610.

Fuller, R., Hill, D. and Tucker, G. (1991) Feeding the birds down on the farm: perspectives from Britain. *Ambio*, 20, 232–237.

Gabunia, L. and Vekua, A. (1995) A Plio-Pleistocene hominid from Dmanisi, East Georgia, Caucasus. *Nature*, 373, 509–512.

Gade, D.W. (1976) Naturalization of plant aliens: the volunteer orange in Paraguay. *Journal of Biogeography*, 3, 269–279.

Gaertner, M., Breeyen, A.D., Hui, C. and Richardson, D.M. (2009) Impact of alien plant invasions on species richness in Mediterranean-type ecosystems: a meta-analysis. *Progress in Physical Geography*, 33, 319–338.

Galay, V.J. (1983) Causes of river bed degradation. *Water Resources Research*, 19, 1057–1090.

Galibert, F., Quignon, P., Hitte, C. and André, C. (2011) Toward understanding dog evolutionary and domestication history. *Comptes Rendus Biologies*, 334, 190–196.

Galloway, J.N. et al. (2008) Transformation of the nitrogen cycle: recent trends, questions, and potential solutions. *Science*, 320, 889–892.

Gameson, A.L.H. and Wheeler, A. (1977) Restoration and recovery of the Thames Estuary. In J. Cairns, K.L. Dickson and E.E. Herricks (eds), *Recovery and Restoration of Damaged Ecosystems*. Charlottesville: University Press of Virginia, 92–101.

Ganguly, D., Rasch, P.J., Wang, H. and Yoon, J.-H. (2012) Climate response of the South Asian monsoon system to anthropogenic aerosols. *Journal of Geophysical Research*, 117, D13209.

Gao, X. and Giorgi, F. (2008) Increased aridity in the Mediterranean region under greenhouse gas forcing estimated from high resolution simulations with a regional climate model. *Global and Planetary Change*, 62, 195–209.

Garcia, X.-F., Schnauder, I. and Pusch, M.T. (2012) Complex hydromorphology of menaders can support benthic invertebrate diversity in rivers. *Hydrobiologia*, 685, 49–68.

García-Ruiz, J.M. (2010) The effects of land uses on soil erosion in Spain: a review. *Catena*, 81, 1–11.

García-Ruiz, J.M., López-Moreno, J.I., Vincente-Serrano, S.M., Lasanta-Martínez, T. and Beguería, S. (2011) Mediterranean water resources in a global change scenario. *Earth-Science Reviews*, 105, 121–139.

Gardelle, J., Bertier, E. and Arnaud, Y. (2012) Slight mass gain of Karakoram glacoiers in the early twenty-first century. *Nature Geoscience*, 5, 322–325.

Gardner, C.M.K., Cooper, D.M. and Hughes, S. (2002) Phosphorous in soils and field drainage water in the Thame catchment, UK. *The Science of the Total Environment*, 282–283, 253–262.

Gates, D.M. (1993) *Climate Change and Its Biological Consequences*. Sunderland, MA: Sinauer.

Gattuso, J.-P., Frankignoulle, M., Bourge, I., Romaine, S. and Buddemeier, R.W. (1998) Effect of calcium carbonate saturation of seawater on coral calcification. *Global and Planetary Change*, 18, 37–46.

Gattuso, J.-P. and Hansson, L. (eds) (2011) *Ocean Acidification*. Oxford: Oxford University Press.

Gaylord, D.R. (1990) Holocene palaeoclimatic fluctuations revealed from dune and interdune strata in Wyoming. *Journal of Arid Environments*, 18, 123–138.

Gedan, K.B., Silliman, B.R. and Bertness, M.D. (2009) Centuries of human-driven change in salt marsh ecosystems. *Annual Review of Marine Science*, 1, 117–141.

Gedan, K.B., Altieri, A.H. and Bertness, M.D. (2011) Uncertain future of New England salt marshes. *Marine Ecology Progress Series*, 434, 229–237.

Gedney, N., Cox, P.M., Betts, R.A., Boucher, O., Huntingford, C. and Stott, P.A. (2006) Detection of a direct carbon dioxide effect in continental river runoff records. *Nature*, 439, 835–838.

Geertz, C. (1963) *Agricultural Involution: The Process of Ecological Change in Indonesia*. Berkeley: University of California Press.

Gehring, C., Denich, M. and Vlek, P.L.G. (2005) Resilience of secondary forest regrowth after slash-and-burn agriculture in central Amazonia. *Journal of Tropical Ecology*, 21, 519–527.

Germer, S., Neill, C., Krusche, A.V. and Elsenbeer, H. (2010) Influence of land-use on near-surface hydrological processes: undisturbed forest to pasture. *Journal of Hydrology*, 380, 473–480.

Gero, A.F., Pitman, A.J., Narisma, G.T., Jacobson, C. and Pielke, R.A. (2006) The impact of land cover change on storms in the Sydney Basin, Australia. *Global and Planetary Change*, 54, 57–78.

GESAMP (IMO/FAO/UNESCO/WMO/IAEA/UN/UNEP Joint Group of Experts on the Scientific Aspects of Marine Pollution) (1990) *The state of the marine environment*. UNEP Regional Seas Reports and Studies, 115, Nairobi.

Ghassemi, F., Jakeman, A.J. and Nix, H.A. (1995) *Salinisation of Land and Water Resources*. Wallingford: CAB International, 536.

Ghersa, C.M., de la Fuente, E., Suarez, S. and Leon, R.J.C. (2002) Woody species invasion in the Rolling Pampa grasslands, Argentina. *Agriculture, Ecosystems & Environment*, 88, 271–278.

Gilbert, G.K. (1917) Hydraulic mining debris in the Sierra Nevada. *United States Geological Survey professional paper*, 105.

Gilbert, O.L. (1975) Effects of air pollution on landscape and land-use around Norwegian aluminium smelters. *Environmental Pollution*, 8, 113–121.

Gill, A.B. (2005) Offshore renewable energy: ecological implications of generating electricity in the coastal zone. *Journal of Applied Ecology*, 42, 605–615.

Gill, T.E. (1996) Eolian sediments generated by anthropogenic disturbance of playas: human impacts on the geomorphic system and geomorphic impacts on the human system. *Geomorphology*, 17, 207–228.

Gillon, D. (1983) The fire problem in tropical savannas. In F. Bourlière (ed.), *Tropical Savannas*. Oxford: Elsevier Scientific, 617–641.

Gil-Romera, G., Turton, D. and Sevilla-Callejo, M. (2011) Landscape change in the lower Omo valley, southwestern Ethiopia: burning patterns and woody encroachment in the savanna. *Journal of Eastern African Studies*, 5, 108–128.

Gimingham, C.H. (1981) Conservation: European heathlands. In R.L. Spect (ed.), *Heathlands and Related Shrublands*. Amsterdam: Elsevier Scientific, 249–259.

Gimingham, C.H. and de Smidt, I.T. (1983) Heaths and natural and semi-natural vegetation. In W. Holzner, M.J.A. Werger and I. Ikusima (eds), *Man's Impact on Vegetation*. Hague: Junk, 185–199.

Ginoux, P., Prospero, J.M., Gill, T.E., Hsu, N.C. and Zhao, M. (2012) Global-scale attribution of anthropogenic and natural dust sources and their emission rates based on MODIS deep blue aerosol products. *Reviews of Geophysics*, 50, RG3005.

Giorgi, F. (2006) Climate change hot-spots. *Geophysical Research Letters*, 33, L08707.

Givati, A. and Rosenfeld, D. (2004) Quantifying precipitation suppression due to air pollution. *Journal of Applied Meteorology*, 43, 1038–1056.

Glacken, C. (1967) *Traces on the Rhodian Shore: Nature and Culture in Western Thought from Ancient Times to the End of the Eighteenth Century*. Berkeley: University of California Press.

Glade, T. (2003) Landslide occurrence as a response to land use change: a review of evidence from New Zealand. *Catena*, 51, 297–314.

Gleick, P.H. (ed.) (1993) *Water in Crisis: A Guide to the World's Freshwater Resources*. New York: Oxford University Press.

Glieck, P.J. (2002) Dams. In A.S. Goudie (ed.), *Encyclopedia of Global Change*. New York: Oxford University Press, 229–234.

Global Native Species Network. http:www.gisin.org?cwis438/Websites/GISNDirectory/SpeciesStatus_Topoinvasives.php? (Accessed December 7th 2012)

Goddard, M.A., Dougill, A.J. and Benton, T.G. (2010) Scaling up from gardens: biodiversity conservation in urban environments. *Trends in Ecology & Evolution*, 25, 90–98.

Goebel, T., Waters, M.R. and O'Rourke, D.H. (2008) The late Pleistocene dispersal of modern humans in the Americas. *Science*, 319, 1497–1502.

Goldberg, E.D., Hodge, V., Koide, M., Griffin, J., Gamble, E., Bicker, O.P., Metisoff, G., Holdren, G.R. and Brown, R. (1978) A pollution history of Chesapeake Bay. *Geochimica et Cosmochimica Acta*, 42, 1413–1425.

Goldewijk, K.K. (2001) Estimating global land use change over the past 300 years: the HYDE database. *Global Biogeochemical Cycles*, 15, 417–433.

Goldsmith, F.B. (1983) Evaluating nature. In A. Warren and F.B. Goldsmith (eds), *Conservation in Perspective*. Chichester: Wiley, 233–246.

Goldsmith, V. (1978) Coastal dunes. In R.A. Davis (ed.), *Coastal Sedimentary Environments*. New York: Springer-Verlag, 420 pp.

Goldstein, M.C., Rosenberg, M. and Cheng, L. (2012) Increased oceanic micrplastic debris enhances oviposition in an endemic pelagic insect. *Biology Letters*, doi:10.1098/rsbl.2012.0298.

Gomez, B. and Smith, C.G. (1984) Atmospheric pollution and fog frequency in Oxford, 1926–1980. *Weather*, 39, 379–384.

Gong, Z.-T. (1983) Pedogenesis of paddy soil and its significance in soil classification. *Soil Science*, 135, 5–10.

Gonzalez, M.A. (2001) Recent formation of arroyos in the Little Missouri Badlands of southwestern Dakota. *Geomorphology*, 38, 63–84.

Goode, J.R., Luce, C.H. and Buffington, J.M. (2012) Enhanced sediment delivery in a changing climate in semi-arid mountain basins: implications for water resource management and aquatic habitat in the northern Rocky Mountains. *Geomorphology*, 139–140, 1–15.

Goodman, D. (1975) The theory of diversity-stability relationships in ecology. *The Quarterly Review of Biology*, 50, 237–266.

Goreau, T.J. and Hayes, R.L. (1994) Coral bleaching and ocean 'hot spots'. *Ambio*, 23, 176–180.

Gorman, M. (1979) *Island Ecology*. London: Chapman & Hall.

Gornitz, V., Rosenzweig, C. and Hillel, D. (1997) Effects of anthropogenic intervention in the land hydrologic cycle on global sea level rise. *Global and Planetary Change*, 14, 147–161.

Gornitz, V., Couch, S. and Hartig, E.K. (2002) Impacts of sea level rise in the New York City metropolitan area. *Global and Planetary Change*, 32, 61–88.

Gosden, C. (2003) *Prehistory: A Very Short Introduction*. Oxford: Oxford University Press.

Gottesfeld, P. and Cherry, C.R. (2011) Lead emissions from solar voltaic energy systems in China and India. *Energy Policy*, 39, 4939–4946.

Gottschalk, L.C. (1945) Effects of soil erosion on navigation in Upper Chesapeake Bay. *Geographical Review*, 35, 219–238.

Goudie, A.S. (1972) Vaughan Cornish: geographer. *Transactions of the Institute of British Geographers*, 55, 1–16.

Goudie, A.S. (1973) *Duricrusts of Tropical and Subtropical Landscapes*. Oxford: Clarendon Press.

Goudie, A.S. (1977) Sodium sulphate weathering and the disintegration of Mohenjo-Daro, Pakistan. *Earth Surface Processes*, 2, 75–86.

Goudie, A.S. (1983) Dust storms in space and time. *Progress in Physical Geography*, 7, 502–530.

Goudie, A.S. (ed.) (1990) *Techniques for Desert Reclamation*. Chichester: Wiley.

Goudie, A.S. (1992) *Environmental Change* (3rd edn). Oxford: Clarendon Press.

Goudie, A.S. (1994) Deserts in a warmer world. In A.C. Millington and K. Pye (eds), *Environmental Change in Drylands: Biogeographical and geomorphOLogical Perspectives*. Chichester: Wiley, 1–24.

Goudie, A.S. (2002) *Great Warm Deserts of the World: Landscape and Evolution*. Oxford: Oxford University Press.

Goudie, A.S. (2006) Global warming and fluvial geomorphology. *Geomorphology*, 79, 384–394.

Goudie, A.S. and Middleton, N.J. (2006) *Desert Dust in the Global System*. Heidelberg: Springer Verlag.

Goudie, A.S. and Middleton, N.S. (1992) The changing frequency of dust storms through time. *Climate Change*, 20, 197–225.

Goudie, A.S. and Viles, H.A. (1997) *Salt Weathering Hazards*. Chichester: Wiley.

Goudie, A.S. and Wilkinson, J.C. (1977) *The Warm Desert Environment*. Cambridge: Cambridge University Press.

Goudie, A.S., Viles, H.A. and Pentecost, A. (1993) The late-Holocene tufa decline in Europe. *The Holocene*, 3, 181–186.

Goudie, A.S., Parker, A.G. and Al-Farraj, A. (2000) Coastal change in Ras Al Khaimah (United Arab Emirates): a cartographic analysis. *The Geographical Journal*, 166, 14–25.

Goulson, D. (2003) Effects of introduced bees on native ecosystems. *Annual Review of Ecology, Evolution, and Systematics*, 34, 1–26.

Gourou, P. (1961) *The Tropical World* (3rd edn). London: Longman.

Govorushko, S.M. (2011) *Natural Processes and Human Impacts*. Heidelberg: Springer.

Gowlett, J.A.J., Harris, J.W.K., Walton, D. and Wood, B.A. (1981) Early archaeological sites, hominid remains and traces of fire from Chesowanja, Kenya. *Nature*, 284, 125–129.

Grabherr, G., Gottfried, M. and Pauli, H. (2010) Climate change impacts in alpine environments. *Geography Compass*, 4, 1133–1153.

Grable, J.L. and Harden, C.P. (2006) Geomorphic response of an Appalachian Valley and Ridge stream to urbanization. *Earth Surface Processes and Landforms*, 31, 1707–1720.

Graedel, T.E. and Crutzen, P.J. (1995) *Atmosphere, Climate and Change*. New York: Scientific American Library.

Graetz, D. (1994) Grasslands. In W.B. Meyer and B.L. Turner II (eds), *Changes in Land Use and Land Cover: A Global Perspective*. Cambridge: Cambridge University Press, 125–147.

Graf, J.B., Webb, R.H. and Hereford, R. (1991) Relation of sediment load and flood-plain formation to climatic variability, Paria River drainage basin, Utah and Arizona. *Geological Society of America Bulletin*, 103, 1405–1415.

Graf, W.K. (1977) Network characteristics in suburbanizing streams. *Water Resources Research*, 13, 459–463.

Graf, W.L. (1978) Fluvial adjustments to the spread of tamarisk in the Colorado Plateau region. *Geological Society of America Bulletin*, 86, 1491–1501.

Graf, W.L. (1981) Channel instability in a sand-river bed. *Water Resources Research*, 17, 1087–1094.

Graf, W.L. (1988) *Fluvial Processes in Dryland Rivers*. Berlin: Springer Verlag.

Graf, W.L. (2001) Damage control: restoring the physical integrity of America's rivers. *Annals of the Association of American Geographers*, 91, 1–27.

Grainger, A. (1990) *The Threatening Desert: Controlling Desertification*. London: Earthscan.

Grainger, A. (1992) *Controlling Tropical Deforestation*. London: Earthscan.

Grainger, A. (2008) Difficulties in tracking the long-term global trend in tropical forest area. *Proceedings of the National Academy of Sciences of the United States of America*, 105, 818–823.

Grant, M.J., Waller, M.P. and Groves, J.A. (2011) The *Tilia* decline: vegetation change in lowland Britain during the mid and late Holocene. *Quaternary Science Reviews*, 30, 394–408.

Grattan, J.P., Gilbertson, D.D. and Hunt, C.O. (2007) The local and global dimensions of metalliferous pollution derived from a reconstruction of an eight thousand year record of copper smelting and mining at a desert-mountain frontier in southern Jordan. *Journal of Archaeological Science*, 34, 83–110.

Gray, R. (1993) Regional meteorology and hurricanes. In G.A. Maul (ed.), *Climatic Change in the Intra-Americas Sea*. London: Edward Arnold, 87–99.

Grayson, D.K. (1977) Pleistocene avifaunas and the overkill hypothesis. *Science*, 195, 691–693.

Grayson, D.K. (1988) Perspectives on the archaeology of the first Americans. In R.C. Carlisle (ed.), *Americans before Columbus: Ice Age Origins*. Pittsburgh: University of Pittsburgh, 107–123.

Green, F.H.W. (1978) Field drainage in Europe. *The Geographical Journal*, 144, 171–174.

Green, P.A., Vörösmarty, C.J., Galloway, J.N., Peterson, B.J. and Boyer, E.W. (2004) Pre-industrial and contemporary fluxes of nitrogen through rivers: a global assessment based on typology. *Biogeochemistry*, 68, 71–105.

Greenfield, H.J. (2010) The secondary products revolution: the past, the present and the future. *World Archaeology*, 42, 29–54.

Greenland, D.J. and Lal, R. (1977) *Soil Conservation and Management in the Humid Tropics*. Chichester: Wiley.

Gregory, J.M. and Oerlemans, J. (1998) Simulated future sea-level rise due to glacier melt based on regionally and seasonally resolved temperature changes. *Nature*, 391, 474–476.

Gregory, J.M. et al. (2001) Comparisons of results from several AOGCMs for global and regional sea level change 1900–100. *Climate Dynamics*, 18, 225–240.

Gregory, J.M., Huybrechts, P. and Raper, S.C.B. (2004a) Threatened loss of the Greenland ice-sheet. *Nature*, 428, 616.

Gregory, K.J. (1976) Drainage networks and climate. In E. Derbyshire (ed.), *Geomorphology and Climate*. Chichester: Wiley, 289–315.

Gregory, K.J. (1985) The impact of river channelization. *The Geographical Journal*, 151, 53–74.

Gregory, K.J. and Walling, D. (1973) *Drainage Basin Form and Process: A Geomorphological Approach*. London: Arnold.

Gregory, R.D. et al. (2007) Population trends of widespread woodland birds in Europe. *The Ibis*, 149 (Suppl. 2), 78–97.

Gregory, R.D., Noble, D.G. and Custance, J. (2004b) The state of play of farmland birds: population trends and conservation status of lowland farmland birds in the United Kingdom. *The Ibis*, 146 (Suppl. 2), 1–13.

Grieve, I.C. (2001) Human impacts on soil properties and their implications for the sensitivity of soil systems in Scotland. *Catena*, 42, 361–374.

Griffiths, J.F. (1976) *Applied Climatology, an Introduction* (2nd edn). Oxford: Oxford University Press.

Grimm, N.B., Chacon, A., Dahm, C.N., Hostetler, S.W., Lind, O.T., Starkweather, P.L. and Wertsbaugh, W.W. (1997) Sensitivity of aquatic ecosystems to climatic and anthropogenic changes: the Basin and Range, American south-west and Mexico. In C.E. Cushing (ed.), *Freshwater Ecosystems and Climate Change in North America: A Regional Assessment*. Chichester: Wiley, 205–223.

Grosjean, M., Núñez, L., Castajena, I. and Messerli, B. (1997) Mid-Holocene climate and culture change in the Atacama Desert, northern Chile. *Quaternary Research*, 48, 239–246.

Gross, M.G. (1972) Geological aspects of waste solids and marine waste deposits, New York Metropolitan region. *Geological Society of America Bulletin*, 83, 3163–3176.

Grove, A.T. and Rackham, O. (2001) *The Nature of Mediterranean Europe: An Ecological History*. New Haven and London: Yale University Press.

Grove, J.M. (1988) *The Little Ice Age*. London: Routledge.

Grove, R.H. (1983) *The Future for Forestry*. Cambridge: British Association of Nature Conservationists.

Grove, R.H. (1997) *Ecology, Climate and Empire: Colonialism and Global Environmental history, 1400–1940*. Cambridge: White Horse Press.

Grove, R.H. and Damodaran, V. (2006) Imperialism, intellectual networks and environmental change. *Economic and Social Weekly*, 14 October, 4345–4354.

Grover, H.D. and Musick, H.B. (1990) Shrubland encroachment in southern New Mexico, USA: an analysis of desertification processes in the American southwest. *Climatic Change*, 17, 305–330.

Gruber, S. and Haeberli, W. (2007) Permafrost in steep bedrock slopes and its temperature-related destabilization following climate change. *Journal of Geophysical Research*, 112, F02S18.

Guha, S.K. (ed.) (2000) *Induced Earthquakes*. Dordrecht: Kluwer.

Guicherit, R. and Roemer, M. (2000) Tropospheric ozone trends. *Chemosphere – Global Change Science*, 2, 167–183.

Guitard, R., Sachana, M., Caloni, F., Croubels, S., Vandenbroucke, V. and Berny, P. (2010) Animal poisoning in Europe. Part 3: wildlife. *Veterinary Journal (London, England: 1997)*, 183, 260–265.

Guo, J.H. et al. (2010) Significant acidification of major Chinese croplands. *Science*, 327, 1008–1010.

Guo, S., Wang, J., Xiang, L., Ying, A. and Li, D. (2002) A macro-scale and semi-distributed monthly water balance model to predict climate change impacts in China. *Journal of Hydrology*, 268, 1–15.

Guo, X. and Zheng, G. (2009) Advances in weather modification from 1997 to 2007 in China. *Advances in Atmospheric Sciences*, 26, 240–252.

Gupta, H.K. (2002) A review of recent studies of triggered earthquakes by artificial water reservoirs with special emphasis on earthquakes in Koyna, India. *Earth-Science Reviews*, 58, 279–310.

Gurnell, A., Lee, M. and Souch, C. (2007) Urban rivers: hydrology, geomorphology, ecology and opportunities for change. *Geography Compass*, 1, 1118–1137.

Guthrie, R.D. (2003) Rapid body size decline in Alaskan Pleistocene horses before extinction. *Nature*, 426, 169–171.

Gutierrez-Elorza, M. (2001) *Geomorfología climática*. Barcelona: Omega.

Haberl, H. et al. (2007) Quantifying and mapping the human appropriation of net primary production in earth's terrestrial ecosystems. *Proceedings of the National Academy of Sciences of the United States of America*, 104, 12942–12947.

Hadley, R.F. (1961) Influence of riparian vegetation on channel shape, northeastern Arizona. *U.S.Geological Survey professional paper 424C*, 30–31.

Haeberli, W. and Burn, C.R. (2002) Natural hazards in forests: glacier and permafrost effects as related to climate change. In R.C. Sidle (ed.), *Environmental Change and Geomorphic Effects in Forests*. Wallingford: CABI, 167–202.

Haff, P.K. (2010) Hillslopes, rivers, plows and trucks: mass transport on Earth's surface by natural and technological processes. *Earth Surface Processes and Landforms*, 35, 1157–1166.

Haigh, M.J. (1978) *Evolution of slopes on artificial landforms – Blaenavon, UK*. Research paper 183, Department of Geography, University of Chicago.

Hails, J.R. (ed.) (1977) *Applied Geomorphology*. Amsterdam: Elsevier.

Hails, R.J. (2002) Assessing the risks associated with new agricultural practices. *Nature*, 418, 685–688.

Hall, S.R. and Mills, E.L. (2000) Exotic species in large lakes of the world. *Aquatic Ecosystem Health and Management*, 3, 105–135.

Halpern, B.S., Selkoe, K.A., Micheli, F. and Kappel, C.V. (2007) Evaluating and ranking the vulnerability of global marine ecosystems to anthropogenic threats. *Conservation Biology*, 21, 1301–1315.

Han, Z., Wang, T., Dong, Z., Hu, Y. and Yao, Z. (2007) Chemical stabilization of mobile dunefields along a highway in the Taklimakan desert of China. *Journal of Arid Environments*, 68, 260–270.

Hannah, L., Lohse, D., Hutchinson, C., Carr, L. and Lankerani, A. (1994) A preliminary inventory of human disturbance of world ecosystems. *Ambio*, 23, 246–250.

Hansen, J., Johnson, D., Lacis, A., Lebedeff, S., Lee, P., Rims, D. and Russell, G. (1981) Climatic impact of increasing atmospheric carbon dioxide. *Science*, 213, 957–966.

Hanson, P.J. and Weltzin, J.F. (2000) Drought disturbance from climate change: response of United States forests. *The Science of the Total Environment*, 262, 205–220.

Happ, S.C. (1944) Effect of sedimentation on floods in the Kickapoo Valley, Wisconsin. *The Journal of Geology*, 52, 53–68.

Harlan, J.R. (1975a) Our vanishing genetic resources. *Science*, 188, 617–622.

Harlan, J.R. (1975b) *Crops and Man*. Madison: American Society of Agronomy.

Harlan, J.R. (1976) The plants and animals that nourish man. *Scientific American*, 235 (3), 88–97.

Harnischmacher, S. (2010) Quantification of mining subsidence in the Ruhr District (Germany). *Géomorphologie*, 3, 261–274.

Harper, K.C. (2008) Climate control: United States weather modification in the cold war and beyond. *Endeavour*, 32, 20–26.

Harris, C., Davies, M.C.R. and Etzelmüller, B. (2001) The assessment of potential geotechnical hazards associated with mountain permafrost in a warming global climate. *Permafrost and Periglacial Processes*, 12, 145–156.

Harris, D.R. (ed.) (1980) *Human Ecology in Savanna Environments*. London: Academic Press.

Harris, D.R. (ed.) (1996) *The Origins and Spread of Agriculture and Pastoralism in Eurasia*. London: UCL Press.

Harris, G., Thirgood, S., Hopcraft, J.G.C., Cromsigt, P.P.G.M. and Berger, J. (2009) Global decline in aggregated migrations of large terrestrial mammals. *Endangered Species Research*, 7, 55–76.

Harris, J.M., Oltmans, S.J., Bodeker, G.E., Stolarski, R., Evans, R.D. and Quincy, D.M. (2003) Long-term variations in total ozone derived from Dobson and satellite data. *Atmospheric Environment*, 37, 3167–3175.

Harris, S.A. (2002) Causes and consequences of rapid thermokarst development in permafrost or glacial terrain. *Permafrost and Periglacial Processes*, 13, 237–242.

Harrison, G.P. and Whittington, H.W. (2002) Susceptibility of the Batoka Gorge hydroelectric scheme to climate change. *Journal of Hydrology*, 264, 230–241.

Harrison, S., Glasser, N., Winchester, V., Haresign, E., Warren, C. and Jansson, K. (2006) A glacial outburst flood associated with recent mountain glacier retreat Patagonian Andes. *The Holocene*, 16, 611–620.

Harvey, A.M. (1989) The occurrence and role of arid zone alluvial fans. In D.S.G. Thomas (ed.), *Arid Zone Geomorphology*. London: Belhaven Press, 136–158.

Harvey, A.M. and Renwick, W.H. (1987) Holocene alluvial fan and terrace formation in the Bowland Fells, Northwest England. *Earth Surface Processes and Landforms*, 12, 249–257.

Harvey, A.M., Oldfield, F., Baron, A.F. and Pearson, G.W. (1981) Dating of post-glacial landforms in the central Howgills. *Earth Surface Processes and Landforms*, 6, 401–412.

Harvey, L.D.D. (2000) *Global Warming. The Hard Science*. Harlow: Prentice Hall.

Hassol, S.J. (2004) *Impacts of a Warming Arctic*. Cambridge: Cambridge University Press.

Hauck, M., Zimmermann, J., Jacob, M., Dulamsuren, C., Bade, C., Ahrends, B. and Leuschner, C. (2012) Rapid recovery of stem increment in Norway spruce at reduced SO$_2$ levls in the Harz Mountains, Germany. *Environmental Pollution*, 164, 132–141.

Hawass, Z. (1993) The Egyptian monuments: problems and solutions. In M.J. Thiel (ed.), *Conservation of Stone and Other Material*. London: Spon, 19–25.

Hawksworth, D.L. (1990) The long-term effects of air pollutants on lichen communities in Europe and North America. In G.M. Woodwell (ed.), *The Earth in Transition: Patterns and Processes of Biotic Impoverishment*. Cambridge: Cambridge University Press, 45–64.

Hay, J. (1973) Salt cedar and salinity on the Upper Rio Grande. In M.T. Farvar and J.P. Milton (eds), *The Careless Technology*. London: Tom Stacey, 288–300.

Hayden, B. (1995) A new review of domestication. In T.D. Price and A.B. Gebauser (eds), *Last Hunters, First Farmers*. Santa Fe: School of American Research, 273–299.

Haynes, C.V. (1991) Geoarchaeological and palaeohydrological evidence for a Clovis-age drought in North America and its bearing on extinction. *Quaternary Research*, 35, 438–450.

Healy, T. (1991) Coastal erosion and sea level rise. *Zeitschrift für Geomorphologie*, Supplementband, 81, 15–29.

Healy, T. (1996) Sea level rise and impacts on nearshore sedimentation. *Geologische Rundschau*, 85, 546–553.

Heathwaite, A.L., Johnes, P.J. and Peters, N.E. (1996) Trends in nutrients. *Hydrological Processes*, 10, 263–293.

Heberger, M., Cooley, H., Herrera, P., Gleick, P.H. and Moore, E. (2011) Potential impacts of increasing coastal flooding in California due to sea-level rise. *Climatic Change*, 109 (Suppl.), S229–S249.

Heever, S.C. and Cotton, W.R. (2007) Urban aerosol impacts on downwind convective storms. *Journal of Applied Meteorology and Climatology*, 46, 828–850.

Heine, K., Niller, H.P., Nuber, T. and Scheibe, R. (2005) Slope and valley sediments as evidence of deforestation and land-use in prehistporic and historic Eastern Bavaria. *Zeitschrift für Geomorphologie*, Supplementband, 139, 147–171.

Helldén, U. (1985) Land degradation and land productivity monitoring – needs for an integrated approach. In A. Hjört (ed.), *Land Management and Survival*. Uppsala: Scandinavian Institute of African Studies, 77–87.

Helliwell, D.R. (1974) The value of vegetation for conservation. II: M1 motorway area. *Journal of Environmental Management*, 2, 75–78.

Henderson, S., Dawson, T.P. and Whittaker, R.J. (2006) Progress in invasive plants research. *Progress in Physical Geography*, 30, 25–46.

Henderson-Sellers, A. and Blong, R. (1989) *The Greenhouse Effect: Living in a Warmer Australia*. Kensington, NSW: New South Wales University Press.

Henderson-Sellers, A. and Gornitz, V. (1984) Possible climatic impacts of land cover transformation with particular emphasis on tropical deforestation. *Climatic Change*, 6, 231–257.

Henderson-Sellers, A. and Robinson, P.J. (1986) *Contemporary Climatology*. London: Longman.

Henry, J. et al. (2012) A common pesticide decreases foraging success and survival in honey bees. *Science*, 336, 348–350.

Herb, W.R., Janke, B., Mohseni, O. and Stefan, H.G. (2008) Thermal pollution of streams by runoff from paved surfaces. *Hydrological Processes*, 22, 987–999.

Hereford, R. (1984) Climate and ephemeral-stream processes: twentieth-century geomorphology and alluvial stratigraphy of the Little Colorado River, Arizona. *Geological Society of America Bulletin*, 95, 654–668.

Hereford, R., Jacoby, G.C. and McCord, V.A.S. (1995) Geomorphic history of the Virgin River in the Zion National Park area, southweset Utah. *US Geological Survey circular*, 95–515.

Hess, W.N. (ed.) (1974) *Weather and Climate Modification*. New York: Wiley.

Hester, E.T. and Doyle, M.W. (2011) Human impacts to river temperature and their effects on biological processes: a quantitative analysis. *Journal of the American Water Resources Association*, 47, 571–587.

Hewlett, J.D., Post, H.E. and Doss, R. (1984) Effect of clear-cut silviculture on dissolved ion export and water yield in the Piedmont. *Water Resources Research*, 20 (7), 1030–1038.

Heywood, V.H. (1989) Patterns, extents and modes of invasions by terrestrial plants. In J.A. Drake (ed.), *Biological Invasions: A Global Perspective*. Chichester: Wiley, 31–55.

Heywood, V.H. and Watson, R.T. (eds) (1995) *Global Biodiversity Assessment*. Cambridge: Cambridge University Press.

Hicke, J.A. et al. (2012) Effects of biotic disturbanvces on forest carbon cycling in the United States and Canada. *Global Change Biology*, 18, 7–34.

Hickey, J.J. and Anderson, O.W. (1968) Chlorinated hydrocarbons and eggshell changes in raptorial and fish-eating birds. *Science*, 162, 271–272.

Hickler, T. et al. (2012) Projecting the future distribution of European potential natural vegetation zones with a generalized, tree species-based dynamic vegetation model. *Global Ecology and Biogeography*, 21, 50–63.

Hilborn, R., Branch, T.A., Ernst, B., Magnusson, A., Minte-Vera, C.V. and Mark, D.S. (2003) State of the world's fisheries. *Annual Review of Environment and Resources*, 28, 359–399.

Hill, A.R. (1975) Ecosystems stability in relation to stresses caused by human activities. *The Canadian Geographer*, 19, 206–220.

Hills, T.L. (1965) Savannas: a review of a major research problem in tropical geography. *The Canadian Geographer*, 9, 216–228.

Hobbs, P.V. and Radke, L.F. (1992) Airborne studies of the smoke from the Kuwait oil fires. *Science*, 256, 987–991.

Hoegh-Guldberg, O. (1999) Climate change, coral bleaching and the future of the world's coral reefs. *Marine and Freshwater Research*, 50, 839–866.

Hoegh-Guldberg, O. (2001) Sizing the impact: coral reef ecosystems as early casualties of climate change. In G.R. Walther, C.A. Burga and P.J. Edwards (eds), *Fingerprints of Climate Change*. New York: Kluwer/Plenum, 203–228.

Hoegh-Guldberg, O. (2011) Coral reef ecosystems and anthropogenic climate change. *Regional Environmental Change*, 11 (Suppl. 1), S215–S227.

Hoegh-Guldberg, O. and Bruno, J.F. (2010) The impact of climate change on the world's marine ecosystems. *Science*, 328, 1523–1528.

Hoegh-Guldberg, O. et al. (2007) Coral reefs under rapid climate change and ocean acidification. *Science*, 318, 1737–1742.

Hoelzle, M. and Trindler, M. (1998) Data management and application. In W. Haeberli et al. (eds), *Into the Second Century of Worldwide Glacier Monitoring: Prospects and Strategies*. Paris: UNESCO, 53–64.

Hoelzmann, P., Keding, B., Berke, H., Kröpelin, S. and Kruse, H.-J. (2001) Environmental change and archaeology: lake evolution and human occupation in the eastern Sahara during the Holocene. *Palaeogeography, Palaeoclimatology, Palaeoecology*, 169, 193–217.

Hoffmann, M. et al. (2011) The changing fate of the world's mammals. *Philosophical Transactions of the Royal Society of London. Series B, Biological Sciences*, 366, 2598–2610.

Hogsden, K.L. and Harding, J.S. (2012) Consequences of acid mine drainage for the structure and function of benthic stream communities: a review. *Freshwater Science*, 31, 108–120.

Holden, J., Chapman, P.J. and Labadz, J.C. (2004) Artificial drainage of peatlands: hydrological and hydrochemical process and wetland restoration. *Progress in Physical Geography*, 28, 95–123.

Holden, J., Evans, M.G., Burt, T.P. and Horton, M. (2006) Impact of land drainage on peatland hydrology. *Journal of Environmental Quality*, 35, 1764–1778.

Holland, G.J., McBridge, J.L. and Nicholls, N. (1988) Australian region tropical cyclones and the greenhouse effect. In G.I. Pearman (ed.), *Greenhouse, Planning for Climate Change*. Leiden: Brill, 438–455.

Holliday, V.T. (2004) *Soils in Archaeological Research*. New York: Oxford University Press.

Hollis, G.E. (1975) The effects of urbanization on floods of different recurrence interval. *Water Resources Research*, 11, 431–435.

Hollis, G.E. (1988) Rain, roads, roofs and runoff: hydrology in cities. *Geography*, 73, 9–18.

Hollis, G.E. and Luckett, J.K. (1976) The response of natural river channels to urbanization: two case studies from Southeast England. *Journal of Hydrology*, 30, 351–363.

Holloway, G. and Sou, T. (2002) Has Arctic sea ice rapidly thinned? *Journal of Climate*, 15, 1691–1701.

Holm, K., Bovis, M. and Jacob, M. (2004) The landslide response of alpine basins to post-Little Ice Age glacial thinning and retreat in southwestern British Columbia. *Geomorphology*, 57, 201–216.

Holtz, W.G. (1983) The influence of vegetation on the swelling and shrinking of clays in the United States of America. *Géotechnique*, 33, 159–163.

Holz, A. and Veblen, T.T. (2011) The amplifying effects of humans on fire regimes in temperate rainforests in western Patagonia. *Palaeogeography, Palaeoclimatology, Palaeoecology*, 311, 82–92.

Holzer, T.L. (1979) Faulting caused by groundwater extraction in South-central Arizona. *Journal of Geophysical Research*, 84, 603–612.

Holzmueller, E.J. and Jose, S. (2010) Invasion success of cogongrass, an alien C4 perennial grass, in the southeastern United States: exploration of the ecological basis. *Biological Invasions*, 13, 435–442.

Hooke, J. (ed.) (1998) *Coastal Defence and Earth Science Conservation*. Bath: Geological Society of London.

Hooke, R.L. (1994) On the efficacy of humans as geomorphic agents. *USA Today*, 4 (217), 224–225.

Hope, G. (1999) Vegetation and fire response to late Holocene human occupation in island and mainland north west Tasmania. *Quaternary International*, 59, 47–60.

Horn, R., van der Akker, J.J.H. and Arvidsson, J. (eds) (2000) *Subsoil Compaction: Distribution, Processes and Consequences*. Reiskirchen: Catena Verlag.

Houghton, J.T. (2009) *Global Warming: The Complete Briefing* (4th edn). Cambridge: Cambridge University Press.

Houghton, J.T., Jenkins, G.J. and Ephraums, J.J. (1990) *Climate Change: The IPCC Scientific Assessment*. Cambridge: Cambridge University Press.

Houghton, J.T., Callander, B.A. and Varney, S.K. (eds) (1992) *Climate Change 1992: The Supplementary Report of the IPCC Scientific Assessment*. Cambridge: Cambridge University Press.

Houghton, R.H. and Skole, D.L. (1990) Carbon. In B.L. Turner II (ed.), *The Earth Transformed by Human Action*. Cambridge: Cambridge University Press, 393–408.

Howard, K.W.F. and Beck, P.J. (1993) Hydrochemical implications of groundwater contamination by road deicing chemicals. *Journal of Contaminant Hydrology*, 12, 245–268.

Howarth, R. et al. (2011) Coupled biogeochemical cycles: eutrophication and hypoxia in temperate estuaries and coastal marine ecosystems. *Frontiers in Ecology and the Environment*, 9, 18–26.

Howden, N.J.K., Burt, T.P., Worralle, F., Whelan, M.J. and Bieroza, M. (2010) Nitrate concentrations and fluxes in the River Thames ovber 140 years (1868–2008): are increases irreversible? *Hydrological Processes*, 24, 2657–2662.

Howe, G.M., Slaymaker, H.O. and Harding, D.M. (1966) Flood hazard in mid-Wales. *Nature*, 212, 584–585.

Hoyos, C.D., Agudelo, P.A., Webster, P.J. and Curry, J.A. (2006) Deconvolution of the factors contributing to the increase in global hurricane intensity. *Science*, 312, 94–97.

Hsu, P.-C., Li, T., Luo, J.-L., Murakami, H., Kitoh, A. and Zhao, M. (2012) Increase of global monsoon area and precipitation under global warming: a robust signal? *Geophysical Research Letters*, 39, L06701.

Hu, G., Dontg, Z., Lu, J. and Yan, C. (2012) Driving forces responsible for aeolian desertification in the source region of the Yangtze River from 1975 to 2005. *Environmental Earth Sciences*, 66, 257–263.

Hu, R. (2006) Urban land subsidence in China. *IAEG2006 Paper 786*, Geological Society of London.

Huang, Q.H. and Cai, Y.L. (2007) Spatial pattern of karst rock desertification in the middle of Guizhou Province, southwestern China. *Environmental Geology*, 52, 1325–1330.

Huang, R. and Chan, L. (2004) Human-induced landslides in China: mechanism study and its implications on slope management. *Chinese Journal of Rock Mechanics and Engineering*, 23, 2766–2777.

Huang, W., Clochon, R., Gu, Y., Larick, R., Fang, Q., Schwartz, H., Yonge, C., de Vos, J. and Rink, W. (1995) Early *homo and* associated artefacts from Asia. *Nature*, 378, 275–278.

Hübner, R., Herbert, R.J.H. and Astin, K.B. (2010) Cadmium release caused by the die-back of the saltmarsh cord grass Spartina anglica in Poole Harbour (UK). *Estuarine, Coastal and Shelf Science*, 87, 553–560.

Hudson, N. (1987) Soil and water conservation in semi-arid areas. *FAO Soils Bulletin*, 55.

Hugenholtz, C.H. and Wolfe, S.A. (2005) Biogeomorphic model of dunefield activation and stabilization on the northern Great Plains. *Geomorphology*, 70, 53–70.

Huggel, C., Clague, J.J. and Korup, O. (2012) Is climate change responsible for changing landslide activity in high mountains? *Earth Surface Processes and Landforms*, 37, 77–91.

Hughes, R.G. and Paramor, O.A.L. (2004) On the loss of saltmarshes in south-east England and methods for their restoration. *Journal of Applied Ecology*, 41, 440–448.

Hughes, R.J., Sullivan, M.E. and Yok, D. (1991) Human-induced erosion in a highlands catchment in Papua New Guinea: the prehistoric and contemporary records. *Zeitschrift für Geomorphologie*, Supplementband, 83, 227–239.

Hughes, T.P. (1994) Catastrophes, phase shifts, and large-scale degradation of a Caribbean coral reef. *Science*, 265, 1547–1551.

Hughes, T.P., Bellwood, D.R., Baird, A.H., Brodie, J., Bruno, J.F. and Pandolfi, J.M. (2011) Shifting base-lines, declining coral cover, and the erosion of reef resilience: comment on Sweatman et al. (2011). *Coral Reefs*, 30, 653–660.

Hull, S.K. and Gibbs, J.N. (1991) Ash dieback: a survey of nonwoodland trees. *Forestry Commission Bulletin*, 93, 32.

Hultine, K.R. and Bush, S.E. (2011) Ecohydrological consequences of non-native riparian vegetation in the southwestern United States: a review from an ecophysiological perspective. *Water Resources Research*, 47, W07542.

Huntington, E. (1914) *The climatic factor as illustrated in and America*. Carnegie Institution of Washington publication, 192.

Huntington, T.G. (2003) Climate warming could reduce runoff significantly in New England, USA. *Agricultural and Forest Meteorology*, 117, 193–210.

Hupp, C.R., Pierce, A.R. and Noe, G.B. (2009) Floodplain geomorphic processes and environmental impacts of human alteration along coastal plain rivers, USA. *Wetlands*, 29, 413–429.

Hurd, L.E., Mellinger, M.W., Wold, L.L. and McNaughton, S.J. (1971) Stability and diversity at three trophic levels in terrestrial successional ecosystems. *Science*, 173, 1134–1136.

Idso, S.B. (1983) Carbon dioxide and global temperature: what the data show. *Journal of Environmental Quality*, 12, 159–163.

Idso, S.B. and Brazel, A.J. (1984) Rising atmospheric carbon dioxide concentrations may increase streamflow. *Nature*, 312, 51–53.

Ikawa-Smith, F. (1982) Current issues in Japanese archaeology. *American Scientist*, 68, 134–145.

Illies, J. (1974) *Introduction to Zoogeography*. London: Macmillan.

Imeson, A. (2012) *Desertification, Land Degradation and Sustainability*. Chichester: Wiley-Blackwell.

Imeson, A. and Emmer, I.M. (1992) Implications of climate change on land degradation in the Mediterranean. In L. Jeftic, J.D. Milliman and G. Sestini (eds), *Climate Change and the Mediterranean*. London: Arnold, 95–128.

Imeson, A.C. (1971) Heather burning and soil erosion on the North Yorkshire Moors. *Journal of Applied Ecology*, 8, 537–541.

Imhoff, M.L., Bounoua, L., Ricketts, T., Loucks, C., Harriss, R. and Lawrence, W.T. (2004) Global patterns in human consumption of net primary productivity. *Nature*, 429, 870–873.

Imhoff, M.L., Zhang, P., Wolfe, R.E. and Bounoua, L. (2010) Remote sensing of the urban heat island effect across biomes in the continental USA. *Remote Sensing of Environment*, 114, 504–513.

Immer Zeel, W.W., van Beek, L.P.H. and Bierkens, M.F.P. (2010) Climate change will affect the Asian water towers. *Science*, 328, 1382–1385.

Indoitu, R., Orlovsky, L. and Orlovsky, N. (2012) Dust storms in Central asia: spatila and temporal variations. *Journal of Arid Environments*, 85, 62–70.

Innes, J., Blackford, J. and Simmons, I. (2010) Woodland disturbance and possible land-use regimes during the Late Mesolithic in the English uplands: pollen, charcoal and non-pollen palynomorph evidence from Bluewath Beck, North York Moors, UK. *Vegetation History and Archaeobotany*, 19, 439–452.

Innes, J.L. (1983) Lichenometric dating of debris-flow deposits in the Scottish Highlands. *Earth Surface Processes and Landforms*, 8, 579–588.

Innes, J.L. (1987) Air pollution and forestry. *Forestry Commission Bulletin*, 70.

Innes, J.L. (1992) Forest decline. *Progress in Physical Geography*, 16, 1–64.

Innes, J.L. and Boswell, R.C. (1990) Monitoring of forest condition in Great Britain 1989. *Forestry Commission Bulletin*, 94, 57.

Institute of Hydrology (1991) *Institute of Hydrology report 1990–91*. Wallingford: Institute of Hydrology.

Intergovernmental Panel on Climate Change (IPCC) (1996) *Climate Change 1995*. Cambridge: Cambridge University Press.

Intergovernmental Panel on Climate Change (IPCC) (2001) *Climate Change 2001: The Scientific Basis*. Cambridge: Cambridge University Press.

Intergovernmental Panel on Climate Change (IPCC) (2007a) *Climate Change 2007: The Physical Science Basis*. Cambridge: Cambridge University Press.

Intergovernmental Panel on Climate Change (IPCC) (2007b) *Climate Change 2007; Impacts, Adaptation and Vulnerability*. Cambridge: Cambridge University Press.

International Tanker Owners Pollution Federation (ITOPF) http://www.itopf.co.uk/information-services/data-and-statistics/statistics/ (accessed 15 November 2012).

Irish, J.L. et al. (2010) Potential implications of global warming and barrier island degradation on future hurricane inundation, property damages, and population impacted. *Ocean and Coastal Management*, 53, 645–657.

Isaac, E. (1970) *Geography of Domestication*. Englewood Cliffs: Prentice Hall.

Ives, A.R. and Carpenter, S.R. (2007) Stability and diversity of ecosystems. *Science*, 317, 58–62.

Ives, J.D. and Messerli, B. (1989) *The Himalayan Dilemma: Reconciling Development and Conservation*. London: Routledge.

Iwase, A., Hashizume, J., Izulo, M., Takahasji, K. and Sato, H. (2011) Timing of megafaunal extinction in the Late Pleistocene on the Japanese Archipelago. *Quaternary International*, doi:10.1016/j.quaint.2011.03.029.

Iwashima, T. and Yamamoto, R. (1993) A statistical analysis of the extreme events: long-term trend of heavy daily precipitation. *Journal of the Meteorological Society of Japan*, 71, 637–640.

Jacks, G.V. and Whyte, R.O. (1939) *The Rape of the Earth: A World Survey of Soil Erosion*. London: Faber & Faber.

Jackson, R.B. and Jobbágy, E.G. (2005) From icy roads to salty streams. *Proceedings of the National Academy of Sciences of the United States of America*, 102, 14487–14488.

Jacobs, J. (1969) *The Economy of Cities*. New York: Random House.

Jacobs, J. (1975) Diversity, stability and maturity in ecosystems influenced by human activities. In W.H. Van Dobben and R.H. Lowe-McConnell (eds), *Unifying Concepts in Ecology*. The Hague: Junk, 187–207.

Jacobsen, T. and Adams, R.M. (1958) Salt and silt in ancient Mesopotamian agriculture. *Science*, 128, 1251–1258.

James, L.A. (1989) Sustained storage and transport of hydraulic gold mining sediment in the Bear River, California. *Annals of the Association of American Geographers*, 79, 570–592.

James, L.A. and Marcus, W.A. (eds) (2006) 37th Binghamton Geomorphology Symposium: the human role in changing fluvial systems. *Geomorphology*, 79, 1–506.

James, L.A. (2011) Contrasting geomorphic impacts of pre- and post-Columbian land-use changes in Anglo America. *Physical Geography*, 32, 399–422.

Jarvis, P.H. (1979) The ecology of plant and animal introductions. *Progress in Physical Geography*, 3, 187–214.

Jeffries, M. (1997) *Biodiversity and Conservation*. London: Routledge.

Jeffries, M.J. (2012) Ponds and the importance of their history: an audit of pond numbers, turnover and the relationship between the origins of ponds and their contemporary plant communities in south-east Northumberland, UK. *Hydrobiologia*, 689, 11–21.

Jenkins, M. (2003) Prospects for biodiversity. *Science*, 302, 1175–1177.

Jenkins, M.E., Davies, T.J. and Stedman, J.R. (2002) The origin and day-of-week dependence of photochemical ozone episodes in the UK. *Atmospheric Environment*, 36, 999–1012.

Jennings, J.N. (1952) *The origin of the Broads*. Royal Geographical Society research series, 2.

Jenny, H. (1941) *Factors of Soil Formation*. New York: McGraw-Hill.

Jernelöv, A. (2010) The threats from oil spills: now, then, and in the future. *Ambio*, 39, 353–366.

Jiang, Y., Luo, Y., Zhao, Z. and Tao, S. (2009) Changes in wind speed over China during 1956–2004. *Theoretical and Applied Climatology*, doi:10.1007/s00704-009-0152-7.

Jickells, T.D., Carpenter, R. and Liss, P.S. (1991) Marine environment. In B.L. Turner, W.C. Clark, R.W. Kates, J.F. Richards, J.T. Matthews and W.B. Meyer (eds), *The Earth as Transformed by Human Action*. Cambridge: Cambridge University Press, 313–334.

Jim, C.Y. (1998) Urban soil characteristics and limitations for landscape planting in Hong Kong. *Landscape and Urban Planning*, 40, 235–249.

Jin, H., Li, S., Cheng, G., Shaoling, W. and Li, X. (2000) Permafrost and climatic change in China. *Global and Planetary Change*, 26, 387–404.

Joern, A. and Keeler, K.H. (eds) (1995) *The Changing Prairie*. New York: Oxford University Press.

Johannessen, C.L. (1963) Savannas of interior Honduras. *Ibero-Americana*, 46, 160 pp.

Johnson, A.I. (ed.) (1991) *Land subsidence*. International Association of Hydrological Sciences publication 200.

Johnson, C. and Wroe, S. (2003) Causes of extinction of vertebrates during the Holocene of mainland Australia: arrival of the dingo, or human impact? *The Holocene*, 13, 941–948.

Johnson, C.N. (2005) What can the date on late survival of Australian megafauna tell us about the cause of their extinction? *Quaternary Science Reviews*, 24, 2167–2172.

Johnson, C.N. (2009) Ecological consequences of Late Quaternary extinctions of megafauna. *Proceedings. Biological Sciences/The Royal Society*, 276, 2509–2519.

Johnson, D.B. and Hallberg, K.B. (2005) Acid mine drainage remediation options: a review. *The Science of the Total Environment*, 338, 3–14.

Johnson, D.L. and Lewis, L.A. (1995) *Land Degradation: Creation and Destruction*. Oxford: Blackwell.

Johnson, N.M. (1979) Acid rain: neutralization within the Hubbard Brook ecosystem and regional implications. *Science*, 204, 497–499.

Johnston, D.W. (1974) Decline of DDT residues in migratory songbirds. *Science*, 186, 841–842.

Johnston, D.W., Turner, J. and Kelly, J.M. (1982) The effects of acid rain on forest nutrient status. *Water Resources Research*, 18, 448–461.

Jomelli, V., Brunstein, D., Déqué, M., Vrac, M. and Grancher, D. (2009) Impacts of future climatic change (2070–2099) on the potential occurrence of debris flows: a case study in the *Massif des Ecrins* (French Alps). *Climatic Change*, 97, 171–191.

Jonard, M. et al. (2012) Deterioration of Norway spruce vitality despite a sharp decline in acid deposition: a long-term integrated study. *Global Change Biology*, 18, 711–725.

Jones, H.P. and Schmitz, O.J. (2009) Rapid recovery of damaged ecosystems. *PLoS ONE*, 4, e5653.

Jones, J.A.A., Liu, C., Woo, M.-K. and Kung, H.-T. (eds) (1996) *Regional Hydrological Response to Climate Change*. Dordrecht: Kluwer.

Jones, J.D.G. (2011) Why genetically modified crops? *Philosophical Transactions. Series A, Mathematical, Physical, and Engineering Sciences*, 369, 1807–1815.

Jones, P.D. and Lister, D.H. (2009) The urban heat island in Central London and urban-related warming trends in Central London since 1900. *Weather*, 64, 323–327.

Jones, P.D. and Reid, P.A. (2001) Assessing future changes in extreme precipitation over Britain using regional climate model integrations. *International Journal of Climatology*, 21, 1337–1356.

Jones, R., Benson-Evans, K. and Chambers, F.M. (1985) Human influence upon sedimentation in Llangorse Lake, Wales. *Earth Surface Processes and Landforms*, 10, 227–235.

Jordà, G., Marbà, N. and Duarte, C.M. (2012) Mediterranean seagrass vulnerable to regional climate warming. *Nature Climate Change*, doi:10.1038/NCLIMATE1533.

Joughlin, I. and Alley, R.B. (2011) Stability of the West Antarctic ice sheet in a warming world. *Nature Geoscience*, 4, 506–513.

Joyce, L., Aber, J., McNulty, S., Dale, V., Hansen, A., Irland, L., Neilson, R. and Skog, K. (2001) Potential consequences of climate variability and change for the forests of the United States. In National Assessment Synthesis Team (ed.), *Climate Change Impacts on the United States: The Potential Consequences of Climate Variability and Change*. Cambridge: Cambridge University Press, 489–521.

Julian, M. and Anthony, E. (1996) Aspects of landslide activity in the Mercantour Massif and the French Riviera, southeastern France. *Geomorphology*, 15, 275–289.

Kabakci, H., Chevalier, P.M. and Papendick, R.I. (1993) Impact of tillage and residue management on dryland spring wheat development. *Soil and Tillage Research*, 26, 127–137.

Kadiri, M., Ahmadian, R., Bockelmann-Evans, B., Rauen, W. and Falconer, R. (2012) A review of the potential water quality impacts of tidal renewable energy systems. *Renewable and Sustainable Energy Reviews*, 16, 329–341.

Kadomura, H. (1983) Some aspects of large-scale land transformation due to urbanization and agricultural development in recent Japan. *Advances in Space Research*, 2 (8), 169–178.

Kadomura, H. (1994) Climatic change, droughts, desertification and land degradation in the Sudano-Sahelian region: a historico-geographical perspective. In H. Kadomura (ed.), *Savannization Processes in Tropical Africa II*. Tokyo: Department of Geography, Tokyo Metropolitan University, 203–228.

Kai, F.M., Tyler, S.C., Randerson, J.T. and Blake, D.R. (2011) Reduced methane growth rate explained by decreased Northern Hemisphere microbial sources. *Nature*, 476, 194–197.

Kang, S.M., Polvani, L.M., Fyfe, J.C. and Sigmond, M. (2011) Impact of polar ozone depletion on subtropical precipitation. *Science*, 332, 951–954.

Kaplan, J.O., Krumhardt, K.M. and Zimmermann, N. (2009) The prehistoric and preindustrial deforestation of Europe. *Quaternary Science Reviews*, 28, 3016–3034.

Kar, S.K., Liou, Y.-A. and Ha, K.-J. (2009) Aerosol effects on the enhancement of cloud-to-ground lightning over major urban areas of South Korea. *Atmospheric Research*, 92, 80–87.

Karakurt, I., Aydin, G. and Aydiner, K. (2012) Sources and mitigation of methane emissions by sectors: a critical review. *Renewable Energy*, 39, 40–48.

Kareiva, P. and Levin, S.A. (eds) (2003) *The Importance of Species*. Princeton: Princeton University Press.

Karim, M.F. and Mimura, N. (2008) Impacts of climate change and sea-level rise on cyclonic storm surge floods in Bangladesh. *Global Environmental Change*, 18, 490–500.

Karl, T.R., Melillo, J.M. and Peterson, T.C. (2009) *Global Climate Change Impacts in the United States*. Cambridge: Cambridge University Press.

Karnes, L.B. (1971) Reclamation of wet and overflow lands. In G.-H. Smith (ed.), *Conservation of Natural Resources*. New York: Wiley, 241–255.

Karnosky, D.F. (2003) Impacts of elevated atmospheric CO_2 on forest trees and forest ecosystems: knowledge gaps. *Environment International*, 29, 161–169.

Karnosky, D.F., Ceulemans, R., Scarascia-Mugnozza, G.E. and Innes, J.L. (2001) *The Impact of Carbon Dioxide and Other Greenhouse gases on Forest Ecosystems*. Wallingford: CABI Publishing.

Karnosky, D.F., Skelly, J.M., Percy, K.E. and Chappelka, A.H. (2007) Perspectives regarding 50 years of research on effects of tropospheric ozone air pollution on US forests. *Environmental Pollution*, 147, 489–506.

Kaser, G. (1999) A review of the modern fluctuations of tropical glaciers. *Global and Planetary Change*, 22, 93–103.

Kasischke, E.S. et al. (2010) Alaska's changing fire regime – implications for the vulnerability of its boreal forests. *Canadian Journal of Forest Research*, 40, 1313–1324.

Kasperson, V.X., Kasperson, R.E. and Turner, B.L. II (1995) *Regions at Risk: Comparisons of Threatened Environments*. Tokyo: United Nations University Press.

Kates, R.W., Turner, B.L.I. and Clark, W.C. (1990) The great transformation. In B.L. Turner, W.C. Clark, R.W. Kates, J.F. Richards, J.T. Matthews and W.B. Meyer (eds), *The Earth as Transformed by Human Action*. Cambridge: Cambridge University Press, 1–17.

Kaufmann, R.K., Seto, K.C., Schneider, A., Liu, Z., Zhou, L. and Wang, W. (2007) Climate response to rapid urban growth: evidence of a human-induced precipitation deficit. *Journal of Climate*, 20, 2299–2306.

Kauppi, P. and Posch, M. (1988) A case study of the effects of CO_2-induced climatic warming on forest growth and the forest sector. In M.L. Parry, T.R. Carter and N.T. Konijn (eds), *The Impact of Climatic Variations in Agriculture* (Vol. 1). Dordrecht: Kluwer, 183–195.

Kaushal, S.S., Groffman, P.M., Likens, G.E., Belt, K.T., Stack, W.P., Kelly, V.R., Band, L.E. and Fisher, G.T. (2005) Increased salinization of fresh water in the northeastern United States. *Proceedings of the National Academy of Sciences of the United States of America*, 102, 13517–13520.

Kean, J.W., Staley, D.M. and Cannon, S.H. (2011) In situ measurements of post-fire debris flows in southern California: comparisons of the timing and magnitude of 24 debris-flow events with rainfall and soil moisture conditions. *Journal of Geophysical Research*, 116, F04019.

Keatley, B.E., Bennett, E.M., MacDonald, G.K., Taranu, Z.E. and Gregory-Eaves, I. (2011) Land-use legacies are important determinants of lake eutrophication in the Anthropocene. *PLoS ONE*, 6, e15913.

Keefer, D.K., de France, S.D., Mosely, M.E., Richardson, J.B., Satterlee, D.R. and Day-Lewis, A. (1998) Early maritime economy and El Niño events at Quebrada Tachuay, Peru. *Science*, 281, 1833–1835.

Keller, E.A. (1976) Channelisation: environmental, geomorphic and engineering aspects. In D.R. Coates (ed.), *Geomorphology and Engineering*. Stroudsburg: Dowden, Hutchinson and Ross, 115–140.

Kemp, D.D. (2004) *Exploring Environmental Issues: An Integrated Approach*. London: Routledge.

Kemp, K., Palmgren, F. and Mancher, O.H. (1998) *The Danish air quality monitoring programme. Annual report for 1997*. NERI Technical report No 245, Roskilde: National Environmental Research Institute.

Kemp, W.M. et al. (2005) Eutrophication of Chesapeake Bay: historical trends and ecological interactions. *Marine Ecology Progress Series*, 303, 1–29.

Kennish, M.J. (2001) Coastal salt marsh systems in the U.S.: a review of anthropogenic impacts. *Journal of Coastal Research*, 17, 731–748.

Kent, M. (1982) Plant growth problems in colliery spoil reclamation. *Applied Geography*, 2, 83–107.

Kernan, M., Battarbee, R.W., Curtis, C.J., Monteith, D.T. and Shilland, E.M. (2010) *UK Acid Waters Monitoring Network 20 year interpretative report*. London: Environmental Change Research centre, University College London.

Keshkamat, S.S., Tsendbazar, N.-E., Zuidgeest, M.H.P., van der Veen, A. and de Leeeuw, J. (2011) The environmental impact of not having paved roads in arid regions: an example from Mongolia. *Ambio*, doi:10.1007/S13280-011-0855-3.

Khalil, M.A.K. and Rasmussen, R.A. (1987) Atmospheric methane: trends over the last 10,000 years. *Atmospheric Environment*, 21, 2445–2452.

Kibler, K., Tullos, D. and Kondolf, M. (2011) Evolving expectations of dam removal outcomes: downstream geomorphic effects following removal of a small, gravel-filled dam. *Journal of the American Water Resources Association*, 47, 408–422.

Kiepe, P. (1996) Cover and barrier effect of Cassia siamea hedgerows on soil conservation in semi-arid Kenya. *Soil Technology*, 9, 161–171.

Kiersch, G.A. (1965) The Vaiont reservoir disaster. *Mineral Information Service*, 18, 129–138.

Kuhlmann, D.H. (1988) The sensitivity of coral reefs to environmental pollution. *Ambio*, 17, 13–21.

Kimura, B. et al. (2011) Ancient DNA from Nubian and Somali wild asses provides insights into donkey ancestry and domestication. *Proceedings. Biological Sciences/The Royal Society*, 278, 50–57.

King, C.A.M. (1975) *Introduction to Physical and Biological Oceanography*. London: Edward Arnold.

Kingston, P.F. (2002) Long-term environmental impact of oil spills. *Spill Science and Technology Bulletin*, 7, 53–61.

Kinouchi, T., Yagi, H. and Miyamoto, M. (2007) Increase in stream temperature related to anthropogenic heat input from urban wastewater. *Journal of Hydrology*, 335, 78–88.

Kinsey, D.W. and Hopley, D. (1991) The significance of coral reefs as global carbon sinks – response to greenhouse. *Palaeogeography, Palaeoclimatology, Palaeoecology*, 89, 363–377.

Kirby, C. (1995) Urban air pollution. *Geography*, 80, 375–392.

Kirch, P.V. (1982) Advances in Polynesian prehistory: three decades in review. *Advances in World Archaeology*, 2, 52–102.

Kirkbride, M.P. and Warren, C.R. (1999) Tasman Glacier, New Zealand: 20th century thinning and predicted calving retreat. *Global and Planetary Change*, 22, 11–28.

Kirkpatrick, J. (1994) *A Continent Transformed: Human Impact on the Natural Vegetation of Australia*. Melbourne: Oxford University Press.

Kirono, D.G.C., Kent, D.M., Hennessy, K.J. and Mpelasoka, F. (2011) Characteristics of Australian droughts under enhanced greenhouse conditions: results from 14 global climate models. *Journal of Arid Environments*, 75, 566–575.

Kirwan, M.L., Murray, A.B., Donnelly, J.P. and Corbett, D.R. (2011) Rapid wetland expansion during European settlement and its implication for marsh survival under modern sediment delivery rates. *Geology*, 39, 507–510.

Kittredge, J.H. (1948) *Forest Influences*. New York: McGraw-Hill.

Klein, R.G. (1983) The stone age prehistory of southern Africa. *Annual Review of Anthropology*, 12, 25–48.

Kleypas, J.A., Buddemeier, R.W., Archer, D., Gattuso, J.-P., Langdon, C. and Opdyke, B.N. (1999) Geochemical consequences of increased atmospheric carbon dioxide on coral reefs. *Science*, 284, 118–120.

Knapp, A.B. (2010) Cyprus's earliest prehistory: seafarers, foragers and settlers. *Journal of World Prehistory*, 23, 79–120.

Knight, M., Thomas, D.S.G. and Wiggs, G.F.S. (2004) Challenges of calculating dunefield mobility over the 21st century. *Geomorphology*, 59, 197–213.

Knight, P.G. (1999) *Glaciers*. Cheltenham: Stanley Thornes.

Knighton, A.D. (1991) Channel bed adjustment along mine-affected rivers of northeast Tasmania. *Geomorphology*, 4, 205–219.

Knox, J.C. (1972) Valley alluviation in southwestern Wisconsin. *Annals of the Association of American Geographers*, 62, 401–410.

Knox, J.C. (1977) Human impacts on Wisconsin stream channels. *Annals of the Association of American Geographers*, 67, 323–342.

Knox, J.C. (1987) Historical valley floor sedimentation in the Upper Mississippi Valley. *Annals of the Association of American Geographers*, 77, 224–244.

Knox, J.C. (1993) Large increase in flood magnitude in response to modest changes in climate. *Nature*, 361, 430–432.

Knox, J.C. (2001) Agricultural influence on landscape sensitivity in the upper Mississippi river valley. *Catena*, 42, 193–224.

Knox, J.C. (2002) Agriculture, erosion and sediment yields. In A.R. Orme (ed.), *The Physical Geography of North America*. Oxford: Oxford University Press, 482–500.

Knutson, T.R. and Tuleya, R.E. (1999) Increased hurricane intensities with CO$_2$-induced warming as simulated using the GFDL hurricane prediction system. *Climate Dynamics*, 15, 503–519.

Knutson, T.R., Tuleya, R.E. and Kurihara, Y. (1998) Simulated increase of hurricane intensities in a CO$_2$-warmed climate. *Science*, 279, 1018–1020.

Knutson, T.R., Sirutis, J.J., Garner, S.T., Vecchi, G.A. and Held, I.M. (2008) Simulated reduction in Atlantic hurricane frequency under twenty-first-century warming conditions. *Nature Geoscience*, 1, 359–364.

Koch, P.L. and Barnosky, A.D. (2006) Late Quaternary extinctions: state of the debate. *Annual Review of Ecology, Evolution, and Systematics*, 37, 215–250.

Kohen, J. (1995) *Aboriginal Environmental Impacts*. Sydney: University of New South Wales Press.

Koide, M. and Goldberg, E.D. (1971) Atmospheric and fossil fuel combustion. *Journal of Geophysical Research*, 76, 6589–6596.

Komar, P.D. and Allan, J.C. (2008) Increasing hurricane-generated wave heights along the U.S. East Coast and their climate controls. *Journal of Coastal Research*, 24, 479–488.

Komar, P.D., McManus, J. and Styllas, M. (2004) Sediment accumulation in Tillamook Bay, Oregon: natural processes versus human impacts. *The Journal of Geology*, 112, 455–469.

Kornis, M.S., Mercado-Silva, N. and Vander Zanden, M.J. (2012) Twenty years of invasion: a review of round goby *Neoglobius melanstomus* biology, spread and ecological implications. *Journal of Fish Biology*, 80, 238–285.

Kort, E.A. et al. (2012) Atmospheric observations of Arctic Ocean metnane emissions up to 82°N. *Nature Geoscience*, doi:10.1038/NGEO1452.

Kotb, T.H.S., Watanabe, T., Ogino, Y. and Tanji, K.K. (2000) Soil salinization in the Nile Delta and related policy issues in Egypt. *Agricultural Water Management*, 43, 239–261.

Kotlyakov, V.M. (1991) The Aral Sea basin: a critical environmental zone. *Moscow Environment*, 33 (1), 4–9, 36–38.

Kotwicki, V. and Isdale, P.J. (1991) Hydrology of Lake Eyre, Australia: El Niño link. *Palaeogeography, Palaeoclimatology and Palaeoecology*, 84, 87–98.

Krabill, W., Frederick, E., Manizade, S., Martin, C., Sonntag, J., Swift, R., Thomas, R., Wright, W. and Yungel, J. (1999) Rapid thinning of parts of the southern Greenland Ice Sheet. *Science*, 283, 1522–1524.

Kratzmann, M.G. and Hapke, C.J. (2012) Quantifying anthropogenically driven morphological changes on a barrier island: Fire Island National Seashore, New York. *Journal of Coastal Research*, 28, 76–88.

Kravtsova, V.I. and Tarasenko, T.V. (2010) Space monitoring of Aral Sea degradation. *Water Resources*, 37, 285–296.

Kriegler, E., Hall, J.W., Held, H., Dawson, R. and Schellnhuber, H.J. (2009) Imprecise probability assessment of tipping points in the climate system. *Proceedings of the National Academy of Sciences of the United States of America*, 106, 5041–5046.

Krug, E.C. and Frink, C.R. (1983) Acid rain on acid soil: a new perspective. *Science*, 221, 520–525.

Kühn, I. and Klotz, S. (2006) Urbanization and homogenization – comparing the floras of urban and rural areas in Germany. *Biological Conservation*, 127, 292–300.

Kullman, L. (2001) 20th century climate warming and tree-limit rise in the southern Scandes of Sweden. *Ambio*, 30, 72–80.

Kumar, S., Merwade, V., Kam, J. and Thurner, K. (2009) Streamflow trends in Indiana: effects of long term persistence, precipitation and subsurface drains. *Journal of Hydrology*, 374, 171–183.

Kunkel, K.E. (2003) North American trends in extreme precipitation. *Natural Hazards*, 29, 291–305.

Kuo, C. (1986) Flooding in Taipeh, Taiwan and coastal drainage. In J.G. Titus (ed.), *Effects of Changes in Stratospheric Ozone and Global Climate*. Washington DC: UNEP/USEPA, 37–46.

Kustu, M.D., Fan, Y. and Robock, A. (2010) Large-scale water cycle perturbation due to irrigation pumping in the US High Plains: a synthesis of observed streamflow changes. *Journal of Hydrology*, 390, 222–244.

Kwong, Y.T.J. and Gan, T.Y. (1994) Northward migration of permafrost along the Mackenzie Highway and climatic warming. *Climatic Change*, 26, 399–419.

Labadz, J.C., Burt, T.P. and Potter, A.W.L. (1991) Sediment yield and delivery in the blanket peat moorlands of the southern Pennines. *Earth Surface Processes and Landforms*, 16, 255–271.

Labat, D., Goddéris, Y., Probst, J.L. and Guyot, J.L. (2004) Evidence for global runoff increase related to climate warming. *Advances in Water Resources*, 27, 631–642.

Lal, R. (2002) Soil carbon dynamics in cropland and range land. *Environmental Pollution*, 116, 353–362.

Lal, R., Kimble, J., Levine, E. and Stewart, B.A. (eds) (1995) *Soil Management and Greenhouse Effect*. Boca Raton: CRC Lewis.

Le Maitre, D.C., Versfeld, D.B. and Chapman, R.A. (2000) The impact of invading alien plants on surface water resources in South Africa: a preliminary assessment. *Water SA*, 26, 397–408.

Lamb, H.H. (1977) *Climate: Present, Past and Future. 2: Climatic History and the Future*. London: Methuen.

Lambeck, K. (1988) *Geological Geodesy*. Oxford: Clarendon Press.

Lambert, J.H., Jennings, J.N., Smith, C.T., Green, C. and Hutchinson, J.N. (1970) *The making of the Broads: a reconsideration of their origin in the light of new evidence*. Royal Geographical Society research series, 3.

Lamprey, H. (1975) The integrated project on arid lands. *Nature and Resources*, 14, 2–11.

Lancaster, N. (1995) *Geomorphology of Desert Duneas*. London: Routledge.

Landsberg, H.E. (1981) *The Urban Climate*. New York: Academic Press.

Landsea, C.W. (2000) El Niño/Southern Oscillation and the seasonal predictability of tropical cyclones. In H.F. Diaz and V. Markgraf (eds), *El Niño and the Southern Oscillation*. Cambridge: Cambridge University Press, 148–181.

Langbein, W.B. and Schumm, S.A. (1958) Yield of sediment in relation to mean annual precipitation. *Transactions of the American Geophysical Union*, 39, 1076–1118.

Langford, T.E.L. (1990) *Ecological Effects of Thermal Discharges*. London: Elsevier Applied Science.

Lanly, J.P., Singh, K.D. and Janz, K. (1991) FAO's 1990 reassessment of tropical forest cover. *Nature and Resources*, 27, 21–26.

Lantuit, H. and Pollard, W.H. (2008) Fifty years of coastal erosion and retrogressive thaw slump activity on Herschel Island, southern Beaufort Sea, Yukon Territory, Canada. *Geomorphology*, 95, 84–102.

Lapp, S., Byrjne, J., Townshend, I. and Kienzle, S. (2005) Climate warming impacts on snowpack accumulation in an alpine watershed. *International Journal of Climatology*, 25, 521–536.

Larick, R. and Ciochon, R.L. (1996) The African emergence and early Asian dispersals of the genus Homo. *American Scientist*, 84, 538–551.

Laris, P. (2011) Humanizing savanna biogeography: linking human practices with ecological patterns in a frequently burned savanna of southern Mali. *Annals of the Association of American Geographers*, 101, 1067–1088.

Larsen, R., Bell, J.N.B., James, P.W., Chimonides, P.J., Rumsey, P.J., Tremper, A. and Purvis, O.W. (2007) Lichen and bryophyte distribution on oak in London in relation to air pollution and bark acidity. *Environmental Pollution*, 146, 332–340.

Larson, E.R. and Kipfmueller, K.F. (2012) Ecological disaster on the limits of observation? Reconciling modern declines with the long-term dynamics of Whitebark pine communities. *Geography Compass*, 6, 189–214.

Larson, F. (1940) The role of bison in maintaining the short grass plains. *Ecology*, 21, 113–121.

Lasanta, T., García-Ruez, J.M., Pérez-Rontomé, C. and Sancho-Marcén, C. (2000) Runoff and sediment yield in a semi-ard environment: the effect of land management after farmland abandonment. *Catena*, 38, 265–278.

Latif, M. and Keenlyside, N.S. (2009) El Niño/Southern oscillation response to global warming. *Proceedings of the National Academy of Sciences of the United States of America*, 106, 20578–20583.

Lawson, D.E. (1986) Response of permafrost terrain to disturbance: a synthesis of observations from northern Alaska, USA. *Arctic and Alpine Research*, 18, 1–17.

Le, C., Zha, Y., Li, Y., Lu, H. and Yin, B. (2010) Eutrophication of lake waters in China: cost, causes and control. *Environmental Management*, 45, 662–668.

Lean, J. and Warrilow, D.A. (1989) Simulation of the regional climatic impact of Amazon deforestation. *Nature*, 342, 126–133.

Learmonth, J.A., Macleod, C.D., Santos, M.B., Pierce, G.J., Crick, H.P.Q. and Robinson, R.A. (2006) Potential effects of climate change on marine mammals. *Oceanography and Marine Biology: An Annual Review*, 44, 431–464.

Leatherman, S.P. (2001) Social and economic costs of sea level rise. In B.C. Douglas, M.S. Kearney and S.P. Leatherman (eds), *Sea Level Rise: History and Consequences*. San Diego: Academic Press, 181–223.

Lecce, S., Pavlowsky, R. and Schlomer, G. (2008) Mercury contamination of active channel sediment and floodplain deposits from historic gold mining at Gold Hill, North Carolina, USA. *Environmental Geology*, 55, 113–121.

Leclercq, N., Gattuso, J.-P. and Jaubert, J. (2000) CO_2 partial pressure controls the calcification rate of a coral community. *Global Change Biology*, 6, 329–334.

Leduc, C., Favreau, G. and Schoreter, P. (2001) Long term rise in a Sahelian water-table: the Continental Terminal in South-West Niger. *Journal of Hydrology*, 243, 43–54.

Lee, D.O. (1992) Urban warming – an analysis of recent trends in London's heat island. *Weather*, 47, 50–56.

Lee, D.S. et al. (2010), Transport impacts on atmosphere and climate: aviation. *Atmospheric Environment*, 44, 4678–4734.

Lee, E., Sacks, W.J., Chase, T.N. and Foley, J.A. (2011) Simulated impacts of irrigation on the atmospheric circulation over Asia. *Journal of Geophysical Research*, 116, D08114.

Lee, R.B. and DeVore, I. (1968) *Man the Hunter*. Chicago: Aldine.

Le Houérou, H.N. (1977) Biological recovery versus desertization. *Economic Geography*, 63, 413–420.

Leifeld, J., Muller, M. and Fuhrer, J. (2011) Peatland subsidence and carbon loss from drainaed temperate fens. *Soil Use and Management*, 27, 170–176.

Lently, A.D. (1994) Agriculture and wildlife: ecological implications of subsurface irrigation drainage. *Journal of Arid Environments*, 28, 85–94.

Lenton, T.M. (2012) Arctic climate tipping points. *Ambio*, 41, 10–22.

Lenton, T.M., Held, H., Kriegler, E., Hall, J.W., Lucht, W., Rahmstorf, S. and Schellnhuber, H.J. (2008) Tipping elements in the Earth's climate system. *Proceedings of the National Academy of Sciences of the United States of America*, 105, 1786–1793.

Lents, J.M. and Kelly, W.J. (1993) Clearing the air in Los Angeles. *Scientific American*, October, 18–25.

Leopold, L.B. (1951) Rainfall frequency: an aspect of climatic variation. *Transactions of the American Geophysical Union*, 32, 347–357.

Leopold, L.B., Wolman, M.G. and Miller, J.P. (1964) *Fluvial Processes in Geomorphology*. San Francisco: Freeman.

Lerner, D. (1990) *Groundwater recharge in urban areas*, IAHS Publication no. 198, 59–65.

Letey, J. (2001) Cases and consequences of fire-induced soil water repellency. *Hydrological Processes*, 15, 2867–2875.

Letourneau, D.K. and Burrows, B.E. (eds) (2001) *Genetically Engineered Organisms: Assessing Environmental and Human Health Effects*. Washington DC: CRC Press.

Leung, D.Y.C. and Yang, Y. (2012) Wind energy development and its environmental impact: a review. *Renewable and Sustainable Energy Reviews*, 16, 1031–1039.

Levine, J.M., Vila, M., D'Antonio, C.M., Dukes, J.S., Grigulis, K. and Lavorel, S. (2003) Mechanisms underlying the impacts of exotic plant invasions. *Proceedings. Biological Sciences/The Royal Society*, 270, 775–781.

Lev-Yadun, S., Gopher, A. and Abbo, S. (2000) Enhanced: the cradle of agriculture. *Science*, 288, 1602–1603.

Lewin, J., Bradley, S.B. and Macklin, M.G. (1983) Historical valley alluviation in mid-Wales. *Geological Journal*, 18, 331–350.

Leys, J. (1999) Wind erosion on agricultural land. In A.S. Goudie, I. Livingstone and S. Stokes (eds), *Aeolian Environments, Sediments and Landforms*. Chichester: Wiley, 143–166.

L'Homer, A. (1992) Sea level changes and impact on the Rhône Delta coastal lowlands. In M.J. Tooley and S. Jelgersma (eds), *Impacts of Sea Level Rise on European Coastal Lowlands*. Oxford: Blackwell, 136–152.

Li, P., Feng, X., Qiu, G., Shang, L. and Wang, S. (2012) Mercury pollution in Wuchuan mercury mining area, Guizhou, southwestern China: the impacts from large scale and artisanal mercury mining. *Environment International*, 42, 59–66.

Li, X. et al. (2008) Cryospheric change in China. *Global and Planetary Change*, 62, 210–218.

Li, Y., Cui, J., Zhang, T., Okuro, T and Drake, S. (2009) Effectiveness of sand-fixing measures on desert land restoration in Kerqin Sandy Land, northern China. *Ecological Engineering*, 35, 118–127.

Li, Z. et al. (2012) Decreasing trend of sunshine hours and related driving forces in southwestern China. *Theoretical and Applied Climatology*, 109, 305–321.

Liddle, M. (1997) *Recreation Ecology*. London: Chapman & Hall.

Liébault, F. and Piégay, H. (2002) Causes of 20th century channel narrowing in mountain and piedmont rivers of southeastern France. *Earth Surface Processes and Landforms*, 27, 425–444.

Liebhold, A.M., Brockerhoff, E.G., Garrett, L.J., Parke, J.L. and Britton, K.O. (2012) Live plant imports: the major pathway for forest insect and pathogen invasions of the US. *Frontiers in Ecology and the Environment*, 10, 135–143.

Liebsch, D., Marques, M.C.M. and Goldenberg, R. (2008) How long does the Atlantic Rain Forest take to recover after a disturbance? Changes in species composition and ecological features during secondary succession. *Biological Conservation*, 141, 1717–1725.

Lienert, J. (2004) Habitat fragmentation effects on fitness of plant populations – a review. *Journal for Nature Conservation*, 12, 53–72.

Likens, G.E. (2010) Acid rain. In *Encyclopedia of Earth* (http://www.eoearth.org/article/Acid_rain?topic=49506), last accessed 05 January 2013.

Likens, G.E. and Bormann, F.H. (1974) Acid rain: a serious regional environmental problem. *Science*, 184, 1176–1179.

Likens, G.E., Wright, R.F., Galloway, J.N. and Butler, T.J. (1979) Acid rain. *Scientific American*, 241 (4), 39–47.

Lin, N., Emanuel, K., Oppenheimer, M. and Vanmarcke, E. (2012) Physically based assessment of hurricane surge threat under climate change. *Nature Climate Change*, doi:10.1038/NCLIMATE1389.

Ling, F. and Zhang, T. (2003) Impact of the timing and duration of seasonal snow cover on the active layer and permafrost in the Alaskan Arctic. *Permafrost and Periglacial Processes*, 14, 141–150.

Ling, S.D. (2008) Range expansion of a habitat-modifying species leads to loss of taxonomic diversity: a new and impoverished reef state. *Oecologia*, 156, 883–894.

Lioy, P.J. and Georgopoulos, P.G. (2011) New Jersey: a case study for the reduction in urban and suburban air pollution for the 1950s to 2010. *Environmental Health Perspectives*, 119, 1351–1355.

List, J.H., Sallenger, A.H., Hansen, M.E. and Jaffe, B.E. (1997) Accelerated sea level rise and rapid coastal erosion: testing a causal relationship for the Louisiana barrier islands. *Marine Geology*, 140, 437–465.

Liu, J. and Diamond, J. (2005) China's environment in a globalizing world. *Nature*, 435, 1179–1186.

Liu, J., Hayakawa, N., Lu, M., Dong, S. and Yuan, J. (2003) Hydrological and geocryological response of winter streamflow to climate warming in Northeast China. *Cold Regions Science and Technology*, 37, 15–24.

Liu, J., Curry, J.A., Wang, H., Song, M. and Horton, R.M. (2012) Impact of declining Arctic sea ice on winter snowfall. *Proceedings of the National Academy of Sciences of the United States of America*, doi/10.1073/pnas.1114910109.

Liu, L., Lee, G.-A., Jiang, L. and Zhang, J. (2007) The earliest rice domestication in China. *Antiquity*, 81, 1–2.

Liu, L., Liu, Z., Ren, X., Fischer, T. and Xu, Y. (2011) Hydrological impacts of climate change in the Yellow River Basin for the 21st century using hydrological model and statistical downscaling model. *Quaternary International*, 244, 211–220.

Liu, S., Zhang, Y. and Ding, Y. (2009) Estimation of glacier runoff and future trends in the Yangtze River source region, China. *Journal of Glaciology*, 55, 353–362.

Liu, X.B., Zhang, X.Y., Wang, Y.X., Sui, Y.Y., Zhang, S.L., Herbert, S.J. and Ding, G. (2010) Soil degradation: a problem threatening the sustainable development of agriculture in northeast China. *Plant, Soil and Environment*, 56, 87–97.

Liu, Y.-P. et al. (2006) Multiple maternal origins of chickens: out of the Asian jungles. *Molecular Phylogenetics and Evolution*, 38, 12–19.

Lloyd, J.W. (1986) A review of aridity and groundwater. *Hydrological Processes*, 1, 63–78.

Loehle, C. and Eschenbach, W. (2012) Historical bird and terrestrial mammal extinction rates and causes. *Diversity & Distributions*, 18, 84–91.

Loidi, J., Biurrun, I., Campos, J.A., García-Mijangos, I. and Herrera, M. (2010) A biogeographical analysis of the European Atlantic lowland heathlands. *Journal of Vegetation Science*, 21, 832–842.

Lomborg, B. (2001) *The Skeptical Environmentalist*. Cambridge: Cambridge University Press.

Loreau, M., Naeem, S., Inchausti, P., Bengtsson, J., Grime, J.P., Hector, A., Hooper, D.U., Huston, M.A., Raffaelli, D., Schmid, B., Tilman, D. and Wardle, D.A. (2001) Biodiversity and ecosystem functioning: current knowledge and future challenges. *Science*, 294, 804–808.

Loreau, M., Naeem, S. and Inchausti, P. (eds) (2002) *Biodiversity and Ecosystem Functioning. Synthesis and Perspectives*. Oxford: Oxford University Press.

Lovelock, J. (2006) *The Revenge of Gaia*. London: Allen Lane.

Lovelock, J. (2008) A geophysiologist's thoughts on geoengineering. *Philosophical Transactions. Series A, Mathematical, physical, and engineering Sciences*, doi:10:1098/rsta.2008.0135.

Lowe, D.J. (2008) Polynesian settlement of New Zealand and the impacts of volcanism on early Maori society: an update. In D.J. Lowe (ed.), *Guidebook for Pre-Conference North Island Field Trip*. New Zealand: Massey University, Palmerston North, New Zealand Society of Soil Science, 142–147.

Lowe, P.D. (1983) Values and institutions in the history of British nature conservation. In A. Warren and F.B. Goldsmith (eds), *Conservation in Perspective*. Chichester: Wiley, 329–352.

Lowe-McConnell, R.H. (1975) Freshwater life on the move. *The Geographical Magazine*, 47, 768–775.

Lowenthal, D. (2000) *George Perkins Marsh, Prophet of Conservation*. Seattle: University of Washington Press.

Loya, Y., Sakai, K., Yamazato, K., Sambali, H. and Van Woesik, R. (2001) Coral bleaching: the winners and the losers. *Ecology Letters*, 4, 122–132.

Lu, H. et al. (2009) Earliest domestication of common millet (Panicum miliaceum) in East Asia extended to 10,000 years ago. *Proceedings of the National Academy of Sciences of the United States of America*, 106, 7367–7372.

Lu, Z., Streets, D.G., Zhang, Q., Wang, S., Carmichael, G.R., Cheng, Y.F., Wei, C., Chin, M., Diehl, T. and Tan, Q. (2010) Sulfur dioxide emissions in China and sulphur trends in East Asia since 2000. *Atmospheric Chemistry and Physics*, 10, 6311–6331.

Lugo, A.E. (2000) Effects and outcomes of Caribbean hurricanes in a climate change scenario. *The Science of the Total Environment*, 262, 243–251.

Lugo, A.E., Cintron, G. and Goenaga, C. (1981) Mangrove ecosystems under stress. In G.W. Barrett and R. Rosenberg (eds), *Stress Effects on Natural Ecosystems*. Chichester: John Wiley, 129–153.

Lund, J.W.G. (1972) Eutrophication. *Proceedings of the Royal Society of London. Series B. Biological Sciences*, 180, 371–382.

Lupton, M.K., Rojas, C., Drohan, P. and Bruns, M.A. (2012). *Restoration Ecology*, doi:10.111/j.1526-100X.2012.00902.x.

Lyell, C. (1835) *Principles of Geology* (4th edn, Vol. III). London: Murray, (12th edn 1875).

Lynas, M. (2011) *The God Species*. London: Fourth Estate.

Lynch, J.A., Rishel, G.B. and Corbett, E.S. (1984) Thermal alteration of streams draining clearcut watersheds: quantification and biological implications. *Hydrobiologia*, 111, 161–169.

Lyons, S.K., Smith, F.A., Wagner, P.J., White, E.P. and Brown, J.H. (2004) Was a 'hyperdisease' responsible for the late Pleistocene megafaunal extinction? *Ecology Letters*, 7, 859–868.

Mabbutt, J.A. (1977) *Desert Landforms*. Cambridge, Massachusetts: MIT Press.

Mabbutt, J.A. (1985) Desertification of the world's rangelands. *Desertification Control Bulletin*, 12, 1–11.

MacCracken, M. et al. (2001) Scenarios for climate variability and change. In National Assessment Synthesis Team (ed.), *Climate Change Impacts on the United States: The Potential Consequences of Climate Variability and Change*. Cambridge: Cambridge University Press, 13–71.

MacDonald, D. and Burnham, D. (2011) *The State of Britain's Mammals 2011*. London: People's Trust for Endangered Mammals.

Macfarlane, M.J. (1976) *Laterite and landscape*. London: Academic Press.

Macias-Fauria, M., Forbes, B.C., Zetterberg, P. and Kumpula, T. (2012) Eurasian Arctic greening reveals teleconnections and the potential for structurally novel ecosystems. *Nature Climate Change*, doi:10.1038/NClimate1558.

MacKay, M. and Seglenieks, F. (2011) On the simulation of Laurentian Great Lakes water levels under projections of global climate change. *Climatic Change*, doi:10.1017/s10584-012-0560-z.

Macklin, M.G. and Lewin, J. (1986) Terraced fills of Pleistocene and Holocene age in the Rheidol Valley, Wales. *Journal of Quaternary Science*, 1, 21–34.

Macklin, M.G., Passmore, D.G., Stevenson, A.C., Colwey, A.C., Edwards, D.N. and O'Brien, C.F. (1991) Holocene alluviation and land-use change on Callaly Moor, Northumberland, England. *Journal of Quaternary Science*, 6, 225–232.

Macklin, M.G., Jones, A.F. and Lewin, J. (2010) River response to rapid Holocene environmental change: evidence and explanation in British catchments. *Quaternary Science Reviews*, 29, 1555–1576.

Mader, H.J. (1984) Animal habitat isolation by roads and agricultural fields. *Biological Conservation*, 29, 81–96.

Magdaleno, F., Fernández, J.A. and Merino, S. (2011) The Ebro River in the 20th century or the ecomorphological transformation of a large and dynamic Mediterranean channel. *Earth Surface Processes and Landforms*, doi:10.1002/esp.2258.

Magilligan, F.J. (1985) Historical floodplain sedimentation in the Galena River basin, Wisconsin and Illinois. *Annals of the Association of American Geographers*, 75, 583–594.

Magilligan, F.J. and Goldstein, P.S. (2001) El Niño floods and culture change: a late Holocene flood history for the Rio Moquegua, Southern Peru. *Geology*, 29, 431–434.

Mahmood, R., Foster, S.A., Keeling, T., Hubbard, K.G., Carlson, C. and Leeper, R. (2006) Impacts of irrigation on 20th century temperature in the northern Great Plains. *Global and Planetary Change*, 54, 1–18.

Mahowald, N.M. et al. (2010) Observed 20th century desert dust variability: impact on climate and biogeochemistry. *Atmospheric Chemistry and Physics*, 10, 10875–10893.

Mahowald, N.M. and Luo, C. (2003) A less dusty future? *Geophysical Research Letters*, doi:10.1029/2003GL017880.

Mahowald, N.M., Muhs, D.R., Levis, S., Rasch, P.J., Yoshioka, M., Zender, C.S. and Luo, C. (2006) Change in atmospheric mineral aerosols in response to climate: last glacial period, preindustrial, modern and doubled carbon dioxide climates. *Journal of Geophysical Research*, doi:10.1029/2005JD006653.

Maignien, R. (1966) *A Review of Research on Laterite*. Paris: UNESCO.

Mainguet, M. (1995) *L'homme et la sécheresse*. Paris: Masson.

Majer, E.L., Baria, R., Start, M., Oates, S., Bommer, J., Smith, B. and Asanuma, H. (2007) Induced seismicity associated with Enhanced Geothermal Systems. *Geothermics*, 36, 185–222.

Major, J. et al. (2008) Initial fluvial response to the removal of Oregon's Marmot Dam. *Eos*, 27, 241–243.

Major, J.J. et al. (2012) Geomorphoic response of the Sandy River, Oregon, to removal of Marmot Dam. *U.S.Geological Survey professional paper*, 1792.

Malamud, B.D., Morein, G. and Turcotte, D.L. (1998) Forest fires: an example of self-organized critical behavior. *Science*, 281, 1840–1842.

Malingreau, J.P., Eva, H.D. and de Miranda, E.E. (2011) Brazilian Amazon: a significant five year drop in deforestation rates but figures are on the rise again. *Ambio*, doi:10.1007/S13280-011-0196-7.

Malkisnon, D., Wittenberg, L., Beeri, O. and Barzilai, R. (2011) Effects of repeated fires on the structure, composition, and dynamics of Mediterranean maquis: short- and long-term perspectives. *Ecosystems*, 14, 478–488.

Malm, W.C., Schichtel, B.A., Ames, R.B. and Gebhart, K.A. (2002) A 10-year spatial and temporal trend of sulfate across the United States. *Journal of Geophysical Research*, 107 (D22), 4627.

Manabe, S. and Stouffer, R.J. (1980) Sensitivity of a global climate model to an increase of CO_2 concentration in the atmosphere. *Journal of the Atmospheric Sciences*, 37, 99–118.

Manabe, S. and Wetherald, R.T. (1986) Reduction in summer soil wetness by an increase in atmospheric carbon dioxide. *Science*, 232, 626–628.

Manney, G.L. et al. (2011) Unprecedented Arctic ozone loss in 2011. *Nature*, 478, 469–475.

Mannion, A.M. (1992) Acidification and eutrophication. In A.M. Mannion and S.R. Bowlby (eds), *Environmental Issues in the 1990s*. Chichester: Wiley, 177–195.

Mannion, A.M. (1995) *Agriculture and Environmental Change*. Chichester: Wiley.

Mannion, A.M. (1997) *Global Environmental Change* (2nd edn). Harlow: Longman.

Mannion, A.M. (2002) *Dynamic World. Land-Cover and Land-Use Change*. London: Arnold.

Marden, M. (2012) Effectiveness of reforestation in erosion mitigation and implications for future sediment yields, East Coast catchments, New Zealand: a review. *New Zealand Geographer*, 68, 24–35.

Marden, M., Arnold, G., Seymour, A. and Hambling, R. (2012) History and distribution of steepland gullies in response to land use change, East Coast Region, North Island, New Zealand. *Geomorphology*, 153–154, 81–90.

Mark, A.F. and McSweeney, G.D. (1990) Patterns of impoverishment in natural communities: case studies in forest ecosystems – New Zealand. In G.M. Woodwell (ed.), *The Earth in Transition: Patterns and Processes of Biotic Impoverishment*. Cambridge: Cambridge University Press, 151–176.

Marker, M.E. (1967) The Dee estuary: its progressive silting and salt marsh development. *Transactions of the Institute of British Geographers*, 41, 65–71.

Marks, P.L. and Bormann, F.H. (1972) Revegetation following forest cutting: mechanisms for return to steady-state nutrient cycling. *Science*, 176, 914–915.

Marlon, J.R. et al. (2012) Long-term perspective on wildfires in the western USA. *Proceedings of the National Academy of Sciences of the United States of America*, doi/10.1073/pnas.1112839109.

Marquiss, M., Newton, I. and Ratcliffe, D.A. (1978) The decline of the raven, *Corvus corax*, in relation to afforestation in southern Scotland and northern England. *Journal of Applied Ecology*, 15, 129–144.

Marsh, G.P. (1864) *Man and Nature*. New York: Scribner.

Marsh, G.P. (1965) *Man and Nature*. D. Lowenthal (ed.). Cambridge, MA: Belknap Press.

Marshall, E., Weinberg, M., Wunder, S. and Kaphengst, T. (2011) Environmental dimensions of bioenergy development. *EuroChoices*, 10, 43–48.

Marshall, L.G. (1984) Who killed Cock Robin? An investigation of the extinction controversy. In P.S. Martin and R.G. Klein (eds), *Quaternary Extinctions*. Tucson: University of Arizona Press, 785–806.

Marston, R.A. and Dolan, L.S. (1999) Effectiveness of sediment control structures relative to spatial patterns of upland soil loss in an arid watershed, Wyoming. *Geomorphology*, 31, 313–323.

Martin, P.S. (1967) Prehistoric overkill. In P.S. Martin and H.E. Wright (eds), *Pleistocene Extinctions*. New Haven: Yale University Press, 75–120.

Martin, P.S. (1974) Palaeolithic players on the American stage: man's impact on the Late Pleistocene megafauna. In J.D. Ives and R.G. Barry (eds), *Arctic and Alpine Environments*. London: Methuen.

Martin, P.S. (1982) The pattern and meaning of Holarctic mammoth extinction. In D.M. Hopkins, J.V. Matthews, C.S. Schweger and S.B. Young (eds), *Paleoecology of Beringia*. New York: Academic Press, 399–408.

Martin, P.S. and Klein, R.G. (1984) *Pleistocene Extinctions*. Tucson: University of Arizona Press.

Martin, P.S. and Wright, H.E. (eds) (1967) *Pleistocene Extinctions*. New Haven: Yale University Press, 75–120.

Martin, Y.E., Johnson, E.A., Gallaway, J.M. and Chaikina, O. (2011) Negligible soil erosion in a burned mountain watershed, Canadian Rockies: field and modelling investigations considering the role of duff. *Earth Surface Processes and Landforms*, 36, 2097–2113.

Martinez Raya, A., Duran Zuazo, V.H. and Francia Martinez, J.R. (2006) Soil erosion and runoff response to plant-cover strips on semiarid slopes (SE Spain). *Land Degradation and Development*, 17, 1–11.

Maskell, L.C., Smart, S.M., Bullock, J.M., Thompson, K. and Stevens, C.J. (2010) Nitrogen deposition causes widespeard loss of species richness in Bbritish habitats. *Global Change Biology*, 16, 671–679.

Maslowski, W., Kinney, J.C., Higgins, M. and Roberts, A. (2012) The future of Arctic Sea Ice. *Annual review of Earth and Planetary Sciences*, 40, 625–654.

Mason, I.M., Guzkowska, M.A.J., Rapley, C.G. and Street-Perrott, F.A. (1994) The response of lake levels and areas to climatic change. *Climatic Change*, 27, 161–197.

Mason, S.J., Waylen, P.R., Mimmack, G.M., Rajaratnam, B. and Harrison, J.M. (1999) Changes in extreme rainfall events in South Africa. *Climatic Change*, 41, 249–257.

Mather, A.S. (1983) Land deterioration in upland Britain. *Progress in Physical Geography*, 7, 210–228.

Mather, A.S. (2007) Recent Asian forest transitions in relation to transition theory. *International Forestry Review*, 9, 491–502.

Mattheus, C.R., Rodriguez, A.B., McKee, B.A. and Curring, C.A. (2010) Impact of land-use change and hard structures on the evolution of fringing marsh shorelines. *Estuarine, Coastal and Shelf Science*, 88, 365–376.

Maugh, T.H. (1979) The Dead Sea is alive and well . . . . *Science*, 205, 178.

Maurer, E.P. and Duffy, P.B. (2005) Uncertainty in projections of streamflow changes due to climate change in California. *Geophysical Research Letters*, 32, L03704.

Mavruk, S. and Avsar, D. (2008) Non-native fishes in the Mediterranean from the Red Sea, by way of the Suez Canal. *Reviews in Fish Biology and Fisheries*, 18, 251–262.

May, L., Defew, L.H., Bennion, H. and Kirika, A. (2012) Historical changes (1905–2005) in external phosphorus loads to Loch Leven, Scotland, UK. *Hydrobiologia*, 681, 11–21.

May, R.M. (1979) Fluctuations in abundance of tropical insects. *Nature*, 278, 505–507.

May, T. (1991) Südspanische matorrales als Kulturofolgevegetation. *Geoökodynamik*, 12, 87–107.

McCann, K.S. (2000) The diversity-stability debate. *Nature*, 405, 228–233.

McCarney-Castle, K., Voulgaris, G. and Kettner, A.J. (2010) Analysis of fluvial suspended sediment load contribution through Anthropoecne history to the South Atlantic Bight coastal zone, U.S.A. *The Journal of Geology*, 118, 399–416.

McCarthy, T.S. (2011) The impact of acid mine drainage in South Africa. *South African journal of science*, doi:10.4102/sajs.v107i5/6.712.

MacCarthy, J., Thistlethwaite, G., Salisbury, E., Pang, Y. and Misselbrook, T. (2012) *Air quality pollutant inventories for England, Scotland, Wales and Northern Ireland: 1990–2010.* Didcot: AEA.

McClanahan, T.R. (2000) Bleaching damage and recovery potential of Maldivian coral reefs. *Marine Pollution Bulletin*, 40, 587–597.

McClelland, J.W., Holmes, R.M., Peterson, B.J. and Stieglitz, M. (2004) Increasing river discharge in the Eurasian Arctic: consideration of dams, permafrost thaw, and fires as potential agents of change. *Journal of Geophysical Research*, 109, D18102.

McConnell, J.R., Aristarain, A.J., Banta, J.R., Edwards, P.R. and Simões, J.C. (2007) 20th-century doubling in dust archived in an Antarctic Peninsula ice core parallels climate change and desertification in South America. *Proceedings of the National Academy of Sciences of the United States of America*, 104, 5743–5748.

McCulloch, M., Fallon, S., Wyndham, T., Hendy, E., Lynch, J. and Barnes, D. (2003) Coral record of increased sediment flux to the inner Great Barrier Reef since European settlement. *Nature*, 421, 727–730.

McGlone, M.S. and Wilmshurst, J.M. (1999) Dating initial Maori environmental impact in New Zealand. *Quaternary International*, 59, 5–16.

McGuire, B. (2010) Climate forcing of geological and geomorphological hazards. *Philosophical Transactions. Series A, Mathematical, Physical, and Engineering Sciences*, 368, 2311–2315.

McKinney, M.L. (2002) Urbanization, biodiversity, and conservation. *Bioscience*, 52, 883–890.

McKinney, M.L. (2008) Effects of urbanization on species richness: a review of plants and animals. *Urban Ecosystems*, 11, 161–176.

McKnight, T.L. (1959) The feral horse in Anglo-America. *Geographical Review*, 49, 506–525.

McKnight, T.L. (1971) Australia's buffalo dilemma. *Annals of the Association of American Geographers*, 61, 759–773.

McLennan, S.M. (1993) Weathering and global denudation. *The Journal of Geology*, 101, 295–303.

McNeall, D., Halloran, P.R., Good, P. and Betts, R.A. (2011) Analyzing abrupt and nonlinear climate changes and their impacts. *Wiley Interdisciplinary Reviews: Climate Change*, 2, 663–686.

McNeill, J.R. (2000) *Something New under the Sun. An Environmental History of the Twentieth Century.* London: Allen Lane.

McNeill, J.R. (2003) Resource exploitation and over-exploitation: a look at the 20th century. In T.S. Benzing and B. Herrmann (eds), *Exploitation and Overexploitation in Societies Past and Present.* Münster: LIT Verlag, 51–60.

McNeill, J.R. (2005) Modern global environmental history. *IHDP Global Update*, 2–7.

McPherron, S.P. et al. (2010) Evidence for stone-tool-assisted consumption of animal tissues before 3.39 million years ago at Dikika, Ethiopia. *Nature*, 466, 857–860.

McWethy, D.B., Whitlock, C., Wilmshurst, J.M., McGlone, M.S. and Li, X. (2009) Rapid deforestation of South Island, New Zealand, by early Polynesian fires. *The Holocene*, 19, 883–897.

Meade, R.H. (1991) Reservoirs and earthquakes. *Engineering Geology*, 30, 245–262.

Meade, R.H. (1996) River-sediment input to major deltas. In J.D. Milliman and B.V. Haq (eds), *Sea-Level Rise and Coastal Subsidence.* Dordrecht: Kluwer, 63–85.

Meade, R.H. and Moody, J.A. (2010) Causes for the decline of suspended-sediment discharge in the Mississippi River system, 1940–2007. *Hydrological Processes*, 24, 35–49.

Meade, R.H. and Parker, R.S. (1985) Sediment in rivers in the United States. *United States Geological Survey water supply paper*, 2276, 49–60.

Meade, R.H. and Trimble, S.W. (1974) Changes in sediment loads in rivers of the Atlantic drainage of the United States since 1900. *Publication of the International Association of Hydrological Science*, 113, 99–104.

Meadows, M.E. and Linder, H.P. (1993) A palaeoecological perspective on the origin of Afromontane grasslands. *Journal of Biogeography*, 20, 345–355.

Meadows, P.S. and Meadows, A. (1999) The environmental impact of the River Indus on the coastal and offshore zones of the Arabian Sea and the north-west Indian Ocean. In A. Meadows and P.S. Meadows (eds), *The Indus River: Biodiversity, Resources, Humankind.* Karachi: Oxford University Press.

Mee, L.D. (1992) The Black Sea in crisis: a need for concerted international action. *Ambio*, 21, 278–286.

Meehl, G.A., Arblaster, J.M. and Collins, W.D. (2008) Effects of black carbon aerosols on the Indian Monsoon. *Journal of Climate*, 21, 2869–2882.

Melillo, J., Janetos, A., Schimel, D. and Kittel, T. (2001) Vegetation and biogeochemical scenarios. In National Assessment Synthesis Team (ed.), *Climate Change Impacts on the United States: The Potential Consequences of Climate Variability and Change.* Cambridge: Cambridge University Press, 74–91.

Mellanby, K. (1967) *Pesticides and Pollution.* London: Fontana.

Menze, B.H. and Ur, J.A. (2012) Mapping patterns of long-term settlement in Northern Mesopotamia at a large scale. *Proceedings of the National Academy of Sciences of the United States of America*, doi:10.1073/pnas.1115472109.

Menzel, L. and Burger, G. (2002) Climate change scenarios and runoff response in the Mulde catchment (Southern Elbe, Germany). *Journal of Hydrology*, 267, 53–64.

Mercer, D.E. and Hamilton, L.S. (1984) Mangrove ecosystems: some economic and natural benefits. *Nature and Resources*, 20, 14–19.

Mercer, J.H. (1978) West Antarctic ice sheet and CO_2 greenhouse effect: a threat of disaster. *Nature*, 271, 321–325.

Meriläinen, J.J., Kustula, V. and Witick, A. (2011) Lead pollution history from 256 BC to AD 2005 inferred from the Pb isotope ratio (206Pb/207Pb) in a varve record of Lake Korttajärvi in Finland. *Journal of Paleolimnology*, 45, 1–8.

Messier, M.S., Shatford, J.P.A. and Hibbs, D.E. (2012) Fire exclusion effects on riparian forest dynamics in southwestern Oregon. *Forest Ecology and Management*, 264, 60–71.

Metcalfe, S. and Derwent, D. (2005) *Atmospheric Pollution and Environmental Change*. London: Hodder Arnold.

Meusberger, K. and Alewell, C. (2008) Impacts of anthropogenic and environmental factors on the occurrence of shallow landslides in an alpine catchment (Urseren Valley, Switzerland). *Natural Hazards and Earth System Sciences*, 8, 509–520.

Meybeck, M. (1979) Concentration des eaux fluviales en éléments majeurs et apports en solution aux océans. *Revue de Géographie Physique et de Géologie Dynamique*, 21a, 215–246.

Meybeck, M. (2001a) Global alteration of riverine geochemistry under human pressure. In E. Ehlers (ed.), *Understanding the Earth System: Compartments, Processes and Interactions*. Heidelbers: Springer, 97–113.

Meybeck, M. (2001b) River basins under anthropocene conditions. In B. von Bodungen and R.K. Turner (eds), *Science and Integrated Coastal Management*. Dahlem: Dahlem University Press, 275–294.

Meyer, G.A. and Pierce, J.L. (2003) Climatic controls on fire-induced sediment pulses in Yellowstone National Park and central Idaho: a long-term perspective. *Forest Ecology and Management*, 178, 89–104.

Meyer, R.S., DuVal, A.E. and Jensen, H.R. (2012) Patterns and processes in crop domestication: an historical review and quantitative analysis of 203 global food crops. *The New Phytologist*, 196, 29–48.

Meyer, W.B. (1996) *Human Impact on the Earth*. Cambridge: Cambridge University Press.

Meyer, W.B. and Turner, B.L. II (eds) (1994) *Changes in Land Use and Land Cover: A Global Perspective*. Cambridge: Cambridge University Press.

Meyfroidt, P. and Lambin, E. (2008) Forest transition in Vietnam and its environmental impacts. *Global Change Biology*, 14, 1319–1336.

Meyfroidt, P. and Lambin, E. (2009) Geographic and historical patterns of reforestation. *Bulletin des séances de l'Academie Royale des Sciences d'outre-mer*, 55, 477–502.

Micklin, P. (1972) Dimensions of the Caspian Sea problem. *Soviet Geography*, 13, 589–603.

Micklin, P. (2010) The past, present and future Aral Sea. *Lakes and Reservoirs: Research and Management*, 15, 193–213.

Middleton, N.J. (2008) *The Global Casino*. London: Hodder Education.

Middleton, N.J. and Thomas, D.S.G. (1997) *World Atlas of Desertification* (2nd edn). London: Edward Arnold.

Midgley, G.F., Hannah, L., Millar, D., Thuiller, W. and Booth, A. (2003) Developing regional and species-level assessments of climate change impacts on biodiversity in the Cape Floristic Region. *Biological Conservation*, 112, 87–97.

Mieck, I. (1990) Reflections on a typology of historical pollution: complementary conceptions. In P. Brimblecombe and C. Pfister (eds), *The Silent Countdown*. Berlin: Springer Verlag, 73–80.

Miettingen, J., Shi, C. and Soo, C.L. (2011) Deforestation rates in insular southeast Asia between 2000 and 2010. *Global Change Biology*, 17, 2261–2270.

Mikesell, M.W. (1969) The deforestation of Mount Lebanon. *Geographical Review*, 59, 1 28.

Millennium Ecosystem Assessment (MA). 2005. *Ecosystems and Human Well-Being: Synthesis*. Washington, DC: Island Press.

Miller, J.D., Safford, H.D., Crimmins, M. and Thode, A.E. (2009) Quantitative evidence for increasing fire severity in the Sierra Nevada and southern Cascade Mountains, California and Nevada, USA. *Ecosystems*, 12, 16–32.

Miller, J.R. and Russell, G.L. (2002) Projected impact of climate change on the energy budget of the Arctic Ocean Global Climate Model. *Journal of Climate*, 15, 3028–3042.

Miller, L. and Douglas, B.C. (2004) Mass and volume contribution to twentieth-century global sea level rise. *Nature*, 428, 406–409.

Miller, R.L. and Tegen, I. (1998) Climate response to soil dust aerosols. *Journal of Climate*, 11, 3247–3267.

Miller, R.S. and Botkin, D.B. (1974) Endangered species: models and predictions. *American Scientist*, 62, 172–181.

Milliman, J.D. (1990) Fluvial sediment in coastal seas: flux and fate. *Nature and Resources*, 26, 12–22.

Milliman, J.D. and Haq, B.U. (eds) (1996) *Sea Level Rise and Coastal Subsidence*. Dordrecht: Kluwer.

Milliman, J.D., Qin, Y.S., Ren, M.E. and Saita, Y. (1987) Man's influence on erosion and transport of sediment by Asian rivers: the Yellow River (Huanghe) example. *The Journal of Geology*, 95, 751–762.

Milliman, J.D., Broadus, J.M. and Gable, F. (1989) Environmental and economic impacts of rising sea level and subsiding deltas: the Nile and Bengal examples. *Ambio*, 18, 340–345.

Milly, P.C.D., Wetherald, R.T., Dunne, K.A. and Delworth, T.L. (2002) Increasing risk of great floods in a changing climate. *Nature*, 415, 514–517.

Milne, W.G. (1976) Induced seismicity. *Engineering Geology*, 10, 83–88.

Min, S.-K., Zhang, X., Zwiers, F.W. and Hegerl, G.C. (2011) Human contribution to more-intense precipitation extremes. *Nature*, 470, 378–381.

Minassian, V.T. (2011) The Green Wall of Africa. *CNRS International Magazine*, 23, 6–7.

Mirza, M.M.Q. (2002) Global warming and changes in the probability of occurrence of floods in Bangladesh and implications. *Global Environmental Change*, 12, 127–138.

Mistry, J. (2000) *World Savannas: Ecology and Human Use*. Harlow: Prentice Hall.

Mithen, S. (2007) Did farming arise from a misapplication of social intelligence? *Philosophical Transactions of the Royal Society of London. Series B, Biological Sciences*, 362, 705–718.

Mitra, A., Chatterjee, C. and Mandal, F.B. (2011) Synthetic chemical pesticides and their effects on birds. *Research Journal of Environmental Toxicology*, 5, 81–96.

Moleele, N.M., Ringrose, S., Matheson, W. and Vanderpost, C. (2002) More woody plants? The status of bush encroachment in Botswana's grazing areas. *Journal of Environmental Management*, 64, 3–11.

Moles, A.T. et al. (2012) Invasions: the trail behind, the pasth ahead, and a test of a disturbing idea. *The Journal of Ecology*, 100, 116–127.

Mölg, T., Georges, C. and Kaser, G. (2003) The contribution of increased incoming shortwave radiation to the retreat of the Rwenzori Glaciers, East Africa, during the 20th century. *International Journal of Climatology*, 23, 291–303.

Moncel, M.-H. (2010) Oldest human expansions in Eurasia: favouring and limiting factors. *Quaternary International*, 223–224, 1–9.

Montgomery, D.R. (1997) What's best on the banks? *Nature*, 388, 328–329.

Moody, J.A. and Martin, D.A. (2001) Initial hydrologic and geomorphic response following a wildfire in the Colorado Front Range. *Earth Surface Processes and Landforms*, 26, 1049–1070.

Mooney, H.A. and Parsons, D.J. (1973) Structure and function of the California Chaparral – an example from San Dimas. *Ecological Studies*, 7, 83–112.

Moore, C.J. (2008) Synthetic polymers in the marine environment: a rapidly increasing, long-term threat. *Environmental Research*, 108, 131–139.

Moore, D.M. (1983) Human impact on island vegetation. W. Holzner, M.J.A. Werger and I. Ikusima (eds), *Man's Impact on Vegetation*. The Hague: Junk, 237–248.

Moore, J. (2000) Forest fire and human interaction in the early Holocene woodlands of Britain. *Palaeogeography, Palaeoclimatology, Palaeoecology*, 164, 125–137.

Moore, N. and Rojstaczer, S. (2001) Irrigation-induced rainfall and the Great Plains. *Journal of Applied Meteorology*, 40, 1297–1309.

Moore, N.W., Hooper, M.D. and Davis, B.N.K. (1967) Hedges, I. Introduction and reconnaissance studies. *Journal of Applied Ecology*, 4, 201–220.

Moore, P.D. (1973) Origin of blanket mires. *Nature*, 256, 267–269.

Moore, P.D. (1986) Unravelling human effects. *Nature*, 321, 204.

Moore, P.D. (2002) The future of cool temperate bogs. *Environmental Conservation*, 29, 3–20.

Moore, T.R. (1979) Land use and erosion in the Machakos Hills. *Annals of the Association of American Geographers*, 69, 419–431.

Morgan, G.S. and Woods, C.A. (1986) Extinction and the zoogeography of West Indian land mammals. *Biological Journal of the Linnean Society*, 28, 167–203.

Morgan, J.A. et al. (2004) Water relations in grassland and desert ecosystems exposed to elevated atmospheric CO_2. *Oecologia*, 140, 11–25.

Morgan, R.P.C. (1977) *Soil erosion in the United Kingdom: field studies in the Silsoe area, 1973–75*. National College of Agricultural Engineering, occasional paper, 4.

Morgan, R.P.C. (1979) *Soil Erosion*. London: Longman.

Morgan, R.P.C. (1995) *Soil Erosion and Conservation* (2nd edn). Harlow: Longman.

Morgan, R.P.C. (2005) *Soil Erosion and Conservation* (3rd edn). Oxford: Blackwell.

Morgan, W.B. and Moss, R.P. (1965) Savanna and forest in Western Nigeria *Africa*, 35, 286–293.

Morris, I. (2010) *Why the West Rules – For Now*. London: Profile Books.

Morton, R.A., Bernier, J.C., Barras, J.A. and Ferina, N.F. (2005) Historical subsidence and wetland loss in the Mississippi delta Plain. *Gulf Coast Association of Geological Societies Transactions*, 55, 555–571.

Morton, R.A., Bernier, J.C. and Barras, J.A. (2006) Evidence of regional subsidence and associated wetland loss induced by hydrocarbon production, Gulf Coast region, USA. *Environmental Geology*, 50, 261–274.

Mote, P.W., Hamlet, A.F., Clark, M.P. and Lettenmaier, D.P. (2005) Declining mountain snowpack in western North America. *Bulletin of the American Meteorological Society*, 86, 39–49.

Motyka, R.J., O'Neal, S., Connor, C.L. and Echelmeyer, K.A. (2002) Twentieth century thinning of Mendenhall Glacier, Alaska, and its relationship to climate, lake calving and glacier run-off. *Global and Planetary Change*, 35, 93–112.

Moulherat, C., Tengberg, M., Haquet, J.-F. and Mille, B. (2002) First evidence of cotton at Neolithic Mehrgarh, Pakistan: analysis of mineralized fibres from a copper bead. *Journal of Archaeological Science*, 29, 1393–1401.

Moulin, C., Lambert, C.E., Dulac, F. and Dayan, U. (1997) Control of atmospheric export of dust from North Africa by the North Atlantic Oscillation. *Nature*, 398, 691–694.

Mousavi, M.E., Irish, J.L., Frey, A.E., Olivera, F. and Edge, B.L. (2011) Global warming and hurricanes: the potential impact of hurricane intensification and sea level rise on coastal flooding. *Climatic Change*, 104, 575–597.

Moyle, P.B. (1976) Fish introductions in California: history and impact on native fishes. *Biological Conservation*, 9, 101–118.

Muhly, J.D. (1997) Artifacts of the Neolithic, Bronze and Iron Ages. In E.M. Myers (ed.), *The Oxford Encyclopaedia of Archaeology in the Near East* (Vol. 4). New York: Oxford University Press, 5–15.

Muhs, D.R. and Holliday, V.T. (1995) Evidence of active dune sand in the Great Plains in the 19th century from accounts of early explorers. *Quaternary Research*, 43, 198–208.

Muhs, D.R. and Maat, P.B. (1993) The potential response of eolian sands to greenhouse warming and precipitation reduction on the Great Plains of the United States. *Journal of Arid Environments*, 25, 351–361.

Mulitza, S. et al. (2010) Increase in African dust flux at the onset of commercial agriculture in the Sahel region. *Nature*, 466, 226–228.

Mullan, D., Favis-Mortlock, D. and Fealy, R. (2012) Addressing key limitations associated with modelling soil erosion under the impacts of future climate change. *Agricultural and Forest Meteorology*, 156, 18–30.

Mulrennan, M.E. and Woodroffe, C.D. (1998) Saltwater intrusion into the coastal plains of the lower Mary River, Northern Territory, Australia. *Journal of Environmental Management*, 54, 169–188.

Munson, S., Belnap, J. and Okin, G.S. (2011) Responses of wind erosion to climate-induced vegetation changes on the Colorado Plateau. *Proceedings of the National Academy of Sciences of the United States of America*, 108, 3854–3859.

Murdoch, W.W. (1975) Diversity, complexity, stability and pest control. *Journal of Applied Ecology*, 12, 795–807.

Murozumi, M., Chow, T.J. and Paterson, C. (1969) Chemical concentrations of pollutant lead aerosols, terrestrial dusts and sea salt in Greenland and Antarctic snow strata. *Geochimica et Cosmochimica Acta*, 33, 1247–1294.

Murton, R.K. (1971) *Man and Birds*. London; Collins.

Musk, L.F. (1991) The fog hazard. In A.H. Perry and L.J. Symons (eds), *Highway Meteorology*. London: Spon, 91–130.

Muturi, G.M., Mohren, G.M.J. and Kimani, J.N. (2009) Prediction of Prosopis species invasion in Kenya using geographical information system techniques. *African Journal of Ecology*, 48, 628–636.

Myers, N. (1979) *The Sinking Ark: A New Look at the Problem of Disappearing Species*. Oxford: Pergamon Press.

Myers, N. (1983) Conversion rates in tropical moist forests. In F.B. Golley (ed.), *Tropical Rain Forest Ecosystems*. Amsterdam: Elsevier Scientific, 289–300.

Myers, N. (1984) *The Primary Source: Tropical Forests and Our Future*. New York: Norton.

Myers, N. (1988) *Natural resource systems and human exploitation systems: physiobiotic and ecological linkages*. World Bank policy planning and research staff, environment department working paper, 12.

Myers, N. (1990) The biodiversity challenge: expanded hotspots analysis. *The Environmentalist*, 10 (4), 243–256.

Myers, N. (1992) Future operational monitoring of tropical forests: an alert strategy. In J.P. Mallingreau, R. da Cunha and C. Justice (eds), *Proceedings World Forest Watch Conference*, Sao Jose dos Campos, Brazil, 9–14.

Myers, N. and Kent, J. (2003) New consumers: the influence of affluence on the environment. *Proceedings of the National Academy of Sciences of the United States of America*, 100, 4963–4968.

Myers, N., Mittermeier, R.A., Mittermeier, C.G., da Fonseca, G.A.B. and Kent, J. (2000) Biodiversity hotspots for conservation priorities. *Nature*, 403, 853–858.

Myles, S. et al. (2011) Genetic structure and domestication history of the grape. *Proceedings of the National Academy of Sciences of the United States of America*, 108, 3530–3535.

Mylne, M.F. and Rowntree, P.R. (1992) Modelling the effects of albedo change associated with tropical deforestation. *Climatic Change*, 21, 317–343.

Mylona, S. (1996) Sulphur dioxide emissions in Europe 1880–1991 and their effect on sulphur concentrations and depositions. *Tellus B*, 48, 662–689.

Nadal-Romero, E., Lasanta, T. and García-Ruiz, J.M. (2012) Runoff and sediment yield from land under various uses in a Mediterranean mountain area: long-term results from an experimental station. *Earth Surface Processes and Landforms*, doi:10.1002/esp.3281.

Naeem, S. (2002) Biodiversity equals instability? *Nature*, 416, 23–24.

Naidoo, V., Wolter, K., Cuthbert, R. and Duncan, N. (2009) Veterinary diclofenac threatens Africa's endangered vulture species. *Regulatory Toxicology and Pharmacology*, 53, 205–208.

Naik, P.K. and Jay, D.A. (2011) Distinguishing human and climate influences on the Columbia River: changes in mean flow and sediment transport. *Journal of Hydrology*, 404, 259–277.

Najjar, R.G. (1999) The water balance of the Susquehanna River Basin and its response to climate change. *Journal of Hydrology*, 219, 7–19.

Nakagawa, K. (1996) Recent trends of urban climatological studies in Japan, with special emphasis on the thermal environments of urban areas. *Geographical review of Japan. Series B*, 69, 206–224.

Nakano, T. and Matsuda, I. (1976) A note on land subsidence in Japan. *Geographical Reports of Tokyo Metropolitan University*, 11, 147–162.

Nanson, G.C. and Tooth, S. (1999) Arid-zone rivers as indicators of climate change. In A.K. Singhvi and E. Derbyshire (eds), *Paleoenvironmental Reconstruction in Arid Lands*. New Delhi and Calcutta: Oxford and IBH, 75–216.

Nash, L.L. and Gleick, P.H. (1991) Sensitivity of streamflow in the Colorado Basin to climatic changes. *Journal of Hydrology*, 125, 221–241.

Nature Conservancy Council (1977) *Nature Conservation and Agriculture*. London: Her Majesty's Stationery Office.

Nature Conservancy Council (1984) *Nature Conservation in Great Britain*. Shrewsbury: Nature Conservancy Council.

Nawaz, M.F., Bourrié, G. and Trolard, F. (2012) Soil compaction impact and modelling. A review. *Agronomy for Sustainable Development*, doi:10.1007/s13593-011-0071-8.

Nearing, M.A. (2001) Potential changes in rainfall erosivity in the US with climate change during the 21st century. *Journal of Soil and Water Conservation*, 56, 229–232.

Nearing, W. et al. (2005) Modeling response of soil erosion and runoff to changes in precipitation and cover. *Catena*, 61, 131–154.

Neave, M., Rayburg, S. and Swan, A. (2009) River channel change following dam removal in an ephemeral stream. *The Australian Geographer*, 40, 235–246.

Neff, J.C. et al. (2008) Increasing eolian dust deposition in the western United States linked to human activity. *Nature Geosciences*, 1, 189–195.

Nelson, F.E. (2002) Climate change and hazard zonation in the Circum-Arctic Permafrost regions. *Natural Hazards*, 26, 203–225.

Nelson, F.E. and Anisimov, O.A. (1993) Permafrost zonation in Russia under anthropogenic climate change. *Permafrost and Periglacial Processes*, 4, 137–148.

Nelson, F.E., Anisimov, O.A. and Shiklomanov, N.I. (2001) Subsidence risk from thawing permafrost. *Nature*, 410, 889–890.

Nepstad, D.C., Stickler, C.M., Soares-Filho, B. and Merry, F. (2008) Interactions among Amazon land use, forests and climate: prospects for a near-term forest tipping point. *Philosophical Transactions of the Royal Society of London. Series B, Biological Sciences*, 363, 1737–1746.

Neris, J., Tejedor, M., Fuentes, J. and Jiménez, C. (2012) Infiltration, runoff and soil loss in Andisols affected by forest fire (Canary Islands, Spain). *Hydrological Processes*, doi:10.1002/hyp.9403.

Neronov, V.M., Lushchekina, A.A., Karimova, T.Y. and Arylova, N.Y. (2012) Population dynamics of a key steppe species in a changing world: the critically endangered saiga antelope. In M.J.A. Werger and M.A. van Staalduinen (eds), *Ecological Problems and Livelihoods in a Changing World*. doi:10.1007/978-94-007-3886_12.

Nesje, A., Lie, O. and Dahl, S.O. (2000) Is the North Atlantic Oscillation reflected in glacier mass balance records? *Journal of Quaternary Science*, 15, 587–601.

Nesje, A. and Dahl, S.O. (2000) *Glaciers and Environmental Change*. London: Arnold.

Nesje, A., Bakke, J., Dahl, S.O., Lie, O. and Matthews, J.A. (2008) Norwegian mountain glaciers in the past, present and future. *Global and Planetary Change*, 60, 10–27.

Neves, E.G., Petersen, J.B., Bartone, R.N. and da Silva, C.A. (2003) Historical and socio-cultural origins of Amazonian Dark earths. In J. Jehmann (ed.), *Amazonian Dark Earths: Origin, Properties, Management*. Dordrecht: Kluwer, 29–50.

New, M., Todd, M., Hulme, M. and Jones, P. (2001) Precipitation measurements and trends in the twentieth century. *International Journal of Climatology*, 21, 1899–1922.

New, M., Liverman, D., Schroeder, H. and Anderson, K. (2011) Four degrees and beyond: the potential for a global temperature increase of four degrees and its implications. *Philosophical Transactions. Series A, Mathematical, Physical, and Engineering Sciences*, 369, 6–19.

Newman, J.R. (1979) Effects of industrial pollution on wildlife. *Biological Conservation*, 15, 181–190.

Newman, W.S. and Fairbridge, R.W. (1986) The management of sealevel rise. *Nature*, 320, 319–321.

Newton, J.G. (1976) Induced and natural sinkholes in Alabama: continuing problem along highway corridors. In F.R. Zwanig (ed.), *Subsidence over Mines and Caverns*. Washington DC: National Academy of Sciences, 9–16.

Nichol, S.L., Augustinus, P.L., Gregory, M.R., Creese, R. and Horrocks, M. (2000) Geomorphic and sedimentary evidence of human impact on the New Zealand landscape. *Physical Geography*, 21, 109–132.

Nicholls, R.J., Hoozemans, F.M.J. and Marchand, M. (1999) Increasing flood risk and wetland losses due to global sea level rise: regional and global analyses. *Global Environmental Change*, 9, S69–S87.

Nicholls, R.J., Wong, P.P., Burkett, V.R., Codignotto, J.O., Hay, J.E., McLean, R.F., Ragoonaden, S. and Woodroffe, C. (2007) Coastal systems and low-lying areas. In M.L. Parry (ed.), *Climate Change 2007: Impacts, Adaptation and Vulnerability*. Cambridge: Cambridge University Press, 315–356.

Nicholls, R.J., Marinova, N., Lowe, J.A., Brown, S., Vellinga, P., de Gusmão, D., Hinkel, J. and Tol, R.S.J. (2011) Sea-level rise and its possible impacts given a 'beyond 4°C world' in the twenty-first century. *Philosophical Transactions. Series A, Mathematical, Physical, and Engineering Sciences*, 369, 161–181.

Nicholson, S.E. (1988) Land surface atmosphere interaction: physical processes and surface changes and their impact. *Progress in Physical Geography*, 12, 36–65.

Nicod, J. (1986) Facteurs physico-chimiques de l'accumulation des formations travertineuses. *Méditerranée*, 10, 161–164.

Nihlgård, B.J. (1997) Forest decline and environmental stress. In D. Brune, D.V. Chapman, M.D. Gwynne and J.M. Pacyna (eds), *The Global Environment*. Weinheim: VCH.

Nijssen, B., O'Donnell, G.M., Hamlet, A.F. and Lettenmaier, D.P. (2001) Hydrologic sensitivity of global rivers to climate change. *Climatic Change*, 50, 143–175.

Nikolskiy, P.A., Sulerzhitsky, L.D. and Pitulko, V.V. (2011) Last straw versus Blitzkrieg overkill: climate-driven changes in the Arctic Siberian mammoth population and the Late Pleistocene extinction problem. *Quaternary Science Reviews*, 30, 2309–2328.

Nikonov, A.A. (1977) Contemporary technogenic movements of the Earth's crust. *International Geology Review*, 19, 1245–1258.

Nitschke, C.R. and Innes, J.R. (2012) Potential effect of climate change on observed fire regimes in the Cordilleran forests of South-Central Interior, British Columbia. *Climatic Change*, doi:10.1007/s10584-012-0522-5.

Niu, F., Li, Z., Li, C., Lee, K.H. and Wang, M. (2010) Increase of wintertime fog in China: potential impacts of weakening of the Eastern Asian monsoon circulation and increased aerosol loading. *Journal of Geophysical Research*, 115, D00k20.

Nobre, C.A., Borma, L.D.S. (2009) 'Tipping points' for the Amazon forest. *Current Opinion in Environmental Sustainability*, 1, 28–36.

Nobre, C.A., Dias, M.A.S., Culf, A.D., Polcher, J., Gash, J.H.C., Marengo, J.A. and Avissar, R. (2004) The Amazonian climate. In P. Kabat, M. Claussen, P.A. Dirmeyer, J.H.C. Gash, L.B. de Guenni, M. Meybeck, R.A. Pielke, C.S. Vörösmarty, R.W.A. Hutjes and S. Lütkemeier (eds), *Vegetation, Water, Humans and the Climate*. New York: Springer, 79–92.

Nordstrom, K.F. (1994) Beaches and dunes of human-altered coasts. *Progress in Physical Geography*, 18, 497–516.

Nordstrom, K.F. and Hotta, S. (2004) Wind erosion from cropland in the USA: a review of problems, solutions and prospects. *Geoderma*, 121, 157–167.

Norris, S. (2001) Thanks for all the fish. *New Scientist*, 29th September, 36–39.

Noss, R.F. (2011) Between the devil and the deep blue sea: Florida's unenviable position with respect to sea level rise. *Climatic Change*, 107, 1–16.

Notebaert, B. and Verstraeten, G. (2010) Sensitivity of West and Central European river systems to environmental changes during the Holocene: a review. *Earth-Science Reviews*, 103, 163–182.

Noti, R., van Leeuwen, J.F.N., Colombaroli, D., Vescovi, E., Pasta, S., La Mantia, T. and Tinner, W. (2009) Mid- and late-Holocene vegetation and fire history at Biviere di Gela, a coastal lake in southern Sicily, Italy. *Vegetation History and Archaeobotany*, 18, 371–387.

Novotny, E.V., Murphy, D. and Stefan, H.G. (2008) Increase of urban lake salinity by road deicing salt. *The Science of the Total Environment*, 406, 131–144.

Nowlis, J.S., Roberts, C.M., Smith, A.H. and Siirila, E. (1997) Human-enhanced impacts of a tropical storm on nearshore coral reefs. *Ambio*, 26, 515–521.

Noy-Meir, I. (1974) Stability in arid ecosystems and effects of men on it. *Proceedings of the 12th International Congress of Ecology*, Wageningen, 220–225.

Nriagu, J.O. (1979) Global inventory of natural and anthropogenic emissions of trace metals in the atmosphere. *Nature*, 279, 409–411.

Nriagu, J.O. and Pacyna, J.M. (1988) Quantitative assessment of worldwide contamination of air, water and soils by trace metals. *Nature*, 337, 134–139.

Nuckolls, A.E., Wurzburger, N., Ford, C.R., Hendrick, R.L., Vose, J.M. and Kloeppel, B.D. (2009) Hemlock declines rapidly with hemlock wooly adelgid infestation: impacts on the carbon cycle of southern Appalachian forests. *Ecosystems*, 12, 179–190.

Nunn, P.D. (1991) *Human and natural impacts on Pacific island environments*. Occasional paper of the East West Environment and Policy Institute, 13. Honolulu.

Nutalaya, P. and Ran, J.L. (1981) Bangkok: the sinking metropolis. *Episodes*, 4, 3–8.

Nutalaya, P., Yong, R.N., Chumnankit, T. and Buapeng, S. (1996) Land subsidence in Bangkok during 1978–1988. In J.D. Milliman and B.U. Haq (eds), *Sea Level Rise and Coastal Subsidence*. Dorcrecht: Kluwer, 105–130.

Nye, P.H. and Greenland, D.J. (1964) Changes in the soil after clearing tropical forest. *Plant and Soil*, 21, 101–112.

Nyssen, J. et al. (2009) Desertification? Northern Ethiopia re-photographed after 140 years. *The Science of the Total Environment*, 407, 2749–2755.

Oba, G., Post, E. and Stenseth, N.C. (2001) Sub-Saharan desertification and productivity are linked to hemispheric climate variability. *Global Change Biology*, 7, 241–246.

Oberle, M. (1969) Forest fires: suppression policy has its ecological drawbacks. *Science*, 165, 568–571.

O'Connor, C.D., Garfin, G.M., Falk, D.A. and Swetnam, T.W. (2011) Human pyrogeography: a new synergy of fire, climate and people is reshaping ecosystems across the globe. *Geography Compass*, 5, 329–350.

Oechel, J.W.C., Hastings, S.J., Vourlitis, G.L., Jenkins, M.A. and Hinkson, C.L. (1995) Direct effects of elevated CO_2 in Chaparral and Mediterranean-type ecosystems. In J.M. Moreno and W.C. Oechel (eds), *Global Change and Mediterranean-Type Ecosystems*. New York: Springer, 58–75.

Oerlemans, J. (1993) Possible changes in the mass balance of the Greenland and Antarctic ice sheets and their effects on sea level. In R.A. Warwick, E.M. Barrows and T.M.L. Wigley (eds), *Climatic and Sea Level Change: Observations, Projections and Implications*. Cambridge: Cambridge University Press, 144–161.

Oerlemans, J. (1994) Quantifying global warming from the retreat of glaciers. *Science*, 264, 243–245.

Oerlemans, J., Anderson, B., Hubbard, A., Huybrechts, P., Jóhannesson, T., Knap, W.H., Schmeits, M., Stroeven, A.P., van de Wal, R.S.W., Wallinga, J. and Zuo, Z. (1998) Modelling the response of glaciers to climate warming. *Climate Dynamics*, 14, 267–274.

Oke, T.R. (1978) *Boundary Layer Climates*. London: Methuen.

Olley, J.M. and Wasson, R.J. (2003) Changes in the flux of sediment in the Upper Murrumbidgee catchment, southeastern Australia, since European settlement. *Hydrological Processes*, 17, 3307–3320.

Olsson, L., Eklundh, L. and Ardö, J. (2005) A recent greening of the Sahel – trends, patterns and potential causes. *Journal of Arid Environments*, 63, 556–566.

Oltmans, S.J. et al. (2006) Long-term changes in tropospheric ozone. *Atmospheric Environment*, 40, 3156–3173.

O'Neal, M.R., Nearing, M.A., Vining, R.C., Southworth, J. and Pfeifer, R.A. (2005) Climate change impacts on soil erosion in Midwest United States with changes in crop management. *Catena*, 61, 165–184.

Oppenheimer, M. (1998) Global warming and the stability of the West Antarctic ice sheet. *Nature*, 393, 325–332.

Oppenheimer, S. (2003) *Out of Eden. The Peopling of the World*. London: Constable.

Orr, J.C., Pantoja, S. and Pörtner, H.-O. (2005a) Introduction to special section: the ocean in a high-CO2 world. *Journal of Geophysical Research*, 110 (C), doi: 10 1029/2005 JC 003086.

Orr, J.C. et al. (2005b) Anthropogenic ocean acidification over the twenty-first century and its impact on organisms. *Nature*, 437, 681–686.

Orts, W.J., Roa-Espinosa, A., Sojka, R.E., Glenn, G.M., Imam, S.H., Erlacher, K. and Pedersen, J.S. (2007) Use of synthetic polymers and biopolymers for soil stabilization in agricultural, construction and military applications. *Journal of Materials in Civil Engineering*, 19, 58–66.

Osborn, T.J., Hulme, M., Jones, P.D. and Basnett, T.A. (2000) Observed trends in the daily intensity of United Kingdom precipitation. *International Journal of Climatology*, 20, 347–364.

Osterkamp, T.E. and Romanovsky, V.E. (1999) Evidence for warming and thawing of discontinuous permafrost in Alaska. *Permafrost and Periglacial Processes*, 10, 17–37.

O'Sullivan, P.E., Coard, M.A. and Pickering, D.A. (1982) The use of laminated lake sediments in the estimation and calibration of erosion rates. *Publication of the International Association of Hydrological Science*, 137, 385–396.

Otterman, J. (1974) Baring high albedo soils by overgrazing: a hypothesised desertification mechanism. *Science*, 186, 531–533.

Otu, M.K., Ramlal, P., Wilkinson, P., Hall, R.I. and Hecky, R.E. (2011) Paleolimnological evidence for the effects of recent cultural eutrophication during the last 200 years in Lake Malawi, East Africa. *Journal of Great Lakes Research*, 37 (Suppl. 1), 61–74.

Outram, A.K. et al. (2009) The earliest horse harnessing and milking. *Science*, 323, 1332–1335.

Overland, J.E., Wood, K.R. and Wang, M. (2011) Warm Arctic – cold continents: climate impacts of the newly open Arctic Sea. *Polar Research*, doi:10.3402/polar.v30i0.15787.

Overpeck, J.T., Rind, D. and Goldberg, R. (1990) Climate-induced changes in forest disturbance and vegetation. *Nature*, 343, 51–53.

Owen, L.A., Kamp, U., Khattak, G.A., Harp, E.L., Keefer, D.K. and Bauer, M.A. (2008) Landslides triggered by the 8 October 2005 Kashmir earthquake. *Geomorphology*, 94, 1–9.

Oxley, D.J., Fenton, M.B. and Carmody, G.R. (1974) The effects of roads on populations of small mammals. *Journal of Applied Ecology*, 11, 51–59.

Ozenda, P. and Borel, J.L. (1990) The possible responses of vegetation to a global climatic change. In M.M. Boer and R.S. de Groot (eds), *Landscape – Ecological Impact of Climatic Change*. Amsterdam: IOS Press, 221–249.

Özkan, H., Willcox, G., Graner, A., Salamini, F. and Kilian, B. (2011) Geographic distribution and domestication of wild emmer wheat (Triticum dicoccoides). *Genetic Resources and Crop Evolution*, 58, 11–53.

Page, H. (1982) Some notes on the geomorphological and vegetational history of the saltings at Brean. *Somerset Archaeology and Natural History*, pp. 120–125.

Page, M.J. and Trustrum, N.A. (1997) A late Holocene lake sediment record of the erosion response to land use change in a steepland catchment, New Zealand. *Zeitschrift für Geomorphologie*, 41, 36992.

Pakeman, R.J., Marrs, R.H., Howard, D.C., Barr, C.J. and Fuller, R.M. (1996) The bracken problem in Great Britain: its present extent and future changes. *Applied Geography*, 16, 65–86.

Palmer, T.N. and Räisänen, J. (2002) Quantifying the risk of extreme seasonal precipitation events in a changing climate. *Nature*, 415, 512–514.

Pandolfi, J.M. et al. (2005) Are U.S.coral reefs on the slippery slope to slime? *Science*, 307, 1725–1726.

Panel on Weather and Climate Modification (1966) *Weather and Climate Modification Problems and Prospects*. Washington DC: National Academy of Sciences.

Pang, J.-F. et al. (2009) mtDNA data indicate a single origin for dogs south of Yangtze River, less than 16,300 years ago, from numerous wolves. *Molecular Biology and Evolution*, 26, 2849–2864.

Parfitt, S.A. et al. (2010) Early Pleistocene human occupance at the edge of the boreal zone in northwest Europe. *Nature*, 466, 229–233.

Parizek, B.R. and Alley, R.B. (2004) Implications of increased Greenland surface melt under global-warming scenarios: ice-sheet simulations. *Quaternary Science Reviews*, 23, 1013–1027.

Park, C.C. (1977) Man-induced changes in stream channel capacity. In K.J. Gregory (ed.), *River Channel Change*. Chichester: Wiley, 121–144.

Park, C.C. (1987) *Acid Rain: Rhetoric and Reality*. London: Methuen.

Park, R.A., Armentano, T.V. and Cloonan, C.L. (1986) Predicting the effects of sea level rise on coastal wetlands. In J.G. Titus (ed.), *Effects of Changes in Stratospheric Ozone and Global Climate* (Vol. 4, *Sea level rise*). Washington DC: UNEP/USEPA, 129–152.

Parker, A.G., Goudie, A.S., Anderson, D.E., Robinson, M.A. and Bonsall, C. (2002) A review of the mid-Holocene elm decline in the British Isles. *Progress in Physical Geography*, 26, 1–45.

Parshall, T. and Foster, D.R. (2002) Fire on the New England landscape: regional and temporal variation, cultural and environmental controls. *Journal of Biogeography*, 29, 1305–1317.

Parson, E.A., Carter, L., Anderson, P., Wang, B. and Weller, G. (2001) Potential consequences of climate variability and change for Alaska. In National Assessment Synthesis Team (ed.), *Climate Change Impacts on the United States: The Potential Consequences of Climate Variability and Change*. Cambridge: Cambridge University Press, 283–312.

Parsons, A.J., Abrahams, A.D. and Wainwright, J. (1996) Responses of interrill runoff and erosion rates to vegetation change in southern Arizona. *Geomorphology*, 14, 311–317.

Parsons, J.J. (1960) Fog drip from coastal stratus. *Weather*, 15, 58.

Pasternack, G.B., Brush, G.S. and Hilgartner, W.B. (2001) Impact of historic land-use change on sediment delivery to a Chesapeake Bay subestuarine delta. *Earth Surface Processes and Landforms*, 26, 409–427.

Paul, K.I., Poglase, P.J., Nyakuengama, J.G. and Khanna, P.K. (2002) Change in soil carbon following afforestation. *Forest Ecology and Management*, 168, 241–257.

Pautasso, M., Döring, T.F., Garbelotto, M., Pellis, L. and Jeger, M.J. (2012) Impact of climate change on plant diseases – opinions and trends. *European Journal of Plant Pathology*, doi: 10.1007/s10658-012-9936-1.

Pearson, A.J., Snyder, N.P. and Collins, M.J. (2011) Rates and processes of channel response to dam removal with a sand-filled impoundment. *Water Resources Research*, 47, W08504.

Pearson, R.G. and Dawson, T.P. (2003) Predicting the impacts of climate change on the distribution of species: are bioclimate envelope models useful? *Global Ecology and Biogeography*, 12, 361–371.

Pechony, O. and Shindell, D.T. (2010) Driving forces of global wildfires over the past millennium and the forthcoming century. *Proceedings of the National Academy of Sciences of the United States of America*, 107, 19167–19170.

Peck, A.J. (1983) Response of groundwater to clearing in western Australia. In *Papers, international conference on groundwater and man*, 327–335.

Peck, A.J. and Halton, T. (2003) Salinity and the discharge of salts from catchments in Australia. *Journal of Hydrology*, 272, 191–202.

Pederson, G.T. et al. (2011) The unusual nature of recent snowpack declines in the North American cordillera. *Science*, 333, 332–335.

Peet, R.K., Glenn-Lewin, D.C. and Wolf, J.W. (1983) Prediction of man's impact on plant species diversity. In W. Holzner, M.J.A. Werger and I. Ikusima (eds), *Man's Impact on Vegetation*. The Hague: Junk, 41–54.

Peglar, S.M. and Birks, H.J.B. (1993) The mid-Holocene Ulmus fall at Diss Mere, Norfolk, south-east England – disease and human impact? *Vegetation History and Archaeology*, 2, 61–68.

Peierls, B.L., Caraco, N.F., Pace, M.L. and Cole, J.J. (1991) Human influence on river nitrogen. *Nature*, 350, 386.

Pelejero, C., Calvo, E. and Hoegh-Guldberg, O. (2010) Paleoperspectives on ocean acidification. *Trends in Ecology & Evolution*, 25, 332–344.

Peng, S. et al. (2012) Surface urban heat island across 419 global big cities. *Environmental Science & Technology*, 46, 696–703.

Pennington, W. (1981) Records of a lake's life in time: the sediments. *Hydrobiologia*, 79, 197–219.

Pereira, H.C. (1973) *Land Use and Water Resources in Temperate and Tropical Climates*. Cambridge: Cambridge University Press.

Perla, R. (1978) Artificial release of avalanches in North America. *Arctic and Alpine Research*, 10, 235–240.

Peters, J.H. (ed.) (1998) *Artificial Recharge of Groundwater*. Amsterdam: Swetzs and Zeitlinger.

Peters, R.L. (1988) The effect of global climatic change on natural communities. In E.O. Wilson (ed.), *Biodiversity*. Washington DC: National Academy Press, 450–461.

Peterson, B.J., Homes, R.M., McClelland, J.W., Vörösmarty, C.J., Lammers, R.B., Shiklomanov, A.I., Shiklomanov, I.A. and Rahmstorf, S. (2002) Increasing river discharge to the Arctic Ocean. *Science*, 298, 2171–2173.

Peterson, C.J. (2000) Catastrophic wind damage to North American forests and the potential impact of climate change. *The Science of the Total Environment*, 262, 287–311.

Pethick, J. (1993) Shoreline adjustments and coastal management: physical and biological processes under accelerated sea level rise. *The Geographical Journal*, 159, 162–168.

Pethick, J. (2001) Coastal management and sea level rise. *Catena*, 42, 307–322.

Petit-Maire, N., Burollet, P.F., Ballais, J.-L., Fontugne, M., Rosso, J.-C. and Lazaar, A. (1999) Paléoclimats Holocènes du Sahara septentionale. Dépôts lacustres et terrasses alluviales en bordure du Grand Erg Oriental à l'extrême – Sud de la Tunisie. *Comptes rendus Académie des Sciences*, Series 2, 312, 1661–1666.

Petley, D.N., Hearn, G.J., Hart, A., Rosser, N., Dunning, S.A., Oven, K. and Mitchell, W.A. (2007) Trends in landslide occurrence in Nepal. *Natural Hazards*, 43, 23–44.

Petts, G.E. (1979) Complex response of river channel morphology subsequent to reservoir construction. *Progress in Physical Geography*, 3, 329–362.

Petts, G.E. (1985) *Impounded Rivers: Perspectives for Ecological Management*. Chichester: Wiley.

Petts, G.E. and Lewin, J. (1979) Physical effects of reservoirs on river systems. In G.E. Hollis (ed.), *Man's Impact on the Hydrological Cycle in the United Kingdom*. Norwich: Geobooks, 79–91.

Pezza, A.B. and Simmonds, I. (2005) The first South Atlantic hurricane: unprecedented blocking, low shear and climate change. *Geophysical Research Letters*, 32, L15712.

Pfisterer, A.B. and Schmid, B. (2002) Diversity-dependent production can decrease the stability of ecosystem functioning. *Nature*, 416, 84–86.

Phien-Wej, N., Giao, P.H. and Nutalaya, P. (2006) Land subsidence in Bangkok, Thailand. *Engineering Geology*, 82, 187–201.

Pickering, C.M. and Hill, W. (2007) Impacts of recreation and tourism on plant biodiversity and vegetation in protected areas of Australia. *Journal of Environmental Management*, 85, 791–800.

Pickersgill, B. (2007) Domestication of plants in the Americas: insights from Mendelian and molecular genetics. *Annals of Botany*, 100, 925–940.

Pierson, F.B., Carlson, D.H. and Spaeth, K.E. (2002) Impacts of wildfire on soil hydrological properties of steep sagebrush-steepe rangeland. *International Journal of Wildland Fire*, 11, 145–151.

Pierson, F.B., Williams, C.J., Hardegree, S.P., Weltz, M.A., Stone, J.J. and Clarke, P.E. (2011) Fire, plant invasions, and erosion events on Western Rangelands. *Rangeland Ecology and Management*, 64, 439–449.

Pimentel, D. (1976) Land degradation: effects on food and energy resources. *Science*, 194, 149–155.

Pimentel, D. (2003) *Biological Invasions*. Washington DC: CRC Press.

Pimentel, D., Harvey, C., Resosuddarmo, P., Sinclair, K., Kurz, D., McNair, M., Crist, S., Shpritz, L., Fitton, L., Saffouri, R. and Blair, R. (1995) Environmental and economic costs of soil erosion and conservation benefits. *Science*, 267, 1117–1122.

Pinker, R.T., Zhang, B. and Dutton, E.G. (2005) Do satellites detect trends is surface solar radiation? *Science*, 308, 850–854.

Piotto, D., Montagnini, F., Thomas, W., Ashton, M. and Oliver, C. (2009) Forest recovery after swidden cultivation across a 40-year chronosequence in the Atlantic forest of southern Bahia, Brazil. *Plant Ecology*, 205, 261–272.

Pirazzoli, P.A. (1996) *Sea Level Changes: The Last 20,000 Years*. Chichester: Wiley.

Pittock, A.B. and Wratt, D. (2001) Australia and New Zealand. In J.J. McCarthy, O.F. Canziani, N.A. Leary, D.J. Dokken and K.S. White (eds), *Climate Change 2001: Impacts, Adaptation and Vulnerability*. Cambridge: Cambridge University Press, 591–639.

Pluhowski, E.J. (1970) Urbanization and its effects on the temperature of the streams on Long Island, New York. *United States Geological Survey professional paper*, 627-D.

Poesen, J.W., Torri, D. and Bunte, K. (1994) Effects of rock fragments on soil erosion by water and different spatial scales: a review. *Catena*, 23, 141–166.

Poesen, J.W.E., Nachtergaele, J., Verstraeten, G. and Valentin, C. (2003) Gully erosion and environmental change: importance and research needs. *Catena*, 50, 91–133.

Pokhrel, Y.N., Hanasaki, N., Yeh, P.J.-F., Yamada, T.J., Kanae, S. and Oki, T. (2012) Model estimates of sea-level change due to anthropogenic impacts on terrestrial water storage. *Nature Geoscience*, 5, 389–392.

Polidoro, B.A. et al. (2010) The loss of species: mangrove extinction risk and geographic areas of global concern. *PLoS ONE*, 5, e10095.

Poloczanska, E.S., Limpus, C.J. and Hays, G.C. (2010) Vulnerability of marine turtles to climate change. *Advances in Marine Biology*, 56, 151–211.

Polvani, L.M., Waugh, D.W., Correa, G.J.P. and Son, S.-W. (2011) Stratospheric ozone depletion: the main driver of twentieth-century atmospheric circulation changes in the Southern Hemisphere. *Journal of Climate*, 24, 795–812.

Ponting, C. (1991) *A Green History of the World*. London: Penguin.

Ponting, C. (2007) *A New Green History of the World*. London: Penguin.

Poole, G.C. and Berman, C.H. (2001) An ecological prespective on in-stream temperature: natural heat dynamics and mechanisms of human-caused thermal degradation. *Environmental Management*, 27, 787–802.

Poore, M.E.D. (1976) The values of tropical moist forest ecosystems. *Unasylva*, 28, 127–143.

Pope, J.C. (1970) Plaggen soils in the Netherlands. *Geoderma*, 4, 229–255.

Post, W.M. and Kwon, K.C. (2000) Soil carbon sequestration and land-use change: processes and potential. *Global Change Biology*, 6, 317–327.

Potter, G.L., Ellsaesser, H.W., MacCracken, M.C. and Luther, F.M. (1975) Possible climatic impact of tropical deforestation. *Nature*, 258, 697–698.

Potter, G.L., Ellsaesser, H.W., MacCracken, M.C. and Ellis, J.C. (1981) Albedo change by man: test of climatic effects. *Nature*, 291, 47–49.

Powell, M. (1985) Salt, seed and yields in Sumerian agriculture: a critique of progressive salinization. *Zeitschrift für Assyrologie und Vorderasiatische Archaologie*, 75, 7–38.

Powlesland, R.G. (2009) Impacts of wind farms on birds: a review. *Science for Conservation*, 289, 1–51.

Prats, J., Val, R., Armengol, J. and Dolz, J. (2010) Temporal variability in the thermal regime of the lower Ebro River (Spain) and alteration due to anthropogenic factors. *Journal of Hydrology*, 387, 105–118.

Price, M. and Reed, D.W. (1989) The influence of mains leakage and urban drainage on groundwater levels beneath conurbations in the United Kingdom. *Proceedings of the Institution of Civil Engineers*, 86, 31–39.

Price, M.F. (1989) Global change: defining the ill-defined. *Environment*, 31 (8), 18–20, 42–44.

Price, S.J., Ford, J.R., Cooper, A.H. and Noal, C. (2011) Humans as major geological and geomorphological agents in the Anthropocene: the significance of artificial ground in Great Britain. *Philosophical Transactions. Series A, Mathematical, Physical, and Engineering Sciences*, 369, 1056–1084.

Prince, H.C. (1959) Parkland in the Chilterns. *Geographical Review*, 49, 18–31.

Prince, H.C. (1962) Pits and ponds in Norfolk. *Erdkunde*, 16, 10–31.

Prince, H.C. (1964) The origin of pits and depressions in Norfolk. *Geography*, 49, 15–32.

Prince, H.C. (1979) Marl pits or dolines of the Dorset Chalklands? *Transactions of the Institute of British Geographers*, 4, 116–117.

Pritchard, H.D. and Vaughan, D.G. (2007) Widespread acceleration of tidewater glaciers on the Antarctic Peninsula. *Journal of Geophysical Research*, 112, F03S29.

Proffitt, M.H., Margitan, J.J., Kelly, K.K., Loewenstein, M., Podolske, J.R. and Chan, K.R. (1990) Ozone loss in the Arctic polar vortex inferred from high-altitude aircraft measurements. *Nature*, 347, 31–33.

Prokopovich, N.P. (1972) Land subsidence and population growth. *24th International Geological Congress Proceedings*, 13, 44–54.

Prowse, T.D. and Beltaos, S. (2002) Climatic control of river-ice hydrology: a review. *Hydrological Processes*, 16, 805–822.

Prudhomme, C. et al. (2012) The drying up of Britain? A national estimate of changes in seasonal river flows from 11 Regional Climatic Model simulations. *Hydrological Processes*, 26, 1115–1118.

Puddu, G., Falucci, A. and Maiorano, L. (2011) Forest changes over a century in Sardinia: implications for conservation in a Mediterranean hotspot. *Agroforestry Systems*, doi:10.1007/s10457-011-9443-y.

Purich, A. and Son, S.-W. (2012) Impact of Antarctic ozone depletion and recovery on Southern Hemisphere precipitation, evaporation, and extreme changes. *Journal of Climate*, 25, 3145–3154.

Pushkina, D. and Raia, P. (2008) Human influence on distribution and extinctions of the late Pleistocene Eurasian megafauna. *Journal of Human Evolution*, 54, 769–782.

Pyne, S.J. (1982) *Fire in America – A Cultural History of Wildland and Rural Fire*. Princeton: Princeton University Press.

Qadir, M., Ghafoor, A. and Murtaza, G. (2000) Amelioration strategies for saline soils: a review. *Land Degradation and Development*, 11, 501–521.

Qadir, M., Noble, A.D., Kureshi, A.S., Gupta, R.K., Yuldashev, T. and Karimov, A. (2009) Salt-induced land and water degradation in the Ara Sea basin: a challenge to sustainable agriculture in Central Asia. *Natural Resources Forum*, 33, 134–149.

Qiu, J. (2012) Evidence mounts for dam-quake link. *Science*, 336, 291.

Quinton, J.N., Edwards, G.M. and Morgan, R.P.C. (1997) The influence of vegetation species and plant properties on runoff and soil erosion: results froma rainfall simulation study in south east Spain. *Soil Use and Management*, 13, 143–148.

Rackham, O. (1980) *Ancient Woodland*. London: Arnold.

Rackham, O. and Moody, J. (1996) *The Making of the Cretan Landscape*. Manchester: Manchester University Press.

Radić, V. and Hock, R. (2011) Regionally differentiated contribution of mountain glaciers and ice caps to future sea-level rise. *Nature Geoscience*, 4, 91–94.

Radivojević, M., Rehren, T., Pernicka, E., Šljivar, D., Brauns, M. and Borić, D. (2010) On the origins of extractive metallurgy: new evidence from Europe. *Journal of Archaeological Science*, 37, 2775–2787.

Radley, J. (1962) Peat erosion on the high moors of Derbyshire and west Yorkshire. *East Midland Geographer*, 3, 40–50.

Rahm, D. (2011) Regulating hydraulic fracturing in shale gas plays: the case of Texas. *Energy Policy*, 39, 2974–2981.

Rahmstorf, S. (2007) A semi-empirical approach to projecting future sea-level rise. *Science*, 315, 368–370.

Raison, R.J. (1979) Modification of the soil environment by vegetation fires with particular reference to nitrogen transformation: a review. *Plant and Soil*, 51, 73–108.

Raji, B.A., Utovbisere, E.O. and Momodu, A.B. (2004) Impact of sand dune stabilization structures on soil and yield of millet in the semi-ard region of NW Nigeria. *Environmental Monitoring and Assessment*, 99, 181–196.

Ramanathan, V. (1988) The greenhouse theory of climate change: a test by an inadvertent global experiment. *Science*, 240, 293–299.

Ramanathan, V. and Carmichael, G. (2008) Global and regional climate changes due to black carbon. *Nature Geoscience*, 1, 221–227.

Ramankutty, N. and Foley, A. (1999) Estimating historical changes in global land cover: croplands from 1700 to 1992. *Global Biogeochemical Cycles*, 13, 997–1027.

Ramankutty, N., Heller, E. and Rhemtulla, J. (2010) Prevailing myths about agricultural abandonment and forest regrowth in the United States. *Annals of the Association of American Geographers*, 100, 502–512.

Ramirez, E. et al. (2001) Small glaciers disappearing in the tropical Andes: a case study of Bolivia: Glacier Chacaltya (16°S). *Journal of Glaciology*, 47, 187–194.

Ramirez-Llodra, E. et al. (2011) Man and the Last Great Wilderness: human impact on the deep sea. *PLoS ONE*, 6 (8), e 22588.

Ranta, P. (2001) Changes in urban lichen diversity after a fall in sulphur dioxide levels in the city of Tampere, SW Finland. *Annales Botanici Fennici*, 38, 295–304.

Ranwell, D.S. (1964) *Spartina* salt marshes in southern England, II: rate and seasonal pattern of sediment accretion. *The Journal of Ecology*, 52, 79–94.

Ranwell, D.S. and Boar, R. (1986) *Coast Dune Management Guide*. Monks Wood: Institute of Terrestrial Ecology.

Rao, K.N., Subraelu, P., Naga Kumar, K.Ch.V., Demudu, G., Hema Malini, B., Rajawat, A.S. and Ajai, (2010) Impacts of sediment retention by dams on delta shoreline recession: evidences from the Krishna and Godavari deltas, India. *Earth Surface Processes and Landforms*, 35, 817–827.

Rapp, A. (1974) *A review of desertization in Africa – water, vegetation and man*. Secretariat for International Ecology, Stockholm, report no. 1.

Rapp, A., Murray-Rust, D.H., Christansson, C. and Berry, L. (1972) Soil erosion and sedimentation in four catchments near Dodoma, Tanzania. *Geografiska Annaler*, 54A, 255–318.

Rapp, A., Le Houérou, H.N. and Lundholm, B. (1976) Can desert encroachment be stopped? *Ecological Bulletin*, 24.

Rasid, H. (1979) The effects of regime regulation by the Gardiner Dam on downstream geomorphic processes in the South Saskatchewan River. *The Canadian Geographer*, 23, 140–158.

Rasmussen, K., Fog, B. and Masden, J.E. (2001) Desertification in reverse? Observations from northern Burkina Faso. *Global Environmental Change*, 11, 271–282.

Ratcliffe, D.A. (1974) Ecological effects of mineral exploitation in the United Kingdom and their significance to nature conservation. *Proceedings of the Royal Society of London*, 339A, 355–372.

Ravi, S., D'Odorico, P., Wang, J., White, C.S., Okin, G.S., Macko, S.A. and Collins, S.L. (2009) Post-fire resource redistribution in desert grasslands: a possible negative feedback on land degradation. *Ecosystems*, 12, 434–444.

Ray, C., Hayden, B.P., Bulger, A.J. and McCormick-Ray, G. (1992) Effects of global warming on the biodiversity of coastal-marine zones. In R.L. Peters and T.E. Lovejoy (eds), *Global Warming and Biological Diversity*. New Haven: Yale University Press, 91–102.

Reale, O. and Dirmeyer, P. (2000) Modelling the effects of vegetation on Mediterranean climate during the Roman Classical Period: part I: climate history and model sensitivity. *Global and Planetary Change*, 25, 163–184.

Reclus, E. (1871) *The Earth* (2 vols). London: Chapman & Hall.

Reclus, E. (1873) *The Ocean, Atmosphere and Life*. New York: Harper and Brothers.

Reed, C.A. (1970) Extinction of mammalian megafauna in the old world late Quaternary. *Bioscience*, 20, 284–288.

Reed, D.J. (1990) The impact of sea level rise on coastal salt marshes. *Progress in Physical Geography*, 14, 465–481.

Reed, D.J. (1995) The response of coastal marshes to sea level rise: survival or submergence? *Earth Surface Processes and Landforms*, 20, 39–48.

Reed, D.J. (2002) Sea-level rise and coastal marsh sustainability: geological and ecological factors in the Mississippi delta plain. *Geomorphology*, 48, 233–243.

Reed, L.A. (1980) Suspended-sediment discharge, in five streams near Harrisburg, Pennsylvania, before, during and after highway construction. *United States Geological Survey water supply paper*, 2072.

Rees, H.G. and Collins, D.N. (2006) Regional differences in response of flow in glacier-fed Himalayan rivers to climatic warming. *Hydrological Processes*, 20, 2157–2169.

Reeve, D., Chadwick, A. and Fleming, C. (2012) *Coastal Engineering* (2nd edn). London: Spon Press.

Regalado, A. (2010) Brazil says rate of deforestation in Amazon continues to plunge. *Science*, 329, 1270–1271.

Reheis, M.C. (1997) Dust deposition downwind of Owens (dry) Lake, 1991–1994: preliminary findings. *Journal of Geophysical Research*, 102, 25998–26008.

Reichert, B.K., Bengston, L. and Oerlemans, J. (2001) Midlatitude forcing mechanism for glacier mass balance investigated using general circulation models. *Journal of Climate*, 14, 3767–3784.

Reij, C., Scoones, I. and Toulmin, C. (eds) (1996) *Sustaining the Soil: Indigenous Soil and Water Conservation in Africa*. London: Earthscan.

Remondo, J., Soto, J., González-Díez, A., Díaz de Terán, J.R. and Cendrero, A. (2005) Human impact on geomorphic processes and hazards in mountain areas in northern Spain. *Geomorphology*, 66, 69–84.

Renard, K.G. and Freid, J.R. (1994) Using monthly precipitation data to estimate the R factor in the revised USLE. *Journal of Hydrology*, 157, 287–306.

Renberg, I. and Hellberg, T. (1982) The pH history of lakes in SW Sweden, as calculated from the subfossil diatom flora of the sediments. *Ambio*, 11, 30–33.

Renfrew, C. (2006) Inception of agriculture and rearing in the Middle East. *Comptes Rendus Palevol*, 5, 395–404.

Revell, D.L., Battalio, R., Spear, B., Ruggiero, P. and Vandever, J. (2011) A methodology for predicting future coastal hazards due to sea-level rise on the California Coast. *Climatic Change*, 109 (Suppl.), S251–S276.

Revelle, R.R. and Waggoner, P.E. (1983) Effect of a carbon dioxide-induced climatic change on water supplies in the western United States. In Carbon Dioxide Assessment Committee (ed.), *Changing Climate*. Washington DC: National Academy Press, 419–432.

Rhemtulla, J.M., Mladenoff, D.J. and Clayton, M.K. (2009) Legacies of historical land use on regional forest composition and structure in Wisconsin, USA (mid-1800s – 1930s – 2000s). *Ecological Applications*, 19, 1061–1078.

Rhoades, J.D. (1990) Soil salinity – causes and controls. In A.S. Goudie (ed.), *Techniques for Desert Reclamation*. Chichester: Wiley, 109–134.

Richards, J.F. (1991) Land transformation. In B.L. Turner, W.C. Clark, R.W. Kates, J.F. Richards, J.T. Matthews and W.B. Meyer (eds), *The Earth as Transformed by Human Action*. Cambridge: Cambridge University Press, 163–178.

Richardson, J.A. (1976) Pit heap into pasture. In J. Lenihan and W.W. Fletcher (eds), *Reclamation*. Glasgow: Blackie, 60–93.

Richardson, S.J. and Smith, J. (1977) Peat wastage in the East Anglian Fens. *Journal of Soil Science*, 28, 485–489.

Richter, A., Burrows, J.P., Nüss, H., Granier, C. and Niemeier, U. (2005) Increase in tropospheric nitrogen dioxide over China observed from space. *Nature*, 437, 129–132.

Richter, D.deB. (2007) Humanity's transformation of Earth's soil: pedology's new frontier. *Soil Science*, 172, 957–967.

Richter, D.O. and Babbar, L.I. (1991) Soil diversity in the tropics. *Advances in Ecological Research*, 21, 315–389.

Rickson, R.J. (2006) Controlling sediment at source: an evaluation of erosion control geotextiles. *Earth Surface Processes and Landforms*, 31, 550–560.

Ridgwell, A.J. (2002) Dust in the Earth system: the biogeochemical linking of land, sea and air. *Philosophical Transactions. Series A, Mathematical, Physical, and Engineering Sciences*, 360, 2905–2924.

Ridgwell, A.J. (2003) Implications of the glacial CO_2 'iron hypothesis' for Quaternary climate change. *Geochemistry, Geophysics, Geosystems*, 4, 1076.

Rignot, E. and Kanagaratnam, P. (2006) Changes in the velocity structure of the Greenland Ice Sheet. *Science*, 311, 986–990.

Rignot, E. and Thomas, R.H. (2002) Mass balance of polar ice sheets. *Science*, 297, 1502–1506.

Rignot, E., Velicogna, I., van der Broejke, M.R., Monaghan, A. and Lenaerts, J. (2011) Acceleration of the contribution of the Greenland and Antarctica ice sheets to sea level rise. *Geophysical Research Letters*, 38, L05503.

Rijnsdorp, A.D., Peck, M.A., Engelhard, G.H., Möllmann, C. and Pinnegar, J.K. (2009) Resolving the effect of climate change on fish populations. *ICES Journal of Marine Science*, 66, 1570–1583.

Rinaldi, M., Wyzga, B. and Surian, N. (2005) Sediment mining in alluvial channels: physical effects and management perspectives. *River Research and Applications*, 21, 805–828.

Rinderer, T.E., Oldroyd, R.P. and Sheppard, W.S. (1993) Africanized bees in the US. *Scientific American*, 269 (6), 52–58.

Ripley, E.A. (1976) Drought in the Sahara: insufficient geophysical feedback? *Science*, 191, 100.

Robb, G.A. and Robinson, J.D.F. (1995) Acid drainage from mines. *The Geographical Journal*, 161, 47–54.

Roberts, B.W., Thornton, C.P. and Pigott, V.C. (2009) Development of metallurgy in Eurasia. *Antiquity*, 83, 1012–1022.

Roberts, N. (1989) *The Holocene: An Enviironmental History*. Oxford: Blackwell.

Roberts, N. (1998) *The Holocene: An Environmental History* (2nd edn). Oxford: Basil Blackwell.

Roberts, N. and Barker, P. (1993) Landscape stability and biogeomorphic response to past and future climatic shifts in intertropical Africa. In D.S.G. Thomas and R.J. Allison (eds), *Landscape Sensitivity*. Chichester: Wiley, 65–82.

Roberts, R.G. et al. (2001) New ages for the least Australian megafauna: continent-wide extinction about 46,000 years ago. *Science*, 292, 1888–1892.

Robinson, M. (1979) The effects of pre-afforestation ditching upon the water and sediment yields of a small upland catchment. *Working paper 252, School of Geography, University of Leeds*.

Robinson, M. (1990) *Impact of improved land drainage on river flows*. Institute of Hydrology, Wallingford, Report 113.

Robinson, M.A. and Lambrick, G.H. (1984) Holocene alluviation and hydrology in the Upper Thames Basin. *Nature*, 308, 809–814.

Rockström, J. et al. (2009) Planetary boundaries: exploring the safe operating space for humanity. *Ecology and Society*, 14 (http://www.ecologyandsociety.org/vol14/iss2/art32/), last accessed 05 January 2013.

Rodda, J.C., Downing, R.A. and Law, F.M. (1976) *Systematic Hydrology*. London: Newnes-Butterworth.

Rodell, M., Velicogna, I. and Famiglietti, J.S. (2009) Satellite-based estimates of groundwater depletion in India. *Nature*, 460, 999–1001.

Rogers, R.D. and Schumm, S. (1991) The effect of sparse vegetative cover on erosion and sediment yield. *Journal of Hydrology*, 123, 19–24.

Rojas-Soto, O.R., Sosa, V. and Ornelas, J.F. (2012) Forecasting cloud forest in eastern and southern Mexico: conservation insights under future climate change. *Biodiversity and Conservation*, 21, 2671–2690.

Rolett, B.V., Zheng, Z. and Yue, Y. (2011) Holocene sea-level change and the emergence of Neolithic seafaring in the Fuzhou Basin (Fujian, China). *Quaternary Science Reviews*, 30, 788–797.

Roman, J. and Palumbi, S.R. (2003) Whales before whaling in the North Atlantic. *Science*, 301, 508–510.

Romero, A., González, I. and Galán, E. (2011) Stream water geochemistry from mine wastes in Peña de Hierro, Riotinto area, SW Spain,: a case of extreme acid mine drainage. *Environmental Earth Sciences*, 62, 645–656.

Romero-Diaz, A., Belmonte-Serrato, F. and Ruiz-Sinoga, J.D. (2010) The geomorphic impact of afforestations on soil erosion in southeast Spain. *Land Degradation and Development*, 21, 188–195.

Romme, W.H. and Despain, D.G. (1989) The Yellowstone fires. *Scientific American*, 261, 21–29.

Roots, C. (1976) *Animal Invaders*. Newton Abbot: David & Charles.

Roques, K.G., O'Connor, T.G. and Watkinson, A.R. (2001) Dynamics of shrub encroachment in an African savanna: relative influences of fire, herbivory, rainfall and density dependence. *Journal of Applied Ecology*, 38, 268–280.

Rose, L.S., Stallins, J.A. and Bentley, M.L. (2008) Concurrent cloud-to-ground lightning and precipitation enhancement in the Atlanta, Georgia,(United States), urban region. *Earth Interactions*, doi:10.1175/2008EI265.1.

Rose, R. (1970) Lichens as pollution indicators. *Your Environment*, 5.

Rosenberg, N.J., Epstein, D.J., Wang, D., Vail, L., Srinivasan, R., Arnold, J.G. (1999) Possible impacts of global warming on the Ogallala Aquifer Region. *Climatic Change*, 42, 677–692.

Rosenzweig, C. and Hillel, D. (1993) The dust bowl of the 1930s: analog of greenhouse effect in the Great Plains? *Journal of Environmental Quality*, 22, 9–22.

Rosenzweig, C. et al. (2009) Mitigating New York City's heat island. *Bulletin of the American Meteorological Society*, 90, 1297–1312.

Rosepiler, M.J. and Reilinger, R. (1977) Land subsidence due to water withdrawal in the vicinity of Pecos, Texas. *Engineering Geology*, 11, 295–304.

Ross, N. et al. (2012) Steep reverse bed slope at the grounding line of the Weddell Sea sector in West Antarctica. *Nature Geoscience*, 5, 393–396.

RoTap (2012) *Review of Transboundary Air Pollution: Acidification, Eutrophication, Ground Level Ozone and Heavy Metals in the UK*. Wallingford: Centre for Ecology and Hydrology.

Rott, H., Müller, F., Nagler, T. and Floricioiu, D. (2010) The imbalance of glaciers after disintegration of Larsen B ice shelf, Antarctic Peninsula. *The Cryosphere Discussions*, 4, 1607–1633.

Rouse, W.R. et al. (1997) Effects of climate change on the freshwaters of Arctic and Subarctic North America. *Hydrological Processes*, 11, 873–902.

Royal Society (2005) *Ocean Acidification due to Increased Atmospheric Carbon Dioxide*. London: Royal Society.

Royal Society (2008) *Ground-Level Ozone in the 21st Century: Future Trends, Impacts and Policy Implications*. London: Royal Society.

Royal Society (2009a) *Geoengineering the Climate*. London: Royal Society.

Royal Society (2009b) *Climate Change: A Summary of the Science*. London: Royal Society.

Royal Society (2010) Climate change: a summary of the science. http://royalsociety.org/policy/publications/2010/climate-change-summary-science/ (last accessed 05 January 2013).

Royal Society Study Group (1983) *The Nitrogen Cycle of the United Kingdom*. London: The Royal Society.

Ruddiman, W.F. (2003) The anthropogenic greenhouse era began thousands of years ago. *Climatic Change*, 61, 261–293.

Ruddiman, W.F. (2005) *Plows, Plagues and Petroleum. How Humans Took Control of Climate*. Princeton and Oxford: Princeton University Press.

Ruddiman, W.F. and Thomsen, J.S. (2001) The case for human causes of increased atmospheric CH_4 over the last 5000 years. *Quaternary Science Reviews*, 20, 1769–1777.

Ruddiman, W.F., Kutzbach, J.E. and Vavrus, S.J. (2011) Can natural or anthropogenic explanations of late-Holocene

CO_2 and CH_4 increases be falsified? *The Holocene*, 21, 865–879.

Rudel, T.K., Coomes, O.T., Moran, E., Achard, F., Angelsen, A., Xu, F. and Lambin, E. (2005) Forest transitions: towards a global understanding of land use change. *Global Environmental Change*, 15, 23–31.

Rudel, T.K., Schneider, L. and Uriarte, M. (2010) Forest transitions: an introduction. *Land Use Policy*, 27, 95–97.

Rule, S., Brook, B.W., Haberle, S.G., Turney, C.S.M., Kershaw, A.P. and Johnson, C.N. (2012) The aftermath of megafaunal extinction: ecosystem transformation in Pleistocene Australia. *Science*, 335, 1483–1486.

Rumschlag, J.H. and Peck, J.A. (2007) Short-term sediment and morphologic response of the Middle Cuyahoga River to the removal of the Munroe Falls Dam, Summit County, Ohio. *Journal of Great Lakes Research*, 33, 142–153.

Russell, J.S. and Isbell, R.F. (eds) (1986) *Australian Soils: The Human Impact*. St Lucia: University of Queensland Press.

Rutherford, I. (2000) Some human impacts on Australian stream channel morphology. In S. Brizga and B. Finlayson (eds), *River Management: The Australian Experience*. Chichester: Wiley, 11–47.

Rutherford, M.C. and Westfall, R.H. (1994) Biomes of southern Africa: an objective characterization. *Mem. Bot. Surv. S. Afr.*, 63, 1–94.

Ryan, P.A. and Blackford, J.J. (2010) Late Mesolithic environmental change at Black Heath, south Pennines, UK: a test of Mesolithic woodland management models using pollen, charcoal and non-pollen palynomorph data. *Vegetation History and Archaeobotany*, 19, 545–558.

Ryder, M.L. (1966) The exploitation of animals by man. *Advancement of Science*, 23, 9–18.

Ryding, S.O. and Rast, R.W. (1989) *The Control of Eutrophication of Lakes and Reservoirs*. Paris: UNESCO.

Saalfield, W.K. and Edwards, G.P. (2010) Distribution and abundance of the feral camel (*Camelus dromedarius*) in Australia. *The Rangeland Journal*, 32, 1–9.

Sabadell, J.E., Risley, E.M., Jorgensen, H.T. and Thornton, B.S. (1982) *Desertification in the United States: Status and Issues*. Washington DC: Bureau of Land Management, Department of the Interior.

Sabadini, R. (2002) Ice sheet collapse and sea level change. *Science*, 295, 2376–2377.

Sahagian, D. (2000) Global physical effects of anthropogenic hydrological alterations: sea level and water redistribution. *Global and Planetary Change*, 25, 38–48.

Saiko, T.A. and Zonn, I.S. (2000) Irrigation expansion and dynamics of desertification in the Circum-Aral region of Central Asia. *Applied Geography*, 20, 349–367.

Sala, O.E. et al. (2000) Global biodiversity scenarios for the years 2100. *Science*, 287, 1770–1774.

Salazar, L.F., Nobre, C.A. and Oyama, M.D. (2007) Climate change consequences on the biome distribution in tropical South America. *Geophysical Research Letters*, 34, L09709.

Salby, M., Titova, E. and Deschamps, L. (2011) Rebound of Antarctic ozone. *Geophysical Research Letters*, 38, L09702.

Sallenger, A.H., Doranm, K.S. and Howd, P.A. (2012) Hotspot of accelerated sea-level rise on the Atlantic coast of North America. *Nature Climate Change*, doi:10.1038/NCLIMATE1597.

Sampaio, G., Nobre, C., Costa, M.H., Satyamurty, P., Soares-Filho, B.S. and Cardoso, M. (2007) Regional climate change over eastern Amazonia caused by pasture and soybean cropland expansion. *Geophysical Research Letters*, 34, L17709.

Sanchez, P.A. and Buol, S.W. (1975) Soils of the tropics and the world food crisis. *Science*, 188, 598–603.

Sanders, W.M. (1972) Nutrients. In R.T. Oglesby, C.A. Carlson and J.A. McCann (eds), *River Ecology and Man*. New York: Academic Press, 389–415.

Sankaran, M., Ratnam, J. and Hanan, N. (2008) Woody cover in African savannas: the role of resources, fire and herbivory. *Global Ecology and Biogeography*, 17, 236–245.

Sankey, J.B., Germino, M.J. and Glenn, N.F. (2009) Relationships of post-fire aeolian transport to soil and atmospheric conditions. *Aeolian Research*, 1, 75–85.

Santer, B.D. et al. (2006) Forced and unforced ocean temperature changes in Atlantic and Pacific tropical cyclogenesis regions. *Proceedings of the National Academy of Sciences of the United States of America*, 103, 13905–13910.

Sapozhnikov, V.V., Mordasova, N.V. and Metreveli, M.P. (2010) Transformations in the Caspian Sea ecosystem under the fall and rise of the sea level. *Oceanology*, 50, 488–497.

Sapp, J. (1999) *What is Natural? Coral Reef Crisis*. New York: Oxford University Press.

Sarmiento, G. and Monasterio, M. (1975) A critical consideration of the environmental conditions associated with the occurrence of savanna ecosystems in tropical America. *Ecological Studies*, 11, 233–250.

Sarre, P. (1978) The diffusion of Dutch elm disease. *Area*, 10, 81–85.

Sauer, C.O. (1938) *Destructive exploitation in modern colonial expansion*. International Geographical Congress, Amsterdam, Vol. III, sect. IIIC, 494–499.

Sauer, C.O. (1952) *Agricultural Origins and dispersals*. New York: American Geographical Society.

Sauer, C.O. (1969) *Seeds, Spades, Hearths and Herds*. Cambridge, Massachusetts: MIT Press.

Saunders, M.A. and Lea, A.S. (2008) Large contribution of sea surface warming to recent increase in Atlantic hurricane activity. *Nature*, 451, 557–560.

Savage, M. (1991) Structural dynamics of a southwestern pine forest under chronic human influence. *Annals of the Association of American Geographers*, 81, 271–289.

Savini, J. and Kammerer, J.C. (1961) Urban growth and the water regime. *United States Geological Survey water supply paper*, 159.

Sax, F.F. and Gaines, S.D. (2008) Species invasions ands extinction: the future of native biodiversity on islands.

Proceedings of the National Academy of Sciences of the United States of America, 105, 11490–11497.

Scanlon, B.R., Faunt, C.C., Longuevergne, L., Reedy, R.C., Alley, W.M., McGuire, V. and McMahon, P.B. (2012) Groundwater depletion and sustainability of irrigation in the US High Plains and Central Valley. *Proceedings of the National Academy of Sciences of the United States of America*, doi:10.1073/pnas.1200311109.

Scarre, C. (ed.) (2005) *The Human Past: World History and the Development of Human Societies*. London: Thames & Hudson.

Scherler, D., Bookhagen, B. and Strecker, M.R. (2011) Spatially variable response of Himalayan glaciers to climate change affected by debris cover. *Nature Geoscience*, 4, 156–159.

Schiermeier, Q. (2007) What we don't know about climate change. *Nature*, 445, 580–581.

Schilling, K.E., Chan, K.-S., Liu, H. and Zhang, Y.-K. (2010) Quantifying the effect of land use land cover change on increasing discharge in the Upper Mississippi River. *Journal of Hydrology*, 387, 343–345.

Schimper, A.F.W. (1903) *Plant-Geography upon a Physiological Basis*. Oxford: Clarendon Press.

Schipper, J. et al. (2008) The status of the world's land and marine mammals: diversity, threat and knowledge. *Science*, 322, 225–230.

Schlünzen, K.H., Hoffmann, P., Rosenhagen, G. and Riecke, W. (2010) Long-term changes and regional differences in temperature and precipitation in the metropolitan area of Hamburg. *International Journal of Climatology*, 30, 1121–1136.

Schmieder, O. (1927a) The Pampa – a natural or culturally induced grassland? *University of California Publications in Geography*, 2, 255–270.

Schmieder, O. (1927b) Alteration of the Argentine Pampa in the colonial period. *University of California Publications in Geography*, 2, 303–321.

Schneider, H., Höfer, D., Irmler, R. and Mäusbacher, R. (2010) Correlation between climate, man and debris flow events – a palynological approach. *Geomorphology*, 120, 48–55.

Schneider, S.H. and Thompson, S.L. (1988) Simulating the effects of nuclear war. *Nature*, 333, 221–227.

Scholz, N.L. et al. (2012) A perspective on modern pesticides, pelagic fish declines, and unknown ecological resilience in highly managed ecosystems. *Bioscience*, 62, 428–434.

Schoner, W., Auer, I. and Bohm, R. (2000) Climate variability and glacier reaction in the Austrian Eastern Alps. *Annals of Glaciology*, 23, 31–38.

Schrieber, B.C. (1986) Arid shorelines and evaporates. In H.G. Reading (ed.), *Sedimentary Environments and Facies*. Oxford: Blackwell Scientific, 189–228.

Schumm, S.A. (1977) *The Fluvial System*. New York: Wiley.

Schumm, S.A., Harvey, M.D. and Watson, C.C. (1984) *Incised Channels: Morphology, Dynamics and Control*. Littleton, Colorado: Water Resources Publications.

Schuur, E.A.G. and Abbott, B. (2011) High risk of permafrost thaw. *Nature*, 480, 32–33.

Schwartz, M.W., Porter, D.J., Randall, J.M. and Lyons, K.E. (1996) Impact of nonindigenous plants. In *Sierra Nevada Ecosystems Project: Final Report for Congress* (Vol. II). Davis: University of California, 1203–1218.

Schwarz, E.H.L. (1923) *The Kalahari or Thirstland Redemption*. Cape Town and Oxford: Oxford University Press.

Schwarz, H.E., Emel, J., Dickens, W.J., Rogers, P. and Thompson, J. (1991) Water quality and flows. In B.L. Turner, W.C. Clark, R.W. Kates, J.F. Richards, J.T. Matthews and W.B. Meyer (eds), *The Earth as Transformed by Human Action*. Cambridge: Cambridge University Press, 253–270.

Scott, D.F. (1997) The contrasting effects of wildfire and clear felling on the hydrology of a small catchment. *Hydrological Processes*, 11, 543–555.

Scott, G.J. (1977) The role of fire in the creation and maintenance of savanna in the Montana of Peru. *Journal of Biogeography*, 4, 143–167.

Seager, R. and Vecchi, G.A. (2010) Greenhouse warming and the 21st century hydroclimate of southwestern North America. *Proceedings of the National Academy of Sciences of the United States of America*, 107, 21277–21282.

Seager, R. et al. (2007) Model projections of an imminent transition to a more arid climate in southwestern North America. *Science*, doi:10.1126/science.1139601.

Searchinger, T. et al. (2008) Use of U.S. croplands for biofuels increase greenhouse gases through emissions from land-use change. *Science*, 319, 1238–1240.

Sears, P.B. (1957) Man the newcomer: the living landscape and a new tenant. In L.H. Russwurm and E. Sommerville (eds), *Man's Natural Environment, a System Approach*. North Scituate: Duxbury, 43–55.

Segall, P. (1989) Earthquakes triggered by fluid extraction. *Geology*, 17, 942–946.

Seidel, K., Ehrler, C. and Martinec, J. (1998) Effects of climate change on water resources and runoff in an alpine basin. *Hydrological Processes*, 12, 1659–1669.

Seifan, N. (2009) Long-term effects of anthropogenic activities on semi-arid sand dunes. *Journal of Arid Environments*, 73, 332–337.

Selby, M.J. (1979) Slopes and weathering. In K.J. Gregory and D.E. Walling (eds), *Man and Environmental Processes*. Folkestone: Dawson, 105–122.

Selman, M., Greenhalgh, S., Diaz, R. and Sugg, Z. (2008) Eutrophication and hypoxia in coastal areas: a global assessment of the state of knowledge. *World Resources Institute policy note*, 1, 6 pp.

Semaw, S., Renne, P., Harris, J.W.K., Feibel, C.S., Bernov, R.L., Fesseha, N. and Mowbray, K. (1997) 2.5-million-year-old stone tools from Gona, Ethiopia. *Nature*, 385, 333–336.

Semtner, A.J. (1984) The climate response of the Arctic Ocean to Soviet river diversions. *Climatic Change*, 6, 109–130.

Seto, S. et al. (2002) Annual and seasonal trends in chemical composition of precipitation in Japan during 1989–1998. *Atmospheric Environment*, 31, 3505–3517.

Setterfield, S., Rossiter-Rachov, N.A., Hutley, L.B., Douglas, M.M. and Williams, R.J. (2010) Turning up the heat: the impacts of *Andropogon gayanus* (gamba grass) invasion on fire behaviour in northern Australian savannas. *Diversity & Distributions*, 16, 854–861.

Shabalova, M.V., van Deursen, W.P.A. and Buishand, T.A. (2003) Assessing future discharge of the river Rhine using regional climate model integrations and a hydrological model. *Climate Research*, 23, 233–246.

Shakesby, R.A. (2011) Post-wildfire soil erosion in the Mediterranean: review and future research directions. *Earth-Science Reviews*, 105, 71–100.

Shakesby, R.A., Doerr, S.H. and Walsh, R.P.D. (2000) The erosional impact of soil hydrophobicity: current problems and future research directions. *Journal of Hydrology*, 231/2, 178–191.

Shakesby, R.A., Wallbrink, P.J., Doerr, S.H., English, P.M., Chafre, C.J., Humphreys, G.S., Blake, W.H. and Tomkins, K.M. (2007) Distinctiveness of wildfire effects on soil erosion in south-east Australian eucalyptus forests assessed in a global context. *Forest Ecology and Management*, 238, 347–364.

Shaler, N.S. (1912) *Man and the Earth*. New York: Duffield.

Shankman, D. and Smith, L.J. (2004) Stream channelization and swamp formation in the U.S. coastal plain. *Physical Geography*, 25, 22–38.

Sharrock, J. T. R. (1976) *The Atlas of Breeding Birds in Britain and Ireland*. Poyser, Berkhamsted and Nature Conservancy Council, 1977, Nature conservation and agriculture. London: Her Majesty's Stationery Office.

Shaw, L.M., Chamberlain, D. and Evans, M. (2008) The House Sparrow *Passer domesticus* in urban areas: reviewing a possible link between post-decline distribution and human socio-economic status. *Journal of Ornithology*, 149, 293–299.

Sheail, J. (1971) *Rabbits and Their History*. Newton Abbot: David & Charles.

Sheffield, A.T., Healy, T.R. and McGlone, M.S. (1995) Infilling rates of a steepland catchment estuary, Whangamata, New Zealand. *Journal of Coastal Research*, 11 (4), 1294–1308.

Sheffield, J. and Wood, E.F. (2008) Projected changes in drought occurrence under future global warming from multi-model, multi-scenario, IPCC AR4 simulations. *Climate Dynamics*, 31, 79–105.

Shehata, W. and Lotfi, H. (1993) Preconstruction solution for groundwater rise in sabkha. *Bulletin of the International Association of Engineering Geology*, 47, 145–150.

Shem, W. and Shepherd, M. (2009) On the impact of urbanization on summertime thunderstorms in Atlanta: two numerical model case studies. *Atmospheric Research*, 92, 172–189.

Sheng, J. and Wilson, J.P. (2009) Watershed urbanization and changing flood behaviour across the Los Angeles metropolitan region. *Natural Hazards*, 48, 41–57.

Shepard, C.C., Agostini, V.N., Gilmer, B., Allem, T., Stone, J., Brooks, W. and Beck, M.W. (2012) Assessing future risk: quantifying the effects of sea level rise on storm surge risk for the southern shores of Long Island, New York. *Natural Hazards*, 60, 727–745.

Sheppard, C.R.C. (2003) Predicted recurrences of mass coral morality in the Indian Ocean. *Nature*, 425, 294–297.

Sheppard, C.R.C., Davy, S.K. and Pilling, G.M. (2009) *The Biology of Coral Reefs*. Oxford: Oxford University Press.

Sherif, M.M. and Singh, V.P. (1999) Effect of climate change on sea water intrusion in coastal aquifers. *Hydrological Processes*, 13, 1277–1287.

Sherlock, R.L. (1922) *Man as a Geological Agent*. London: Witherby.

Sherratt, A. (1981) Plough and pastoralism: aspects of the secondary products revolution. In I. Hodder, G. Isaac and N. Hammond (eds), *Pattern of the Past*. Cambridge: Cambridge University Press, 261–305.

Sherratt, A. (1997) Climatic cycles and behaviour revolutions: the emergence of modern humans and the beginning of farming. *Antiquity*, 71, 271–287.

Sherriff, R.L. and Veblen, T.T. (2006) Ecological effects of changes in fire regimes in *Pinus ponderosa* ecosystems in the Colorado Front Range. *Journal of Vegetation Science*, 17, 705–718.

Sherwood, B., Cutler, D. and Burton, J. (eds) (2002) *Wildlife and Roads. The Ecological Impact*. London: Imperial College Press.

Shi, P., Yan, P., Yuan, Y. and Nearing, M.A. (2004) Wind erosion research in China: past, present and future. *Progress in Physical Geography*, 28, 366–386.

Shi, Y.F. and Liu, S.Y. (2000) Estimation on the response of glaciers in China to the global warming in the 21st century. *Chinese Science Bulletin*, 45, 668–672.

Shiklomanov, I.A. (1985) Large scale water transfers. In J.C. Rodda (ed.), *Facets of Hydrology II*. Chichester: Wiley, 345–387.

Shiklomanov, I.A. (1999) Climate change, hydrology and water resources: the work of the IPCC, 1988–1994. In J.C. van Dam (ed.), *Impacts of Climate Change and Climate Variability on Hydrological Regimes*. Cambridge: Cambridge University Press, 8–20.

Shindell, D. and Faluvegi, G. (2009) Climate response to regional radiative forcing during the twentieth century. *Nature Geoscience*, 2, 294–300.

Shirahata, H., Elias, R.W., Patterson, C.C. and Koide, M. (1980) Chronological variations in concentrations and isotopic composition of anthropogenic atmospheric lead in sediments of a remote subalpine pond. *Geochimica et Cosmochimica Acta*, 44, 149–162.

Short, F.T. et al. (2011) Extinction risk assessment of the world's seagrass species. *Biological Conservation*, 144, 1961–1971.

Shriner, D.S. and Street, R.B. (1998) North America. In R.T. Watson (ed.), *The Regional Impacts of Climate Change*. Cambridge: Cambridge University Press.

Shuman, J.K., Shugart, H.H. and O'Halloran, T.L. (2011) Sensitivity of Siberian larch forests to climate change. *Global Change Biology*, 17, 2370–2384.

Siakeu, J., Oguchi, T., Aoki, T., Esaki, Y. and Jarvie, H.P. (2004) Change in riverine suspended sediment concentration in central Japan in response to late 20th century human activities. *Catena*, 55, 231–254.

Sidle, R.C. and Dhakal, A.S. (2002) Potential effect of environmental change on landslide hazards in forest environments. In R.C. Sidle (ed.), *Environmental Change and Geomorphic Hazards in Forests*. Wallingford: CABI, 123–165.

Sidle, R.C., Kamil, I., Sharma, A. and Yamashita, S. (2000) Stream response to subsidence from underground coal mining in central Utah. *Environmental Geology*, 39, 279–291.

Sidle, R.C., Ziegler, A.D., Negishi, J.N., Nik, A.R., Siew, R. and Turkelboom, F. (2006) Erosion processes in steep terrain – truths, myths, and uncreatinties related to forest management in Southeast Asia. *Forest Ecology and Management*, 224, 199–225.

Sidle, R.C., Furuichi, T. and Kono, Y. (2011) Unprecedented rates of landslidre and surface erosion along a newly constructed road in Yunnan, China. *Natural Hazards*, 57, 31–26.

Sidorchuk, A.Y. and Golosov, V.N. (2003) Erosion and sedimentation on the Russian Plain, II: the history of erosion and sedimentation during the period of intensive agriculture. *Hydrological Processes*, 17, 3347–3358.

Siebert, S., Burke, J., Faures, J.M., Frenken, K., Hoogeven, J., Döll, P. and Portman, F.T. (2010) Groundwater use for irrigation – a global inventory. *Hydrology and Earth System Sciences Discussions*, 7, 3977–4021.

Simas, T., Nunes, J.P. and Ferreira, J.G. (2001) Effects of global climate change on coastal salt marshes. *Ecological Modelling*, 139, 1–15.

Simberloff, D. (2000) Global climate change and introduced species in United States forests. *The Science of the Total Environment*, 262, 253–261.

Simberloff, D. et al. (2010) Spread and impact of introduced conifers in South America: lessons from other southern hemisphere regions. *Austral Ecology*, 35, 489–504.

Simmonds, N.W. (1976) *Evolution of Crop Plants*. London: Longman.

Simmons, I.G. (1993) *Environmental History: A Concise Introduction*. Oxford: Blackwell.

Simmons, I.G. (1996) *Changing the Face of the Earth: Culture, Environment and History* (2nd edn). Oxford: Blackwell.

Simon, J.L. (1996) *The Ultimate Resource 2*. Princeton: Princeton University Press.

Sinclair, A.R.E. and Fryxell, J.M. (1985) The Sahel of Africa: ecology of a disaster. *Canadian Journal of Zoology*, 63, 987–994.

Siriwardena, L., Finlayson, B.L. and McMahon, T.A. (2006) The impact of land use change on catchment hydrology in large catchments: the Comet River, Central Queensland, Australia. *Journal of Hydrology*, 326, 199–214.

Sitoki, L., Gichuki, J., Ezekiel, C., Wanda, F., Mkumbo, O.C. and Marshall, B.E. (2010) The environment of Lake Victoria (East Africa): current status and historical changes. *International Review of Hydrobiology*, 95, 209–223.

Six, D., Reynaud, L. and Letreguilly, A. (2001) Silans de masse des glaciers alpines et scandinaves, leurs relations avec l'oscillation due climat de l'Atlantique nord. *Comptes rendus Academie des Sciences, science de la terre et des planètes*, 333, 693–698.

Slaymaker, O. and Kelly, R.E.J. (2007) *The Cryosphere and Global Environmental Change*. Oxford: Blackwell.

Slaymaker, O., Spencer, T. and Embleton-Hamann, C. (eds) (2009) *Geomorphology and Global Environmental Change*. Cambridge: Cambridge University Press.

Smil, V. (2011) Harvesting the biosphere: the human impact. *Population and Development Review*, 37, 613–636.

Smith, B.J., McCabe, S., McAllister, D., Adamson, C., Viles, H.A. and Curran, J.M. (2010) A commentary on climate change, stone decay dynamics and the 'greening' of natural stone buildings: new perspectives on 'deep wetting'. *Environmental Earth Sciences*, doi:10.1007/s12665-010-0766-1.

Smith, G.D., Coughland, K.J., Yule, D.F., Laryea, K.B., Srivastava, K.L., Thomas, N.P. and Cogle, A.L. (1992a) Soil management options to reduce runoff and erosion on a hardsetting Alfisol in the semi-arid tropics. *Soil and Tillage Research*, 25, 195–215.

Smith, J.B. and Tirpak, D.A. (eds) (1990) *The Potential Effects of Global Climate Change on the United States*. New York: Hemisphere.

Smith, J.B., Richels, R. and Muller, B. (2001a) Potential consequences of climate variability and change for the western United States. In National Assessment Synthesis Team (ed.), *Climate Change Impacts on the United States: The Potential Consequences of Climate Variability and Change*. Cambridge: Cambridge University Press, 219–245.

Smith, J.E. (ed.) (1968) *Torrey Canyon Pollution and Marine Life*. Cambridge: Cambridge University Press.

Smith, K. (1975) *Principles of Applied Climatology*. London: McGraw-Hill.

Smith, M.W. (1993) Climate change and permafrost. In H.M. French and O. Slaymaker (eds), *Canada's Cold Environments*. Montreal and Kingston: McGill-Queens University Press, 292–311.

Smith, N. (1976) *Man and Water*. London: Davies.

Smith, S.J., Pitcher, H. and Wigley, T.M.L. (2001b) Global and regional anthropogenic sulfur dioxide emissions. *Global and Planetary Change*, 29, 99–119.

Smith, T.M., Shugart, H.H., Bonan, G.B. and Smith, J.B. (1992b) Modelling the potential response of vegetation to global climate change. *Advances in Ecological Research*, 22, 93–116.

Smolikowski, B., Puig, H. and Roose, E. (2001) Influence of soil protection techniques on runoff, erosion and plant production on semi-arid hillslopes of Cabo verde. *Agriculture, Ecosystems & Environment*, 87, 67–80.

Snedaker, S.C. (1993) Impact on mangroves. In G. Maul (ed.), *Climatic Change in the Intra-Americas Sea*. London: Edward Arnold, 282–305.

Snedaker, S.C. (1995) Mangroves and climate change in the Florida and Caribbean region: scenarios and hypotheses. *Hydrobiologia*, 295, 43–49.

Snelder, D.J. and Bryan, R.B. (1995) The use of rainfall simulation tests to assess the influence of vegetation density on soil loss on degraded rangelands in the Baringo District, Kenya. *Catena*, 25, 105–116.

Snover, A. (1997) Impacts of global climate change on the Pacific Northwest. Preparatory white paper for OSTP/USGCRP regional workshop on the *Impacts of global climate change on the Pacific Northwest*, July 1997.

So, C.L. (1971) Mass movements associated with the rainstorm of 1966 in Hong Kong. *Transactions of the Institute of British Geographers*, 53, 55–65.

Sodhi, N.S. and Ehrlich, P.R. (eds) (2010) *Conservation Biology for All*. Oxford: Oxford University Press.

Soja, A.J. et al. (2007) Climate-induced boreal forest change: predictions versus current observations. *Global and Planetary Change*, 56, 274–296.

Solomon, S. (1999) Stratospheric ozone depletion: a review of concepts and history. *Reviews of Geophysics*, 37, 275–316.

Somerville, M. (1858) *Physical Geography* (4th edn). London: Murray.

Son, S.-W., Tandon, N.F., Polvani, L.M. and Waugh, D.M. (2009) Ozone hole and Southern Hemisphere climate change. *Geophysical Research Letters*, 36, L15705.

Sophocleous, M. (2010) Review: groundwater management practices, challenges, and innovations in the High Plains aquifer, USA – lessons and recommended actions. *Hydrogeology Journal*, 18, 559–575.

Sopper, W.E. (1975) Effects of timber harvesting and related management practices on water quality in forested watersheds. *Journal of Environmental Quality*, 4, 24–29.

Sorkin, A.J. (1982) *Economic Aspects of Natural Hazards*. Lexington: Lexington Books.

Souch, C. and Grimmond, S. (2006) Applied climatology: urban climate. *Progress in Physical Geography*, 30, 270–279.

Souter, D.W. and Linden, O. (2000) The health and future of coral reef systems. *Ocean and Coastal Management*, 43, 657–688.

Southon, G.E., Green, E.R., Jones, A.G., Barker, C.G. and Power, S.A. (2012) Long-term nitrogen additions increase likelihood of climate stress and affect recovery from wildfire in a lowland heath. *Global Change Biology*, doi: 10.1111/j.1365-2486.2012.02732.x.

Spanier, E. and Galil, B.S. (1991) Lessepsian migration: a continuous biogeographical process. *Endeavour*, 15, 102–106.

Spate, O.H.K. and Learmonth, A.T.A. (1967) *India and Pakistan*. London: Methuen.

Speight, M.C.D. (1973) Oudoor recreation and its ecological effects. A bibliography and review. *Discussion papers in conservation*, 4, *University College, London*.

Spencer, T. (1995) Potentialities, uncertainties and complexities in the response of coral reefs to future sea level rise. *Earth Surface Processes and Landforms*, 20, 49–64.

Spencer, T. and Douglas, I. (1985) The significance of environmental change: diversity, disturbance and tropical ecosystems. In I. Douglas and T. Spencer (eds), *Environmental Change and Tropical Geomorphology*. London: Allen & Unwin, 13–33.

Spencer, T., Teleki, K.A., Bradshaw, C. and Spalding, M.D. (2000) Coral bleaching in the Southern Seychelles during the 1997–1998 Indian Ocean warm event. *Marine Pollution Bulletin*, 40, 569–586.

Sperling, C.H.B., Goudie, A.S., Stoddart, D.R. and Poole, G.C. (1979) Origin of the Dorset dolines. *Transactions of the Institute of British Geographers*, 4, 121–124.

Sperna Weiland, F.C., van Beek, L.P.H., Kwadijk, K.C.J. and Blerkens, M.F.P. (2011) Global patterns of change in discharge rehgimes for 2100. *Hydrology and Earth System Sciences Discussions*, 8, 10973–11014.

Spriggs, M. (2010) Geomorphic and archaeological consequences of human arrival and agricultural expansion on Pacific islands: a reconsideration after 30 years of debate. In S.G. Haberle, J. Stevenson and M. Prebble (eds), *Altered Ecologies: Fire, Climate and Human Influence on Terrestrial Landscapes*. Canberra: ANU E Press.

Squire, G.R. et al. (2003) On the rationale and interpretation of the farm scale evaluations of genetically modified herbicide-tolerant crops. *Philosophical Transactions of the Royal Society of London. Series B, Biological Sciences*, 358, 1779–1799.

Staehelin, J., Harris, N.R.P., Appenzeller, C. and Ebeshard, J. (2001) Ozone trends: a review. *Reviews of Geophysics*, 39, 231–290.

Stanhill, G. and Cohen, S. (2001) Global dimming: a review of the evidence for a widespread and significant reduction in global radiation with discussion of its probable causes and possible agricultural consequences. *Agricultural and Forest Meteorology*, 107, 255–278.

Stanley, D.J. (1996) Nile delta: extreme case of sediment entrapment on a delta plain and consequent coastal land loss. *Marine Geology*, 129, 189–195.

Stanley, D.J. and Chen, Z. (1993) Yangtze delta, eastern China: I. Geometry and subsidence of Holocene depocenter. *Marine Geology*, 112, 1–11.

State of the Birds, United States of America (2009) *State of the Birds United States of America, 2009*. Washington DC: US Department of the Interior.

Staver, A.C., Archibald, S. and Levin, S. (2011) Tree cover in sub-Saharan Africa: rainfall and fire constrain forest and savanna alternative stable states. *Ecology*, 92, 1063–1072.

Steadman, D.W., Stafford, T.W., Donahue, D.J. and Jull, A.J.T. (1991) Chronology of Holocene vertebrate extinction in the Galápagos Islands. *Quaternary Research*, 36, 126–133.

Steffen, W. (2010) Observed trends in Earth System behaviour. *Wiley Interdisciplinary Reviews: Climate Change*, 1, 428–449.

Steffen, W., Crutzen, P.J. and McNeill, J.R. (2007) The Anthropocene: are humans now overwhelming the great forces of nature? *Ambio*, 36, 614–621.

Steffen, W. et al. (2004) *Global Change and the Earth System*. Berlin: Springer.

Stendel, M. and Christensen, J.H. (2002) Impact of global warming on permafrost conditions in a coupled GCM. *Geophysical Research Letters*, 29, 10–11.

Stephens, J.C. (1956) Subsidence of organic soils in the Florida Everglades. *Proceedings of the Soil Science Society of America*, 20, 77–80.

Sterk, G. (2003) Causes, consequences and control of wind erosion in sahelian Africa: a review. *Land Degradation and Development*, 14, 95–108.

Sternberg, H.O.R. (1968) Man and environmental change in South America. *Monographiae Biologicae*, 18, 413–445.

Stetler, L.L. and Gaylord, D.R. (1996) Evaluating eolian-climate interactions using a regional climate model from Hanford, Washington (USA). *Geomorphology*, 17, 99–113.

Stevens, C.J. et al. (2010) Nitrogen deposition threatens species richness of grasslands across Europe. *Environmental Pollution*, 158, 2940–2945.

Stevens, C.J. et al. (2011) Addressing the impact of atmospheric nitrogen deposition on Western European grasslands. *Environmental Management*, 48, 885–894.

Stevenson, A.C., Jones, V.J. and Battarbee, R.W. (1990) The cause of peat erosion: a palaeolimnological approach. *The New Phytologist*, 114, 727–735.

Stewart, D.C. (1956) Fire as the first great force employed by man. In W.L. Thomas (ed), *Man's role in Changing the Face of the Earth*. Chicago: University of Chicago Press, 115–133.

Stewart, I.T. (2009) Changes in snowpack and snowmelt runoff for key mountain regions. *Hydrological Processes*, 23, 78–94.

Stiles, D. (1995) An overview of desertification as dryland degradation. In D. Stiles (ed.), *Social Aspects of Sustainable Dryland Management*. Chichester: Wiley.

Stocker, T.F. (2001) Physical climate processes and feedbacks. In J.T. Houghton (ed.), *Climate Change 2001: The Scientific Basis*. Cambridge: Cambridge University Press, 417–470.

Stoddart, D.R. (1968) Catastrophic human interference with coral atoll ecosystems. *Geography*, 53, 25–40.

Stoddart, D.R. (1971) Coral reefs and islands and catastrophic storms. In J.A. Steers (ed.), *Applied Coastal Geomorphology*. London: Macmillan, 154–197.

Stoddart, J.L. et al. (1999) Regional trends in aquatic recovery from acidification in North America and Europe. *Nature*, 401, 575–578.

Stoffel, M. and Beniston, M. (2006) On the incidence of debris flows from the early Little Ice Age to a future greenhouse climate: a case study from the Swiss Alps. *Geophysical Research Letters*, 33, L16404.

Stokes, S., Thomas, D.S.G. and Washington, R. (1997) Multiple episodes of aridity in southern Africa since the last interglacial period. *Nature*, 388, 154–158.

Stone, B. (2007) Urban and rural temperature trends in mproximity to large US cities: 1951–2000. *International Journal of Climatology*, 27, 1801–1807.

Stork, N.E. (2010) Re-assessing current extinction rates. *Biodiversity and Conservation*, 19, 357–371.

Stott, T. and Marks, S. (2000) Effects of plantation forest clear-felling on stream temperatures in the Plynlimon experimental catchments, mid-Wales. *Hydrology and Earth System Sciences*, 4, 95–104.

Strandberg, C.H. (1971) Water pollution. In G.H. Smith (ed.), *Conservation of Natural Resources* (4th edn). New York: Wiley, 189–219.

Street, F.A. and Grove, A.T. (1979) Global maps of lake level fluctuation since 30,000 years ago. *Quaternary Research*, 12, 83–118.

Streets, D.G., Wu, Y. and Chin, M. (2006) Two-decadal aerosol trends as a likely explanation of the global dimming/brightening transition. *Geophysical Research Letters*, 33, L15806.

Stringer, C. (2003) Out of Africa. *Nature*, 423, 692–699.

Strong, D.R. and Ayres, D.A. (2009) Spartina introduction and consequences in salt marshes. In B.R. Silliman, E.D. Grosholz and M.D. Bertness (eds), *Human Impacts on Salt Marshes: A Global Perspective*. Vancouver: University of British Columbia Press, 3–22.

Sturrock, F. and Cathie, J. (1980) Farm modernisation and the countryside. *Occasional paper 12, Department of Land Economy, University of Cambridge*.

Su, Z. and Shi, Y. (2002) Response of monsoonal temperate glaciers to global warming science The Little Ice Age. *Quaternary International*, 97–98, 123–131.

Suchodoletz, H., von Oberhänsli, H., Faust, D., Fuchs, M., Blanchet, C., Goldhammer, T. and Zöller, L. (2010) The evolution of Saharan dust input on Lanzarote (Canary Islands) – influenced by human activity during the early Holocene? *The Holocene*, 20, 169–179.

Suckale, J. (2010) Moderate-to-large seismicity induced by hydrocarbon production. *The Leading Edge*, 29, 310–319.

Sun, G.E., McNulty, S.G., Moore, J., Bunch, C. and Ni, J. (2002) Potential impacts of climate change on rainfall erosivity and water availability in China in the next 100 years. *Proceedings of the 12th International Soil Conservation Conference*, Beijing.

Suppiah, R. and Hennessy, K.J. (1998) Trends in total rainfall, heavy rain events and number of dry days in Australia. *International Journal of Climatology*, 18, 1141–1164.

Sutton, J.E.G. (1965) Sirikwa holes, stone houses and their makers in the western highlands of Kenya. *Man*, 65, 113–115.

Swank, W.T. and Douglass, J.E. (1974) Streamflow greatly reduced by converting deciduous hardwood stands to pine. *Science*, 18, 857–859.

Swann, A.L.S., Fung, I.Y. and Chiang, J.C.H. (2012) Mid-latitude afforestation shifts general circulation and tropical precipitation. *Proceedings of the National Academy of Sciences of the United States of America*, 109, 712–716.

Swanston, D.N. and Swanson, F.J. (1976) Timber harvesting, mass erosion and steepland forest geomorphology in the Pacific north-west. In D.R. Coates (ed.), *Geomorphology and Engineering*. Stroudsburg: Dowden, Hutchinson and Ross, 199–221.

Sweatman, H., Delean, S. and Syms, C. (2011) Assessing loss of coral cover on Australia's Great Barrier Reef over two decades, with implications for longer term trends. *Coral Reefs*, 30, 521–531.

Swift, L.W. and Messer, J.B. (1971) Forest cuttings raise temperatures of small streams in the southern Appalachians. *Journal of Soil and Water Conservation*, 26, 111–116.

Swift, M.J. and Sanchez, P.A. (1984) Biological management of tropical soil fertility for sustained productivity. *Nature and Resources*, 20, 2–10.

Swisher, C.C., Curtis, G.H., Jacob, T., Getty, A.G. and Suprijo, A. (1994) Age of the earliest known hominids in Java, Indonesia. *Science*, 263, 1118–1121.

Sylla, M.B., Gaye, A.T., Jenkins, G.S., Pal, J.S. and Giorgi, F. (2010) Consistency of projected drought over the Sahel with changes in the monsoon circulation and extremes in a regional climate model projections. *Journal of Geophysical Research*, 115, D16108.

Syvitski, J.P.M. (2008) Deltas at risk. *Sustainability Science*, 3, 23–32.

Syvitski, J.P.M. and Kettner, A. (2011) Sediment flux and the Anthropocene. *Philosophical Transactions. Series A, Mathematical, Physical, and Engineering Sciences*, 369, 957–975.

Syvitski, J.P.M. and Milliman, J.D. (2007) Geology, geography and humans battle for dominance over the delivery of fluvial sediment to the coastal ocean. *The Journal of Geology*, 115, 1–19.

Syvitski, J.P.M. and Saito, Y. (2007) Morphodynamics of deltas under the influence of humans. *Global and Planetary Change*, 57, 261–282.

Syvitski, J.P.M., Vörösmarty, C.J., Kettner, A.J. and Green, P. (2005) Impact of humans on the flux of terrestrial sediment to the global coastal ocean. *Science*, 308, 376–380.

Syvitski, J.P.M. et al. (2009) Sinking deltas due to human activities. *Nature Geosciences*, 2, 681–686.

Szabó, J., Dávid, L. and Lóczy, D. (eds) (2010) *Anthropogenic Geomorphology*. Dordrecht: Springer.

Szabolcs, I. (1994) State and perspectives on soil salinity in Europe. *European Society for Soil Conservation Newsletter*, 3, 17–24.

Ta, W., Xiao, H. and Dong, Z. (2008) Long-term morphodynamic changes of a desert reach of the Yellow River following upstream large reservoirs' operation. *Geomorphology*, 97, 249–259.

Takacs-Santa, A. (2004) The major transitions in the history of human transformation of the biosphere. *Human Ecology Review*, 11, 51–66.

Tallis, J.H. (1965) Studies on southern Pennine peats, IV: evidence of recent erosion. *The Journal of Ecology*, 53, 509–520.

Tallis, J.H. (1985) Erosion of blanket peat in the southern Pennines: new light on an old problem. In R.H. Johnson (ed.), *The Geomorphology of North-West England*. Manchester: Manchester University Press, 313–336.

Talwani, P. (1997) On the nature of reservoir-induced seismicity. *Pure and Applied Geophysics*, 150, 473–492.

Tan, J. et al. (2010) The urban heat island and its impact on heat waves and human health in Shanghai. *International Journal of Biometeorology*, 54, 75–84.

Tang, D.L., Di, B.P., Wei, G., Ni, I.H., Oh, I.S. and Wang, S.F. (2006) Spatial, seasonal and species variations of hartmful algal blooms in the South Yellow Sea and East China Sea. *Hydrobiologia*, 568, 245–253.

Tang, J., Xu, X.B. and Wang, S.F. (2010) Trends of the precipitation acidity over China during 1992–2006. *Chinese Science Bulletin*, 55, 1800–1807.

Tao, X., Wu, P., Tang, C., Liu, H. and Sun, J. (2012) Effect of acid mine drainage on a karst basin: a case study on the high-As coal mining area in Guizhou province, China. *Environmental Earth Sciences*, 65, 631–638.

Taylor, A.H. (2010) Fire disturbance and forest structure in an old-growth *Pinus ponderosa* forest, southern Cascades, USA. *Journal of Vegetation Science*, 21, 561–572.

Taylor, C.M., Lambin, E.F., Stephenne, N., Harding, R.J. and Essery, L.H. (2002) The influence of land use change on climate in the Sahel. *Journal of Climate*, 15, 3615–3629.

Taylor, J.A. (1985) Bracken encroachment rates in Britain. *Soil Use and Management*, 1, 53–56.

Taylor, K.E. and Penner, J.E. (1994) Response of the climatic system to atmospheric aerosols and greenhouse gases. *Nature*, 369, 734–737.

Tegen, I., Lacis, A.A. and Fung, I. (1996) The influence on climate forcing of mineral aerosols from disturbed soils. *Nature*, 380, 419–422.

Teneva, L., Karnauskas, M., Logan, C.A., Bianucci, L., Currie, J.C. and Kleypas, J.A. (2012) Predicting coral bleaching hotspots: the role of regional variability in thermal stress and potential adaptation rates. *Coral Reefs*, 31, 1–12.

Tengberg, M. (2012) Beginnings and early history of date palm garden cultivation in the Middle East. *Journal of Arid Environments*, 86, 139–147.

Terborgh, J. (1992) *Diversity and the Tropical Rain Forest*. New York: Freeman.

Ternan, J.L., Williams, A.G., Elmes, A. and Fitzjohn, C. (1996) The effectiveness of bench-terracing and afforestation for erosion control on Rana sediments in central Spain. *Land Degradation and Development*, 7, 337–351.

Theodoropoulos, D.I. (2003) *Invasion Biology. Critique of a Pseudoscience*. Blythe, California: Avvar Books.

Thirgood, J.V. (1981) *Man and the Mediterranean Forest – A History of Resource Depletion*. London: Academic Press.

Thomas, A.D., Walsh, R.P.D. and Shakesby, R.A. (2000) Solutes in overland flow following fire in eucalyptus and pine forests, northern Portugal. *Hydrological Processes*, 14, 971–985.

Thomas, C.D. et al. (2004) Extinction risk from climate change. *Nature*, 427, 145–148.

Thomas, D.S.G. and Middleton, N.J. (1993) Salinisation: new perspectives on a major desertification issue. *Journal of Arid Environments*, 24, 95–105.

Thomas, D.S.G. and Middleton, N.J. (1994) *Desertification: Exploding the Myth*. Chichester: Wiley.

Thomas, D.S.G. and Wiggs, G.F.S. (2008) Aeolian system responses to global change: challenges of scale, process and temporal integration. *Earth Surface Processes and Landforms*, 33, 1396–1418.

Thomas, D.S.G., Knight, M. and Wiggs, G.F.S. (2005) Remobilization of southern African desert dune systems by twenty-first century global warming. *Nature*, 435, 1218–1221.

Thomas, R., Frederick, E., Krabill, W., Manizade, S. and Martin, C. (2006) Progressive increase in ice loss from Greenland. *Geophysical Research Letters*, 33, L10503.

Thomas, R.H., Sanderson, T.J.O. and Rose, K.E. (1979) Effect of climatic warming on the West Antarctic ice sheet. *Nature*, 277, 355–358.

Thomas, W.L. (ed.) (1956) *Man's Role in Changing the Face of the Earth*. Chicago: University of Chicago Press.

Thompson, J.R. (1970) Soil erosion in the Detroit metropolitan area. *Journal of Soil and Water Conservation*, 25, 8–10.

Thompson, L.G. (2000) Ice core evidence for climate change in the Tropics: implications for our future. *Quaternary Science Reviews*, 19, 19–35.

Thompson, L.G., Brecher, H.H., Mosley-Thompson, E., Hardy, D.R. and Mark, B.G. (2009) Glacier loss on Kilimanjaro continues unabated. *Proceedings of the National Academy of Sciences of the United States of America*, 106, 19770–19775.

Thomson, D.P., Shaffe, G.P. and McCorquodale, J.A. (2002) A potential interaction between sea level rise and global warming: implications for coastal stability on the Mississippi River Deltaic Plain. *Global and Planetary Change*, 32, 49–59.

Thomson, A.M., Brown, R.A., Rosenberg, N.J., Srinivasan, R., Izaurralde, R.C. (2005) Climate change impacts for the conterminous USA: an integrated assessment. Part 4: Water resources. *Climatic Change*, 69, 67–88.

Thornthwaite, C.W. (1956) Modification of the rural microclimates. In W.L. Thomas (ed.), *Man's Role in Changing the Face of the Earth*. Chicago: University of Chicago Press, 567–583.

Thuiller, W., Broennimann, O., Hughes, G., Alkemades, J.R.M., Midgley, G.F. and Corsi, P. (2006) Vulnerability of African mammals to anthropogenic climate change under conservative land transformation assumptions. *Global Change Biology*, 12, 424–440.

Tickner, D.P., Angold, P.G., Gurnell, A.M. and Mountford, J.O. (2001) Riparian plant invasions: hydrogeomorphological control and ecological impacts. *Progress in Physical Geography*, 25, 22–52.

Tie, X. and Cao, J. (2009) Aerosol pollution in China: present and future impact on emnvironment. *Particuology*, 7, 426–431.

Tiffen, M., Mortimore, M. and Gichuki, F.N. (1994) *More People, Less Erosion: Environmental Recovery in Kenya*. London: Wiley.

Tilman, D., Reich, P.B. and Knops, M.H. (2006) Biodiversity and ecosystem stability in a decade-long grassland experiment. *Nature*, 441, 629–632.

Tipping, E. et al. (2000) Reversal of acidification in tributaries of the River Duddon (English Lake District) between 1970 and 1998. *Environmental Pollution*, 109, 183–191.

Tipping, R. (2008) Blanket peat in the Scottish Highlands: timing, cause, spread and the myth of environmental determinism. *Biodiversity and Conservation*, 17, 2097–2113.

Titus, J.G. (1990) Greenhouse effect, sea level rise, and barrier islands: case study of Long Beach Island, New Jersey. *Coastal Management*, 18, 65–90.

Titus, J.G. and Seidel, S. (1986) Overview of the effects of changing the atmosphere. J.G. Titus (ed.), *Effects of Changes in Stratospheric Ozone and Global Climate*. Washington DC: UNEP/USEPA, 3–19.

Tivy, J. (1971) *Biogeography. A Study of Plants in the Ecosphere*. Edinburgh: Oliver and Boyd.

Tiwari, V.M., Wahr, J. and Swenson, S. (2009) Dwindling groundwater resources in northern India, from satellite gravity observations. *Geophysical Research Letters*, 36, L18401.

Tixier-Boichard, M., Bed'hom, B. and Rognon, X. (2011) Chicken domestication: from archaeology to genomics. *Comptes Rendus Biologies*, 334, 197–204.

Tockner, K. and Stanford, J.A. (2002) Riverine flood plains: present state and future trends. *Environmental Conservation*, 29, 308–330.

Todhunter, P.E. and Chihacek, L.J. (1999) Historical reduction of airborne dust in the Red River Valley of the North. *Journal of Soil and Water Conservation*, 54, 543–551.

Tolba, M.K. and El-Kholy, O.A. (1992) *The World Environment, 1972–1992*. London: UNEP/Chapman & Hall.

Tomaselli, R. (1977) Degradation of the Mediterranean maquis. *UNESCO, MAB technical note*, 2, 33–72.

Tomkins, D.M., White, A.R. and Boots, M. (2003) Ecological replacement of native red squirrels by invasive greys driven by disease. *Ecology Letters*, 6, 189–196.

Toon, O.B. (2003) African dust in Florida clouds. *Nature*, 424, 623–624.

Tornquist, T.E. et al. (2008) Mississippi Delta subsidence primarily caused by compaction of Holocene strata. *Nature Geoscience*, 11, 173–176.

Trimble, S.W. (1974) *Man-Induced Soil Erosion on the Southern Piedmont*. Ankeny, Iowa: Soil Conservation Society of America.

Trimble, S.W. (1976) Modern stream and valley sedimentation in the Driftless Area, Wisconsin, USA. *23rd International Geographical Congress*, sect. 1, 228–231.

Trimble, S.W. (1988) The impact of organisms on overall erosion rates within catchments in temperate regions. In H.A. Viles (ed.), *Biogeomorphology*. Oxford: Basil Blackwell, 83–142.

Trimble, S.W. (1997) Stream channel erosion and change resulting from riparian forests. *Geology*, 25, 467–469.

Trimble, S.W. (2003) Historical hydrographic and hydrologic changes in the San Diego Creek watershed, Newport Bay, California. *Journal of Historical Geography*, 29, 422–444.

Trimble, S.W. (2004) Effects of riparian vegetation on stream channel stability and sediment budgets. *Water Science and Application*, 8, 153–169.

Trimble, S.W. (2008) *Man-Induced Soil Erosion on the Southern Piedmont* (2nd edn). Ankeny, Iowa: Soil Conservation Society of America.

Trimble, S.W. (2011) The historical decrease of soil erosion in the eastern United States – the role of geography and engineering. In S.D. Brunn (ed.), *Engineering Earth*. Dordrecht: Springer, 1383–1393.

Trimble, S.W. and Crosson, S. (2000) US soil erosion rates – myth and reality. *Science*, 289, 248–250.

Trimble, S.W. and Lund, S.W. (1982) Soil conservation and the reduction of erosion and sedimentation in the Coon Creek Basin, Wisconsin. *US Geological Survey professional paper*, 1234.

Trimble, S.W. and Mendel, A.C. (1995) The cow as a geomorphic agent: a critical review. *Geomorphology*, 13, 233–253.

Tripathy, D.P. (2010) Environmentally sound,management of e-wastes. *Ecology, Environment and Conservation*, 16, 641–649.

Troels-Smith, J. (1956) Neolithic period in Switzerland and Denmark. *Science*, 124, 876–879.

Tubbs, C. (1984) Spartina on the south coast: an introduction. In P. Doody (ed.), *Spartina Anglica in Great Britain*. Shrewsbury: Nature Conservancy Council, 3–4.

Tucker, C.J., Dregne, H.E. and Newcomb, W.W. (1991) Expansion and contraction of the Sahara Desert from 1980 to 1990. *Science*, 253, 299–301.

Turco, R.P., Toon, O.B., Ackermann, T.P., Pollack, J.B. and Sagan, C. (1983) Nuclear winter: global consequences of multiple nuclear explosions. *Science*, 222, 1283–1292.

Turnbull, L., Wainwright, J. and Brazier, R.E. (2010) Changes in hydrology and erosion over a transition from grassland to shrubland. *Hydrological Processes*, 24, 393–414.

Turner, B.L., Kasperson, R.E., Meyer, W.B., Dow, K.M., Golding, D., Kasperson, J.X., Mitchell, R.C. and Ratick, S.J. (1990) Two types of global environmental change: definitional and spatial-scale issues in their human dimensions. *Global Environmental Change*, 1, 14–22.

Turner, I.M. (1996) Species loss in fragments of tropical rain forest: a review of the evidence. *Journal of Applied Ecology*, 33, 200–209.

Turvey, S.T. (ed.) (2009) *Holocene Extinctions*. Oxford: Oxford University Press.

Turvey, S.T. et al. (2007) First human-caused extinction of a cetacean species? *Biology Letters*, 3, 537–540.

Tweel, A.W. and Turner, R.E. (2012) Watershed land use and river eigineering drive wetland formation and loss in the Mississippi birdfoot delta. *Limnology and Oceanography*, 57, 18–28.

Tyldesley, J.A. and Bahn, P.G. (1983) The use of plants in the European palaeolithic: a review of the evidence. *Quaternary Science Reviews*, 2, 53–81.

Tyler, S.W., Kranz, S., Parlange, M.B., Albertson, J., Katul, G.G., Cochram, G.F., Lyles, B.A. and Holder, G. (1997) Estimation of groundwater evaporation and salt flux from Owens Lake, California, USA. *Journal of Hydrology*, 200, 110–135.

UK Climate Impacts Programme (2001) *Climate Change and Nature Conservation in Britain and Ireland*. Oxford: UKCIP.

UNEP (1991) *United Nations Environment Programme Environmental data Report* (3rd edn). Oxford: Basil Blackwell.

UNEP (2007) *Global Environment Outlook GEO4*. Nairobi: UNEP.

Unger, P.W., Stewart, B.A., Parr, J.F. and Singh, R.P. (1991) Crop residue management and tillage methods for conserving soil and water in semi-arid regions. *Soil and Tillage Research*, 20, 219–240.

US Department of Agriculture, Soil Conservation Service (1972) (http://www.plant-materials.nrcs.usda.gov/pubs/ndpmcrnmama37midw.pdf), last accessed 14 December 2012.

US Environmental Protection Agency (1994) *Technical Document: Acid Mine Drainage Prediction*. Washington DC: US Environmental Protection Agency.

US General Accounting Office (2000) *Acid Rain. Emissions Trends and Effects in the Eastern United States*. Washington DC: US General Accounting office.

Usher, M.B. (1973) *Biological Management and Conservation*. London: Chapman & Hall.

Utset, A. and Borroto, M. (2001) A modelling-GIS approach for assessing irrigation effects of soil salinisation under global warming conditions. *Agricultural Water Management*, 50, 53–63.

Vale, T.R. and Vale, G.R. (1976) Suburban bird population in westcentral California. *Journal of Biogeography*, 3, 157–165.

Valentin, C., Rajot, J.-L. and Mitja, D. (2004) Response of soil crusting, runoff and erosion to fallowing in the sub-humid and semi-arid regions of West Africa. *Agriculture, Ecosystems & Environment*, 104, 287–302.

van Auken, O.W. (2000) Shrub invasion of North American semiarid grasslands. *Annual Review of Ecology and Systematics*, 31, 197–215.

van den Broeke, M. et al. (2009) Partitioning recent Greenland mass loss. *Science*, 326, 984–986.

Vandermeulen, J.H. and Hrudey, S.E. (eds) (1987) *Oil in Freshwater: Chemistry, Biology, Countermeasure Technology*. New York: Pergamon.

van der Ween, C.J. (2002) Polar ice sheets and global sea level: how well can we predict the future? *Global and Planetary Change*, 32, 165–194.

Vanguelov, E.I. et al. (2010) Chemical fluxes in time through forest ecosystems in the UK – soil response to pollution recovery. *Environmental Pollution*, 158, 1857–1869.

Van Herk, C.M., Mathijssen-Spiekman, E.A.M. and de Zwart, D. (2003) Long distance nitrogen effects on lichens in Europe. *Lichenologist*, 35, 347–359.

Van Oost, K., Govers, G., de Alba, S. and Quine, T.A. (2006) Tillage erosion: a review of controlling factors and implications for soil quality. *Progress in Physical Geography*, 30, 443–466.

Vaudour, J. (1986) Travertins holocènes et pression anthropique. *Méditerranée*, 10, 168–173.

Vaughan, D.G. and Doake, C.S.M. (1996) Recent atmospheric warming and retreat of ice shelves on the Antarctic Peninsula. *Nature*, 379, 328–331.

Vaughan, D.G. and Spouge, J.R. (2002) Risk estimation of collapse of the west Antarctic ice sheet. *Climate Change*, 52, 65–91.

Vaughan, N.E. and Lenton, T.M. (2011) A review of climate geoengineering proposals. *Climate Change*, 109, 745–790.

Veblen, T.T. and Stewart, G.H. (1982) The effects of introduced wild animals on New Zealand forests. *Annals of the Association of American Geographers*, 72, 372–397.

Vendrov, S.L. (1965) A forecast of changes in natural conditions in the northern Ob' basin in case of construction of the lower Ob' Hydro Project. *Soviet Geography*, 6, 3–18.

Vennemo, H., Aunan, K., Lindhjem, H. and Seip, H.M. (2009) Environmental pollution in China: status and trends. *Review of Environmental Economics and Policy*, 3, 209–230.

Venteris, E.R. (1999) Rapid tidewater glacier retreat: a comparison between Columbia Glacier, Alaska and Patagonian calving glaciers. *Global and Planetary Change*, 22, 131–138.

Vermeer, M. and Rahmstorf, S. (2009) Global sea level linked to global temperature. *Proceedings of the National Academy of Sciences of the United States of America*, 106, 21527–21532.

Verstraeten, G., Rommens, T., Peeters, I., Poesen, J., Govers, G. and Lang, A. (2009) A temporarily changing Holocene sediment budget for a loess-covered catchment (central Belgium). *Geomorphology*, 108, 24–34.

Vesely, J., Majer, V. and Norton, S.A. (2002) Heterogeneous response of central European streams to decreased acid atmospheric deposition. *Environmental Pollution*, 120, 275–281.

Vice, R.B., Guy, H.P. and Ferguson, G.E. (1969) Sediment movement in an area of suburban highway construction, Scott Run Basin, Fairfax, County, Virginia, 1961–64. *United States Geological Survey water supply paper*, 1591-E.

Viessman, W., Knapp, J.W., Lewis, G.L. and Harbaugh, T.E. (1977) *Introduction to Hydrology* (2nd edn). New York: IEP.

Viets, F.G. (1971) Water quality in relation to farm use of fertilizer. *Bioscience*, 21, 460–467.

Viles, H.A. (2002) Implications of future climate change for stone deterioration. In S. Siegesmund, T. Weiss and J.A. Vollbrecht (eds), *Natural Stone, Weathering Phenomena, Conservation straTegies and Case Studies*. Geological Society of London special publication 205, 407–418.

Viles, H.A. (2003) Conceptual modelling of the impacts of climate change on karst geomorphology in the UK and Ireland. *Journal for Nature Conservation*, 11, 59–66.

Viles, H.A. and Goudie, A.S. (2003) Interannual decadal and multidecadal scale climatic variability and geomorphology. *Earth-Science Reviews*, 61, 105–131.

Viles, H.A. and Spencer, T. (1995) *Coastal Problems: Geomorphology, Ecology and Society at the Coast*. London: Edward Arnold.

Vilímek, V., Zapata, M.L., Klimeš, J., Patzelt, Z. and Santillán, N. (2005) Influences of glacial retreat on natural hazards of the Palcacocha Lake area, Peru. *Landslides*, 2, 107–115.

Vine, H. (1968) Developments in the study of soils and shifting agriculture in tropical Africa. In R.P. Moss (ed.), *The Soil Resources of Tropical Africa*. Cambridge: Cambridge University Press, 89–119.

Vinnikov, K.Y., Robock, A., Stouffer, R.J., Walsh, J.E., Parkinson, C.L., Cavalieri, D.J., Mitchell, J.F.B., Garrett, D. and Zakharov, V.F. (1999) Global warming and Northern Hemisphere sea ice extent. *Science*, 286, 1934–1937.

Vita-Finzi, C. (1969) *The Mediterranean Valleys*. Cambridge: Cambridge University Press.

Vitousek, P.M., Gosz, J.R., Gruer, C.C., Melillo, J.M., Reiners, W.A. and Todd, R.L. (1979) Nitrate losses from disturbed ecosystems. *Science*, 204, 469–473.

Vitousek, P.M., Mooney, H.A., Lubchenco, J. and Melillo, J.M. (1997) Human Domination of Earth's Ecosystems. *Science*, 277, 494–499.

Vogl, R.J. (1977) Fire: a destructive menace or a rational process. In J. Cairns, K.L. Dickson and E.E. Herricks (eds), *Recovery and Restoration of Damaged Ecosystems*. Charlottesville: University Press of Virginia, 261–289.

von Broembsen, S.L. (1989) Invasions of natural ecosystems by plant pathogens. In J.A. Drake (ed.), *Biological Invasions: A Global Perspective*. Chichester: Wiley, 77–83.

Vörösmarty, C.J., Meybeck, M., Fekete, B., Sharma, K., Green, P. and Syvitski, J.P.M. (2003) Anthropogenic sediment retention: major global impact from registered river impoundments. *Global and Planetary Change*, 39, 169–190.

Vuille, M., Francou, B., Wagnon, P., Juen, I., Kaser, G., Mark, B.G. and Bradley, R.S. (2008) Climate change and tropical Andean glaciers: past, present and future. *Earth-Science Reviews*, 89, 79–96.

Vuorenmaa, J., Forsius, M. (2008) Recovery of acidified Finnish lakes: trends, patterns and dependence of catchment characteristics. *Hydrology and Earth System Sciences*, 12, 465–478.

Wackernagel, M. and Rees, W. (1995) *Our Ecological Footprint. Reducing Human Impact on the Earth*. Gabriola Island: New Society Publishers.

Wada, Y., van Beek, L.P.H., van Kempen, C.M., Reckman, J.W.T.M., Vasak, S. and Bierkens, M.F.P. (2010) Global depletion of groundwater resources. *Geophysical Research Letters*, 37, L20402.

Wada, Y., van Beek, L.P.H. and Bierkens, M.F.P. (2012a) Non-sustainable groundwater sustaining irrigation: a global assessment. *Water Resources Research*, 48, W00L06.

Wada, Y., van Beek, L.P.H., Sperna Weiland, F.C., Chao, B.F., Wu, Y.-H. and Bierkens, M.F.P. (2012b) Past and future contribution of global groundwater depletion to sea-level rise. *Geophysical Research Letters*, 39, L09402.

Wagner, T.P. (2009) Shared responsibility for managing electronic waste: a case study of Maine, USA. *Waste Management*, 29, 3014–3021.

Waithaka, J.M. (1996) Elephants: a keystone species. In T.R. McClanahan and T. Young (eds), *East African Ecosystems and Their Conservation*. New York: Oxford University Press, 284–285.

Wakindiki, I.I.C. and Ben-Hur, M. (2002) Indigenous soil and water conservation techniques: effects on runoff, erosion, and crop yields under semi-arid conditions. *Australian Journal of Soil Research*, 40, 367–379.

Walker, G. (2006) The tipping point of the iceberg. *Nature*, 441, 802–805.

Walker, H.J. (1988) *Artificial Structures and Shorelines*. Dordrecht: Kluwer.

Walker, H.J., Coleman, J.M., Roberts, H.H. and Tye, R.S. (1987) Wetland loss in Louisiana. *Geografiska Annaler*, 69A, 189–200.

Walker, M.D., Gould, W.A. and Chapin, F.S. (2001) Scenarios of biodiversity changes in Arctic and alpine tundra. In F.S. Chapin, O.E. Sala and E. Huber-Sannwald (eds), *Global Biodiversity in a Changing Environment*. New York: Springer, 83–100.

Walling, D.E. (2006) Human impact on land-ocean sediment transfer by the world's rivers. *Geomorphology*, 79, 192–216.

Walling, D.E. and Gregory, K.J. (1970) The measurement of the effects of building construction on drainage basin dynamics. *Journal of Hydrology*, 11, 129–144.

Walling, D.E. and Quine, T.A. (1991) *Recent rates of soil loss from areas of arable cultivation in the UK*. IAHS publication 203, 123–131.

Wallwork, K.L. (1956) Subsidence in the mid-Cheshire industrial area. *The Geographical Journal*, 122, 40–53.

Wallwork, K.L. (1974) *Derelict Land*. Newton Abbot: David & Charles.

Walsh, K. and Pittock, B. (1998) Potential changes in tropical storms, hurricanes, and extreme rainfall events as a result of climate change. *Climatic Change*, 39, 199–213.

Walsh, R.P.D. et al. (2006) Changes in the spatial distribution of erosion within a selectively logged rainforest in Borneo 1988–2003. In P.N. Owens and A.J. Collins (eds), *Soil Erosion and Sediment Redistribution in River Catchments*. Wallingford: CAB International, 239–253.

Walter, H. (1984) *Vegetation and the Earth* (3rd edn). Berlin: Springer Verlag.

Walter, R.C. and Merritts, D.J. (2008) Natural streams and the legacy of water-powered mills. *Science*, 319, 299–304.

Wang, H., Fu, L., Du, X. and Ge, W. (2010) Trends in vehicular emissions in China's mega cities from 1995–2005. *Environmental Pollution*, 158, 394–400.

Wang, X., Eerdun, H., Zhou, Z. and Liu, X. (2007) Significance of variations in the wind energy environment over the past 50 years with respect to dune activity and desertification in arid and semiarid northern China. *Geomorphology*, 86, 252–266.

Wang, Y., Yang, Y., Zhao, N., Liu, C. and Wang, Q. (2012) The magnitude of the effect of air pollution on sunshine hours in China. *Journal of Geophysical Research*, 117, D00V14.

Wanger, T.C. (2011) The lithium future – resources, recycling and the environment. *Conservation Letters*, 4, 202–206.

Ward, D. (2005) Do we understand the causes of bush encroachment in African savannas? *African Journal of Range and Forage Science*, 22, 101–105.

Ward, P.J., Marfai, M.A., Yulianto, F.F., Hizbaron, D.R. and Aerts, J.C.J.H. (2011) Coastal inundation and damage exposure estimation: a case study for Jakarta. *Natural Hazards*, 56, 899–916.

Ward, R.C. (1978) *Floods – A Geographical Perspective*. London: Macmillan.

Ward, S.D. (1979) Limestone pavements – a biologist's view. *Earth Science Conservation*, 16, 16–18.

Warner, R.C. and Budd, W.F. (1990) Modelling the long-term response of the Antarctic Ice Sheet to global warming. *Annals of Glaciology*, 27, 161–168.

Warren, A. (ed.) (2002) *Wind Erosion on Agricultural Land in Europe*. Brussels: European Commission.

Warren, A. and Maizels, J.K. (1976) *Ecological Change and Desertification*. London: University College.

Warrick, J.A., Hatten, J.A., Pasternack, G.B., Gray, A.B., Goni, M.A. and Wheatcroft, R.A. (2012) The effects of wildfire on the sediment yield of a coastal California watershed. *Geological Society of America Bulletin*, doi:10.1130/B30451.1.

Warrick, R.A. and Ahmad, Q.K. (eds) (1996) *The Implications of Climate and Sea Level Change for Bangladesh*. Dordrecht: Kluwer.

Wasson, R.J. (2012) Geomorphic histories for river and catchment management. *Philosophical Transactions. Series A, Mathematical, Physical, and Engineering Sciences*, 370, 2240–2263.

Waters, M.R. and Haynes, C.V. (2001) Late Quaternary arroyo formation and climate change in the American southwest. *Geology*, 29, 399–402.

Watkins, T. (2010) New light on Neolithic revolution in south-west Asia. *Antiquity*, 84, 621–634.

Watson, A. (1976) The origin and distribution of closed depressions in south-west Lancashire and north-west Cheshire. Unpublished BA dissertation, University of Oxford.

Watson, A., Price-Williams, D. and Goudie, A.S. (1984) The palaeoenvironmental interpretation of colluvial sediments and palaeosols of the Late Pleistocene hypothermal in southern Africa. *Palaeogeography, Palaeoclimatology, Palaeoecology*, 5, 225–249.

Watts, J. (2005) China: the air pollution capital of the world. *Lancet*, 366, 1761–1762.

Waycott, M. et al. (2009) Accelerating loss of seagrasses across the globe threatens coastal ecosystems. *Proceedings of the National Academy of Sciences of the United States of America*, 106, 12377–12381.

Weart, S.R. (2003) *The Discovery of Global Warming*. Cambridge, Massachusetts: Harvard University Press.

Weaver, J.E. (1954) *North American Prairie*. Lincoln: Johnsen.

Webb, R.H. 1982. Off-road motorcycle effects on a desert soil. *Environmental Conservation*, 9, 197–208.

Weber, E., Sun, S.-G. and Li, B. (2008) Invasive alien plants in China: diversity and ecological insights. *Biological Invasions*, 10, 1411–1429.

Weber, P. (1993) Reviving coral reefs. In L.R. Brown (ed.), *State of the World 1993*. London: Earthscan, 42–60.

Webster, P.J., Holland, G.J., Curry, J.A. and Chang, H.-R. (2005) Changes in tropical cyclone number, duration, and intensity in a warming environment. *Science*, 309, 1844–1846.

Wehrli, M.N., Mitchell, E.A.D., van der Knaap, W.O., Ammann, B. and Tinner, W. (2010) Effects of climatic change and bog development on Holocene tufa formation in the Lorze Valley (central Switzerland). *The Holocene*, 20, 325–336.

Wein, R.W. and Maclean, D.A. (eds) (1983) *The Role of Fire in Northern Circumpolar Ecosystems*. Chichester: Wiley.

Weisrock, A. (1986) Variations climatiques et periodes de sedimentation carbonatée a l'Holocene – l'age des depôts. *Mediterranée*, 10, 165–167.

Wellburn, A. (1988) *Air Pollution and Acid Rain: The Biological Impact*. London: Longman.

Wells, J.T. (1995) Effects of sea level rise on coastal sedimentation and erosion. In D. Eisma (ed.), *Climate Change Impact on Coastal Habitation*. Boca Raton: Lewis, 111–136.

Wells, J.T. (1996) Subsidence, sea level rise, and wetland loss in the lower Mississippi River Delta. In J.D. Milliman and B.V. Haq (eds), *Sea Level Rise and Coastal Subsidence*. Dordrecht: Kluwer.

Wells, N.A. and Andriamihaja, B. (1993) The initiation and growth of gullies in Madagascar: are humans to blame? *Geomorphology*, 8, 1–46.

Wells, P.V. (1965) Scarp woodlands, transported grass soils and concept of grassland climate in the Great Plains region. *Science*, 148, 246–249.

Werritty, A. (2002) Living with uncertainty: climate change, river flows and water resource management in Scotland. *The Science of the Total Environment*, 294, 29–40.

Werritty, A. and Lees, K.F. (2001) The sensitivity of Scottish rivers and upland valley floors to recent environmental change. *Catena*, 42, 251–273.

Werth, D. and Avissar, R. (2002) The local and global effects of Amazon deforestation. *Journal of Geophysical Research – Atmospheres*, 107 (D20), 8087.

Westhoff, V. (1983) Man's attitude towards vegetation. In W. Holzner, M.J.A. Werger and I. Ikusima (eds), *Man's Impact on Vegetation*. The Hague: Junk, 7–24.

Westing, A. and Pfeiffer, E.W. (1972) The cratering of Indochina. *Scientific American*, 226 (5), 21–29.

Westing, A.H. (2013) Nuclear war: its environmental impact. *Springer briefs on pioneers in science and practice*, 1, 80–113.

Wetherald, R.T. and Manabe, S. (2002) Simulation of hydrologic changes associated with global warming. *Journal of Geophysical Research*, 107, D19.

Wheaton, E.E. (1990) Frequency and severity of drought and dust storms. *Canadian Journal of Agricultural Economics*, 38, 695–700.

Whitaker, J.R. (1940) World view of destruction and conservation of natural resources. *Annals of the Association of American Geographers*, 30, 143–162.

White, T., Asfaw, B., De Gusta, D., Gilbert, H., Richards, G.D., Suwa, G. and Howell, F. (2003) Pleistocene *Homo sapiens* from Middle Awash, Ethiopia. *Nature*, 423, 742–747.

Whitehorn, P.R., O'Connor, S., Wackers, F.L. and Goulson, D. (2012) Neonicotinoid pesticide reduces bumble bee colony growth and queen production. *Science*, doi:10.1126/science.1215025.

Whitlock, C., Shafer, S.L. and Marlon, J. (2003) The role of climate and vegetation change in shaping past and future fire regimes in to northwestern US and the implications for ecosystem management. *Forest Ecology and Management*, 178, 5–21.

Whitmore, T.M., Turner, B.L., Johnson, D.L., Kates, R.W. and Gottschang, T.R. (1990) Long term population change. In B.L. Turner, W.C. Clark, R.W. Kates, J.T. Matthews and W.B. Meyer (eds), *The Earth as Transformed by Human Action*. Cambridge: Cambridge University Press, 26–39.

Whitney, G.G. (1994) *From Coastal Wilderness to Fruited Plain*. Cambridge: Cambridge University Press.

Whitten, A.J., Damanik, S.J., Anwar, J. and Nazaruddin, H. (1987) *The Ecology of Sumatra*. Yogyukarta: Gadjah Mada University Press.

Whyte, A.V.I. (1977) Guidelines for field studies in environmental perception. *UNESCO, MAB technical note 5*.

Wicke, B., Sikkema, R., Dornburg, V. and Faaij, A. (2011) Exploring land use changes and the role of palm oil production in Indonesia and Malaysia. *Land Use Policy*, 28, 193–206.

Wigley, B.J., Bond, W.J. and Hoffmann, M.T. (2009) Bush encroachment under three contrasting land-use practices in a mesic South African savanna. *African Journal of Ecology*, 47 (Suppl. 1), 62–70.

Wigmosta, M.S. and Leung, R. (2002) Potential impacts of climate change on streamflow and flooding in snow-dominated forested basin. In R.C. Sidle (ed.), *Environmental Change and Geomorphic Hazards in Forests*. Wallingford: CABI, 7–23.

Wilby, R.L. (2003) Past and projected trends in London's urban heat island. *Weather*, 58, 251–260.

Wilby, R.L. and Gell, P.A. (1994) The impact of forest harvesting on water yield: modelling hydrological changes detected by pollen analysis. *Hydrological Sciences Journal*, 39, 471–486.

Wilby, R.L., Dalgleish, H.Y. and Foster, I.D.L. (1997) The impact of weather patterns on historic and contemporary catchment sediment yields. *Earth Surface Processes and Landforms*, 22, 353–363.

Wild, M. (2009) Global dimming and brightening: a review. *Journal of Geophysical Research*, 114, D00D16.

Wild, M. (2012) Enlightening global dimming and brightening. *Bulletin of the American Meteorological Society*, 93, 27–37.

Wild, M. et al. (2005) From dimming to brightening: decadal changes in solar radiation at Earth's surface. *Science*, 308, 847–850.

Wild, M., Ohmura, A. and Makowski, K. (2007) Impact of global dimming and brightening on global warming. *Geophysical Research Letters*, 34, L04702.

Wildman, L.A.S. and Macbroom, J.G. (2005) The evolution of gravel bed channels after dam removal: case study of the Anaconda and Union City dam removals. *Geomorphology*, 71, 245–262.

Wilken, G.C. (1972) Microclimate management by traditional farmers. *Geographical Review*, 62, 544–560.

Wilkinson, B.H. and McElroy, B.J. (2007) The impact of humans on continental erosion and sedimentation. *Geological Society of America Bulletin*, 119, 140–156.

Wilkinson, C., Linden, O., Cesar, H., Hodgson, G., Rubens, J. and Strong, A.G. (1999) Ecological and socio-economic impacts of 1998 coral mortality in the Indian Ocean: an ENSO impact and a warning of future change? *Ambio*, 28, 188–196.

Wilkinson, W.B. and Brassington, F.C. (1991) Rising groundwater levels – an international problem. In R.A. Downing and W.B. Wilkinson (eds), *Applied Groundwater Hydrology – A British Perspective*. Oxford: Clarendon Press, 35–53.

Williams, E.H. and Bunkley-Williams, L. (1990) The worldwide coral bleaching cycle and related sources of coral mortality. *Atoll Research Bulletin*, pp. 1–71.

Williams, G.P. (1978) The case of the shrinking channels – the North Platte and Platte rivers in Nebraska. *United States Geological Survey circular*, 781.

Williams, M. (1970) *The Draining of the Somerset Levels*. Cambridge: Cambridge University Press.

Williams, M. (1988) The death and rebirth of the American forest: clearing and reversion in the United States, 1900–1980. In J.F. Richards and R. Tucker (eds), *World Deforestation in the Twentieth Century*. Durham, North Carolina and London: Duke University Press, 211–229.

Williams, M. (1989) *Americans and Their Forests*. Cambridge: Cambridge University Press.

Williams, M. (1990) *Wetlands: A Threatened Landscape*. Oxford: Basil Blackwell.

Williams, M. (1994) Forests and tree cover. In W.B. Meyer and B.L. Turner II (eds), *Changes in Land Use and Land Cover: A Global Perspective*. Cambridge: Cambridge University Press.

Williams, M. (2000) Deforestation: general debates explored through local studies. *Progress in Environmental Science*, 2, 229–251.

Williams, M. (2003) *Deforesting the Earth. From Prehistory to Global Crisis*. Chicago: The University of Chicago Press.

Williams, P., Biggs, J., Crowe, A., Murphy, J., Nicolet, P., Waetherby, A. and Dunbar, M. (2010) Ponds report from 2007. Countryside survey technical report No. 7/07.

Williams, P.W. (ed.) (1993) Karst terrains: environmental changes and human impact. *Catena Supplement*, 25, 1–20.

Williams, R.S. and Moore, J.G. (1973) Iceland chills lava flow. *Geotimes*, 18, 14–17.

Williams, W.D. (1999) Salinisation: a major threat to water resources in the arid and semi-arid regions of the world. *Lakes and Reservoirs: Research and Management*, 4, 85–91.

Williamson, M. (1996) *Biological Invasions*. London: Chapman & Hall.

Willis, C.M. and Griggs, G.B. (2003) Reductions in fluvial sediment discharge by coastal dams in California and implications for beach sustainability. *The Journal of Geology*, 111, 167–182.

Willis, K.J., Gillson, L. and Brncic, T.M. (2004) How 'virgin' is virgin rainforest? *Science*, 304, 402–403.

Wilshire, H.G. (1980) Human causes of accelerated wind erosion in California's deserts. In D.R. Coates and J.D. Vitek (eds), *Geomorphic Thresholds*. Stroudsburg: Dowden, Hutchinson & Ross, 415–433.

Wilshire, H.G., Nakata, J.K. and Hallet, B. (1981) Field observations of the December 1977 wind storm, San Joaquin Valley, California. In T.L. Péwé (ed.), *Desert Dust: Origin, Characteristics and Effects on Man*. Special Paper, Geological Society of America. Denver, CO: Geological Society of America, 233–251.

Wilson, C.J. (1999) Effects of logging and fire on runoff and erosion on highly erodible granitic soils in Tasmania. *Water Resources Research*, 35, 3531–3546.

Wilson, K.V. (1967) A preliminary study of the effect of urbanization on floods in Jackson, Mississippi. *United States Geological Survey professional paper*, 575-D, 259–261.

Winkler, E.M. (1970) The importance of air pollution in the corrosion of stone and metals. *Engineering Geology*, 4, 327–334.

Wilson, E.O. (1992) *The Diversity of Life*. Cambridge, Massachusetts: Harvard/Belknap.

Wishart, D. and Warburton, J. (2001) An assessment of blanket shire degradation and peatland gully development in the Cheviot Hills, Northumberland. *Scottish Geographical Magazine*, 117, 185–206.

Wolfe, S.A. and Hugenholtz, C.H. (2009) Barchan dunes stabilized under recent climate warming on the northern Great Plains. *Geology*, 37, 1039–1042.

Wolman, M.G. (1967) A cycle of sedimentation and erosion in urban river channels. *Geografiska Annaler*, 49A, 385–395.

Wolman, M.G. and Schick, A.P. (1967) Effects of construction on fluvial sediment, urban and suburban areas of Maryland. *Water Resources Research*, 3, 451–464.

Wondzell, S.M. and King, J.G. (2003) Postfire erosional processes in the Pacific Northwest and Rocky Mountain regions. *Forest Ecology and Management*, 178, 75–87.

Wong, S., Dessler, A.E., Mahowald, N., Colarco, P.R. and Silva, A. (2008) Long-term variability in Saharan dust transport and its link to North Atlantic sea surface temperature. *Geophysical Research Letters*, 35, L07812.

Woo, M.-K. (1996) Hydrology of northern North America under global warming. In J.A.A. Jones et al. (eds), *Regional Hydrological Response to Climate Change*. Dordrecht: Kluwer, 73–86.

Woo, M.-K., Lewkowicz, A.G. and Rouse, W.R. (1992) Response of the Canadian permafrost environment to climate change. *Physical Geography*, 13, 287–317.

Wood, B. (2002) Hominid revelations from Chad. *Nature*, 418, 133–135.

Woodroffe, C.D. (1990) The impact of sea-level rise on mangrove shorelines. *Progress in Physical Geography*, 14, 483–520.

Woodroffe, C.D. (2008) Reef-island topography and the vulnerability of atolls to sea-level rise. *Global and Planetary Change*, 62, 77–96.

Woodwell, G.M. (1992) The role of forests in climatic change. In N.P. Sharman (ed.), *Managing the World's Forests*. Dubuque, Iowa: Kendall/Hunt, 75–91.

Wooster, W.S. (1969) The ocean and man. *Scientific American*, 221, 218–223.

World Commission on Dams (2000) *Dams and Development*. London: Earthscan.

World Conservation Monitoring Centre (1992) *Global Biodiversity*. London: Chapman & Hall.

World Glacier Monitoring Service (2008) Global glacier changes: facts and figures. (http://www.grid.unep.ch/glaciers/), last accessed 05 January 2013.

World Meteorological Organization (1995) *Climate system review*. Geneva: WMO.

World Resources Institute (1986) *World Resources 1986–7*. New York: Basic Books.

World Resources Institute (1988) *World Resources 1988–9*. New York: Basic Books.

World Resources Institute (1992) *World Resources 1990–91*. New York and Oxford: Oxford University Press.

World Resources Institute (1998) *World Resources 1996–7*. New York and Oxford: Oxford University Press.

Woth, K., Weisse, R. and von Storch, H. (2006) Climate change and North Sea storm surge extremes: an ensemble study of storm surge extremes expected in a changed climate projected by four different regional climate models. *Ocean Dynamics*, 56, 3–15.

Wright, L.W. and Wanstall, P.J. (1977) The vegetation of Mediterranean France: a review. Occasional paper 9, Department of Geography, Queen Mary College, University of London.

Wright, S.A. and Schoellhamer, D.H. (2004) Trends in the sediment yield of the Sacramento River, California, 1957–2001. *San Francisco Estuary and Watershed Science* (on line serial), 2 (2), Article 2.

Wroe, S. and Field, J. (2006) A review of the evidence for a human role in the extinction of Australian megafauna and an alternative interpretation. *Quaternary Science Reviews*, 25, 2692–2703.

Wu, P., Wood, R. and Stott, P. (2005) Human influence on increasing Arctic river discharges. *Geophysical Research Letters*, 32, L02703.

Wu, Y., Wang, S., Streets, D.G., Hao, J., Chan, M. and Jiang, J. (2006) Trends in anthropogenic mercury emissions in China from 1995 to 2003. *Environmental Science & Technology*, 40, 5312–5318.

Wullschleger, S.D., Gunderson, C.A., Hanson, P.J., Wilson, K.B. and Norby, R.J. (2002) Sensitivity of stomatal and canopy conductance to elevated CO_2 concentrations – interacting variables and perspective on scale. *The New Phytologist*, 153, 485–496.

WWF (2010) *Living Planet Report*.

Xing, Y. et al. (2005) A spatial temporal assessment of pollution from PCBs in China. *Chemosphere*, 60, 731–739.

Xu, K. and Milliman, J.D. (2009) Seasonal variations of sediment discharge from the Yangtze River before and after impoundment of the Three Gorges Dam. *Geomorphology*, 104, 276–283.

Xu, K., Milliman, J.D., Yang, Z. and Xu, H. (2007) Climatic and anthropogenic impacts on water and sediment discharges from the Yangtze River (Changjiang), 1950–2005. In A. Gupta (ed.), *Large Rivers: Geomorphology and Management*. Chichester: Wiley, 609–626.

Xu, M., Chang, C.-P., Fu, C., Qi, Y., Robock, A., Robinson, D. and Zhang, H.-M. (2006) Steady decline of east Asian monsoon winds, 1969–2000: evidence from direct ground measuremenrs of wind speed. *Journal of Geophysical Research*, 111, D24111.

Xu, Y.S., Zhang, D.X., Shen, S.L. and Chen, L.Z. (2009) Geohazards with characteristics and prevention measures along the coastal regions of China. *Natural Hazards*, 49, 479–500.

Xue, Y.Q., Zhang, Y., Ye, S.J., Wu, J.C. and Li, Q.F. (2005) Land subsidence in China. *Environmental Geology*, 48, 713–720.

Yaalon, D.H. and Yaron, B. (1966) Framework for man-made soil changes – an outline of metapedogenesis. *Soil Science*, 102, 272–277.

Yamano, H., Sugihara, K. and Nomura, K. (2011) Rapid poleward range expansion of tropical reef corals in response to rising sea surface temperatures. *Geophysical Research Letters*, 38, L04601.

Yang, D., Kanae, S., Oki, T., Koike, T. and Musiake, K. (2003) Global potential soil erosion with reference to land use and climate changes. *Hydrological Processes*, 17, 2913–2928.

Yang, D.Y., Liu, L., Chen, X. and Speller, C.F. (2008) Wild or domesticated: DNA analysis of ancient water buffalo remains from north China. *Journal of Archaeological Science*, 35, 2778–2785.

Yang, M., Nelson, F.E., Shiklomanov, N.I., Guo, D. and Wan, G. (2010a) Permafrost degradation and its environmental effects on the Tibetan Plateau. *Earth-Science Reviews*, 103, 31–44.

Yang, P., Hong, G., Dessler, A.E., Ou, S.S.C., Liou, K.-N., Minnis, P. and Vardhan, H. (2010b) Contrails and induced cirrus. *Bulletin of the American Meteorological Society*, 91, 473–478.

Yang, Q., Wang, K., Zhang, C., Yue, Y., Tian, R. and Fan, F. (2011a) Spatio-temporal evolution of rocky desertification and its driving forces in karst areas of Northwestern Guangxi, China. *Environmental Earth Sciences*, 64, 383–393.

Yang, S.L., Milliman, J.D., Li, P. and Xu, K. (2011b) 50,000 dams later: erosion of the Yangtze River and its delta. *Global and Planetary Change*, 75, 14–20.

Yao, T. et al. (2012) Different glacier status with atmospheric circulations in Tibetan Plateau and surroundings. *Nature Climate Change*, doi:10.1038/NCLIMATE1580.

Yasunari, T.J. et al. (2012) Estimated range of black carbon dry deposition and the related snow albedo reduction over Himalayan glaciers during pre-monsoon periods. *Atmospheric Environment* doi:10.1016/j.atmsoenv.2012.03.031.

Yechieli, Y., Abelson, M., Bein, A., Crouvi, O. and Shtivelman, V. (2006) Sinkhole 'swarms' along the Dead Sea coast: reflection of disturbance of lake and adjacent groundwater systems. *Geological Society of America Bulletin*, 118, 1075–1087.

Yesner, D.R. (2001) Human dispersal into interior Alaska: antecedent conditions, mode of colonization, and adaptations. *Quaternary Science Reviews*, 20, 315–327.

Yi, H., Hao, J. and Tang, X. (2007) Atmospheric environmental protection in China: current status, development trend and research emphasis. *Energy Policy*, 35, 907–915.

Yin, Y., Zhang, K. and Li, X. (2006) Urbainzation and land subsidence in China. *IAEG2006 Paper* 31, Geological Society of London.

Yizhaq, H., Ashkenazy, Y. and Tsoar, H. (2007) Why do active and stabilized dunes coexist under the same climatic conditions? *Physical Review Letters*, doi:10.1103/PhysRevLett.98.188001.

Yizhaq, H., Ashkenazy, Y. and Tsoar, H. (2009) Sand dune dynamics and climate change: a modelling approach. *Journal of Geophysical Research*, 114, F01023.

Yoo, J.-C. and D'Odorico, P. (2002) Trends and fluctuations in the dates of ice break-up of lakes and rivers in Northern Europe: the effect of the North Atlantic Oscillation. *Journal of Hydrology*, 268, 100–112.

Yorke, T.H. and Herb, W.J. (1978) Effects of urbanization on streamflow and sediment transport in the Rock Creek and Anacostia basins, Montgomery County, Maryland, 1962–74. *United States Geological Survey professional paper* 1003.

Yoshikawa, K. and Hinzman, L.D. (2003) Shrinking thermokarst ponds and groundwater dynamics in discontinuous permafrost near Council, Alaska. *Permafrost and Periglacial Processes*, 14, 151–160.

Young, J.E. (1992) *Mining the earth*. Worldwatch Paper 109, 1–53.

Yuill, B., Lavoie, D. and Reed, D. (2009) Understanding subsidence processes in coastal Louisiana. *Journal of Coastal Research*, SI (54), 23–36.

Yule, J.V. (2009) North American Late Pleistocene megafaunal extinctions: overkill, climate change or both? *Evolutionary Anthropology*, 18, 159–160.

Yunus, M. and Iqbal, M. (eds) (1996) *Plant Response to Air Pollution*. Chichester: Wiley.

Zabinski, C. and Davis, M.B. (1989) Hard times ahead for Great Lakes forests: a climate threshold model predicts responses to CO_2-induced climate change. In J.B. Smith and D. Tirpak (eds), *The Potential Effects of Global Climate Change on the United States*. Washington DC: US Environmental Protection Agency, Appendix D, 5–1–5–19.

Zeder, M.A. (2008) Domestication and early agriculture in the Mediterranean Basin: origins, diffusion, and impact. *Proceedings of the National Academy of Sciences of the United States of America*, 105, 11597–11604.

Zedler, J.B. and Kercher, S. (2005) Wetland resources: status, trends, ecosystem services, and restorability. *Annual Review of Environment and Resources*, 30, 39–74.

Zeeberg, J. and Forman, S.L. (2001) Changes in glacier extent of north Novaya Zemlya in the twentieth century. *The Holocene*, 11, 161–175.

Zelazowski, P., Malhi, Y., Huntingford, C., Sitch, S. and Fisher, B. (2011) Changes in the potential distribution of humid tropical forests on a warmer planet. *Philosophical Transactions. Series A, Mathematical, Physical, and Engineering Sciences*, 369, 137–160.

Zeng, N. and Yoon, J. (2009) Expansion of the world's deserts due to vegetation-albedo feedback under global warming. *Geophysical Research Letters*, 36, L17401.

Zhang, D.Q., Tan, S.K. and Gersberg, R.M. (2010a) Municipal solid waste management in China: status, problems, and challenges. *Journal of Environmental Management*, 91, 1623–1633.

Zhang, G.L. and Gong, Z.-T. (2003) Pedogenic evolution of paddy soils in different soil landscapes. *Geoderma*, 115, 15–29.

Zhang, J., Ma, K. and Fu, B. (2010b) Wetland loss under the impact of agricultural development in the Sanjiang Plain, NE China. *Environmental Monitoring and Assessment*, 166, 139–148.

Zhang, J., Steele, M. and Schweifer, A. (2010c) Arctic sea ice response to atmospheric forcings with varying levels of anthropogenic warming and climate variability. *Geophysical Research Letters*, 37, L20505.

Zhang, J. et al. (2010d) Natural and human-induced hypoxia and consequences for coastal areas: synthesis and future development. *Biogeosciences*, 7, 1443–1467.

Zhang, K., Yu, Z., Li, X., Zhou, W. and Zhang, D. (2007) Land use change and land degradation in China from 1991–2001. *Land Degradation and Development*, 18, 209–219.

Zhang, T.-H., Zhao, H.-L., Li, S.-G., Li, F.-R., Shirato, Y., Ohkuro, T. and Taniyama, I. (2004) A comparison of different measures for stabilizing moving sand dunes in the Horqin Sandy Land of Inner Mongolia, China. *Journal of Arid Environments*, 58, 203–214.

Zhang, Y. and Song, C. (2006) Impacts of afforestation, deforestation and reforestation on forest cover in China from 1949 to 2003. *Journal of Forestry*, 104, 383–387.

Zhao, C., Wang, Y. and Zeng, T. (2009) East China Plains: a 'basin' of ozone pollution. *Environmental Science & Technology*, 43, 1911–1915.

Zheng, F.L. (2005) Effects of accelerated soil erosion on soil nutrient loss after deforestation on the loess plateau. *Pedospkere*, 15, 707–715.

Zhong, L., Deng, J., Song, Z. and Ding, P. (2011) Research on environmental impacts of tourism in China: progress and prospect. *Journal of Environmental Management*, 92, 2972–2983.

Zhu, A., Ramanathan, V., Li, F. and Kim, D. (2007) Dust plumes over the Pacific, Indian, and Atlantic Oceans: climatology and radiative impact. *Journal of Geophysical Research*, 112, D16208.

Zhu, C., Wang, B. and Qian, W. (2008) Why do dust storms decrease in northern China concurrently with the recent global warming? *Geophysical Research Letters*, 35, L18702.

Zhu, R.X. et al. (2008) Early evidence of the genus Homo in East Asia. *Journal of Human Evolution*, 55, 1075–1085.

Zizumbo-Villarreal, D. and Colunga-GarcíaMarín, P. (2010) Origin of agriculture and plant domestication in West Mesoamerica. *Genetic Resources and Crop Evolution*, 57, 813–825.

Zohary, D., Hopf, M. and Weiss, E. (2012) *Domestication of Plants in the Old World* (4th edn). Oxford: Oxford University Press.

Zwally, H.J., Abdalafi, W., Herring, T., Larson, K., Saba, J. and Steffen, K. (2002) Surface melt-induced acceleration of Greenland ice-sheet flow. *Science*, 297, 218–221.

INDEX

The Human Impact on the Natural Environment: Past, Present and Future, Seventh Edition. Andrew S. Goudie.
© 2013 John Wiley & Sons, Ltd. Published 2013 by John Wiley & Sons, Ltd.